OPERATIONS RESEARCH AND SYSTEMS ENGINEERING

Operations Research AND *Systems Engineering*

---

edited by CHARLES D. FLAGLE

WILLIAM H. HUGGINS

ROBERT H. ROY

THE JOHNS HOPKINS PRESS : *Baltimore*

Distributed in Great Britain by Oxford University Press, London

Printed in the United States of America
by The Colonial Press Inc., Clinton, Mass.

Library of Congress Catalog Card Number 60–8280

FOREWORD

While there has been no editorial attempt to do so, the pages of this book trace the history of the growing involvement of a university in the day-to-day affairs of government and industry. The chapters of the book are a set of lectures delivered by the authors in The Johns Hopkins University's annual two-week course for management, a course that bears the same title as the book. The lectures assembled for the course at the time that these particular lectures were given came from many academic departments of The Johns Hopkins University and from its affiliated research branches, the Applied Physics Laboratory and the Operations Research Office.

Many and diverse disciplines are represented—physics, economics, statistics, psychology, several branches of engineering, and mathematics. Despite this range of backgrounds, the authors share a common interest that unites them to each other perhaps more than to some colleagues in their own disciplines. This binding interest is the concern for the operation of total systems—human organizations, man-machine systems such as factories or military units, or complex physical systems.

Common terms have developed and these appear intermingled with older, specialized terminologies. We are interested in the "objectives" of an organization, their "feasibility" within the "constraints" of limited resources, and "measures of effectiveness" with which objectives are achieved. The comparison of alternative strategies and tactics for achieving feasible objectives leads hopefully to the choice of an "optimum alternative," which choice is in itself a "decision process."

The initial chapters are devoted to the philosophical and historical aspects of systems engineering and operations research. These are followed by chapters on specific methodologies that have developed or have been adapted for the field. A set of case histories concludes the volume. The editors have footnoted cross references to relate the methodologies to one another and to identify their role in the case studies. This in itself has been a considerable task since the application of mathematical analysis in operations research is often less direct than in conventional engineering. The models and formulae lend insight into system behavior and the functional relationship between behavior and the factors that affect it; but rather rarely do they yield precise numerical predictions like those that tell the design engineer the deflection of a beam or the pressure on an airfoil.

The reader who survives the early chapters must sense the authors' concern for the effects of rapid growth, in our times, of knowledge of the physical world. In our social, political, and commercial organizations the expanding application of new knowledge promotes rapid obsolescence—not only in physical equipment but in the adequacy and capability of humans. In the case of physical equipment little can be done other than to replace it under the economic pressure of obsolescence. We humans, on the other hand, can forestall our own obsolescence by continuous new learning. Each author offers here a central interest of his professional life, presenting it in brief form for a reader who is assumed to be mature and interested, but not a specialist in the field. It is for those to whom this is new knowledge that the editors have taken it upon themselves to bring this material together and publish it.

The editors are grateful to the authors for their continuing co-operation and to The Johns Hopkins Press for its support in assembling this volume. Illustrations of consistently high quality and imagination have been prepared by Mr. John Spurbeck, Director of the Illustration Division of the University, and his assistants.

Baltimore, Maryland
January, 1960

Charles D. Flagle
William H. Huggins
Robert H. Roy

CONTENTS

AUTHORS

ALPHONSE CHAPANIS, Ph. D., is Professor of Psychology and Industrial Engineering, The Johns Hopkins University. Dr. Chapanis was elected President of the Society of Experimental Psychologists for 1959-1960.

NASLI H. CHOKSY, Ph. D., is Assistant Professor of Electrical Engineering, The Johns Hopkins University.

WILLIAM G. COCHRAN, M. A., is Professor of Statistics, Harvard University. Mr. Cochran was formerly Professor of Biostatistics, The Johns Hopkins University School of Hygiene and Public Health. He has served as President of the Institute of Mathematical Statistics, the American Statistical Association, and the Biometric Society.

WALTER E. CUSHEN, Ph. D., is Group Chairman at the Operations Research Office, The Johns Hopkins University.

SIDNEY DAVIDSON, Ph. D., is Professor of Accounting in the School of Business, University of Chicago. Dr. Davidson was formerly Professor of Accounting, The Johns Hopkins University.

ACHESON J. DUNCAN, Ph. D., is Associate Professor of Statistics, The Johns Hopkins University.

CHARLES D. FLAGLE, Dr. Engr., is Associate Professor of Industrial Engineering, The Johns Hopkins University, and Director of Operations Research, The Johns Hopkins Hospital.

RALPH E. GIBSON, Ph. D., is Director of the Applied Physics Laboratory, The Johns Hopkins University. Dr. Gibson was given the Navy Distinguished Public Service Award by the Secretary of the Navy in January, 1958 for his "outstanding contributions to the Department of the Navy in the fields of scientific research and development."

WILLIS C. GORE, Dr. Engr., is Associate Professor of Electrical Engineering, The Johns Hopkins University.

MARVIN A. GRIFFIN, M. S. E., is Instructor in Industrial Engineering, The Johns Hopkins University.

WILLIAM H. HUGGINS, Sc. D., is Professor of Electrical Engineering, The Johns Hopkins University. In 1954 Dr. Huggins was awarded the Air Force Decoration for Exceptional Civilian Service.

ELLIS A. JOHNSON, D. Sc., is Director of the Operations Research Office, The Johns Hopkins University. The Distinguished Civilian Service Medal was awarded to Dr. Johnson in 1958 by the Department of Defense.

RICHARD B. KERSHNER, Ph. D., is a member of the Principal Professional Staff of the Applied Physics Laboratory, The Johns Hopkins University. Dr. Kershner received the Distinguished Public Service Award from the United States Navy for his work in development of the Terrier missile.

ALEXANDER KOSSIAKOFF, Ph. D., is Assistant Director for Technical Operations of the Applied Physics Laboratory, The Johns Hopkins University. Dr. Kossiakoff holds the Presidential Certificate of Merit and the Navy Distinguished Public Service Award.

P. STEWART MACAULAY, A. B., is Executive Vice-President, The Johns Hopkins University and Chairman of the Board of Trustees, Associated Universities, Inc.

VINCENT V. MC RAE, Ph. D., is a Staff Member of the Operations Research Office, The Johns Hopkins University. In 1957 Dr. McRae served on the Technical Staff of the Gaither Committee.

ELIEZER NADDOR, Ph. D., is Associate Professor of Industrial Engineering, The Johns Hopkins University. Dr. Naddor was Director of the Intensive Courses in Operations Research and Systems Engineering from which the papers in this volume have been drawn.

THORNTON L. PAGE, Ph. D., is Professor of Astronomy, Wesleyan University. Dr. Page was formerly Deputy Director of the Operations Research Office, The Johns Hopkins University.

ROBERT H. ROY, B. E., is Professor of Industrial Engineering and Dean of the School of Engineering, The Johns Hopkins University.

RICHARD E. ZIMMERMAN, M. S., is Group Chairman at the Operations Research Office, The Johns Hopkins University. Mr. Zimmerman was awarded the Lanchester Prize in 1957.

PART I

Perspectives

One

Introduction—THE MARKET PLACE AND THE IVORY TOWER

P. STEWART MACAULAY

Before launching upon my subject I should, perhaps, attempt to qualify as an expert. My claim is based on some dozen years of active association with the market place, and an even greater span of years in the ivory tower atmosphere of a university. I have always asserted that this duality of experience gives me special competence to understand where, when, and under what circumstances the twain shall meet.

This claim, however, probably will be challenged on both sides. I must confess that my non-academic experience has been in newspaper work and that it has been limited to the areas of writing and editing. Anyone who knows anything about a newspaper will recognize at once that, to the business office, and especially to the composing room, writers and editors are not regarded as real people living in a real world. Conversely, in my academic career I have been concerned with such sordid matters as budgets and the economics of plant operation. Obviously, therefore, I am not fully accepted by the denizens of the academic ivory tower.

This, it seems, leaves me somewhere in the middle, and perhaps that is a good vantage point from which to launch a few observations on the relationships which exist, or should exist, between the world of affairs and the "out-of-this-world" environment commonly attributed to colleges and universities.

Let me say at the outset that historically the separation between town and gown has never been complete. Many of America's ivory towers were actually created by hard-bitten business men and industrialists. Our great privately established universities reek of oil, of steel, of railroads and tobacco—yes, even of good corn whiskey.

Do you think I stretch a point when I mention whiskey? I am sure you identify the others—Rockefeller's oil and Carnegie's steel helped to build large segments of our system of higher education; railroad money built Stanford University; and profits from the sale of tobacco produced Duke. But what university owes its origin to whiskey? Let me quote a passage from the biography of Johns Hopkins by his great-niece Helen Hopkins Thom: ". . . he decided to go into business for himself and, taking his three brothers, Philip, Gerard and Mahlon as salesmen, he formed the wholesale Provision House of 'Hopkins Brothers.' This house soon did a large business, especially through North Carolina and the valley of Virginia, where they had important connections. The new firm took whiskey in return for goods and sold it under the brand of 'Hopkins' Best.'

"This action on the part of Johns Hopkins offended the Society of Friends and he was temporarily turned out of Meeting. He continued to sell whiskey, nevertheless; but he went regularly to meeting, continued to contribute and was later reinstated. In his later life, however, he felt that he had been wrong in the stand he had taken; and he told his nephew Joseph Hopkins, that he wished he had never sold liquor, and that in so doing he had made the greatest mistake of his life."

Those of us who are concerned with the current finances of the institution which bears his name often wonder whether Johns Hopkins' greatest mistake was not in giving up the trade in liquor. Perhaps many of our current woes would be eased if we could still count upon the profits from "Hopkins' Best." But I have deviated from my main theme.

The men who founded these universities did not expect to collect dividends for themselves or their enterprises in terms of specific contributions by the institutions to business and industry. They felt, perhaps rather vaguely, that education was a good thing. They enjoyed the sensation of having created something of value to society in general. They expected to claim their rewards in heaven or in Gothic architecture bearing their names. The idea that a professor might emerge from the ivory tower with a practical down-to-earth idea certainly never occurred to them.

Nor did such a ghastly thought often occur to the professors. Having had a fairly comfortable ivory tower provided for them, they were content to reside therein, pursuing endlessly and in relative seclusion the conventional academic objective of learning more and more about less and less. The humanists were immersed in their books—ancient books containing the wisdom of the ages, and their own books in which this wisdom, presumably, was distilled. Scientists worked in solitary grandeur in their laboratories, creating curious odors and sounds, or perhaps contriving candid microscopic exposés of life among the amoebae. Political scientists observed the functions of government from the safety of their cloisters but seldom, if ever, ventured into the political arena themselves. Economists with jaundiced eyes surveyed the systems which they generally disapproved, emerged occasionally to bite the hand that fed them, then scuttled back to the protecting walls of tenure and academic freedom.

I have not done enough research to establish without question the *exact* moment at which the first breakthrough occurred—the first uneasy and perhaps questionable marriage between the market place and the ivory tower. It is reasonably certain, however, that this was a union between a chemist and a promoter.

Certainly chemistry was the first of the academic disciplines to produce things of value to the population at large and therefore of interest to industry and commerce. At about the turn of the century the market place began to cast sidelong glances at that part of the ivory tower which encompassed the chemistry laboratories. A few early commercially profitable products of these laboratories burgeoned into many, and an entire new and rapidly growing industry established a debt to academic education and research which it probably will never completely repay.

In this process not many academic chemists got rich. A few, however, were stubborn enough to insist on a share of the profits which their discoveries had created, and so was established the first regular exchange between the market place and the ivory tower.

Gradually the areas of exchange broadened. Engineering, of course, came into the picture quite early, for engineering training was directed specifically at the needs of industry. Engineering research was closely related to the discovery of new products and better ways of producing them.

Perhaps the most dramatic outpouring of academic personnel into the world of affairs occurred during the great depression of the early thirties. Then the economists, the experts in government, the psychologists, and various mixed breeds from within the ivory towers descended upon Washington for the purpose of creating a brave new world. There are some who say that they did not do a very good job, that the change in environment was too much for them, that the world would have been better off if they had stayed where they were. I shall not take sides on this issue, for my purpose is not to evaluate but simply to describe—and there is no doubt that the emergence of social scientists into active relationships to government and even business and industry at that time was a major development.

The next exodus came, of course, with the threat of World War II and a general recognition for the first time that those scientists who had been most securely protected by the ivory tower had something to contribute to the national defense. Thus the physicists first, and then even the mathematicians, joined the chemists and the engineers in an all-out effort to win what was generally recognized as a scientific war.

I find up to this point that I have not mentioned another very important area of academic research and activity generally—biology. I should not wish to overlook this field, for it does enter very completely into the processes I have just attempted to describe. Early in the game, however, it was recognized that a good deal of biological research had immediate and practical implications, mainly, perhaps, in agriculture and in medicine. The tribe as a whole, therefore, never has been completely insulated from the world of affairs. Biology has not followed the gradual and evolutionary pattern which I have described with respect to other academic activities, simply because the practical potentialities of biological studies were recognized in many cases as soon as the discoveries were made. Nevertheless, a limited number of biologists found a new role in the World War II period, joining their fellow scientists in explorations of methods of attack and defense which have been described as "biological warfare."

The ultimate development of this widespread and sporadic emergence from the seclusion of academic halls to the arena in which real problems are attacked and perhaps solved was inevitable as we look backward to it. If the physicists, the chemists, the mathematicians, and the engineers could combine to build an atomic bomb, why could not the same kinds of groups, working in concert, solve other major problems, both military and civil? The concept of the multi-disciplinary approach was utilized in a number of areas during World War II, and it was only natural that the techniques thus devised should carry over. It was equally natural that these techniques should be given a name—the name most commonly accepted is operations research.

I am not competent to give you a precise definition of operations research. Many have been brought forward and no two seem to be in agreement. There is one school, I believe, which seems to feel that practically all problems can be solved by appropriate application of the principles of physics, mathematics, and statistics. Another school, whose leader is Ellis Johnson of The Johns Hopkins Operations Research Office, believes that the solution of many problems requires also the intervention of such specialists as economists, political scientists, historians, and philosophers. Regardless of these differences, one thing emerges clearly—the ivory tower no longer is inviolable. The academician of whatever stripe, if he be so inclined, has the possibility of contributing something to the solution of real problems, and the people who have to contend with real problems are recognizing more and more the resources which lie behind these formerly impregnable walls. This dramatic development is certain to produce changes which will be felt on both sides and which will have a profound effect on both the academic community and business and industry.

The effect may be generally beneficial, but I see some dangers in it. I hope the pressure will never become so great upon the truly academic person that he will be forced by public opinion or economic necessity to participate in activities which are uncongenial to him. Even though the ivory tower as a sanctuary may be crumbling, I believe that strong measures should be taken to preserve within it at least a few cells which are still inviolate. There are in many of the conventional disciplines people of high caliber and high devotion who should be protected against all pressures designed to force them into action situations. They must be permitted to do their thinking without concern for what happens to the results of their thinking. They should be nurtured as men of ideas, not of action. And the greatest defenders of these remaining bastions of pure scholarly or scientific thought should be the businessmen and the industrialists who have most to gain ultimately through the preservation, somewhere in our social and educational system, of that kind of an intellectual environment from which the world's greatest discoveries have come, and from which they will continue to come in the future.

Two

THE DEVELOPMENT AND FUTURE OF OPERATIONS RESEARCH AND SYSTEMS ENGINEERING

ROBERT H. ROY

"When *I* use a word," Humpty Dumpty said, in a rather scornful tone, "it means just what I choose it to mean—neither more nor less."

"The question is," said Alice, "whether you *can* make words mean so many different things."

"The question is," said Humpty Dumpty, "which is to be master—that's all."

Lewis Carroll: *Through the Looking Glass*

"What's in a name? That which we call a rose
By any other name would smell as sweet."

William Shakespeare: *Romeo and Juliet*

Within the past two decades, something called "operations research" in the United States and "operational research" in Great Britain has been born, has flourished and multiplied, has attracted and continues to attract much attention, and appears to hold great promise for the future. More recently, something else called "systems engineering," or "system engineering," has had analogous birth, development, attention, and promise. This chapter deals with what these things are, how they have developed, and what may be expected of them in time to come.

That there have been such births and such developments cannot be denied. Nor can it be said that operations research and systems engineering have failed to attract and hold attention; nor that they lack promise for the future. Operations research and systems engineering *have* been born, *have* developed, *have* commanded notice, and *have* promise for the future. But what they *are* has been the cause of much dispute.

There are some who claim that operations research and systems engineering are new intellectual disciplines in their own right, deserving rank and prestige with mathematics, physics, chemistry, and philosophy. There are others, perhaps less aggressive or pretentious, who say that these are at the very least new professions, that the assemblage of multi-discipline teams—a characteristic of both operations research and systems engineering—yields a whole that is different from, and at the same time greater than, the sum of its parts. Bring together, it is said, a mathematician, a physicist, a statistician, an economist, and a mechanical engineer, let them study an operation with the "operations research point of view," and the result will be different—and very much better than could be achieved otherwise. Assign a similar combination to the creation of a complex man and machine system, give them the "systems point of view," and like benefits will accrue.

Quite naturally, most claims of this kind come from those already identified with operations research and systems engineering, who need the status and security that go with recognition. A physicist who makes a career of operations research may not get much recognition as a physicist, and can win prestige in operations research only if operations research itself is recognized. Hence, operations research to him may be a new discipline. This is exaggerated, to be sure; physicists *have* won fame as operations researchers, but there is some truth in the assertion: these fields *need* to be claimed as new disciplines partly for the sake of those in them.

Those outside operations research and systems engineering have been quite skeptical of these claims and much less kind in their criticisms. Industrial engineers, while busily engaged in learning to use the newer tools of operations research, at the same time allege that it is "nothing but industrial engineering," an assertion that is particularly galling to operations researchers trained in the traditional disciplines, who look down upon industrial engineering as narrow and pragmatic, and want no part of identification with it. Those who have worked in marketing research have made the same claim—operations research is "nothing but marketing research"—and it has been almost as galling.

J. Bronowski, in a review[1] of one of the early books on operations research,[2] agrees in part with this "nothing but" point of view and is not sanguine about the future of operations research:

> Is there then a future for operations research today, either in industry or in war? I doubt it; at least, I doubt whether it can again be useful in the simple sense of the last war and of this book. The heroic age is over; and, dropping with a sigh the glamour and the heady sense of power, we have to face the recognition that the field of opportunity will never again be quite so blank, so simple and so lavish. What was new and speculative on the battlefield turns out, in the practical affairs of industry, to become only a painstaking combination of cost accounting, job analysis, time and motion study and the general integration of plant flow. There is an extension of this to the larger economics of whole industries and nations, but it is hardly likely to be rewarding to first-rate scientists, and calls at bottom for the immense educational task of interesting economists and administrators in the mathematics of differentials and of prediction.
>
> Nor is the art of war likely to offer again such a creamy surface to skim. The easy successes have been scored; the simple mistakes which they put right are understood and should not be made again. The analysis of operations must expect now to go a good deal deeper and to be a great deal more like research. Operations research has done its major work, and it turns out to have been a piece of education—the education of scientists and warriors in a new empiricism.

That was in 1951. In 1957, in his review[3] of a much more recent work,[4] Robert E. Machol adds fuel to the critical fire:

> . . . The O.R. man has stretched his triumphs for a decade by invading industry, but operations research as a separate discipline seems foredoomed unless something more can be synthesized than is apparent in this book. The techniques described in this book are valid and they will be used, but perhaps not most efficiently by professional O.R. men. If O.R. is to continue to grow, it will need to take on a more constructive and coherent aspect.

[1] J. Bronowski, *Scientific American,* Vol. 185 (October, 1951), 75-77.

[2] George E. Kimball and Philip M. Morse, *Methods of Operations Research* (New York: John Wiley and Sons, and Cambridge, Mass.: The Technology Press, M. I. T., 1951).

[3] Robert E. Machol, *Mechanical Engineering,* Vol. 79, No. 9 (September, 1957), 890-91.

[4] C. West Churchman, Russell L. Ackoff, and E. Leonard Arnoff, *Introduction to Operations Research* (New York: John Wiley and Sons, 1957).

It is not without significance, perhaps, that the reviewer himself is co-author of a recent book on systems engineering,[5] for internecine strife is not unknown between those in operations research and those identified with systems engineering. Each individual has his own idea of what these terms mean and what they ought to mean, and each is concerned with what they mean to the world at large. To paraphrase Humpty Dumpty, "The question is, *who* is to be master—that's all."

SCIENTIFIC MANAGEMENT

Forty-eight years ago, in January, 1912, a Special Committee of the House of Representatives of the Congress was similarly concerned about the meaning of a then new and even more controversial term: *Scientific Management.*[6] Their concern derived from the work of Frederick W. Taylor, who had begun his career in 1874 as an apprentice machinist and pattern-maker. As a machinist Taylor perceived that workers restricted output by what he later called "systematic soldiering." As a worker he conformed to this practice, but when he was made foreman he sought to break it up by punitive measures, deeming that it was his job as foreman to "get output." He resorted to fines and to firing, and his men retaliated by violence and by deliberately breaking their machines.[7]

Taylor never forgot these years of bitter conflict. They provided motivation for his later work and for his philosophy of "initiative and incentive," which took expression in two famous papers: "Shop Management"[8] and "The Principles of Scientific Management."[9] In the second of these, Taylor reported his famous experiments on pig-iron

[5] Harry H. Goode and Robert E. Machol, *System Engineering* (New York: McGraw-Hill, 1957).

[6] "Taylor's Testimony Before the Special House Committee," *Scientific Management* (New York: Harper and Brothers, 1947), pp. 145-287.

[7] Frank Barker Copley, *Frederick W. Taylor* (New York: Harper and Brothers, 1923), pp. 3-6.

[8] Frederick W. Taylor, "Shop Management" (New York: American Society of Mechanical Engineers, 1903, and New York: Harper and Brothers, 1911). Republished in *Scientific Management* (New York: Harper and Brothers, 1947).

[9] Taylor, "The Principles of Scientific Management" (New York: Harper and Brothers, 1911). Republished in *Scientific Management* (New York: Harper and Brothers, 1947).

loading and shoveling.[10] These are worth recounting here for their relevance to this chapter.

Pig-Iron Loading

At the Bethlehem Steel Company, yard workers were engaged in the performance of various manual jobs, among which was the loading of pigs of iron into freight cars. It would be difficult to imagine a more elementary, common labor job: all that each worker did was to pick up a 92-pound pig, carry it up an inclined ramp, drop it into the car, and return for another and another. For this unskilled work each man was paid $1.15 per day, for which the average quantity loaded was 12½ long tons.

A study of the operation was made with a stop watch, and calculations from these data were so astonishing that the observations were made again with the same result: a pig-iron loader should be able to handle from 47 to 48 long tons per day, almost four times as much as the going rate.

There followed the selection of the little Pennsylvania Dutchman "Schmidt," long famous in the story, who agreed that he wanted to be a "high-priced man" and who thereupon loaded 47½ long tons each day, motivated by the incentive of a new wage of $1.85.

Shoveling

A similar attack was made upon shoveling, another kind of common labor job performed throughout the yard. Each man hired by the company for this work had to bring his own shovel, and, as might be expected, a varied array of different handles and shapes and sizes of scoop was the rule. In this case Taylor and his cohorts first conducted empirical studies to determine the optimum weight of material per shovelful for the different items handled. Their conclusion was that shovels designed to hold 21 pounds of each kind of material would result in the greatest weight moved per day.

Shovels thereupon were provided by the company for each significantly different density of material: sand, slag, rice coal, iron ore, etc., and then the operation was timed in the manner of the pig-iron study. Again, a measured standard performance was determined and a money incentive offered for attainment of the task level. Results were

[10] *Ibid., Scientific Management,* pp. 41-72.

even more spectacular than in the handling of pig iron: the number of yard laborers was reduced from 400-600 to 140; the average number of tons per man per day rose from 16 to 59; the earnings of each worker rose from $1.15 to $1.88 per day; and the cost per long ton of material moved dropped from $.072 to $.033.[11]

Metal Cutting

During these years Taylor and his associates carried on another investigation, almost monumental in its scope, culminating in 1907 in the publication of his famous paper, "On the Art of Cutting Metals." [12] This was an attempt to measure, by controlled experiment, the effect of each of twelve variables upon the operation of metal cutting tools. His objective was to determine that setting of the machine tool which would be optimum for any combination of hardness, diameter, length, depth of cut, etc.

During the course of these experiments, which stretched over a period of twenty-six years, Taylor and a metallurgist, Maunsel White, devised a technique for tempering tool steel, patented the method, and sold the rights to the Bethlehem Steel Company.[13] Tools treated by the Taylor-White method "would do from two to four times as much work as other tools." [14] Taylor also successfully introduced a heavy stream of cooling water at the nose of the tool which "produced a gain in the cutting speed of *self-hardening tools* of about 33 per cent." [15] The Taylor-White process for tempering tool steel has long since passed into limbo, although the use of liquid coolants is commonplace today on metal cutting machines throughout the world.

Criticism

Taylor's work has been subjected to a great deal of criticism and even strong vilification, some of it deserved. Today's industrial society would not tolerate a fourfold increase in labor output for a 60 per cent rise in wages, even when accompanied by the rationalization that the

[11] *Ibid.*, p. 71.

[12] Frederick W. Taylor, "On the Art of Cutting Metals," *Transactions of the American Society of Mechanical Engineers,* Vol. 28, 31-350.

[13] Copley, *op. cit.*, pp. 79-118.

[14] Taylor, "On the Art of Cutting Metals," p. 39.

[15] *Ibid.*

economic gain should be divided among workers, owners, and customers in some equitable proportion.

This circumstance alone would have aroused resentment. When coupled with statements like the following, strong antagonism could be the only result:

> Now one of the very first requirements for a man who is fit to handle pig iron as a regular occupation is that he shall be so stupid and phlegmatic that he more nearly resembles in his make-up the ox than any other type.[16]
>
> When, on the other hand, they receive much more than a 60 per cent increase in wages, many of them will tend to work irregularly and tend to become more or less shiftless, extravagant, and dissipated. Our experiments showed, in other words, that it does not do for most men to get rich too fast.[17]
>
> The full possibilities of functional foremanship, however, will not have been realized until almost all of the machines in the shop are run by men who are of small calibre and attainments, and who are therefore cheaper than those required under the old system.[18]

To academicians, Taylor's greatest offenses were perhaps his pretentious claims to science. With respect to "that class of work in which the limit of a man's capacity is reached because he is tired out," Taylor propounded a "law of heavy laboring" in which he dogmatically declared that "a first-class workman can only be under load 43 per cent of the day. He must be entirely free from load during 57 per cent of the day."[19]

In his work on metal cutting, Taylor and his associates did not foresee the development of multi-variable experimental design and painstakingly followed the classical technique of investigating each of the twelve variables one at a time. The complex relationships emerging from this work[20] do not inspire confidence in its scientific validity and tend to cast doubt upon its conclusions.

Most of all, Taylor has been criticized for arguing that his methods of analysis and measurement, coupled with wage incentives, with his ideas on employee selection, with functional foremanship, and with a unique and undefinable philosophy, resulted in a new kind of "scientific management." Even with the then more prevalent notion that

[16] Taylor, *Scientific Management,* p. 59.

[17] *Ibid.,* p. 74.

[18] *Ibid.,* p. 105.

[19] *Ibid.,* p. 57.

[20] Taylor, "On the Art of Cutting Metals," pp. 107, 159, 163, 195, 205, etc.

"scientific" was synonymous with "systematic," this claim is difficult to accept today.

SCIENTIFIC MANAGEMENT, OPERATIONS RESEARCH, AND SYSTEMS ENGINEERING

Despite these and other criticisms, Taylor's severest critics would admit that his work constituted a great industrial breakthrough. Just as many of the effects upon workers of mass production manufacturing may be decried, so Taylor's system of measurement and incentive may be decried—while we continue to enjoy the economic benefits of both. It is no exaggeration to say that the work of Frederick W. Taylor has had consequences equal in importance to the invention of the spinning jenny, the power loom, and the steam engine, the primogenitors of the Industrial Revolution.

But Taylor's work had importance beyond this, in ways directly germane to operations research and systems engineering. It has been said that the great textile inventions of the Industrial Revolution were not so important intrinsically as they were in creating the idea in men's minds of transfer of skill and intelligence from men to machines. In exactly the same way, one can say that Taylor's contributions, great as they were intrinsically, were even more valuable in revealing, perhaps for the first time, the merit of creating elements of organization whose missions are not the performance of operations but the analysis of them.

There is too strong a tendency to split hairs over physical scientists versus social scientists, to emphasize the importance of the mixed team, the "whole system" approach, the operations research and systems "points of view," and to ignore the significance of the first basic step: the formation of organization for research on operations. It is this which is transcendently important, far more so than the backgrounds of the particular individuals who are to do the work. In this respect, the organization relationship and the organization mission of Taylor and his associates was exactly the same as for those in operations research today, and closely analogous for those in systems engineering as well.

A second attribute of Taylor's work is even less understood. When he studied pig-iron handling and shoveling he emerged with the claim that his measurements defined a "fair day's work" and that his incentive reward was pegged at just the right level to secure optimum output. When all of the twelve variables of metal cutting had been studied and the results synthesized on a slide rule, that instrument was purported to give the *best* setting for any given set of conditions.

These claims are now rejected as being pretentious. Thus, there is a tendency to forget an ever-present managerial requirement: to engage a pig-iron handler or a shoveler, or to turn a job on a lathe, decisions *must* be made on *all* of these things. Twelve and one-half long tons of pig iron *already* had been determined as a fair day's work, not by managerial decision but by tacit acceptance, and $1.15 had been determined as a fair recompense in the same way. Jobs *already* were being run on lathes at feeds and speeds which had to be decided by the operator or by his foreman in order to do the work at all.

The criterion for evaluating Taylor's results, then, should not be the *perfection* he claimed, but only the degree to which his work led to *better* decisions than those which were possible and necessary before. The margin of superiority in these terms of relative goodness was very great indeed.

This point has relevance for both operations research and systems engineering. Elsewhere in this book are chapters on inventory theory, linear programming, and operational gaming, to mention only three of the topics covered. To use the models developed in inventory theory, for example, in order to resolve a practical inventory problem, one must make decisions on such things as holding costs and costs of shortage. What does it cost to store one more item? What does it cost to be without an item in time of demand? Obviously, these are not easy questions to answer. If there is abundant room in the warehouse, the marginal cost of storing additional items will be trivial; if the warehouse is jammed, this cost may be high. If the cupboard is bare, like Mother Hubbard's, the cost of shortage may be zero if the customer is complacently willing to wait, or catastrophically high if he decides to take his business elsewhere.

Similarly, in the development of war gaming, decisions had to be made and expressed quantitatively for movement, firepower, terrain, etc., in order to play the game at all. Obviously, these could not be perfect decisions, any more than those for holding costs and shortage costs could be perfect decisions—or any more than Taylor's decisions on the setting up of a lathe could be perfect.

Does this mean, then, that inventory theory and operational gaming should be criticized and denied a place in the sun, because they fall short of perfection? Not at all. One cannot operate a factory, or a store, or any other enterprise without an inventory. Stores and stock may be kept at minimal levels on a hand-to-mouth basis or they may be provided in abundance; quantities may be regulated carefully or casually but, whether by decision or default, there must be an inventory. And, in exactly the same way, there must be decisions on weapons in order to have a military force at all. The question is: do inventory models and operational gaming provide *better* answers for the decisions which *must* be made? The answer is: they do. Moreover, they provide *very much better* answers. The margin of superiority for these newer techniques can be very great, just as it was for Taylor's techniques in time study, wage incentives, and metal cutting.

There are other points of resemblance between Taylor's work and operations research and systems engineering. Mixed or multi-discipline teams are said to be an innovation important to both operations research and systems engineering. The concept is important, but one cannot be sure that it is new. Taylor had at least twelve colleagues in his metal working studies and had this to say about three of them:

> Mr. White (Maunsel White) is undoubtedly a much more accomplished metallurgist than any of the rest of us; Mr. Gantt (H. L. Gantt) is a better all around manager, and the writer of this paper has perhaps the faculty of holding on tighter with his teeth than any of the others. . . . Mr. Barth (Carl G. Barth), who is a very much better mathematician than any of the rest of us, has devoted a large part of his time . . . to carrying on the mathematical work.[21]

Surely, this is a mixed team, additionally blessed by the leadership of one capable of "holding on tighter with his teeth."

Operations research also espouses the development of mathematical models, representative of the operation under study. Although the notion may be fanciful and the point strained in the making, the slide rule developed by Carl Barth (Figure 1) appears to be a mathematical model, too, representative of the operations it portrays.

These points of resemblance, however, are of lesser significance. The important things are the creation of organization whose mission is research on operations and understanding of the fact that such organization elements can provide much better answers for decisions which must be made.

[21] Taylor, "On the Art of Cutting Metals," p. 35.

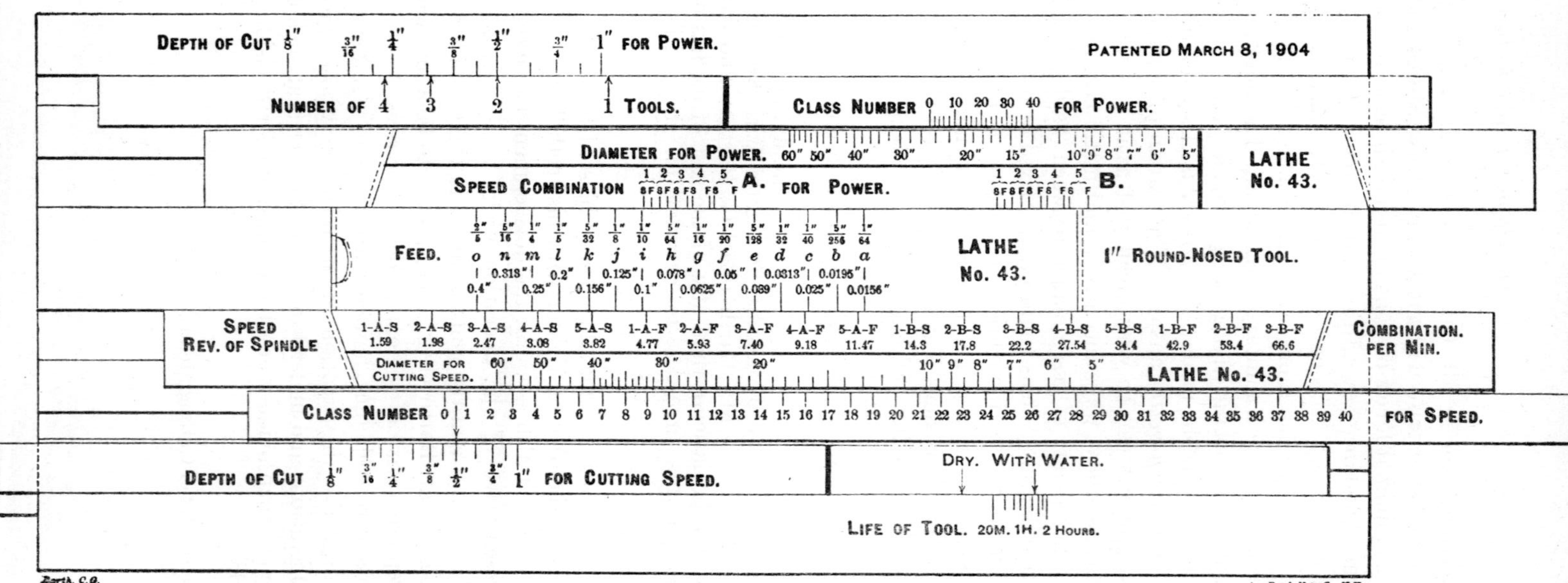

Figure 1. Barth slide rule.

THE DEVELOPMENT OF OPERATIONS RESEARCH

Taylor's work has been followed by a number of other impressive aids to managerial decision-making: by the development of accounting techniques, by Gantt's chart,[22] Gilbreth's micromotion studies,[23] Shewhart's statistical quality control,[24] by the emergence of aptitude tests, and, in a somewhat more qualitative sense, by the dramatic Hawthorn Experiments.[25] World War II brought about a further contribution, born in Great Britain of military necessity, and there called operational research.[26]

Like most things of this kind, a precise beginning is hard to define and so also is the moment of conscious realization that a new and different concept is at work. "In 1939, at the outbreak of the war in Europe, there was a nucleus of a British operational research organization already in existence,"[27] and their contributions quickly were followed and augmented in various important ways: in improving the early-warning radar system, in anti-aircraft gunnery, in antisubmarine warfare, in civilian defense, in the determination of convoy size, and in the conduct of bombing raids upon Germany.[28]

For some of this work Professor P. M. S. Blackett of the University of Manchester, a Fellow of the Royal Society and a Nobel Laureate, assembled a group which ". . . included three physiologists, two mathematical physicists, one astrophysicist, one Army officer, one surveyor, one general physicist, and two mathematicians. 'Blackett's circus,' as the group was called, was soon able to demonstrate the value of the mixed-team approach to operational problems."[29] Thus was born the mixed or multi-discipline team which has become a

[22] H. L. Gantt, *Organizing for Work* (New York: Harcourt, Brace and Howe, 1919).

[23] F. B. Gilbreth, *Motion Study* (New York: D. van Nostrand, 1911).

[24] W. A. Shewhart, *Economic Control of Manufactured Product* (New York: D. van Nostrand, 1931).

[25] F. J. Roethlisberger and William J. Dickson, *Management and the Worker* (Cambridge: Harvard University Press, 1939).

[26] Florence N. Trefethen, "A History of Operations Research," *Operations Research for Management,* eds. Joseph F. McCloskey and Florence N. Trefethen (Baltimore: The Johns Hopkins Press, 1954), Vol. I, 3-35.

[27] *Ibid.,* p. 5.

[28] *Ibid.,* pp. 5-10.

[29] *Ibid.,* p. 6.

distinguishin[illegible]racteristic of both operations research and systems engineering.

These notable accomplishments in the British military had their counterparts during the war in our own armed forces, in contributing to the effectiveness of the Air Force, in antisubmarine warfare, and in the sea mining of Japan, to mention but a few.[30] The success of these endeavors is evidenced today by the flourishing continuance of operations research organizations in all departments of our armed forces: the Operations Analysis Group (OAG) with the Air Force, the Operations Evaluation Group (OEG) with the Navy, and the Operations Research Office (ORO) with the Army.

More recently, again with leadership shown by the British, operations research has spread into business, industry, and civil government. Operational research groups exist in Great Britain for the iron and steel industry, for coal, road and rail transport, textiles, agriculture, brickmaking, shoes, etc., most but not all of them under government sponsorship, in the civil service.

In the United States, operations research groups have worked upon problems in hospitals, department stores, supermarkets, railroads, newspapers, toll bridges, electric utilities, and petroleum refining, to mention but a few. Results have varied from notable to nothing, but there is a growing and impressive record of accomplishment.

Other developments also are worthy of mention. In Great Britain the Operational Research Society of the United Kingdom has been formed and in 1957 served as hosts to the First International Conference on Operational Research at Oxford, England. Co-sponsors in this enterprise were the Operations Research Society of America and the Institute of the Management Sciences. Delegates to the International Conference represented twenty-one nations. Of these, seven reported comprehensive operations research activity, with eleven indicating more modest development.[31] Clearly, whatever operations research may be, it is flourishing.

Two other developments in the United States have not been duplicated in Great Britain, at least not in the same way nor to the same degree. These are the growth of academic progress in the universities

[30] *Ibid.*, pp. 12-20.

[31] *Proceedings of the First International Conference on Operational Research* (Baltimore: Operations Research Society of America, 1957), pp. 400-513 and 522-26.

and the formation of operations research groups by at least four large consulting firms. Courses germane to operations research have been offered at various British universities (e.g., Birmingham), but there is no counterpart of the advanced degree programs offered by American institutions such as Case, M.I.T., Michigan, Johns Hopkins, and others. British consulting firms, unless the development has been recent, do not have counterparts of the operations research groups in such American firms as Arthur D. Little, Inc.; Booz, Allen and Hamilton; Haskins and Sells, etc. Perhaps this is because of the strength of the British groups in the civil service.

SYSTEMS ENGINEERING

Now that operations research has been brought up to date, so to speak, it is fair to speak of systems engineering. Doing so is more difficult than for operations research, because systems engineering does not appear to have had as well defined a birth and development as operations research, at least not under the name systems engineering. It has, however, aroused much the same kind of controversy.

Just as some have declared that operations research is "nothing but industrial engineering," so there are those who feel that systems engineering is "nothing but people who can draw block diagrams," or, more specifically, it is the "engineering process" dressed up in a new name. One such comment is worth quoting:

> The Curtis Report, which is the blue print being followed by the Air Modernization Board, defined systems engineering. The specific elements of the systems engineering process are:
>
> 1. Defining the problem.
> 2. A synthesis of postulated solution.
> 3. Test of solutions.
> 4. Development of components.
> 5. Test of the system with the developed components.
> 6. Establishment of the system.
> 7. Operation of the system.
> 8. Analysis of the system performance.[32]

Commenting upon this in a private communication, Dr. Willis Gore

[32] *Electronic News,* May 12, 1958, p. 18.

says, "This represents, I believe, the understanding that the greatest number of people have when they use the term. I believe this definition amounts to a statement of the 'engineering' process and that the term 'systems' is redundant. This is not what some people are trying to make it mean."

That the term "systems" is redundant will be sharply disputed by some, who are sensitive to the distinction between solving *component* problems and *systems* problems and who believe in something called the "systems point of view." This, of course, is exactly analogous to those who believe in that nebulous thing, the "operations research point of view." Whether or not these beliefs have real substance and meaning will be considered in a moment.

Still others take a much more grandiose view of systems engineering *vis à vis* operations research, comprehending the latter merely as part of the former. One member of the faculty at U. C. L. A. has stated that their catalog gives the subhead "Systems Engineering (Operations Research)" and has declared that operations research does not yield verifiable results, while systems engineering does.[33] On this same theme Goode and Machol have this to say:

> . . . operations research is a very broad field. It is not like the other tools of system design, because it is not well defined and because it includes many of these tools. In fact, there has been a recent tendency to broaden the definition of operations research so that it is practically synonymous with system design. However, there is a fundamental difference in approach: The operations analyst is primarily interested in making procedural changes, while the system engineer is primarily interested in making equipment changes.[34]

This, it seems, gets at the nub of the difference. The operations research team is concerned with operations *per se,* and is more likely to be concerned with operations in being than with operations in prospect. Systems engineers are concerned with operations, too, but are more likely to refer to them as man and machine systems and are much more likely to emphasize the machines than the procedures by which the machines are used. Furthermore, systems engineers are more likely to be engaged in the design of systems yet to be, rather than the operation of systems in being.

In a certain sense, then, operations research and systems engineer-

[33] Comment by Dr. Alexander W. Boldyreff at a Conference on Education for Operations Research, Cornell University, April, 1958.

[34] Goode and Machol, *System Engineering,* p. 130.

ing *are* the same. Both engage in the analysis of complex man and machine systems or, one may also say, man and machine operations; both utilize multi-discipline teams; both employ the scientific method; both emphasize the "whole system" rather than the component approach; and, above all, both are staff elements of organization whose mission is the analysis of operations.

The differences are much less important than the similarities. Historically, those who have created the proximity fuse, a large part of the telephone system, and the guided missile have come to call themselves, and to be called, systems engineers. They have come from many fields but, very generally, tend to have an electronics-communications-servomechanisms-human engineering-design quality. On the other hand, those who planned the sea mining of Japan, who created the massive ship convoy, and proposed the integration of Negro soldiers in the armed forces have come to call themselves, and to be called, operations researchers. They, too, have come from many fields but, again very generally, have tended to have a somewhat different aura, this time of mathematical models, stochastic processes, statistics, probability, economics, and behavioral science.

These are hardly adequate connotations, but they do make a point. The differences between operations research and systems engineering lie more in the people who do the work than in concept, philosophy, or procedure.

EVALUATION

If operations research and systems engineering are conceded to be similar to each other, does this still mean that they are at the same time different and new? Or are they, in their resemblance, "nothing but" industrial engineering, marketing research, those who can draw block diagrams, or merely the "engineering process" in a new dress?

There are two answers to these questions: (1) operations research and systems engineering, beyond industrial engineering and marketing research, beyond the drawing of block diagrams, have attracted capable scientists and scholars to the study of "action problems" in the "real world," in the world of affairs; (2) there *is* something new and different in the multi-discipline team and the system, as opposed to the component approach.

This attraction of scientists and scholars to the world of action and decision is an important, not a little thing, and it will have consequences just as significant as the attraction of engineers to the study of industrial production after the work of Taylor, just as significant as the principles of transfer of skill and intelligence, consciously realized and exploited in the eighteenth and nineteenth centuries.

The assertion that there is something different and better in the multi-discipline team and the whole system approach is equally important but impossible to prove. Creation of a guided missile will require expertness in thermodynamics, hydraulics, communications, servomechanisms, metallurgy, structures, fluid mechanics, chemistry, for its component elements: engine, fuel, fuel system, radio, shape, structure, and launching mechanism. Is the assemblage of such experts into a *team* for the purpose of making a superior *missile* better than independent work aimed at making superior *component elements?* It is, and those who have worked in such a way will so testify. There is something more in the team idea than merely the "engineering process."

HAZARDS

The statement, just made, that the attraction of scholars and scientists to the world of affairs will have profound consequences augurs well for the future of operations research and systems engineering, contrary to the opinions of Bronowski and Machol already stated. Despite this bright future, however, difficulties lie ahead, some of which may lead to disrepute and some to frustration.

Taylor's work in methods analysis, time study, and wage incentives quickly was followed by the emergence of "efficiency experts," who, lacking professional competence, had no real understanding of what Taylor had done and applied his "system" willy-nilly as a cure-all. Reputable industrial engineers have paid dearly for the opprobrium brought to the profession by these quacks.

Operations research faces precisely this same danger. Just as methods study, time study, and wage incentives were interpreted as the essence of Taylor's work, rather than the broader concept of staff organization for the analysis of operations, so linear programming, inventory theory, queuing theory, game theory, are very likely to be interpreted as the *meaning* of operations research, rather than merely as the tools they are.

Let a designated operations research group look at a traffic problem and see only the queue, to the exclusion of broader strategic possibilities, and they will be the exact counterparts of the efficiency expert who applies time study to every operation he sees. The danger lies in preoccupation with mathematical models, in forgetting that numbers often come from value judgments, in mistaking methods and techniques for substance. The danger is a very real one; operations research and systems engineering are likely to have their efficiency experts, too.

To the extent that operations research seeks to change procedures, as stated by Goode and Machol, and to the extent that these procedural changes alter the habits of people, operations research workers face hostility in the conduct of their studies and in the implementation of their findings and recommendations. The notion that a group of outsiders, themselves incapable of performing an operation, can tell veteran, expert operators how to do better is, as stated elsewhere, preposterously contrary to our own notions about ourselves.[35]

Hostility from this source is augmented by the history of Taylor and his less perceptive followers and by the imputation of criticism which is often involved in the act of selecting, from among several, a particular operation to be studied. If operations research groups are to avoid frustration and a sense of futility; if they desire to see their proposals carried out, they must understand the hostility against which they sometimes work and learn to practice the art of persuasion to the n-th degree.[36]

SUMMARY

In the last half of the eighteenth century the invention of textile machines, accompanied by successful invention of the steam engine, gave rise to what has been known since as the Industrial Revolution. Other, equally notable inventions, such as the art of printing from

[35] Robert H. Roy, *The Administrative Process* (Baltimore: The Johns Hopkins Press, 1958), p. 74.

[36] *Ibid.*, Chapter VIII.

movable types, had long antedated these innovations but had consequences of a different kind. The mold and matrix for casting type and the press for printing it involved the transfer of the skill and intelligence of the scribe to the machine, in exactly the same way as the spinning jenny and the power loom later transferred the skill and intelligence of the handicraft spinner and weaver to these newer machines. Yet printing, while it helped to foster an intellectual revolution, the Renaissance, did not bring with its invention an understanding of these important principles. Three centuries were to pass before the great textile inventions spread conscious knowledge that the transfer of skill and intelligence from man to machine could have important consequences.

In a comparable way, Eli Whitney's demonstration of the principle of interchangeability, shown to be feasible in the manufacture of muskets, required a century before three Cadillacs could be disassembled and reassembled without selective fitting, before there was widespread understanding of the value of the concept.

The purpose in drawing these analogies is to argue that the most important concept revealed by the work of Frederick W. Taylor is that of assigning to specific elements of organization the mission, not of performing operations, but of analyzing them. It is this which is the mission of both operations research and systems engineering. It is this, *as a concept,* which is transcendently important. It is my belief that executives in organizations everywhere are now on the threshold of widespread understanding of the value of organization for operations analysis. It is this, more than any other single thing, which gives high promise for the future of operations research and systems engineering.

Within the framework of this organization concept, both operations research and systems engineering utilize multi-discipline teams, who employ the scientific method and are, in the ideal state, imbued with the spirit of teamwork and the system, as opposed to the component, approach. Both have attracted scholars and scientists to problems in the world of affairs, to the enrichment of scholarly learning and experience, and to the attainment of better knowledge for decisions which must be made.

Between operations research and systems engineering there presently are differences in the kinds of people who comprise the respective fields, in the kinds of problems they attack, in the methods and tools they employ, in the ways they think about themselves, and in the objectives they pursue. But concern over procedure versus concern over equipment, devotion to operations versus devotion to man and machine systems, study of operations and systems in being versus study of systems and operations in prospect, all these are less important than the main identity: the organization assignment of multi-discipline teams of scientists and scholars to the mission of operations and systems analysis by the methods of science. If this is the "education of scientists . . . in a new empiricism," so be it. It is still new, it is different, and it is important.

Three

OPERATIONS RESEARCH IN THE WORLD CRISIS IN SCIENCE AND TECHNOLOGY[1]

ELLIS A. JOHNSON

INTRODUCTION

The title of this talk was chosen a year ago, because of deep concern with the extraordinarily fast progress of the Soviet Union in the development of superior weapons systems. It was clear then that they had already surpassed us in some areas of military development and would inevitably surpass us in all other areas if the trends that existed at that time continued.

As it now stands, the USSR is clearly our equal in the quality of atomic weapons, two years ahead of us in missiles, and generally our equal in other areas. This is disappointing and dangerous to us, but is not necessarily crucial. What will be fatal to us within a decade, unless the President acts forcefully now, is the fact that Soviet technical progress is 50 per cent faster than ours because they can develop a new weapon in five years while it takes us ten years. The Western World in particular, but all of humanity in the long run, is faced with a crisis which will become increasingly severe in the management of man's increasing knowledge in the physical sciences and life sciences, in peace as well as in military affairs.

[1] Reprinted, with some additions, by permission from *Operations Research*, Vol. 6, No. 1 (January-February, 1958).

This chapter will discuss the relation of science to war and to human welfare, the new instability that increasing physical knowledge has brought to the management and control of war and of human welfare, and the difficulty of choosing in peace as well as in war the methods of managing development and its supporting applied research. It also will outline the need for long-range planning, not the rigid planning of communism or socialism, but a new kind of long-range planning suited to a democratic culture. Beginning with this decade, long-range planning becomes increasingly important to the individual and to the family, to the company in industry, to the city and state, as well as to the nation, and indeed to all of humanity. Finally, the role of operations research in long-range planning will be presented.

It is first necessary to establish certain premises, with some supporting evidence, to lay the foundation for conclusions which are intended to be interesting and convincing.

The first of these is the assumption that there is an important relationship between the per capita energy consumption in a given country and the per capita national income, i.e., the national welfare of the country. The support for this assumption is shown in Figure 1. The source is from United States statistics. Although there is a considerable scatter, there is no doubt that per capita energy consumption is correlated with per capita national income. The hypothesis is that there is a causal connection and that it is the most important factor in improving the welfare in each country. Upon examination of the countries with high living standards, it becomes clear that there is no direct correlation with national resources. The relationship appears to be rather with the organization of society towards use of the maximum amount of energy per person. This seems reasonable. The pursuit of high per capita levels of national income is deeply dependent on man's control over energy, over space, and over time; that is, over the use of power for peaceful and military purposes, the use of air, land, and sea vehicles for transport between the nations of the world, and the development of all of the methods of communications between the peoples of the world, as well as within each society.

The second hypothesis is that in the future man's ability to use increasingly greater amounts of energy in order to increase his welfare and his control over the environment depends on his effort in research and development.

In Figure 2 there is shown the relation between the percentage of moneys spent by the governments of various countries on research and

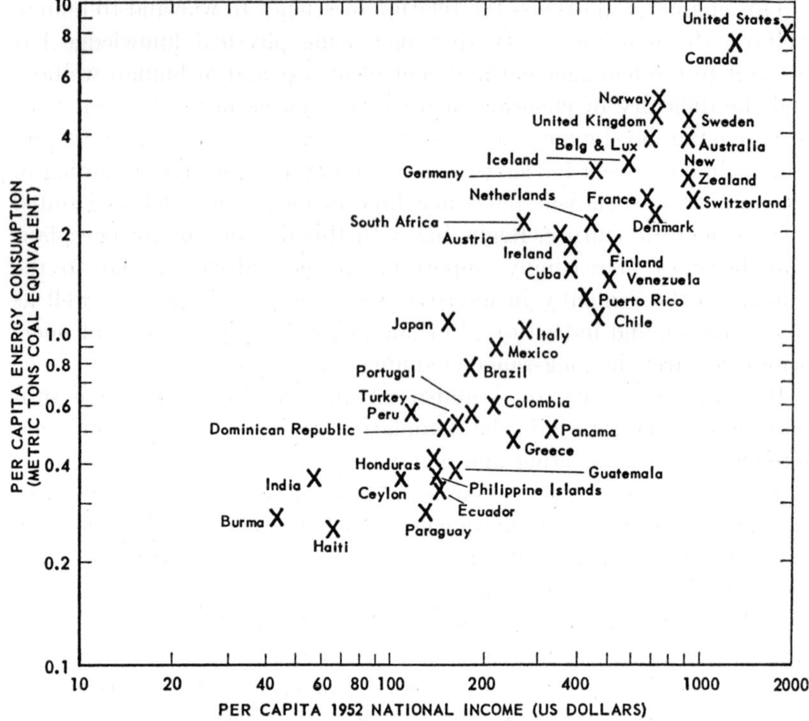

Figure 1. Per capita national income and energy consumption.

development and the per capita national income of each. It is clear that there is a direct relation, and the hypothesis is made that the relation is a causal one and that high expenditure on research and development is one of the important ingredients that brings into being a high per capita national income.

It seems sensible; and, in fact, if the ability of man to produce to his own benefit depends on the energy he commands, in turn there can be no doubt that to bring into being and to command atomic power, electric power, Diesel generated power, or modern steam generated power—as well as the many machines, tools, vehicles and motors that use power—requires increasingly greater effort on the part of scientists and engineers.

In early days, perhaps up to a decade or two ago, many great inventions could be made without much direct relation to science and by

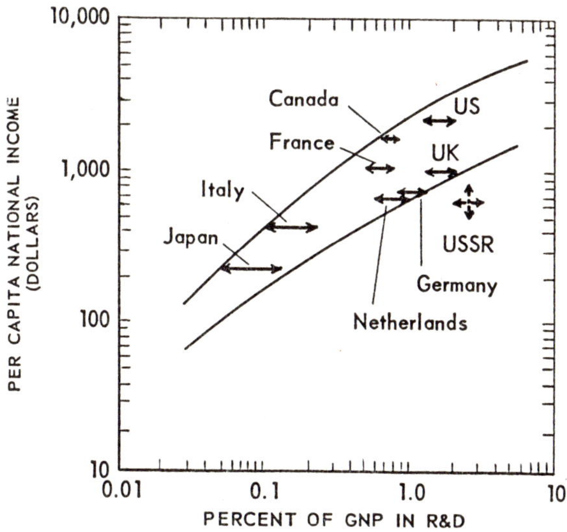

Figure 2. Per capita national income versus per cent of gross national product devoted to research and development.

people without scientific or engineering education. Thus our command over natural resources did not depend uniquely upon science. But because this was so in the past, it does not follow that we can extrapolate this pleasant and easy pattern into the future. For example, Figure 3 shows the increasing consumption of electric energy in the United States from 1900 and extrapolated to the year 2000. In this century the amount of power used in the United States will have increased by a factor of about 15,000 and by nearly as much per capita. It is true that the steam engine could be invented, coal mined, and easily located petroleum found and used without scientific aid, but there is no possibility that our continued development and control of these large amounts of energy can be provided without a sophisticated use of science. Seventy-five per cent of the electrical energy in the year 2000 will need to come from nuclear fuels. There are 1500 Q units of nuclear fuels available compared to only about 30 Q units of other kinds of fossil fuels in the United States. The nuclear fuels require technologies which can be understood and placed under the command of man by only the most skilled and well-educated scientists and engineers. It is possible that the utilization of solar energy may require

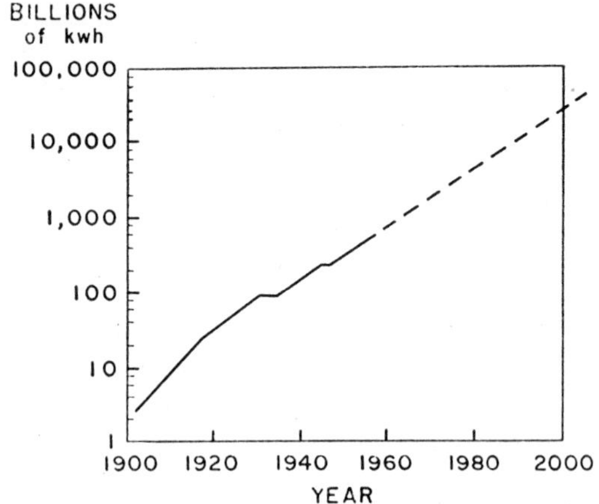

Figure 3. Generation of power by U.S. utilities industry.

less scientific sophistication, but one doubts that this will be the case. It may be that the biological sciences will be more important in the use of solar energy than the physical sciences.

Thus the relations of the physical sciences and of engineering have changed with respect to invention, and in a direction that now requires a long and difficult apprenticeship before the individual can contribute to man's control over energy, space, and time.

It is not only science and its inventions that are important to man; it is the rate at which these occur that leads to crisis. As an example from the military art, we may note that in earlier years weapons systems had a very long life. This was a period in which development was slow and the same kind of weapon remained useful over many decades. Until about 1000 A.D., weapons had a useful life of about 400 years; from 1000 to 1500 A.D., about 100 years; from 1500 A.D. until about the beginning of the twentieth century, a useful life of about 50 years. But today, weapons systems have a life span in the range of about three to seven years. In fact, today, all weapons systems tend to be obsolete by the time the first units come off of the production line.

It is clear that our increased well-being requires technical change, but the problem of dealing with this high rate of technical progress becomes incredibly difficult. A similar acceleration has occurred in our own lives during the more normal times of peace. Many of us began our lives when the horse provided the primary motive power for local transportation, when there were no airplanes, radio, TV, mechanical washing machines, dish washers, or electric lights, and indeed this is hardly believable to our own children. The questions we need to ask, then, are "How has this come about, and will it continue, and if it does, can we deal with it?"

THE INCREASE IN PHYSICAL KNOWLEDGE

For the last 250 years each generation has met with a much greater effect, in its time, of growth of knowledge in the physical sciences. The sociological impact of this physical knowledge has been far more striking than the more muted problems posed by the nature of man. Difficulties of a purely behavioral or social nature have at no time presented any serious surprise, even though the specific problems have been serious. We are more likely to be surprised if such difficulties, including war, do *not* occur. The fact that there are increasing difficulties in management has therefore been more natural, overall, than the particular difficulties connected with the digestion and use of physical knowledge.

Exponential Growth of Knowledge

There is adequate evidence that knowledge in the physical sciences has been growing at an exponential rate since 1700. Such knowledge appears to be doubling every fifteen years, in contrast with knowledge in the social and management sciences which appears to be doubling at the much slower rate of once every fifty years, or at about the same rate as the population of the world. The slow rate of growth in social science is due in part to its meager support. For example, only about 2 per cent of US research and development funds are spent on social sciences.

Figure 4 gives a crude measure of the growth of world-wide scientific knowledge, which is primarily in the physical sciences, and shows the growth in the number of scientific journals, including abstract journals. There are now about 100,000 scientific journals, accompanied by about 300 abstract journals which attempt to enumerate the great

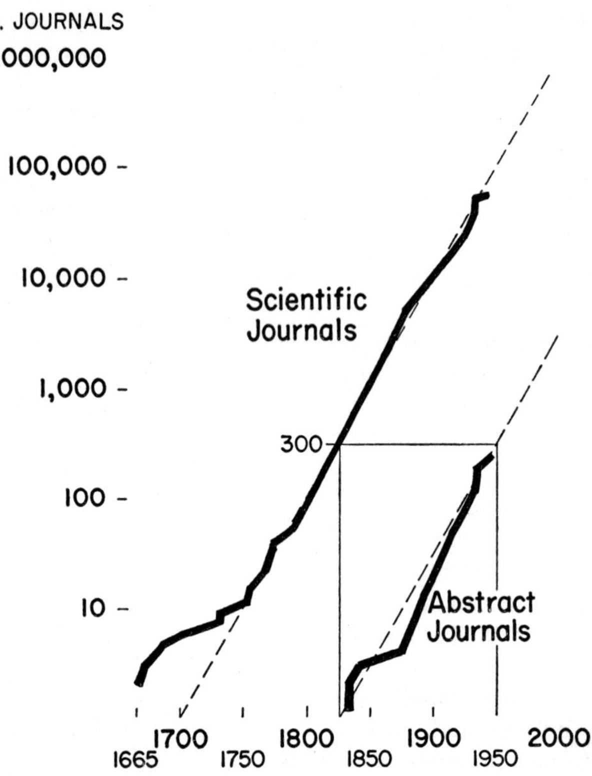

Figure 4. Growth of scientific and abstract journals.

amount of knowledge continuously generated in the original articles.

The practical effect of this rapid growth of physical knowledge has been of great importance in every country. In the US, for example, the amount of money spent annually on research and development has increased from about 166 million dollars a year in 1930 to more than five billion dollars a year in 1953, and twelve billion in 1960, or fifteen-fold in 23 years in constant dollars. Thus, the cost of research and development in the US has doubled every six years. The number of active scientists in the US has increased from approximately 46,000 in 1930 to 250,000 in 1956, a five-fold increase.

THE EFFECT OF TECHNOLOGY ON INDUSTRY AND ITS PRODUCTS

The increase in physical knowledge has practical effects on our society in both peacetime and in war. In peacetime there is no longer any product stability. The "one horse shay" remained good modern practical transportation for a century. Can you imagine using a 1915 automobile as good practical work-a-day transportation in 2015? Nowadays the things we use in our daily life change every year and, in fact, almost every day and every week. For example, the General Electric Company, where "progress is our most important product," estimates that 40 per cent of the electronics products being produced ten years from now have yet to be invented. Large companies such as G.E. can survive in this unstable production world because of their size and diversification and their command over the advertising media that persuade us to buy new products.

The problem is much more difficult for small companies whose size does not permit extensive diversification. They must seek another kind of production flexibility, for, if they produce one product today, it may be either out of fashion tomorrow or the market may be saturated by competition. A small company, then, must be fast on its feet, and must be able to switch rapidly from one product to another. This makes survival difficult and puts a premium on flexibility and ability to change to new products, rather than on the development of long-time skills in a single "stable" product.

There are, of course, cultural constraints which ameliorate the peacetime effects of new developments. Monopoly industry and cartels have tried to enforce a kind of cultural stability so far as production is concerned. For example, sewing machines remained essentially stable and unimproved until tariff changes permitted the importation of foreign sewing machines, and this in turn brought about tremendous improvements in US machines.

Another kind of constraint reflects the repercussions when massive investments have to be "thrown away" with the introduction of great improvements. General improvements in railroad transport, for example, have been constrained on this account. In spite of these constraints, new developments usually cannot be put off too long. Otherwise our physical knowledge is used to improve some other form of competition. Modern air transport, for example, has taken away half of the railroad passenger traffic, and buses another part. The effect of new products thus is strong on each of us as an individual, and the value of our investments is uncertain, even if we choose very carefully.

For young people there is a serious problem in education, because it is no longer possible to choose a profession, art, or craft and to feel that one can be educated for a whole lifetime. The question is—education for what? For the physical sciences or engineering? For the biological sciences? For those who enter into the crafts, the problem is equally difficult. As products change, skills must change with them, and training to service new products becomes increasingly difficult.

The effect of this rapid change is to make our lives uncertain in work, in earning capacity, and in prestige. This effect will become more and more severe.

THE EFFECT ON MILITARY DEVELOPMENT

It is in our military competition with the Soviet Union that we have a forecast of the crisis that will be an everyday problem, even if we were to have a world at peace.

To begin with, we must accept the observed fact that the Soviet Union can fully develop a new weapons system in five years but that the same task takes us ten years.

The effect of increasing physical knowledge on the cost of weapons in a weapons system has been very great in terms of money and complexity. For example, the increase in cost of aircraft with time is shown in Figure 5. It can be seen that this cost has increased ten-fold from 1945 to 1955. As would be expected, the over-all increase is a function of the cost of important elements of the weapon.

The cost of electronics in fighter aircraft has increased from $3,000 in 1939 to $300,000 in 1954. Other factors than cost are increased by technical advance. In bombers, the weight of the bombsight has increased from 125 pounds in 1940 to 2,000 pounds in 1955. Each pound added to equipment increases the gross weight of the bomber as much as ten pounds overall. Affected also are complexity and reliability. In aircraft gas turbines the number of parts has increased from 9,000 in 1946 to 20,000 in 1957. Of precious engineering hours, 17,000 were required to produce a fighter aircraft in 1940 and 1,400,000 in 1955.

All of these technically caused increases, rising at a tremendous rate

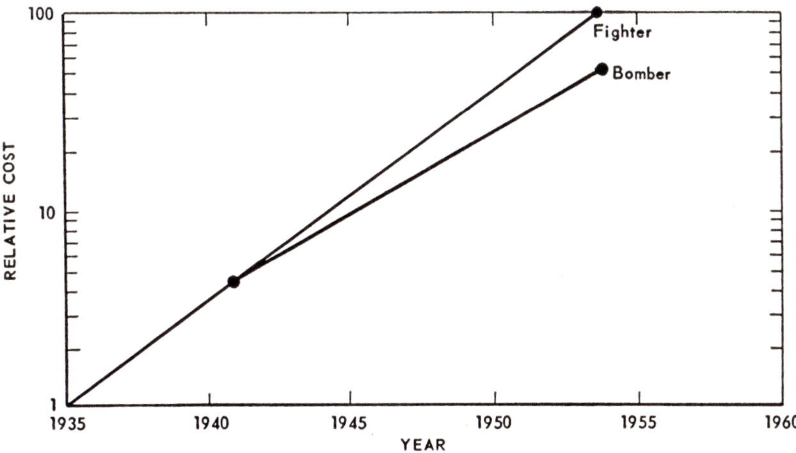

Figure 5. Increased cost of aircraft with time.

as a function of time, stem from the basic fact that knowledge in the physical sciences is itself increasing exponentially. This is very well illustrated by the number of choices available to the designer in 1935 in the design of a major bomber weapons system. Then there were on the order of two major choices available to the designer; in 1955 there were more than 360 choices available.

If we assume that the number of technical choices is proportional, *on the average,* to the amount of physical knowledge, one can hypothesize that the number of choices available to weapons designers will be a Gamma or factorial function of the amount of knowledge existing at a particular time. If the number of choices is equal to some function of the knowledge existing in the physical sciences and is represented by two in 1935, then we can calculate the number of choices which are equal to factorial N where N doubles every fifteen years as a function of time. This has been done in Figure 6. It can be seen that the calculated curve corresponds reasonably well with the observed value for 1955. We can also deduce from this curve that our situation is indeed unstable in attempting to find a solution to the problem of attack or defense, when the actual number of choices is now multiplying at such a tremendous rate.

The explanation for the observed increase in cost and complexity in attack and defensive systems is that they result from the very existence

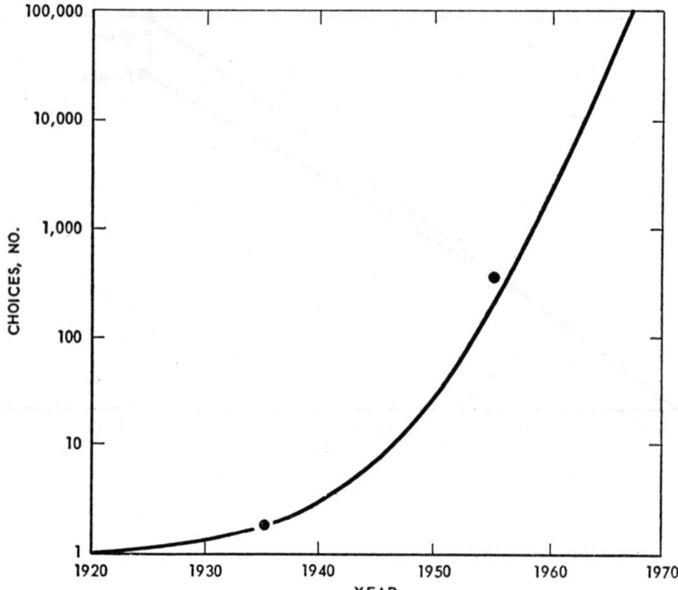

Figure 6. Increase in technical choices with time.

of such a large number of choices and upon the regenerative interaction between attack and defense. When an enemy attack system appears, the defense reacts by using more modern technology to reduce the effectiveness of the attack system. Some years after the original attack system is first established, the attack system itself reacts to the improved defense by choosing, from amongst the many new alternative possible improvements that have appeared in the interim, one or more alternatives that will produce a new attack system of increased effectiveness, inevitably of increased complexity and cost.

In a second cycle, the defensive system in its turn reacts again from an even greater increase in the number of choices available. The cycle repeats itself continuously, and the attack and defense systems acquire the actually observed increase in complexity and cost needed to maintain parity between offense and defense. The executive is often almost a helpless observer in this process, which results from technological competition and the executive's expressed desire for the "best" or "most effective" weapons system. Once loose, such forces of technology are not easily turned off or reharnessed.

Table 1. Doubling Periods in Technical Systems

	Years
Scientific Knowledge	15
Management Knowledge	50
Cost of US R & D	6
Cost of US Fighter Electronics	3
Man-hours of Labor for US Fighter Aircraft	3
Pounds of Aircraft Electronics	3
Complexity	10

Table 1 gives the estimated doubling rates for various factors in weapons systems. These questions arise: Do we necessarily have to let ourselves be driven to more and more expensive choices? Can we not find more timeless solutions in cost in which the military budget does not have to double every few years? Let us next examine these questions.

This argument and the supporting data have been concerned with a particular weapons system, that of air warfare. However, the same mechanism operates in the grosser aspects of competition between the development of weapons systems which serve competing major tactics and strategy. Thus, the development of nuclear weapons with which we are now so vitally concerned is equally the result of the exponential increase in knowledge. Such weapons replace not only conventional high explosives but many other forms of strategic and tactical weaponry, and have led to new strategic forms of attack.

We can postulate that such gross improvements in weaponry require a certain number of "bits" of knowledge. For example, the overall knowledge required to develop nuclear weapons was generated primarily in the period from about 1900 to 1940; we will postulate that this period generated about one "bit" of knowledge. Accepting this postulate, another "bit" of knowledge was generated between 1940 and 1955. Concerned primarily with the mastery over space and time and continued improvements in the mastery over energy, the revolutionary effect of this new knowledge, generated in about fifteen years, has certainly been about the equivalent of the "bit" generated in the preceding forty years.

Extending our observations, we can expect two "bits" of knowledge to be generated in the period from 1955 to 1970, thus exacerbating the instability which has resulted from our inability to manage competitively the rapid increase of knowledge in the physical sciences.

The Effect of Lead-Times and Obsolescence Policy

From Figure 6 it seems clear that the conservative management policies of the last half of the nineteenth and the first half of the twentieth century were generally adequate to deal with the small number of choices available to the scientist and the engineer. For example, in aircraft systems up to 1935 there were probably not more than one or two different choices available at any particular time, and management did not have to do much head-scratching. During these three decades, new weapons systems choices in aircraft developed at such a slow rate that management could cope with *technical* progress without too much difficulty, since there was adequate time to argue, coordinate, and choose. Then a difference of a year or two, or even a factor of two, in lag time between two enemies was of comparatively less importance than the eventual use of large masses of materiel, since with reasonable diligence there was not likely to be any tremendous difference in military effectiveness that could not be compensated for by overwhelming production.

Because war can now be decided in days instead of years, it is argued that this is no longer the case; that relative military power now depends more importantly upon making choices in a minimum time and in achieving the minimum possible lead-time between decisions and the availability of a weapons system for use. At present it appears that Soviet lead-time between concept and use may be about five years, on the average, while the US lead-time may be about ten years.

In examining the effect of the disparity in lead-times in the competition between a Soviet air attack system and a US air defense system, let us agree to use the curve of Figure 6 as an accepted measure of the technological choices available to the two sides. Let us now assume that the Soviet Union decided upon a particular weapons system in 1955. At that time there were about 200 choices available from Figure 6. It is reasonable to assume that the Soviet Union is able to keep its technical choices secret from us from one to two years. Under these circumstances, it would be 1957 before we become aware of the probable characteristics of the forthcoming Soviet weapons system, which is scheduled by them to become fully operational in 1960. Then suppose that we react by making a decision in 1957 to counter this new Soviet attack system by an adequate defense system within our economic capability. At that time we will have 500 choices available, which gives us an improved situation with respect to the Soviet system generated two years before, despite the fact that the Soviets will be able to incorporate at least some of the improvements generated in the interim. There are, however, serious limitations as to how many improvements can be actually incorporated by the USSR if the original Soviet schedule is held to. However, with a ten-year lead-time, the US defense system will not come into being as an operationally effective system until 1967—seven years after the Soviet capability first became operational.

We now have to consider obsolescence policy on the part of the Soviet Union. The meager data bearing on this problem indicate that the Soviet Union, and the US in air-attack systems, have policies such that weapons systems have an operational life of between five to seven years. In our case, therefore, the Soviet weapons system, which becomes operational in 1960, will have been phased out sometime between 1965 and 1967 and will have been replaced by a new weapons system based upon the technology of about 1960 and 1962, and the US defensive technology will thus come into being *after* the Soviet attack system which it was designed to counter is superseded by a superior Soviet attack system. Furthermore, the new Soviet attack system which will come into being between 1965 and 1967 will be based upon the technology of 1960 to 1962, at which time there will have been available roughly 5,000 choices, as compared to the 200 choices available to them at the time of their design of the original system. Under these circumstances, the US defense system will come into operational being against a technologically superior attack system, against which it may have a negligible or very reduced effectiveness and indeed may be completely ineffective. Again, this is an argument in detail.

It is perhaps obvious that a particular weapon cannot usually be improved indefinitely. Instead it approaches a technological limit in effectiveness after a period of reasonably intensive development and needs marginal improvement thereafter primarily to maintain production feasibility (in the use of new or currently available raw materials) and efficiency. Sometimes the increase in scientific knowledge permits a bigger spurt in the improvement of the weapon, but this is rare (the rifle is an example of an essential, good, but stable weapon). Weapons systems, often built around a major weapon or tactical concept (the manned bomber is an example) also evolve in this sigmoid cycle. This does not for long limit the practical application of increased scientific knowledge which turns to alternative weapons systems (to the ICBM as a replacement for the bomber).

If we now return to the grosser argument involving "bits" of knowledge, we can see that the designers of the 1960-1962 attack systems of the Soviet Union will have the advantage of about one-half an additional "bit" of knowledge which should be sufficient to permit them to generate a system which will make obsolete defensive systems conceptually generated prior to 1957, at the time the US made its defensive decisions.

The arguments given above make it crystal clear that if physical knowledge is advancing as fast as appears to be the case, then, assuming equivalent technological capability within each culture as is indicated by current evidence, the struggle between offense and defense can be maintained at parity only if there are no more than a few months' lag between Soviet decisions on their attack system and US decisions on its defense system and if US lead-time is equal to or shorter than the Soviet lead-time. To maintain such parity on the basis of equal lead-time requires the additional factor of a US forecast that provides a fairly accurate estimate of the enemy's course of action. By the very nature of the situation the forecast rather than exact knowledge must also be the basis of the actual decisions on defensive systems even with maximum effort expended to determine Soviet decisions at the earliest possible dates. It is as yet not certain as to whether forecasts can be made well enough when there are so many alternative technical choices available to both sides. We can state with confidence, however, that on the average a US policy relying appreciably on a defensive strategic posture will surely fail unless lead-times are equal to or less than Soviet lead-times. Therefore, it is argued that the most critical single factor involved in the defense against Soviet nuclear attack is the reduction of US weapons systems lead-times. These must be reduced from ten to about five years, or to about the same performance as that achieved in the US systems laboratories of World War II.

The Dilemma of the Optimum

If there are thousands of choices from which particular weapons systems can be chosen, the executive is faced with the cruel dilemma of having to choose from amongst the tempting array which he must pay for with limited resources in raw materials, manpower, time, and energy. Presumably he should choose the "best" buy. But this can hardly be done by the non-technical executive *only* with the aid of intuition, as in earlier times. What he is usually forced to do is to have "staff" studies, or "systems analysis," or "operations research," etc., made, which will narrow down the choices to a manageable few. The trouble with such studies is that they often take from one to four years to make, if they are to have a high reliability, and thus add to the already long lead-time. They may often, because of long study time, be concerned with choices already conceptually obsolete at the time they are completed.

The need to choose the "optimum" out of complexity is thus antagonistic to the need for speed in going from choice to operation. A "quick" weapon, technically inferior because the optimum was not chosen, may be just as fatal when battle is joined as a weapon that is technically inferior due to obsolescence because too long was taken to bring it into being. It is clear that the executive must be both quick and wise. He must choose from amongst hundreds of alternatives the "optimum" weapons system with "reasonable" accuracy and he must do so quickly after new knowledge appears. An executive concerned with these two requirements must possess in his own right both the skill of executives trained primarily in operations and technical skill of an equally high order.[2] But quick the executive must be.

This is practicable only if each major executive is skilled in the profession he controls. In turn, this implies that the major functions and professions of research, development, production, and combat must be *separately* organized in the Department of Defense.

Effect on Military-Economic Competition

It has been shown that prior to the initiation of overt military hostilities the primary effect of the development of increased physical knowledge is to make possible technological maneuver, i.e., the kaleidoscopic development and redevelopment of a complex of attack and defense weapons systems, whose primary mission is to affect the military-economic competition in order to provide an advantage to the attacker or the attacked at the inception of hostilities.

In consonance with the preceding discussion, the effect of new tech-

[2] One scientist has paraphrased this in the vernacular: "In Washington we need executives with more savvy, more guts, and fewer committees."

nological choices and improvements forces the side with the longer lead-time into the development of military defense systems that are never able to match the opposing attack systems. The side defending against an attack system will be able to meet the gambits of the attacking side on time, although perhaps not economically, *only* if its lead-time is short in comparison with the other side, and if lag in intelligence on enemy decisions is short. In fact, the sum of intelligence lags plus technological lag time must be shorter for the defense system than for the attack system.

Furthermore, "bits" of knowledge are used primarily to find attack systems and strategies that are of a sufficiently new character to outmode existing and imaginable defense systems, especially from an economic point of view. Again we cannot help but conclude that either the problem is one of detail in the race between attack and defense or one involving the race in the use of new knowledge to design weapons suitable for quite new and different strategies. Other things being equal, the side with a grave disadvantage in lead-time, as is now the case of the US, runs an unreasonable risk of being outgamed politically, strategically, and economically in both offense and defense strategies and of losing the military-economic race.

But what about the size of the military budget? It appears from our analysis that the increase in the rate of generation of knowledge has greatly increased rather than decreased costs. The hope of the taxpayer, the economist, and the treasury has been that military effectiveness is so increased out of proportion to cost that only a few weapons are needed at less over-all cost to the nation. Instead, we have seen that new knowledge is used to counter the magic of old knowledge and that new and increasingly expensive solutions must be used, if the USSR chooses to threaten us with increasingly modern weapons systems. Thus, if we are to protect the nation, we are forced as in earlier days to match military budget to budget, or if the military-economic exchange rate is adverse, as is the likely case in defense versus attack, we may need to spend *more* for defense than the enemy spends on attack systems. In the long run, therefore, the effect of improved technology is to exacerbate the requirement for the *maximum* possible military budget acceptable to the nation. New technology increases rather than decreases the cost of war.

The only effect of improved technology is to replace production maneuver by technological maneuver.

In summary, therefore, it is stated as a requirement that the greatest need of the US is to achieve parity or superiority to the USSR in the lead-time between the concept of a weapons system and its operational readiness, and to accept a military budget larger than the "effective" USSR budget.

THE INDUSTRIAL-MILITARY SYSTEM AND LEAD-TIME

Let us turn to a consideration of the factors that make our present lead-time so long. Military research and development in the United States is now achieved in a system whose components are primarily industry and the Department of Defense. Although there are a few great in-house military laboratories that do creative work in their own right, most of the creative weapons development and design is done by defense industry. In theory the guidance and management of this effort comes from the military officers of the Department of Defense. It is of the greatest importance to note that civilians with military research and development experience have, except for a small and relatively unimportant residue, been organized out of all of the positions of management responsibility and authority for military research and development programs in the Department of Defense. This was a deliberate action taken by the three services, acting in concert, beginning shortly after the end of World War II. The concentrated effort followed the dissolution of the Office of Scientific Research and Development and culminated in the dissolution of the Research and Development Board.

The direct influence of science and engineering within the Department of Defense and at higher echelons of the government now is channeled through the shadowy façade of scientific advisory committees that actually have no responsibility or authority for the development programs, have negligible influence upon these programs, and are relatively unacquainted with the substantive content of the military-technical problems involved. Almost the entire creative burden of weapons development, therefore, falls upon industry, yet there is no formal and systematic method to keep industry informed of the tactical and strategic military requirements. The industrial contractor can have formal access to tactical and strategic military requirements only on a need-to-know basis, and after he has a contract. This has led, and is leading, to more and more study contracts and to extensive informal discussions. These usually result in proposals *by industry*, either formally or informally, for new military weapons and weapons systems. After such proposals are processed within the Department of Defense, in a time-consuming cycle in which their compatibility with fiscal, tactical, and strategic constraints are considered, requirements are fed back to industry for competitive bids. Often the industry originating the requirements is the successful bidder. It is obvious that this random and grotesque procedure not only extends the time from concept to initiation of development, but it operates under the most severe handicap possible with respect to any high assurance that the development is best suited to meet the strategic requirements of the United States.

Assuming, however, that the development is a desirable one, a number of requirements must be met by industry. First, in order to survive at all, industry must make a profit—a profit adequate to assure the survival of the company—survival that stems primarily from production contracts that follow successful development. The second requirement is that steady employment for scientists and engineers must be assured within the industrial development laboratories. It takes approximately two years for a new employee in a company to become fully productive. This makes it impracticable to hire and fire research and development people on a short-term basis and still retain the competence necessary for successful development. A steady work load can be assured only by having a large backlog of development projects. In this case, the development time for each project is stretched out, increasing the lead-time, while each project awaits its turn on the priority list. On the other hand, if there is no backlog of development contracts, the work program on the current contract is rigidly scheduled on a stretch-out basis while the sales engineers of the company work desperately to get the new contracts that will insure survival of the development laboratory. In this case, development time is also prolonged, again increasing the lead-time.

There are two other unfortunate effects in this industry-defense system as it is now designed. The company must make the profit required for survival on the production of the new weapon. This means that the company will tend to choose and process developments that have a very high probability of being successful. This leads to development that is heavily weighted in the direction of product improvement, since high-profit, high-risk development programs are too uncertain to provide assurance that the production facilities of the company will be employed. Thus, our present industrial-military partnership tends to result in slow, steady, but mediocre progress.

Another unfortunate factor is use of cost-plus-fixed-fee contracts by the Department of Defense. These contracts have two effects. In development, they tend to result in an exorbitant use of technical personnel in each development, since the fee is always tied to the total cost; the greater the cost (that is, the greater the number of scientists and engineers working on a weapons development), the higher the profit to the industry. There is thus no incentive towards economical use of scientific and engineering personnel. A second unfortunate effect of the cost-plus-fixed-fee contract is that there is no incentive to take a high risk, which might result in a high profit, because there is no particular immediate reward for success or punishment for failure. In an older time, the competition in weapons development led to a situation where success was accompanied by high profits. This provided a desirable and effective incentive to industry.

There are a number of additional factors that greatly increase weapons development lead-time. The tremendous desire of the Department of Defense is for an immediate payoff in useful end item weapon systems. This strong desire for immediate results has led to a tremendous neglect of applied research, and component subsystems development for weapons. In practice, the prior and independent development of good components (a rocket of high thrust, for example, is a *component* of a missile or satellite launcher) has been relatively neglected. Such development is time-consuming and, because there is little prestige for the office-manager within his tour of duty, there is little interest among the operating group. Far too small a portion of defense funds has been allocated to component and subsystem development. Thus, when a weapons system development is determined upon, the applied research and the component and subsystems development must be initiated if it is critical to the system development, or else the development must be frozen at the existing state of the art.

Before leaving the subject of the difficulties of industrial weapons development, another most serious problem must be mentioned: the effect of security regulations involving need-to-know. These are rigidly enforced, so that not only is industry unable to make proposals based upon the most forward-looking aspects of tactical and strategic requirements, but also industry does not, in general, have adequate access to the tremendous amount of prior or parallel work of other industry or military laboratories. As a result, the classified libraries of industry are inadequate relative to their need for classified information. The fault of this does not lie in industry. It lies squarely with the military managers, who do not understand the need for "tautomeral" information in research and development, and who lack appreciation of the fact that need-to-know cannot be rigidly limited or interpreted when the problem is one of search for information, the value of which cannot be appreciated until it is found. Thus, the extraordinary lack of access of defense industry to classified information and the limited communications within the industry-defense system again tremendously extend the development time for US weapons systems. Industry attempts to compensate for this by frequent and extensive informal visits to defense and industrial installations. This is a poor method of obtaining the necessary classified information, which wastes technical manpower and again extends development lead-time.

Another effect of the need-to-know restriction is that this so limits communications within the Department of Defense that there is unnecessary replication of work already done. This does not mean competition in development, but the actual re-solving of old problems. In no sense is this deliberate on the part of the officers in the Department of Defense; it is the effect of the system involving need-to-know. The military managers are the first to deplore duplication, but they have no way of controlling this, especially when interservice barriers must be crossed. The effect is again to increase development time, because, if problems have already been solved, they should be adopted instead of wasting technical manpower to solve them again.

The need-to-know crisis is the result of lack of experience on the part of the higher-ranking military personnel, especially with respect to the requirement of "tautomeral" information on the part of research and development personnel. We actually are keeping our secrets so closely that this has aided the Soviet Union to draw ahead of us. Thus the primary purpose of the need-to-know requirement has been made ineffectual.

The serious communication block, with an original basis honestly designed to maintain security at a time when we were ahead of the rest of the world in weapons development, has now become one of the most vicious tools for interservice rivalry, as well as within each service.

Indeed, it is used by industry itself to limit the advantages of a competitor. Thus, again lead-time is lengthened and our technical and scientific manpower is wasted.

Let us now turn to the situation in the United States in World War II. During that period, the United States, as well as other countries, was tremendously productive in developing many new weapons. A variety of management methods were used in different laboratories but there are two points that stand out, whether or not a functional, a project, or a hybrid type of internal management was employed in a particular laboratory.

First, most of the great weapons systems were developed in systems laboratories independent of production control. This was true in Germany and in Great Britain as well. Second, and most important, every laboratory was treated as though it were a facility. There was never a question as to whether there would be enough work to do; there was never the need to stretch out the development time, because every laboratory had more work than it could do during the entire war. The laboratories worked at maximum speed and concentrated on each weapons system, usually one by one. There was no difficult tie-in between development and production from the viewpoint of profit. There was an acceptance of high-risk, high-profit development tasks and, if there was a failure, this could be forgiven, if the effort to make the development successful had been a great and honest one. There was in general a high and sophisticated support of component and subsystem development, as well as of all the necessary applied research. This was particularly true in OSRD. Thus the treatment of industrial, military, university, and OSRD laboratories as facilities, integrated into systems laboratories, led to the short lead-times of one to five years that were achieved in American weapons development systems during World War II. There were, of course, some other factors.

The best scientists in the country joined in weapons development during the war. Most of them returned to universities or other research centers at the end of the war. And finally, the motivations were high for all of us. We not only had a high goal, but we were sure that we could achieve our goal. While these factors should not be dismissed, they do not have an importance comparable to the treatment of the development laboratories as facilities in a consolidated weapons systems concept.

Research in general was managed by civilians, since most of the military personnel were then engaged in military operations or in training the military cadres. This was important. Another factor that was striking during World War II was the very free exchange of classified information within and between industry and Department of Defense management. We had then not achieved our pre-eminence in weapons systems. The present excessively severe security rules had not been formulated or designed and thus communications were free. This undoubtedly speeded up development time.

Examination of the current methods used by the Soviet Union in weapons development indicates that they treat their laboratories as facilities. There is never a problem of running out of work; there is steady employment for all research and development personnel. Did they learn this lesson from our own methods of World War II? The second striking aspect of their system is that everything indicates that they provide a heavy support to applied research, to component development, and to subsystem development prior to or in parallel and independent of the specific end item weapons systems. This is just the method we used during the war, especially in OSRD. Did the Soviet Union learn these leasons from our techniques, the ones we have forgotten?

To summarize the principal factors which contributed to success on the part of the Free World during World War II in weapons development, and which appear to lead to important successes now on the part of the Soviet Union: First, development laboratories were treated as facilities, were organized as weapons systems laboratories, and their development programs were not tied on a one-to-one basis to production. Second, the communication of classified information was provided on an adequately free basis, permitting maximum utilization of progress made, regardless of where results were obtained in the over-all system. Third, management was primarily by competent civilians trained in research and development. Fourth, all of the motivations were designed to obtain maximum technical progress, rather than an absolute assurance of production.

We need to return to the methods we used during wartime, if we are to compete with the Soviet Union in weapons design.

LONG-RANGE PLANNING

The advantages of planning are self-evident to all who look to future satisfactions and have responsibility for budgeting toward their achievement. Systematic planning must proceed from a clear concept of objectives and from basic policies for achieving those objectives. There are, for example, certain cultural traits that direct us in determining policies—human rights, legitimate competition, etc. What makes planning so difficult—difficult above all else—is the uncertainty of the future amidst revolutionary developments resulting from man's able conquests in science and technology. We cannot plan on the basis of evolutionary changes when our world is moving forward in revolution. Today, more than at any other time in our history, Americans view our defense program as one that has been too slow in keeping up with the times. The USSR's spectacular strides forward in science and development over the last decade, dramatized recently for all to see, have alerted our own people to the requirement for more effective planning for defense of the United States.

To identify the requirements of planning, here is a philosophy that applies to all parts of our military system and can be translated to cover industrial and other large-scale operations. To be fully effective such a philosophy must permeate an entire organization and guide and direct *all* the planning in that organization.

The planning philosophy or system suggested incorporates fundamental characteristics that stem from a scientific approach to the problem of planning. The goal is a planning system that provides in an orderly, logical way for a *continuous and up-to-date search for solutions* based upon sound and timely decision-making, instead of blind adherence to "five-year plans." This proposal is not tied to any specific organization; rather it focuses upon the functioning of a logical *decision-making* system within which long-range plans are developed. Implied is a feed-around of guidance, information, and partial solutions, in a ceaseless search for the flexible, accommodating, over-all solution that adjusts continuously to the dynamics of technology. Figure 7 shows schematically this system.

It is logical to start with the forecast—the basis for rational action.

The forecast provides the first and fundamental part of a rational basis for future operational planning. The forecast for any particular time frame is based upon existing trends in their relation to the values and objectives chosen, and makes a prognosis of situations which may arise in a future period if these trends continue. With skill and experience such formal forecasts can be made with sufficient certainty to provide a far better basis for rational action than the present informal and almost random decision-making system. Without a rather explicit

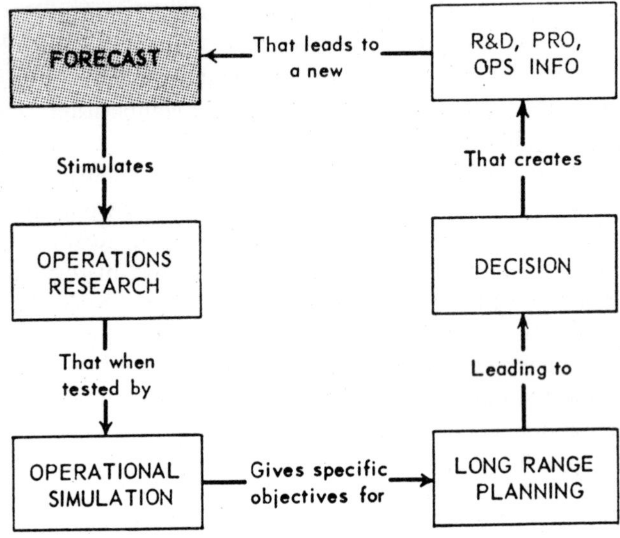

Figure 7. Forecast and planning cycle.

forecast based upon trends, the US has no basis for long-range planning.

There are three specific areas of information that provide data for the formulation of a US defense forecast: autointelligence that provides information about ourselves; foreign intelligence that provides information about our enemies and allies; and historical intelligence that brings together and analyzes the lessons of history. From intelligence studies in these three areas it is possible to develop a probabilistic forecast of the US defense situation. On these foundations, the forecast represents a weighed and balanced analysis, rather than an intuitive impression of current trends and their possible effect upon developments in the future. Such a forecast must not be regarded as a prediction of what *will* happen in the future, but a prediction of what will *probably* happen if present trends are allowed to continue.

Innovations and Operations Research

In view of the predictions, the focus of the search is to determine possible new technical, tactical, and strategic courses of action that will enhance the US military position—or an industrial situation.

Possibilities lie in the area of problem-solving innovations. There is no actual lack of *general* ideas for such solutions. The practical difficulty is in identifying and gaining support for a truly novel proposal that will provide a significant advance—not just marginal betterment. It is the role of operations research to identify and compare the most promising alternatives.

Wise and timely decisions on acceptability of proposed innovations are of the utmost importance in the competition for short lead-time. Decisions on weapons systems are difficult and complex in an unstable technological situation, involving high risk in cost and effort. But the effectiveness of the end result depends in large part on the timeliness of the decision itself. It is in the realm of decision-making that the US is paying its greatest penalty in the race with the USSR in military technology. In the USSR hierarchy, decisions are made at the combined high level of government, military, and scientific organization. Decisions are made quickly throughout the cycle from research to use. The USSR learned its techniques from the methods developed and used by the Free World during World War II but, alas, abandoned by the Free World since that time.

Long-Range Planning Group

Operations research identifies from available forecasts the possibility of initiating new trends that may turn disadvantages into advantages. Innovations are directed to provide specific ways and means to achieve success. Theories that support innovation represent basic elements in long-range planning. They are the building blocks for long-range planners.

It is, then, the task of a group of highly professional, hard-thinking, imaginative planners to screen these ideas and proposals and to identify specific means and feasible plans to meet objectives.

Decision-Making

The basic problem of US peacetime military management is balance in the programs and choices that will provide readiness in the short-, mid-, and long-range futures of the US. This requires a military posture of sufficient flexibility to meet a wide range of possible USSR political and military actions. Good decisions to achieve such a posture are aided in the making and in their implementation by good philosophy and good organizational structure, but they are determined more by that intangible human factor of outstanding and forceful leadership than by any other single factor. The good judgment, strength, and determination of able leadership in the democratic tradition are necessary to engender forward-looking philosophy, efficient organization, and advantageous use of our resources to bring into being the force necessary to assure the survival of the republic.

The decision-maker requires all the aid possible from the advanced problem-solving procedures now available for decision-making, such as operations research. As yet, the philosophy of scientific aid for the decision-maker has not been fully accepted. Too many decision-makers rely fully on the intuitive methods that were sufficient in the past, when the problems of an enterprise were comprehensible to a single human being. They have not yet recognized that this is outmoded in truly complex situations, that the utilization of the available tools of science must be combined with intuition developed through experience in the different professions.

The executive should have the benefit of the results of all of the sophisticated management techniques when individual plans and sets of objectives covering proposed strategies, tactics, and/or weapons systems are submitted to him. The executive has the responsibility for the final decision that determines whether a proposal is to be subjected to the operational cycle of research, development, production, training, and operations.

The Cycle

The full cycle of planning has been depicted in Figure 7 and new cycles are continuous. There is a ceaseless search for a solution, a continuously fresh approach. Figure 8 shows the cycle in more detail.

The US military establishment faces a multitude of decisions in determining and achieving the best offensive-defensive mix to assure national safety and the support of the Free World. Time is of the essence—there should be no procrastination in establishing the best possible operational pattern directed to these national aims and safeguards. Objectives have to be set. The jobs to be done have to be identified. The responsibility for getting those jobs done has to be delegated to persons, laboratories, and operational groups—professionally qualified persons and groups who comprehend the missions and are capable of carrying them, as they did in the systems labora-

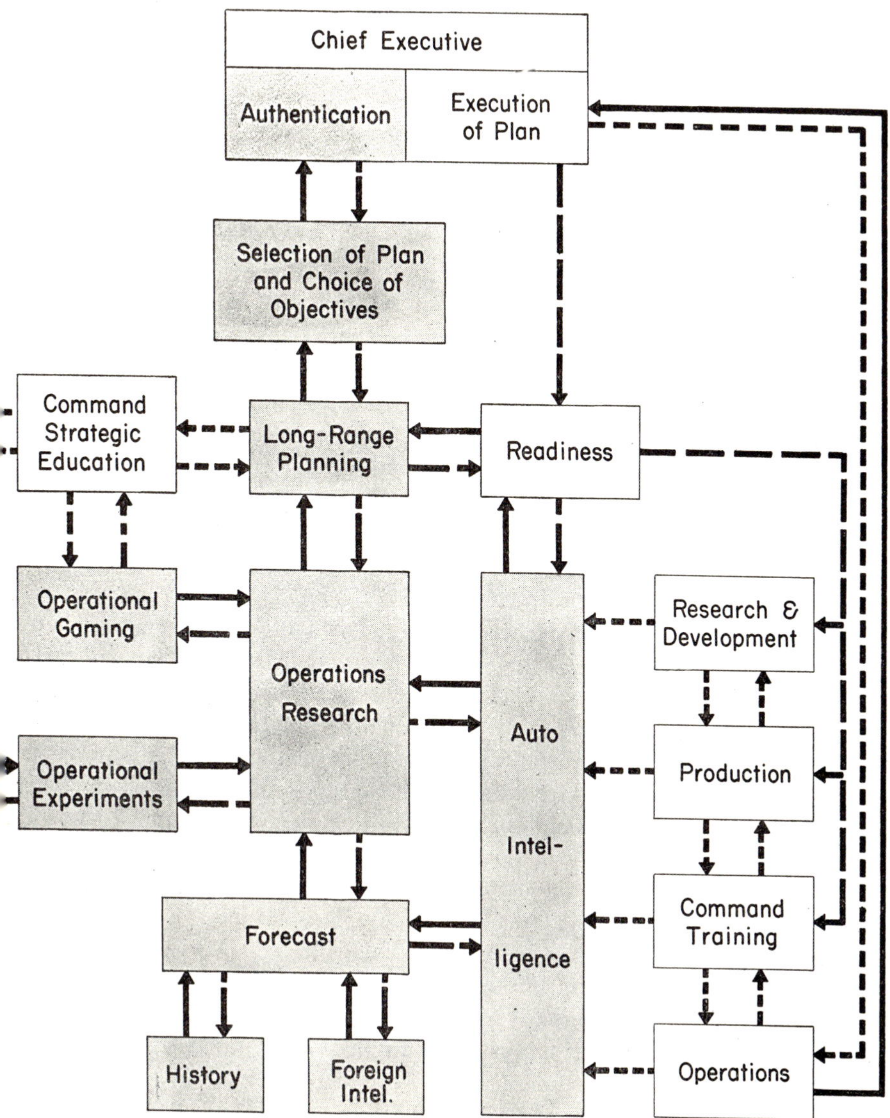

Figure 8. Schematic planning system for innovated warfare.

tories that developed the atomic bomb, radar, VT fuzes, and the other innovations of World War II.

CONCLUSIONS

Preceding pages have outlined the effect of the increase in physical knowledge upon our peacetime society, and the crucial and decisive effect upon our military competition with the Soviet Union. This crisis in technology has produced a crisis in the management of our military establishment. It is becoming equally severe for all of the rest of humanity; it has increased the tempo of the second industrial revolution now sweeping the world. In turn, this tempo, which has been quickening for centuries, rose above the threshold of crisis in this decade, i.e., about 1950. It will become increasingly severe and our primary problem is to find a way to manage our affairs in this new situation. The solutions must be found by far more conscious long-range planning and this should be aided by a very great increase in our support of the social sciences.

If the increase in physical knowledge has made the future so uncertain, then we must plan much further ahead in a way that will provide much greater flexibility, whether this be in peaceful or military affairs, whether it be for the individual or for the country. We will need to adapt our culture to more rapid change, and accept such adaptation much more readily than in the past, instead of conducting our lives, our actions, and our affairs as though there continued to be a stable and known way of doing new things. We will need to estimate well ahead of time the cost and values of doing things in new and different ways, so that when the time comes to make a decision we will already have thought the problem through. We will have to examine our individual, group, and national values to see what it is we want to do in a rapidly changing world, and to see what we can consciously do to manipulate in our favor the real and perhaps hostile physical world environment.

Such long-range planning is compatible with the ideals of our Republic. It is not only shrewd but honorable to figure things out so as to conserve and increase our resources. It is more dignified to control the population of a country than to have ten million people die of famine because of a drought. It is more sensible to spend the needed seventy billion dollars a year now for defense, in order to prevent a world-wide thermonuclear war, than to increase our standards of living by 5 per cent each year. It is more dignified to find orderly ways of preventing cruel results than to permit ignorance and uncontrolled chance to destroy the purposefulness of our actions. Therefore, we must meet the crisis brought about by our increase in physical knowledge by improving management of ourselves, our country, and, joining with the rest of the peoples of the world, of all humanity, without, at the same time, sacrificing dignity and freedom of the individual.

We will not be able to compete with the Soviet Union in military technology until research and development are completely separated functionally in management and execution from production and combat operations. Furthermore, this can be done with competence only if the entire control is placed under the direct authority of professional civilian research and development executives, trained in the research profession.

Military personnel should use their great talents in military roles of combat, combat training, and the development of war plans.

It is high time for the Department of Defense to recognize the critical importance of a civilian military partnership in which authority and leadership are directly correlated with professional competence, that is to say, with professional experience.

Finally, the incentive for speed and quality in military research by industry must be vitalized by:

Adequate access to classified information,
Elimination of cost-plus-fixed-fee-contracts,
Three-year development contracts that can be extended each year if performance is good,
Separate support of applied research, component development, and subsystem development,
Removal of the tie-in between development and production contracts, and
Increased profit for speed and quality in research.

The methods of operations research offer guidance in solving these difficult problems in this time of crisis.

Four

THE RECOGNITION OF SYSTEMS ENGINEERING

RALPH E. GIBSON

Systems engineering is already being recognized in the practice of engineering as a branch of the art with problems, methods, and objectives peculiar to itself. It is becoming recognized in the academic world as an association of disciplines, modes of thought and operation, which is worthwhile presenting systematically to the student as preparation for intellectual leadership in solving an important class of problems likely to confront his generation. Recognition is also being given to the fact that certain patterns of viewpoint, standards of value, background and education, aptitudes and inclinations of mind, mark a man as likely to succeed in the development and application of systems engineering.

This chapter, in introducing a series of discussions by authorities on systems engineering, will handle the subject in a very simple way, thereby bringing into the open some of the fundamental ideas involved.

To begin with a definition, a system is *an integrated assembly of interacting elements, designed to carry out co-operatively a predetermined function.* The living organism, the animal, or above all, the human body with its central nervous system, is an example *par excellence* of a system. Indeed, the human body is the primary source of our fundamental ideas about systems. St. Paul used the body to illustrate a social system, the Church, "But now are they many members, yet but one body." Following the human habit of making anthropomorphic models, we have used ideas gained intuitively from personal experience to understand new phenomena or to devise new creations.

PRINCIPLES OF SYSTEM DESIGN

Example

The human body has been systems-engineered by Nature through processes that have taken millions of years and are still going on. Although these processes sometimes seem wasteful, dependent for fundamental improvement on chance breakthroughs in the form of new species with greater survival qualities, the results command our utmost admiration. Economy of materials and flexibility of design characterize the structure which houses and implements a brain and nervous system capable of receiving and interpreting intelligence from the external world and controlling the reactions of the body with delicacy and precision. The whole is powered by an alimentary system where energy from chemical sources is supplied upon demand through a fascinating chain of chemical processes. It is an integrated assembly of thousands of elements all co-operating to carry out a certain function, namely, survival in a competitive and changing world. In this system, the elements are essentially simple but so numerous and interdependent that the system as a whole presents a very complex picture to the observer. It is so complex that for centuries patient empirical studies in the healing arts, and later scientific research in medical sciences, have been necessary to support the activities of the practitioners of medicine and surgery who really act as repair men for this system.

Definition of Function

Even a superficial study of the human system brings out several principles which are fundamental in the design and development of all systems. The first has been already alluded to; it is the definition of the function the system is to perform. It is essential that this function (or these functions) be selected with great care and defined clearly and, if possible, simply. Above all, the functions selected for the system must be stable (i.e., have a high degree of permanence) in a competitive world. (There is possibly a real connection between the permanence of a function and the simplicity of its formulation.)

It is here that operations analysis enters the picture. The purpose that the proposed system is to achieve must be analyzed in detail with realism not only reflecting the present jobs to be done but also anticipating by sound extrapolation the exigencies of the future. The capabilities of the proposed system, and variations of it to achieve this purpose, must be assessed quantitatively and compared with potentially competitive systems, the ultimate basis of comparison being the ratio of effectiveness to cost.

Here is an example. Suppose we set out to build a guided missile system designed to protect an installation of some sort against attack by aircraft. From a knowledge of the state of the art, we can put together on paper an assembly of components such as engines, airframes, control and guidance mechanisms, payloads, and so forth, for the missile, and equipment such as launchers, radars, and computers for supplying it with information and starting it on its mission. We can assemble these in different ways, producing several preliminary concepts. We now have a problem of selection—what is the best way to put this system together, which is the best concept to adopt? To make this selection, we must first analyze the present and future tactical situations. The threat must be estimated—what is the speed of the attacking bombers, at what ranges can they bomb accurately, how many will attack at once, what tactics are they likely to employ? These questions must be answered, not only for the present threat, but for the future threat as forecast by sound extrapolations from present tactical and technical trends. The capability of our proposed system to function with high effectiveness in the tactical situation so developed is then assessed and compared with other means of combatting aircraft attack, such as guns or interceptors. Progressive approximations to realistic estimates of effectiveness per unit cost for different systems can then be made, and from the whole an optimized system concept may be distilled. Interplay of scientific and engineering considerations on the one hand, and the estimates of requirements in terms of present and future threats on the other, determine at what requirement level the development of the first prototype will be aimed. Technical analysis also permits prediction of the growth potential of the proposed system. The growth potential capabilities of different systems are given great weight in the selection of the systems concept to be developed. On the basis of such analysis, a sound systems objective can be established and the long haul of development and engineering undertaken with confidence. This discussion is somewhat in-

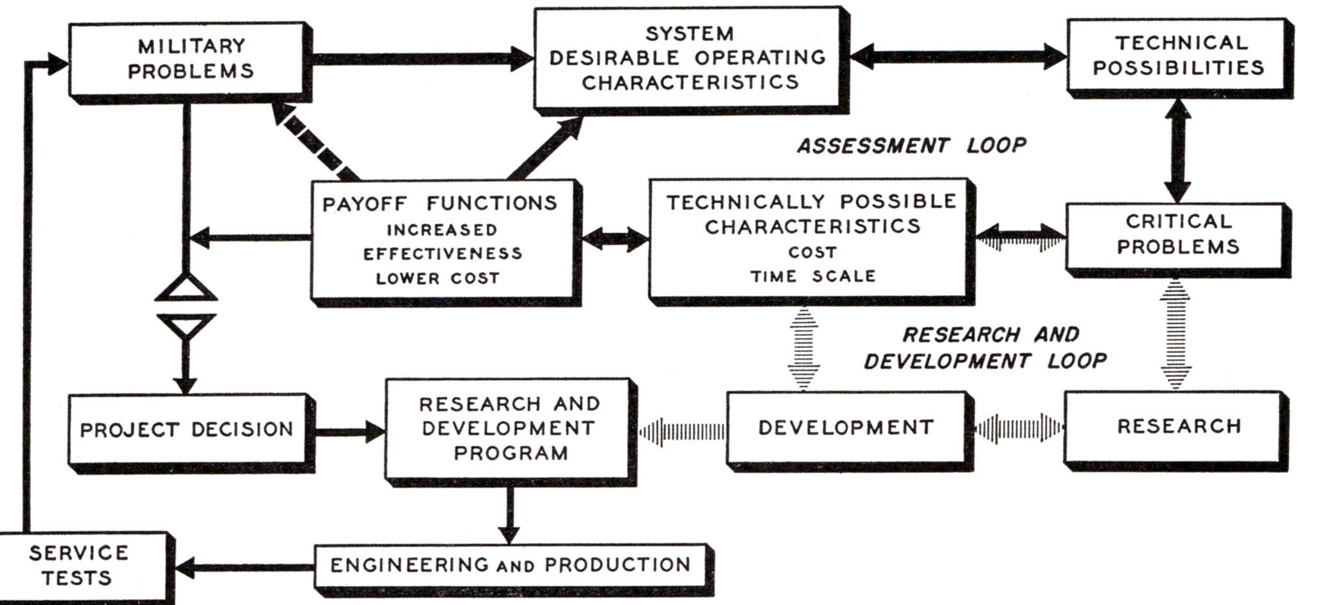

Figure 1. Functions in planning and execution of a project.

volved but the process is not a simple or direct one. The diagram in Figure 1 may help in visualizing it.

The stimulus for engineering the system is a set of military problems (upper left) which generates objectives and the need to assess the alternative forms of systems which may satisfy these objectives. The assessment of alternatives may raise questions whose answers come from research. Out of assessment and research emerges a basis for choice among alternatives—a project decision. The development of the system ultimately impinges on the military problems which inspired it, producing modifications to the system and the development of new systems.

Feedback Control

A second principle obvious in the human system is that of *feedback control.* This principle is a simple one that is illustrated by the action of the governor on a steam engine or the control of room temperature by a thermostat. Feedback in its simplest form consists of the use of a fraction of the output of a machine to control its power input. If, as in the governor, *increase* in the machine's output brings about a decrease in the power input, we have negative feedback, and stability of operation results. If, on the other hand, increase in output calls directly or indirectly for increase in input, instability results, or the system oscillates as in the tube circuit of a radio transmitter.

Closed-loop control by feedback is to be contrasted with open-loop control where special design efforts and strict adherence to close tolerances are required. Suppose we wish to design a steam engine that will work always at constant speed and maintain the speed within very close limits. We can do so by controlling the steam pressure in the boiler very carefully, by calibrating the throttle valve so that a given opening corresponds exactly to a given rate of entry of steam into the engine, and finally by keeping the load on the engine always constant. This involves accurate calibrations of many components of the system, and, if any depart from pre-set tolerances the engine will slow down or speed up by an unknown amount. This is the open-loop method of control. The closed-loop counterpart is given by a governor which *measures* the speed directly, or rather, measures the difference between the actual speed and the desired speed, and *signals* back to the throttle valve commands to open or close in order to bring the error to zero. It does this rapidly enough to make the response effective. Thus the speed of the engine is maintained without the requirement for tight tolerances in the boiler pressure, the load, or the throttle valve. Three requirements are inherent in a feedback control system—first, the quantity to be controlled must be susceptible of precise measurement so that a difference between its actual and required values can generate a definite error signal; second, an adequate system of communications must exist between the element sensing the error and the primary source of motive power for the device (regulator); third, that the controller and the regulator and the communications between them act in a timely manner. The timeliness of such actions and interactions is a matter of great importance. Lags in response can produce instability, and in certain cases feedback control is almost impossible. For example, the player of an instrument such as an organ *must* operate by an open-loop control. He has a feedback mechanism in that he can hear the results of the causes he produces but by the time he hears a wrong note or a faulty expression, it is too late to make a correction.

The principle of control by feedback is a very old one; indeed, it probably played a most important evolutionary role in the very first stages of animal life. As soon as the prototypes of animal life started to move around and encounter unexpected conditions, the organism that could sense directly new circumstances and regulate itself accordingly gained an overwhelming advantage over its competitors operating on inflexible programs. The process of evolution toward the higher animals and man has seen steady progress in the utilization of feedback control. Indeed, Sherrington showed more than fifty years ago that in man between one-quarter and one-half of the nerve fibers supplying the muscles are concerned with feedback functions. It is hardly an exaggeration to say that scientific advances that extended the possibility of feedback control to a wide variety of systems inaugurated the era of modern systems engineering.

Protection of the Critical Function

A third principle discernible in the human system is another concerned with selection and this may be called the *protection of the critical function.* In the design and development of a system, components play an essential part, but some play such critical parts that without them the system would not achieve its objective at all. The ascendancy of man over the other forms of plant and animal life is due to the power of his brain to reason, to imagine, and to create. The higher brain is the critical subassembly in the human organism from the viewpoint of its fulfilling its prime function as a system. A complex network of feedback loops protects this mechanism by providing a stable environment in which it can realize its fullest effectiveness. This refers to the mechanisms producing the condition of *homeostasis* or constancy of internal environment. We are endowed with a series of mechanisms largely under the control of the lower part of the brain that controls the environment in which the brain itself operates. Factors determining this environment are the body temperature, hydrogen ion concentration, sugar content, oxygen content, water content of body fluids, etc. These are kept constant within narrow limits. As an example, take body temperature, whose control within narrow limits is designated *thermostasis.* In one of her experiments, Nature endowed the dinosaurs with spectacular size, fertility, defensive and offensive armour unparalleled in any other creatures. They became extinct because, among other things, they lacked a mechanism for thermostasis. They were "cold-blooded creatures"; that is to say, they assumed the temperature of their environment and their central nervous system reflected these changes. In cold weather their nervous activity was reduced to a very low value; they reverted to a state of torpor or sleep. In warm weather the temperature of their bodies rose, bringing their nervous activity to a highly excited state similar to convulsions accompanying fever. As Gray Walter says,

> . . . The acquisition of internal temperature control, thermostasis, was the supreme event in evolution and, indeed, in all natural history. It made possible the survival of mammals on a cooling globe. That was its general importance in evolution. Its particular importance was that it completed, in one section of the brain, an automatic system of stabilization for the vital functions of the organism —the condition known as homeostasis. With this arrangement, other parts of the brain are left free for functions not immediately related to the vital engine or senses, for functions surpassing the wonders of homeostasis itself.

To summarize the third principle, the human system contains a large number of feedback control loops whose primary function is to provide automatically a constant internal environment in which the brain can perform its higher functions in comparative peace. An important role of these loops is to keep down the *internal noise level* so that weak signals from the brain may be passed on to their appropriate destinations conveying clear, definite, and complete information with the minimum contamination by noise. The body as a whole is an engine to provide energy and nourishment for the brain, to provide mechanisms for implementing signals from the brain, and to provide a low noise level environment for its functioning and communications.

Ultrastability

Closely related to the third principle is a fourth, the principle of *ultrastability,* which covers mechanisms that are often referred to as adaptive servomechanisms. The basis of this principle lies in the fact that the dynamic response range of any control loop is limited. If an ultrastable system encounters conditions that push the primary loop controlling a certain quantity beyond its limits, it falls back on a second control mechanism which attempts to adjust matters so that the first one is again able to perform its control function. Thus, for example, if the human body finds itself in an environment so cold that the thermostatic mechanism no longer keeps it stable, it resorts to reactions such as drawing the knees up to the chin so as to minimize the surface exposed, or moving toward a fire, or putting on more clothes. If it gets too close to the fire, the control mechanism induces a withdrawal, and so forth.

This principle reflects the numerous resources which may be called in automatically to prevent changes in conditions causing catastrophic failure of the system to perform its function. It is a principle which one may suspect is very difficult to implement in artificial systems.

Redundancy

A fifth principle is that of *redundancy.* In many key places nature has engineered this system with ample excess capacity. Large sections of the brain and nearly three-quarters of the kidneys may be removed without disastrous effects on the system. Furthermore, numerous alternate routes may be used for communications, loss of one merely requiring the learning to use another. This principle is well recognized in systems engineering but for very practical reasons is practiced by man much less than by nature.

Reliability

The sixth principle to be observed in biological systems may be stated as follows—complexity does not necessarily produce unreliability. It is estimated that there are of the order of 10^{10} neurons (the basic switching mechanism) in the central nervous system which, by and large, functions with great reliability without intermission over the whole life of the machine, an average of more than fifty years. Our largest digital machines contain between 10,000 and 100,000 basic switching elements. Their complexity is only 1/100,000 that of the central nervous system, and yet no one can claim that our computing machines are 100,000 times more reliable than the central nervous system.

The design and arrangement of the elements, for example, whether they work in series or in parallel with each other, and the width or narrowness of the limits in which they can work are probably more important than the mere number of interacting elements in determining the reliability of a system. Redundancy does promote reliability, even if it involves added complexity.

SOCIAL CONSIDERATIONS

Thus far, this chapter has been devoted to some of the thoughts suggested by a study of a well-engineered system, the human body, in order to form a background for a closer study of what is involved in the term "systems engineering." Before this study proceeds, a few words should be said about the social foundation of the subject. Systems engineering is an activity forced on us by the demands for more and more automatic devices. These demands are fairly deep seated, having their roots in an expanding population, a widespread increase in the standards of living, and an expanding economy. This calls for the production of more and more sophisticated commodities in greater quantities. Somewhat over a hundred years ago, the steam engine and the coal that supplied its energy started the replacement of animal and human labor as the motive power for machines and transport. This was not only a replacement; it led to machines that would have been impossible otherwise, and to an economic utilization of power on a scale absolutely impossible when its chief sources were wind, water wheels, animals, and men. It is true that the use of steam and other sources of power did relieve men of much hard work and not too elevating toil and freed them for more productive jobs, but there are not enough men and women in the country to turn out by pre-industrial revolution methods the quality and quantity of goods we now consume. It would be quite a problem to handle the million horses that would be needed to supply Baltimore with the power it now consumes. Twenty thousand horses or their equivalent in eagles would not propel a modern passenger airplane at its cruising speed.

The same consideration holds today as regards the rapid extension of the use of more and more sophisticated machines. At the outset this movement starts by replacing human elements in operating systems that automatically generate intelligence or information which can be used in the operation of control systems. Sooner or later, however, the movement goes much beyond this stage and automatic systems develop capacities which could be duplicated only by the efforts of a prohibitive number of human beings, permitting the undertaking of jobs on a scale or at a price that would be impossible without them. Thus, in its initial phases, automation may relieve men and women from routine work and permit them to exercise higher skills; it may undermine the training investment of some people of limited intellectual means and outlook, but its capacity for growth makes it an essential instrument in the extension to a larger fraction of the world's population of the standard of living now enjoyed by a small fraction. The implications of this in the promotion of world peace are obvious.

Large-scale use of power from fossil fuels has added its bit to the debit side of the ledger of human resources and the same will be true of the widespread use of automatic equipment. Necessity develops the use of physical and mental powers that may atrophy when circumstances no longer place vital demands on their use.

It seems that already the automobile and other power equipment have reduced the capacity of American youth to walk, climb, or carry loads. Games that exercise the body are a remedy for this situation. Likewise we may expect the widespread introduction of automatic machinery to have a deleterious effect on a number of talents and skills that have been widely cultivated in the past. Games that exercise such skills as counting, sorting, steering, and so forth, together with more encouragement of handicrafts and "do it yourself" activities, will tend to prevent automatic machinery from destroying these human resources.

SYSTEMS ENGINEERING

In general terms, the job of a systems engineer may be stated as follows. Given a function to be performed, he (1) formulates an optimized concept of a system that will perform this function automatically (i.e., with the minimum participation of human operators), (2) translates the concept into reliable practical hardware giving the desired performance, and (3) accomplishes the above with timeliness and economy. Stated in these terms the task sounds quite familiar, the only suspicion of novelty arising from the word, "automatically." The construction of a spinning machine or a motor bicycle answers to the same description. However, modern systems engineering implies considerably more than appears on the surface of the above description. It implies that the function to be performed be a fairly sophisticated one involving selection, discrimination, and requiring some degree of adaptability on the part of the system, combined with a high degree of automaticity.

Modern systems engineering also implies that the elements of the system may be chosen from widely different areas of science and engineering. It also implies that the system will require the application of a number of the principles outlined in the earlier part of this chapter—in short that the modern system is going to take on more and more of the characteristics of an organism. These implications and the job itself lead us to consider that systems engineering does pose some unique requirements on its practitioners, requirements similar to those that distinguish medical science and its practice from the greater body of biological sciences. In particular, it is a field where action must be backed up by scientific understanding—it is too complicated for purely empirical approaches. This point needs further development.

Figure 2 shows a division of systems engineering into a number of areas, each area containing problems unique to itself. These areas are not necessarily identified with any one branch of science or engineering but may combine knowledge and techniques from a number of these branches. The outer circle represents areas of technical activity containing problems whose solutions lead to the design of components and equipment. The inner circle deals with the interaction of the components and equipment in the *system* to be built. It is to be emphasized that the actual equipments selected, both as regards their external characteristics, expressed as inputs and outputs, and in their internal characteristics of design and reliability, are the products of interactions around each circle and between each circle. The selection and integration of these equipments is one of the systems engineer's most important jobs.

Consider, for example, the design and development of an anti-

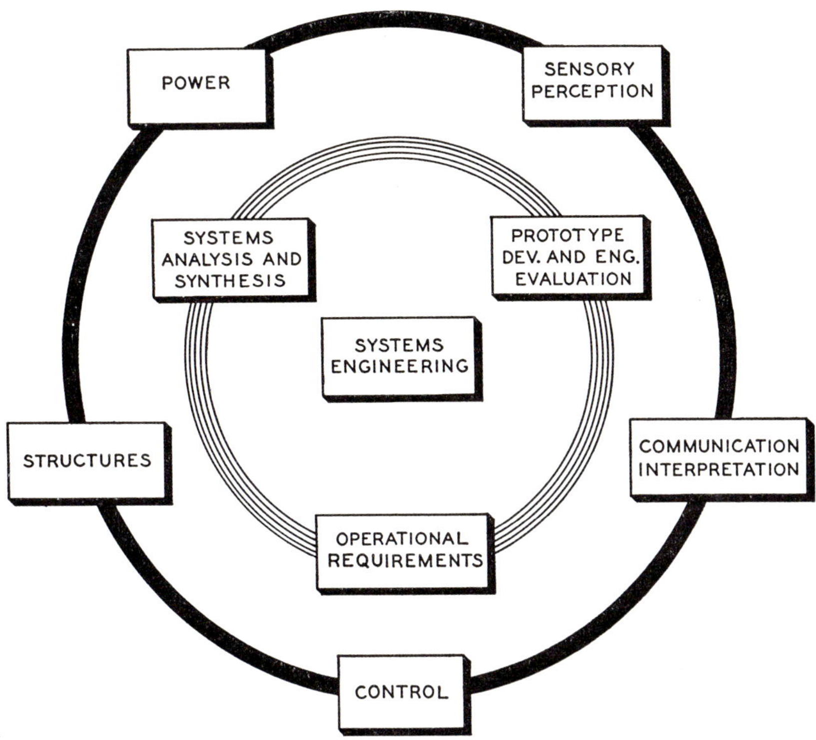

Figure 2. Systems engineering.

aircraft beam-riding guided missile. Through analysis of the type already described, it is decided that the missile must have a certain range, speed, accuracy, maneuverability, and warhead effectiveness. In the power area, these requirements call for engines to launch and sustain the missile at cruising speed over a given range, radars to be effective over that range, and power to supply the missile servos for a given time. In the sensory perception area, we must perceive the presence of hostile targets sufficiently accurately that they can be located at the end of the beam of a tracking radar which determines their position in space and follows them automatically. At the same time, we must place in our missile apparatus to sense its distance from the axis of the tracking radar beam according to *information communicated* along the beam from its source. The beam-rider receiver senses and interprets this information and sends the results to the control system where steering orders are generated and implemented through servomechanisms that deflect rudders. The functions which the control system must perform with lightning rapidity are quite complex. Not only must the control system deflect the steering surfaces rapidly and precisely by amounts that depend on the speed and aerodynamics of the missile, it must also sense the response of the missile and make adjustments in the rudder settings to prevent overshoot or instability of flight path and missile attitude. An elaborate network of feedback loops enables the control system to perform these functions through sensory perception elements of its own, such as gyros and accelerometers, that sense the response of the missile to orders and, through computers, communicate back to the servos commands that prevent wide oscillations in flight path or attitude. Similarly, the tracking radar has a control system which enables it to follow the target with the minimum of oscillations. The information received by the radar and the missile is always contaminated with noise. For example, the target plane does not act as a point source; the center of reflection of the radar beam wanders from place to place over its surface giving the impression of rapidly changing speeds and directions of its motion. This is one source of noise that confuses the information signal and, along with others, places the ultimate limits on the performance we can obtain. For highest performance, intelligence and control systems must be able to distinguish between the true signal and the noise, even when their relative amplitudes are comparable.

In the structures field problems arise concerning the design of a body with optimum aerodynamic properties which will house the equipment mentioned and that combines lightness of weight with strength to withstand the stresses set up by high-speed flight and rapid maneuver, including the temperature changes generated by frictional heating. The weight of the structure and its contents determines the size of the engines. We have made one complete tour around the circle.

Figure 2 suggests that the responsibilities of a systems engineer may be summed up by four words—*concept, selection, analysis, synthesis.*

Having determined the functions that the system is to perform, the first step to be taken is the formulation of a systems *concept* in which the structure of the *whole system* is outlined, its general methods of operation determined, and the external functions (inputs and outputs) of the various elements defined. This calls for creativity, imagination, and broad background.

Selection enters at many points in the evolution of a system. It must be exercised in the choice of one concept versus another, in the choice of equipments (components), and in the choice of details of equipment. Since many components interact in a system, their selection depends on an over-all knowledge of the system as well as on detailed knowledge of the components themselves. Many compromises must be made—some components have much more leeway in their performance limits than others. This knowledge must be dealt into the selection process. Critical thinking based on extensive knowledge is the basis of successful selection.

Analysis reduces opinions to quantitative terms and follows the scientific logic of cause and effect throughout the maze of the entire system—it provides the ultimate scientific basis for selection. It predicts ideal performance and the departures from the ideal caused by the limitations of practically realizable equipment. It provides a scientific basis for design criteria, specifications, and tests.

Synthesis means, first, the integration of equipment into a theoretical system, and, second, the translation of this result into practical hardware that gives the required performance, and can be produced, maintained, and operated by human agents.

THE SYSTEMS ENGINEER—RECOGNITION OF EDUCATIONAL BACKGROUND

A catalogue of all the branches of science and engineering that feed into the fields of technical activity shown on the diagram in Figure 2 would run the entire gamut of mathematics, the physical sciences, and engineering. Such a listing will not be given here because it is absurd to suggest that a systems engineer should be expert in all these branches of technology. It is suggested, however, that he must be conversant in all these branches, the term "conversant" meaning that the systems engineer should be able to communicate clearly and unambiguously with men who *are* expert in all these fields.

To be conversant with one another, men must share a common background and a common set of standards of value. A man's background and standards of value are largely a product of his education. We turn, therefore, to a general consideration of requirements in the education of a systems engineer. To bring out some important points, a simplified picture of modern technology follows.

There are two main components of modern technology, scientific research and the practice of the "useful arts." Fundamentally, they have different incentives, objectives, and methods which are illustrated in Figure 3. On the right we see a circuit (heavy solid line) representing the "useful arts" or its modern counterpart in our discussion today, engineering. On the left-hand side of the diagram is another circuit representing scientific research, whose incentive is curiosity inspired by observation of natural phenomena and whose objective is understanding. The methods of scientific research deserve a few words. Scientific research itself may be regarded as a system in which a number of human activities cooperate to carry out one function, the development of an understanding of man and his environment. Let us examine these activities.

Figure 4 gives in block diagram form a picture of these activities interacting in a series of circuits to produce the results characteristic of scientific research. Starting from the observation of phenomena or events, the results of exploration and description, we proceed to establish facts by careful experiment and to confirm them by communicating them to others for independent verification. This chain is shown by diagonal arrows. Simultaneously, a very important circuit involving the upper three left blocks is activated. This circuit represents what Claud Bernard, the French physiologist, called the "interplay of experimental theory and experimental practice." From the observation and careful study of phenomena or events, facts are obtained which may then be fitted together in an experimental pattern, i.e., a working hypothesis. If the facts fit well into a generalization, the latter immediately suggests new subjects for observation or new experiments from which come new facts, and so the activity in the circuit builds up, and with it *confidence in the validity of the facts and the consistency of the theory.* On the other hand, if the facts do not fit into a recognized pattern, one must first make a further study of their validity to insure that they have not been vitiated by some error; and errors may arise in very subtle ways. At the same time, it may be necessary to re-examine the pattern or theory and, if necessary, modify it to accommodate the new facts. The process is a cyclic one, and only when the *facts and the experimental theory fit together* can we be content with either. The product of this circuit is a satisfy-

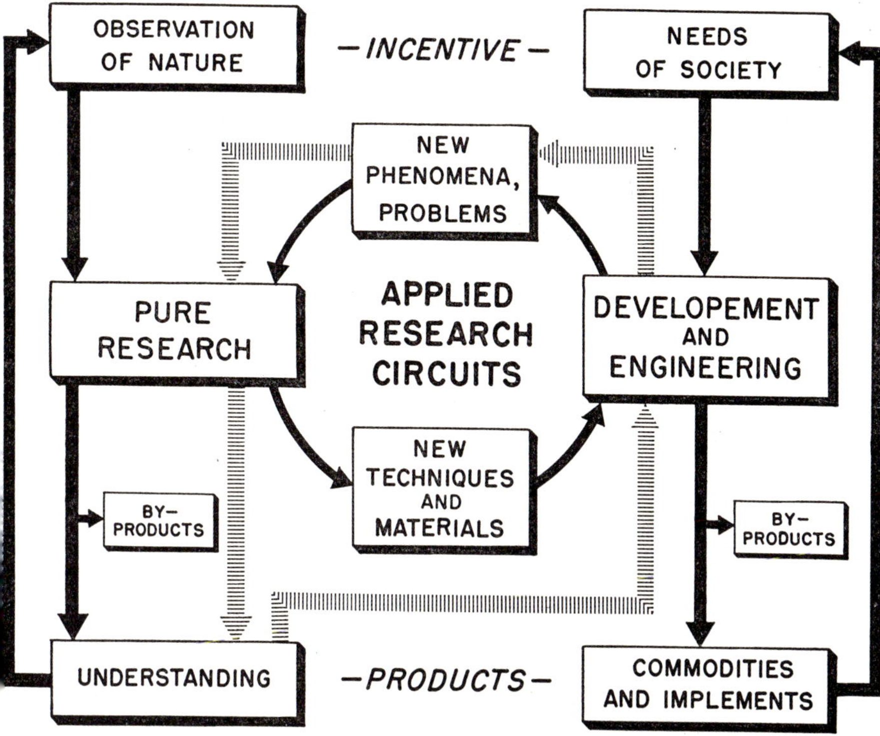

Figure 3. Regenerative circuits in science and technology.

ing pattern or general theory (lower left box) which enables us to understand the phenomena or events in the field of study, which comprehends all the facts, links them with facts from other fields, and enables us to predict verifiable new phenomena or events. It represents the major theories or patterns which accommodate large bodies of facts such as the Laws of Thermodynamics, the Laws of Motion, the Theory of Relativity, the Quantum Theory, Maxwell's Electromagnetic Equations, the Mendelian Laws, and so forth.

The thin, solid arrows represent the circuit in which facts and theories are subjected to scrutiny and verification by others. It is also the circuit in which new minds with fresh experiences can extend valid facts and amplify consistent theories.

The diagram brings out two very important feedback circuits. In order to extend and integrate the patterns and to assay their consist-

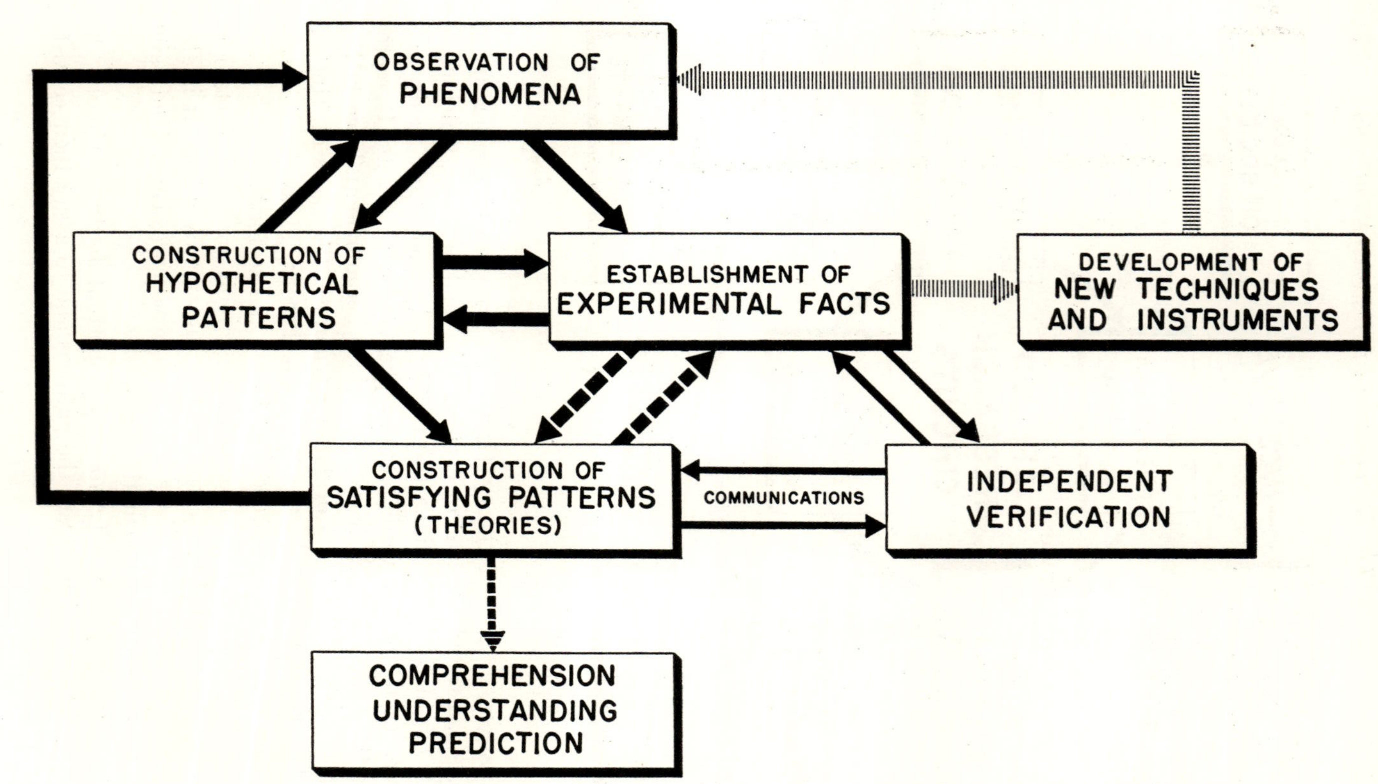

Figure 4. Regenerative circuits in scientific research.

ency over wide ranges of facts, it has been found necessary to seek for facts in every region susceptible to precise observation. The new substances, instruments, and techniques—we may even include concepts—discovered and developed in the course of one series of researches may be used to explore into new regions for more facts. The build-up in the circuit due to this positive feedback has been most spectacular; indeed, the history of natural philosophy is marked by milestones, each indicating the discovery of a new device or technique which opened up to human experience regions that were hitherto inaccessible. These devices were means to an end, but the end would never have been achieved without the means. Telescopes, microscopes, X-rays, radioactivity, cyclotrons, chemical analysis, electronics, high-speed computing machines have all been means of opening up new continents for valid experiences.

The arrow leading into the top box (observations of phenomena and events) also indicates a positive feedback, the autocatalytic effect of *understanding*. A satisfactory theory or pattern of facts broadens and deepens understanding, pointing the way to new fruitful fields where facts of significance, interest, and potential for application are likely to be discovered. It helps research men to make more intelligently the most important decision of all, namely the choice of problems in which to invest years of their lives. With the aid of new instruments, techniques, and methods, both experimental and theoretical, these decisions may be implemented and the investigators may pursue their researches into new and more complex fields with increasing facility and confidence.

There are regions of interest in science where it is not possible to make precise observations or accumulate facts under completely controlled conditions. In such cases the system works in a deductive mode through the feedback from "satisfying patterns" to observation. In cosmogony, for example, it is not practical to build up a theory of the origin of the universe from reproducible facts obtained from direct observation of the processes concerned. However, starting from a comprehensive pattern of facts from physics and chemistry and certain assumptions, it is possible to draw a theoretical picture of the origin of the universe in *sufficient* detail that certain critical consequences which are susceptible to observation may be deduced. Facts extracted from observation may then be compared with those deduced from theory. The history of the sciences mentioned shows clearly that as our satisfying patterns grow in depth and breadth, the deductions drawn lead to more and more pertinent and refined observations and our confidence in them grows accordingly. This circuit has found wide application in attacks on complicated problems or those dealing with past or future events. Its power depends on the existence of broadly based established patterns of facts, a condition which is sometimes not fully appreciated in attempts to apply "scientific methods" in new or complicated fields.

There is no loop in the diagram where negative feedback exists. This is a condition for rapid, even explosive build-up of activity in the system. Indeed, we may reasonably conclude that there is *built into* scientific research itself, a mechanism that can produce a *continually accelerating expansion,* a mechanism that has the power of breaking down the barriers that at any given time prevent exploration into particular regions of nature, a mechanism that makes undefinable the limitations imposed on the system.

This description of the methods of scientific research has been treated at length for two reasons: first, it gives an interesting generalization of the concept of a system and brings out one property of a system, namely, the important part played by systematic flow of inputs and outputs from one element to another; second, an ingrained conscious or unconscious faith in this method is an essential part of the background of any forward-looking technologist of tomorrow. In view of the complexity of his subject and the variety of knowledge it comprises, this is especially true of the systems engineer.

Returning for a moment to Figure 3, observe the domain of "applied research" which lies in between the extremes of pure research on the one hand and the practice of the useful arts on the other. This is the region in which nearly all of our larger research institutions now operate; it is the region where new systems are born and nurtured. Applied research has not been represented by a block with special incentives and objectives, because it shares these in common with the two end circuits. Instead, two circuits have been used to typify its functions and show its interrelations with the end members.

The hatched arrow indicates the demands of development and engineering for understanding. In the course of a development, new phenomena may be encountered or problems arise that require elucidation from a broad point of view. If brought to his attention, these stimulate the interest of the scientist, and the understanding resulting from his researches feeds back to broaden the basis on which the development rests, to predict promising modifications, or suggest remedies for troubles. The development of high-performance jet engines furnishes an excellent example of the working of the hatched circuit. These engines depend on combustion reactions in gases moving at relatively high speeds, and in the course of their development many significant problems have been brought to light in chemical kinetics, fluid dynamics, and thermodynamics—the need for a fundamental theory of flames has been accentuated. These problems have challenged the research physicists and chemists to develop understanding, and already their results have been fed back into the design of practical engines.

The circular arrow indicates a circuit energized by what are essentially by-products in the quest for understanding, new substances, techniques, or principles. The outstanding example of this circuit is the use of atomic energy, which applies on a large scale substances and techniques that were completely in the domain of pure research only a few years ago. Another current example is a by-product of solid-state physics (one of the more academic subjects in modern physics), namely the transistors. The circuit of circular arrows has already started to oscillate in the transistor field and revolution in electronics is in process.

Figure 3 presents a simplified and unified picture of the system of interlocking technical thought and action we call modern technology—the world in which our technical men work and for which they must be educated. Special emphasis should be placed on one feature brought out by the diagram, namely the essential place of scientific research in modern technology. Pure research is the source of understanding and understanding is the catalyst of technological progress. Research is a necessity, not a luxury, in modern life.

EDUCATION FOR SYSTEMS ENGINEERING

In the light of the foregoing discussion, it is suggested that an important goal in setting up a curriculum for the embryo systems engineer is to provide him with ability and confidence to operate in these applied research circuits. This may be approached from two directions. In the first place, we may require that the student specialize in one of the pertinent sciences, such as physics, chemistry, mathematics, or even some parts of biology, carrying his studies far enough to turn out a major piece of research. Whatever field he works in, he should acquire reasonable facility with mathematics. In this way he could get a sound appreciation of the methods and standards of scientific research. To this must now be added knowledge of certain branches of engineering, such as mechanical and electrical engineering, together with a familiarity with electronics and at least an appreciation of the problems of process engineering (general design for and methods of fabrication). Whether this is done concurrently with the scientific training or as graduate work in connection with an advanced degree is a matter which is left to those who have had more experience in teaching. Approaching the problem from the other direction, we might establish the initial requirements for systems engineering as courses in mechanical, electrical, communications, and process engineering with the basic science and mathematics courses required to back them up, followed by a year or two of research in a strictly scientific research atmosphere, i.e., an atmosphere where understanding is the prime objective and not commodities. Either approach would lead toward the goal outlined and could be supplemented by courses in cybernetics, information theory, servomechanisms, operations research, and human engineering that cover the subjects of special interest in the field.

Figure 3 suggests that there is a method of combining these two approaches that deserves special attention. The region designated "applied research circuits" has an old and well-established analogue, namely a university hospital. In the hospital professors from the medical school provide direct contact with scientific research in the biological sciences, including biochemistry and biophysics. This is the equivalent of the left-hand solid circuit. On the clinical side, corresponding to the right-hand solid circuit, the hospital is in contact with the urgent needs of society for relief from suffering, and its objective is to use all the resources of science and art to provide services to meet these needs. In the hospital itself, research in the medical and surgical fields covers the whole region from research for understanding to applied work to develop a new instrument, a new technique, or to explore new clinical applications of drugs, techniques, or methods of treatment. The intern may enter with a medical education that has laid stress on the scientific side, biochemistry, biophysics, physiology, and so forth, or with one that has emphasized the empirical side of the healing arts. In the hospital he has a chance to see both in operation and to come to understand the problems, the methods, and the standards of both.

The establishment, in association with the universities, of organizations similar to the hospitals, with analogous general functions but with technical objectives in the systems engineering field rather than the medical field, is worth careful and serious attention. It would provide for a well-rounded training of the systems engineer in a way that is difficult to visualize with our present broad gulf between faculties of philosophy and engineering.

The hospital analogy suggests another point. The overwhelming urgency of the practical problems, the saving of life and the relief of pain, brings before the intern the need for quick decisions and prompt action, a sense of the importance of timely action, a feeling of competition with the forces of nature, and a realization of the importance of all that is covered by the term "good human relations." In such an atmosphere, the man who has had a thorough grounding in the scientific and research aspects of medicine has an advantage—he is already conversant with the scientific side; the pressures of the surroundings force him to learn the empirical lore of his profession and cultivate the virtues just mentioned. The man whose previous training has been more practical or empirical, unless his inclinations lead him strongly towards the fundamental basis of his subject, will not be likely to acquire the same breadth of outlook. He will have the opportunity to do so but not the same environmental drive. The fragility of this generalization is admitted—factors such as the predilections of those who command prestige in an organization play a large role in determining environmental pressures—but, on the whole, urgent immediate human needs have a powerful appeal. This suggests that better preparation for a systems engineer would be a fundamental training in the physical sciences and mathematics with a minor in electronics and mechanical engineering. This would be followed by pure research work in fields fundamental to systems engineering, then by internship in an organization where pure research, applied research, and development are well balanced, and where the exigencies of time scales, specifications, economy, competition, and other practical problems bring out in him the capability of sound decisions and prompt actions.

SUMMARY

The design and engineering of systems is a laborious enterprise which mankind has brought on itself in its tireless striving for a better standard of living. It is another in a long series of ascending steps man has taken to use his mind to invent machines which can do in a professional way that which he himself has done in an amateur way. Systems engineering introduces us to a new order of complexity in that a system depends for the performance of its assigned functions on the intimate co-operation of a number of devices, each a complicated mechanism in itself. It, therefore, accelerates the obsolescence of the cut and dried empirical methods that formerly characterized all the useful arts and which still play a significant role in them, engineering and medicine not excepted. In systems engineering even more than in other arts, understanding, the product of scientific research, is the catalyst of technological progress. It is an essential for survival in a competitive world.

Underlying our design of systems are many principles derived from living organisms and especially from the human body. It is interesting to note that in applying these principles to systems of our own making, where our errors confront us squarely and no kindly Nature hides our mistakes, we must understand them in *quantitative* terms. This understanding may then be fed back to sharpen our knowledge and broaden our understanding of the source of the principle, the living organism itself. A picture of the interplay of the practical virtues of prompt decisions and timely and effective action on the one hand and the intellectual virtues of critical thinking and imaginative understanding on the other in the development of modern technology leads to some suggestions about the education of systems engineers. In particular, the recognition of systems engineering induces the recognition of the need for an engineering analogue of the hospital in a university structure.

If asked to comment on the attributes needed in a successful systems engineer, one might do well to quote Francis Bacon.

> . . . There is one principal and as it were radical distinction between different minds, in respect of philosophy and the sciences; which is this: that some minds are stronger and apter to mark the differences of things, others to mark their resemblances. The steady and acute mind can fix its contemplations and dwell and fasten on the subtlest distinctions; the lofty and discursive mind recognizes and puts together the finest and most general resemblances. Both kinds, however, easily err in excess, by catching the one at gradations, the other at shadows.

The successful systems engineer so balances the attributes of discrimination and association that he recognizes real from trivial differences, substantive from illusory resemblances; he is lofty in concept; knowledgeable and wise in selection; critical, alert, and scientifically exact in analysis; and in synthesis, imaginative in plan, meticulous in operation, and practical in execution.

Five

THE SYSTEMS ENGINEERING PROCESS

ALEXANDER KOSSIAKOFF

INTRODUCTION

The purpose of this chapter is to examine the problems associated with the engineering of a new complex system. The growing magnitude of the tasks being undertaken in systems engineering has created problems in organization and execution which severely tax the ingenuity and capacity of many of our major industries, government-sponsored laboratories, and even government itself. As has been pointed out previously, the lead time for developing and putting into use a new weapons system requires approximately ten years. In the competitive military situation in which we find ourselves, this long lead time puts us at a very real disadvantage in providing a defensive and deterrent strength adequate to prevent future war. Thus, the payoff for reducing lead time, and with it the colossal cost associated with producing new systems, is of enormous importance. In commercial systems the time factor is not as vital in a national sense, but is no less urgent to the particular industry concerned. The purpose of talking about the systems engineering process is to try to find some of the general principles which govern it, and then to see if the logical consequences of those principles lead to improved methods of attacking the problem. It is much easier to explain why the process is so long and costly than to find practical ways of improving it significantly. However, the attempt to understand the process is worth making.

Orders of Automation

It is of some interest to examine a recent classification system for the order of automaticity for machines based on the kinds of human faculties embodied in the machines.[1] Table 1 gives the characteristics of each proposed class and examples of devices and machines having this order of automaticity. Hand tools which replace no human energy or control functions are classified as machines of zero order. Power machines such as bench saws, pneumatic drills and the like, which supply energy to do the basic machine function, are defined as first-order machines. The second order introduces the function of automatic feed, but requires the operator to do the component setup, and on-off control. Third order introduces the first true "automatic" machines, which have the function of open-loop control. Examples are turret lathes, screw machines, and many others. These automatic machines were fundamental to the advent of mass production.

We now come to the first basic elements of systems with the introduction of closed-loop feedback control. This brings in the functions of perception and response, which distinguish animate organisms from inanimate matter. This is defined as the fourth order of automaticity. The fifth order introduces the function of performing calculations, in which machine control is based on automatic solution of control equations. Finally come machines obeying complex logic by performing "deductive reasoning." Examples of such machines are telephone circuits, elevator controls, and the like. This is the order of automaticity which is characteristic of our present day complex systems. With these faculties it is possible to combine in one ordered entity a multitude of individual components, supplied with power, connected by control links, matched to one another by computing elements, and coordinated by switching mechanisms which exercise the processes of selection to adapt the system to varied input conditions.

As a matter of interest we might look at the remaining classifications which conceive of orders of automaticity beyond those which we have been able to realize up to now. The seventh order is defined as machines able to learn by experience—a machine which learns from its mistakes and attempts different modes of operation as necessary. Some

[1] George H. and Paul S. Amber, "A Yardstick for Automation," *Instruments and Automation,* April, 1957.

Table 1. Orders of Automation

Order of automaticity	Human faculty accomplished by machine	Characteristics of this order of automaticity	Examples of devices and machines having this order of automaticity
Zero Order	None. Replaces no human energy or control function.	Includes all hand tools. They increase worker's efficiency, but replace no human function.	Shovel, knife, hammer, pliers.
		Includes all hand energized machines. They give mechanical advantages, but replace no human function.	Block & tackle, pencil sharpener, bow & arrow, pump, can opener.
First Order	Energy required to do basic machine function.	Uses mechanical power source, but machine feed and control completely dependent upon operator.	Cement troweling machine; bench saws; pneumatic drill; portable electric tools; floor polisher; wood lathe.
Second Order	All energy required to do basic machine function. Also automatic feed.	Uses 100% mechanical power, but operator must do complete set-up, and on-off and control.	Radial drill, bench lathe, power hack saw, pipe threading machine.
Third Order	Controlling the machine.	Completely self acting—the first true "automatics." "Open loop" performance.	Turret lathe; transfer machine; grinding machines; gear hobbers; screw machines; special manufacturing machines, such as for cigarettes, lamp bulbs, and bottling.
Fourth Order	Monitoring machine performance.	Measures machine performance, compares to a standard, corrects machine as necessary. "Closed loop" feedback control performance.	Ruling engine, honing machines, cold reduction mill, chemical process plants, oil refineries, generators, speed regulators, level controls.
Fifth Order	Performing calculations.	Machine control is based on automatic solution of control equations.	Machines that use small special purpose computers such as machine-mean computer; standard deviation computer.
Sixth Order	Obeying complex logic, performs "deductive reasoning."	Machine control is based on automatic solution of complex formal logic conditions (A_5 and A_6 overlap somewhat).	Telephone circuits, bowling pin spotter, elevator controls, cypac machine controls, "Robots."
Seventh Order	Learning by experience.	Machine learns from its mistakes, and attempts different modes of operation as necessary, and improves its techniques.	A paper talk machine so far, which is theoretically possible, but has not yet been attempted.
Eighth Order	Performs inductive reasoning. Has intuition and executive type judgment.	Machine extrapolates from its experience, and forms modes of operation beyond actual experience, resembling intuitive operation and judgment.	This could be an automatic Operations Research machine that sets up much of its own local programming.
Ninth Order	Creativeness—originality.	Machine comes up with original, creative concepts, and can work beyond its programming.	As there is no agreement what constitutes creativeness for men, there will be no agreement on creative machines. Do we want a machine of this type?
Tenth Order	Dominance.	Machines give the orders to the operators and designers.	Science fiction writers take over from here. A dictatorship by machines!

very interesting work is being done to endow simple machines with the capability to learn. A maze follower which learns how to go through a maze by trial and error, and has a memory which remembers fruitless turns and minimizes their recurrence, is an example of these first attempts and the so-called "Perceptron" is another. The eighth order of automaticity introduces the faculty of inductive reasoning where the machine extrapolates from its experience and has functions resembling intuition and judgment. The ninth order introduces creativeness and originality, and the tenth and final order is the logical culmination of mental power, namely, dominance. As it is said, "Science fiction writers take over from here." With the difficulties we are experiencing with systems possessing lower orders of automaticity, the higher orders proposed sound fantastic. But with the rapid and accelerating growth of technology the fantastic has a way of becoming real.

Forces which Generate New Systems

Much has been said in other chapters regarding the forces which bring new systems into being. At this point it is only necessary to emphasize one aspect which strongly affects the environment in which new systems are developed: the pressures generated by these forces can be extremely high, particularly in the creation of military systems and in commercial fields where competition is strong. This puts a strain on the systems engineering process which often strongly colors the decisions made. The manner in which these pressures operate will become more obvious later.

Problems of Systems Engineering

The process of engineering a new system consists of the solution of a series of problems involved in the assembly of a large number of diverse elements into an ordered whole. The solutions of these problems are basically different from those encountered in science. The terms of reference of each problem cannot be stated precisely; they are not solvable in closed form. Each problem has an infinity of solutions. The selection of the proper solution rests on judgment of the relative importance of incommensurate quantities. The problems must be solved by an iterative process of successive approximations. The parameters in the problems are time-dependent. To carry out the process requires a large number of highly trained persons. It is time-consuming and costly. The final result is a compromise between the initially desired objectives and the capacity for realizing them with the manpower, time, and funds available. These are some of the problems characteristic of systems engineering.

Method of Presentation

It is difficult to present the elements of the systems engineering process in a logical and ordered fashion. This difficulty stems from the very nature of the systems concept, namely, the complex interaction of the individual elements of the system. As in the system itself, the various elements of the systems engineering process are inextricably connected with one another by a series of interlocking feedback loops. There is no unique way of arranging these elements one after another.

Fortunately, chronological presentation is not subject to the feedback problem. Therefore, this chapter will examine the problems encountered in the engineering of a new system as they might occur sequentially, during the course of the development and engineering of the system. Using the chronological approach, then, the systems engineering process will be discussed in three sections: first, the initiation of a program to engineer a new system, which culminates in a decision to proceed with a full-scale development and engineering program; second, the engineering of a so-called systems prototype, which culminates in the construction and test of a full-scale system having all of the functional characteristics of the final product and which provides a realistic demonstration of the potential effectiveness of the system; third, the engineering of the final system, which culminates in finished drawings suitable for production of the requisite number of complete systems for operational use. The three steps of Figure 1 represent successive approximations to the final system, starting with a preliminary design, proceeding to a functional prototype, and ending with the final article. It should be pointed out that in complex systems there may be additional steps, such as, for example, the construction of several successively improved prototypes in case the initial step falls too far short of the desired goals and cannot be used as a basis for final design.

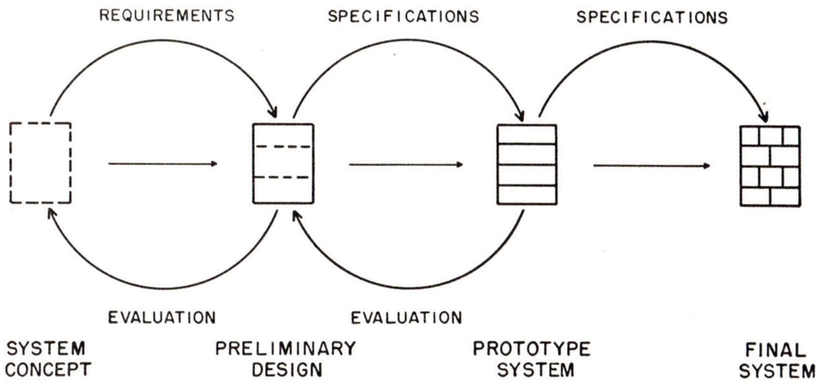

Figure 1. Phases of systems engineering.

Differences in Function of Systems Engineers in System Development

At this point attention should be drawn to differences in possible descriptions of the systems engineering process that stem from a difference in selection of the type of system chosen for primary consideration. In guided missile systems, for example, systems engineering continues through the development phase and well into the production phase in an active capacity. This is because it has been found impractical to design a guided missile system without building a hardware prototype and demonstrating in flight that it fulfills the functional requirements for which it has been designed. The functional interactions between the various elements of the missile and its environment are just too multifold and complex to be adequately analyzed on paper or in the laboratory. In actual practice the design and optimization of the system proceeds throughout the prototype design, construction, and testing phase.

A possible distinction may be made between the systems engineer and the development engineer—for the case in which the particular component for which the development engineer is responsible can be accurately specified by the systems designer and developed successfully within this specification. Otherwise, the systems engineer can phase out only when the components of the system become "stable." When this is not the case, it becomes necessary to alter components during development in such a way that they affect the design of other components, and the operation must be under the control of the systems engineer. He alone can determine the degree to which such interactions can occur without violating the integrity of the system as a whole. The role of the systems engineer in the development of a new system therefore depends on the degree of system stability which can be achieved in the various phases of the development. For systems whose elements can be prescribed in the early design phase, the problem can be turned over to the development engineers, with participation by systems engineers on a follow-through basis only. In systems where the components interact strongly, and particularly where there are environmental factors which cannot be fully assessed prior to the construction and test of the full-scale models, the systems engineer must remain in control of the technical program until the final stages of prototype development or even final production design.

INITIATION OF SYSTEMS ENGINEERING PROGRAM

Before proceeding with the full-scale development and engineering of a new system, extensive planning and exploratory studies are often needed. This is a process that requires a thorough and objective appraisal of the need for the new system, of the concepts applicable to fulfilling this need, and of the technology available for bringing the system into being. In selecting the most promising approaches, it is necessary to assess the uncertainties that might be encountered and to take all of these factors into account in appraising the time, manpower, and cost of implementing the system development. Obviously, the more thoroughly and expertly this is done, the more certain will be the outcome of the entire program. This section will present several factors pertinent to this phase of systems engineering which are considered to be especially important in the development of military systems.

Uncertainty of Program Objectives

In the design of a military weapons system a basic difficulty arises in the definition of the objective which the system must be designed to meet. Every weapons system must be matched against a presumed enemy capability to defeat it. This capability can only be guessed, using fragmentary intelligence estimates of the enemy's weapons development and the technological state of the art, and by comparison with our own achievements in similar systems. Furthermore, these capabilities must be estimated many years in advance of the time at which our own system will become operational. In many ways the best criterion of enemy capability comes from an assessment of our own, but this process somewhat resembles a dog chasing his own tail, by effectively making us compete with ourselves. This is a job for operations analysis. Thus, at the outset the military weapons system designer is posed with a problem whose solution is complicated and uncertain. This factor leads to a basic instability in the weapons system concept and often results in changes in objectives caused by revised estimates of enemy capabilities.

Conflict Between Time Scale and Performance

Fundamental in the competitive world is the factor of time. In order to realize an advantage from a new superior system, it must be produced before a competitor can produce one equally good or better. Thus, an essential objective of systems engineering is the production of a superior system in *minimum time*. Furthermore, even if the race is won, the advantage gained is always temporary—eventually the competitor will produce a system still better, which will "capture the market" unless advanced versions of the first system can be produced early enough to maintain the initial superiority. In a very real sense one must run in order to stand still.

This situation brings about a crucial problem in systems engineering—the conflict between performance objectives and time scales. The underlying factors can be described by a series of statements:

1. A new system is of no value until it is put into use.
2. Development of a new system characteristically requires a long period of time from conception to use.
3. The length of time required cannot be predicted accurately because it depends on the successful solution of many problems, some of which require varying degrees of invention.
4. Technology is advancing rapidly, thus tending to make obsolescent any long-term development which is based on techniques now in hand.
5. To survive competition, the system must simultaneously exceed in capability and precede in time any probable competitive system.

The above statements describe the awesome dilemma which confronts any organization which undertakes a new system development. The choices, therefore, are these:

1. A new system can be developed in a relatively short time if it is based on proven techniques. If this is done its margin of superiority over existing systems is not likely to be large—possibly not sufficient to offset the costs of development, engineering, tooling, and the host of other costs incident to the introduction of a new item. It will probably be obsolete in a relatively short time.
2. A more advanced new system can *probably* be developed by utilizing promising, but yet unproven, techniques. The time and cost required to bring such a system into being will certainly be greater. Neither can be predicted with assurance—perhaps within 50 per cent or 100 per cent. During the same time the competitor may bring in a new system of his own, which may capture the market.

This situation is illustrated in Figure 2. In this figure is plotted the relationship between the performance sought for and the probability of achievement of a hypothetical system. The three curves represent three different approaches based, respectively, on a very short, an intermediate, and a relatively long time scale. The curve for the very short time scale represents an approach in which a large risk is taken on the attainment of the desired solution quickly, with a correspondingly high probability of failure. The intermediate curve represents a

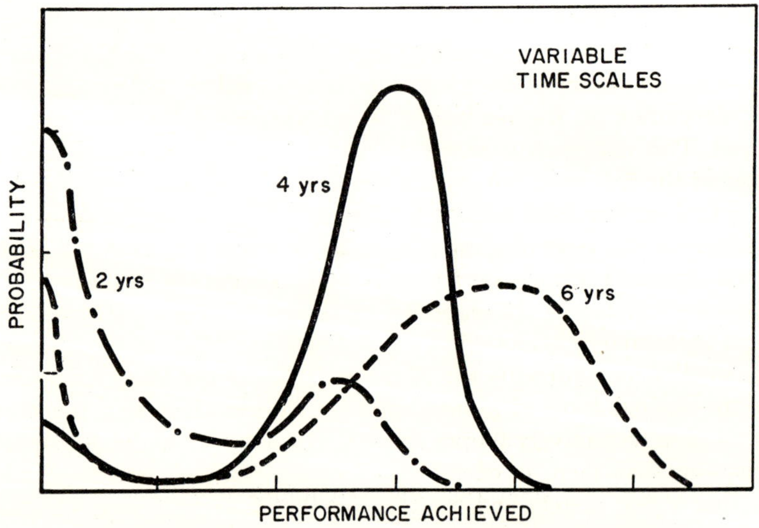

Figure 2. Probability of attaining given performance.

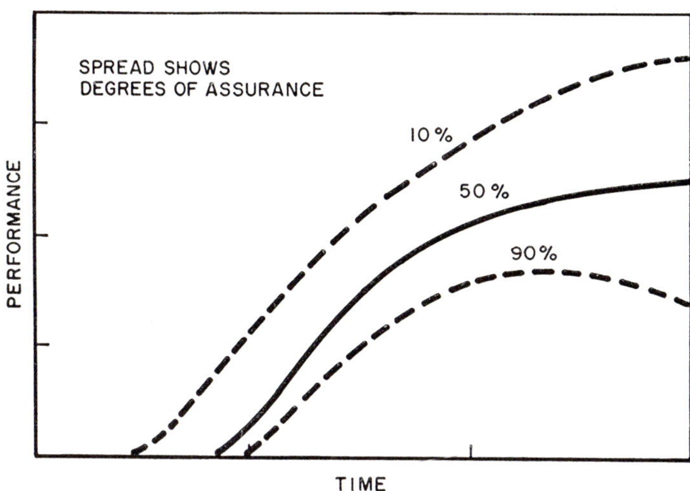

Figure 3. Time of attaining given performance.

relatively conservative approach, both from the standpoint of time and expected performance. It is shown to have a correspondingly high probability of success and a relatively narrow margin of probable departure of the final performance from that sought. The third curve represents an approach designed to achieve a higher performance system, depending on the realization of advances in technology not yet proved. This approach is shown to attain a higher performance than either of the first two, but with a considerably greater uncertainty in the actual performance achieved. Likewise, it has a significant chance of failure if the technological advances do not materialize and if the system approach cannot be altered to use existing techniques.

Another version of the same set of curves is given in Figure 3, in which performance is plotted as the ordinate and time, as the abscissa. The solid curve represents the 50 per cent confidence level for achieving the desired level of performance in a given time, and the other curves give respectively 90 per cent and 10 per cent confidence levels. It is seen that there is effectively a minimum time associated with creating a new system and that the likelihood of achieving a certain performance in a given time tends to level off. This presupposes that in each case a certain approach is taken and held to throughout the program. Actually, of course, the approach can be altered during the process, with the result that the short term approach is spread out with a commensurate increase in probability of success. Likewise, in most cases the probability of failure in the long term approach can be ameliorated.

The obvious consequence of the above illustrations is that the time factor must be used as an explicit variable in selecting the method of approach to the problem. The performance sought for and the time allowed must be *optimized* with regard to the over-all profit expected. The method of approach cannot be selected on the basis of philosophy alone—an organization which operates on an unswerving philosophy of sticking to proven techniques will be overtaken by a more daring concern. One which operates on blue-sky optimism will go bankrupt through failure. Such philosophy must be replaced by a very careful and thorough analysis, based on the best obtainable intelligence of the probable action of the competition, of the state of the market, and of the technological prospects of all major components of the system. A management decision can then be made which still involves a gambling risk, but in which the odds are heavily weighted on the side of success.

Conflict of Approaches

It was mentioned earlier that one of the fundamentals in designing complex systems is the effect of strong interaction between the diverse system elements. Another chapter illustrates the relations between the principal components of a guided missile. To elaborate a little on that picture, the guidance and control systems of the missile must bring it within lethal range of its target before the time that the target can release its destructive payload. The accuracy of the guidance system is dependent on the characteristics and tactics of the target, on the intelligence concerning target location from the ground, and on the characteristics of the air-borne guidance equipment. The closeness of miss is therefore determined by the combination of the accuracy of guidance intelligence, missile maneuverability, and target evasive tactics. The closeness of approach to the target which is required of the guidance system is dictated by the lethality of the warhead and the vulnerability of the target. The size and power of the propulsion system is dictated by the weight of equipment that the missile has to carry, its speed, and the maximum range of target intercept. The weight of structure is likewise a function of the equipment size, power plant size, and maneuverability requirements on the missile.

All of these factors interact closely with one another. Guidance accuracy can be relaxed as the warhead is made larger and more lethal. This in turn would require an increase in missile weight and a consequent decrease in speed and range, or an increase in power plant size and an even greater increase in total weight. Power plant requirements could be relieved by use of lighter structure and decrease in the weight of guidance equipment. However, lightening the structure may make it so flexible that body vibrations will couple into the control system and produce instability. This would require lowering the gain of the control system with consequent loss in maneuverability and decrease of effectiveness. These examples can be multiplied many times to illustrate the kinds of compromises which are forced on the systems designer.

The result of over-optimizing a guided missile in a particular respect is illustrated by Figure 4. This picture shows the necessity for the mixed team approach for the engineering of a balanced system.

This factor can be illustrated more fully by looking at a particular subsystem of a missile, namely the control system. This is a problem of special complexity because of the diverse elements involved in its solution. The missile control system represents a closed loop in which the control forces are generated by aerodynamic lift; the control surfaces which produce them are actuated by hydraulic servos; the response is affected by structural deflections and vibrations of the missile body; the motion is sensed by electromechanical instruments such as gyros and accelerometers; and the appropriate signals to the servos to correct and damp the flight path are generated by electronic circuits. The control equations are complicated by non-linearities in aerodynamic forces as a function of control surface deflection, by variations of these with flight speed, by induced weathercock oscillation of the vehicle as a whole, and by changes in aerodynamic moments produced by distortion of the control surfaces under aerodynamic load. Furthermore, the air through which the vehicle travels is not still but turbulent, and the vehicle must tolerate sporadic gusts as well as prevailing winds. The design and analysis of the closed-loop behavior for a missile control system requires an aerodynamicist, a servo engineer, an instrumentation engineer, an electronic engineer,

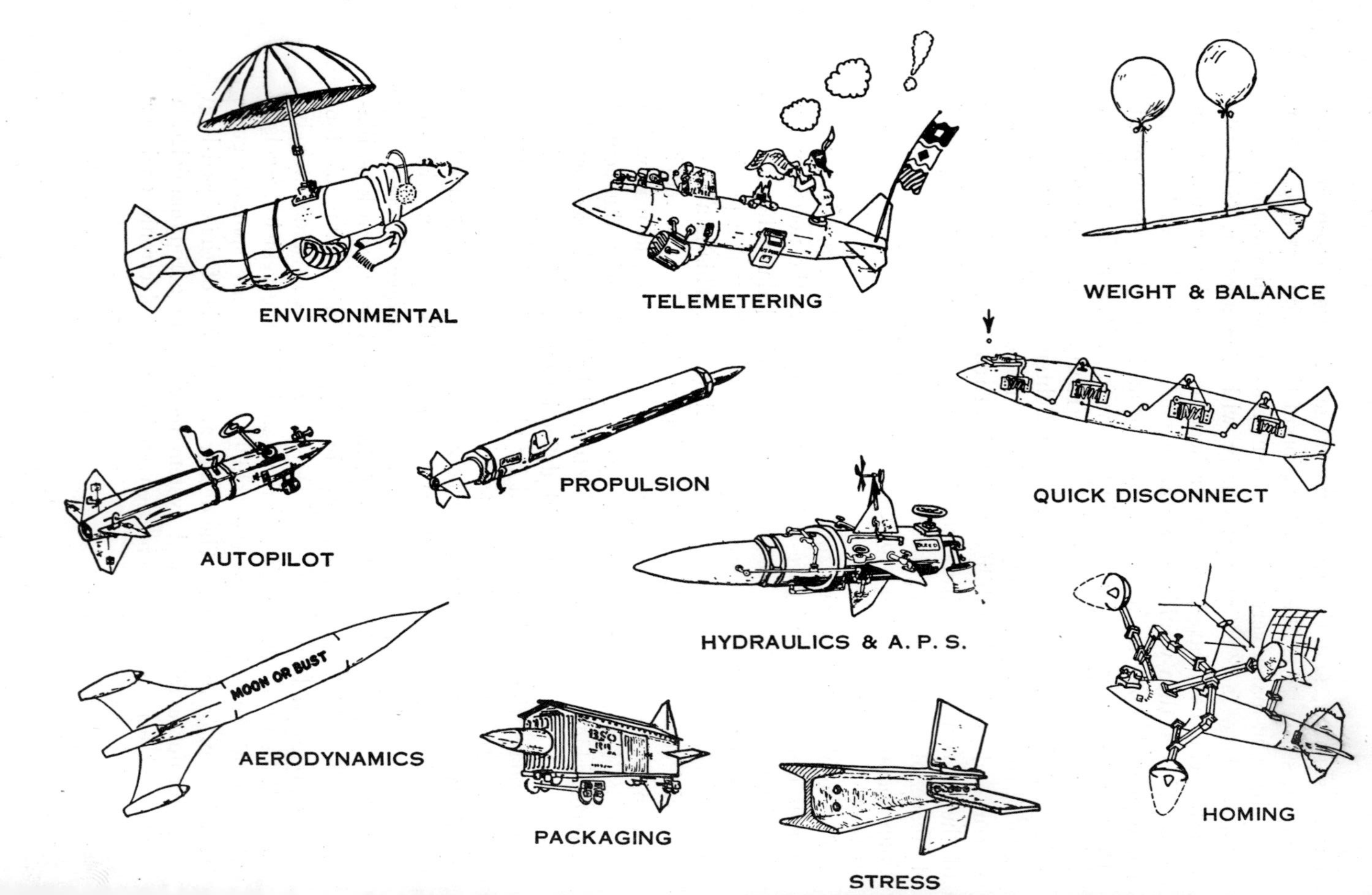
ENVIRONMENTAL
TELEMETERING
WEIGHT & BALANCE
QUICK DISCONNECT
AUTOPILOT
PROPULSION
HYDRAULICS & A. P. S.
MOON OR BUST
AERODYNAMICS
PACKAGING
STRESS
HOMING

and a dynamicist aided by data from wind tunnels, simulators or analog computers, structural tests, and component tests.

In designing a control system it is not enough to assemble a mixed team representing all of the above specialities. Each of the individuals trained in these fields speaks a different language from the others, which is almost totally unintelligible to anyone not expert in the particular field. This is an area which urgently requires engineers trained in control system design who can understand the language of all of the specialists concerned. With the support of the mixed team (which has the necessary depth of understanding in each of the requisite fields), a control system engineer can optimize the systems and meet the over-all requirements with due regard for the most effective compromise which can be made among the techniques employed.

Appraisal of Chances for Success

At the conclusion of the preliminary design study for a new system, the systems engineer has to present management with an objective appraisal of the chances for success in undertaking the system developed. Figure 5 is a symbolic representation of some of the factors operating for and against the successful development of a new system. These factors in themselves are obvious. It is necessary to point out, however, that the very idea of weighing the chances for success involves a quantitative measure of the factors concerned. Even though it may be extremely difficult to appraise some of these factors quantitatively, it is nevertheless necessary to attempt to do so, using the best information available. Of particular importance is the appraisal of the technical risks and of the probable time and effort required to carry the program through to completion. It is here that the problem of performance objectives versus time scale must be examined most critically in order to select the method of approach that is most likely to provide a successful result in the minimum time and for least cost.

ENGINEERING OF SYSTEMS PROTOTYPE

Subdivision of Problem

The decision having been made to undertake active development of a system, though not necessarily to put the item into production and use, the problem is to define, organize, and carry out the development

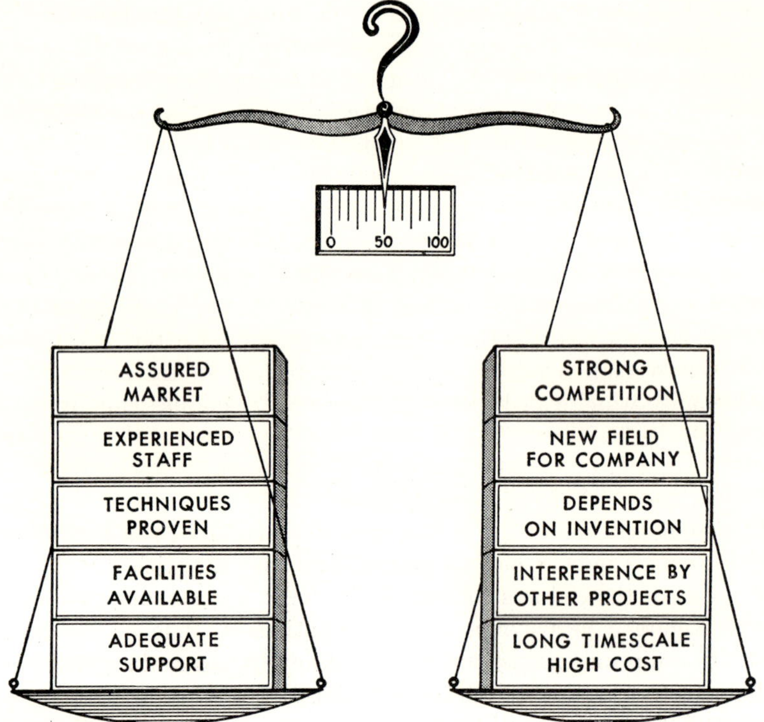

Figure 5. Balance of chances for success.

of all the functional elements of the system, to assemble them into a prototype, and to evaluate the performance of the prototype in terms of the objectives which it has been decided to meet. The general development plan has already been prepared in the initial phase of the program for the purpose of estimating the time and effort required for its completion. However, it is necessary now to prepare a detailed plan in terms of definite assignments of tasks to specific organizational groups who will be requested to carry out the numerous jobs of component analysis, development, and test.

It has been mentioned that one of the basic problems in systems development is the complex interaction between the various components which make up a system. If these interactions are not minimized to the greatest possible extent, the conduct of the program will require so much communication and control of the various groups engaged that the organization will break down and produce a system that will not work as a whole. *The technique which must be used to minimize component interaction is to organize the components into subsystems in such a way that most of the interactions occur within the subsystems, while the individual subsystems interact as little as possible with one another.* It is important to do this not only functionally but also physically if the full benefits of such a procedure are to be realized. This corresponds to the collection of closely interacting components into physically separable sections which can be treated as separate entities during development and preliminary evaluation. Proper and improper sectionalization are illustrated schematically in Figure 6.

This concept of sectionalization has been found to be particularly important in the guided missile field where interactions between functional elements are extremely strong. In many of the early designs of missiles, components were located mainly where they could be tucked into a minimum volume. If a hole existed somewhere in the structure, and if there was a black box of the right size and shape to fit into this hole, it was frequently put there. This not only precluded the possibility of building and testing functionally unified subsystems but also made the missile a maze of cables and hydraulic lines. This arrangement made the manufacture and testing of missiles unwieldy and promoted unreliability due to the incompatibility of components located in close proximity to one another. Modern missile design has

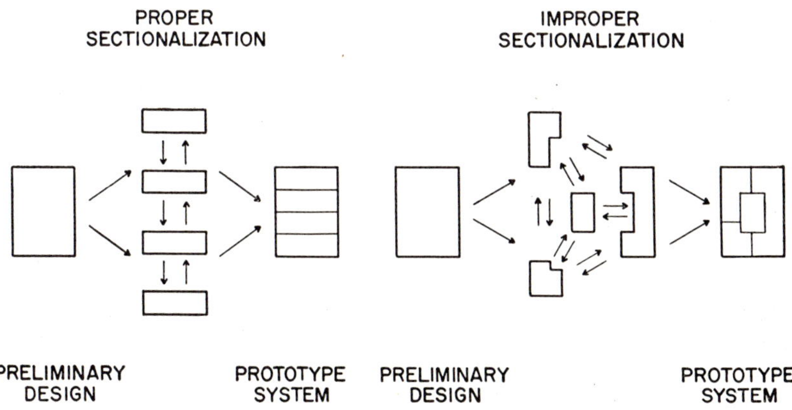

Figure 6. System sectionalization.

recognized the advantage of sectionalization and uses it as a dominant factor in determining the physical arrangements of components in the missile. Missile components, such as the guidance intelligence system, autopilot electronics, servo and actuator system, warhead, propulsion system, and primary power system, are generally built as physical entities which can be removed, tested, and replaced. This is a similar concept to that of modular packaging, which is becoming standard in military electronics.

The organization of the system into subsystems and sections culminates in the preparation of specifications for each subsystem so that when assembled they will operate as a system. The problem of their development can now be assigned and a program undertaken in earnest.

Organization of Program

When a detailed technical plan has been established for attacking the various portions of the system, it is necessary to organize the effort which will be required to carry through the prototype development. This is an area that should be given particular attention by any organization undertaking a major system development program.

The matter of organization is one where opinions differ widely among persons engaged in development and engineering activities. It is often said that the course of a program is determined much more by the individuals who are responsible for it than by the formal organization in which they work. With the wrong kind of staff, the job will not be done well regardless of how the organization is laid out. On the other hand, the same persons who claim philosophically that organization is to a large extent irrelevant often hold strong ideas as to what the "right" form of organization is. While there are a number of different schools of thought, there are two which predominate in industrial engineering organizations. These might be called the *departmental form* and the *task-force form* of organization.

In the departmental form of organization, the engineering department consists of a number of semipermanent groups staffed by specialists in the various fields of engineering. For example, the groups might be aerodynamics, structures, dynamics, microwave, circuit design, electronic packaging, etc., depending on the fields in which the engineering organization operates. In addition, there are groups responsible for providing services and facilities for the departments, such as drafting, environmental test, field test, experimental shop, etc. In this type of organization, projects are handled by assignment of a project engineer, who reports either to the chief engineer or to a chief project engineer. The project engineer has the responsibility of co-ordinating the engineering activities carried out within the departments. To carry out this function he is assisted by a small staff responsible for technical liaison, planning, scheduling, and monitoring of program costs.

In a task type organization the engineering department is organized entirely in terms of projects, except for service groups of the type mentioned for the departmental organization. Each project has reporting to it a number of groups staffed according to the magnitude and needs of the project.

In trying to assess whether either of these or some other form of organization is best suited to carry out a major system development, it is necessary to examine the requirements which the nature of the problem places upon the organization. The following considerations appear to be pertinent to this question.

1. MAGNITUDE OF FORCE ENGAGED

In a typical large-scale systems development program the total number of persons involved during the hardware development phase of the program may include hundreds of engineers and a total force numbering a thousand or more. This is particularly true if one includes not only the systems development organization itself but its subcontractors. This can hardly be called a technical team; it is more like a technical army.

2. NECESSITY FOR CONTROL

The magnitude of the effort requires that close control be exercised on the allocation of manpower to various tasks, on the design of components which interact closely with one another, on the adherence of all portions of the effort to the program schedule, and on the running costs of the program. Without such control there is likely to be considerable unco-ordinated activity which, if allowed to get beyond reasonable limits, can throw the program completely off schedule.

3. NECESSITY FOR DECISIONS

Whenever any portion of the program uncovers a new problem, either an unforeseen difficulty or a breakthrough, it is necessary to make a prompt decision as to how this problem should be dealt with. This decision cannot always be made at the section level if other parts of the system are affected. It is clear, therefore, that the magnitude and diversity of the effort and the pressures under which it is conducted require a high degree of organizational control over the program throughout its expensive and critical phases.

4. ANALOGY TO A MILITARY CAMPAIGN

Men have long experimented with organization to find methods best suited to carry out a particular type of task. The question one asks himself in considering an organization for a relatively new type of operation, such as engineering of a large-scale system, is whether or not there are other large-scale operations which have significant similarities to the systems engineering process whose organization may give a clue to this problem. One is impressed by the resemblance of the process of engineering a complex system to that of waging a military campaign. A military campaign requires a large, highly trained, organized body of persons (i.e., an army) conducting a connected series of operations (i.e., a campaign) to bring about the desired result.

In this sense, the direct objective is not waging war in the violent or hostile sense, but in the more restricted sense of a contest or struggle. The contest may be with a potential enemy, as in weapons systems development, or with a potential competitor, as in the case of commercial systems. In either case intelligence must be obtained regarding the probable tactics of the adversary, and the campaign laid out to overwhelm him. The campaign must be carefully planned by the equivalent of a general staff, making the best use of the available strength and weapons. The execution of the campaign must be carried out under strict discipline to make certain that each individual does his part in the over-all effort. The entire operation may take three to ten years to complete—a time directly comparable to a full-scale war.

At first sight the comparison of the systems engineering operation to a military campaign may meet with intuitive objections. Scientific progress has been mainly achieved by the creative work of individuals —not a disciplined mass of technicians. As a concession to the requirement for joint action by diversely trained people, the concept of a technical "team" has been generated. Men trained as professional scientists, engineers, or businessmen learn to think as individuals and find military discipline distasteful.

Most of these objections tend to disappear upon closer scrutiny of the situation. In the actual process of engineering a system, new principles and knowledge are not being sought—existing knowledge is being applied to specific problems. The undertaking is too large and costly to depend on discoveries being made along the way. True, discoveries are sometimes made during the process, but they are the by-products, not the necessary ingredients. Again, since the process is so long, advances in technology often affect its outcome. The same is true in a military campaign. It must be waged on the basis of forces and weapons in hand, although new weapons often are introduced to great advantage during the operation.

The team concept applies in any co-operative operation. The systems engineering process utilizes a large number of teams to attack the various problems encountered. In a military campaign, the individual small units also operate to a large extent as teams. But the team is inherently an organization of a small number of coequal individuals and cannot be extended to as vast and diverse an enterprise as the engineering of a large-scale system. The systems engineering operation must be executed in accordance with an established time schedule and within a prescribed cost—otherwise the success of the entire operation may be jeopardized. Such an operation calls for a highly organized effort, with well-defined lines of responsibility and authority. The diversity of skills required means that only a few can participate in all decisions; thereafter the scope must be circumscribed to areas within the capabilities of the individuals concerned.

The decision process in systems engineering is very similar to that in military situations. The basic problem is too complex for the effect of all factors to be known with reliance. Thus, decisions always involve an element of risk. Most problems can be attacked in several ways, but the manpower available is never sufficient to attack in all directions; a decision must be made on the one most likely to succeed. Unforeseen difficulties arise often, and, unless dealt with decisively, can throw the entire effort into confusion. Someone has to decide quickly how to reorient effort to maintain the momentum. The basic plan may itself change because of new external circumstances dictated by policy—the effort must be quickly adapted to the new conditions. The decision-making process in systems engineering may be described aptly by the term "technical generalship." The systems engineer himself can be thought of as a technical general. The term "general" well describes both the breadth of judgment and the level of decision required of the systems engineer.

5. APPLICATION TO SYSTEMS ENGINEERING ORGANIZATION

The analogy can be carried further to see how it might relate to the type of organization which is appropriate to systems engineering. In this case it is necessary, as with any analogy, to pick out only those features where real similarities exist, rather than be carried away with the idea and push it well beyond its usefulness. It would be foolish to suggest that all of the persons engaged in the systems engineering program be put into uniform and given military rank.

It seems that the following general principles of organization which are used in the prosecution of a military campaign can well be applied to the organization of systems development.

(a) *Mobilization.* When a military campaign is undertaken, resources necessary for its prosecution are assembled and put under the command of a general. In the same manner, major systems development requires the mobilization of manpower selected to carry out the specific job at hand. Such an effort cannot be absorbed into a general purpose organization, with authority diffused among a large number of semi-independent supervisors. This is one difficulty encountered in attempting to utilize a departmental organization for systems engineering.

(b) *Authority.* It is necessary that a single individual have charge of the systems development program as a whole. He still must operate within the framework established by management for the conduct of the program. However, he must have authority for making technical and program decisions within this framework. By virtue of the magnitude of the job he has to supervise, he must be highly placed in the organizational structure, so that his decisions will be followed by the organization at large. He must be a program supervisor, not simply a program co-ordinator.

(c) *Staff.* No single person can make all the decisions necessary to wage a successful campaign. In a similar manner no single individual can make all the decisions involved in a large-scale technical development. Nor can the decisions affecting the entire system be delegated to low levels in the organization. This calls for existence of a strong systems engineering staff possessing the necessary depth of knowledge in all of the component fields and working as a mixed team to formulate the decisions required to keep the program moving.

(d) *Task forces.* The general waging a campaign has at his direct command the forces required to engage the enemy in the field. In the midst of battle there is seldom time to call for assistance from home bases. In like manner the man in charge of the systems development program should have under his direct supervision those development groups which must work most closely together throughout the course of the program. This organization should not be static, but should be built up as the program proceeds and demobilized when it is completed.

(e) *Support forces.* The supply of the field armies in terms of ammunition, new weapons, routine services, and specialist functions is handled by centralized supporting forces which serve all of the armies. Similarly, the systems development groups should be supported by specialist and service groups which are not under the supervision of the program supervisor. This type of organization serves to make the best use of common facilities and specialists who can best work with each other away from the fighting front.

This analysis calls for an organization which is a compromise between the departmental and the task force types of organization prevalent in most industrial companies. Of the two types it is closer to the task force type of organization, in terms of the authority vested in the man in charge of the program. It calls for considerable imagination from management to put such an organization into effect and make the necessary adjustments within and among programs. However, the payoff which can be realized by wise management of the organizational structure and assignments can have a major payoff in the over-all effectiveness of the organization. Therefore, it is worthwhile for management to take a vital part in seeing that the organization is properly tailored for the large-scale jobs which they undertake.

Control of Program Effort

Already emphasized is the necessity for exercising close control of the manpower engaged in a program. This is obvious from the large numbers of persons engaged, which connotes that improper utilization of effort increases the cost and time scales of the program. The main resource available to the development organization is its technical staff, and its productivity is directly related to the effectiveness with which this staff is used.

There is one particular factor in the control of program effort which seems highly significant. This factor relates to the dangers of overstaffing. In these days of general shortages of experienced engineers throughout the country, there is a tendency to overcome lack of quality by quantity. This often makes matters considerably worse instead of better. The productivity of an organization does not, as it is sometimes supposed, increase with the total number of men engaged in a given task. On the contrary, it tends to follow a relationship of the general sort that is shown in Figure 7. For a given task there is an effective minimum of people required to make significant progress. After this minimum is exceeded, productivity rises rapidly, levels off, and actually eventually decreases. This fact has been recognized by many authorities who have studied the productivity of organizations. The saturation effect arises from the saturation of supervision. Problems of maintaining a co-ordinated effort among a group of people multiply at a rate faster than the number of people engaged. The time comes when the supervisors have too little time for co-ordinating the work of their staff with that in other sections of the program. In other words, they are left with no time to think. The saturation of supervision has some regenerative properties. In particular, the morale of the working staff can suffer through lack of guidance and through lack of recognition and utilization of their efforts. The communication between individuals working on the same problem also gets out of hand if too many are engaged.

Figure 7 also shows the cost of accomplishing a given job as a function of the manpower engaged. The time to accomplish the job is inversely proportional to the rate of productivity. The cost, which is proportional to the manpower engaged times the time, is proportional to the manpower divided by the productivity. It is seen that the minimum of this curve actually comes before the peak of the productivity manpower curve. Thus, the most efficient effort is one for which the peak of productivity has not quite been reached.

This situation brings about an interesting regenerative effect on estimates of program cost. If an organization ever starts operating at manpower levels beyond which the program cost is a minimum, they will find that their program costs exceed original estimates by sub-

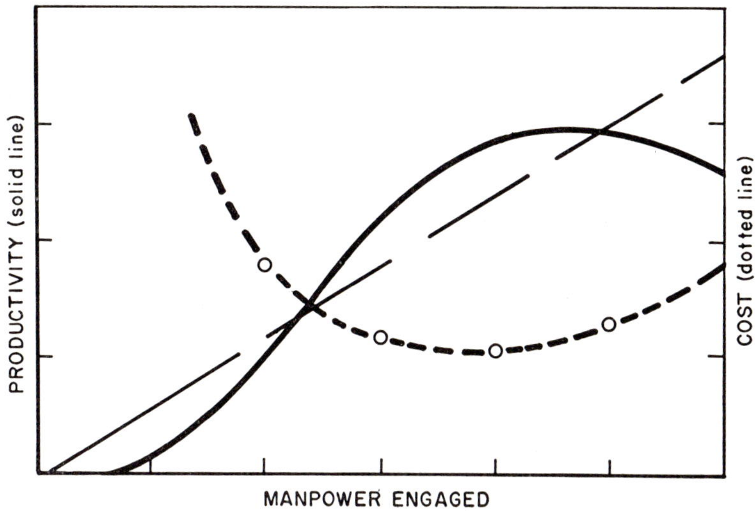

Figure 7. Productivity and cost versus manpower.

stantial amounts. Using this experience as a guide for future cost estimates, they will tend progressively to increase them with successive jobs. Furthermore, since the productivity realized will also fall short of predictions there will be a tendency to devote even greater manpower to successive jobs, thereby further throwing the effort out of balance. The result will be that the organization will simply slide farther and farther along the curve of decreasing productivity and increasing cost without realizing what is happening. Their cost predictions will increase, and their cost will likewise, proving them right. This is not a fanciful situation but one which occurs all too often.

Control of Component Design

In a system whose components interact strongly and whose development requires that all of its parts be designed soundly and expeditiously, the control of component design throughout the program is of utmost importance. In development work it is necessary to give the component designer considerable latitude in the manner in which he attacks his problems and arrives at solutions within the specifications that he is given. However, there is often a tendency to let this latitude go too far. Discussed briefly here are three aspects of this problem to which special attention should be given during the systems development program.

1. ALTERNATIVE DESIGN APPROACHES

Inherent in the engineering of a system is the fact that each problem has a variety of possible solutions. For each component there are several different design approaches, each of which possesses certain potential advantages and disadvantages. In this situation there is a tendency to pursue several of these in parallel to determine which is best before making a final selection. This process is warranted under two conditions:

(a) In the exploratory stages where the work is confined to paper analysis or small-scale laboratory testing; and

(b) Where it is a major conflict between performance achievable and potential risk between two alternative approaches.

In the first case, exploration on a small scale can lead to significant improvement in the approach at low cost; in the second case, an insurance program may be required in case the more ambitious approach fails. However, there are many more instances in which a given approach is almost certain to provide an adequate answer and where alternative solutions offer small gains in return for the effort expended. In these cases, pursuing alternative approaches is "gilding the lily." It must be remembered that if two approaches are pursued and come out with very similar performance, the job of evaluating each one to determine which should be adopted requires additional effort beyond that initially expended. It is much better in these cases to make an arbitrary decision in selecting the method of approach and utilize the manpower saved on more difficult parts of the problem.

2. OVER-DESIGN

A problem related to the first is the natural tendency of engineers to invent new means of squeezing the last bit of performance out of a given type of component. This tendency, of course, is what results in technological progress, but in a full-scale systems development program it might well prove disastrous. It not only results in a greater expenditure of effort to accomplish the task, but also subjects the outcome to a greater degree of risk by dependence on untried methods. Here it is necessary to distinguish really good sound ideas from mere gadgeteering.

3. DESIGN JEALOUSY

A new scientific hypothesis or discovery is considered valid only after it has been subjected to scrutiny and after its truth has been verified. A new design, however, is seldom exposed to such treatment. Since design is more of an art than it is a science, its evaluation by others tends to be subjective. Nevertheless, it is extremely important that well-defined procedures for design review be established in order to provide collective judgment before the design is accepted as a basis for hardware. This can be provided by the systems engineering staff, as well as by individuals specially competent in various parts of the system.

In this connection, it might be mentioned that there is a tendency for program supervisors to confine their activities to administration and stay out of design problems. This tendency should be counteracted as strongly as possible. It is most essential that the man in charge of the program be familiar with its details in order to have a proper perspective of its chance of success and of needs for making decisions forestalling possible difficulties.

Control of Program Schedules

It goes without saying that control of the program schedule is essential to a successful development operation. However, the problems inherent in doing so are more complex than most people realize. In particular there is a paradox in program scheduling which poses a difficult dilemma for management. This is the principle of the slipping time schedule.

Since time is a most essential ingredient in the definition of system objectives, a time table is always constructed at the beginning of each systems engineering program to guide the various phases of the task, and to establish a completion date on which the system will be ready for exploitation. There have been programs where the schedule has been met, and these are properly a great source of pride to the organizations who have accomplished this enviable aim. However, it has been the almost universal experience that the initial time table is not met by a substantial margin. Furthermore, there is a remarkable uniformity in the degree to which the schedule has been missed. A study made for the Defense Department of projected versus actual schedules in the early phases of various guided missile programs showed that the slippage was very nearly the same for all, and amounted to about 50 per cent. That means that a job scheduled to require twelve months actually took eighteen, and one scheduled for two years took three. The effect is pictured schematically in Figure 8.

It is not difficult to find the reason for this phenomenon. It comes directly from the conflict between desire for high performance and the need for producing a timely solution. This results in the formulation of a program which attempts to achieve the maximum goals in a minimum time, in which contingencies for unforeseen problems are squeezed out. Such problems inevitably arise when one is working on the frontiers of technology and producing a complex device never before assembled in a similar form. A problem in any one critical area will slow down the entire program. The result is slippage.

One might expect that knowledge that this situation prevails would permit the establishment of what are termed *realistic* schedules. Obviously slippages are a nuisance—they do not enhance the reputation of the organization, they upset management planning, and they generally cause lost motion during the progress of the task. Why not simply add on a 50 per cent safety factor in the first place?

The answer is complicated, but amounts, in most cases, to the conclusion that such action would not be wise. In the first place, extension of the schedule would not insure that it would now be met. It is usually discovered that the extra time provided can be put to good use in several ways. A more ambitious approach is now possible, which has promise of even greater performance capabilities in the product. Also, it is possible to examine a number of alternative solutions to problems for which a single best-bet approach had been selected. In this way the time allowed for *unforeseen* contingencies will be eaten up by attempts for greater performance (at greater risk) and by "insurance" programs of which only a few may actually forestall a potential crisis. The net result is that the new "realistic" schedule will also slip, probably by a similar percentage.

In the second place, the cost of development and engineering is a direct function of the time schedule. Except for so-called "crash programs" the longer the time allowed, the higher will be the cost. Unless this is compensated by major gains in performance, it is a net loss.

It is seen, therefore, that the established schedule plays a dynamic role in determining and guiding the technical approach and cannot be

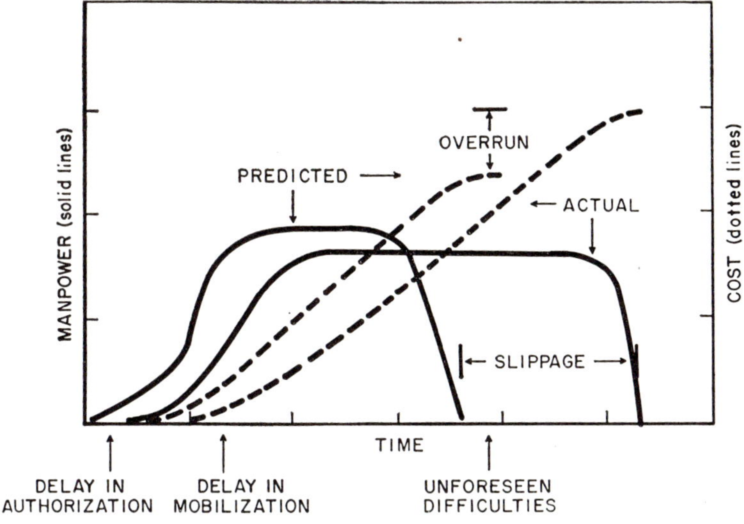

Figure 8. Manpower and cost history.

altered without a feedback into the entire program. Extending it may actually increase the amount of slippage and will certainly increase costs. It must be kept tight to achieve the desired results.

From the above discussion it might be concluded that management must become resigned to missing schedules. While this is not necessarily so, it is clear that scheduling remains a major unsolved problem in systems engineering, one worthy of the most serious study.

In practice there are three directions in which savings in time usually can be made. The first is to simplify the objectives which the system was initially intended to meet. It is often found that a number of these, which turn out to require major effort, are frosting on the cake and can be removed without any real decrease in the effectiveness of the system. The second is to simplify the method of attack by eliminating duplicate approaches whose purpose is largely to optimize further an already fully adequate portion of the system. The third is to gear the organization specifically to the job in hand, and to alter it as necessary as the job progresses. It cannot be overemphasized that the timing problem is crucial throughout the systems engineering process and must be handled with wisdom and vigor.

Control of Program Cost

Relatively little need be said about the control of cost during the prototype development program. Mentioned below are three aspects which are not always properly handled.

1. MAINTENANCE OF CURRENT BUDGETS

Most organizations maintain a fairly elaborate cost accounting system in which people charge time against the various projects in which they are engaged. However, this represents an a posteriori analysis of program costs, since the time taken to tabulate and organize the information is often so long as to have little influence on the course of the program. In other words, the information in terms of unexpectedly large costs often comes too late to do much about it. The result is an overrun which causes embarrassment to all concerned. What is needed here is not so much an accurate-to-the-penny accounting of what has been spent two to three months before but rather an up-to-date approximate cost estimate and an enlightened comparison of this with the budgeted rate of expenditure. If this information is properly organized and presented to the supervisors responsible, they could quickly note which phases of the program are not proceeding according to plan, and take early action either to correct the situation or to adjust the over-all budget accordingly. It is not likely that such a service can be provided by the accounting department of an organization, but rather might better be included as part of the project staff or elsewhere in the engineering department itself.

2. AUTHORITY FOR EXPENDITURE OF FUNDS

It has frequently been found true that persons having a high degree of responsibility for making program decisions have little authority to expend funds outside the organization. This stems from the usual chain of administrative approvals required for the purchase of anything from a hammer upwards. Thus elaborate controls are established to avoid misspending a dollar where hundreds of thousands of dollars in terms of engineering effort are often at stake. As a solution to this problem, it is suggested that persons in the organization responsible for making major program decisions be given commensurate authority to expend funds on purchases and subcontracts, provided they are within the established program plan and budget. Thereafter any necessary administrative approvals should be exercised with discretion as a veto power.

3. COST CONSCIOUSNESS

It has been observed that in many industries there is a notable lack of cost consciousness among the engineers responsible for designing systems. This is perhaps more true of the systems *design* engineer than of the systems engineer. In any event, it seems that the striving for high performance appears to be a more challenging and interesting goal than striving for low cost. This is particularly true of the current crop of young engineers who have not come up the hard way through production and consequently have little personal knowledge of the costs incident to manufacturing a given article. Accordingly, too little attention is often given in the prototype design stage to the use of design methods which will result in a producible article. The difficulty with this is that once the basic design has been established the system is so complex that production engineers have little latitude to alter component design to ease production. This combination of circumstances contributes to the enormous cost of systems engineering and of the systems themselves.

To bring the system-cost problem under control will require a major reorientation in thinking in many organizations, both among management and among engineers, particularly the latter. Some way must be found to give design engineers a thorough grounding in production problems, and to provide an incentive for ingenuity to reduce complexity and cost comparable to that now existing for improving performance. Only by a radical change in viewpoint is progress likely to be made in this direction.

Prototype Evaluation

The culmination of the prototype development program is the assembly and testing of the prototype model against the specifications for which it was designed. It is only when the prototype is first assembled and operated as a unit that the compatibility of the components with one another can be assessed realistically. In a system such as a guided missile, there are a large number of compatibility conditions, including not only functional but electrical and structural, which are only fully assessed when the first prototype missile is assembled as a complete unit. At this time one might well discover that the center of gravity of the missile is not quite in the position predicted, despite careful estimates of weights and moments of the components. If not corrected, this would require changes in the autopilot because of the alteration of aerodynamic moments about the center of gravity. Thus, one often finds prototype missiles and even final tactical missiles containing a hundred pounds of ballast located in the nose or in the tail. This is only one trivial example of what can happen despite painstaking efforts to eliminate all incompatibilities in component design.

The evaluation of the prototype system is a painstaking task, particularly if severe environmental conditions are involved. In the case of guided missiles this evaluation consists in a long series of ground tests using signal generators to simulate guidance inputs, large-scale shake testers, drop towers, and many other items of special test equipment to detect any system defects in the laboratory. Following this operation it is necessary to carry out a series of prototype flight tests to demonstrate that the system operates in the flight environment. The analysis and interpretation of ground test and flight test data is a science in itself and requires a highly trained analysis group to derive the proper conclusions from the data.

The end result of the prototype evaluation phase must lay the groundwork for the final design of the system. If the results show that the original specifications are not fully met, the specifications must either be altered or a program laid out for redesign to meet the specifications. If such redesign is extensive, it may require the construction of additional prototypes incorporating the necessary improvements before a decision to engineer the system for production is made. In addition there is assembled a voluminous amount of collateral data bearing on items of marginal reliability, production tolerances, test methods, and other factors which will be useful in the final design stage. In addition, the final decisions on the design of the system test equipment are made during this stage.

ENGINEERING OF FINAL SYSTEM

The phase referred to as the engineering of the final system overlaps and is inextricably tied up with the engineering and evaluation of the system prototype. Properly speaking, the engineering of the final system begins when a commitment is made to produce the initial quantity of systems. Depending on the size of advance attempted in the new system, this commitment may be made well before the completion of the prototype, or sometimes it may be deferred until the completion of prototype evaluation. This phasing is directly dependent upon the degree of risk involved in undertaking the large-scale effort required in final engineering compared to the pressure for an early completion of the task. In the case of a major system advance the first prototype may fall so short of the desired goal that a second and third prototype stage may be required.

The relation of the final system engineering phase to the prototype system is shown in Figure 9. These three points should be noted in the figure.

1. The final engineering process represents detailed design of the individual system components or sections. This results in finished engineering drawings for production.
2. The starting point for this process is the individual components of the system prototype. At this level of detail a systems approach

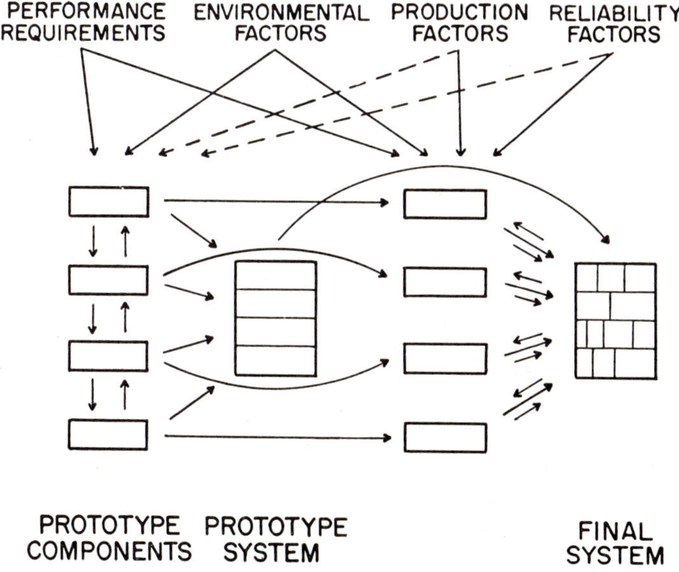

Figure 9. Final engineering process.

is impractical—the components must be kept stable during the final stage.

3. The principal inputs are environmental factors, production factors, and reliability factors. These have already been considered in the prototype design, but not fully met, particularly the last two.

The engineering of the final prototype is the least glamorous phase of systems engineering. It is also frequently the most hazardous. It is important to recognize the most common pitfalls.

Weakness of Detailed Design

The apparently straightforward and unglamorous nature of component final design often results in its assignment to the least experienced and least competent engineers in the organization. This is particularly true in mechanical design, due to the general scarcity of really good mechanical design engineers. The result can be a multitude of minor design faults, which can pyramid into a major crisis when the system is assembled. The fact is that it takes very smart people to do *all* of the design because of the many subtle interactions between system components. Few organizations have an adequate number of engineers to handle the entire system. The only solution is to keep a substantial number of top engineers throughout the final design phase and maintain tight design control in all respects.

Departure from Prototype Design

The temptation for redesign of the prototype components during the final design stage is hard to resist. With the passage of time since the prototype design was frozen, new, potentially better techniques have become available. Who can prove that these new techniques will not work? This temptation must be resisted at all costs. In the final design stage there are enough mandatory changes required so that any further departures from the proven prototype design are risky in the extreme. The burden of proof is not on the one who says that a new design will not work but on the one who claims it will. The solution to this problem is to make explicit provisions for incorporating such changes after the start of production. In this way the equipment can be kept modern in an orderly manner without undue risks.

The Premature Freeze

A departure from the prototype design may be unintentionally provoked by a premature prototype freeze. Such a design will inevitably require mandatory alteration to correct deficiencies. The process of final design is not well suited to handling changes. Thousands of drawings are usually involved. A single change may affect dozens of drawings. If it affects other components, a chain reaction can start throughout the whole structure. In the worst cases the entire engineering process may become paralyzed through saturation. The decision as to when final design can begin requires the highest level of technical insight and judgment.

Production Factors

The problem of system cost is met head on in the final engineering phase. Most factors which affect the cost of the system can be broken down into the following categories:

(1) Number of units to be produced.
(2) Number of parts.
(3) Special precision components.

1. PRODUCTION RATES

For complex equipment the unit cost is high and the market is small. In addition the rate of obsolescence is high. For these reasons the total number of systems which are produced to a given design is usually in the tens or possibly hundreds. This is a far cry from mass production and requires a job-shop type of operation. The operation does not warrant a high degree of tooling. Accordingly, it is costly.

In order to keep costs down in a job-shop operation it is necessary to devise special tools which do a number of operations without changing the set-up. The newly developed machines which can be automatically programmed by using a tape input are ideally suited to this purpose. The "merry-go-round" method of electronic assembly is another special technique developed for small production runs. In this method, each operator does a few simple operations, as in a mass production assembly line, but after wiring perhaps thirty to fifty assemblies, all operators change over to a new set of operations and the units are recirculated to them. This system retains many of the good features of the mass assembly method but can be carried out with a small number of assembly technicians.

2. NUMBER OF PARTS

The number of parts used in a system is a direct function of system design. It is often possible to simplify the system considerably in the final engineering phase by eliminating features which contribute relatively little to the system performance. In eliminating parts care must be taken not to place the remaining components under conditions exceeding their normal design limits. This again is a job for top engineers because of the possible feedback of the resulting changes on other components.

3. SPECIAL PRECISION COMPONENTS

Frequently a very considerable part of the system cost is tied up directly or indirectly in precision components. Such components include gyros, accelerometers, servo valves, complex switches, radar tubes, radar plumbing, and the like. Most of these, unfortunately, are custom-built items and are very costly. A considerable portion of their cost comes from the degree of testing they must undergo. They are often delicate and particularly subject to environmental effects. A highly profitable field exists in standardizing such components to the point where large volume can reduce their cost. The largest obstacle is the tendency of each system designer to impose special requirements which force the use of a custom-built component. Both industry and government could make major gains in this area.

Complexity and Operability

Throughout the preceding discussion numerous references have been made to the complexity of systems. Some remarks have also been made that the tendency is to make systems too complex. The fact remains, however, that most systems, no matter how well-designed, must inherently be complex in order to perform the functions for which they are designed. This follows directly from the fact that systems attempt to reproduce mechanically and electronically some of the functions of

the human brain. It has been estimated that an electronic model of the human brain would require of the order of 10^{10} vacuum tubes. Whereas present systems do not attempt to achieve such a goal, it is nevertheless true that to reproduce even the most elementary operation of the human brain requires a very high order of complexity.

The consequences of complexity are many and one of the most important has to do with marginal reliability. The proper functioning of a system depends on the proper functioning of virtually every component vacuum tube, relay, resistor, and wire. A short to ground anywhere in the system is likely to put it completely out of operation. The reliability of a system composed of a large number of elements can be expressed as the product of the individual reliabilities of these elements. If the system contains 1,000 elements, each of which has a reliability of 99.5 per cent during a given time interval, the entire system will have a reliability of 1 per cent. It is clear that enormous pains must be taken with each seemingly trivial part in order to insure that the over-all system will be operative to an acceptable degree.

When a failure does occur, the complexity of the system, together with its closed-loop characteristics, makes it extremely difficult to localize the failure so that corrective action can be taken. The recognition of this fact has brought about radically new concepts in system design to provide facility for maintenance. In other words, since it is not humanly possible to design systems to be 100 per cent reliable, they must be designed in such a way as to make them operable for the largest possible percentage of the time. If the time for repair can be kept to sufficiently low levels by clever design, practically realizable failure rates can be accepted.

At this point it is necessary to remember that every system contains human beings as an integral part. In a guided missile system, the ground equipment is monitored by human observers who continuously watch the signals received by the radar and the correctness of the operation of all the equipment. In the case of the guided missile itself, its operation is periodically checked by automatic test equipment and necessary repairs made by maintenance personnel. The missile, its test equipment, and the maintenance man constitute a subsystem which keeps the missile in "shape." In these terms the system has self-healing properties. The efficiency with which self-healing maintains a system operable is dependent on the frequency with which it gets sick, the rapidity with which the exact cause of the sickness can be diagnosed, and the ease with which it can be cured by replacement of the defective part. The more complex the system, the more care must go into these features of its design which permit it to be healed quickly. Modular design, diagnostic testing, and human engineering are important tools used to solve this problem.

Six

A SURVEY OF OPERATIONS RESEARCH TOOLS AND TECHNIQUES

THORNTON L. PAGE

To make this "Survey of OR Tools and Techniques" complete is literally an impossible task, for there is no limited set of tools and techniques in this activity we describe by the term "operations research." It would be as difficult to enumerate all the techniques of scientific research in general. The requirement of operations research is to solve a practical problem, using any techniques that can be applied, and new techniques are being developed all the time, probably without limit. Nevertheless, an attempt will be made to enumerate broad classes of techniques, with some examples of the less common ones.

GENERAL METHODS OF SCIENTIFIC RESEARCH

Very generally, the recognized techniques of scientific research apply in this field, including the general one of taxonomy, or *classification.* Many of our major sciences have started simply through classification of a number of objects or activities. Most obvious is biology, where the species of all living things were classified. From this classification came understanding and eventually the theory of evolution. In astronomy the same thing happened in the classification of stellar spectra, which led to an understanding of the life cycle of a star. Astronomers gained an understanding of stars in general from a taxonomy—describing all the types of stars that were observed and putting these types into logical order.

Another general scientific technique is *model-building*. Two broad types of models can be distinguished: (1) *analogical models*, where there is a one-to-one correspondence between the construct and the activity to be explained or understood, and (2) *abstract models*, which are usually entirely mathematical. As one of the sciences, operations research can be expected to depend on techniques of these types, as will be discussed in more detail later.

CHARACTERISTICS OF OPERATIONS RESEARCH

The major point in this discussion is that operations research, by its very nature, is the application of *all* forms of human knowledge to the solution of a *whole* problem. In this it differs from many of the other scientific disciplines; its tools and techniques are very diverse. Anything "goes" in an operations research study as long as it leads to better understanding of the problem. By contrast, the other scientific disciplines have tended to specialize—to narrow down the boundaries of the problems they consider—and in a university today one finds very detailed specialists. In medicine, for instance, one finds few, if any, general practitioners; there are, instead, radiologists, endocrinologists, dermatologists, and the like. This has led, of course, to rapid advances in the science—certainly a good thing—but when it comes to solving *broad* problems, the specialist tends to try only those techniques most familiar to him. Insofar as specific techniques are developed in operations research—insofar as they narrow down or put a boundary on the considerations of any one problem—they do not help operations research in its broadest aspects. Limiting the techniques to be used is antithetical to the idea of solving whole problems.

The point about useful techniques, then, is that the operations analyst never should be bound by them. He never should say, "I can only solve this problem with linear programming." (In such a case he is not doing operations research—he is doing linear programming.) Here lies the basis for using *teams* in operations research; only a group of men can provide competence in a variety of scientific disciplines and techniques.

Operations research is best considered as the essence of applied science, as illustrated in Figure 1, underlying the major divisions of human knowledge generally recognized in a university. Four of these are represented (although they can be broken down into many more), ranging from the physical sciences through the biological and social sciences to the humanities. They are put down in this order be-

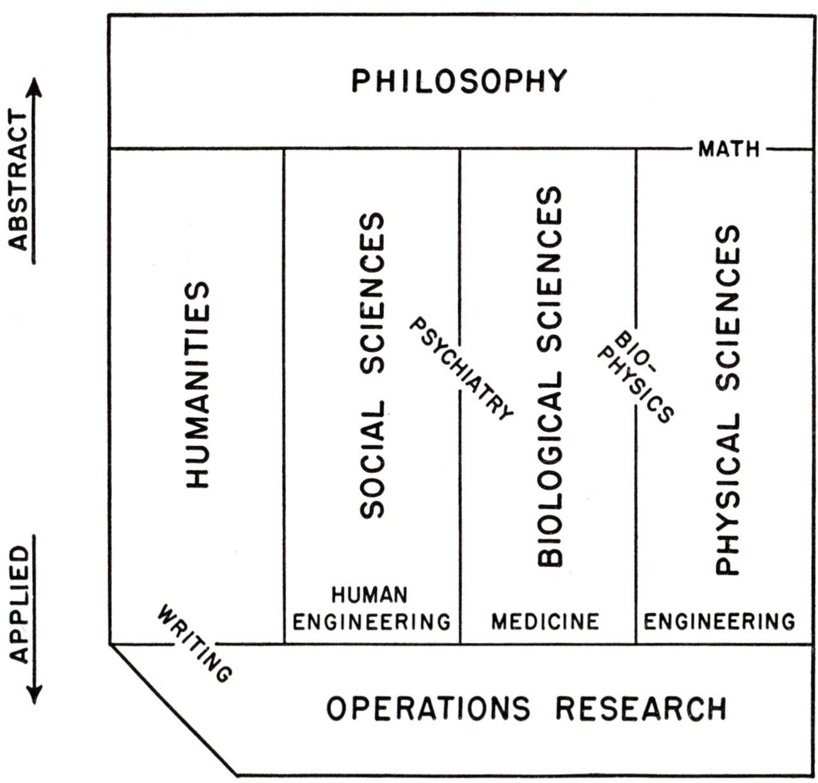

Figure 1. Relationship of operations research to the broad divisions of human knowledge.

cause it is convenient to represent inter-disciplinary research areas such as biophysics as lying on the border of the physical and biological sciences, or psychiatry as lying between the biological and social sciences.

The meaning of the vertical dimension of Figure 1 is shown, the more abstract sciences being near the top, and the more applied sciences being near the bottom. At the top of the physical science block, for instance, theoretical physics and mathematics appear the most abstract part of that block, and the most removed from the real world. At the lower end go all the branches of engineering, closest to the real world. Philosophy is placed across the top of the diagram as a discipline which binds together the other blocks of human knowledge in the abstract. Across the bottom is operations research, in a way binding the others together and *using* all of them in its essential concern with the real world. The operations research block is tapered off at the left because no one as yet has used the humanities specifically in operations research. However, the reporting of an operations research study is of recognized importance, and since it involves persuasion and good writing, it may well be considered an application of the humanities.

The diagram is simply intended to illustrate that the techniques and tools of operations research should be drawn from all these blocks of human knowledge. There should be no artificial boundaries—no limitations—on the techniques that are used in the solution of a problem. It is characteristic of operations research that it draws on techniques from a variety of the disciplines, and a problem that can be solved simply by engineering techniques, for instance, is not an operations research problem. (The engineers would agree—they prefer to call such a problem an engineering problem!) It is only when *more* than simple engineering techniques are involved that the thing becomes operations research.

MODEL-BUILDING—FORMULATION OF A THEORY

Basic to the solution of any operations research problem is the constructing of a model. Model-building is stressed because, in order to get anywhere in a scientific study, one has to have some means of bringing together facts and data of different kinds. This is the primary function of a model, which can be highly mathematical—and either deterministic or probabilistic—or which can be mostly qualitative (non-mathematical). A "deterministic mathematical model" means something like the laws of mechanics, Newton's laws, which describe the motions, for instance, of the planets moving around the sun. There is nothing indefinite about these laws. One of them is $F = ma$, force equals mass times acceleration; another is that in the vicinity of the sun the force of the gravitational attraction toward the sun is $F = G\ mM/r^2$, a constant times the mass of the planet times the mass of the sun divided by the square of the distance between them. With these equations it is possible to determine exactly where a planet such as the earth will move, given the initial conditions of where it is and how fast it is moving at the start.[1]

A deterministic model used in operations research is the *Lanchester model* of warfare, now standard for reasoning about requirements for military forces and the effectiveness of certain improvements. Two forces, say M and N, are considered to "interact"; that is, they fight. One side shoots at the other, the second shoots back, and the differential expressions $dN/dt = -AM$ and $dM/dt = -BN$ have been shown (1) to describe the course of several actual battles. In these Lanchester equations, dN/dt and dM/dt represent the time rate of change of the two forces, each proportional to the strength of the opposing force. That is, the number of casualties per day in force N depends on the size of the opposing force M, and vice versa. The constants A and B represent the efficiency of the forces M and N. These Lanchester equations have been widely used with many modifications to represent battle between ground forces, with and without supporting weapons, and between air forces.

This kind of deterministic mathematical model is perhaps the most useful, *if* one can achieve it, but in many of the problems of operations research such a model will not work—the relationships are not inherently deterministic. In fact, by and large, operations research is concerned with problems where the relationships are stochastic or probabilistic, as distinct from deterministic. An illustrative analogy is the modern theory of atoms and their internal workings, which does not attempt to predict exactly where an electron would be in some future time. The techniques for handling such problems have been highly developed over the last century, and statistical methods are now available for devising models where the factors are not precisely determined, as they can be in the case of the planets. Note that Newtonian mechanics and atomic theory are used as analogical models in this

[1] The physicist reader will recognize that perturbations can also be included in this highly simplified statement.

brief discussion of what models are. A number of examples of probabilistic mathematical models used in operations research are given later.

It is not necessary for a model to be mathematical in operations research studies and, even if it is mathematical, it may be much simpler than the examples cited—involving no elaborate equations. A very simple *analogical model* is the "pipeline" used in studies of military supply systems, where one thinks of the full military supply moving forward to an army as if it were moving through a big pipe. Among other things, such a model takes account of inventories of stocks in transit, which can be a very large part of the total inventory; it is also an effective model of the delays in delivery when the pipeline is first filled. In other words, this simple idea of a pipeline replaces all the complex of air delivery, shipping, and rail nets between the source of supply and field army depots.

Another analogical model was once proposed in jest at ORO (2) for solving traffic problems, for which no satisfactory mathematical model has been devised. The kinds of traffic problems in mind were such as the best distribution of traffic over a road net to supply an army in some theater of war, or the optimum design of a highway network. In each of these one would like to represent by equations or by some other means the amount of traffic that can go from one point to another over a network of roads between them. The proposed model was called the "Antiac" in true computer terminology, the "ac" at the end standing for automatic computer. "Ant" refers to the use of ants.

As proposed, the idea is to string up a road net represented by strings between two tables. The strings are tied together, some possibly of larger dimension (ropes), in order to represent the larger capacity of a dual highway. On one table is a source of ants, an ant hill, and on the other a source of food, probably sugar. Ants are well known for their tendency to go after sugar and bring it to their ant hill. Moreover, a large number of ants can try a great many different alternative routes, ending up with the best one for their purpose. Use of a Geiger counter and radioactive sugar would simplify counting the ants as they go by on each route.

After the counts settle down, the equilibrium traffic pattern of ants along the various strings and ropes may bear some relationship to the real problem of trucks with men as drivers—and if one wants to find the effect of taking out one of the routes, just cut the string and see how the ants reshape their traffic pattern to make maximum use of the modified "road net"! The Antiac has never been used, but might be worth trying for exhibition purposes, if nothing else. It is a probabilistic analogical type of model, the movement of ants along strings representing movement of traffic on roads better than analogical models based on simple electrical currents, for instance.

As you can see, there are a very large number of meanings that the word "model" can have, varying from the almost qualitative pipeline concept to a sophisticated mathematical theory. It may be defined as any concept that gives insight or facilitates reasoning about the problem at hand.

SIMULATION

The Antiac goes further than the idea of a model in that it might be used to try the effect of changing parameters (such as the capacity of roads, as represented by the diameter of strings) in repeated "plays." Such a use is called operational gaming or simulation, and offers great promise in operations research. Any type of model can be used in gaming, although it generally involves probabilistic models in research studies, so that there is some point to repeated plays.

One of the earliest forms of simulation was *war gaming,* which was used extensively by the German General Staff before World War II, and in which a campaign was conducted battle by battle, on a map. Military judgment, or probabilistic calculations, or other calculations based on mathematical models of battle, can be used to determine the outcome of each battle, but the decisions of the players on each side usually are so variable that the game does not necessarily come out the same way if it is replayed. One of the most important aspects of such gaming is the *training* it provides for the players.

The technique of simulation is described in greater detail in other parts of this book. Also a good expository treatment of war gaming has been written by Thomas and Deemer (3). Therefore, only the application of simulation to industrial operations will be discussed here.

One of the simplest concerned the design of a railroad classification yard (4) where the problem was to see if a new hump and track layout could better handle the load of freight trains arriving to be broken up and reassembled for new destinations. The make-up of incoming trains, car by car, had been recorded for a period of a year or more. Therefore, it was possible to make up stacks of cards representing trains, each card representing an individual freight car, with destination indicated. On a schematic track layout these simulated trains were brought in, sorted on the simulated hump at a realistic rate, and the new stacks of cards collected on schematic sidings. The "game" was played by the yard men themselves, who corroborated the "rules" and began to take an interest in the problem. The relative congestion and delay involved in each track layout tried was readily observed, and the most desirable design of the yard was established to everyone's satisfaction.

A more complex example of simulation is the management game developed for the American Management Association (5), in which the players represent several companies in a single industry. The operations of these companies are represented by a suitable mathematical model involving costs of production, investment in new plant, research, inventories, price and demand for the products, effectiveness of promotion, taxes, etc., and the players can decide, during play, how effort should be allocated. A single game represents several years on the market, during which players may go bankrupt, or make large profits.

Whether the outcomes of such elaborate simulation are significant in themselves is open to question, and the operations analysts working on simulation are very cautious in using them. If the model is in error, or if the numbers used in the model are not all correct, the outcomes of the game might be entirely misleading. However, it is often possible to use simulation in a differential or relative manner, comparing a spread of many outcomes under one set of conditions with the spread obtained when one condition is changed.[2] In this use, errors in the numbers used may be less critical.

USEFUL SCIENTIFIC DISCIPLINES

If one goes through published operations research studies, such as those in the *Journal of the Operations Research Society* since its start in 1951—a sample of operations research studies—one gains the impression that mathematics is the most widely used discipline; and within mathematics, the theories of statistics and probability. However,

[2] See, for example, chapter 15.

this is a severely biased view, due to a selection that arises primarily from the effects of security.

Both in industry and in the government, operations research studies that are any good are important enough to be kept from the competitor or opponent. In almost every case, the operations analyst who wants to exchange ideas with his colleagues is forbidden to give away company secrets or national secrets. Under such limitation a large number of papers are devoted simply to the mathematics of the solution, leaving out all the numbers. It is a most unfortunate result that, from the published papers, one gets an impression of operations research as just a lot of mathematics. Papers consisting of definitions and mathematics that come to no practical conclusions are *not* representative of operations research, although, of course, the mathematical formalism may be very useful in some operations research problem.

After mathematics, the physical and biological sciences are next in importance, particularly as good preparation for operations research. Research experience in these sciences seems to be the most useful background for operations research, probably because those disciplines are heavily dependent on theoretical constructs, or models. They are both also dependent upon experiments, experimentation is an important potential tool of operations research.

More than that, experiments and observation provide the facts that keep operations research tied to reality. It is very easy to be carried away by a model; for instance, to forget that the Lanchester equations are an imperfect representation of men in action, omitting psychological factors such as poor morale, good leadership, etc.—or even to use them through a phase with negative forces, as was done in one computation several years ago! Although the model is essential for progress in the solution of a problem, it can be grossly misused.

By contrast, the experimental approach to solving a problem—trying it out under controlled conditions—certainly avoids or corrects such mistakes or errors in theory. Unfortunately, it is usually very difficult to perform significant experiments in operations research. This does not mean that data from field experience are not used, but rather that, as in astronomy, data are obtained primarily from passive observation. Until recently, there was no possibility of conducting experiments in astronomy; astronomers could only look at what was going on, record all the facts, and build a theory of what is "behind" these facts, or what "caused" them.

In the same way, operations analysts must generally be content to observe passively, and experimentation has been notable by its absence. There just are not many cases where it is possible to make a controlled experiment. Among the few exceptions was an experiment carried out in Korea during a study of artillery effectiveness in 1953—possibly the only controlled experiment made during wartime on the battlefield. The problem was to determine what the effect of counterbattery fire might be; that is, just how useful it was for our troops to shoot back when an enemy battery was shooting at them. A model had been devised, involving the accuracy of artillery fire, so that it was possible to calculate the probability of damage to the enemy artillery in terms of the number of rounds fired at it. As usual, this model ignored psychological effects.

There were plenty of enemy batteries firing at our troops, and it was possible to observe every round from each enemy gun. So, the experimenters waited until they spotted one of these enemy batteries firing and then timed the rounds. At some later time our artillery opened up and gave the enemy battery various "doses" while his rounds were still being counted to see whether our troops' return fire changed his rate of fire. Typically, again, the results of that study are classified, but it can be reported that the experiment, at least, was successful.

Experiments have also been performed in industrial operations research, although generally the interference with the operation of a big industry is serious if it must deviate in a significant way from normal, as is necessary to perform an experiment. Moreover, it is difficult to provide a control. If two variants on an operation are to be compared experimentally, other conditions must be held constant—a difficult matter outside the laboratory. Thus it can be said that experimentation is *not* characteristic of operations research as yet, though its close counterpart, *simulation,* is, and though more efforts at controlled overall system experimentation are to be expected in the future. In the meantime, operations research is primarily dependent on passive observation for its basic data.

After mathematics and the physical and biological sciences come the social sciences, from which operations research also draws (per Figure 1). The concepts and techniques of the social sciences are somewhat different from those discussed thus far, in being much broader and less precise. However, these concepts take account of human factors that are generally ignored, particularly in the physical sciences. These human factors are mainly psychological ones. The most obvious illustration of their use is in human engineering, where the design of equipment is related to the reactions and feelings of the men that will use the equipment, as well as to the purely physical efficiency. Another example is taking labor relations into account in industrial operations, and a third is the integration of Negroes in the Army (6). Marketing studies also are generally dependent on these sociological techniques. One specific technique is the use of questionnaires or interviews to estimate human attitudes.

LEVELS OF OPERATIONS RESEARCH

Operations research can be undertaken at various levels in an organization, and this affects the tools and techniques that may best be used. Most of this discussion will involve "suboptimization," a concept worth a good deal of thought. The word was introduced to operations research by Charles J. Hitch (7) in the first volume of the *Journal of the Operations Research Society*. In this excellent and non-mathematical article, Hitch discusses the question of the level of an operations research study and the effect of optimizing at various levels.

In fact, it is the endeavor in all operations research to optimize at the right level, whereas in engineering (under our definition) the engineer is engaged in optimizing within the narrower boundaries of his specialty. An engineering design may be optimum, for instance, in terms of the immediate cost of production, but may not be optimum with respect to some other goal of the whole system, such as maximizing long-term profits of a corporation, or the stability of an industry. There may have been a case of suboptimization in the US automobile industry over the past ten years, as automobiles were designed to satisfy certain immediate desires of consumers in this country, but also have caused such congestion on the roads and in parking spaces, and have become so expensive and cumbersome that consumers are turning to foreign makes. For the US automobile industry as a whole, a different design—or variety of designs—might have been far better in the long run.

The question of the proper level of optimization is undoubtedly the most important single factor in setting up an operations research study; the requirement is to optimize at a level appropriate to the problem. In a given company this optimization usually has to take place with respect to the whole activity of the company, and even this often leads to results different from optimization for the whole industry. In national problems, proper optimization has to take into account the economy of the whole country and its national policy, a scope that is beyond the capabilities of a small research group. Here is the dilemma: in order to avoid suboptimization entirely one must take on so big a problem that one may never finish it. But a more limited problem, one that can be solved, is apt to lead to a wrong answer because some significant considerations were omitted.

One clear example of suboptimization dates back to the beginnings of operations research during World War II when the Navy was concerned about protecting large ships against mines. The mines that the Germans introduced at the beginning of the war were magnetically operated by the passage of a ship, and one way of defeating them was to "degauss" the ship—that is, to reduce the ship's magnetic field by putting coils of wire around it and running the proper electric current through them. At the Naval Ordnance Laboratory in Washington some very elaborate coils were designed for various naval ships, which succeeded in reducing the magnetic fields so that the ship would not fire a mine except very close to it, and in very shallow water where mine sweepers could easily be used. The probability of magnetic mines sinking ships was reduced by a large factor, and it looked as if the problem of protecting ships against mines had been solved.

But it turned out that suboptimization had occurred. An engineering problem had been solved, a human problem ignored. The magnetic field changed with the heading of the ship, and thus the current through these coils had to be adjusted as the ship turned to different headings. Curves had been provided showing exactly how these adjustments should be made, but no one in the Navy ever succeeded in following these elaborate instructions as a routine while ships maneuvered. In the end, much simpler systems of coils were adopted as a compromise on the engineering problem that allowed a better solution of the human problem. (The alternative engineering solution—automatic adjustment—proved too complex.)

The various levels where operations research can take place are well illustrated in the US Army; for instance, the highest level is in Washington with the Chief of Staff, and is appropriate to broad problems of logistics, long-range planning, research, development of new equipment, organization, communications, selection, training, and the like, many of which are being studied in the Operations Research Office. At lower levels there are problems of equal interest that are of importance to theater and field army commanders—more limited in scope, as appropriate to the authority of the commander for whom the study is done. For instance, the problem may be to make the best allocation of available communications facilities, and, as far as the local commander is concerned, this is not solved for the immediate future by recommendations for a major change in the whole Army's system of communications. That is, the solution *must be* a suboptimization with respect to the over-all Army problem, and properly so.

Another example of operations research at a subordinate level in the US Army is Richard E. Zimmerman's study of the effectiveness of tanks. This is of great value for the tank design people, though it leads to a suboptimization since it does not consider whether tanks might better be replaced by, say, helicopters—a question outside the scope of his study.

Thus, there is a hierarchy of operations research studies that can be done at different levels, and, ideally, these should feed into one another in order to achieve optimization at all levels. That is, the results of the tank study should be used in a higher-level study to learn what number of tanks may be needed in a future war. Fitting together a set of studies that covers all levels in an organization is a goal that has not yet been effectively achieved. There is hope that a systematic set of operational games will facilitate such a synthesis.

SPECIFIC MATHEMATICAL TECHNIQUES

Some of the mathematical techniques that have been used in operations research and that have recognized names will be discussed briefly. Several will be covered in detail in other chapters.

Game theory (8), not to be confused with gaming or operational simulation, deals with strategies used in repeated competition, such as small-unit battles in warfare, bidding on the market, or marketing itself. In each case one must first set down the decisions that can be made consistently by both sides in the competition; the goal of game theory is to derive simple rules for making these decisions so that the player using these rules—a strategy—will win most often in *repeated* situations. The word "repeated" is emphasized because there is a danger in applying game theory to very large, unique decisions, such as the military decision to invade Europe during World War II, or a company decision to move its industrial operation from one part of the country to another. If one is dealing with something that happens just once he may not have enough understanding of the decision to take account of all the factors. Game theory should therefore only be applied to repeated contests.

One of the most difficult steps in applying game theory is to decide how to measure "payoff," which is another way of expressing the problem of suboptimization. In a simple betting game like poker the payoff is simply the excess of bets won over bets lost, although even this can be complicated by local rules or by the desire to ingratiate oneself by losing. In many industrial operations the goal is to make as much money as possible, but this can be complicated by the interval of time involved; large dividends right away may mean smaller dividends later on. In small military actions it is often extremely difficult to see just what the payoff function should be—that is, how winning ground pays off against the loss of lives.

Linear programming (9) is sometimes confused with game theory or gaming, and will be covered fully in another part of this book. It is a way of optimizing outputs depending on a number of different inputs that are subject to control. A good example is maximizing the profit of an oil refinery, where the cracking process allows various outputs of gasoline, kerosene, fuel oil, and other products. Given the market prices of all these products, the costs of raw materials, and the volume relationships, what is the optimum output mix? The method of solution also applies to the transportation problem—the minimum-cost pattern of deliveries of goods from several factories to several distribution points, given the unit transportation costs between each pair of end points.

When large matrices are involved (e.g., scores of factories and distribution points) the solution is laborious and may require the use of a large computing machine. If the equations are non-linear (as would be the case for reduced unit transportation cost in large shipments), or vary with time ("dynamic" programming), their solution becomes very complex.

Queuing theory (10) is a statistical method of estimating the delays and the waiting lines that occur whenever service has to be provided in sequence for "customers" arriving at a random rate. The same mathematical equations can represent a large variety of things and actions that get delayed when they "line up" in some order for a "service." For instance, queuing theory can be applied to the aircraft that are stacked up over an airfield, or to messages awaiting transmission in a communications system, or to automobiles waiting at a toll gate. It has also been extended to more complex cases, introducing priorities, for instance, such that a priority "customer" goes to the head of the line (or a priority message is handled first in a communications center, or a jet airliner is landed ahead of propeller aircraft, etc.).

Search theory (11), another highly developed technique, first referred to the allocation of effort in conducting a search for a submarine known to be in a general area. Since many operations involve a search of some sort (e.g., a search for the solution to a problem) and an allocation of effort to various parts of this search, the method may have broad application. For instance, marketing may be considered as a search for a customer.

Symbolic logic is a shorthand representation of thought processes such as might lead to a decision. It has proved useful in formulating instructions to large electronic computers. Together with *information theory* (concerning the amount of information in a message or in a file cabinet, and the rate at which it can be transmitted over communications channels), *organization theory* (concerning the way in which individuals interact in an organization), and *servo-mechanism theory* (concerning the control function), a field of *decision theory* is developing—a comprehensive model of all the factors affecting repeated decisions by those who must make them in a large organization. One of the immediate problems is: How much information is needed for a decision, and how timely must it be?

AN EMPIRICAL STUDY OF DECISION

One approach to these questions about decision is to study decisions in the making. In the US Army, this has been done (12) by recording all the decisions made by a field army command and all the information available as the decisions were made during a large Army maneuver. The basic reason for this study was to learn how communications and organization might be improved in the Army. Similar observations could be made in industrial organizations.

The maneuver took place in Europe in 1954 and was observed by a team of ten operations analysts, deployed by pairs on the "battlefield" to try to see what went on. This was a difficult problem in itself—to keep up with what went on in four days of mock battle during which a four-division force attacked ground held by a three-division force, each division involving over 10,000 men. The battle was not entirely scheduled; there were umpires with every unit who decided on the spot who won a local engagement, what actions were feasible, how many casualties there were, etc. Of course, a maneuver like this is never exactly like a battle; it is carried out primarily for purposes of instruction and training; therefore, one has to be careful in interpreting the data. Nevertheless, it seemed clear that the types of decisions made by the commanders on both sides of this simulated battle were fairly typical.

The first thing determined was that the commander never has enough information for the decisions he would like to make. There were twenty-six major decisions clearly made during the four days at field army level—such as a decision to commit a reserve division, or to use an atomic weapon on the enemy, or to change a plan of attack because an enemy atomic weapon had been used.

The operations analysts collected all the information reported in the form of messages logged in—about 300 messages—and the question was how the messages affected the decisions. First, the messages were classified roughly, and were scored by how many had a bearing on each class of decisions. This taxonomy developed ten classes of information, based largely on the types of decisions made. For instance, in deciding to use an atomic bomb, the commander had to be sure it would not hurt his own troops, and in such cases he seldom had good information on where his own troops were—he only knew where they were several hours ago. Thus one class of information concerned location of the front lines; another concerned location of the enemy; another, logistics information, etc.

The amounts of information available in these classes were measured by the numbers of messages received—a very rough measure of information. In information theory the accepted unit is the "bit," or binary digit, which can be conceptually defined by the game of "twenty questions." If it takes ten questions, answered yes or no, to determine what is being thought of, this information is worth roughly ten bits. These messages referred to such things as strong enemy resistance in a specified sector, or the burst of an enemy atomic weapon, or the shortage of a division's supplies. After attempting to convert a few messages into bits, the analysts found the job almost impossible, and proceeded by assuming that each message had approximately the same informational content—the same number of bits. Comparison between the relative amounts of information in each class, and the relative *use* of each class in the twenty-six decisions, showed that certain classes generally needed increasing more than others.

But the usefulness of information was affected by the delay in reporting. For instance, front line information one hour old was obviously much more valuable than if it were ten hours old; a measure of sufficiency of the information with respect to these actual decisions was derived from the average delay in reporting, measured from the time of action or observation recorded on each, to the time it was received at Army headquarters. These delay times were distributed to a log-normal distribution, and it is an interesting fact that this distribution turns up in every study of delays involving human action. The explanation undoubtedly lies in the random addition of a large number of small delays along the line.

From these measured delays the probability can be inferred that information would be received at various times after the event. From this, and estimates of the rapidity of action and delay in implementing decisions, the operations analysts were able to devise an index of how valuable the information was for these decisions. Each of the classes of information was weighted by the number of times it was actually used. In this manner the analysts were able to show that the communications system serving the Army would have twice the value in decisions of this type if it were speeded up—if the delays were reduced by a specified amount. The rest of the study had to do with means of accomplishing this speed-up.

NON-MATHEMATICAL TECHNIQUES

As a final example of operations research techniques, a study (6) has been chosen that involves almost no mathematics, this to counteract the over-emphasis of mathematical techniques by the selection process mentioned earlier. Also, the study illustrates all the characteristics of operations research—the solution of a whole problem leading to definite recommendations that could be implemented by the customer.

The *problem* was one referred to the Operations Research Office in 1951: Should the US Army adopt a policy of integrating Negroes in all-white Army units or not?

The *decision criteria,* on which this matter would be resolved, might have included many considerations of an ethical and political nature. The criterion actually used was simply the practical one of achieving maximum over-all efficiency in the use of manpower resources in the operation of the US Army for the foreseeable future. Negroes now account for over 10 per cent of manpower available to the Army.

Restraints: As is proper, and in accordance with the spirit of the Constitution, the Army must be responsive to public opinion and national policies in a matter of this nature. This political requirement is strongly enforced through congressional committees with powers of investigation, through congressional action on the Army's budget and on laws concerning the Army, and through the president's acting as commander-in-chief. Thus it might have been that a policy matter of this nature would have been entirely settled by presidential decree under congressional pressure, in which case a scientific study would have been inconsequential.

However, no formal restraints of a political nature were imposed or assumed in the course of the study, although it was noted that the Congress, by the Army Reorganization Act of 1950, repealed an act dating from 1869 which required the Army to maintain four all-Negro regiments, and that the President's Executive Order of 1948 directed that equality of opportunity and treatment should be given in the Armed Forces regardless of race, color, or creed. A more serious restraint was at first felt to be the deep-seated emotional attitudes and prejudices of some individuals within the Army, primarily those of southern background and culture. It was alleged further that the Army would no longer appeal as a career to southern whites if Negroes were integrated with whites in its organization. The fact that these restraints did not exist is one of the major conclusions of the study which, in one way of speaking, consisted of an examination of all such restraints.

Data used in the study included Army General Classification Test scores, and other indicators of intelligence and educational background. These showed that the differences between whites and Negroes are so great at present that it is impossible to maintain Army standards in selecting leaders, primarily non-commissioned officers, in an all-Negro unit. The test data also showed, on the other hand, that there were too limited opportunities in the small number of all-Negro units to take full advantage of the talents of really outstanding Negroes. Here were the practical disadvantages of segregation in the Army: It resulted in units of sub-standard military capability, and it did not allow full use of Negro talents available.

The practical value of integration inferred from this reasoning was confirmed by data collected in the Korean War, where Negroes had been assigned to all-white units because no white replacements were available. A study of the battle records, interview data from senior commanding officers, and interview-questionnaire data from all ranks in the integrated units, established that the Negro makes a good soldier under such conditions, whereas similar data from World War II had shown poor performance in all-Negro units.

The *techniques* used in the study were primarily drawn from the social sciences, the "critical-incidents" technique employed in the opinion surveys proving very effective. Junior officers and men were asked to write about incidents they observed first-hand in battle and particularly remembered as examples of good or bad performance by co-equals. The results showed little race bias in integrated units; both Negroes and whites featured in outstandingly good—sometimes heroic —performance (as judged by their peers), and both were guilty of poor performance—such as neglecting equipment, responding slowly, or failing to carry out orders.

Interestingly enough, interview data showed that white commanders' opinions of Negroes' performance as soldiers changed markedly after the respondents commanded an integrated unit—from predominantly disparaging to predominantly favorable. In fact, this turned out to be the most significant correlation between opinion on Negroes and background, far more so than the correlation with geographic origin or culture (i.e., Southerners' bias).

A great deal of effort was devoted to demographic analysis of the Negro population in the US, and its changes, to a survey of all-white military units to determine how many Negroes could be absorbed, to a survey of all-Negro units to determine the feasibility of "reverse-integration," and to a sociological study of related problems—mostly problems of integration on Army posts. It was concluded from these that integration was feasible.

The research team comprised six to seven ORO analysts, assisted by nine consultants and three sub-contracting groups, and undertook the study on a "crash" basis in a period of about four months, during which the interview and questionnaire data were obtained in Korea, the historical data assembled, analysis made, and a preliminary report written.

Results achieved: The US Army announced integration of Negroes with whites in several of its major commands, including the Far East Command, US Army in Europe, and the Military District of Washington, shortly after the results of this study were presented. The advance recognition of sociological problems to be expected from integration on Army posts (from the sharing of such facilities as service clubs, swimming pools, and dance halls) assisted the Army to take corrective action before these problems arose.

CONCLUSION

This brief survey attempts to show the wide variety of techniques that have been used in operations research studies, and the unlimited nature of the subject. Nevertheless, several specific conclusions emerge:

1. Some form of *model* is necessary in an operations research study, although the type and specificity of this theoretical construct can range from the broadly analogical to the strictly mathematical.
2. Many models used in operations research are stochastic or probabilistic, and in these cases *simulation* is often an effective technique for reaching a solution. Such simulation varies from crude trials with human players to highly sophisticated calculations with electronic computing machines.
3. The *levels of optimization* in an operations research study must be selected with care to avoid the dangers of suboptimization on the one hand and, on the other, the impossibility of completing a study of too broad a scope.
4. The broader the problem, the more difficult it is to obtain reliable *data* on which to base a theoretical model. More attempts should be made to perform controlled experiments on whole systems.
5. The characteristic of a true operations research study is that it provide realistic answers to an *actual practical problem*. In this context the tools and techniques used should never be limiting; the goal is to select techniques that allow all significant factors of the actual problem to be considered.

REFERENCES

(1) Engel, J. H. "A Verification of Lanchester's Law," *Operations Research* Vol. 2 (1954), 163-71.
(2) "The Antiac, a Computer of Hymenopterous Design," *Operations Research* Vol. 1 (1953), 254.

(3) Thomas, C. J., and Deemer, W. L., Jr. "The Role of Operational Gaming," *Operations Research,* Vol. 5 (1957), 1-27.

(4) Crane, Roger R. "Analysis of a Railroad Classification Yard," *Operations Research for Management.* Vol. II. Edited by Joseph F. McCloskey and John F. Coppinger. Baltimore: The Johns Hopkins Press, 1956.

(5) Ricciardi, Franc M., *et al.* "Top Management Decision Simulation." New York: American Management Association, 1957.

(6) Hausrath, A. H. "Utilization of Negro Manpower in the Army," *Operations Research,* Vol. 2 (1954), 17-30.

(7) Hitch, C. "Sub-Optimization in Operations Problems," *Operations Research,* Vol. 1 (1953), 87-99.

(8) von Neumann, J., and Morganstern, O. *The Theory of Games and Economic Behavior.* Princeton: Princeton University Press, 1955.

(9) Dantzig, G. B. "Concepts, Origins and Use of Linear Programming," *Proceedings of the First International Conference on Operational Research.* Baltimore: Operations Research Society of America, 1957.

(10) Morse, P. M. *Queues, Inventories, and Maintenance.* New York: John Wiley and Sons, 1958.

(11) Koopman, B. O. "The Theory of Search. III. The Optimum Distribution of Searching Effort," *Operations Research,* Vol. 5 (1957), 613-27.

(12) Page, T. "The Value of Information in Decision Making," *Proceedings of the First International Conference on Operational Research.* Baltimore: Operations Research Society of America, 1957.

Seven

A SURVEY OF SYSTEMS ENGINEERING TOOLS AND TECHNIQUES

RICHARD B. KERSHNER

DEFINITION OF A SYSTEM

Attempting to define a *system,* in the sense in which the word is used in such phrases as "systems engineering" or "systems optimization," is a very ambitious and perhaps foolhardy undertaking. In fact, many feel such an attempt is doomed to failure for the reason that the word *system* is so general and inclusive that any attempt to define it necessarily requires the use of other words and these, being even more general and inclusive, are less well understood by the reader than the word system itself. Those who feel this way would probably be best satisfied by a statement that a *system* is the sort of thing discussed in this book. And, from a purely logical point of view, such an attitude is completely justified. For it is indeed impossible to define all words in a language without circularity[1] and, perhaps, the word *system* could quite properly be classed as one which should be left undefined with its meaning acquired by observing it in use. However, it is possible to make statements about the sorts of things considered in systems engineering which are at least suggestive if not actually informative.

One reason for making this attempt is that the word "system" is

[1] Those interested can find a complete discussion of this point in Richard B. Kershner and L. R. Wilcox, *The Anatomy of Mathematics* (New York: The Ronald Press Co., 1950), Chapter 2.

used in common language with at least two different meanings which do not agree with the use implied in the phrase "systems engineering." In common language the word is sometimes used to refer simply to a collection or aggregation of similar or interrelated things as "a system of thunderheads," "a system of rivers," "the system of fixed stars," "the solar system." On the other hand, the word is also used to mean a collection of rules for procedure such as "a system for winning at roulette." In this chapter the word system will be used to apply to a situation in which both elements are present. Thus a system will consist of a collection of interrelated things *together with* a set of rules for procedure or behavior. The collection of things is called the *system composition,* and the set of rules is called the *system operation.* Finally, for a system to come within our purview, there is a third requirement: either the composition or the operation of the system must be under human control. The solar system is not subject to systems engineering, although an artificial earth satellite is.

A system is a collection of entities or things (animate or inanimate) which receives certain inputs and is constrained to act concertedly upon them to produce certain outputs, with the objective of maximizing some function of the inputs and outputs. It will be noticed that this definition uses some very general words indeed—entities, inputs, outputs, function. Nonetheless, some of the important distinguishing aspects of the sort of things treated in systems engineering are suggested by this collection of words.

First, the words "act upon" are very important. A system is something dynamic. A completely static object is not a system. Thus, for most people, a stone is not a system, although for a geologist who sees a stone as something that used to be mud and is in the process of becoming sand, a stone might be at least dynamic. A building, as such, is not a system, although a hotel, together with its staff and operating rules, which receives inputs (food, fuel, guests, complaints, water, bills, etc.) and operates on them to produce outputs (garbage, hot air, comfort, bills, etc.) with the objective of maximizing profits, is a large and complex system and contains many subsystems (a heating system, a plumbing system, a billing system, etc.). A system is dynamic, and, indeed, dynamics is one of the mathematical disciplines often invoked in systems engineering.

Next, the words "with the objective of" are suggestive of an important fact. For our purposes a system exists only when somebody has something in mind. There must be an intent. A thunderstorm receives inputs and produces outputs—it is undoubtedly dynamic. The dynamics of a thunderstorm can be and have been subjected to mathematical analysis. And, in common language, a thunderstorm is often called a system. However, the lack of human intent or control of the dynamics involved makes most of the techniques of modern systems engineering not applicable. It does not seem appropriate for "systems engineering" to subsume all dynamics, or indeed all of physics, chemistry, and biology, simply because a dynamic situation is involved. Accordingly the words "systems engineering" are usually restricted to situations when the interaction between human intention and actions and the performance of the system are being considered.

Finally, a few words are in order about the way in which the "objective" of a system was stated, i.e., "maximizing some function of the inputs and outputs." It may be felt that this is rather special and restrictive. However, a little thought should make it clear that it is not at all restrictive. Any imaginable end result which depends on the values of the inputs and outputs can be expressed in this way. The word "function" is very general indeed and simply means something with a value (valuation) which depends on the inputs and outputs. The function to be maximized may be thought of as some measure of worth or value of the system. The process of systems optimization, with which this book is primarily concerned, consists really of two parts; first, the formulation of a (value) function which it is desired to maximize, and, second, the variation of the systems composition or operation in such a way as to accomplish this maximization. The first of these problems is often the more difficult of the two.

The process of "optimization" always takes place with a number of constraints with respect to what variables of systems composition or operation are allowed to be varied. There are always some variables which are not varied. Hence every "optimization" is really a "suboptimization." For example, consider a heating system in a building which is thermostatically controlled. A heating engineer called in to adjust an existing system will confine his attention to modifications of minor variables (voltages, contact spacing, etc.). His optimization will consist of minimizing the difference between the temperature called for, T_c, and the temperature actually provided, T_p. A heating engineer designing a system for a new building will certainly not attempt to minimize $T_c - T_p$. This difference can always be reduced by installing more elaborate and expensive equipment. Hence he must create some more elaborate function involving system cost and reliability as well as $T_c - T_p$ as his system value. An engineer developing a heating system for general public sale must also consider customer appeal and will have to include more complex factors (esthetics, noise, etc.) in his value function. Finally, from a sufficiently broad point of view, the "optimum" solution of the heating problem might be to breed a new race of man impervious to temperature variations and so eliminate all the mechanisms. It is unlikely that any systems engineer will be given so broad an assignment as this last. This is what was meant by saying that, in the real world, only suboptimizations are ever performed.

It has been said that our attention would be confined to situations where either system composition or system operation is under human control and subject to variation. When primary attention is being placed on variation of the system composition, it is customary to use the words "systems engineering," while if the primary emphasis is on variation of the rules of procedure or system operation then the appropriate words are "operations analysis." Obviously these activities are strongly interrelated. Usually a variation in system composition requires some change in operation and conversely. There is a great need for a word for the over-all process of seeking an optimum (or appropriate suboptimum) composition and operation of a system to accomplish a desired objective. There is a growing tendency for systems engineers (operations analysts) to use the words "systems engineering" (operations analysis) for this over-all optimization process and hence to subsume operations analysis (systems engineering) as simply a special subdoctrine.

CLASSIFICATION OF SYSTEMS

It was stated above that "systems engineering" applies to situations in which one is considering the interaction between human intentions or actions and the performance of the system. It is convenient to classify systems according to the way in which this interaction enters the problem.

The minimum human involvement occurs in a system whose components are purely mechanical (chemical, electrical, hydraulic, pneumatic) devices and whose operation is determined by the laws of physics (and chemistry). A guided missile or an atomic bomb is such a system. Here the human element is confined to the choice (design) of the system composition. This activity is pure systems engineering. Such a system will be called *mechanistic.*

Many systems contain humans as operating elements or components acting in such a way that they would be (at least in principle) replaceable by machines. Whenever the humans are simply lifting, moving, pointing or switching things in specific response to dial readings, lights, or other optical signals, noises (including verbal commands) and the like, the function performed can be considered as essentially mechanical. The pilot or gunner of a combat airplane and the operator of a telephone switchboard perform such functions and have been successfully replaced by machines (guided missiles, dial phones). On the other hand, the pilot and stewardess on a commercial airplane and the chief operator of a telephone system perform many essentially non-mechanical functions (public relations, morale, repair, etc.). A system in which the humans involved in the operation are performing essentially mechanical functions will be called *quasi-mechanistic.* It is necessary to distinguish a quasi-mechanistic system from a mechanistic system because different techniques are necessary for their analysis. However much the systems designer intends the human elements of the system to behave like mechanisms, they will, of course, fail to do so. Their functioning is far less predictable, and they are subject to different types of malfunction. They will also insist on performing extra functions, some of which are very desirable, such as giving notice of impending malfunction either of themselves ("Gee, Boss, I feel sick.") or of the rest of the system ("That bearing over there is smoking.").

Many systems are designed to operate with a set of important inputs which are so numerous that it has not been found possible to create a set of operating rules which will cover all possible variations of all input parameters. For example, a manufacturing organization must be prepared to cope with variations in raw material supply and cost, in consumer demand, in the tax structure and a host of other governmental regulations, in population, wages, labor unions, transportation, etc. For this purpose such a system uses humans in an essentially non-mechanical way to make decisions which modify the system operation. Since the purpose of the modification is to adapt the system operation to a new set of conditions, such a system is often called *adaptive.* Since the nature of the adaptation is usually to modify the operation or utilization of the system, these systems are also called *systems with variable utilization.* The decision-making process involved is, or should be, operations analysis. It used to be onsidered that the adaptive capability provided by humans in an daptive system was essentially non-mechanical and no machine could e designed which would exhibit such capabilities. However, in recent ears, the development of computing machines has progressed to the oint that adaptive performance of a sort is provided by mechanisms. ome persons are beginning to believe that there is no essential differnce in kind between the type of adaptive behavior exhibited by such achines and the operation of the human brain.[2] However, for our urposes, it is enough to recognize that the techniques for analysis of n adaptive system are quite different from those applicable to a ystem which is mechanistic in the old-fashioned sense.

Finally, there is one specific type of variation of input which poses roblems of such a special nature that it is necessary to distinguish is situation and develop special techniques for its analysis. This tuation arises in case the system must operate in the presence of one r more other systems which have as a part of their objective the estruction or defeat of the given system (minimizing the function hich the given system is attempting to maximize). Commercial competition and war are the two most common areas in which such situtions arise, and such systems are called *competitive.*

ETERMINISM AND STATISTICS

It was indicated above that, as humans take a more prominent role the operation of a system, prediction of the performance of the ystem in a given circumstance becomes more difficult. Humans, in a articular instance, have a way of acting unpredictably. This does not ean, however, that all analysis must be abandoned. For, while the erformance of a particular human in a particular situation is never recisely predictable, it is often possible to make very accurate statements about the average behavior of a human in a particular situation. s the performance of a particular system becomes less completely redictable, either because of the introduction of human factors or for ny other reason (e.g., the introduction of "acts of God"), then the chnique of analysis shifts from mechanics to statistics. This is such n important point that it seems wise to devote some discussion to the nderlying philosophy.

The tremendous success of Newtonian mechanics in making possible

[2] This subject is discussed at length in W. Ross Ashby, *Design for a Brain* New York: John Wiley and Sons, 1952).

the accurate prediction of the results of countless experiments involving objects ranging from suns to molecules led many nineteentl
century physicists and philosophers to assume that mechanics coul
explain everything if only we were informed enough to know all th
initial conditions and capable enough to do the sums. These peopl
saw the universe as a giant machine whose every action was pre
scribed by its present configuration and past history. This was th
philosophy of determinism and its adherents were legion.

During this Newtonian period the theory of probability and th
techniques of statistics were being developed—at first to analyz
games of chance (dice throwing, card drawing, etc.). Later, statistic
(statistical mechanics) was used to analyze the behavior of large dy
namical systems in which the particles were so numerous (as th
molecules of a gas) that it was hopeless to solve the complete New
tonian equations for each of the interacting particles. However, th
determinists thought of probabilities as being only a statement abou
the degree of information they had on a subject and statistics as bein
a trick to avoid an excessively tedious set of calculations—and no
that there was anything basically different (inherently unpredictable
about the situations in which these doctrines were applied. When
die is cast, they argued, if we could accurately measure all the initia
values of position, orientation, and linear and angular velocities, an
also knew the elasticity of the floor and die, then Newtonian mechanic
could calculate which side would wind up on top. For most purpose
it is more useful and more important to know without doing this fear
some calculation even once that if a die is cast a large number c
times then the five will land on top about one-sixth of the time. How
ever, in this particular case there is some justification for the deter
ministic viewpoint.

But science had some surprises in store for the determinists. Th
first bad blow was the discovery of the laws governing the behavio
of radioactive atoms. These substances follow a law that is, from th
standpoint of a determinist, completely preposterous. A radioactiv
atom has the ability to emit some form of radioactivity and simul
taneously change into a different atom. Associated with each radic
active species there is a length of time (called the half-life) with th
following property: if a bucket full of a particular type of radioactiv
atoms is kept in any environment for a half-life of time, then at th
end of this time it will be found that just half of the atoms have emit
ted radiation and changed their species. Whether or not any particula

atom has changed seems not to depend in any way on the environment o which it has been subjected. Nothing (short of nuclear bombard-nent) seems even to affect the probability of a particular atom chang-ng in the next minute or century. The future behavior of a radioactive atom appears to be inherently unpredictable. Yet very accurate sta-istical statements can be made: just half of the atoms will change in he period of a half-life.

If this were not enough to finish off the most ardent determinist, the liscovery of the Heisenberg uncertainty principle came along to ad-minister the *coup de grace.* This principle counters the determinist's tatement, that he could calculate the universe if he only knew exactly ll the initial conditions, with a demonstration that the initial condi-ions are *inherently* unknowable. Even for a single elementary particle ne cannot know both the position and velocity (momentum) with bsolute precision. This principle essentially reverses the roles assigned o mechanics and statistics by the nineteenth century determinist. tatistics becomes the fundamental technique for discussing the uni-erse. Mechanics is simply a lazy man's way to treat the motion of a arge number of atoms by discussing only gross average motions of statistical fiction called the center of gravity.

Independently of all this philosophy, what the systems engineer has o know is that mechanics and statistics are two powerful techniques or predicting the behavior of systems and that he must use the echnique that is appropriate to the extent of knowledge that he has nd needs.

At the beginning of this section it was indicated that statistical echniques were most likely to be used when unpredictable human actors influenced the behavior of the system. It should be clear by ow, however, that this is by no means the only reason for the use of statistical approach. The design of an atomic bomb is based on atistical techniques. If one were to design a meteorite warning sys-em for a spaceship, then times of arrival of meteorites and their eights would have to be treated as statistical inputs in spite of the act that any individual meteorite is following Newtonian mechanics ith classic precision.

There is a tendency to attach too much significance to the fact that eterministic calculations predict performance absolutely while sta-stical calculations predict only probabilities. For, on the one hand, deterministic calculation only accurately predicts system behavior *the absence of failure* or malfunction. When the probability of malfunction is introduced, as it always must be, then only a probability statement can be made about even the most completely deterministic system. And, on the other hand, when statistics is used, as in calculations of radioactivity or in statistical mechanics, to encompass the behavior of a tremendously large number of individual events, then the probability of a particular answer can be so near to one as to constitute certainty for any practical purposes.

SYSTEM ORGANIZATION

In the next few sections attention will be concentrated on mechanistic systems, since they are the special province of systems engineering as distinguished from operations analysis. Some of the techniques used in dealing with problems of system optimization for such mechanistic systems will be indicated.

The first step in preparing to analyze system performance is to write down the composition and operation of the system in some suggestive and useful form. The most usual form is some variety of *block diagram.* In the block diagram, various components or subsystems of the system are represented by properly labeled blocks (rectangles), and the interrelations are suggested by arrows or connecting lines which also may be labeled when appropriate. For example, a particular type of guided missile might be represented by the block diagram shown in Figure 1.

A block diagram, such as that in Figure 1, is very useful but generally only for certain very specific purposes. A large number of such diagrams, drawn with different objectives, is generally needed. In fact most of the blocks shown in Figure 1 are themselves relatively large complex systems which must in turn be analyzed and diagrammed. Frequently, at more detailed levels, the featureless blocks of the typical block diagram are replaced either by special standard symbols (e.g., a triangle for an electronic amplifier) or by symbolic sketches which indicate the functioning of the device under consideration. For example, the block marked "servo" in Figure 1 might be an electro-hydraulic servo which functions in a manner indicated in Figure 2. A still greater level of detail would replace the triangle indicating the "servo amplifier" by a diagram showing the individual resistors, condensers, vacuum tubes, etc., and their interconnections. At this level for electronic equipment, the block diagram is called a *circuit diagram*

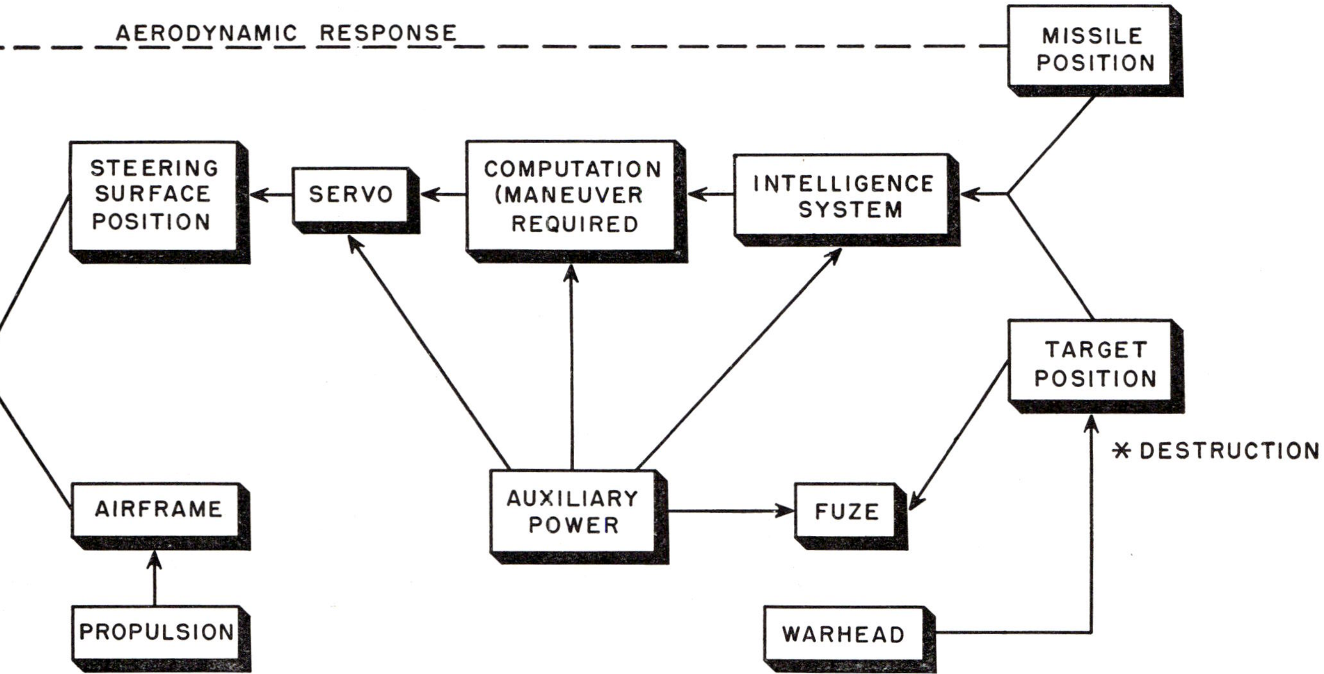

Figure 1. Block diagram representing a guided missile.

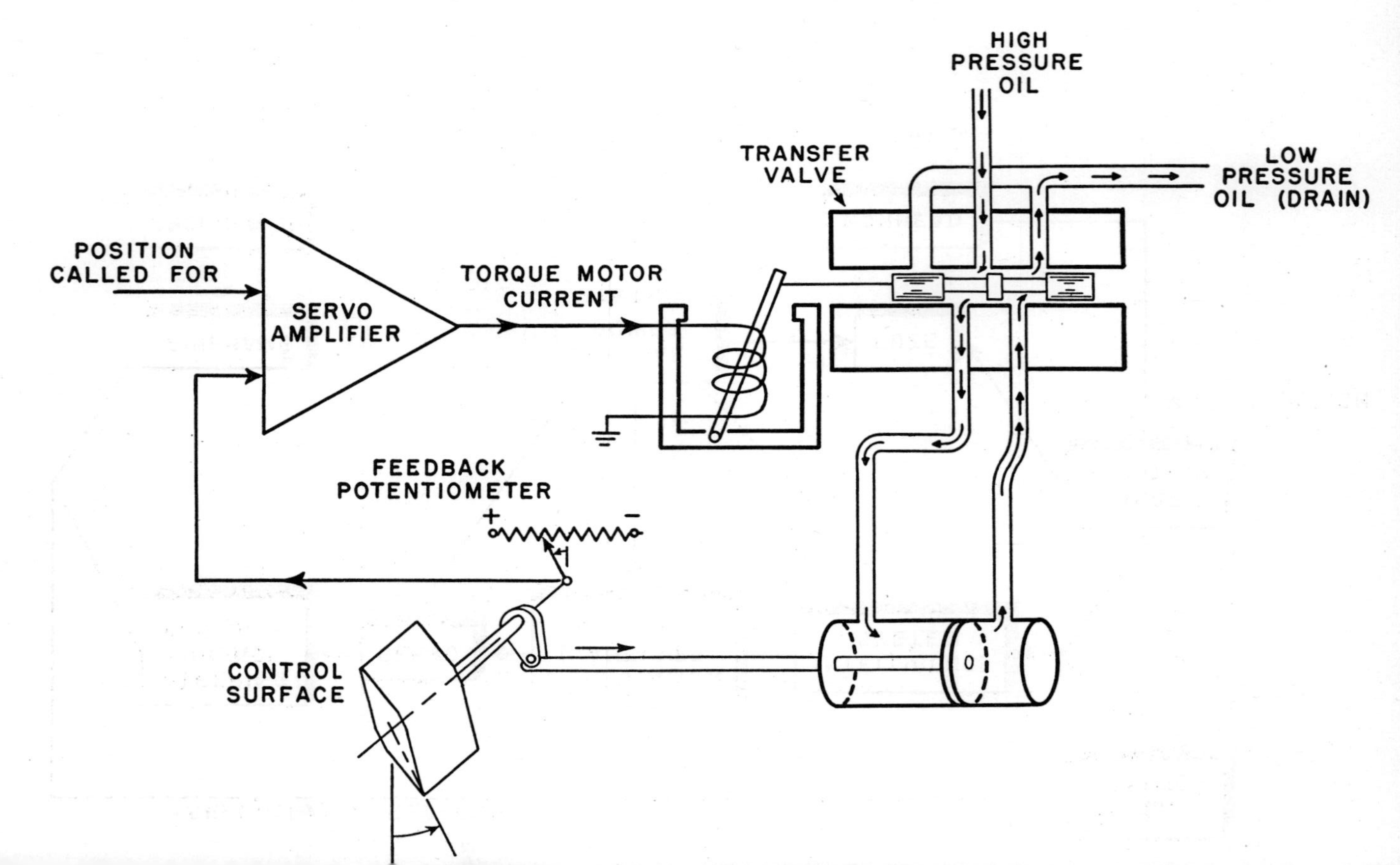
HIGH PRESSURE OIL
TRANSFER VALVE
LOW PRESSURE OIL (DRAIN)
POSITION CALLED FOR
SERVO AMPLIFIER
TORQUE MOTOR CURRENT
FEEDBACK POTENTIOMETER
+
−
CONTROL SURFACE

The way in which a system is organized into interacting subsystems depends entirely on the purpose of the subdivision. The breakdown of a guided missile into functionally distinct subsystems, shown in Figure 1, is very useful in aiding the analysis of system performance. A quite different breakdown might be used to guide composite design or packaging studies. In choosing the location and arrangement of parts in a guided missile, it is very wise to group together parts that require a similar technology. For example, it is very desirable to group all electronic packages together in an "electronic section" of the missile. For, although the system function of different electronic packages may be quite different (e.g., the fuze and the servo amplifier), nonetheless all the electronic packages require similar assembly techniques, handling equipment, repair and maintenance tools and test equipment, spare parts, and personnel training. Design of a complex piece of equipment in such a way that it is broken into separate sections, each of which consists of components all embodying a similar technology, is sometimes called the "echelon design" or the "sectionalized design" approach. Properly carried out, each section is built to specification insuring interchangeability, so that repair of a faulty device can be carried out by first identifying the particular section in which the fault lies and then replacing the faulty section with a spare section, without the need for any adjustment to assure operation of the over-all device. Isolation of the fault to a given section is made easy by the different nature of the sections, e.g., it is usually easy to distinguish a hydraulic failure from an electrical failure. After the over-all unit is repaired by the use of a replacement section the lower level repair of the faulty section can be carried out by specialists in the specific technology embodied in that section. Figure 3 shows a guided missile design which was based on the sectionalized philosophy.

SUBSYSTEM INTERACTIONS

A block diagram, such as Figure 1, seldom shows all the interconnections between subsystems, for if it did the diagram would become an unreadable maze of lines. To illustrate the complexity of the interrelations between the components of such a system as the guided missile of Figure 1, let us consider (in oversimplified form) a specific systems design problem that actually arose in the development of such a guided missile. In a specific design of such a missile it was found

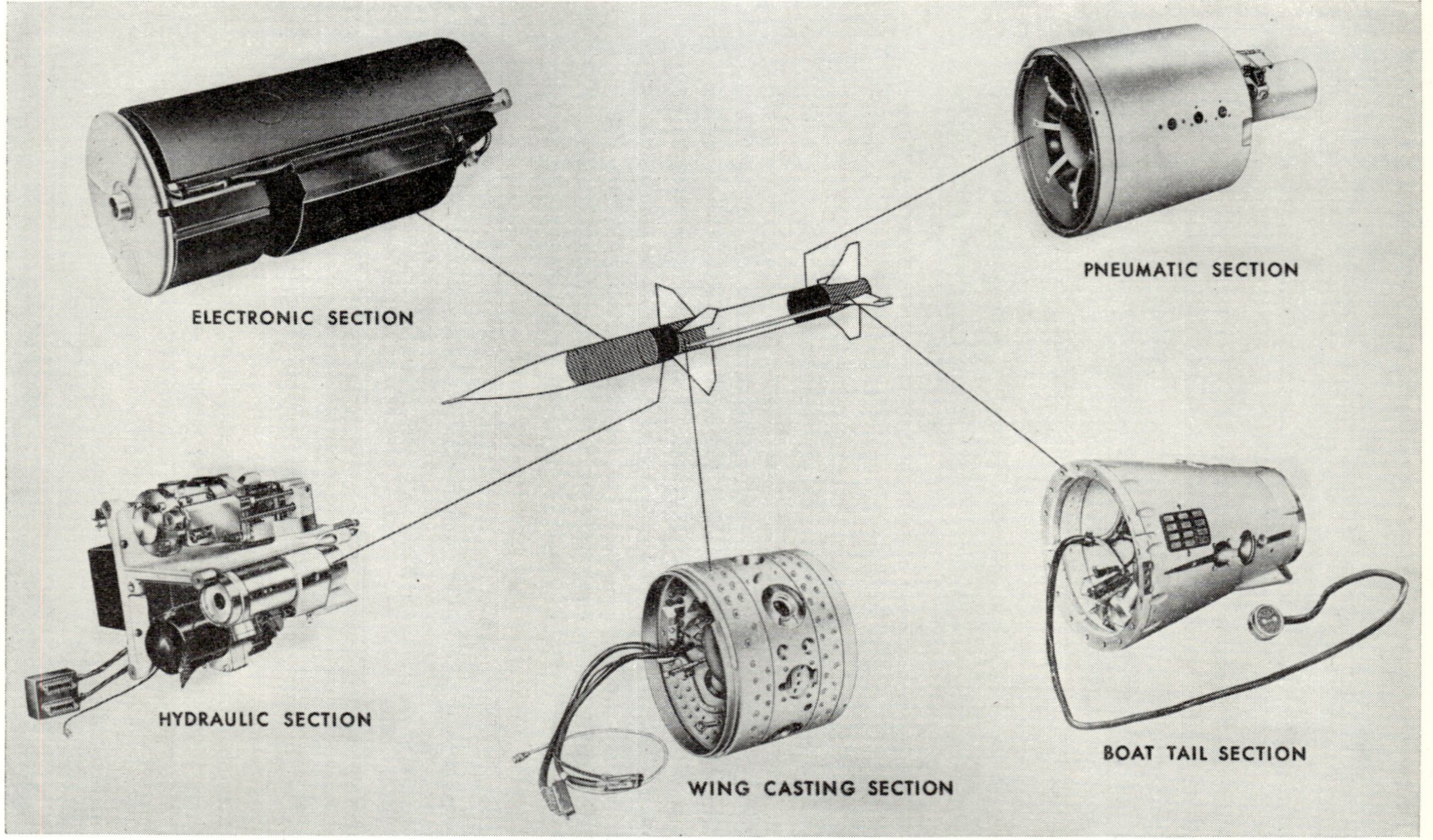

Figure 3. Terrier 1B sectionalization.

that the auxiliary power system was exhausted somewhat before target intercept was achieved for targets at maximum range. The missile did not achieve its intended maximum range capability. The solution would seem to be obvious: the power supply is inadequate; hence increase it. But it is not that simple. Increasing the power supply requires increasing the weight and size of the over-all missile; this in turn reduces the speed (for a fixed propulsion system) so that it takes longer to reach the target, thus requiring more auxiliary power and aggravating the problem. And even if a simple increase in power supply would solve the problem it might not be the best solution. Actually all of the following possibilities separately and in combination (together with a good many more too complicated to describe here) were seriously considered in this case.

(1) Increase the size of the auxiliary power supply system.
(2) Improve the efficiency of the auxiliary power supply system.
(3) Redesign the fuze (intelligence system, computation system, servo system) to require less electrical power and hence make the existing supply adequate.
(4) Redesign the computation system to reduce the maneuver requirements and hence reduce the servo power required.
(5) Redesign the wings to reduce aerodynamic hinge moment and hence reduce the servo power required.
(6) Increase the average speed to arrive at target more quickly, before power is exhausted, by
  (a) Increasing propulsion,
  (b) Reducing warhead weight,
  (c) Reducing drag, by
    (i) Aerodynamic redesign,
    (ii) Computer redesign to reduce the maneuver requirements and hence reduce the drag due to maneuver.

In the specific case of reference a combination of (4) and (6.c.ii) proved to be the optimum solution of the problem.

The main lesson to be learned from the above is that a problem which was stated to be a power supply problem could be solved by various approaches involving literally any major subsystem of the missile including the aerodynamics or the warhead. The old-fashioned engineering approach of assigning a power supply problem to the power supply specialist and leaving him alone does not get the best answers in a guided missile design job. Somebody has to look at all of the possibilities—this somebody is the systems engineer. He can, of course, call on specialists to provide the detailed analysis of any one approach; the electronics expert for (3), the dynamicist for (4), the propulsion expert for (6.a), the aerodynamicist for (6.c), etc., but the systems engineer has to create the valuation technique by means of which the various possible solutions are compared on a common basis.

SYSTEMS ENGINEERING ESTIMATES

In comparing the various possible solutions to the power supply problem discussed in the previous section, the systems engineer must base his decision on comparison of the volume, weight, cost, reliability, performance, etc., of each of the proposed solutions. Obviously the only certain way to make a completely factual comparison is to carry out the redesign in detail in each of the proposed manners and then evaluate the finished products. Equally obvious is that such an approach is generally not consistent with the cost and timescale requirements of the project. Accordingly, it is usually necessary to make the design decisions on the basis of *a priori* estimates of volume, weight, cost, reliability, performance, etc. Procedures for arriving at such estimates are among the most commonly used tools of the systems engineer, and these procedures run the gamut from very crude empirical "rule of thumb" approaches to detailed analysis utilizing large and complex digital computing machines. The detailed analysis techniques, both deterministic and statistical, are discussed at length in this and other books. But little is found in print concerning the use of empirical techniques, yet very frequently the systems engineer must make use of an empirical approach.

To illustrate the necessity for such an approach, let us discuss the problem of estimating volumes and weights in a guided missile. There are detailed and very elaborate methods for computing the weight of the air frame structure based on analysis of the loads, vibration, flutter, aerodynamic heating, etc., which are taught in courses on aeronautical engineering. Such problems are assigned to a well-trained stress analyst, and the answers can generally be relied on (except in the areas of flutter and fatigue where present theory is not completely adequate). In a similar way the weight of the propulsion system, the hydraulic components, and the auxiliary power supply can be calculated by methods taught in engineering schools and based on reasonably well-developed theory.

The warhead is in a different category. The damage created by a warhead of a given size is such an intricate function of the nature of the warhead design, of the target, and of the geometry (spacing and attitude) of the intercept that direct computation of the results is generally not feasible. Accordingly, a large number of experiments have been performed and the results embodied in empirical formulas which are used to make warhead effectiveness calculations. Still another type of problem is posed by the electronic equipment. Even when the detailed circuit diagram on an electronic component is known, the volume and weight of the final device is not subject to accurate computation —for most of the volume of an electronic package consists of the complex shaped piece of empty space between the components and much of the weight is attributable to the metal package used to wrap up this complex piece of space. An electronic package in which the actual components (including wiring) occupies 40 per cent of the total package volume is considered exceptionally dense. Variations in packaging techniques—the use of printed wiring instead of cables, the types of connections used (plugs or solder terminals)—all can make a major difference in the weight and volume of the final package. Accordingly, estimates of the weight and volume of electronics packages are always based on empirical equations embodying the results of experience which is chosen to be as relevant as possible.

There are good reasons why such equations are not taught in the engineering schools and included in the books. Their number is legion and their validity is transitory. An equation for the packaging density of commercial radio equipment is not applicable to guided missile components because the great importance of compactness in the missile has forced the development of techniques for miniaturization in packaging. For the same reason, a formula for guided missile packaging which was valid two or three years ago is no longer valid.

Because of the transitory validity of empirical relationships of this nature, such formulas are often called "state of the art" equations. They represent current practice. One of the important assignments of a systems engineer is to maintain in a currently valid form a file of such "state of the art" empirical equations applicable to the systems under his cognizance. This should be done in a very explicit and objective form and embody the results of other organizations rather than simply the past experience of his own organization. In this way such relationships serve as a means for helping to assure that the engineering of his own organization is maintained at a competitive level. There is a tendency, because of the relatively unscientific basis and uncertain validity of state of the art equations, to avoid their use. Unfortunately an attempt to avoid explicit use of such relations generally amounts to replacing them by vague, unformulated "seat of the pants" or "engineering judgment" considerations which are still less scientific and far less objective.

A particular area in which reliance on empirical formulas is almost always necessary is the calculation of reliability. There are a few cases where failure can be expected due to a calculable process (e.g., automobile batteries), but in most cases failure will result from a collection of causes too subtle or intricate to admit of accurate mechanistic calculation. As a result, reliability is calculated by the use of statistical techniques and the various probabilities involved are determined by experiment or experience. Considerable work has been done, especially in recent years, in developing the statistical techniques and establishing the necessary experimental numbers. There are today available numbers for the expected life or failure rate of a vast collection of devices including such things as vacuum tubes, relays, resistors, solder joints, pumps, motors, etc. However, it is necessary to observe very great caution in the use of these numbers to calculate the reliability of a complex device.

A device of the order of complexity of a guided missile contains thousands or tens of thousands of components such as resistors or solder joints. Assuming, what is nearly true for a guided missile, that a single failure of any of these components will lead to a failure of the missile, then the probability of successful operation of the missile is the product of all the probabilities of successful operation of each component. Multiplying together a very large number of quantities which are each less than unity gives a value near zero unless most of the individual numbers are *extremely* near to unity. In other words it makes quite a difference whether the probability of success of most of the small components is assumed to be 0.999990 or 0.999999. And, quite clearly, establishing the difference between 0.999990 and 0.999999 experimentally is a hopeless undertaking since the number of experiments which would have to be performed would run into the billions. Historically, most of the early estimates of guided-missile reliability were very discouraging for essentially this reason. In practice, guided missiles, after the completion of development, have turned out to have a quite acceptable reliability. The lesson to be learned from the foregoing discussion is that reliability estimates should be based on a breakdown of the device into only a relatively small number of major components or subassemblies, and experimental numbers or realistic estimates must be obtained for these subassemblies as complete operating units.

Cost estimates are still another area where empirical relations must be used. Such relations are frequently expressed in terms of a cost per pound for a given type of equipment. And, rather surprisingly, for many types of equipment the cost is roughly proportional to the weight over large factors. However, there are a number of complicating factors involved in the estimating of costs. An obvious point is that cost of a unit is very strongly dependent on the number of units being built. It is also a strong function of the number of units that have been built, the degree of tooling, the extent to which the tooling has been paid off (amortized), the state of competition, etc. An obvious point is that cost is inextricably related to reliability. This comes about in several ways. First, unreliable equipment poses additional costs in the form of inspection, repair, spares, and the like. Second, and often more important, it may require an increase in the number of units necessary to accomplish a given objective. For example, in the case of an antiaircraft guided missile the cost per missile is not nearly as significant as the cost per target kill.

SYSTEMS ENGINEERING PERFORMANCE ANALYSIS

In contrast to the approximate and empirically based calculations that are characteristic of cost and reliability estimates, the calculation of *performance* of a mechanistic system is frequently based on a highly accurate solution of deterministic laws of physics (or chemistry) that characterize the system. It is fortunate that more accurate methods are available for performance calculations since these calculations are frequently far more critical than the calculation of cost or reliability. In a guided missile, for example, the difference between a steering system which is stable and one which is unstable may depend on a minor change of one individual gain setting, but the effect is to change a very valuable device into a totally useless one.

For a mechanical system the relevant laws of physics are expressed in terms of relationships connecting positions (in space), velocities, and accelerations of significant reference points. If the system is electrical, then the applicable laws express relations involving electric charge and current (amperage). In a chemical system the relations generally involve the quantity or concentration of specific chemicals and the rate at which the chemicals are transformed into other chemical compounds or other states. If the system is hydraulic or pneumatic, the laws connect pressure, temperature, or volume, and the rate of change of these quantities as the fluid and heat flows through the system. The important fact to note is that in all cases the physical laws describing the system involve connections between certain quantities and the rate at which these quantities change under the dynamic constraints of the system. Realization of this point in the mechanical case is assisted by recalling that velocity is simply the rate of change of position, and acceleration is defined as the rate of change of velocity. In electrical systems, current is the rate of change of the quantity of electricity, and the basic laws governing alternating current circuits, for example, are expressible in terms of these two quantities and the rate of change of the current.

In calculus the rate of change of a quantity is called its derivative (with respect to time), and an equation connecting quantities with their derivatives (and perhaps, the derivatives [rate of change] of these derivatives) is called a differential equation. The deterministic calculations of the performance of a mechanistic system, then, consists generally of the solution of differential equations.

The subject of differential equations is treated at length in mathematics courses and textbooks. In particular the type of equations specifically applicable to servo systems, such as that represented by Figure 2, have been the subject of extensive development in recent years so that a special subject called servo dynamics or servo mechanism theory is now recognized and taught.

Unfortunately, in spite of the extensive theories mentioned above, the complete hand-computation of solutions of differential equations generally falls somewhere in the range from tedious to impossible, with a tendency to lie at the impossible end of the range. Fortunately there is a way, though a somewhat expensive way, out of this dilemma. It was mentioned that the performance of electric circuits could be described in terms of differential equations. To a certain extent the converse is true; that is, for very broad classes of differential equations it is possible to construct electric circuits which are described by the given differential equations. When this is possible one can construct such a circuit and let the electricity solve the differential equations. Large and flexible machines in which a vast variety of such circuits can be set up conveniently by plug board techniques have been developed and are widely used for the solution of differential equations. Such machines are called analogue computers since the behavior of the circuit can be considered to be analogous to the behavior of the system which the differential equations under consideration were intended to describe. It is not too strong a statement to say that the development of guided missiles and other modern systems of a comparable complexity would have been impossible (in any reasonable timescale) without the prior development of such powerful and flexible computing machines. In fact, historically, the development of guided missiles began before such machines were available and through this period (during and just after World War II) progress was very slow and painful.

In mathematics courses it is shown that "in principle" differential equations can always be solved *approximately* by straightforward numerical methods involving nothing but the ordinary arithmetic processes of addition, subtraction, multiplication and division. It is further shown that the approximate answer obtained in this way can be made as accurate as desired at the expense of increasing the amount of arithmetic required. Until quite recently this fact had a certain theoretical and philosophical interest but relatively little practical significance, since, in practical cases, the number of arithmetical operations necessary to achieve an accurate approximation to the solution of a complex equation is so large that the time required to perform these computations by hand is measured in years or centuries. Recently, however, a change with truly revolutionary implications has been brought about by the introduction of machines which can perform arithmetic operations quite rapidly. These machines are called digital computers—since, basically, all they can do is add digits.[3] But they can add digits in less than a millionth of a second. This one simple (though remarkable) fact means that the whole technique of numerical approximation to the solution of differential equations becomes a practical matter. Calculations that would have required years by hand methods are accomplished in minutes. Digital computers of great power and flexibility are more recent than analogue machines, but already their impact on the development of technology has been almost incalculable.

Because both analogue and digital computers are available to the modern systems engineer, he is frequently faced with a choice as to which to use for a particular problem. There are no simple rules to

[3] See Chapters 11 and 22.

answer this problem. Basically he must determine which technique gives him a satisfactory answer at a lower price. But there are some general rules that help in reaching a decision. The analogue machine is limited in accuracy (in general, the accuracy is about 10 per cent) while digital machines can operate to any desired degree of accuracy. On the other hand, the analogue machine operates conveniently "in real time." This means that it is frequently possible to operate the analogue machine in such a way that one second of machine operation corresponds to one second of operation of the real-life system under investigation. This "real time" operation is of great importance in connection with the next subject—simulation.

SIMULATION

During the earliest stages of the development of a complex deterministic system the performance of any presumed design is normally computed by the use of general-purpose analogue or digital machines. But as the development progresses, certain portions or subsystems become available in actual "hardware" form. At this stage it is possible to substitute this actual device for the analogous circuits or equations in the computation and, in a sense, let this device compute its own performance. The resulting compromise between a computation and a test is called a simulation, and a special purpose computer designed to incorporate a certain number of actual "hardware" system components is called a simulator. For dynamic calculations it is clearly necessary that the calculation operate in real time, in the sense of the preceding paragraph. For this reason simulators are usually based on analogue computations (although digital computations in real time have been performed in a few cases).

In the case of the guided missile of Figure 1, the ultimate simulation uses the entire guided missile in hardware form and uses computation techniques only for those portions of the system operation which require actual firing of the missile to bring them into operation. In such a final simulation a signal is fed into the intelligence system of a missile in the laboratory which simulates the signal that would be received in flight for a given target and missile position. The missile responds to this signal in completely realistic fashion and, in particular, moves its control surfaces in response to the servo action. At this point, since the missile is not actually in flight, computation must take over. The simulator is arranged to measure the control surface motion and calculate the change in missile position (maneuver) that would have resulted in flight. Then the change in relative target and missile position due to this maneuver and the target motion during the same time is calculated, and finally the change in signal received by the intelligence system is computed. This signal is provided by a signal generator, and operation proceeds ("the loop is closed"). Such simulation methods make it possible to run the equivalent of thousands of flight tests with a vast variety of presumed target situations (various speeds, altitudes, ranges, evasive maneuver, multiple targets, etc.) without expending the actual missile equipment. Of course, any computation, including this type of simulation, necessarily has elements of unrealism and cannot replace the need for actual flights or eliminate the occurrence of unexpected phenomena when the flights occur. A calculation is no better than the assumptions made in establishing the equations, and it is always possible (or even likely, in the case of guided missiles) that a significant phenomenon was overlooked when the equations were established. When this occurs, as evidenced by flight behavior not in agreement with the simulator performance, the neglected phenomenon is introduced into the simulator by an appropriate modification. In fact, it is fair to state that the development program is complete when the actual flights and the simulations agree in performance. The simulation technique is, of course, particularly important in the development of those systems which (like atomic bombs or guided missiles) are expended or destroyed in actual operation so that repeated final tests are extremely expensive.

STATISTICAL METHODS

In the last section deterministic calculations of system performance were discussed. As mentioned earlier, such methods are applicable, at most, to mechanistic systems, and as soon as human participation in the system operation occurs it is necessary to introduce arguments of probability and calculate, not how the system will certainly and invariably perform, but only how it will perform a certain fraction of the times it is tested. Sometimes, as in the case of the atomic bomb already mentioned, statistical methods must be used even for a mechanistic system when the laws of physics that apply (like those for radioactivity) are inherently statistical in nature.[4]

When the performance of a given experiment (or the operation of a given system) does not invariably yield the same result, but instead has a particular result a certain fraction, p, of the time, then the given result is said to have probability p. Probabilities can often be determined experimentally simply by performing the relevant experiment a large number of times and tabulating the results. Sometimes probabilities can be computed *a priori* without the necessity of performing the experiment.

The possibility of *a priori* calculation exists when it is possible to catalogue the various outcomes of the experiment as consisting of a number of "equally likely" possibilities. To see how this process works, consider, for example, the outcomes of throwing dice, illustrated in Tables 6 and 7 of Chapter 9.

A number of techniques for the computation of probabilities have been developed and are taught in courses in probability and statistics; these techniques, though equivalent, are often more convenient than the direct catalogue of possibilities. Somewhat as in the case of differential equations, in particular cases it is very often impracticable or impossible to carry out a hand computation of probabilities. For, in practical cases, there are often very complex interdependencies that make the number of individual cases that need consideration almost astronomically large. The type of interdependencies that frequently occur can be illustrated by considering a rather dull but very informative solitaire game.

The game to be described requires three boxes, each containing a certain number of pieces or counters which are indistinguishable except by color, as, for example, poker chips. Specifically, the game starts with the following situation:

The first box contains	1,000	red and 3,000 white chips;
the second box contains	100	red and 50 white chips;
the third box contains	2	white and 1 blue chip.

The game consists of drawing successive chips (blind) from the boxes according to the following rules:

Draw from the first box for the first three moves and thereafter unless the last three chips drawn are all red.

[4] See the discussion of the Poisson process in Chapter 14.

If the last three or four chips drawn are all red, draw from the second box.

If the last five chips drawn are all red draw from the third box.

If the blue chip is drawn, the game is won. (If a white chip is drawn at any point go back to the first box.) If the first or second box is found to be empty in attempting to draw from it, the game is lost.

Problem: What is the probability of winning the game?

Notice that the probabilities at any stage depend, in a reasonably complicated way, on the entire past history of the game. Even on the second move the probability of drawing a red chip is 999/3,999 if the first chip was red and 1,000/3,999 if the first chip was white. Of course, later on the probabilities become grossly different when different boxes are to be drawn from.

It would, in principle, be possible to calculate the probability of winning the game by the use of the elementary rules for combining probabilities or, equivalently, by cataloguing all possible sequences of moves. The computational labor involved in this approach, however, makes it a very impractical matter. Indeed, it would be far easier to determine the probability experimentally by simply playing the game a large number of times and noting the percentage of times the game is won. Even the experimental approach is quite a lengthy process since an individual game might require drawing over 4,000 chips successively and a rather large number of games would have to be played to give a reasonable approximation to the answer. Once again modern computing machines can come to the rescue.

It is reasonably easy to set up an electronic digital computer to play a game such as that under discussion. All that is necessary is to provide the machine with some way of doing the equivalent of drawing a chip and noting its color. This requires having the machine make a decision (choose a number, for example) which is based on a random process which has the same probability as that applying to the color of the chip at any given stage of the game.

Before describing how a digital machine can be made to make choices based on any prescribed probability, it is desirable first to describe a convenient standardized way for a human to do the same thing—by the use of random digits. A digit is, of course, any of the ten symbols, 0, 1, 2, 3, 4, 5, 6, 7, 8, 9. A sequence of random digits is a sequence of digits that might be obtained by successive throws of a "ten-sided die." Unfortunately, there is no regular solid with ten faces (so a "ten-sided die" does not exist) but there is a regular twenty-sided solid—the icosahedron—and if the twenty faces are numbered in pairs, with two faces marked 0, two faces marked 1, etc., then casting this "die" will produce each of the ten digits with equal likelihood. Such icosahedra have been constructed and used for generating random digits. Now suppose a person wishes to make a selection with a probability of 999/3,999 (as in the second draw of the poker chip game when the first draw was red). Of course, this could be done, as in the game, by counting out 999 red chips and 3,000 white chips, mixing them, and drawing a chip. But there is another approximate way that requires only random digits. Express the probability decimally, to the degree of accuracy desired, e.g., to four places 999/3,999 = .2498. Then select four random digits by rolling an icosahedron or by choosing them from a random digit table (which may be thought of as containing the results of someone else's having rolled the icosahedron) and write them down as a four-digit number. If the number obtained is between 0,000 and 2,497, inclusive, the trial is called a success (a red chip in the game), and if it is between 2,498 and 9,999, inclusive, it is called a failure (a white chip in the game). The principle is that, since all digits are equally likely, the 10,000 four-digit numbers (0,000 to 9,999) are all equally likely. Thus, there is a probability .2498 of a four-digit number formed from random digits being less than 2,498.

The technique outlined above shows how a person could play the game of the three boxes without bothering to acquire a few thousand poker chips. He need only acquire an icosahedral die to roll, or, still better, a table of random digits, and decide the color of the chip he would have drawn if he really had the poker chips by observing the size of the number he "draws" from the random digit table. After each move he must, of course, calculate the probabilities governing his next move; however, this is quite trivial since, at each stage, he knows exactly what decisions were reached (what chips were drawn) in all previous moves, and thus he knows what remains in each box and which box he must draw from next. This process may be thought of as a simulated way of playing the poker chip game.

It is easy to state how a digital machine can play the poker-chip game. It is inconvenient to provide an electronic machine with actual poker chips or icosahedral dice, but it is easy to provide it with a random digit table. It is also possible to cause it to generate random digits for itself in quite a number of ways. Given the random digits, all the operations that the human must perform in this "simulated" poker-chip game are of the type that a digital machine performs with ease—comparing numbers to see which is larger and performing elementary arithmetic. Now emerges the point of all this. The tremendous speed of the digital machine is such that it can complete a game in the time a human player would require for the first few moves. This fact is of great importance for the reason that completing the game *once* gives almost no information—only that the particular outcome is a possible one. Playing the game a very large number of times, however, determines the ratio of wins to the total plays of the game, i.e., the probability of a win. The high speed of a modern digital machine makes a large number of repetitions of playing the game a practicable matter.

The process performed by the computing machine as described in the last paragraph may be considered a simulated experimental method of determining complex probabilities. It is generally referred to as the *Monte Carlo method* or the Monte Carlo technique. Its use has made possible the solution of many important systems engineering or operations analysis problems.

A typical practical case where the Monte Carlo method has been used is in the study of traffic problems. At an intersection where control by a traffic light is contemplated it is desired to allot the time for the red, yellow, and green in each direction to minimize the impediment to traffic flow. This is a very complex problem, particularly because of the occurrence of turns. The basic probabilities influencing the problem—the probability of a car arriving in a given direction within any particular time interval, the probability of a left turn or of a right turn, etc.—can be readily determined by observation (usually with the help of automatic traffic counters). However, the interactions are so complicated (a car intending to turn left will be held up if there are cars crossing in the opposite direction, etc.) that direct calculation of the final results (how long a line of cars will form—with what probability) is impossible. Direct experiment is expensive and otherwise undesirable since bad choices will cause rather serious complaints from taxpayers and voters. The simulated experiments of the Monte Carlo method provide an excellent solution.

Another application of the Monte Carlo method which is beginning to be widely used is to the determination of production tolerances for the parts of a complex device such as a guided missile. Clearly, loose tolerances for parts increase the probability of a failure to pass final tests (it may not even "go together" if the parts are near the extremes of a loose tolerance). Failure to pass final tests costs money since rework is necessary. On the other hand, tight tolerances are expensive also; 1 per cent resistors cost much more than 10 per cent resistors. By the use of a Monte Carlo process it is actually possible to determine the tolerances which will yield a given quality of end product at the lowest manufacturing cost.

The availability of truly high speed digital machines with high capacity is so recent that the capabilities of these machines and the power of some of the techniques, including the Monte Carlo method, which these machines make practical, have not yet been widely recognized by industry. It is hoped, however, that enough has been said here to indicate at least the reason for the statement, made earlier, that these machines will have and are having a truly revolutionary impact on technology.

HUMAN ENGINEERING

All of the techniques described above can be and have been applied to problems in the systems engineering of the purely mechanistic system. The traffic problem, mentioned above, is more nearly an operations analysis problem, but the manufacturing tolerance problem is a *design* control problem and hence is pure systems engineering according to the definitions of this chapter.

Discussed briefly here is a field that is required only when humans are introduced as operating elements or components of the system. It has already been mentioned that the performance of a system incorporating humans, even a quasi-mechanistic system, cannot be computed deterministically because human behavior is not invariable. Hence statistical or probabilistic methods are used. For example, the Monte Carlo method might well be used. To use this, however, or any other method of computing the probabilities of system performance, the basic probabilities of component performance must be determined in some way. When the component is a human, the technique of determining the performance probabilities is a task of *human engineering.* Human engineering applies primarily to quasi-mechanistic systems, where the human is operating in a mechanical (non-intellectual) fashion, since it has not yet proved possible to compute the probability of sound judgment. But it is possible, by properly controlled and documented experiments, to determine the probability of performance of a relatively mechanical action by a human.

Suppose that a human is assigned the task of pushing one of two buttons in response to the signal provided by a light being turned on. Then the human engineer can determine (has determined) the following sorts of things:

The average time between light on and button push.
The effect of the position, size, shape, color of the light, and button on this average time.
The probability of pushing the wrong button.
The probability of not noticing the light for twenty seconds, two minutes, all night.
The effect of fatigue, drugs, etc., on the previous numbers.

Clearly, the systems engineer for a quasi-mechanistic system must have such human engineering data available to him in order to carry out a performance analysis of the system.

COMPETITION

In a non-mechanistic system, where humans enter into the operation of the system in a non-mechanical way—making judgments—the analysis of system performance becomes much more difficult. In fact, it might be supposed that system analysis would become totally impossible because of the non-predictable nature of human judgment, and, indeed, for the most part this is true. However, it is not completely true, for if the human judgments are reached by a moderately predictable process, such as the use of operations analysis techniques, then it is quite possible that another operations analyst could reach the same conclusions and hence, in a sense, predict the nature of the judgments.

An attempt to predict the performance of a non-mechanistic system becomes of great importance in a competitive situation. In particular it is important to be able to estimate the performance of a competitor's system. Fortunately it is not usually necessary to know exactly how the competitor will perform but only how well he might perform (if he does worse, so much the better). For this purpose it is reasonable to assume that he makes his decisions in the optimum way (for his own advantage) and that his decisions can be predicted by "putting yourself in his place" and performing the same operations analysis that he presumably is doing.

As long as one is attempting to compete with a system which is unaware of or uninterested in one's existence (a very small business trying to establish itself may be in this position), the type of analysis described above can be carried out straightforwardly. In a competition between near equals, however, it can often be assumed that the decision-making of Competitor A is based on an attempt to deduce or predict the behavior of Competitor B. In this case the analysis acquires a new dimension of difficulty, and a new mathematical technique known as *game theory* is being developed to deal with such situations.

The fundamentals of game theory were established by the late great von Neumann[5] to provide a means for analyzing competitive systems, which can be typified by a game in which the players simultaneously and secretly make decisions (for example, choose digits) and then reveal their choices and transfer score (or money) from one to the other on the basis of some predetermined payoff function which is a function of the choices made by all players.

The technique of game theory makes it possible, in solvable cases, to determine a *strategy;* that is, a set of choices with a probability (or frequency) for each choice which is optimum in some sense for a given player. It is not always clear in what sense to "optimize" a game theory strategy. A common criterion is called the minimax criterion in which a strategy is determined which minimizes the maximum losses (or maximizes the minimum gains) for the player. This and other aspects of game theory are discussed in Chapter 16.

Game theory has been applied to economic problems (commercial competition) and even more extensively to problems of warfare. Of course, in very complex practical problems of this nature it is necessary to make many simplifying assumptions to reduce the situation to one that is amenable to analysis. This fact has led some to question the validity of the results of such an approach to complex problems. Such skeptics will point out that even such a well-specified game as chess cannot be analyzed in detail by present computers, and, although digital computers have been programmed to play chess, they play at best a rather poor game and any expert can readily beat them. This is true, but it should be recalled that chess is apparently a nearly fair game; that is, either player has nearly an equal chance to win. Most warfare problems are not fair games and, frequently, one side has a very considerable advantage. When this is the case, even a greatly

[5] J. von Neumann and O. Morgenstern, *The Theory of Games and Economic Behavior* (Princeton: Princeton University Press, 1955).

simplified analysis may reveal the unfairness of the situation and reliably indicate which party has the advantage. On the other hand the fact of such a disparity is frequently not at all obvious without a game theory analysis.

Competitive problems, particularly warfare problems, are sometimes treated in a simulated experimental fashion that is an extension of the Monte Carlo technique. In this approach the decision-making part of the problem is simulated, quite directly, by having people make decisions in the same manner that would be required in the real life situation. This approach uses opposing teams of players who make the command decisions (to attack or retreat, open fire, call for air support, etc.), a team of referees who monitor the game, and a computer which calculates the results. The underlying probabilities are introduced as in a Monte Carlo calculation. For example, a decision to fire is made by the appropriate team, but the decision as to whether or not the firing results in a kill is reached by choosing random digits and comparing with the (known) kill probability of the weapon. This technique, known as *war gaming* since it has been applied most extensively to warfare, is exemplified by the simulation of tactical war games described in Chapter 24.

DECISION

The foregoing sections have covered briefly a number of the techniques that are used to analyze the size, weight, cost, reliability, performance, etc. of a system. The reason for performing such analyses is, normally, to assist in reaching some decision—to choose one of two systems or to modify a given system. In order to reach a decision it is not enough to have available the results of such analyses; it is also necessary to have a criterion on the basis of which the decision will be made. We must not only know how the system will perform—we must know whether that performance is "good" or "bad." Wrong decisions have been reached many times in the presence of excellent knowledge of all the facts simply by using the wrong criterion of value.

In judging artillery weapons it is customary to put great stress on accuracy. This fact was very significant in delaying the introduction of artillery rockets as against guns due to the generally lower accuracy of rockets. But, by the end of World War II it was beginning to be realized that, when used for barrage purposes, the lower accuracy of rockets was actually a slight advantage since a broader barrage could be laid down without retraining the rocket launchers. There is little point in being accurate when one does not know exactly where the target is. More strikingly, there was a situation in World War II in which an antiaircraft gun was used in connection with a gun-direction system that contained inherent unpredictable biases. Purposely degrading the accuracy of the gun resulted in increased kills, since the highly accurate gun consistently shot at the slightly incorrect spot calculated by the gun director. In weapons systems the normal criterion of value is cost per kill, and great care must be taken to avoid using some other criterion (accuracy) which is assumed to be equivalent but is actually not. Great care must also be taken to be sure of including all the hidden costs (training, repair, supply, transportation, etc.) as well as the more obvious direct costs.

In developing a decision criterion it is commonly necessary to do what one is told in grade school must never be done—compare apples and oranges. Military planners are frequently faced with deciding whether to proceed with the procurement of a particular defense system or wait for a more effective system which will be available at a later date. Here one must make a judgment which compares time with effectiveness. Such a decision is reached by finding a large enough framework (e.g., national defense) to encompass both possibilities as alternate subsystems and determine what function one really desires to maximize (e.g., probability of survival) in the over-all system and then attempt to estimate this function for the two alternate cases.

In its most unpleasant form the comparison of unlike things may require a choice between dollars and human lives. For example, in comparing the effectiveness of a guided missile and a fighter airplane as antiaircraft devices it is necessary to establish some equivalent dollar value for the lives of the pilots which are lost in the use of the fighter plane. Many people find this emotionally disturbing to the point that they refuse to write down such a value. The important thing to remember here is that, assuming a decision must be made, making the decision on arbitrary or intuitive grounds always implicitly assigns such a value whether one likes it or not and using the intuitive approach may well implicitly assign a ridiculously small value. It is always better to be aware of the implications of decisions; the most reliable way to assure this is to make decisions by a reasoned, calculated approach rather than by intuition.

The process of reaching decisions has been studied extensively in recent years, to the degree that there is beginning to be recognized a special subject known as *decision theory* with which the systems engineer should be familiar.

CONCLUSION

It is worth repeating that the process of system optimization involves two distinct problems: first, determining the function which is to be maximized (establishing the valuation technique or the decision criterion), and second, performing the maximization. In the development of a new system this process is repeated a large number of times with increasing realism and detail.

The first step in this process, carried out entirely on paper with no hardware fabrication, is usually called the *preliminary design* or *composite design phase*. In this phase the system performance may be calculated in a reasonably sophisticated way by use of the techniques described above (including the use of analogue or digital computers), but the weights, volumes, costs, reliability, etc., are generally estimated quite roughly by the empirical methods mentioned earlier. Evaluation of a number of competitive composite designs culminates in a definition of the system in the form of a block diagram. In a complex system many of the "blocks" may themselves represent relatively complicated subsystems which must be the subject of further preliminary design studies on a more specific level. After a sufficiently specific block diagram is chosen the actual hardware development begins. Throughout the hardware development the system performance is recomputed (simulated) repeatedly with the actual hardware elements operating in the system where possible. When the entire system is available in hardware form a complete evaluation test program is conducted. This final program must include a determination of system behavior under the complete gamut of environmental inputs. If the system is to be produced in quantity, the job of the systems engineer is still not through at this point. The final job should be the determination of production tolerances to assure reproduction of acceptable performance at minimum cost. As mentioned above, even for this last job, modern techniques and modern computing machines can give far better answers than were available in the past.

From all of this it should be clear that the job of the systems engineer is not new. The decisions the systems engineer must make are decisions that always had to be made in the development or modification of any system. But, until recent years, such decisions were all too often made more or less intuitively, on the basis of "experience" or "engineering judgment" with totally inadequate analysis of the possibilities. What is new is a realization of the power of various analysis techniques in providing a greatly superior basis for engineering decisions and the availability of computing machines of great speed, flexibility, and capacity that make these analysis techniques practicable.

PART II

Methodologies

Eight

SIMPLIFIED MODELS IN OPERATIONS RESEARCH

ELIEZER NADDOR

INTRODUCTION

Models in General

A model is a representation of a system under study. The representation may be in the form of a photograph, a blueprint, or a three-dimensional scale reproduction. These few examples are characteristic of models when one is interested in studying geometric or physical relationships of the components of systems. Other representations of systems may be in the schematic form of organization charts, flow charts, or graphs of various kinds. These can be used to depict or describe functional relationships. Still other representations are feasible: financial statements, sales forecasts, inventory records, etc., are not usually referred to as models, but in the sense of this discussion they may be regarded as such.

In economics and many types of research a model of a system is usually given in symbolic terms. An area S of a rectangle may be represented in terms of its length, a, and width, b, thus

(1) . . . $$S = ab$$

This mathematical model thus relates areas of rectangles to their lengths and widths.

The distance x travelled by a freely falling body for some time t is

(2) . . . $$x = \frac{g}{2} t^2$$

where g is the acceleration due to gravity. This model describes distances of freely falling bodies in terms of the time of fall and in terms of a constant g.

The cost C of buying n units can be given by

$$(3) \ldots \qquad C = np$$

where p is the cost of buying one unit. Here, the model describes the total cost of n units in terms of the number of units n and in terms of the cost of one unit p.

These models are symbolic abstractions of pertinent attributes of the systems under study. They are very useful for such studies since it is generally easier to examine and manipulate the model than the system itself.

Models in Operations Research

In operations research, models are used to help make decisions, since operations research problems are generally decision problems. There is some ultimate objective that the decision-maker wants to maximize (or minimize). The model is then used to find the effect of various alternatives on the objective, thus eventually helping in making the decision.

ALTERNATIVES

For a meaningful decision problem to exist there must be more than one way to perform the operations studied. If there is only one way, there is no decision problem. The various ways to perform operations are referred to as *alternatives subject to control* or, in short, as *alternatives*. They are denoted symbolically by A_1, $A_2 \ldots A_n$, where n is the number of possible alternatives. The decision problem is to find the *best* alternative.

In many cases n may be very large and it is either impractical or technically impossible to consider all alternatives. In these cases the decision problem may be to find an alternative which is best in a specified set of alternatives (where n is reasonably small). Usually, in these cases, the specified set of alternatives includes the alternative currently in operation and one is trying to find another alternative which is *better*.

CONTROLLABLE FACTORS

To describe an alternative one must recognize the parts of which it is made up. These are called *controllable factors,* or, in short, *factors.* That the alternative is a function of its factors $x_1, x_2, \ldots x_m$ is denoted symbolically by

(4) . . . $$A = A(x_1, x_2, \ldots, x_m)$$

Consider, for example, an inventory problem in which one is interested in finding the inventory stock level and the frequency of raising inventory to this level. Suppose that three possible stock levels may be considered: 100 lb., 200 lb. and 300 lb. and that two frequencies are possible: 1 month and 2 months. The alternatives and their factors may be displayed as in Table 1.

Table 1. Alternatives and Factors in an Inventory Problem

Alternatives	A	A_1	A_2	A_3	A_4	A_5	A_6
Factors	x_1	100 lb.	200 lb.	300 lb.	100 lb.	200 lb.	300 lb.
	x_2	1 month	1 month	1 month	2 months	2 months	2 months

In this example there are six alternatives. For each alternative there are two factors: a stock level and a frequency. Since there are three possible stock levels and two possible frequencies the total number of alternatives is six. And in general, if there are m factors, and if the ith factor can have n_i possible values, then the number of possible alternatives is

(5) . . . $$n = n_1 \cdot n_2 \cdot n_3 \ldots . n_m$$

We distinguish between factors that can be measured quantitatively and factors that can only be described qualitatively. To describe a specific alternative one must specify the *magnitude* of the corresponding quantifiable factors and the specific *attributes* of the corresponding qualitative factors.

The inventory example above illustrates a case where the factors are quantifiable, and Table 1 described the alternatives in terms of the magnitudes of the corresponding factors.

In the case study presented in Chapter 27, the reader will find two qualitative factors: (i) the system of reports and (ii) their preparation method. There, the first factor is the system contents and four attributes, S_1, S_2, S_3, and S_4, are considered. The second factor is the preparation method and three attributes, P_1, P_2, and P_3, are considered.

The alternatives and their factors may be displayed as in Table 2.

Table 2. Alternatives and Factors in a System with Qualitative Factors

Alternatives	A	A_1	A_2	A_3	A_4	A_5	A_6	A_7	A_8	A_9	A_{10}	A_{11}	A_{12}
Factors	x_1	S_1	S_2	S_3	S_4	S_1	S_2	S_3	S_4	S_1	S_2	S_3	S_4
	x_2	P_1	P_1	P_1	P_1	P_2	P_2	P_2	P_2	P_3	P_3	P_3	P_3

In formulating an operations research problem a great deal of effort is usually devoted to the identification of the factors and to the selection of alternatives to be studied. When all factors are quantifiable, the problem of selection of alternatives for study is usually not too difficult. But when some of the factors are qualitative the selection of alternatives becomes a difficult problem. As of now, rigid scientific methods for selection of alternatives for study are not available, when some of the factors are qualitative. Selection by judgment is necessary in such cases.

CRITERIA

In order to find the best (or a better) alternative, one must have some *criteria* by which to evaluate the various alternatives. In the inventory problem mentioned above the criteria are:

(a) The average amounts in inventory during the year should be as small as possible. The maximum amount in inventory should never exceed 1200 cu. ft., the available storage space.
(b) There should be as few shortages as possible, and they should not occur more frequently than 10 per cent of the time.
(c) The number of setups per year should be as small as possible.

These criteria may be classified in two groups: the *restriction criteria* and the *minimization* (or *maximization*) *criteria*. The restriction criteria in our case are:

(i) The maximum amount in inventory must not exceed 1200 cu. ft.
(ii) The frequency of shortages must not exceed 10 per cent.

The minimization criteria are:

(1) The average amount in inventory during the year should be minimized.
(2) The number of shortages per year should be minimized.
(3) The number of setups per year should be minimized.

If each pound of the material stocked occupies 5 cu. ft. and if the demand is assumed to be uniform at a rate of 110 lb. per month, the amounts in inventory may be represented graphically as in Figures 1 and 2. Figure 1 describes the amounts in inventory during a year for monthly frequencies of raising stock levels to 100 lb., 200 lb., and 300 lb. Figure 2 describes the amounts in inventory during a year for bi-monthly frequencies of raising stock levels to 100 lb., 200 lb., and 300 lb.

The criteria may now be displayed as in Table 3.

Table 3. Evaluation of Alternatives in an Inventory System

Alternatives	A	A_1	A_2	A_3	A_4	A_5	A_6
Factors Inventory Level	x_1	100 lb.	200 lb.	300 lb.	100 lb.	200 lb.	300 lb.
Frequency	x_2	1 month	1 month	1 month	2 months	2 months	2 months
Criteria Restriction Space (1200 cu. ft.)	R_1	500 cu. ft.	1000 cu. ft.	1500 cu. ft.	500 cu. ft.	1000 cu. ft.	1500 cu. ft.
Frequency of Shortages (10%)	R_2	9.1%	0%	0%	54.5%	9.1%	0%
Minimization Average Inventory	M_1	45 lb.	145 lb.	245 lb.	23 lb.	91 lb.	190 lb.
Shortages	M_2	120 lb.	0	0	720 lb.	120 lb.	0
Setups	M_3	12	12	12	6	6	6

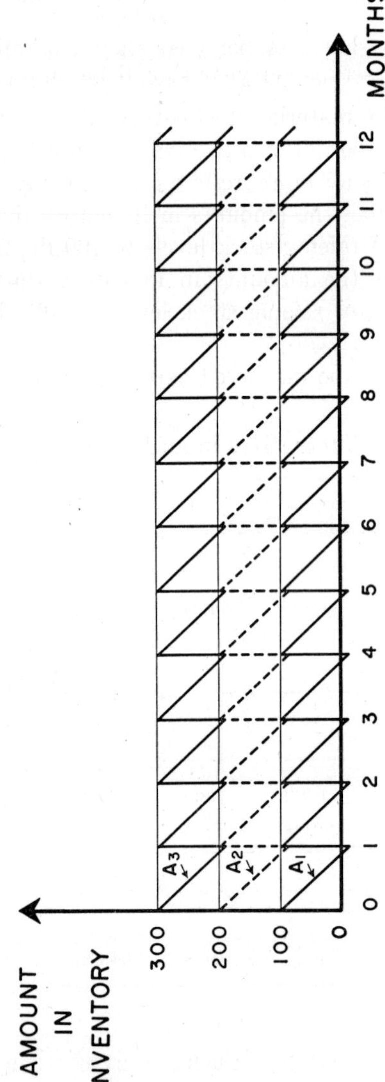

Figure 1. Amounts in inventory for alternatives A_1, A_2, A_3.

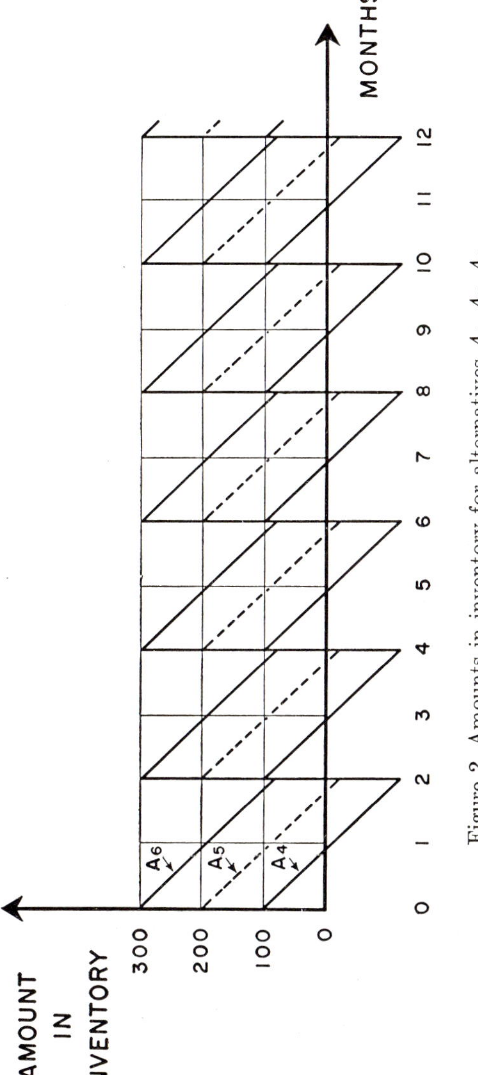

Figure 2. Amounts in inventory for alternatives A_4, A_5, A_6.

If we examine the restriction criteria we can immediately eliminate alternatives A_3, A_4, and A_6. Alternatives A_3 and A_6 would require 1500 cu. ft., whereas the first restriction criterion requires that no more than 1200 cu. ft. be used. Alternative A_4 would cause 54.5 per cent shortages which is in excess of the maximum allowable 10 per cent. Thus alternatives A_1, A_2, A_5 remain to be considered.

If the first criterion were the most important (minimizing inventory), then the best alternative would be A_1, since it would result in only 45 lb. average inventory. If the second criterion (minimizing shortages) were the most important, alternative A_2 would be the best, since no shortages would occur. And, finally, if the third criterion were the most important (minimizing setups), then alternative A_5 would be chosen, since it requires only six setups per year.

TOTAL EFFECTIVENESS

We have seen how alternatives may be evaluated in terms of criteria. It remains to be shown how the best alternative can be found.

Two approaches are possible. Both lead to a relation between an alternative and its total effectiveness. We designate the total effectiveness by E. It is a function of the alternative, or the factors of the alternative.

$$E = E(A) = E(x_1, x_2, \ldots, x_m) \quad (6)$$

In using the first approach one must be able to define a *unit of effectiveness* by which the minimization (or maximization) criteria can be measured. In this case a convenient measure of total effectiveness is given by

$$E = c_1M_1 + c_2M_2 + \ldots c_rM_r \quad (7)$$

where

M_k is a measure of the kth criterion
c_k is a unit effectiveness of the kth criterion

and where all c_kM_k have the same unit of effectiveness.

In the inventory problem we might have

c_1 = \$0.40 per pound inventory carried per year
c_2 = \$2.00 per pound short
c_3 = \$50.00 per setup

This leads to Table 4.

Table 4. Total Effectiveness (First Approach)

Alternative	A	A_1	A_2	A_3	A_4	A_5	A_6
Factor 1	x_1	100	200	300	100	200	300
Factor 2	x_2	1	1	1	2	2	2
Criteria Restriction 1	R_1	500	1000	1500	500	1000	1500
Restriction 2	R_2	9.1%	0%	0%	54.5%	9.1%	0%
Minimize 1	M_1	45 lb.	145 lb.	245 lb.	23 lb.	91 lb.	190 lb.
Minimize 2	M_2	120	0	0	720	120	0
Minimize 3	M_3	12	12	12	6	6	6
Effectiveness Criteria M_1	c_1M_1	$18.0	58.0	—	—	36.4	—
Criteria M_2	c_2M_2	$240.0	0	—	—	240.0	—
Criteria M_3	c_3M_3	$600.0	600.0	—	—	300.0	—
Total	E	$858.0	658.0	—	—	576.4	—

Here, then, the minimum total effectiveness is achieved for alternative A_5. The best alternative, thus, is alternative A_5.

In using the second approach to obtain the total effectiveness one must be able to obtain the following measures:

(i) The efficiency of each alternative for each criterion;
(ii) The relative importance of each criterion.

In this case the total effectiveness is given by:

$$E = w_1F_1 + w_2F_2 + \ldots + w_rF_r \quad (8)$$

where

F_k is the efficiency of the kth criterion, where $-1 \leqq F_k \leqq 1$
w_k is the relative importance of the kth criterion, where $w_k \geqq 0$ and $\Sigma\, w_k = 1$

In the inventory problem we might have:

$$w_1 = 0.2$$
$$w_2 = 0.5$$
$$w_3 = 0.3$$

and the efficiencies might be as in Table 5.

Table 5. Efficiencies of Criteria

Average Inventory	M_1	600	410	320	270	220	175	135	100	65	30	0
Efficiency	F_1	−1.00	−0.80	−0.60	−0.40	−0.20	0	+0.20	+0.40	+0.60	+0.80	+1.00
Shortages	M_2	132	71	55	44	35	27	21	15	9	4	0
Efficiency	F_2	−1.00	−0.80	−0.60	−0.40	−0.20	0	+0.20	+0.40	+0.60	+0.80	+1.00
Setups	M_3	20	18	16	14	12	10	8	6	4	2	0
Efficiency	F_3	−1.00	−0.80	−0.60	−0.40	−0.20	0	+0.20	+0.40	+0.60	+0.80	+1.00

This would lead to Table 6.

Table 6. Total Effectiveness (Second Approach)

Alternative	A	A_1	A_2	A_3	A_4	A_5	A_6
Factor 1	x_1	100	200	300	100	200	300
Factor 2	x_2	1	1	1	2	2	2
Criteria							
Restriction 1	R_1	500	1000	1500	500	1000	1500
Restriction 2	R_2	9.1%	0%	0%	54.5%	9.1%	0%
Minimize 1	M_1	45	145	—	—	91	—
Minimize 2	M_2	120	0	—	—	120	—
Minimize 3	M_3	12	12	—	—	6	—
Efficiency 1	F_1	0.71	0.15	—	—	0.43	—
Efficiency 2	F_2	−0.99	1.00	—	—	−0.99	—
Efficiency 3	F_3	−0.20	−0.20	—	—	0.40	—
Effectiveness							
Criteria M_1	w_1F_1	0.144	0.030	—	—	0.086	—
Criteria M_2	w_2F_2	−0.495	0	—	—	−0.495	—
Criteria M_3	w_3F_3	−0.060	−0.060	—	—	0.120	—
TOTAL	E	−0.411	−0.030	—	—	−0.289	—

In this case, the maximum effectiveness is achieved for alternative A_2.

ILLUSTRATIONS

Seven illustrations of operations research models are given below. All are rather simple examples of more complex problems. However, they all contain important elements of the following corresponding seven classical operations research models: allocation, queueing, inventory, sequencing, routing, replacement, and competition. Each illustration describes a typical numerical operations research problem. A model is constructed in each case, and a specific numerical solution is obtained. Comments are generally given on the relation of the simple illustration to the more general operations research models.

Stewardess Assignment

An airline maintains ten flights daily between cities A and B, as follows:

Flight No.	Leaves	City	Arrives	City
101	8:00 a.m.	A	9:30 a.m.	B
103	10:00 a.m.	A	11:30 a.m.	B
105	1:00 p.m.	A	2:30 p.m.	B
107	5:00 p.m.	A	6:30 p.m.	B
109	8:00 p.m.	A	9:30 p.m.	B
102	7:00 a.m.	B	9:00 a.m.	A
104	10:00 a.m.	B	12:00 a.m.	A
106	12:00 noon	B	2:00 p.m.	A
108	5:00 p.m.	B	7:00 p.m.	A
110	10:00 p.m.	B	12:00 midn.	A

The flying time from A to B is 90 minutes, whereas from B to A it is 120 minutes, due to the prevailing winds between the cities.

A team of two stewardesses is assigned to each flight. Five teams are therefore necessary to provide stewardess service to all flights. Each team has its base either in city A or in city B. If a team is based in one city, then the time it spends in the other city, before taking a return flight, is called *layover time*. This is unproductive time and is equally undesirable for the airline and the stewardesses.

The problems faced by the airline are as follows:

1. Where should the five teams of stewardesses be based?
2. How should the teams be assigned to the ten flights?

We may consider each of the two problems above as a factor subject to control. The first factor may be shown to have 2^5, or 32, different combinations. The second factor may be shown to have 5! or 120 different combinations, so that the total number of feasible alternatives is 32 × 120 or 3840. In spite of this large number of alternatives this problem can be solved rather quickly by trial and error methods. For a larger number of flights this is no longer an easy problem.

A general method for solving the problem which is equally applicable to any number of flights will be presented.

THE LAYOVER TIME MATRICES

The unit of effectiveness in this problem is evidently the layover hour. It is therefore instructive to tabulate the total layover hours for any assignment of teams to flights in any city. This is done in Tables 7 and 8.

Table 7. Layover Time at *B* for Teams Based at *A*

		Arrive at *B* on Flight No.				
		101	103	105	107	109
Leave from *B* on Flight No.	102	21½	19½	16½	12½	9½
	104	½	22½	19½	15½	12½
	106	2½	½	21½	17½	14½
	108	7½	5½	2½	22½	19½
	110	12½	10½	7½	3½	½

Table 8. Layover Time at *A* for Teams Based at *B*

		Leave from *A* on Flight No.				
		101	103	105	107	109
Arrive at *A* on Flight No.	102	23	1	4	8	11
	104	20	22	1	5	8
	106	18	20	23	3	6
	108	13	15	18	22	1
	110	8	10	13	17	20

We might check a few entries in these tables. For example, a team based in *A* which leaves *A* on Flight 105 arrives at *B* at 2:30 P.M. If it returns to *A* on Flight, say, 108, it will leave *B* at 5:00 P.M. Thus, it lays over 2½ hours at *B*. Hence in Table 7 we should have in the intersection of Flights 105 and 108 the layover time of $2\frac{1}{2}$. This checks with the entry in the table.

Now suppose the same pair of flights, 105 and 108, were assigned to a team which is based at B. They would leave B on Flight 108, arriving at A at 7:00 P.M. They would return to B on Flight 105, leaving A at 1:00 P.M., the following day only. Thus, they spend 18 hours at A.

(A further check on the accuracy of calculations is the following. For any pair of flights, the sum of the layover time at A and B is either $20\frac{1}{2}$ hours or $44\frac{1}{2}$ hours. These numbers represent the difference between the number of hours in one or two days and the total number of hours in flight, i.e., $24 - (1\frac{1}{2} + 2) = 20\frac{1}{2}$ and $48 - (1\frac{1}{2} + 2) = 44\frac{1}{2}$.)

If we now examine each specific pair of flights, we can determine immediately whether the team taking this pair should be based in A or in B. For example, the team taking the pair 105, 108 should be based at A since their layover time would be $2\frac{1}{2}$ hours compared to 18 hours if the team were based at B. Similarly, the team taking the pair 101, 102 should be based at A. But the team taking the pair 104, 105 should be based at B.

We can now prepare Table 9, which will represent the least layover time and the corresponding base for each possible pair of flights.

Table 9. Minimal Layover Time and Preferred Base for All Possible Pairs of Flights

	101	103	105	107	109
102	$21\frac{1}{2}$ *A*	1 *B*	4 *B*	8 *B*	$9\frac{1}{2}$ *A*
104	$\frac{1}{2}$ *A*	22 *B*	1 *B*	5 *B*	8 *B*
106	$2\frac{1}{2}$ *A*	$\frac{1}{2}$ *A*	$21\frac{1}{2}$ *A*	3 *B*	6 *B*
108	$7\frac{1}{2}$ *A*	$5\frac{1}{2}$ *A*	$2\frac{1}{2}$ *A*	22 *B*	1 *B*
110	8 *B*	10 *B*	$7\frac{1}{2}$ *A*	$3\frac{1}{2}$ *A*	$\frac{1}{2}$ *A*

Now each flight has to be paired with some other flight and no flight can be paired off with more than one flight.

The Stewardess Assignment Problem may now be stated as follows. In Table 9 find five layover times, corresponding to five pairs of flights, so that each flight is represented and so that the sum of the layover times is minimized.

A MATHEMATICAL MODEL

Let x_{ij} be a variable with the following properties:

$x_{ij} = 1$ means that flight i will be paired with flight j;

$x_{ij} = 0$ means that flight i will *not* be paired with flight j

where

$$i = 102, 104, 106, 108, 110$$
$$j = 101, 103, 105, 107, 109.$$

Since Flight 102 has to be paired with one and only one other flight we must have

$$(9) \ldots \quad x_{102,101} + x_{102,103} + x_{102,105} + x_{102,107} + x_{102,109} = 1$$

In a similar manner we get equations (10) to (18) for Flights 104, 106, 108, 110, 101, 103, 105, 107, 109:

$$(10) \ldots \quad x_{104,101} + x_{104,103} + \ldots + x_{104,109} = 1$$
$$\vdots$$
$$(13) \ldots \quad x_{110,101} + x_{110,103} + \ldots + x_{110,109} = 1$$
$$(14) \ldots \quad x_{102,101} + x_{104,101} + \ldots + x_{110,101} = 1$$
$$\vdots$$
$$(18) \ldots \quad x_{102,109} + x_{104,109} + \ldots + x_{110,109} = 1$$

The total layover time, E, will be:

$$\begin{aligned}(19) \ldots \quad E = {} & 21\tfrac{1}{2}x_{102,101} + 1x_{102,103} + \ldots + 9\tfrac{1}{2}x_{102,109} \\ & + \tfrac{1}{2}x_{104,101} + 22x_{104,103} + \ldots + 8x_{104,109} \\ & + \ldots \\ & + 8x_{110,101} + 10x_{110,103} + \ldots + \tfrac{1}{2}x_{110,109}\end{aligned}$$

We thus have 25 variables (x_{ij}), 10 equations, which these variables have to satisfy (equations 9 to 18), and we have to find a specific value (0 or 1) for each variable so that the total layover time (equation 19) will be minimized.

AN ITERATIVE SOLUTION

The mathematical model just given is a special case of *linear programming,* and can be solved by iterative methods, some of which are described in Chapter 13 and in reference (12).

Applying one of these methods (*the reduced matrices method*) we proceed as follows:

From Table 9 we obtain Table 10 by reducing each row by the minimum layover time in that row.

Table 10. The Reduced Layover Times for Flights 102, 104, 106, 108, 110

	101	103	105	107	109
102	$20\frac{1}{2}$	0	3	7	$8\frac{1}{2}$
104	0	$21\frac{1}{2}$	$\frac{1}{2}$	$4\frac{1}{2}$	$7\frac{1}{2}$
106	2	0	21	$2\frac{1}{2}$	$5\frac{1}{2}$
108	$6\frac{1}{2}$	$4\frac{1}{2}$	$1\frac{1}{2}$	21	0
110	$7\frac{1}{2}$	$9\frac{1}{2}$	7	3	0

For example, in the fourth row the minimum layover time is one hour. Therefore, all layover times corresponding to Flight 108 have been reduced by one hour.

From Table 10 we obtain Table 11 by reducing each column by the minimum layover time in that column.

Table 11. The Reduced Layover Time for Flights 105, 107

	101	103	105	107	109	
102	$20\frac{1}{2}$	0	$2\frac{1}{2}$	$4\frac{1}{2}$	$8\frac{1}{2}$	
104	0	$21\frac{1}{2}$	0	2	$7\frac{1}{2}$	1
106	2	0	$20\frac{1}{2}$	0	$5\frac{1}{2}$	
108	$6\frac{1}{2}$	$4\frac{1}{2}$	1	$18\frac{1}{2}$	0	
110	$7\frac{1}{2}$	$9\frac{1}{2}$	$6\frac{1}{2}$	$\frac{1}{2}$	0	
		2		3	4	

The first two columns and the last column are not affected since the minimum layover times in Table 10 are all zero hours. The minima of the third and fourth columns are $\frac{1}{2}$ hour and $2\frac{1}{2}$ hours respectively and the corresponding layover times of Flights 105 and 107 have been reduced accordingly.

In Table 11 we now draw a minimum number of vertical and horizontal lines through all zero hours layover times (lines 1, 2, 3, 4). Then we find the minimum layover time through which no lines have been drawn (one hour, in Table 11). This minimum is now subtracted from the rows through which no horizontal lines were drawn and added to the columns through which vertical lines were drawn. Table 12 is thus obtained.

Table 12. The Reduced Layover Times for Flights 102, 106, 108, 110, and Increased Layover Times for Flights 103, 107, and 109

	101	103	105	107	109
102	$19\frac{1}{2}$	0*	$1\frac{1}{2}$	$4\frac{1}{2}$	$8\frac{1}{2}$
104	0*	$22\frac{1}{2}$	0	3	$8\frac{1}{2}$
106	1	0	$19\frac{1}{2}$	0*	$5\frac{1}{2}$
108	$5\frac{1}{2}$	$4\frac{1}{2}$	0*	$18\frac{1}{2}$	0
110	$6\frac{1}{2}$	$9\frac{1}{2}$	$5\frac{1}{2}$	$\frac{1}{2}$	0*

In Table 12 we again attempt to draw a minimum number of vertical and horizontal lines through all the zero hours layover time. We find that the minimum number is five. Hence we know that an optimal assignment is available with a total of zero hours reduced layover time.

This assignment is composed of the following pairings (marked * in Table 12):

102-103, 104-101, 106-107, 108-105, 110-109

Referring back to Table 9 and using these pairings the solution of the Stewardess Assignment Problem is obtained:

Team	*Base*	*Leave on Flight*	*Return on Flight*	*Layover Time*
1	*B*	102	103	1 hour
2	*A*	101	104	½ hour
3	*B*	106	107	3 hours
4	*A*	105	108	2½ hours
5	*A*	109	110	½ hour
			Minimum Total Layover Time	7½ hours

Simulation of a Production Queue

A simplified production process is represented in Figure 3.

Figure 3. A simplified production process.

On the average, it takes 12½ minutes to produce one semi-finished item on Machine *A* and 10 minutes to make it into a finished item on Machine *B*. The exact frequencies of production times are as follows:

Machine *A*: 10 minutes per item for 50 per cent of the items;
15 minutes per item for 50 per cent of the items;
Machine *B*: 5 minutes per item for 25 per cent of the items;
10 minutes per item for 50 per cent of the items;
15 minutes per item for 25 per cent of the items.

The production time of any item on either Machine *A* or Machine *B* is random. That is, it may be either 10 or 15 minutes on Machine *A*, and 5, 10, or 15 minutes on Machine *B*.

Machine *A* is in continuous operation, producing a semi-finished item every 12½ minutes on the average. Since the average production time of Machine *B* is only 10 minutes, it will sometimes be idle (20 per cent of the time, to be exact). Yet, because of the random times of production, semi-finished parts may be held up before they can be processed on Machine *B*.

Semi-finished items may thus have to "wait" before being processed on Machine *B*. While they wait, they are said to be in a waiting line, or a "queue." We propose to find out how long on the average each semi-finished item is waiting in the queue.

SYMBOLIC REPRESENTATION

The waiting time problem can be reformulated as follows:

Semi-finished items arrive at random at Machine B every t_a minutes where

$$(20) \ldots \qquad t_a = \begin{cases} 10 \text{ minutes with a probability of } \frac{1}{2} \\ 15 \text{ minutes with a probability of } \frac{1}{2} \end{cases}$$

It takes t_s minutes to produce a finished part where

$$(21) \ldots \qquad t_s = \begin{cases} 5 \text{ minutes with a probability of } \frac{1}{4} \\ 10 \text{ minutes with a probability of } \frac{1}{2} \\ 15 \text{ minutes with a probability of } \frac{1}{4} \end{cases}$$

When a semi-finished item arrives at Machine B any one of three situations may occur:

(a) Machine B is in operation and the semi-finished item is kept waiting t_w minutes before its production can be started;
(b) Machine B may have been idle t_i minutes since the previous item was produced;
(c) Machine B has just finished producing an item and can therefore start immediately producing the semi-finished item.

In case (*a*), $\quad t_w > 0$ and $t_i = 0$

In case (*b*), $\quad t_w = 0$ and $t_i > 0$

In case (*c*), $\quad t_w = 0$ and $t_i = 0$

Let $\bar{t}_a$, $\bar{t}_s$, $\bar{t}_w$ and $\bar{t}_i$, designate the average times of t_a, t_s, t_w and t_i respectively. We know that

$$(22) \ldots \qquad \bar{t}_a = 12\tfrac{1}{2} \text{ minutes per item}$$

$$(23) \ldots \qquad \bar{t}_s = 10 \text{ minutes per item}$$

and therefore

$$(24) \ldots \qquad \bar{t}_i = \bar{t}_a - \bar{t}_s = 2\tfrac{1}{2} \text{ minutes per item}$$

The problem is to find $\bar{t}_w$, the average waiting time per item.

This is one of the problems with which queueing theory is concerned. Although we have simplified it considerably, it is still a very difficult problem to solve if one wants to use straightforward mathematical techniques (see Chapter 14).

A SIMULATION MODEL

Assume that production on Machine A is started at 7:00 A.M. and that semi-finished items are neither waiting for Machine B nor is one being processed on it at that time. The first semi-finished item will thus be ready at either 7:10 A.M. or 7:15 A.M. To be specific, let us say that it is ready at 7:10. At that time it will be put in Machine B and the finished item will be available at 7:15, 7:20 or 7:25 A.M. To be specific, again, let us say that it is available at 7:25 A.M.

Thus, for the first item we have:

$$t_a = 10 \text{ minutes}, t_s = 15 \text{ minutes}, t_w = 0 \text{ and } t_i = 10 \text{ minutes}$$

That is, it was 10 minutes on Machine A and 15 minutes on Machine B; it did not wait to get on Machine B but Machine B was idle 10 minutes before the first item reached it.

Let us now further assume that the second item took again 10 minutes on Machine A and only 5 minutes on Machine B. It leaves Machine A at 7:20 A.M. (10 minutes after the first item left it). It gets on Machine B at 7:25 A.M. (only after the first item has left it) and leaves Machine B at 7:30 A.M. Hence for the second item,

$$t_a = 10 \text{ minutes}, t_s = 5 \text{ minutes}, t_w = 5 \text{ minutes and } t_i = 0 \text{ minutes}$$

All the above information may be recorded as in Table 13.

Table 13. Simulation of a Production Queue

Item No.	Time on Machine A	Time of Arrival at Machine B	Time of Starting on Machine B	Time on Machine B	Time of Leaving Machine B	Waiting Time	Idle Time
N	t_a	T_a	T_b	t_s	T_e	t_w	t_i
1	10	7:10	7:10	15	7:25	0	10
2	10	7:20	7:25	5	7:30	5	0
3	15	7:35	7:35	10	7:45	0	5
4	10	7:45	7:45	5	7:50	0	0
5	15	8:00	8:00	10	8:10	0	10
6	15	8:15	8:15	5	8:20	0	5
7	10	8:25	8:25	15	8:40	0	5
8	10	8:35	8:40	15	8:55	5	0
9	10	8:45	8:55	15	9:10	10	0
10	10	8:55	9:10	10	9:20	15	0
11	10	9:05	9:20	5	9:25	15	0

In the table we recorded information for eleven items. The first two items are the ones discussed earlier. The others have been added to give a more complete picture of the queue of the semi-finished items. The number of items in the queue at time T is given in Figure 4.

Table 13 is a schedule of the production process. Similarly, Figure 4 represents it. These representations are therefore models of the production queue. We refer to such models as *simulation models.*

SOLVING BY THE MONTE CARLO METHOD

The simulation model provides the framework by which we can find the average waiting time per item. It remains to be shown how one simulates the times on Machines A and B (Columns t_a and t_s in Table 13) and how one actually computes the average waiting time. The *Monte Carlo Method* is used for this purpose.

Consider first the times t_a on Machine A. By equation (20) we know that $t_a = 10$ minutes with a probability of $\frac{1}{2}$ and $t_a = 15$ minutes with a probability of $\frac{1}{2}$. One way, then, to simulate these things is to toss a coin. Whenever a "head" appears we let $t_a = 10$ minutes, and whenever a "tail" appears let $t_a = 15$ minutes. Another way of simulating t_a is to use a table of random numbers. In using the table the following rule may be adopted: whenever a digit is even, assume $t_a = 10$ minutes, and whenever the digit is odd, assume $t_a = 15$ minutes. Thus reading down a column we may find the digits: 8, 2, 6, 0, 5, 3, etc. Hence, the corresponding t_a are 10, 10, 10, 10, 15, 15 minutes, etc.

Now consider the times t_s on Machine B. As given, $t_s = 5$ minutes with a probability of $\frac{1}{4}$, $t_s = 10$ minutes with a probability of $\frac{1}{2}$, and $t_s = 15$ minutes with a probability of $\frac{1}{4}$. One way to simulate these times is to toss two coins. Whenever two "heads" appear we let $t_s = 5$ minutes. Whenever a "head" and a "tail" appear, we let $t_s = 10$ minutes. And whenever two "tails" appear we let $t_s = 15$ minutes. Another way of simulating is, again, to use a table of random numbers. The rule for finding t_s may be the following: whenever a two-digit number is between 01 and 25 assume $t_s = 5$ minutes; whenever the two-digit number is between 26 and 75 assume $t_s = 10$ minutes; and whenever the number is between 76 and 00 assume that $t_s = 15$ minutes. For our example, a sequence of two-digit numbers may be 92, 91,

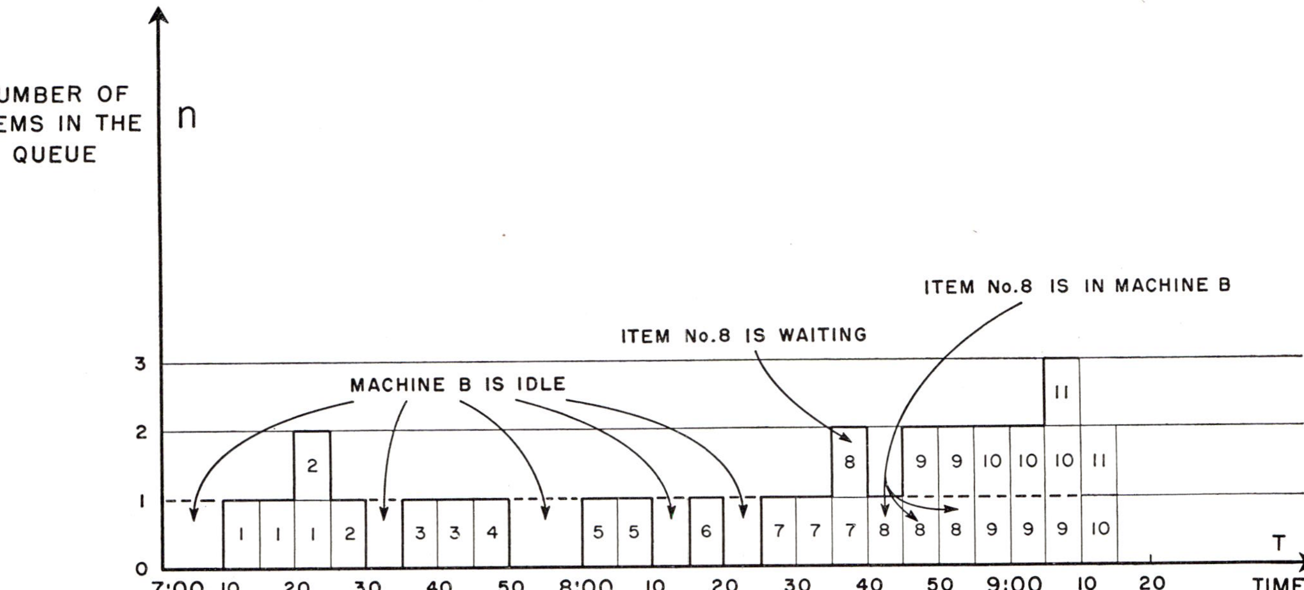

Figure 4. Items in a queue.

01, 49, 43, 99, etc. Therefore, the corresponding t_s are: 15, 15, 5, 10, 10, 15 minutes, etc.

Using this method for 50 items, Table 14 would result.

Table 14. A Monte Carlo Run of Size 50

N	t_a	T_a	T_b	t_s	T_e	t_w	t_i
0	—	0	—	—	0	—	—
1	10	10	10	15	25	0	10
2	10	20	25	15	40	5	0
3	10	30	40	5	45	10	0
4	10	40	45	10	55	5	0
5	15	55	55	10	65	0	0
6	15	70	70	15	85	0	5
7	15	85	85	15	100	0	0
⋮	⋮	⋮	⋮	⋮	⋮	⋮	⋮
50	15	620	620	15	635	0	0
Total	620			525		85	110

On the average, for this Monte Carlo run of size 50, the following results are obtained: $\bar{t}_a = 620/50 = 12.4$ minutes; $\bar{t}_s = 525/50 = 10.5$ minutes; $\bar{t}_w = 85/50 = 1.7$ minutes; $\bar{t}_i = 110/50 = 2.2$ minutes.

Repeating this for another and similar Monte Carlo run of size 50, the following results are obtained for comparison: $\bar{t}_a = 12.3$ minutes; $\bar{t}_s = 9.4$ minutes; $\bar{t}_w = 1.0$ minutes; $\bar{t}_i = 2.7$ minutes.

In ten Monte Carlo runs of size 1,000, simulated on a digital computer, $\bar{t}_w$ ranged between 1.165 and 1.915 minutes, with a mean about 1.5 minutes.

This essentially is the Monte Carlo Method. We use random numbers to simulate some of the parameters of the system under study. We then apply them in the simulation model of the system to obtain estimates of a solution.

Some questions, however, still remain unanswered:

(a) How large a Monte Carlo run should we use?
(b) How many times should the run be repeated?
(c) What confidence limits can one specify for the parameter being estimated?

The discussion of these three questions is beyond the scope of this chapter. For further information the reader is referred to reference (9) and Chapter 15.

The Newspaper Boy Inventory

A newspaper boy is required to buy his papers at 7¢ each; he sells them at 10¢ and gets 1¢ for each unsold paper. He has found by experience that on the average three customers want to buy papers, and that they appear at random.

How many papers should he buy?[1]

Before proceeding with the solution of this simplified inventory problem we should say a few words about the statistical distribution of customers. We know that on the average there are three potential customers and that they arrive at random. One can therefore assume that the number of customers on any day has the Poisson distribution. That is, the probability, $P(x)$, that there will be x customers is:

(25) . . . $$P(x) = e^{-3}3^x/x!$$

Table 15 gives rounded off values of $P(x)$ for various values of x. The table also gives the cumulative values of the distribution $P(x)$:

Table 15. Probability of x Customers Arriving

x	0	1	2	3	4	5	6	7	8
$P(x)$	0.05	0.15	0.22	0.22	0.17	0.11	0.05	0.02	0.01
$\sum_{x=0} P(x)$	0.05	0.20	0.42	0.64	0.81	0.92	0.97	0.99	1.00

[1] For another version of this problem see reference (11).

THE MODEL OF THE SYSTEM

The newspaper boy problem may be restated as follows:

Papers are bought for b cents, they are sold for s cents, and unsold papers are returned for r cents. The probability that there are x customers on any day is $P(x)$ where $x = 0, 1, 2, \ldots, x_{\max}$.

How many papers S should be bought?

We can now develop the model of the system. If the boy buys S papers he will pay for them bS cents no matter how many customers arrive. Thus the cost of buying the papers $C(S)$ is

$$(26) \ldots \qquad C(S) = bS$$

To find his expected revenue R he must consider two possibilities: (i) The number x of customers arriving is smaller than the number of papers S bought, i.e., $x < S$; (ii) the number x of customers arriving is equal or it is larger than the number of papers S bought, i.e., $x \geqq S$.

In case (i) the boy will sell x papers and get for them sx cents. He will return $S - x$ papers and get for them $r(S - x)$ cents. Thus the expected revenue in case (i), R_i, is:

$$(27) \ldots \qquad R_i = sx + r(S - x) \quad \text{where} \quad x < S$$

In case (*ii*) the boy will sell all his papers S and get for them sS. That is, the revenue in case (*ii*), R_{ii}, is:

$$(28) \ldots \qquad R_{ii} = sS \quad \text{when} \quad S \leqq x \leqq x_{\max}$$

Therefore, if the boy buys S papers his total expected revenue R is:

$$(29) \ldots \quad R(S) = \sum_{x=0}^{S-1} R_i\, P(x) + \sum_{x=S}^{x=x_{\max}} R_{ii}\, P(x)$$

$$= \sum_{x=0}^{S-1} [sx + r(S - x)]\, P(x) + \sum_{x=S}^{x=x_{\max}} sS\, P(x)$$

Finally, then, if the boy buys S papers he will be expected to make $M(S)$ cents, where

(30) . . .

$$M(S) = R(S) - C(S) = \sum_{x=0}^{S-1} [sx + r(S - x)]\, P(x) + \sum_{x=S}^{x=x_{\max}} sSP(x) - bS$$

SOLUTION OF THE MODEL

Equation (30) represents the model of the system studied. The problem now is to find what value of S will maximize M.

Since S can take only discrete values, i.e., $S = 0, 1, 2, \ldots$, the differential calculus cannot be used to find the optimal value of S.

One way of solving the model is the following. Let S_o be the optimal value of S. Then the following relations must hold:

$$(31) \ldots \qquad M(S_o) \geqq M(S_o - 1)$$

$$(32) \ldots \qquad M(S_o) \geqq M(S_o + 1)$$

Equations (31) and (32) actually state that if S_o is the number of papers the boy should buy, then he could not possibly make more by buying either $S_o - 1$ papers or $S_o + 1$ papers.

Substitution of (30) in (31) and (32) and rearranging the terms eventually leads to the following solution:

$$(33) \ldots \qquad \sum_{x=0}^{x=S_o-1} P(x) \leqq \frac{s-b}{s-r}$$

$$(34) \ldots \qquad \sum_{x=0}^{x=S_o} P(x) \geqq \frac{s-b}{s-r}$$

Any value S_o satisfying *both* equations (33) and (34) gives the solution of the newspaper boy inventory problem.

APPLYING THE SOLUTION

In our specific problem we have $s = 10$ cents, $b = 7$ cents, $r = 1$ cent. Hence

$$(35) \ldots \qquad \frac{s-b}{s-r} = (10-7)/(10-1) = \tfrac{1}{3} = 0.333 \ldots$$

The values of $P(x)$ and $\sum_{x=0}^{x=S} P(x)$ are given in Table 15.

We note that equation (33) holds for value of $S = 0, 1, 2$. Equation (34) holds for $S = 2, 3, \ldots, 8$. Thus the only value of S satisfying both equations is $S = 2$. This, then, is the solution of our problem. The boy should buy two papers.

$$S_o = 2 \quad (36)$$

As a check we give Table 16 which tells us how much the newspaper boy can be expected to make $M(S)$, as a function of the number of papers bought S.

Table 16. The Boy's Expected Profit

Papers Bought	S	0	1	2	3	4	5	6	7	8
Expected Profit	$M(S)$	0	2.55¢	3.75¢	2.97¢	0.21¢	−4.08¢	−9.36¢	−15.09¢	−21.00¢

From the table we see that if the boy buys more than four papers he is certain to lose money. We also confirm that his best policy is to buy two papers. This will give him an expected daily profit of 3.75¢.

Sequencing Jobs on Machines

Five jobs, a, b, c, d, and e, have to be processed on two machines, A and B. Each job first goes on Machine A and then on Machine B. Also, the same order of processing is maintained on both machines, i.e., if the jobs go on Machine A in a sequence, say, b, e, c, d, a, they also go on Machine B in this sequence.

The processing times of the jobs on the machines are given in Table 17.

Table 17. Processing Times of Jobs on Machines

	a	b	c	d	e
A	7	5	1	9	2
B	3	2	4	3	6

The total time that elapses from the beginning of processing of the first job on Machine *A* to the end of processing of the last job on Machine *B* depends on the sequence of the jobs.

The problem is to find the sequence which gives the minimum total elapsed time.

SOLUTION WITH GANTT CHARTS

The sequence *b, e, c, d, a* was just mentioned. The total elapsed time for this sequence can be found by the use of a Gantt Chart as in Figure 5.

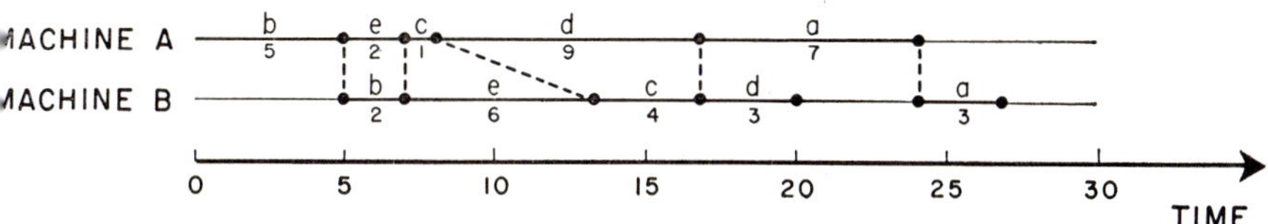

Figure 5. Gantt Chart for the sequence *b, e, c, d, a.*

The information in the Gantt Chart can also be given in tabular form as in Table 18.

Table 18. Times of Completion of Jobs for the Sequence *b, e, c, d, a*

		Job				
		b	*e*	*c*	*d*	*a*
Time Job Is Completed on Machine	*A*	5	7	8	17	24
	B	7	13	17	20	27

Job *b* is the first job to be processed. It is completed at time 5 on Machine *A* and then placed immediately on Machine *B*, coming off it at time 7. The second job in the sequence, Job *e*, is placed on Machine *A* at time 5, just after Job *b* has been completed on it. Job *e* is completed at time 7 on Machine *A* and can be put immediately on Machine *B* since by that time Job *b* has been completed on Machine *B*. The next job, Job *c*, is completed at time 8 on Machine *A*, but cannot be put on Machine B, since the previous job, Job e, is completed on B only at time 17. Similar reasoning has been applied in the preparation of the Gantt Chart in Figure 5 and of Table 18.

The total elapsed time is therefore 27 for the sequence b, e, c, d, a.

This method can be used to evaluate the total elapsed time for other sequences. If all sequences are examined and their total elapsed times computed, then the sequence giving the least total elapsed time can be spotted and is the solution of the problem.

In the above example there are 120 different sequences. This can be shown as follows:

As a first job in the sequence we can choose any one of the five jobs. After the first job has been picked, only four jobs can be chosen for the second job. Then three jobs can be chosen for the third job in the sequence, two jobs for the fourth job, and finally there will remain only one job for the fifth position. The total number of sequences, N, is therefore:

$$(37) \ldots \qquad N = 5 \times 4 \times 3 \times 2 \times 1 = 120$$

And in general, if there are n jobs to be sequenced, the total number of sequences is:

$$(38) \ldots \qquad N = n(n-1)(n-2) \ldots 3 \cdot 2 \cdot 1 = n!$$

Let us consider now the practicability of examining all sequences using Gantt Charts.

When five jobs are to be sequenced there are 120 different sequences to be examined. This may not seem to be a large number of sequences. However, for six jobs the number of sequences is 720 and for seven jobs it is 5040. It thus seems that it would be impractical to examine by Gantt Charts all sequences when the number of jobs is, say, over six. The reader may then suggest that an electronic computer be programmed to examine *all* sequences by tabulating information as in Table 18. However, this, too, is impractical even with our fastest computers. Let us estimate, for example, how long it would take to examine all possible sequences for 20 jobs. The number of different sequences is 2.43×10^{18}. Assuming that a computer can examine a million sequences per second and that it works 24 hours a day, 365 days a year, it would take

$$(39) \ldots \qquad \frac{2.43 \times 10^{18}}{10^6 \times 60 \times 60 \times 24 \times 365} = 77{,}000 \text{ years}$$

to examine all sequences.

SOLVING WITH JOHNSON'S RULE

The sequencing problem stated at the beginning of this section can be solved by using a simple rule developed by S. M. Johnson (6):

(i) In Table 17 find the smallest processing time. If the time is on Machine A, then the corresponding job should be the first in the sequence. If the smallest time is on Machine B, then the corresponding job should be the last in the sequence.

(ii) After the job has been placed (it is either first or last), do not consider its times and repeat step (i) above.

(iii) Steps (i) and (ii) are repeated until all jobs have been placed.

In our problem the smallest processing time is 1, it is on Machine A, and corresponds to Job c. Hence, c should be the first job in the sequence.

Ignoring Job c, the smallest time is 2 and it belongs to Jobs b and e. For Job b, 2 is on Machine B, hence b should be the last job. For Job e, 2 is on Machine A, hence e should be the first job. Thus far, then, we have placed jobs c, b, e:

	1st	*2nd*	*3rd*	*4th*	*5th*
The optimal sequence	c	e			b

Ignoring now Jobs c, b, e the smallest time is 3, and it belongs to the remaining Jobs a, d. Since the 3 is on Machine B for both, it is immaterial which of these should be placed last. That is, there are two optimal sequences:

	1st	*2nd*	*3rd*	*4th*	*5th*
Optimal Sequence I	c	e	a	d	b
Optimal Sequence II	c	e	d	a	b

For both sequences we find that the total elapsed time is 26 as in Table 19.

Table 19. Times of Completion of Jobs for Optimal Sequences

		Optimal Sequence I					Optimal Sequence II				
		c	e	a	d	b	c	e	d	a	b
Time Job Is Completed on Machine	A	1	3	10	19	24	1	3	12	19	24
	B	5	11	14	22	26	5	11	15	22	26

THE n-JOBS AND m-MACHINES PROBLEM

The problem tackled so far pertains to five jobs and two machines. The general sequencing problem has n jobs and m machines.

When $m = 2$, i.e., for two machines, and any number of jobs n, Johnson's Rule for finding the optimal sequence holds. But it does not hold when $m = 3$. The problem of finding an optimal sequence for n jobs on m machines where m is 3 or larger is as yet an unsolved problem. Some of our best mathematicians have tackled the problem but no general solution to it has yet been found.

The number of different sequences is still $n(n-1) \ldots 3 \cdot 2 \cdot 1 = n!$ hence the Gantt Chart method is not a practical method for a large number of jobs. The best we can do is to guess by trial and error at a finite number of possible "good" sequences and select from these the best sequence by computing for each sequence its total elapsed time.

The Travelling Salesman Problem

A school bus has to stop and pick up children daily at locations B, C, D in Figure 6. The bus leaves the school at A and returns to the school after it has picked up the children.

What route should the bus take so that it travels the shortest distance?

Problems of this type have come to be known as "travelling salesman" problems. The travelling salesman's home may be at A. He is supposed to visit cities B, C, D and return home. What is his best route?

In our problem avenues run in the north-south direction. The odd numbered avenues are one-way from south to north and the even numbered avenues are one-way from north to south. The streets run in the west-east direction. The odd numbered streets are one-way from east to west and the even numbered streets are one-way from west to east, with the exception of 8th Street which is a two-way street.

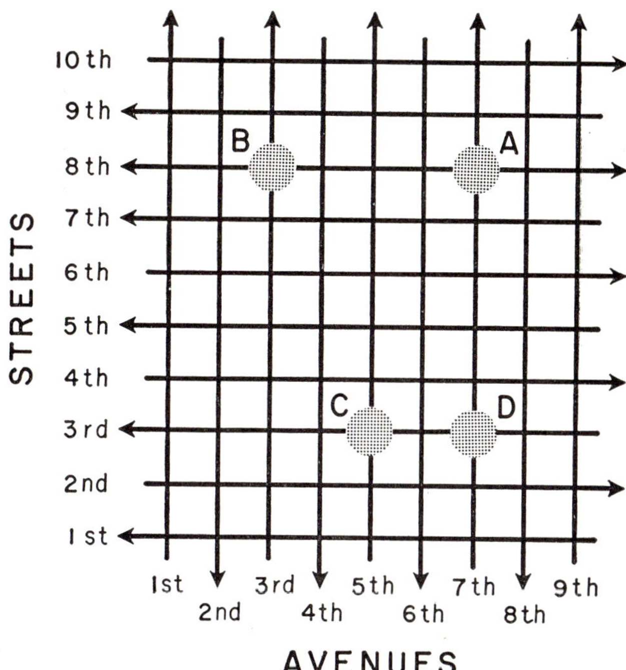

Figure 6. Location of School A and Bus Stations B, C, D, in a city with one-way streets.

THE DISTANCE MATRIX

Our problem can be, of course, solved by inspection. This is possible because of the small number of stops. The solution is the route A to D to C to B to A. From A to D travel along 8th Street, 8th Avenue, 3rd Street, a total of 7 city blocks. From D to C, along 3rd Street, 2 city blocks. From C to B, along 3rd Street and 3rd Avenue, 7 city blocks. And, finally, from B to A, along 8th Street, 4 city blocks. The total length of the route is thus $7 + 2 + 7 + 4 = 20$ city blocks.

For a large number of stops it is rather difficult to obtain a solution by inspection. It is quite helpful in such cases to prepare a *distance matrix* which gives the distance between any two stops, as Table 20.

Table 20. The Distance Matrix

		To			
		A	*B*	*C*	*D*
From	*A*	—	4	7	7*
	B	4*	—	9	11
	C	7	7*	—	6
	D	5	9	2*	—

The distances for the solution, *A* to *D* to *C* to *B* to *A*, are marked with asterisks in the table. We can use the table to find the total distances for all possible routes, thus:

Number	Route	Total Distance
1	*A* to *B* to *C* to *D* to *A*	24 city blocks
2	*A* to *B* to *D* to *C* to *A*	24 city blocks
3	*A* to *C* to *B* to *D* to *A*	30 city blocks
4	*A* to *C* to *D* to *B* to *A*	26 city blocks
5	*A* to *D* to *B* to *C* to *A*	32 city blocks
6	*A* to *D* to *C* to *B* to *A*	20 city blocks

For three stops we find that there are six possible routes. The first stop can be chosen in three different ways. After it has been selected, the second stop can be chosen in two ways. For the third stop there is only one choice, after the first two have been selected. Thus the total number of choices is $3 \times 2 \times 1 = 6$.

And in general for n stops the total number of different routes N is

$$(40) \ldots \quad N = n(n-1)(n-2)\ldots 3\cdot 2\cdot 1 = n!$$

The distance matrix in Table 20 is not symmetric, since the distance between two stops is not always the same, but depends on the direction of streets and avenues. Thus to go from *C* to *D* only 2 blocks are travelled, but to go from *D* to *C* the distance is 6 blocks.

When the distance matrix is symmetric then each route has a corresponding identical route with an opposite direction. For problems of this kind the number of different routes N is

$$(41)\ \ldots \qquad N = (\tfrac{1}{2})n(n-1)(n-2)\ldots 3\cdot 2\cdot 1 = (\tfrac{1}{2})n!$$

The distance matrix need not always be in terms of geometric distances. One could equally use as a measure of effectiveness the distance in time, or even the "distance" in dollars. In this latter case the distance matrix could be in terms of cost of going from one stop to another.

RELATION TO THE ASSIGNMENT PROBLEM

Suppose the distance matrix of Table 20 is marked for each possible route as in Table 21.

Table 21. Marked Distances for All Possible Routes

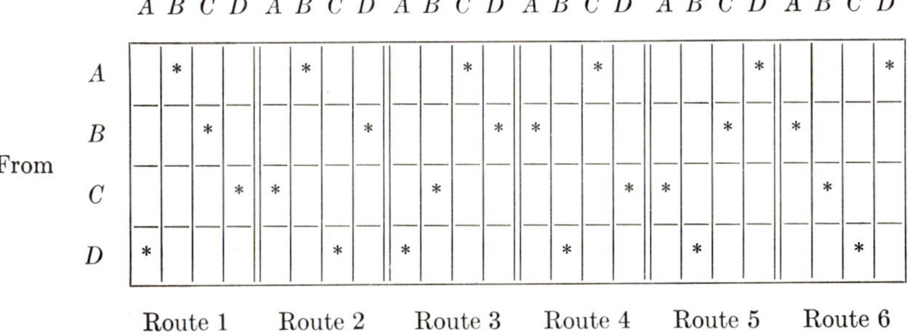

		Route 1				Route 2				Route 3				Route 4				Route 5				Route 6			
		A	*B*	*C*	*D*	*A*	*B*	*C*	*D*	*A*	*B*	*C*	*D*	*A*	*B*	*C*	*D*	*A*	*B*	*C*	*D*	*A*	*B*	*C*	*D*
From	*A*		*				*					*				*					*				*
	B			*					*				*	*						*		*			
	C				*	*					*						*	*					*		
	D	*						*		*					*				*					*	

We note that in every distance matrix each row and each column is marked once and only once. This is obvious since every route has to arrive at a stop and has to leave the stop.

We recall that a similar situation occurred in solving the Stewardesses' Assignment Problem. There, too, for a feasible solution each row and each column had to be represented once and only once. It would then appear that one could use the method of reduced matrices to solve the Travelling Salesman Problem.

One difficulty immediately arises. A feasible solution of the Assignment Problem may involve an element on the diagonal of the distance matrix. Such an assignment is meaningless in the Travelling Salesman Problem. To ensure that this not occur, the distances for the diagonal elements in the distance matrix are given relatively large numbers, say, 100 as in Table 22.

Table 22. Preparing the Distance Matrix for Solution with the Method of Reduced Matrices

From \ To	*A*	*B*	*C*	*D*
A	100	4	7	7
B	4	100	9	11
C	7	7	100	6
D	5	9	2	100

The method of reduced matrices can now be used. It leads to Table 23.

Table 23. The Reduced Distance Matrix

From \ To	*A*	*B*	*C*	*D*
A	96	0*	3	3
B	0*	96	5	7
C	1	1	94	0*
D	3	7	0*	98

The solution of the Assignment Problem having the matrix of Table 22 is marked in Table 23. However, applying it to the Travelling Salesman Problem we get the meaningless route A to B to A and C to D to C.

THE GENERAL SOLUTION

No general method has yet been found to solve Travelling Salesman Problems. One can always try the method of reduced matrices of the Assignment Problem. This may sometimes lead to a solution provided that the Assignment Problem solution is a feasible solution of the Travelling Salesman Problem. But if the solution is not feasible, there is no way of getting a solution to the Travelling Salesman Problem. This problem, as the $m \times n$ Sequencing Problem, has been tackled by some of our best mathematicians, but so far they have not succeeded in solving it.

The interested reader may find additional information on the Travelling Salesman Problem in References (3) and (4).

A Replacement Policy

A piece of equipment costs initially $60,000. Its maintenance costs in the first and succeeding years are $10,000, $22,000, $34,000, etc., increasing by $12,000 per year.

How frequently should this piece of equipment be replaced, assuming that it has no salvage value, that it performs equally well as long as maintenance is provided, and that only the above mentioned cost figures have to be considered?

OPTIMAL REPLACEMENT PERIODS

Suppose that the equipment is replaced every year. One would then spend $60,000 every year for purchasing and $10,000 for maintenance, a total of $70,000 per year.

Now suppose that it is replaced every two years. Every two years one would spend $60,000 for purchasing and $32,000 for maintenance ($10,000 plus $22,000), a total of $92,000, or $46,000 per year. Similar reasoning leads to Table 24.

From the table we see that it is best to replace the equipment every three years. This would result in a total equipment and maintenance cost of $42,000 per year.

Table 24. Yearly Costs as a Function of Frequency of Replacement

Years	Maintenance Cost in Year	Total Equipment and Maintenance Costs	Cost Per Year
1	$10,000	$ 70,000	$70,000
2	22,000	92,000	46,000
3	34,000	126,000	42,000
4	46,000	172,000	43,000
5	58,000	230,000	46,000
6	70,000	300,000	50,000

CHOICE OF EQUIPMENT

We can use the method just described to solve problems of choice of equipment.

Suppose, for example, that a company owns the equipment described above and another piece of equipment is offered which has the following characteristics:

Initial cost: $84,000; maintenance cost in first and succeeding years: $12,000, $18,000, $24,000, etc., increasing by $6,000 per year.

Both pieces of equipment perform equally well as long as maintenance is provided.

Which piece of equipment should be preferred?

A table similar to Table 24 can be prepared for the second piece of equipment (Table 25).

Table 25. Yearly Costs for the Second Piece of Equipment

Years	Maintenance Cost in Year	Total Equipment and Maintenance Costs	Cost Per Year
1	$12,000	$ 96,000	$96,000
2	18,000	114,000	57,000
3	24,000	138,000	46,000
4	30,000	168,000	42,000
5	36,000	204,000	40,800
6	42,000	246,000	41,000
7	48,000	294,000	42,000

The second piece of equipment should therefore be replaced every five years. This would cost a total of $40,800 per year.

Since the first piece of equipment costs at best $42,000 per year, the second piece should therefore be preferred.

A further variation of the problem may be now stated. Assume that the company has had the first piece of equipment for three years and is just about ready to replace it with similar equipment. It now receives information that the second piece of equipment would be available in another year.

What should its policy be?

The company has essentially four alternatives:

I. Replace the first piece of equipment with similar equipment and then replace it again with the second piece of equipment after one year.

II. As (I) but replace with the second piece of equipment after two years.

III. As (I) but replace with the second piece of equipment after three years.

IV. Keep the first piece of equipment and replace it with the second piece after one year.

The costs of the various alternatives are given in Table 26.

Table 26. Total Costs for the Four Alternatives

Alternative	Year 1	Year 2	Year 3	Year 4	Year 5
I	$70,000	$40,800	$40,800	$40,800	$40,800
II	70,000	22,000	40,800	40,800	40,800
III	70,000	22,000	34,000	40,800	40,800
IV	46,000	40,800	40,800	40,800	40,800

For the first piece of equipment we entered in the table the actual costs of purchasing and maintenance as they occur from year to year. For the second piece of equipment we entered the average cost per year, $40,800, assuming that this equipment will be replaced optimally every five years as per Table 25.

To compare the four alternatives we must now select a comparable period of time. This period is evidently composed of the first three years. The total costs during this period for the three years are given in Table 27.

Table 27. Total Costs During Three Years

Alternative	Costs in Years 1, 2 and 3
I	$151,600
II	132,800
III	126,000
IV	127,600

The best policy, therefore, is to replace the first piece of equipment with similar equipment; keep this new equipment for another three years and only then replace it with the second piece of equipment.

COMMENTS ON DISCOUNTING

In problems involving expenditures of large sums of money over relatively long periods of time, one sometimes uses the *present value* of future expenditures as a measure of effectiveness.

Suppose, for example, that we want to find present values for our first replacement problem. In that problem the piece of equipment costs $60,000 and the maintenance costs in first and succeeding years are $10,000, $22,000, $34,000, increasing by $12,000 every year. Assume that money is discounted at an annual rate of 6 per cent; i.e., assume that one dollar spent a year from today has a present value of only $1/1.06 = 94¢$. Let us now find the present value of equipment and maintenance costs as a function of the frequency of replacement.

If the equipment were replaced every year the total present value of the costs would be:

$$(42) \quad \ldots \quad 70{,}000 + 70{,}000/1.06 + 70{,}000/1.06^2 + \ldots$$

$$= 70{,}000\,(1 + 1/1.06 + 1/1.06^2 + \ldots)$$

$$= 70{,}000\left(\frac{1}{1 - 1/1.06}\right) = \$1{,}236{,}000$$

The first term of equation (42) gives the present value of the equipment bought in the first year and the maintenance cost in that year.

The second term gives the corresponding present value of costs in the second year. And so on to infinity. The terms of equation (42) form an infinite series whose sum is \$1,236,000. Thus, the present value of *all* costs of equipment and maintenance, if replacement occurs every year, is \$1,236,000.

Similarly, the present value of all costs for replacements every two years is as in equation (43):

$$(43) \ldots \quad 70{,}000 + 22{,}000/1.06 + 70{,}000/1.06^2 + 22{,}000/1.06^3 + \ldots$$
$$= (70{,}000 + 22{,}000/1.06)(1 + 1/1.06^2 + 1/1.06^4 + \ldots)$$
$$= (70{,}000 + 22{,}000/1.06)\left[\frac{1}{1 - 1/1.106^2}\right]$$
$$= \$825{,}000$$

And in the same way we find in (44) and (45) the present value of all costs for replacements every three and four years.

$$(44) \ldots$$
$$(70{,}000 + 22{,}000/1.06 + 34{,}000/1.06^2)[1/(1 - 1/1.06^3)] = \$755{,}000$$

$$(45) \ldots$$
$$(70{,}000 + 22{,}000/1.06 + 34{,}000/1.06^2 + 46{,}000/1.06^3)[1/(1 - 1/1.06^4)]$$
$$= \$768{,}000$$

On the basis of the present value of these discounted costs the best policy is, again, to replace every three years. Here, then, the decision as to how frequently to replace is the same whether costs are discounted or not.

Matching Pennies, Nickels, and Dimes

Two players, P_1 and P_2, simultaneously put a coin each on a table. They only use three coins: a penny—1¢, a nickel—5¢, and a dime —10¢.

The total amount of money represented by the two coins is then examined. If the total is divisible by 3, the first player, P_1, collects both coins. If the total is divisible by 10, the second player, P_2, collects both coins. Otherwise each player collects the coin which he put on the table.

If you were P_1, how would you play this game? On the average, how much can you expect to win or lose, if you played this game many times?

Player P_1 seems to have an advantage in this game. He can ensure that he will not lose at all. For, if he continuously puts a penny on the table, Player P_2 cannot force a win, although he knows what P_1 is doing. The best P_2 can do, in this case, is to put either a penny or a dime (for totals of 2 and 11) and the game will end in a draw.

On the other hand Player P_2 is not in a similar comfortable position. If Player P_1 knows what P_2 is putting down, P_2 will *always* lose. For, if P_2 puts a penny, P_1 will put a nickel (for a total of 6) and will win P_2's penny. If P_2 puts a nickel, P_1 will put a penny (for a total of 6) and will win P_2's nickel. If P_2 puts a dime, P_1 will put a nickel (for a total of 15) and will win P_2's dime.

Player P_1 can ensure that he will never lose the game. Player P_2 can ensure that he will not lose more than 1 penny (by putting a penny each time). Hence it seems that Player P_1 should be able to win on the average some amount between 0 and 1 penny. This amount is referred to as the *value of the game.*

The game just described is a very special case of a class of games with which the Theory of Games deals. We shall presently use some of the methods of the Theory of Games to find how Player P_1 should play this game and what the value of the game is.

THE PAYOFF MATRIX

All possible outcomes of playing our game can best be displayed as in Table 28.

Table 28. Total Amount on Table

		P_2 puts		
		Penny	Nickel	Dime
P_1 puts	Penny	1¢	6¢	11¢
	Nickel	6¢	10¢	15¢
	Dime	11¢	15¢	20¢

Each player has three alternatives for each play of the game: putting a penny, a nickel, or a dime. Hence nine outcomes are possible as in Table 28.

When the total is divisible by 3, P_1 wins. This occurs four times. When the total is divisible by 10 P_2 wins. This occurs only twice. In all other cases no player wins. This occurs three times.

The corresponding wins and losses for the various alternatives are given in Table 29.

Table 29. Payoff Matrices to Players P_1 and P_2

		P_1			P_2		
		Penny	Nickel	Dime	Penny	Nickel	Dime
P_1	Penny	0¢	5¢	0¢	0¢	−5¢	0¢
	Nickel	1¢	−5¢	10¢	−1¢	5¢	−10¢
	Dime	0¢	5¢	−10¢	0¢	−5¢	10¢
		P_1's Payoff			P_2's Payoff		

When both players put a penny on the table the total amount is 2 and no player wins. Therefore the payoff 0¢ has been entered in the corresponding square in P_1's payoff matrix and in P_2's payoff matrix. When P_1 puts a penny and P_2 puts a nickel, the total is 6¢. It is divisible by 3 and therefore P_1 wins 5¢. Therefore, the payoff 5¢ has been entered in the corresponding square in P_1's payoff matrix and −5¢ (to indicate a loss) has been entered in the corresponding square in P_2's payoff matrix.

In a similar manner all other entries have been made.

Whenever P_1 wins an amount of money, P_2 always loses that amount. And, conversely, whenever P_1 loses an amount P_2 always wins it. The games in which this condition holds are called *Zero-Sum Games*. And indeed, the sum of the corresponding squares in P_1's and P_2's payoff matrices is always zero. It is therefore customary in zero-sum games to consider only P_1's payoff matrix, since P_2's matrix is always the same with all signs reversed.

MIXED STRATEGIES

Let us now, for a moment, put ourselves in the role of player P_1, and let us consider Table 29.

P_1 can compare the relative advantages of putting a penny or a dime on the table. In the first case his corresponding payoffs are 0¢, 5¢, and 0¢, and in the second they are 0¢, 5¢ and −10¢. It therefore seems obvious to P_1 that putting a penny on the table is at least as good, and sometimes even better, *no matter* what P_2 does. Player P_1 can therefore completely disregard the third alternative, that of putting a dime down. Stated in other words, P_1's optimal strategy should not include the putting of a dime on the table.

Now let us consider player P_2. He is aware of the payoffs, just as P_1 is, and can therefore deduce that P_1 will never put a dime on the table. P_2 can now compare the relative advantages of putting down a penny or a dime. In the first case his corresponding payoffs are 0¢ and −1¢ and in the second they are 0¢ and −10¢. (The payoffs corresponding to P_1's putting down a dime are not considered since, by previous reasoning, P_2 knows that P_1 will not put a dime down.) It therefore seems obvious to P_2 that putting a penny on the table is at least as good, and sometimes even better, no matter whether P_1 puts down a penny or a nickel. Thus, P_2's optimal strategy, too, should not include the putting of a dime on the table.

This leads to the reduced payoff matrix in Table 30.

Table 30. The Reduced Payoff Matrix

		P_2	
		Penny	Nickel
P_1	Penny	0¢	5¢
	Nickel	1¢	−5¢

The table gives P_1's payoff. The same matrix with all signs reversed would give P_2's payoff.

From the table we see that if P_1 always puts a penny down, P_2 wil also put a penny down, and the payoff to P_1 would be 0¢. If he always puts a nickel down, P_2 will also put a nickel down and P_2 would lose 5¢.

It therefore seems that P_1 should sometimes put down a penny and sometimes a nickel. We refer to such a way of playing the game as a *mixed strategy*. Similarly, one can quickly reason out that P_2 should also have a mixed strategy; otherwise he would either lose 1¢ or 5¢.

SOLUTION OF GAME

There are several methods for finding optimal mixed strategies. Some of these are described in Chapter 16, "Introduction to the Theory of Games."

Using William's Oddments Method, for example, we find the following odds:

Player P_1: Odds for Penny: $1 - (-5) = 6$
Odds for Nickel: $5 - 0 = 5$
Player P_2: Odds for Penny: $5 - (-5) = 10$
Odds for Nickel: $1 - 0 = 1$

In other words the optimal strategy for P_1 is to put down a penny 6/11 of the time and a nickel 5/11 of the time. The optimal strategy for P_2 is to put down a penny 10/11 of the time and the nickel 1/11 of the time.

We can now compute how much P_1 can expect to win in this game if both he and P_2 use their optimal strategies and when they play the game many times. There will be four different outcomes. Both players will put down pennies $6/11 \times 10/11 = 60/121$ of the time. P_1 will put a penny and P_2 will put a nickel $6/11 \times 1/11 = 6/121$ of the time. P_1 will put a nickel and P_2 will put a penny $5/11 \times 10/11 = 50/121$ of the time. And finally, both will put nickels $5/11 \times 1/11 = 5/121$ of the time.

P_1 can therefore expect to win in this game:

$$(60/121) \times 0 + (6/121) \times 5 + (50/121) \times 1 + (5/121) \times (-5) = 5/11 \text{ cents}$$

The complete solution of the problem is therefore as follows:

P_1 should put down a penny 6/11 of the time, a nickel 5/11 of the time, and he should never put down a dime. P_2 should put down a penny 10/11 of the time, a nickel 1/11 of the time and he, too, should never put down a dime.

If both players follow these optimal mixed strategies, then P_1 will be expected to win on the average 5/11¢ and P_2 will be expected to lose on the average 5/11¢.

n-PERSON GAMES AND NON-ZERO-SUM GAMES

We have just solved a simple game. This game involved two players and it was a zero-sum game. It now seems in order to say a few words about games involving more than two players and games which are not zero-sum. Such games are referred to as n-*person* and *non-zero-sum games.*

In a previous section we pointed out that the sequencing problem for n jobs and m machines and the travelling salesman problem have not yet been solved. The situation is even sadder for n-person and non-zero-sum games. Not only have most such games not been solved as yet, but there does not even seem to be a satisfactory way to say what a solution of such games should be. We have succeeded in giving a solution of a two-person zero-sum game, and the reader must have intuitively agreed with us as to the meaning of the solution (as well as with the solution itself). As of now, unfortunately, it is impossible to define a solution of an n-person game or a zero-sum game with which everyone will agree.

We cannot say more about this topic here. It is mentioned to illustrate a case for which no acceptable model can as yet be constructed.

CONCLUSION

The purpose of this chapter has been to introduce the reader to models in operations research. The author elected to do this by using simple illustrations. These illustrations are simplifications of systems which may be encountered in practice, yet they contain the basic elements inherent in the systems. The *Stewardess Assignment* illustration exemplifies systems in which allocation of resources (Stewardesses) is subject to constraints (flights). Many of these systems can be solved by the methods of *Linear Programming* (see Chapter 13). The *Production Queue* illustration exemplifies systems involving *Waiting Lines* or *Queueing* (see Chapter 14) and the important technique of *Simulation* (see Chapter 15). The *Newspaper Boy* illustration exemplifies systems in which one attempts to balance the costs of shortages (lost sales) with the costs of surpluses (loss on returns). These systems are solved using the methods of *Inventory Control* (see Chapter 12). The *Sequencing of Jobs* and the *Travelling Salesman* illustrations exemplify numerous systems involving ordering or sequencing of events. Such systems are frequently encountered in production and business environments, but no general methods have yet been developed for their solutions. The *Replacement Policy* illustration exemplifies systems in which one attempts to balance maintenance costs with capital investment costs. The *Matching* illustration exemplifies systems in which two or more decision makers are in competition. Such systems are studied using the methods of *Theory of Games* (see Chapter 16).

Many difficulties can usually be expected in attempting to construct models of *real world* situations. There are the difficulties of describing physical phenomena which are familiar to the physicist, the chemist, and the biologist. But in addition, there are additional difficulties arising out of the complexity of social organizations, the human elements involved, the problems of measurement of intangibles, etc.

Operations researchers have been devoting much of their time to the solution of various kinds of models and some very powerful mathematical tools have been used for this purpose. Operations researchers have also been devoting time to construction of models to represent specific systems under study.

Relatively little attention has been paid the following problems:

(a) How good is a model of a system under study? How closely does it represent the system? How can one measure the goodness of fit of the representation?

(b) What is the sensitivity of the model? Would a "simpler" model lead to "wrong" decisions?

(c) What data should be used in the model? How accurate should the data be? What is the effect of estimates and predictions on decisions?

These and related questions should be always borne in mind in the construction and solution of models. More research is needed before we can say that we have foolproof methods to utilize operations research models for decision-making,

REFERENCES

(1) Russell L. Ackoff, "The Development of Operations Research as a Science," *Operations Research,* Vol. 4, No. 3, pp. 271-286, June, 1956.

(2) C. West Churchman, Russell L. Ackoff and E. Leonard Arnoff, Editors, *Introduction to Operations Research,* John Wiley and Sons, New York 1957, pp. 157-194.

(3) G. A. Cross, "A Method for Solving Travelling-Salesman Problems," *Operations Research,* Vol. 6, No. 6, pp. 791-812, November-December, 1958.

(4) Merrill M. Flood, "The Travelling Salesman Problem," *Operations Research for Management,* Vol. II, pp. 340-357, The Johns Hopkins Press Baltimore, 1956.

(5) Basilio Giardina, "La Programmazione dell'ottima Sequenza di Produzione," *Bolletino del Centro per la Ricerca Operativa,* No. 1, pp. 6-27, 1955 and No. 2, pp. 38-42, 1956.

(6) S. M. Johnson, "Optimal Two- and Three-Stage Production Schedules with Setup Time Included," *Naval Research Logistics Quarterly,* Vol. I, No 1, pp. 61-68, March, 1954.

(7) Gilbert W. King, "Applied Mathematics in Operations Research," Chapter 10 in *Modern Mathematics for the Engineer,* pp. 211-242, McGraw-Hill New York, 1956.

(8) George Kozemetsky and Paul Kircher, *Electronic Computers and Management Control,* pp. 131-138, McGraw-Hill, New York, 1956.

(9) H. A. Meyer (Editor), *Symposium on Monte Carlo Methods,* John Wiley and Sons and Chapman and Hall, New York and London, 1956.

(10) J. Sayer Minas, "Formalism, Realism and Management Science," *Management Science,* Vol. 3, No. 1, pp. 9-14, October, 1956.

(11) Philip M. Morse and George E. Kimball, *Methods of Operations Research* pp. 31-32, The Technology Press of M. I. T. and John Wiley, New York 1951.

(12) Joseph F. McCloskey, Fred Hausmann and Lawrence Friedman, Arthur Yaspan, "An Analysis of Stewardess Requirements for a Major Domestic Airline," *Naval Research Logistics Quarterly,* Vol. 4, No. 3, pp. 183-20 September, 1957.

(13) Eliezer Naddor, "Some Models of Inventory and an Application," Chapter 24 in *Analyses of Industrial Operations,* pp. 429-443, Richard D. Irwin Inc., Homewood, Illinois, 1959.

(14) J. D. Williams, *The Compleat Strategyst,* McGraw-Hill, New York, 195

Nine

REVIEW OF SOME BASIC STATISTICS [1]

ACHESON J. DUNCAN

INTRODUCTION

The word "statistics" comes from the word "state." Many decades ago "statistics" were facts pertaining to the state. Gradually the term was applied primarily to numerical facts pertaining to the state and then to numerical facts in general. In the course of time, techniques were developed for analyzing and interpreting numerical facts, and the word "statistics" came to include the techniques as well as the data to which the techniques were applied. Today this dual usage of the word still persists, and its meaning has to be gleaned from the context. Here it will be used as a set of procedures for dealing with and thinking about data. In this broader sense statistics covers the laying out or "design" of experiments and the planning of surveys for the gathering of numerical data, as well as the analysis and interpretation of the data themselves. Statistical thinking has also been extended to the decision-making related to the analysis and interpretation of the data.

Statistics is closely related to the theory of probability, since the interpretation of numerical data and the decisions recommended are generally restricted to situations involving the play of chance forces. It is the stability inherent in *random, mass* phenomena that makes it possible for the statistician to use measures of past variability to make predictions about future variability. Numerical data that have not

[1] In writing this chapter the author has drawn heavily from his book *Quality Control and Industrial Statistics* (2nd ed.; Homewood, Ill.: Richard D. Irwin, Inc., 1959).

been produced in some way or other by the action of chance forces are not proper material for the application of the principles of statistical inference. Such data may be *described* by various statistical measures, but they are not suitable for drawing probabilistic conclusions about any larger set of data of which they may be a part.

FREQUENCY DISTRIBUTIONS

The basic tool in the kit of the statistician is the frequency distribution. This is a list of the magnitudes assumed by a given variable together with the frequency with which it takes on these magnitudes (7).

Table 1. Number of Defects per Sheet for Fifty Sheets of a Given Material

Sheet No.	Number of Defects	Sheet No.	Number of Defects
1	1	26	0
2	0	27	1
3	0	28	3
4	1	29	2
5	0	30	1
6	2	31	0
7	0	32	0
8	2	33	1
9	1	34	1
10	1	35	0
11	0	36	1
12	0	37	1
13	1	38	1
14	0	39	0
15	1	40	0
16	1	41	0
17	1	42	0
18	1	43	0
19	0	44	1
20	0	45	0
21	1	46	0
22	0	47	4
23	1	48	2
24	0	49	1
25	1	50	0

Distributions of Discrete Variables

In Table 1 is listed the number of defects found in sheets of a given material when the sheets were inspected in order of production. If we count the number of sheets that have 0 defects, 1 defect, 2 defects, etc., we get the results shown in Table 2. This is a frequency distribution of

Table 2. Frequency Distribution of the Data of Table 1

Number of Defects (x)	Number of Sheets (F)
0	23
1	21
2	4
3	1
4	1
	50

the number of defects per sheet. It is a frequency distribution of a *discrete* variable. A graph is shown in Figure 1.

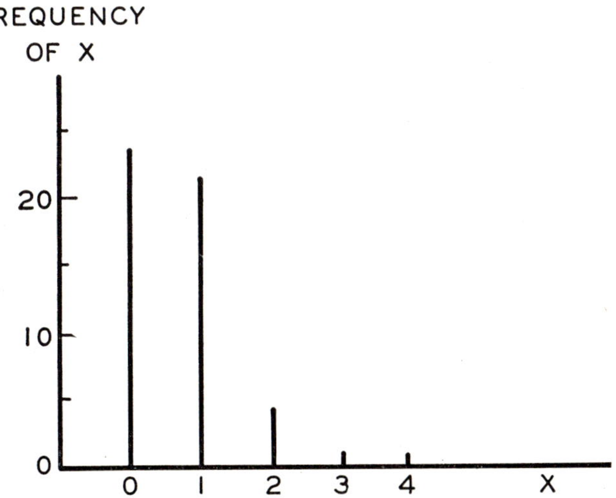

Figure 1. A frequency distribution of a discrete variable. (Data from Table 1)

Distributions of Continuous Variables

Table 3 gives data on the mean roughness of samples taken from

Table 3. Measurements of Mean Roughness of Fifty Rolls of Newsprint (Courtesy of the American Newspaper Publishers Association Research Institute, Inc.)

Roll No.	Roughness	Roll No.	Roughness	Roll No.	Roughness
1	92	18	93	35	77
2	95	19	84	36	83
3	107	20	100	37	79
4	122	21	87	38	61
5	107	22	80	39	79
6	100	23	111	40	114
7	47	24	111	41	104
8	89	25	80	42	87
9	64	26	83	43	100
10	153	27	67	44	124
11	56	28	106	45	59
12	99	29	90	46	68
13	89	30	97	47	57
14	86	31	121	48	84
15	72	32	78	49	89
16	87	33	133	50	73
17	97	34	103		

different rolls of newsprint. If we group the measurements by intervals of 20 starting[2] at 29.5, we get the results shown in Table 4. This is a frequency distribution of measurements of roughness of newsprint. It is a frequency distribution of a *continuous* variable.

A graph of Table 4 is shown in Figure 2; this is called a "histogram." In constructing a histogram it is sometimes preferable to show relative frequencies instead of actual frequencies. Also in certain instances (particularly if the grouping interval is not uniform in size) it is convenient to represent the relative frequency by the area of the rectangles instead of their height.

[2] Since the data have been rounded off to the nearest half unit, we start at 29.5 instead of 30. See (5).

Table 4. A Frequency Distribution of the Data of Table 3

Roughness Measurements (x)	Number of Rolls Falling in Specified Intervals (F)
equal to 29.5 but less than 49.5	1
equal to 49.5 but less than 69.5	7
equal to 69.5 but less than 89.5	19
equal to 89.5 but less than 109.5	15
equal to 109.5 but less than 129.5	6
equal to 129.5 but less than 149.5	1
equal to 149.5 but less than 169.5	1
	50

With continuous data, such as the measurements of roughness above, it is possible to conceive of a *frequency curve.* If we take a very large number of measurements and make an *area histogram of relative fre-*

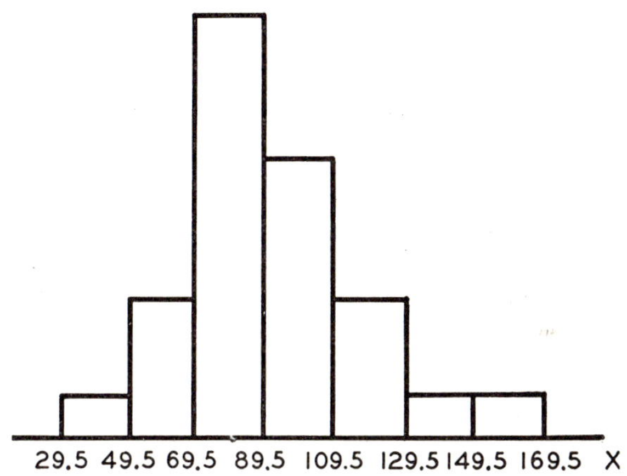

Figure 2. A histogram. (Data from Table 4)

quencies with very small grouping intervals, the tops of the rectangles will tend to trace out a curve. The frequency curve is thus the limit of an area histogram of relative frequency when the number of cases distributed is infinitely large and the grouping intervals are infinitesimally small (1). As a limit the frequency curve is a pure abstraction, but like many abstractions it is found to be very useful in statistical analysis.

Characteristics of Frequency Distributions

Statisticians early came to recognize four major characteristics of frequency distributions (7). One was the location of the distribution on the scale of the variable. Was it located in the region of high values, low values, intermediate values? A second recognized characteristic was the general spread of the distribution. Were the values widely dispersed or closely confined to a given region? A third recognized characteristic was the symmetry or lack of symmetry of the distribution, the "skewness" as it is called, and the fourth recognized characteristic was the peakedness or flatness of the distribution, called "kurtosis."[3] Here we shall consider only one measure of location, the *arithmetic mean*, and one measure of general variability, the *standard deviation.*

For a discrete variable, the mean is computed by multiplying each value of the variable x by its frequency, adding and dividing by the total frequency (7). Thus, with respect to Table 2, we have:

$$\text{Mean } x = \frac{\Sigma Fx}{n}$$

$$= \frac{23(0) + 21(1) + 4(2) + 1(3) + 1(4)}{50} = 0.72$$

With a histogram of a continuous variable the mean is computed by multiplying the *mid-point* of each grouping interval by the frequency of that interval, adding and dividing by the total frequency[4] (7). Thus with respect to Table 4 we have:

$$\text{Mean } x = \Sigma \frac{Fx}{n}$$

$$= \frac{1}{50}[1(39.5) + 7(59.5) + 19(79.5) + 15(99.5) + 6(119.5) + 1(139.5) + 1(159.5)]$$

$$= 89.5$$

[3] Actually kurtosis is an excess of cases in the middle of the distribution and out on the tails with a deficiency in the middle ranges (5).

[4] This introduces a "grouping error." For the mean, the maximum grouping error is half a group. Usually it is much less than this. When the data are used to make inferences about the mean of the population from which the data are a random sample, the grouping error is added independently to the sampling error (6).

For a frequency curve the ordinate of which is $\phi(x)$ we have (10):

$$\text{Mean } x = \int_{-\infty}^{\infty} x\phi(x)\,dx$$

since by definition $\int_{-\infty}^{\infty} \phi(x)\,dx = 1$.

The standard deviation is the weighted root mean square deviation from the mean of the distribution (7). We shall represent it by a Greek sigma. Thus with respect to the data of Table 2, we have:[5]

$$\sigma = \sqrt{\frac{23(0-0.72)^2+21(1-0.72)^2+4(2-0.72)^2+1(3-0.72)^2+1(4-0.72)^2}{50}}$$

$$= \sqrt{7.216} = 2.69$$

With respect to the data of Table 4, we have:

$$\sigma = \Big(\frac{1}{50}\,[1(39.5-89.5)^2 + 7(59.5-89.5)^2 + 19(79.5-89.5)^2 + 15(99.5-89.5)^2 + 6(119.5-89.5)^2 + 1(139.5-89.5)^2 + 1(159.5-89.5)^2]\Big)^{\frac{1}{2}}$$

$$= (500)^{\frac{1}{2}} = 22.4$$

For a continuous frequency curve we would have (10):

$$\sigma = \left(\int_{-\infty}^{\infty} (x - \text{mean})^2\,\phi(x)\,dx\right)^{\frac{1}{2}}$$

The square of the standard deviation is called the *variance*.

The above are simply intended to be explanations of the mean and standard deviation. In practice, the computations can be greatly facilitated by coding and using "arbitrary origins" (7).

The Normal Frequency Distribution

It can be said without qualification that in statistics the most important of all frequency distributions is the "normal" distribution. Many data in real life appear to follow closely the form of a normal distribution. Typical examples are heights of men, intelligence test scores, errors of measurement, and lengths of life of electric light bulbs.

[5] In computing the standard deviation of a finite set of data, many statisticians prefer to divide by n-1 instead of n, where n is the number of cases. This eliminates some bias and simplifies certain calculations.

If the original data are not normal, often a logarithmic or other transformation can be found that will yield a normal distribution. This frequent occurrence of normal distributions in real life is what statistical theory would lead us to expect, since it can be shown on logical grounds that the normal distribution is the appropriate form for many natural phenomena.

In addition to its appearance in real life, the normal distribution is very important in statistical theory itself. Much of the exact theory of sampling that has been developed in recent years assumes that the items in the population follow a normal distribution. Indeed, little theory has been developed for populations that are not normal.

A normal distribution can be quite precisely defined. If the distribution of a variable (x) is such that the relative frequency with which x takes on the values less than or equal to x_0 is given by

$$\int_{-\infty}^{x_0} \frac{1}{\sigma\sqrt{2\pi}} e^{-(x-\mu)^2/2\sigma^2}\, dx,$$

then x is said to be normally distributed (10). Here μ is the mean of the distribution and σ is its standard deviation. A picture of a normal frequency curve is shown in Figure 3. The relative frequency of x equal

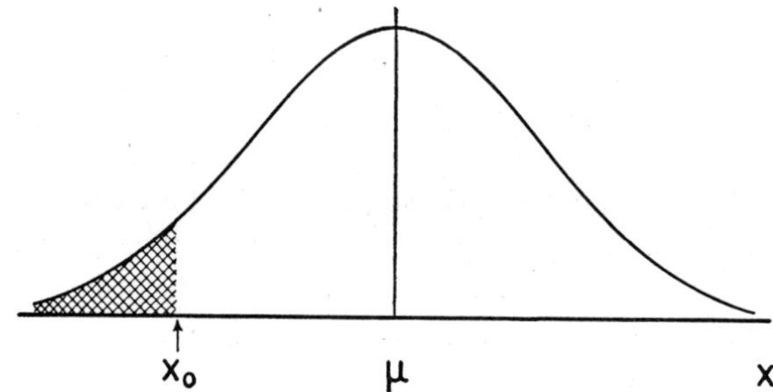

Figure 3. The normal frequency distribution.

to or less than x_0 is given by the area under the curve from $-\infty$ to x_0. This is the graphic counterpart of the above integral.

If we make a transformation of x, viz., $z = \frac{x - \mu}{\sigma}$, we shall get a new variable that has a zero mean and a unit standard deviation. The distribution of z is

$$\text{Relative frequency of } z \leq z_0 = \int_{-\infty}^{z_0} \frac{1}{\sqrt{2\pi}} e^{-z^2/2}\, dz,$$

which is independent of the mean μ and standard deviation σ. The variable z is called the *standard normal deviate* and the frequency distribution of z is the *standard normal frequency distribution.* All normal distributions can be reduced to this standard form by transforming the variable from x to z (i.e., by measuring x from its mean and expressing it in standard deviation units). Since all normal distributions can be reduced to standard form, it is possible to construct a single table of relative frequencies that can be used for all normal distributions. Table 5 is a condensed form of such a table. An example will explain its use.

Suppose x is known to be normally distributed with a mean equal to 100 and a standard deviation equal to 10. Then to find the relative fre-

Table 5. A Condensed Table of the Standard Normal Frequency Distribution*

z	Relative Frequency less than or equal to z	z	Relative Frequency less than or equal to z
0	0.5000	1.7	0.9554
0.1	0.5398	1.8	0.9641
0.2	0.5793	1.9	0.9713
0.3	0.6179	1.960	0.9750
0.4	0.6554	2.0	0.9772
0.5	0.6915	2.1	0.9821
0.6	0.7257	2.2	0.9861
0.7	0.7580	2.3	0.9893
0.8	0.7881	2.326	0.9900
0.9	0.8159	2.4	0.9918
1.0	0.8413	2.5	0.9938
1.1	0.8643	2.6	0.9953
1.2	0.8849	2.7	0.9965
1.282	0.9000	2.8	0.9974
1.3	0.9032	2.9	0.9981
1.4	0.9192	3.0	0.9987
1.5	0.9332	3.090	0.9990
1.6	0.9452	3.291	0.9995
1.645	0.9500	3.7190	0.9999

* Reproduced with permission from A. M. Mood, *Introduction to the Theory of Statistics* (New York: McGraw-Hill, 1950), p. 423.

quency of x equal to or less than 110, we computed $z_{110} = \frac{x - \mu}{\sigma} = \frac{110 - 100}{10} = 1$, and note from Table 5 that the relative frequency of z equal to or less than 1 is 0.8413. Owing to the symmetry of the distribution the relative frequency of x equal to or less than 90 would be $1.000 - 0.8413 = 0.1587$. Of course, the relative frequency of x between 90 and 110 is $0.8413 - 0.1587 = 0.6826$.

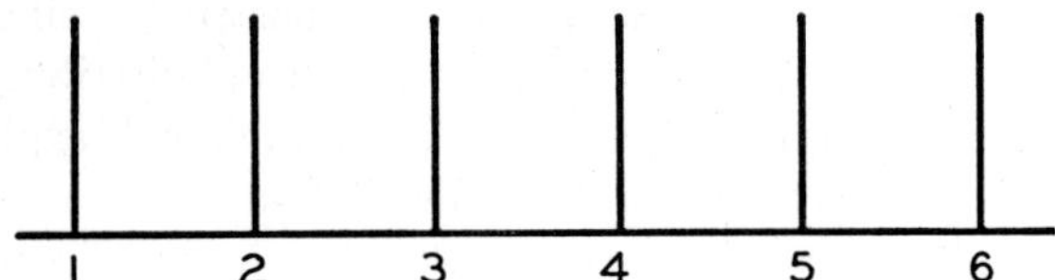

Figure 4. Distribution of results to be obtained with the roll of a single die. (Each ordinate $= \frac{1}{6}$)

The explanation of why the distribution of so many variables is normal or approximately normal in form lies in a very important statistical theorem, known as the *central limit theorem* (2, 10). This says

Table 6. Distribution of Results To Be Obtained with the Roll of Two Dice

Sum	Combination Yielding Specified Sum	Frequency	Relative Frequency
2	11	1	.028
3	12, 21	2	.056
4	13, 22, 31	3	.083
5	14, 23, 32, 41	4	.111
6	15, 24, 33, 42, 51	5	.139
7	16, 25, 34, 43, 52, 61	6	.167
8	26, 35, 44, 53, 62	5	.139
9	36, 45, 54, 63	4	.111
10	46, 55, 64	3	.083
11	56, 65	2	.056
12	66	1	.028
		36	1.000

that if x is the sum of a number of independent variables $y_1, y_2, \ldots, y_n$, then whatever the distributions of the individual y's (subject to certain very general restrictions), the distribution of x approaches the normal form as the number of variables n gets larger and larger. This means that when in the real world a variable is the physical sum of a large number of other independent variables, the distribution of the component variable will be approximately normal.

In some cases, a sum approaches normality very quickly. This may be seen by an examination of the results obtained with rolling dice. If we roll a single die, a 1, or 2, or 3, or 4, or 5, or 6 are all equally likely. The distribution of results obtained with a single die thus follows the rectangular-like distribution shown in Figure 4. When we roll two dice, however, the sum of the two results follows a triangular-like distribution (see Table 6 and Figure 5), and when we roll three dice, the distribution of the sum follows very closely the shape of a normal distribution. (See Table 7 and Figure 6.)

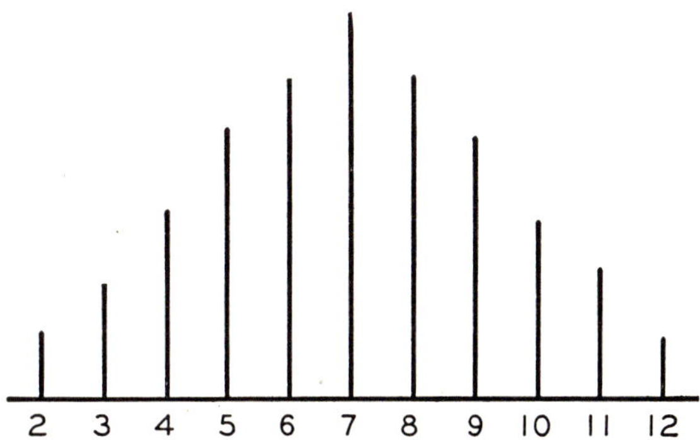

Figure 5. Distribution of results to be obtained with the roll of two dice. (Data from Table 6)

DISTRIBUTIONS OF RESULTS OF RANDOM SAMPLING

The distributions of results of random sampling are of great interest since many decisions are based on sample data. Sometimes the results of sampling can be discovered only by actually taking many samples

Table 7. Distribution of Results To Be Obtained with the Roll of Three Dice

Sum	Combination Yielding Specified Sum	Possible Arrangements of Given Combinations	Frequency of Sum	Relative Frequency of Sum
3	111	1	1	.005
4	112	3	3	.014
5	113	3	6	.028
	122	3		
6	114	3	10	.046
	123	6		
	222	1		
7	115	3	15	.069
	124	6		
	133	3		
	223	3		
8	116	3	21	.097
	125	6		
	134	6		
	224	3		
	233	3		
9	126	6	25	.116
	135	6		
	144	3		
	225	3		
	234	6		
	333	1		
10	136	6	27	.125
	145	6		
	226	3		
	235	6		
	244	3		
	334	3		
11	146	6	27	.125
	155	3		
	236	6		
	245	6		
	335	3		
	344	3		

Table 7. *Continued*

Sum	Combination Yielding Specified Sum	Possible Arrangements of Given Combinations	Frequency of Sum	Relative Frequency of Sum
12	156	6	25	.116
	246	6		
	255	3		
	345	6		
	336	3		
	444	1		
13	166	3	21	.097
	256	6		
	355	3		
	346	6		
	445	3		
14	266	3	15	.069
	356	6		
	455	3		
	446	3		
15	366	3	10	.046
	456	6		
	555	1		
16	466	3	6	.028
	556	3		
17	566	3	3	.014
18	666	1	1	.005
			216	1.000

(the Monte Carlo method), but in many cases the distribution can be derived by mathematical analysis. We shall illustrate this by deriving the sampling distribution of a proportion or fraction.

The Binomial and Hypergeometric Distributions

Let us sample a process in which defective items occur *at random*. Let our samples consist of five items and let the probability of a defec-

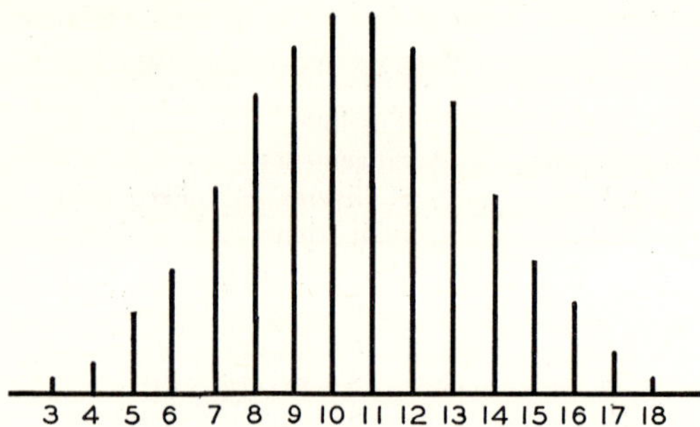

Figure 6. Distribution of results to be obtained with the roll of three dice. (Data from Table 7)

tive item occurring equal p. Then the relative frequency with which we will get (i.e., the probability[6] of getting) 0, 1, 2, 3, 4, or 5 defective items in our samples may be derived as follows. The chance of getting no defective items will be $(1-p)(1-p)(1-p)(1-p)(1-p) = (1-p)^5$ since the defective items are presumed to occur at random. The chance that the first item in a sample will be defective and the rest effective will be $p(1-p)^4$. Altogether there will be five different ways in which one defective item may appear in our sample so that the total probability of getting a sample (any sample) with just one defective item will be $5p(1-p)^4$. Similarly, the probability that the first two items in a sample will be defective and the rest effective will be $p^2(1-p)^3$. There will be $(5 \times 4)/2 = 10$ different ways, however, in which we can get two defective items out of five, so that the total probability of getting a sample with just two defective items will be $10p^2(1-p)^3$. Continuation of this argument leads to the results given in Table 8.

In general, if we take a random sample of n from a process (or infinite population) in which the probability of an item with characteristic A is p, then the probability of getting exactly x out of n sample items with characteristic A is (5):

$$\text{Probability of } x \text{ out of } n = \frac{n!}{x!(n-x)!}\, p^x(1-p)^{n-x}.$$

[6] When we deal with random phenomena and use the relative frequencies in a population to predict relative frequencies in a large number of random samples from the population, we shall identify relative frequency with "probability" (5).

Table 8. Probability of Getting 0, 1, 2, 3, 4, or 5 Defective Items in a Random Sample of 5 from a Process Running 100p Per Cent Defective

No. of Defective Items in Sample	Probability
0	$p^0 (1 - p)^5$
1	$5p (1 - p)^4$
2	$10p^2 (1 - p)^3$
3	$10p^3 (1 - p)^2$
4	$5p^4 (1 - p)^1$
5	$p^5 (1 - p)^0$

This is called the *binomial distribution,* since it is the general term in the expansion of the binominal $(p + q)^n$ where $q = 1 - p$. If we take x/n as the variable, the mean x/n is p and the standard deviation of x/n is $\sqrt{p(1 - p)/n}$. The National Bureau of Standards has published *Tables of the Binomial Probability Distribution* (12) for values of n from 2 to 49 and values of p from 0.01 to 0.90 in steps of 0.01.

Although the binomial distribution is the distribution of a sample proportion when sampling from an infinite population, it can for most practical purposes be taken as the distribution of a sample proportion when sampling from a relatively large finite population. A good practical rule suggests the use of the binomial when the sample is equal to or less than 10 per cent of the population (5).

Exact probabilities for random sampling from a finite population are given by the *hypergeometric distribution* (5). If C_r^N is the number of combinations that can be made of N things in sets of r and if a population of N items contains M defective items, then the probability that a random sample of n items from this population will contain exactly x defective items is:

$$\text{Probability of } x \text{ out of } \quad n = \frac{C_x^M \, C_{n-x}^{N-M}}{C_n^N}.$$

The probabilities of both the binomial and hypergeometric distributions are difficult to compute. Fortunately, good approximate values can be obtained by using other distributions for which computation of probabilities is easier or facilitated by readily available tables. One of these is the normal distribution (5). If $p > 0.10$ and n is such that $np > 5$, the normal distribution does not do badly in computing a sum of binomial or hypergeometric probabilities. The approximation is better, the closer p is to 0.5. The mathematical procedure is to use a normal curve with the same mean and same standard deviation as the binomial or hypergeometric distribution, and when summing for the probabilities of x or less, to compute the area under the normal curve below $x + 0.5$. If $p \leqq 0.10$ and n is large, a good approximation is given by the Poisson distribution, which is discussed in the next section.

The Poisson Distribution

The binomial distribution can be used to explain another distribution that is very important in industrial applications of statistics and in operations research. This is the Poisson distribution, which is named after the French mathematician who developed it. The Poisson distribution is given by the formula

$$\text{Probability of} \quad x = \frac{\lambda^x e^{-\lambda}}{x!}.$$

The Poisson distribution is the mathematical limit of the binomial distribution when in the binomial formula p approaches 0, and n approaches ∞, but the product pn remains equal to a constant λ. This may be proved as follows. We have:

$$\frac{n!}{x!(n-x)!}\,p^x(1-p)^{n-x} = \frac{n(n-1)(n-2) \text{ to } x \text{ factors}}{x!}\,\frac{\lambda^x}{n}\left(1-\frac{\lambda}{n}\right)^{n-x}$$

$$= \frac{1(1-1/n)(1-2/n)\cdots(1-(x-1)/n)}{x!}\,\lambda^x(1-\lambda/n)^{-x}[1-\lambda/n]^{\left(\frac{-n}{\lambda}\right)(\ \lambda)}.$$

If now n approaches ∞, p approaches 0, but λ remains constant, the above approaches

$$\frac{\lambda^x e^{-\lambda}}{x!}.$$

The mean of a Poisson distribution is λ and its standard deviation is $\sqrt{\lambda}$.

Tables of the Poisson distribution have been computed by E. C. Molina of the Bell Telephone Laboratories and are published under the title *Poisson's Exponential Binomial Limit* (8). For values of λ from 0.001 to 100 these tables[7] give the probability of an individual value of x and also the cumulative probability that x will equal or

[7] Molina's tables use a instead of λ.

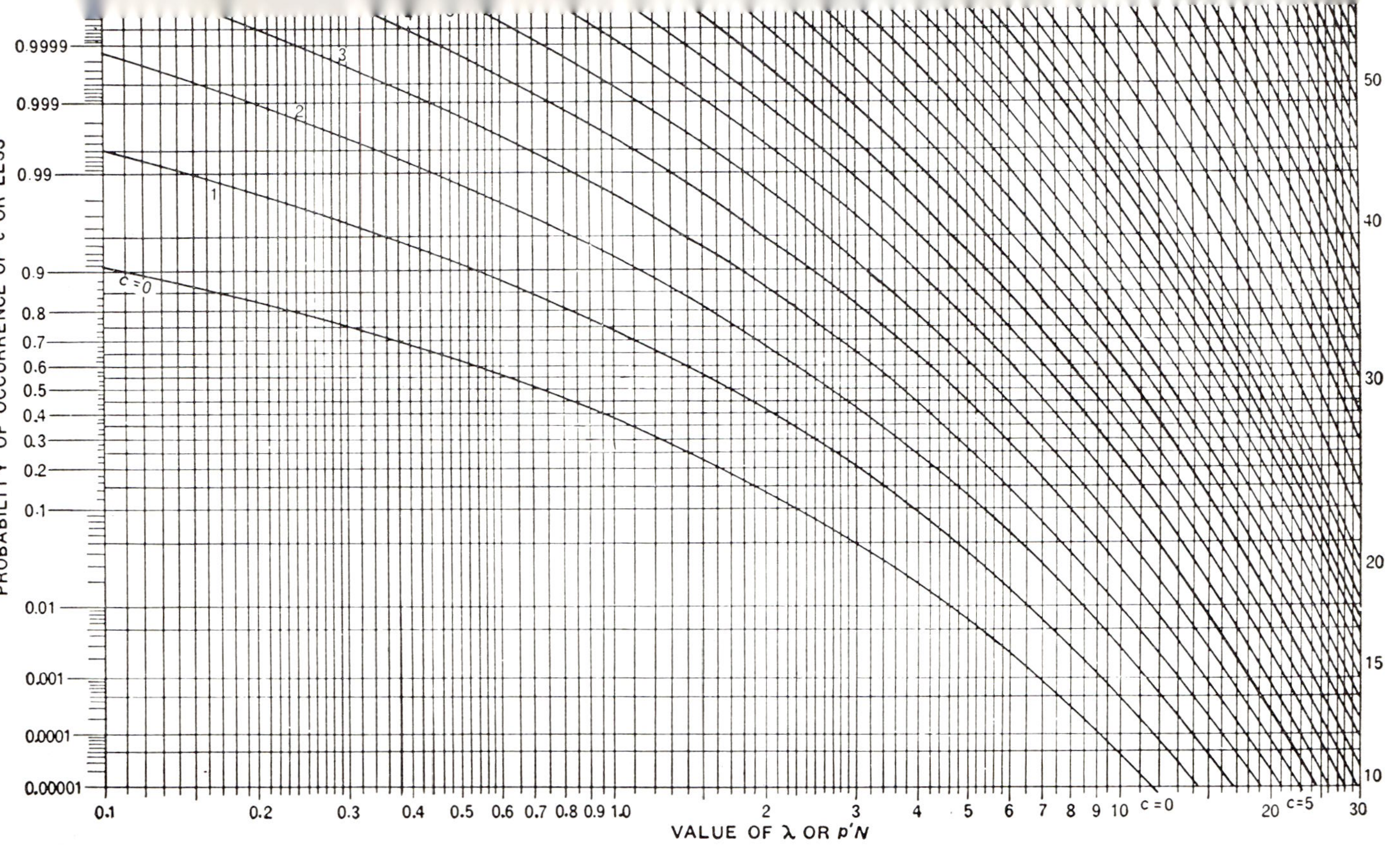

Figure 7. Cumulative curves for Poisson distribution. Reproduced with permission from H. F. Dodge and H. G. Romig, *Sampling Inspection Tables* (New York: John Wiley and Sons, 1944).

exceed a given value c. Figure 7, which is a reproduction of a chart by Miss F. Thorndike, also of the Bell Telephone Laboratories, offers a quick means of determining cumulative Poisson probabilities.

As noted in the previous section, the Poisson distribution is often used as a means of approximating binomial and hypergeometric probabilities when the chance of occurrence of a given event is relatively small (for instance, p is at least less than 10 per cent) and the opportunities for occurrence are large. For example, if we took a random sample of 100 from a large lot in which the proportion of defective items was about 5 per cent, then we could use the Poisson distribution to estimate the probability of getting 2 or less defective items, say, in our sample. (The Thorndike chart gives this immediately as equal to 0.13.) More exact probabilities would be given by the binomial in this instance and the true probabilities would be given by the hypergeometric. The Poisson distribution will nevertheless yield a figure close enough for practical purposes.

The Poisson distribution, however, is more than simply an approximation to the binomial and hypergeometric distributions. Many data in real life tend to follow a Poisson distribution. In target shooting, the number of hits per square inch (or square foot on a larger target) will tend to be distributed in the form of a Poisson distribution. So also will the number of blemishes per table top, the number of defects per television set, and the like. In his chapter on queueing theory, C. D. Flagle shows[8] that if events occur at random over time, the number of events per fixed unit of time will tend to have a Poisson distribution. Thus, if cars arrive randomly at a toll booth at an average rate a, the number arriving per minute, for instance, will fluctuate in accordance with a Poisson distribution with $\lambda = a$.

The Sampling Distribution of a Mean

It will be noted that the mean of a sample of n items is $1/n$ times the sum of these items. Since a mean is thus a constant times a sum and since we have seen above that a sum of random variables tends to be normally distributed, it turns out that sample means tend to be normally distributed. If the population itself is normally distributed, the means of samples therefrom will be exactly normally distributed, but even if the population is nonnormal, sample means will come close to having a normal distribution unless the sample size is small and

[8] See pp. 417, 418.

the population deviates extremely from normal. It can also be shown (1) that the mean of the distribution of sample means is the mean of the population from which the samples are drawn at random, and the standard deviation of this distribution of sample means is the standard deviation of the population divided by the *square root* of the sample size. Thus, the distribution of means of samples of four will have a standard deviation that is one-half that of the individual values. Compare Figure 8.

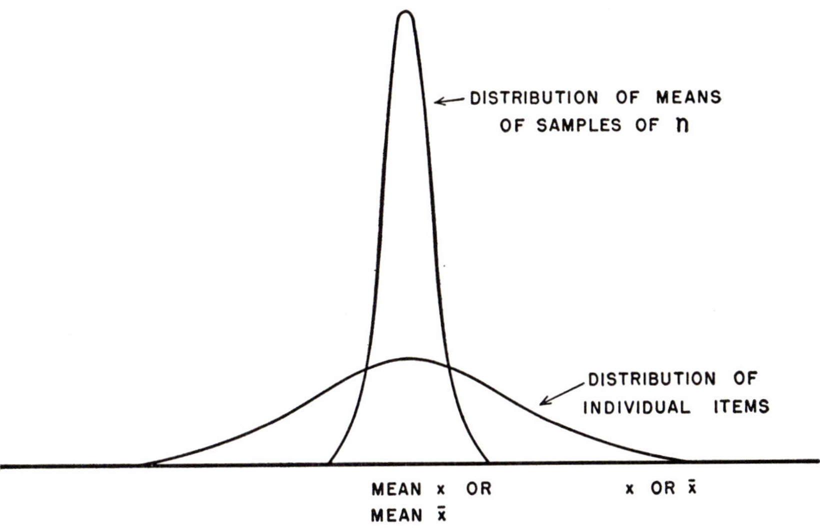

Figure 8. Distribution of individual items and distribution of sample means.

The standard deviation of a distribution of sample means is called the *standard error of the mean.* In general, the standard deviation of the sample values of any specified "statistic" is called the standard error of that statistic.

STATISTICAL INFERENCE

One of the basic problems in statistics is concerned with the inferences that can be made from sample data about the population from which the sample was taken.

It is to be noted, first of all, that the theory of statistical inference is based on *random sampling* and does not apply to samples selected so as to satisfy some individual's notion of representativeness. A random sample may or may not be representative of the population in the sense that it may have the same characteristics as the population.[9] The basic fact about a random sample is that the statistician can usually determine its reliability in the sense that he can compute how much variation may reasonably be attributed to the sampling process. In other words, a statistician can normally compute the standard error of a statistic computed from a random sample. He cannot do the same for a sample whose elements have been selected according to some plan or purpose.

To assure himself that a sample is a random one, the statistician usually makes use of random numbers. Several tables of random numbers have been constructed, one of the largest being the Rand Corporation's *A Million Random Digits with 100,000 Normal Deviates* (11). Random numbers can also be generated by large computing machines. Random numbers should be used to draw random samples whenever it is in any way possible to do so. Considerable ingenuity may sometimes be required to accomplish this, but the effort and expense are usually worth it. The use of random numbers has been recognized by the American courts, whereas other methods of procuring random samples may have to be justified by special pleading.

Confidence Intervals

One of the basic notions in statistical inference is that of a confidence interval. A confidence interval is a range of values that is stated to contain the value of some population parameter, the interval in any given instance being computed from a specific sample taken at random from the population (5, 7). The probability that the statement is true is the *confidence coefficient* associated with the given confidence interval.

The simplest of all examples of confidence intervals is that for the mean of a normal population of which the standard deviation is known. Thus, suppose we make measurements with a given gage the precision of which is known from past use. To be specific, suppose that the measurements made by the given gage are known to have a standard deviation of 0.002 (regardless of what is being measured). Then if we take four

[9] Of course, the larger the sample, the more representative it is likely to be.

measurements of the same thing, e.g., 3.153, 3.159, 3.152, and 3.156 and compute the mean $\left(\frac{3.153 + 3.159 + 3.152 \text{ and } 3.156}{4} = 3.155\right)$, we can make the statement that the interval $3.155 \pm 1.960 \frac{(0.002)}{\sqrt{4}} = \begin{cases} 3.15696 \\ 3.15304 \end{cases}$ will have a 0.95 chance of including the "true value" (i.e., the mean of an infinite set of such measurements). It is assumed here that the "errors" of measurement are randomly and normally distributed.

In general, if the mean of a random sample of n values from a normal population is $\bar{x}$ and if the population is known to have a standard deviation of σ, then the 0.95 confidence interval for the mean of the population will be given by $\bar{x} \pm 1.960\, \sigma/\sqrt{n}$. The proof of this runs as follows.

It was noted in an earlier section that if we take many random samples from a normal population with mean μ and standard deviation σ, the means of these samples will be distributed in the form of a normal distribution with a mean of μ and a standard deviation (standard error) of $\sigma/\sqrt{n}$. Thus from Table 5 we have:

$$\text{I. Probability}\left(-1.960 \frac{\sigma}{\sqrt{n}} \leq \bar{x} - \mu \leq 1.960 \frac{\sigma}{\sqrt{n}}\right) = 0.95.$$

This probability statement can be written in two parts, however. Thus

$$\text{II. Probability}\left(\mu \leq \bar{x} + 1.960 \frac{\sigma}{\sqrt{n}} \textit{ and } \bar{x} - 1.960 \frac{\sigma}{\sqrt{n}} \leq \mu\right) = 0.95.$$

But this is the same as saying that

$$\text{III. Probability}\left(\bar{x} - 1.960 \frac{\sigma}{\sqrt{n}} \leq \mu \leq \bar{x} + 1.960 \frac{\sigma}{\sqrt{n}}\right) = 0.95.$$

Statement III is the confidence interval statement. The quantity 0.95 is the confidence coefficient which yields the factor 1.960 and the sample result

$\bar{x} - 1.960 \frac{\sigma}{\sqrt{n}}$ is the *lower 0.95 confidence limit* and the sample result

$\bar{x} + 1.960 \frac{\sigma}{\sqrt{n}}$ is the *upper 0.95 confidence limit.*

If we do not know the standard deviation of the population, but take a random sample of over thirty items, we can set up a confidence interval for μ as in the known σ case by substituting the sample standard deviation for σ. If samples are smaller than thirty, we must make use of other factors than 1.960. These factors will be found in any statistics text that lists the percentage points of the t-distribution (also known as Student's distribution).

Confidence intervals can be computed for most population parameters. If samples are large, 0.95 confidence intervals will be given approximately by the corresponding sample statistic plus and minus twice the estimated standard error of that statistic. For more precise methods the reader is referred to (10). It will suffice here to point out that if a sample is large (and the population at least ten times larger), 0.95 confidence limits for the proportion (p) of a population having a specified characteristic may be computed (5) from the formula: Confidence limits for p will be given approximately by

$$\frac{x}{n} \pm 2\sqrt{\frac{\left(\frac{x}{n}\right)\left(1-\frac{x}{n}\right)}{n}}$$

If the population is not at least ten times the sample, a better formula is

$$\frac{x}{n} \pm 2\sqrt{\frac{\left(\frac{x}{n}\right)\left(1-\frac{x}{n}\right)}{n}\left(1-\frac{n}{N}\right)}$$

where n is the size of the sample and N is the size of the population. For valid use of these formulas, the sample must of course be random and p must not be too small. A good rule is that pn should not be less than 5. For moderate values of p, somewhat better results will be given by Figure 9. An excellent set of tables for confidence intervals for p is given in (9).

Statistical Tests of Hypotheses

Another important aspect of statistical inference is concerned with tests of hypotheses about a population. The procedure is to assume the hypothesis is true and to lay off reasonable limits for chance variation. If a specified sample exceeds these limits, the hypothesis is rejected (5). The limits are so set that the risk of rejecting the hypothesis when it is true is limited to a predetermined quantity α.

If a hypothesis pertains to a specific parameter of a population, it

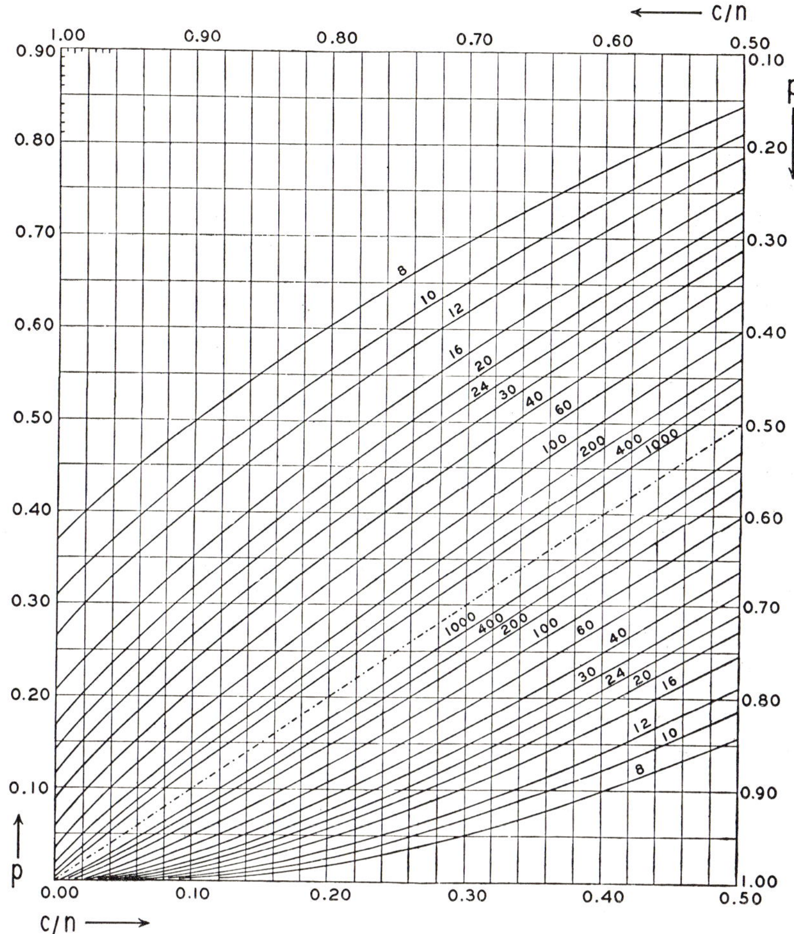

Figure 9. Chart for determining 0.95 confidence limits for a universe proportion for samples of size 10 to 1000. Reproduced with permission from E. S. Pearson and H. O. Hartley, *Biometrika Tables for Statisticians* (Cambridge: Cambridge University Press, 1958), p. 204.

Note: The numbers printed along the curves indicate the sample size n. If for a given value of the abscissa c/n, p_A and p_B are the ordinates read from (or interpolated between) the appropriate lower and upper curves, then

$$\Pr\{p_A \leq p \leq p_B\} \leq 1 - 2\alpha.$$

can usually be tested by computing the confidence interval for that parameter. For example, a hypothesis pertaining to a population mean or a population proportion can be tested by computing the confidence interval for this mean or proportion and noting whether it includes the hypothetical value. If it does not, the hypothesis is rejected. If the confidence coefficient for this confidence interval is $1 - \alpha$, the risk of falsely rejecting the hypothesis in these instances is α.

Nonparametric tests can also be made. A type of hypothesis that is often of interest in operations research is whether a given population is distributed in a particular way. We may want to know, for example, whether a certain variable has a binomial distribution; or another, a Poisson distribution; or a third variable, a normal distribution. Such hypotheses are usually tested by carrying out a χ^2-test of goodness of fit (5, 7). By way of illustration we shall apply a χ^2-test to determine whether the variable of Table 2 could reasonably be believed to follow a Poisson distribution.

A χ^2-test of goodness of fit compares the actual frequencies obtained with those expected on the basis of the assumed distributional form. The mean of the data of Table 2 is 0.72. If the variable in general follows the Poisson form with $\lambda = 0.72$, we would have expected the frequencies to have been those given in Table 9. The χ^2-test consists

Table 8. Poisson Frequencies[1] for $\lambda = 0.72$ and $n = 50$

Number of Defects	Number of Sheets Expected on Basis of Hypothesis that Distribution is Poisson in Form
0	24.4
1	17.5
2	6.3
3 or more	1.8

$$\sum \frac{(F_i - f_i)^2}{f_i} = 1.64$$

$\chi^2_{0.05}$ (for $d.f. = 2$) $= 5.99$.

Poisson hypothesis not rejected since $1.64 < 5.99$.

[1] Derived from Molina's tables with a taken equal to 0.72. The theoretical frequencies are expressed as decimals since they are the means around which individual sample values would fluctuate.

in comparing these theoretical frequencies with those actually shown in Table 2. The comparison index is $\sum \frac{(F_i - f_i)^2}{f_i}$ where F_i is the actual frequency for the ith class and f_i is the theoretical frequency. This is computed in Table 10 for the data of Table 2. The actual test consists

Table 10. Illustration of a χ^2 Test of Goodness of Fit: Comparison of Poisson Expected Frequencies with the Actual Frequencies of Table 2.

No. of Defects	F_i	f_i	$(F_i - f_i)$	$(F_i - f_i)^2$	$(f_i - F_i)^2/f_i$
0	23	24.4	−1.4	1.96	0.08
1	21	17.5	3.5	12.25	0.70
2	4	6.3	−2.3	5.29	0.89
3 or more[1]	2	1.8	0.2	0.04	0.02
Totals	50	50.0			1.64

[1] The last class is expressed as 3 or more, since the Poisson distribution runs to ∞.

in noting how the computed value of $\sum_i \frac{(F_i - f_i)}{f_i}$ stands in relation to a special critical value that comes from a table of the χ^2 distribution. (Hence the name χ^2 test.) The reader is referred to (1) or any statistics textbook that discusses χ^2 tests. It will suffice here to note that for the data of Table 10, the critical χ^2 is 5.99. Since our sample value of $\sum_i \frac{(F_i - f_i)^2}{f_i} < 5.99$, we do not reject the hypothesis that the variable of Table 2 has a Poisson distribution. The critical χ^2 was so chosen in this case that the probability of falsely rejecting the hypothesis is 0.05.

DECISION THEORY

Historically the theory of testing hypotheses led to decision theory. When we test hypotheses we usually decide either to accept or reject the hypothesis. (In some cases we may not accept immediately but may reserve judgment pending the acquisition of more data.) In the chapter on "Statistical Quality Control" we shall discuss sampling inspection in which decisions are made to accept or reject lots on the basis of random samples. Decision theory is primarily devoted to finding optimum procedures for making decisions involving variables subject to random variation. For details, the reader is referred to *Statistical Decision Functions* by A. Wald (13).

REGRESSION ANALYSIS

This chapter will be concluded with a brief account of two statistical procedures that have been found very useful in industrial research. These are regression analysis and analysis of variance.

Two Variable Linear Regression

Figure 10 shows a scatter diagram of measures of printability of newsprint and its smoothness.[10] Although there is considerable scatter, the reader will note a general tendency for printability to improve with smoothness. He will also note that this relationship appears to be a linear one. The line or curve that shows how the *mean value* of one variable (y) changes with values of another variable (x) is called the line or curve of regression of y on x. When this takes the form of a straight line, the regression is said to be linear.

The word regression stems from the particular problem that the regression technique was invented to solve. In the latter part of the nineteenth century the English scientist Sir Francis Galton became interested in the theory of evolution and especially its relationship to human characteristics (5). The question he posed was this. If two tall parents tend to have still taller children and two short parents tend to have still shorter children, why has not the human race tended over the centuries to breed giants on the one hand and dwarfs on the other? He sought an answer to this question by collecting data on the average heights of parents and the average heights of their fully grown children. What he found was that the line showing the relationship between the average height of the parents and the average height of the children was sloped not at 45 degrees, but at a smaller angle. In other words, heights of the offspring of tall parents were on the average less than the average heights of their parents, while the heights of offspring of short parents were on the average taller than the average heights of their parents. Galton thus concluded that the height of

[10] Smoothness is the negative of the measure of roughness.

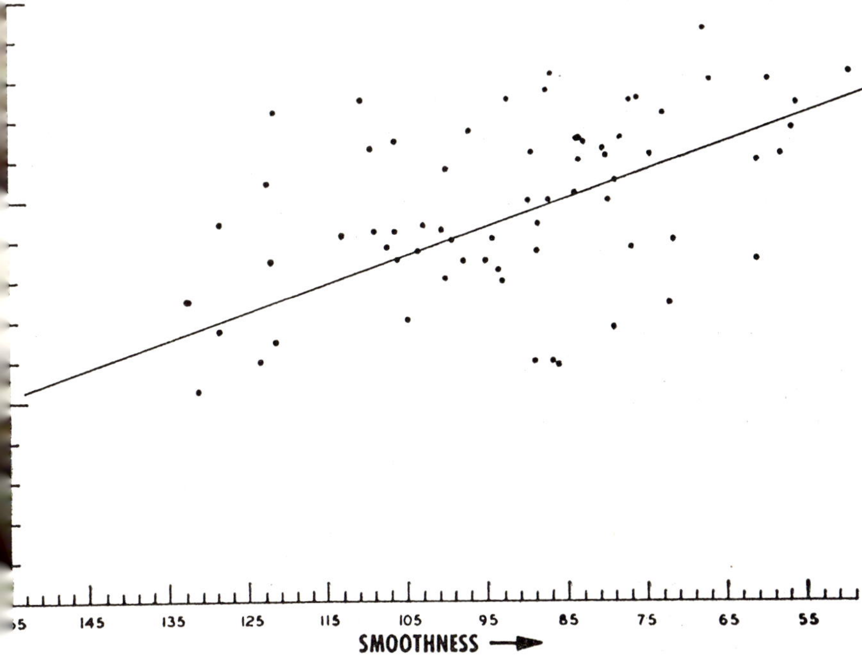

Figure 10. Effect of smoothness on printing quality.

offspring tended "to regress" back to the average height of the race. Ever since this ground breaking work of Galton, lines of average relationships in statistics are called lines of regression.

USE OF A LINE OF REGRESSION

Before a discussion of the problem of estimating a line of regression from sample data, consider what kind of analysis is made possible by knowledge of a line of regression. To be concrete, suppose that printability of newsprint is related to its smoothness by the equation:

Printability = 60 + 0.07Smoothness + a random variable (ϵ) that is independent of Smoothness.

Let the mean of ϵ be zero and let its standard deviation be σ. Under these conditions we would have:

Mean Printability for a Given Smoothness = 60 + 0.07Smoothness.

This last is the equation for the line of regression of Printability on Smoothness.

The first use that can be made of the above equation is to determine mean Printability from Smoothness. Thus, if a given sheet of newsprint had a Smoothness measure of 100, we would "estimate" its Printability to be 67. This is an "estimate" in that it refers to the mean Printability of sheets that have Smoothness measures of 100. If, however, we know σ, the standard deviation of the random variable ϵ, and if we feel justified in making the additional assumption that ϵ is normally distributed, we can determine limits for the Printability of *the individual sheet* of newsprint for which the Smoothness measured is 100. These limits are given by adding and subtracting 1.96σ from the mean Printability for Smoothness of 100. In other words, if Smoothness measured 100 for a given sheet of newsprint, we could estimate the Printability of that sheet as lying between $67 - 1.96\sigma$ and $67 + 1.96\sigma$, and we would have a 0.95 chance of being right. Estimation of the dependent variable from the independent variable is thus the first use of a line of regression.

Being an instrument for estimation, a line of regression can also be a means of adjusting for the effect of a "nuisance variable." Suppose, for example, we wished to study the effect of certain factors on the performance of a cutting tool. The tool has a limited life and wears more or less uniformly over time. In studying variations in performance, therefore, the wearing of the tool over time should be allowed for by computing the regression of cutting performance on time.

The regression equation also enables us to translate the variability of the independent variable into the variability "induced" thereby in the dependent variable. Thus with reference to the equation given above for the regression of Printability of newsprint on Smoothness, if Smoothness could be improved by 30 points, Printability would be raised on the average by about 2 points ($30 \times 0.07 = 2.1$). The ability of the mills to improve Printability by improving Smoothness is thus quantitatively determined.[11]

If a regression equation has held valid over a past period, we can also tell from it and from supplementary knowledge of σ about what per cent of the variation in the dependent variable has been "statistically" accounted for by the independent variable. Thus if over the

[11] The equation does not, of course, tell us anything about the cost to the mills of improving Smoothness.

past the standard deviation of Smoothness has been 20 units and the standard deviation of ϵ has been 2.5 units, it will turn out that the variance (standard deviation squared) of Printability will be approximately[12]

$$(0.07)^2(20)^2 + (2.5)^2 = 1.96 + 6.25 = 8.21.$$

From this it may be concluded that over the period under review variation in Smoothness accounted for $1.96/8.21 = 23.8$ per cent of the variance of Printability. In other words, if the mills could turn out newsprint that showed no variation in Smoothness whatsoever and conditions otherwise remained the same as in the past, variance of Printability would still be 76.2 per cent of its former value.

ESTIMATION OF REGRESSION PARAMETERS

In practice we do not know the line of regression of one variable on another but have to estimate it from sample data. This immediately gives rise to the statistical problem of how best to effect the estimation. Mathematically the problem of estimating a line of regression is this: Given an equation of the form $y_r = \alpha + \beta x$, how can we best estimate α and β from a given sample of data? Here y_r stands for the regression or mean value of y for a given x.

One way of estimating α and β is to plot the sample data on a diagram like Figure 10 and draw in a line freehand. We could then take the intercept and slope of this freehand line as estimates of α and β. The criterion of goodness of fit is here a visual one and rather subjective.

Another way of estimating α and β is to guess them and then see how good the guess is by comparing the actual values of y for the sample data with the estimates yielded by the regression equation that has the guessed values of α and β. If the comparison is not good, we could guess them again and make a second comparison. Eventually, by this trial and error procedure we would find a pair of values for α and β that would give a good comparison between estimated and actual values of y. This method would take considerable time, however, and

[12] It is assumed that the Smoothness component, viz., (0.07)Smoothness, and the chance variable (ϵ) are independent of each other and that the period in question is sufficiently long that 0.07Σ [(Smoothness − Mean Smoothness) *times* (ϵ − Mean ϵ)] is approximately 0.

in the end the question would still go unanswered as to whether the values of α and β finally selected were the best that could have been obtained by this method. We would also have to decide, possibly arbitrarily, in what sense they would be best.

A method of making good estimates of α and β that has found great favor with statisticians is the *method of least squares.* This says that those values of α and β are best that, for the given set of sample data, minimize the sum of the squares of the differences between the values of y in the sample and those given by the regression equation with the selected values of α and β. Hence the name "method of least squares."

The use of the least squares criterion for estimating α and β leads to rather simple algebraic results. Thus, suppose we have the following:

Actual y's	Estimated y's $= y_r = \alpha + \beta x$
y_1	$y_{r1} = \alpha + \beta x_1$
y_2	$y_{r2} = \alpha + \beta x_2$
.	. .
.	. .
.	. .
y_n	$y_{rn} = \alpha + \beta x_n$

Then the sum of the squared differences may be written

$$(y_1 - \alpha - \beta x_1)^2 + (y_2 - \alpha - \beta x_2)^2 + \cdots + (y_n - \alpha - \beta x_n)^2$$

or more succinctly $\sum_i (y_i - \alpha - \beta x_i)^2$. If we differentiate this with respect to α and β and set the derivatives equal to 0, we get the two simultaneous equations:

$$\sum_i y_i = na + b \sum_i x_i$$

$$\sum_i x_i y_i = a \sum_i x_i + b \sum_i x_i^2$$

where a and b are the values of α and β that satisfy the equations. The quantities a and b are known as the least squares estimates of α and β.

Least squares estimates are good in that they are unbiased (1). This means that over many samples the various estimates of a and b would have mean values equal to the true values, α and β. It can also be shown that in many samples, all of which have the same set of x values, but different random components (ϵ), the distribution of a and b will be normal in form and have standard deviations (i.e., "standard errors") given by:

$$\sigma_a = \sigma \sqrt{\frac{1}{n} + \frac{\bar{x}^2}{n\sigma_x^2}}$$

$$\sigma_b = \frac{\sigma}{\sigma_x \sqrt{n}}$$

where $\bar{x}$ and σ_x are the mean and standard deviation of the specified "fixed" set of x's. (See (10).) For a single sample, the x's are, of course, the x's for that sample. The value of σ can be estimated[13] by calculating $\frac{\Sigma\,(y - a - bx)^2}{n}$. From estimates of the standard errors, approximate 0.95 confidence limits for α and β can be computed from:

$$a \pm 2\sigma \sqrt{\frac{1}{n} + \frac{\bar{x}^2}{n\sigma_x^2}}$$

$$b \pm 2\sigma / \sigma_x \sqrt{n}$$

where σ is the sample estimate of σ.

The least squares method thus yields both "point" estimates of α and β, and confidence limits for them. It is customary when writing the equation for an estimated line of regression to put the standard errors immediately under the coefficients to which they apply. Thus, in general we would write:

$$\begin{array}{ccccc} y_r & = & a & + & bx \\ & & (\pm\sigma_a) & & (\pm\sigma_b) \end{array}$$

Of course, when we use an estimated regression equation to predict an individual value of y, we must not only allow for the deviation of the individual from the mean of such individuals (a factor similar to the $\pm 1.96\sigma$ factor used when α, β and σ were known), but must also allow for the sampling error in the estimates of α and β. In general, for a sample size of 30 or more, 0.95 *prediction limits* for an individual y will be given (5) approximately by:

$$a + bx_1 \pm 2\sigma \sqrt{\frac{1}{n} + \frac{(x_1 - \bar{x})^2}{n\sigma_x^2} + 1}$$

where x_1 is the x from which the y value is being predicted.

[13] A shorter procedure is to estimate σ^2 from $\sigma^2 = \frac{1}{n}\,[\Sigma\, y^2 - (\Sigma\, y)^2/n - b\,\Sigma\, xy]$. It is assumed throughout the discussion that n is reasonably large, say, greater than 30. For smaller n a division of $n - 2$ would be used in estimating σ^2.

Over the data from which the values of α and β are estimated, the fraction of the variance in y that is statistically accounted for by its linear relation with x is given (1) by:

$$r^2 = \frac{[\Sigma (x - \bar{x})(y - \bar{y})]^2}{n\sigma_x^2\sigma_y^2}.$$

The statistic r is the sample coefficient of correlation.

Regressions of More Than Two Variables

When a regression equation contains more than two variables, the analysis becomes more complicated, but the principles remain the same. A regression equation in which the dependent variable y is expressed as a linear function of several variables $x_1, x_2, \ldots x_k$ can be used to predict y from the x's. Over past data it can also be used to determine how much of the variation in y can be statistically accounted for by each x individually and jointly with each of the other x's and also by all the x's acting in combination. If the sample regressions are estimated by least squares, we can compute standard errors for each of the estimates of the regression parameters and also prediction limits for individual y's estimated from the sample regression equation. We can also calculate a sample *multiple correlation coefficient,* which indicates (when squared) how much of the variation in y in the past has been statistically accounted for by the x's as a group. We can also calculate sample β *coefficients* which can be used to estimate the influence of the x's individually and in pairs. For details, the reader is referred to (1, 4, and 5).

In recent years considerable interest has developed in the use of regression equations involving second degree terms to map out response surfaces. See (3) and (5).

ANALYSIS OF VARIANCE

Analysis of variance may be viewed as a special variation of regression analysis (1). In standard regression analysis, the dependent variable x is measurable. It is possible, however, to view "values of x" as simply different qualitative categories of a specified classification. For example, suppose we are interested in a measurable characteristic (y) of the output of c different machines. Let the machines have certain biases that contribute differential effects $\theta_1, \theta_2, \ldots \theta_c$ to the y values of their outputs. Then we can set up a regression equation:

$$y_{ij} = \mu + \theta_1 + \theta_2 + \cdots + \theta_c + \epsilon_{ij} \qquad \begin{matrix} i = 1, 2, \cdots r \\ j = 1, 2, \cdots c \end{matrix}$$

in which μ is an over-all constant (corresponding to α in the previous section), and ϵ is a random variable representing effects on y of factors independent of the machine differentials $\theta_1, \theta_2, \ldots \theta_c$. In this formulation of the relationship, the differential θ_j only enters into the equation when the output of machine j is under consideration. If we fit such an equation by least squares, we can estimate all of the parameters μ, θ_1, $\theta_2, \ldots \theta_c$ and also the standard deviation of ϵ. If the ϵ_{ij} are assumed to be independently normally distributed, we can test the hypothesis that the θ_j's are all individually 0, which, in other language, is the hypothesis that the machines all operate alike with respect to their average effects on the y characteristic.

Analysis of variance equations can include other class differentials $\gamma_1, \gamma_2, \ldots \gamma_d, \tau_1, \tau_2, \ldots \tau_e$, and the like, as well as the θ differentials, and it is possible to test for the existence of each set of class differentials independently. It is also possible that an analysis of variance equation will contain *interaction* elements that allow for the joint effects of two or more categories.

Analysis of variance is closely tied to the design of experiments in that a particular experimental design has associated with it a particular analysis of variance model. A sizeable block of statistical theory is devoted to estimating and testing hypotheses about the parameters in the analysis of variance models associated with different experimental designs. See Chapter 18; also see (1) and (3).

REFERENCES

(1) Anderson, R. L., and Bancroft, T. A. *Statistical Theory in Research.* New York: McGraw-Hill, 1952.

(2) Cramer, Harold. *Mathematical Methods of Statistics.* Princeton: Princeton University Press, 1946.

(3) Davies, Owen L. (ed.). *The Design and Analysis of Industrial Experiments.* London: Oliver and Boyd, 1954.

(4) Davies, Owen L. (ed.). *Statistical Methods in Research and Production, with Special Reference to the Chemical Industry.* London: Oliver and Boyd, 1957.

(5) Duncan, Acheson J. *Quality Control and Industrial Statistics*. Homewood, Ill.: Richard D. Irwin, Inc., 1959.

(6) Statistical Research Group, Columbia University. *Techniques of Statistical Analysis*. New York: McGraw-Hill, 1947.

(7) Freund, Richard D. *Modern Elementary Statistics*. New York: Prentice-Hall, 1952. (A non-mathematical text for beginners.)

(8) Molina, E. C. *Poisson's Exponential Binomial Limit*. New York: D. Van Nostrand, 1949.

(9) Mainland, D., Herrera, L., and Sutcliffe, M. I. *Tables for Use with Binomial Samples*. New York: New York University College of Medicine, 1956.

(10) Mood, Alexander M. *Introduction to the Theory of Statistics*. New York: McGraw-Hill, 1950.

(11) The Rand Corporation. *A Million Random Digits with 100,000 Normal Deviates*. Glencoe, Ill.: The Free Press, 1955.

(12) U. S. Department of Commerce, National Bureau of Standards. *Tables of the Binomial Probability Distribution*. Washington, D. C.: U. S. Government Printing Office, 1949.

(13) Wald, A. *Statistical Decision Functions*. New York: John Wiley and Sons, 1950.

Ten

STATISTICAL QUALITY CONTROL [1]

ACHESON J. DUNCAN

HISTORY

One of the interesting developments in recent decades has been the application of statistical techniques to problems of industrial management. A prominent instance of this has been the growth of statistical quality control.

Statistical quality control in the United States may be said to have begun in 1924 when Dr. Walter A. Shewhart, of the Bell Telephone Laboratories, made his first sketch of a "control chart" (15). The new technique was developed in several articles and memoranda, and in 1931 Dr. Shewhart published his book on the *Economic Control of Quality of Manufactured Product*. This was the period when, through public lectures in the United States and England, Dr. Shewhart sowed the seeds of subsequent interest. In this same period Dr. Harold F. Dodge and Dr. Harry G. Romig, also of the Bell Telephone Laboratories, developed their ideas on sampling inspection. The work of these three men, together with that of Leslie E. Simon in Army Ordnance, may be said to have laid the foundations for modern statistical quality control.

Like all new developments, the use of statistical quality control spread very slowly at first. The major force leading to more ready adoption was World War II. American entry into the conflict presented

[1] In writing this chapter the author has drawn heavily from his book, *Quality Control and Industrial Statistics* (2nd ed.; Homewood, Ill.: Richard D. Irwin, Inc., 1959).

the federal government with a vast logistics problem. Large quantities of war materiel were needed in a hurry and it was necessary that it be of good quality. This pressure led the federal government to bring to Washington a number of men who had been successfully applying statistical quality control in the Bell System. These men set up scientific sampling inspection plans, which in turn put pressure on suppliers to adopt statistical quality control procedures in their manufacturing operations. This economic pressure was supplemented by an educational program under which short courses in statistical quality control were conducted throughout the country. At the request of the War Department, the American Standards Association initiated a project that eventuated in the publication of the *American War Standards* Z1.1-1941 and Z1.2-1941, "Guide for Quality Control and Control Chart Method of Analyzing Data," and *American War Standard* Z1.3-1942, "Control Chart Method of Controlling Quality during Production." This wartime pressure led to a very rapid growth in the United States in the use of statistical quality control.

Toward the end of World War II and shortly thereafter, interested groups of men in different cities established local "quality control" societies, and in Buffalo the Society of Quality Control Engineers, in co-operation with the University of Buffalo, began in 1944 the publications of *Industrial Quality Control.* In 1946 these local quality control societies were brought together to form the American Society for Quality Control. The American Society took over the publication of *Industrial Quality Control* and became the primary agency in the western hemisphere for training and interchange of knowledge in the new statistical techniques. By 1958 it had over 10,000 members.

The federal government has continued its interest in statistical quality control by promoting theoretical research in new methods and in the issuance of government standards and handbooks. The latter have not only been directives for use by government inspectors but have also become models for many private inspection procedures.

SAMPLING INSPECTION

There may be said to be three major aspects of statistical quality control: sampling inspection, process control, and research. Each of these will be discussed briefly in turn.

Criticisms of Old Methods

Until the rise of the new statistical procedures, industrial product was either inspected 100 per cent or sample inspected on a proportional basis. Proportional inspection usually consisted of taking a flat percentage sample and accepting the lot only if the sample contained a very small (possibly zero) percentage of defective items. Under the new procedures, the aims and objectives of inspection are clearly defined and methods are employed under which the risks of various decisions can be mathematically determined.

Criticism of the old method of 100 per cent inspection runs as follows: (1) It is pointed out that 100 per cent inspection is slow and costly; (2) investigation has shown that as frequently conducted, 100 per cent inspection is far from perfect inspection. Because of the large volume of inspection required under this system, the work becomes monotonous and boring and the temptation to use short cuts is strong. As a result there is little assurance that 100 per cent inspection prevents all defective material from getting by; (3) finally, it is argued that reliance by management and production departments on final 100 per cent inspection may lead to careless production attitudes and procedures. This leads to production of many defective items which are not caught by final inspection and hence get out to customers. It also means a large percentage of scrap and rework in connection with the defective items that are caught.

Criticism of the flat percentage method of sampling arises primarily from the fact that it does not give constant risks. Samples of 500 from a lot of 5,000 will give much better protection against accepting bad material than samples of 50 from lots of 500. The new statistical procedures employ plans that give relatively constant risks for lots of the same quality whatever the lot size; or, if variation in sample size with lot size is permitted (to make for more stable inspection costs), the variation is such as to divide the increase or decrease in risk between the supplier and the consumer. Whatever procedure is adopted, the point of view is that risks be intelligently appraised.

Lot-by-Lot Acceptance Sampling

Modern sampling inspection plans are basically of two types. Under one type, a decision is made simply to accept or reject. Such a scheme is usually applied to material submitted in lots and is known as lot-by-lot acceptance sampling. If a lot is rejected it is returned to the supplier; or, if the customer's inspection is carried out in the supplier's plant, the lot is simply not shipped to the customer.

The second type of sampling inspection plan includes, in addition to a decision to accept or reject, provision for further inspection of rejected material. The cost of this extra inspection may be borne by the supplier, directly or indirectly, through a price concession. This second type of sampling inspection plan is usually called rectifying inspection. Rectifying inspection plans are discussed below.[2]

SINGLE SAMPLING PLANS

Lot-by-lot acceptance sampling is most easily explained with reference to single sampling plans. A single sampling plan is defined by a sample size n and an acceptance number c. For example, we may speak of the sampling plan $n = 100$, $c = 2$. This means that as lots are submitted for inspection a *random* sample of 100 is taken from each lot. If 2 or less of these 100 items are defective, the lot is accepted; if 3 or more are defective, the lot is rejected. Although tables of single sampling plans may recommend a change of sampling plan with wide variations in lot size, there is no provision in the sampling plan itself for varying the sample size with the lot size. Thus, under the plan $n = 100$, $c = 2$, a sample of 100 items would be taken whether a lot numbered 800 or 1,200.

All acceptance sampling plans have associated with them an *operating characteristic curve* or OC curve, as it is called. This gives the risk of accepting a lot as a function of the quality of the lot being submitted for inspection. The OC curve for the single sampling plan $n = 100$, $c = 2$ is shown in Figure 1. This shows, for example, that the probability of accepting a lot that is 1 per cent defective is about 0.92, while the probability of accepting a lot that is 5 per cent defective is about 0.13. This means that if a stream of lots, all 1 per cent defective, are submitted for inspection under this plan, about 92 per cent of them will be accepted and 8 per cent rejected. On the other hand, if a stream of lots, all 5 per cent defective, are submitted for inspection, about only 13 per cent will be accepted and 87 per cent rejected.

The OC curve of Figure 1 was derived on the assumption that lots were "infinite" in size. As pointed out in Chapter 9, however, the proba-

[2] See p. 264.

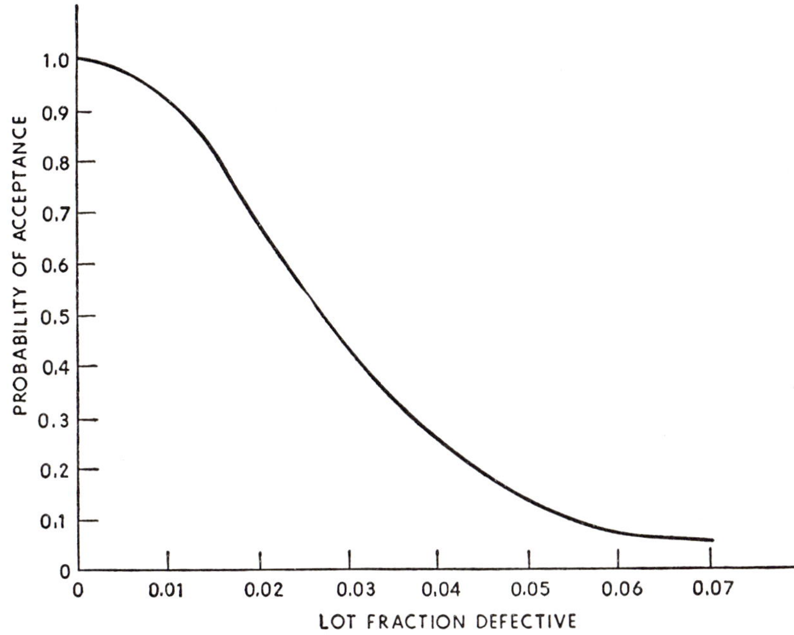

Figure 1. Operating characteristic curve for the sampling inspection plan $N = 100$, $c = 2$. Reproduced with permission from Acheson J. Duncan, *Quality Control and Industrial Statistics*, 2nd ed., (Homewood, Ill.: Richard D. Irwin, Inc., 1959), p. 134.

bilities will be approximately the same for lots that are at least ten times the sample size, or in this case, 1,000 or more. In making the computations, the Poisson distribution was used since the lot per cent defectives were small. It would be a good exercise for the reader to check a few points in Figure 1 by using Figure 7 of Chapter 9.

The OC curve of a single sampling plan is changed as n and c are changed. If c is increased, but n held constant, the plan is relaxed and the OC curve shifts to the right; i.e., there is greater chance of accepting lots of specified quality. If c is reduced, but n held constant, the plan is tightened and the OC curve shifts to the left; i.e., there is less chance of accepting lots of specified quality. (See Figure 2.) Usually, if a company wishes to reduce its risks of acceptance without changing its sampling costs, it reduces c, leaving n unchanged.

If the sample size n and acceptance number c are both changed so as

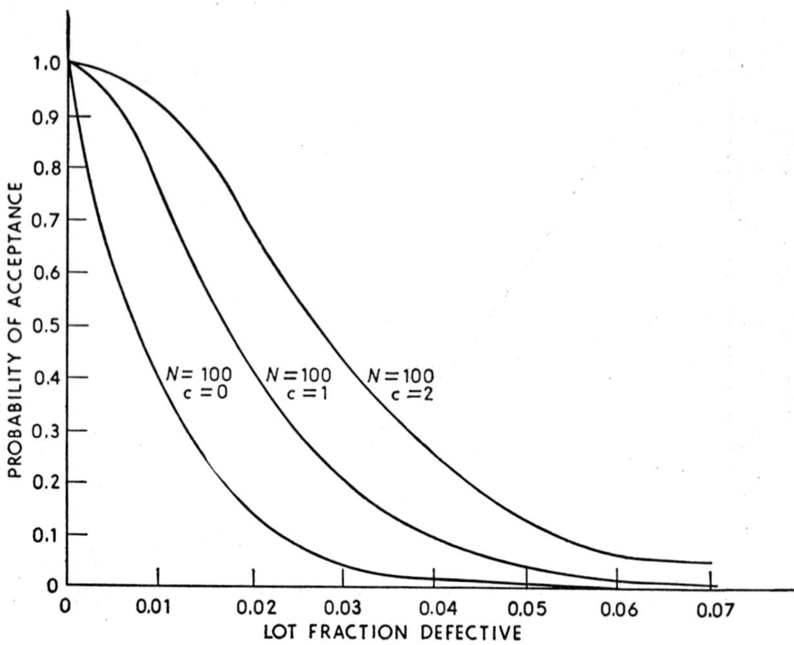

Figure 2. Operating characteristic curves for different acceptance numbers. Reproduced with permission from Acheson J. Duncan, *Quality Control and Industrial Statistics,* 2nd ed. (Homewood, Ill.: Richard D. Irwin, Inc., 1959), p. 136.

to maintain a constant proportion, the OC curve becomes steeper the larger the value of n. (See Figure 3.) A steep OC curve indicates a discriminating sampling plan in that it has a high probability of accepting lots of good quality (low per cent defective) and a small probability of accepting lots of bad quality (high per cent defective). Only a plan with perfect 100 per cent inspection would discriminate completely between good and bad lots. (See Figure 3.) All sampling plans have some slope to their OC curves, which means that there is some chance of rejecting "good" lots and some chance of accepting "bad" ones. These are the risks of sampling. Under modern sampling plans, these risks are known and carefully considered.

Since the OC curves of all sampling plans have some slope, there is always a range of lot qualities over which the chance of acceptance and rejection are both sizeable quantities. With reference to Figure 1, for example, lots the quality of which varies from 1 per cent to 5 per cent

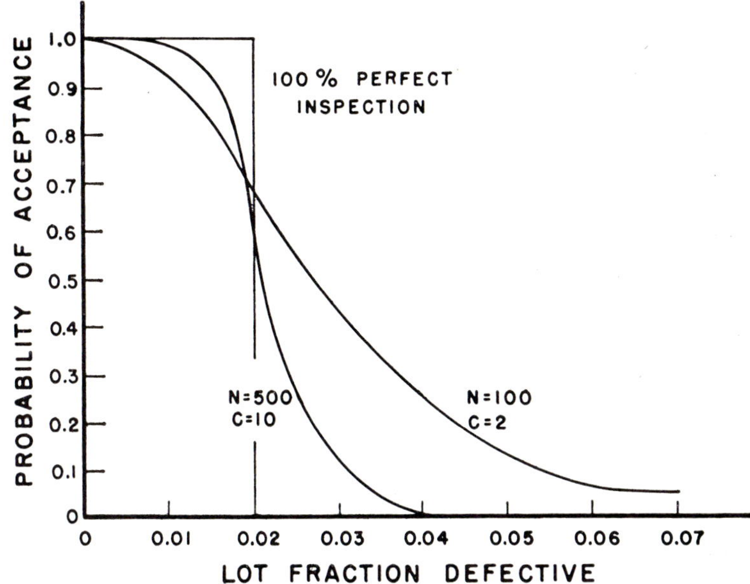

Figure 3. Operating characteristic curves for samples of different size.

have a chance of acceptance that varies from 92 per cent to 13 per cent and a chance of rejection that varies from 8 per cent to 87 per cent. Lots which are 2.5 per cent defective have about a 50-50 chance of acceptance and rejection.

In describing a sampling plan, there has grown up a practice of distinguishing this range of lot qualities that are more or less acceptable by giving special names to particular points. In the region of good lots for which the probability of acceptance is high, of perhaps 95 per cent, it is customary to designate a particular lot quality as "the acceptable quality level," abbreviated as AQL. Likewise, in the region of lots that are on the borderline of being definitely bad, and which under the given sampling plan have a low probability of acceptance, it is customary to designate a particular lot quality as the "lot tolerance per cent defective," or LTPD. The latter is also often represented by the symbol p_t. Thus, if we know the AQL and LTPD of a sampling plan, we know the general range of lot qualities over which lots are more or less acceptable.

Furthermore, it is customary to give special names to the risks associated with lots of AQL and LTPD quality. Since the AQL of a sampling plan serves as the target at which the producer is encouraged to aim, the risk of having lots of AQL quality rejected is called the *producer's risk*. Similarly, since the LTPD indicates the lower tolerance of quality acceptable to the consumer, the risk of accepting lots of LTPD quality is called the *consumer's risk*. It sets an upper limit for acceptance of lots of worse quality than the LTPD. The producer's risk is usually designated by α and the consumer's risk by β. Hence, if we know the AQL, LTPD, α and β of a sampling plan, we have a good idea of the nature of its OC curve.

One of the problems in statistical quality control is to design a sampling inspection plan that has specified values for the AQL, LTPD, α and β. This can be most easily accomplished with the help of special tables. Here, however, it will suffice to describe a graphic method for designing a single-sampling plan that can be used in connection with Figure 7 of Chapter 9. The procedure is to cut out an L-shaped piece of cardboard so that the two legs form a right angle. Then the ratio or logarithmic scale of Figure 7 of Chapter 9 (i.e., the abscissa scale of Figure 7) is marked off on the bottom inside edge of the L, the inner corner of the L being given the value 1. The point on this scale corresponding to the ratio of the LTPD to the AQL is specifically marked. Call this point A. Next the cardboard L is laid on Figure 7 so that the bottom inner edge coincides with the horizontal line for which the probability of occurrence equals the consumer's risk β. Then, while the L is in the above position, the point corresponding to a probability of occurrence equal to $1 - \alpha$ is laid off on the vertical leg of the L. Call this point B. To find the single sampling plan that will have the specified AQL, LTPD, α and β, we move the L horizontally to the right from the position just described until it is found that one of the c-lines of Figure 4 passes through both A and B simultaneously. It will generally be impossible to do this precisely, so we choose one of the c lines that comes close to both A and B simultaneously. The c of the line so chosen will be the acceptance number for the plan. The n for the plan is found by noting the value of λ (or np) at which the ordinate of the c-line is just $1 - \alpha$ and then dividing this by the AQL. As an alternative, we could find the value of λ at which the ordinate of the c-line is just β and then divide this by the LTPD. This graphical procedure is illustrated in Figure 4.

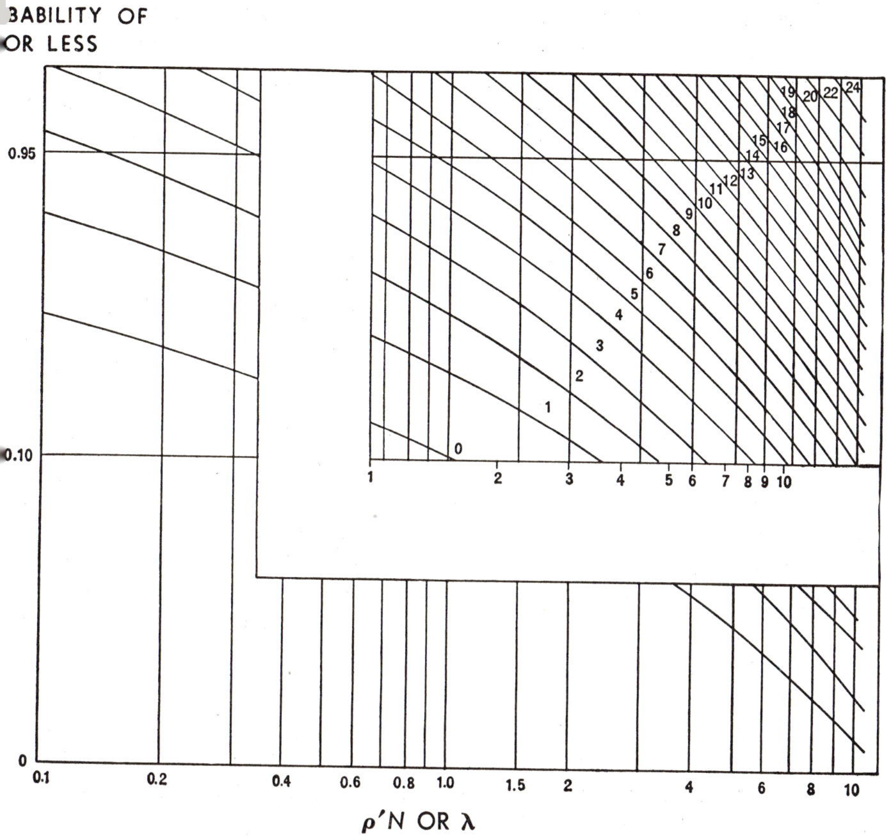

Figure 4. Illustration of a graphic method of finding a single-sampling plan. Reproduced with permission from A. R. Burgess, "A Graphical Method of Determining a Single Sampling Plan," *Industrial Quality Control,* May, 1948, pp. 25-27.

OTHER TYPES OF ACCEPTANCE SAMPLING PLANS

Besides single sampling plans, statisticians have evolved double, multiple and item-by-item sequential sampling plans and also plans that make use of the sample mean and standard deviation to estimate the per cent defective, known as "variables sampling plans." An OC curve with specified AQL, LTPD, α and β can be obtained (approximately) by any one of these sampling plans, so that there is no difference between the various types of plans as to the protection given the producer and consumer. The types of plans do differ, however, as to the amount of inspection that is required to accomplish a given degree of risk. The choice of one type of plan or another is thus a matter of simply finding which is the most economical for a given situation.

EFFECT OF ACCEPTANCE SAMPLING ON QUALITY

It is to be noted that sampling plans that simply accept or reject lots are first of all a means of protecting the consumer from accepting a definitely bad lot or protecting the producer from having a definitely good lot rejected. An acceptance sampling plan *directly* affects the quality of the material passed by the plan only by averaging up the quality of the material accepted. Thus it accepts a higher percentage of lots of good quality than it does of lots of bad quality so that the quality of the lots accepted is on the average higher than the quality of the lots rejected. If, however, lots were all of the same quality, good or bad, acceptance sampling would accept some and reject others, but the quality of the lots accepted would be no higher[3] than the quality of the lots rejected.

The *indirect effects* of acceptance sampling, however, are much more important than the direct effects. When a supplier finds many of his lots rejected, he is under great pressure to improve the quality of his production. If he does not take such action, the consumer is likely to find himself a new supplier. Some sampling programs of the federal government reinforce this normal pressure on the supplier by shifting to "tightened" inspection when records show that the quality of the lots submitted by the supplier averages especially low for the last ten lots.

Rectifying Inspection

When rejected lots are submitted to further inspection, the quality of the material passed by the inspection procedure may be significantly increased by the discovery and elimination or replacement of defective items. Such an inspection program is called *rectifying inspection.* A commonly employed procedure for rectifying inspection is to attach to a regular acceptance sampling plan the additional provision that all

[3] Except for the few bad ones thrown out in the inspection of the sample.

"rejected lots" should be retained by the customer and inspected $100p$ per cent.

When a rider for 100 per cent inspection of rejected lots has been added to an acceptance sampling plan, the quality of lots coming out of inspection will vary, even though they have been of identical quality going in. If a lot of size N is accepted after a sample of n has been inspected, the $N - n$ uninspected items will have essentially the same quality as the original lot. If a lot is rejected, however, its defective items will presumably be removed or replaced in the subsequent screening of the lot and it will leave inspection without any defective items.[4] If a stream of lots is submitted for inspection, each of which is $100\,p$ per cent defective and if the sampling plan is such that it accepts on the average $100P_a$ per cent of lots of this quality (the ordinate of the OC curve for $100p$ per cent quality), then of the many lots submitted, $100\,P_a$ per cent will be passed without further inspection and $100(1 - P_a)$ per cent will be screened and go out with 0 per cent defective. The average quality of the material coming out of inspection, the Average Outgoing Quality (or AOQ as it is called), will be

$$\text{AOQ} = P_a p\left(1 - \frac{n}{N}\right);$$

or if n is small relative to N, we will have approximately

$$\text{AOQ} = P_a p.$$

For if there are 1,000 lots of N items each, all of which are say $100p$ per cent defective, then about $1{,}000P_a$ will be accepted and the number of defective items remaining in the uninspected parts of the lots will be approximately $p(1{,}000P_a)(N - n)$. The samples and the remainders of the rejected lots will all be inspected 100 per cent, and if we assume that the defective items are replaced, the fraction of defective items in the $1{,}000N$ items coming out of inspection will be $\dfrac{p(1{,}000\,P_a)(N - n)}{1{,}000N}$, which equals $pP_a\left(1 - \frac{n}{N}\right)$.

The AOQ of a sampling plan is a function of p. If p is small, most lots will be passed without screening, but the AOQ will be low because p was small in the beginning. If p is large, most lots will be screened and the AOQ will be low because of the 100 per cent inspection. For intermediate values of p, however, the AOQ will be higher than for either

[4] It is assumed here that the 100 per cent inspection is perfect.

relatively low or relatively high values of p. The AOQ curve for the plan $n = 100$, $c = 2$ is shown in Figure 5.

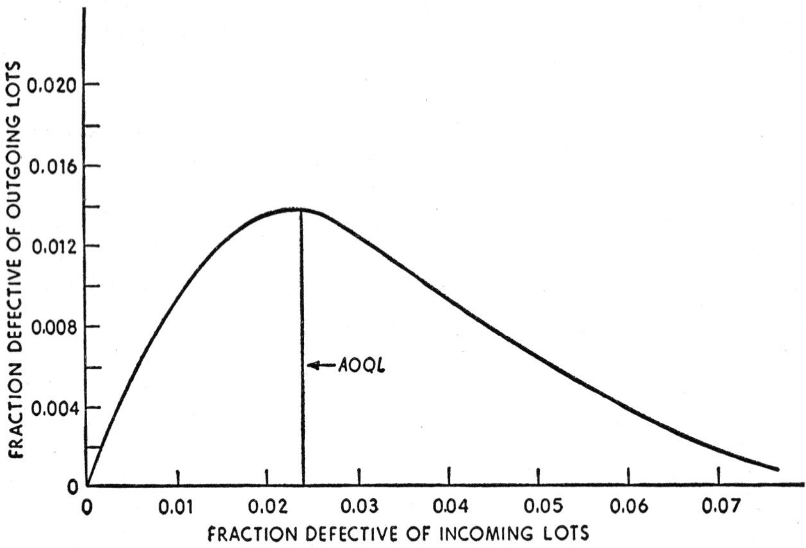

Figure 5. Average outgoing quality curve for the sampling inspection plan $N = 100$, $c = 2$. Reproduced with permission from Acheson J. Duncan, *Quality Control and Industrial Statistics,* 2nd ed. (Homewood, Ill.: Richard D. Irwin, Inc., 1959), p. 280.

As the AOQ curve rises and then falls, it passes through a maximum point. This maximum value of the AOQ is known as the Average Outgoing Quality Limit and is represented by the abbreviation AOQL. It will be noted that the AOQL is the *height* of the AOQ at its maximum point and not the value of p at which the maximum occurs. It is to be especially emphasized that the AOQL pertains to an average over many lots. The quality of a *single lot,* for which the AOQL is an average, will be either about $100p$ per cent defective or 0 per cent defective. The guarantee offered by the AOQL is then valid only for a large series of lots and not for a single lot.

It is possible to design a rectifying inspection plan that guarantees a given AOQL and at the same time minimizes average total inspection for lots of the quality usually inspected. Such plans will be found in the Dodge-Romig *Sampling Inspection Tables* (9).

Continuous Sampling Plans

When items are produced on a continuous assembly line and not in lots or batches, lot-by-lot acceptance sampling, with or without rectification, leads to certain difficulties. Under continuous production, for example, lotting may require that items be allowed to accumulate at given points in production. This creates banks of items at various production points, requires extra space, increases inventories and, in the case of explosive material, produces additional safety hazards. These and other difficulties with lotting may interfere with efficient production operations.

For the above reasons, special sampling procedures have been applied to continuous production. These are commonly called "continuous sampling plans" in contrast to "lot-by-lot" plans and are all rectifying in character in that through partial screening the quality of the outgoing material is better than that of the incoming material.

Continuous sampling plans were first suggested by Harold F. Dodge in 1943 (8), and most of the plans in use today are of the Dodge type or some variation thereof. Other types of continuous sampling plans have been suggested but will not be described here. (See (11).)

The original Dodge plan for sampling inspection of continuous production is known as CSP-1 and runs as follows. It starts with 100 per cent inspection of the items as they pass a given production point.[5] When i consecutive items are found to be free of defects, 100 per cent inspection ceases and the items are sample inspected at the rate of $100f$ per cent. This continues until a defective item has been found. Then reversion is immediately made to 100 per cent inspection and the cycle starts over again. In sampling, the items should be selected at random. Two procedures are used. In one, each sample item is selected at random from a block of $1/f$ items; in the other, items are sample inspected with probability f. In the former instance, no block of $1/f$ items goes by without at least one item being inspected. Once the item has been selected, however, none of the remaining items in the block will be inspected, and, if the operators know this, it may lead to less careful workmanship. If the second method is employed, there is no fixed limit of items that may be allowed to go by without any inspection. To avoid this, some plans call for inspecting at least one item in a block of specified size.

[5] It is, however, possible to start with sampling.

With each CSP-1 plan, there is associated an AOQL which represents the limit of the average quality turned out by the plan over a long (theoretically infinite) run of production. The Dodge plan will guarantee an AOQL if the process to which it is applied is in statistical control,[6] i.e., if the defective items occur at random in the production process with a constant probability p. It may be contended, however, that in all but a few very unlikely situations, this theoretical AOQL will in fact be the true AOQL even if the process is not strictly in control (11). One extreme situation where this would not be true would be that in which there was a perfect synchronization between inspection and production in which the quality of output was $100p$ per cent defective when inspection was on a sampling basis and 0 per cent defective when on a 100 per cent inspection basis. In most realistic situations, it is believed the actual AOQL is close to the theoretical AOQL.

Dodge's chart (8) for selecting continuous sampling plans with specified AOQL's is reproduced as Figure 6. It will be noted that a plan with a given AOQL can be obtained with different values of i and f. Dodge recommends, however, that it is best not to have f too small (for instance, less than 0.02) because this may not give sufficient protection against spotty material.

A number of interesting variations in Dodge's CSP-1 plan have been adopted. The belief existed in many quarters that to require an immediate return to 100 per cent inspection on the finding of a single defect was too conservative a procedure. Dodge and Torrey (10) thus offered CSP-2 which required that *another* defective item be found in the next i items sample inspected before return is made to 100 per cent inspection. In CSP-3 Dodge and Torrey (10) added the additional requirement that the next four successive items be inspected following the finding of a defect, with return to 100 per cent inspection if one of these was defective. This variation was aimed to increase the protection against spotty material while, at the same time, lightening up in the requirement for immediate return to 100 per cent inspection.

A third variation has been offered by Lieberman and Solomon (14). This permits several levels of sampling inspection. Usually a basic sampling rate f is adopted and then this is decreased in the geometric progression $f^2, f^3 \ldots f^k$. The rule is to proceed to the next reduced level of inspection whenever i consecutive items have been inspected at the current rate of sampling and found free of defects. When a defect is found, return is made to the next higher rate of sampling. In practice,

[6] See pp. 273 ff.

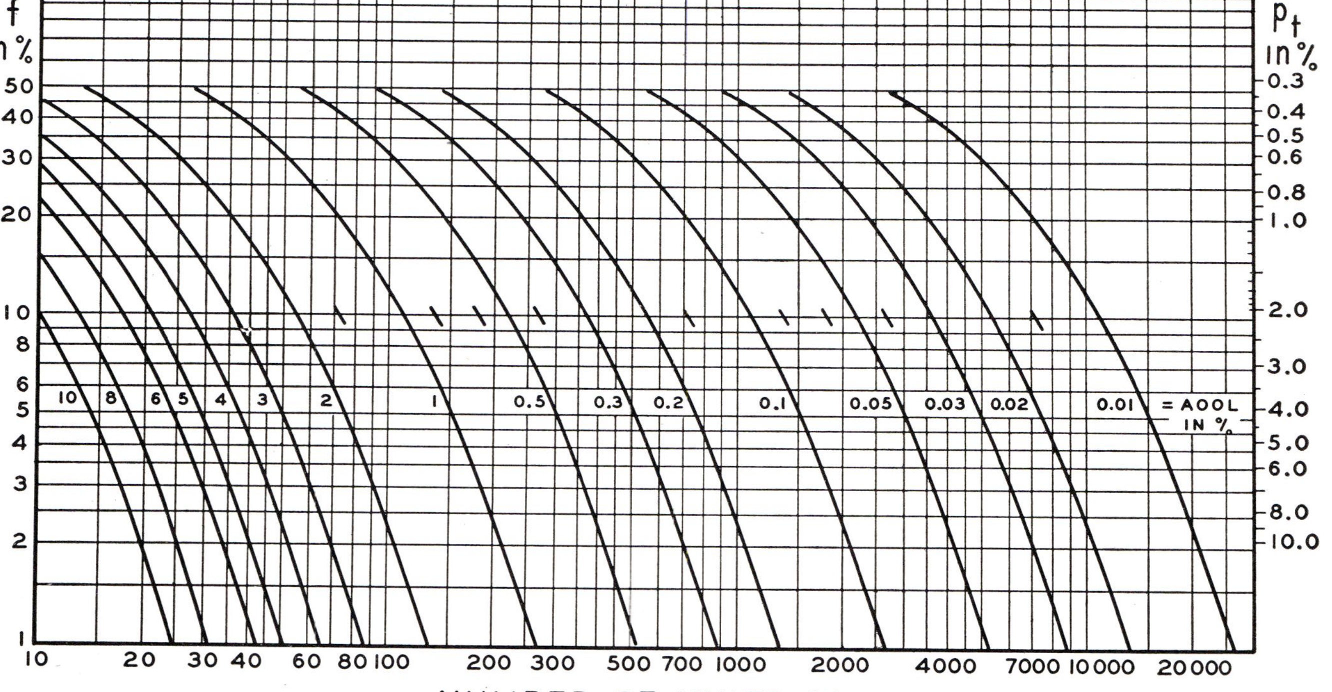

Figure 6. Curves for determining values of f and i of a Continuous-Sampling (CSP-1) Plan for a given AOQL. Reproduced from the *Annals of Mathematical Statistics,* Vol. XIV, p. 274.

the number of levels does not usually exceed five and may be less. This feature gives a plan greater flexibility in increasing the rate of inspection when quality of production gets poorer and decreasing the rate of inspection when quality of production gets better, without causing abrupt changes from a low rate of sampling to 100 per cent inspection. Such plans are called multi-level continuous sampling plans or MLP plans.

Federal Government Standards and Handbooks

LOT-BY-LOT ACCEPTANCE SAMPLING

There are several federal government standards and handbooks providing sampling inspection plans for government use. The best known is the Department of Defense Military Standard 105B, "Sampling Procedures and Tables for Inspection by Attributes," issued in 1958.[7] This is a set of lot-by-lot single, double, and multiple sampling plans together with OC curves. All that is needed to select a plan from the standard is a decision as to the AQL, lot size, and type of plan desired. The AQL is the key decision that is usually made at a high level by negotiation between the government and the supplier. The choice of lot size and type of plan is left to the government inspector.

Table 1 shows how in the single sampling plans of Military Standard 105B the sample size is varied with the lot size. It also gives the risks of rejecting lots of 0.025 quality and the risk of accepting lots of 0.08 quality for each plan.

Military Standard 105B provides for shifts to tightened inspection (by reducing the acceptance number) when the "process average," as estimated from the previous five samples, exceeds a stipulated allowance in excess of the AQL. It also provides for a shift to reduced inspection, whenever the sample results average significantly below the AQL. This is provided for by a shift to a sample size that is one-fifth of the normal sample size. The object of the shift to tightened inspection is to put pressure on the supplier to bring his production back to the agreed upon AQL. The object of the second is to save inspection costs whenever a supplier is doing exceptionally well. Provision is made

[7] This is a minor revision of Mil. Std. 105A, which was originally issued in 1950. An appendix to 105A was issued in 1955.

Table 1. Variation of Sample Size with Lot Size in Military Standard 105B

Lot Size	Sample Size	Acceptance Number for an AQL = 0.025	Approximate Risk of Rejecting 0.025 Quality	Approximate Risk of Accepting 0.08 Quality
2-8	2	*	—	—
9-15	3	*	—	—
16-25	5	*	—	—
26-40	7	0	.20	.53
41-65	10	*	—	—
66-110	15	†	—	—
111-180	25	1	.15	.38
181-300	35	2	.07	.35
301-500	50	3	.05	.30
501-800	75	4	.05	.26
801-1,300	110	6	.03	.22
1,301-3,200	150	8	.03	.16
3,201-8,000	225	11	.02	.07
8,001-22,000	300	14	.01	.02
22,001-110,000	450	20	.01	.00
110,001-550,000	750	31	.00	.00
550,001 and over	1500	56	.00	.00

* Take $n = 7$ (or whole lot if necessary) and $c = 0$.
† Take $n = 25$ and $c = 1$.

for return to normal inspection when a reverse change in conditions appears to warrant it.

A very recent government standard is the Department of Defense publication, *Military Standard 414*, "Sampling Procedures and Tables for Inspection by Variables for Percent Defective." This is a collection of lot-by-lot single sampling plans that will give essentially the same OC curves as plans provided for in Military Standard 105B. The sample sizes required are much less, however, than those required in Military Standard 105B. How the two standards match up for selected plans is shown in Table 2. It will be seen that the savings run up to 86.7 per cent. The standard also provides for plans that make use of the range instead of the standard deviation in estimating the per cent defective. The saving in sample size is not as great with the range as with the standard deviation, but its simplicity and facility of use may

Table 2. Comparison of Sample Sizes for Military Standard 414 with Those for Military Standard 105B, Single Sampling

Lot Sizes	Mil. Std. 105B	Mil. Std. 414	Approximate Percentage Saving*
3-8	2	3	—
9-15	3	3	none
16-25	5	4	20.0
26-40	7	5	28.6
41-65	10	7	30.0
66-110	15	10	33.3
111-180	25	15	40.0
181-300	35	20	42.9
301-500	50	25	50.0
501-800	75	30	60.0
801-1,300	110	35	68.2
1,301-3,200	150	40	73.3
3,201-8,000	225	50	77.8
8,001-22,000	300	75	75.0
22,001-110,000	450	100	77.8
110,001-550,000	750	150	80.0
550,001 and over	1,500	200	86.7

* The comparisons are not precise since the OC curves for the attributes and variables plans are only roughly equivalent.

make it more economical in some cases than the standard deviation.

It is to be noted that the OC curves given for the variables plans in the standard are based squarely on the assumption of random sampling from a normally distributed lot. When a lot has a severely asymmetric (shewed) distribution, the OC curves will underestimate the true probability of acceptance.

CONTINUOUS SAMPLING PLANS

In 1959 the Department of Defense issued a handbook (H 106) for continuous sampling plans (21). These are multi-level plans of the Lieberman-Solomon type with provision for operation up to five levels. Plans are given with basic f's of ½ and ⅓ and with AOQL's ranging from 0.10 to 15 per cent. The standard Average Fraction Inspected and Average Outgoing Quality Curves are given, as well as tables and curves for selecting the number of levels to use.

PROCESS CONTROL

Sampling inspection plans yield quality assurance. Real quality *control* is process control, and this is usually accomplished through control charts or similar devices.

A control chart is a statistical device for informing management as to which variations in the quality of output may be viewed as random and which should be attributed to assignable causes. In the former instance, minor changes in the process to adjust for the cause of variation are not recommended, since such changes will in this case aggravate rather than mitigate the fluctuations in quality. In the latter instance, the management is directed to search out the assignable cause of variation and take appropriate corrective measures. The control chart is thus a statistical index telling management when and when not to act on the process. Explained here are what are known as Fraction Defective Charts (p-Charts), Defects-per-Unit Charts (c-Charts), and Mean and Range Charts ($\overline{X}$- and R-Charts).

p-*Charts*

A p-Chart shows the fraction defective of various segments of production. It may be based on the whole output of given subperiods or on random samples from these subgroups of output. The key is to select subgroups of output for which conditions of production are as homogeneous as possible, so that effects of changes in conditions (assignable causes) will show up primarily as differences between subgroups rather than differences within subgroups. The procedure is to average the fraction defectives of the various subgroups and then, on the assumption that all of the items come from a homogeneous set of production conditions, to set up control limits that are appropriate boundaries for the operation of chance forces. If any points exceed these limits, or if there are long runs within the limits, say a run of seven or more above or below average (11), it becomes a sign that the process is subject to other than chance forces and that assignable causes should be looked for.

To illustrate, suppose one tosses a coin 100 times and gets 45 heads; some one else tosses it 100 times and gets 60 heads. Tom, Dick, and Harry each toss it 100 times and get 54, 39, and 62 heads. If we average these results, we have $\frac{45 + 60 + 54 + 39 + 62}{500} = 0.52$ or 52 per cent as the average percentage of heads. If we consider all the tossings to be random, with the probability of a head equal to 0.52 (this is an empirical estimate of the actual probability of a head with the given coin), then we can expect (See Chapter 9, p. 234) that the variations from one sample of 100 tossings to another will follow a binomial distribution with $n = 100$ and $p = .52$. Of course, most anything could happen by chance, but engineers have decided that in setting up control limits for the operation of chance forces, limits of ± 3 standard deviations (i.e., $\pm 3\sigma$) around the average work well. Since the standard deviation of a binomial distribution is $\sqrt{\frac{p(1-p)}{n}}$ (compare Chapter 9, p. 235), then 3σ limits in our case will be given by

$$0.52 \pm 3\sqrt{\frac{(.52)(.48)}{100}} = \begin{cases} 0.67 \\ 0.37 \end{cases}$$

These are shown in Figure 7 with individual sample results. In this case all samples fall within the control limits and one would conclude that only chance forces were operating to cause the variations in sample results. No assignable causes are present.

Let us extend the above limits for judgment of another set of toss-

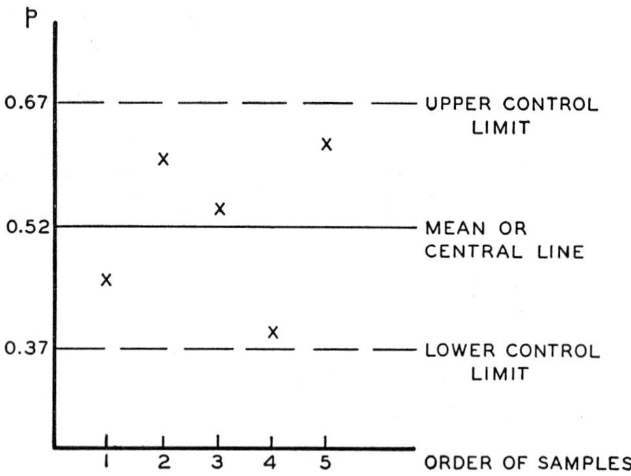

Figure 7. Illustration of a p-Chart. In practice the p-Chart would be constructed from more than five samples, perhaps from twenty or more.

ings by John, Fred, and Jim. Let John get 42 heads and Fred 59 heads, both of which are still within the limits, but let Jim get 69 heads. This last result is outside the "control limits," and we would decide that some assignable cause was present. We would probably look to see if Jim had substituted another coin for the one that had been used previously or whether he was using a special method of tossing.

c-*Charts*

The c-Chart is used in industry when the inspection unit is a constant length or area of output; for instance, a given number of feet of wire or square feet of newsprint, or a given complex unit such as a television set. The inspection consists in determining the number of defects per unit. A similar chart could be used in the study of queueing problems to determine whether the variable in question (for instance, the number of arrivals per unit of time) was behaving in a random manner or whether there had been a change in the average rate of occurrence.

To illustrate a c-Chart consider the data of Table 3. Here are listed the number of entries of elements of combat intelligence in successive

Table 3. Entries of Elements of Combat Intelligence in Successive Two-Hour Periods

Period	Number of Entries	Period	Number of Entries
1	4	13	1
2	1	14	3
3	4	15	5
4	4	16	1
5	2	17	4
6	0	18	1
7	1	19	4
8	2	20	4
9	4	21	6
10	2	22	3
11	4	23	3
12	3		—
		Total	66
		Mean =	2.87
		Standard Deviation = $\sqrt{2.87}$ =	1.69

two-hour periods. The mean is 2.87 entries per period. If we assume that entries are made at random over time with an average rate of 2.87 per period, then we can conclude that variation in the number of entries per period will follow a Poisson distribution, the mean of which will be 2.87 and the standard deviation $\sqrt{2.87}$. (See Chapter 9, pp. 236-8.) Hence a *c*-Chart for these data will have a central line equal to 2.87 and control limits at $2.87 \pm 3\sqrt{2.87}$. Since we cannot have a negative limit, however, we take only the upper limit, which will be 7.94. The *c*-Chart with this central line and upper control limit are shown in Figure 8. The chart shows no points outside the limits and

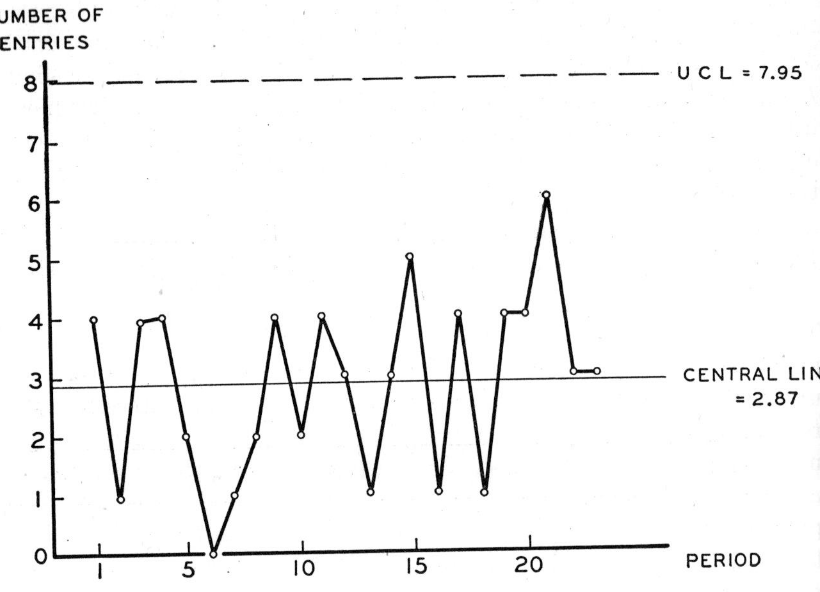

Figure 8. A *c*-Chart for the data of Table 3.

no runs longer than 5. The chart thus verifies the hypothesis of random entries at a stable rate. If in the future we should get 8 or more entries in a two-hour period or there should be a run of 7 or more periods in each of which entries *equaled or exceeded* 3 (or in each case were *less* than 3), then we would conclude that some assignable cause had occurred to modify the average rate of entry.

$\overline{X}$- *and* R-*Charts*

Mean and range charts, or $\overline{X}$- and R-Charts as they are usually called, are based on measurement data instead of data that arise simply from classification and counting. They are suitable for study and control of a single measurable quality characteristic. If it is important to control more than a single quality characteristic, it would be necessary to have $\overline{X}$- and R-Charts for each characteristic. Although more of them may be required, $\overline{X}$- and R-Charts have certain advantages over p-Charts and c-Charts. Since they are based on measurement, much smaller samples can be used, say samples of 4 or 5 instead of samples of 30 or 40. Also they give more information. The $\overline{X}$-Chart will indicate shifts in the average of the process, and the R-Charts will indicate changes in the variability of the process (or uniformity of product). This more precise knowledge of the change or changes that occur will help the engineer greatly in tracking down the assignable cause. It also helps in general in giving a greater insight into the nature of the process itself. Of all control charts, $\overline{X}$- and R Charts are used the most.

We will explain $\overline{X}$ and R-Charts with reference to an example. In Table 4A are data on the "basis weight" of various rolls of newsprint produced by manufacturer A. They are the means and ranges of measurements of basis weight for ten samples from each of five different rolls, the rolls being taken from the process at different times. To set up an R-Chart for these data, the ranges are averaged and 3σ limits are laid off from this average. The results are shown in Figure 9 (supplier A). It is seen that the sample ranges all fall nicely within the limits and there are no runs or other patterns that suggest non-randomness. It will be noted that five rolls are too few for the analysis of runs. (For this purpose, twenty or more would be desirable.) The chart suggests that the variations in variability are solely the result of chance forces. (For a definite conclusion of "control," it would be desirable to have at least twenty samples within limits without peculiar runs or other evidence of non-randomness.)

The average range is a measure of the chance variability *within rolls*. On the assumption of the non-existence of assignable causes it becomes a yardstick for determining the "allowable chance variation"

Table 4A. Means and Ranges of Ten Measurements of Basis Weight for Each of Five Rolls of Newsprint: Supplier *A* (Courtesy of the American Newspaper Publishers Association Research Institute)

Roll Number	Mean of Ten Measurements	Range of Ten Measurements
1	33.15	1.1
2	31.35	1.4
3	31.83	0.8
4	30.59	1.0
5	31.52	1.1

Grand Mean $\overline{\overline{X}} = 31.69$

Upper Control Limit
for $\overline{X}$-Chart* $= 31.69 + 0.308(1.08)$

Lower Control Limit
for $\overline{X}$-Chart $= 31.70 - 0.308(1.08)$
$= 31.36$

Mean Range $\overline{R} = 1.08$

Upper Control Limit
for R-Chart** $= 1.777(1.08)$

Lower Control Limit
for R-Chart** $= 0.223(1.08)$
$= 0.24$

* 0.308 is the A_2 factor for $n = 10$ and equals $3/d_2\sqrt{n}$.

The factor d_2 is the mean value of R/σ where σ is the standard deviation of subgroup measurements. (See (12).)

** 1.777 and 0.223 are the D_4 and D_3 factors. These equal

$$\left(1 + \frac{3d_3}{d_2}\right) \quad \text{and} \quad \left(1 - \frac{3d_3}{d_2}\right)$$

respectively where d_3 is the standard deviation of R/σ. (See (12).)

between rolls. It is thus the basis for setting up the 3σ limits for the $\overline{X}$-Chart. The results are also shown in Figure 9 (supplier A). It will be seen in this case that the variation *between rolls* is not in control in the sense that it can be explained as the result of the same random influences that cause variation in basis weight *within rolls*. In other words, the kinds of forces that cause variations in this mill's newsprint from time to time have a greater effect on basis weight than the forces that cause variation over a short period of time. These longer range forces could thus be viewed as "assignable causes" worthy of investigation. For example, it would be worth finding out why sample No. 1 had such a high basis weight and sample No. 4 such a low basis weight.

Table 4B presents a similar set of data for another newsprint manufacturer. The $\overline{X}$- and R-Charts for these data are shown in Figure 9

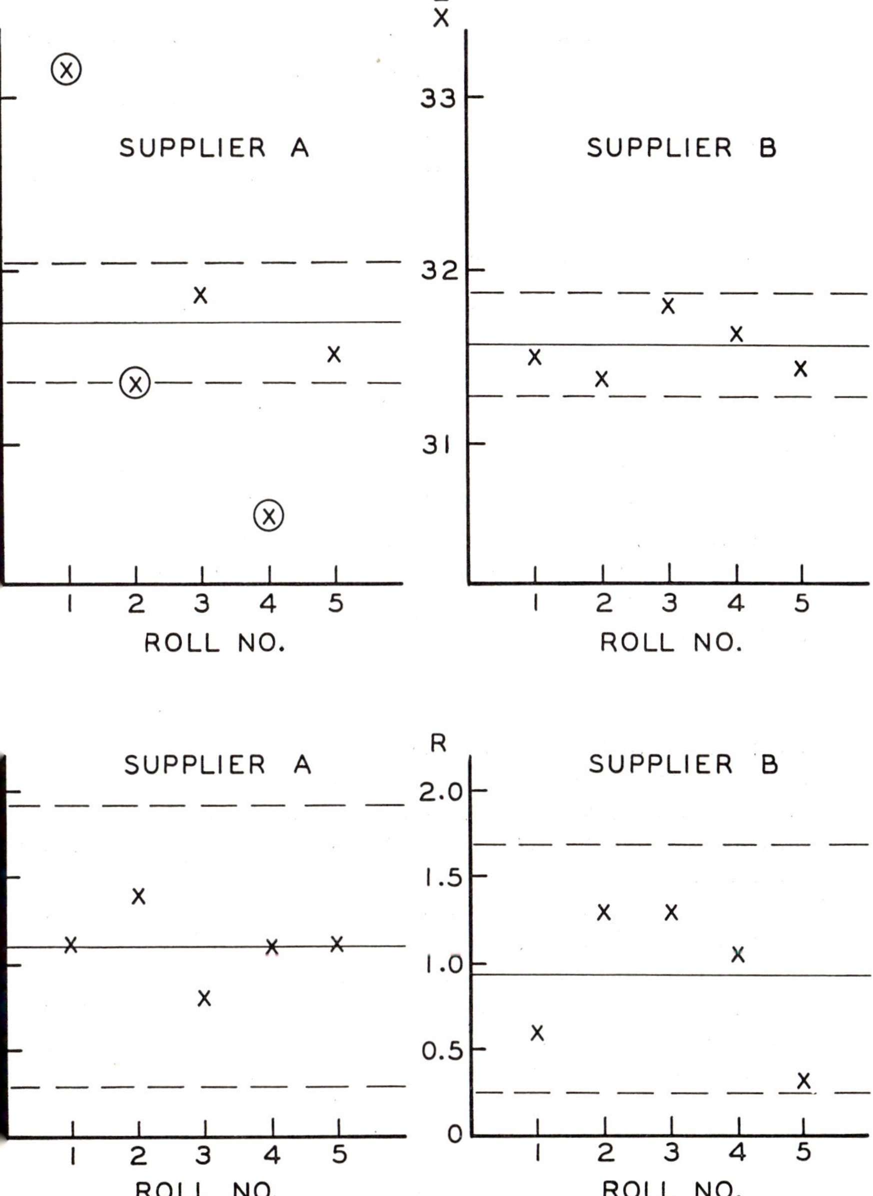

Figure 9. $\overline{X}$ and R-Charts for Table 4A and 4B.

Table 4B. Means and Ranges of Ten Measurements of Basis Weight for Each of Five Rolls of Newsprint: Supplier *B* (Courtesy of the American Newspaper Publishers Association Research Institute)

Roll Number	Mean of Ten Measurements	Range of Ten Measurements
1	31.50	0.6
2	31.38	1.3
3	31.82	1.3
4	31.62	1.1
5	31.43	0.3

Grand Mean $\overline{\overline{X}} = 31.55$

Mean Range $\overline{R} = 0.92$

Upper Control Limit
for $\overline{X}$-Chart $= 31.55 + 0.308(0.92)$
$= 31.83$

Upper Control Limit
for R-Chart $= 1.777(0.92)$
$= 1.63$

Lower Control Limit
for $\overline{X}$-Chart $= 31.55 - 0.308(0.92)$
$= 31.27$

Lower Control Limit
for R-Chart $= 0.223(0.92)$
$= 0.21$

(For source of factors 0.308, 1.777 and 0.223, see Table 4A.)

(supplier B). It will be noted that the average basis weight is about the same for the two manufacturers and the average variability is also about the same. In both cases, the variation in basis weight *within rolls* appears to be in control. With the second manufacturer, however, the average basis weight also seems to be in control in the sense that the forces that cause variation over a long time appear to be the same as the chance forces that cause variation over a short time. There appears to be no assignable cause to investigate in this case.

In general, the quality control engineer prefers to base his conclusions upon a larger set of data than five samples, but the principles he would follow would be the same as those developed above.

RESEARCH

The third facet of statistical quality control is research. This aspect will not be developed here. It will suffice to note that considerable use has been made of *regression analysis* and *analysis of variance* (see Chapter 9, pp. 246-53) and of statistical theories relating to *design* of *experiments.* (See Chapter 18.) Also in recent years there has been a rapidly growing interest in the Chemical Division of the American Society for Quality Control in statistical procedures for finding optimum operating conditions and mapping response surfaces (see Chapter 18, p. 526). Similar principles have been applied directly to plant operations where the statistical approach has been called *evolutionary operation* (5).

QUALITY CONTROL AND OPERATIONS RESEARCH

Many of the engineers engaged in quality control are also interested in operations research, and some of the operations analysts are interested in quality control. The quality control men point out that industrial variables must be brought into control, i.e., be made simple chance variables, before they become suitable for much of the statistical and mathematical analysis of the operations analyst. The quality control men thus contend that their work as engineers should precede or be joined with that of the operations research men. The feeling in some quarters is that operations research men are too prone to assume that the variables with which they deal in industry are already in a "state of statistical control."

The operations research men who have become interested in quality control point out to the quality control engineers that their sampling plans, control charts, and research should often be designed with respect to a more comprehensive set of conditions. In other words, the operations research men contend that the quality control men are prone to suboptimize over too small a segment of activity.

As evidence of the tie-in of quality control and operations research, it may be noted that the American Society on Quality Control has a committee on the application of operations research to quality control problems. The journal of the society, *Industrial Quality Control,* has printed a number of expository articles on the techniques of operations research and papers in these subjects are frequently presented at quality control conventions. On the other hand, the Operations Research Society of America reviews articles in *Industrial Quality Control* for possible candidates for the Lanchester prize. There is thus a growing realization by both groups of the usefulness of the work of the other.

REFERENCES

(1) American Standards Association. "Guide for Quality Control" and "Control Chart Method of Analyzing Data," *American Standards Z1.1-1958 and Z1.2-1958*. These are revisions of *American War Standards Z1.1-1941* and *Z1.2-1941*.

(2) American Standards Association. "Control Chart Method of Controlling Quality during Production," *American Standard Z1.3-1958*. This is a revision of *American War Standard Z1.3-1942*.

(3) American Statistical Association. *Acceptance Sampling—A Symposium*. Washington, D. C.

(4) Box, G. E. P. "Evolutionary Operation: A Method for Increasing Industrial Productivity," *Applied Statistics,* Vol. VI (1957), 81-101.

(5) Burr, Irving W. *Engineering Statistics and Quality Control.* New York: McGraw-Hill, 1953.

(6) Cowden, Dudley J. *Statistical Methods in Quality Control.* New York: Prentice-Hall, 1957.

(7) Davidson, Harold O. "OR Concepts Useful in QC," *Industrial Quality Control,* Vol. XII, No. 4 (October, 1955), 10 ff.

(8) Dodge, Harold F. "A Sampling Inspection Plan for Continuous Production," *Annals of Mathematical Statistics,* Vol. XIV (1943), 264-79.

(9) Dodge, Harold F., and Romig, Harry G. *Sampling Inspection Tables–Single and Double Sampling.* 2nd ed. New York: John Wiley and Sons, 1959.

(10) Dodge, Harold F., and Torrey, Mary N. "Additional Continuous Sampling Inspection Plans," *Industrial Quality Control,* Vol. VII, No. 5 (1951), 5-9.

(11) Duncan, Acheson J. *Quality Control and Industrial Statistics*. 2nd ed. Homewood, Ill.: Richard D. Irwin, Inc., 1959.

(12) Grant, Eugene L. *Statistical Quality Control.* New York: McGraw-Hill, 1952.

(13) *Industrial Quality Control,* the journal of the American Society for Quality Control, Milwaukee, Wis.

(14) Lieberman, G. J., and Solomon, H. "Multi-Level Continuous Sampling Plans," *Annals of Mathematical Statistics,* Vol. XXVI (1955), 686-704.

(15) Littauer, S. B. "The Development of Statistical Quality Control in the United States," *American Statistician,* Vol. IV, No. 5 (October, 1950), 14-50.

(16) Olmstead, Paul S. "QC Concepts Useful in OR," *Industrial Quality Control,* Vol. XII, No. 4 (October, 1955), 11 ff.

(17) Shewhart, W. S. *Economic Control of Quality of Manufactured Product.* New York: D. Van Nostrand, 1931.

(18) Simon, Leslie E. *An Engineer's Manual of Statistical Methods.* New York: John Wiley and Sons, 1941.

(19) United States Department of Defense. *Sampling Procedures and Tables for Inspection by Attributes* (Military Standard 105B). Washington, D. C.: U. S. Government Printing Office, 1958.

(20) United States Department of Defense. *Sampling Procedures and Tables for Inspection by Variables for Percent Defective* (Military Standard 414). Washington, D. C.: U. S. Government Printing Office, 1957.

(21) United States Department of Defense. *Multi-Level Continuous Sampling Procedures and Tables by Attributes* (Handbook 106).

Eleven

ELECTRONIC DIGITAL COMPUTERS

WILLIS C. GORE

INTRODUCTION

Because modern high-speed electronic digital computers are destined to play an ever-increasing role in business and industrial operations, up-to-date management must have some awareness of the important applications, capabilities, and limitations of these devices. This chapter will not consider applications in any detail, but rather will attempt to acquaint the computer layman with the basic operations and principles that are used in nearly all such machines, large and small. Acquiring some familiarity with how such machines operate is necessary if management is to develop working insight into where and how computers may (or may *not*) be applied effectively.

Basically, these machines do the ordinary arithmetic operations that a desk computer can perform. These include addition, subtraction, multiplication, and division. The advantage of these machines, however, does not result in the fact that they can perform some tasks that other machines cannot do, but simply that they can accomplish these very much faster than an ordinary desk computer. While a desk calculator can multiply two ten-digit numbers in approximately six seconds, a modern high-speed computer can perform a similar kind of computation in approximately six-millionths of a second. Thus, this modern machine could perform as much arithmetic in one hour as the desk computer could accomplish in one million hours. Since there are approximately eight working hours in a day, the one million hours is 125,000 days or 25,000 working weeks. Allowing for fifty weeks of work during a year, one arrives at the conclusion that the modern machine does as much arithmetic in one hour as a desk computer could accomplish in five hundred years!

Problems which in principle could have been solved a long time ago, but which in practice were never reduced to answers because of the tedious and time-consuming process of doing the ordinary arithmetic involved, may now be programmed and solved by means of these high-speed machines. Thus, the charge of several hundred dollars normally made per hour of machine time seems relatively modest when one realizes that it is equivalent to approximately five hundred years of work by one man and one desk computer.

NUMBER SYSTEMS

To continue, how do these machines operate basically? How do they perform their arithmetic? Before a discussion of the operation of these machines is begun, it might be well to review some fundamentals which are so commonplace as to seem almost insignificant.

We first realize that the basis of all arithmetic is simply counting. There is no arithmetic operation that could not be performed by a simple counting procedure. Thus, the basis of all of our arithmetic is counting, and the basis of our counting is a decimal counting system. A study of numbers and counting systems originating in many places of the world shows that they are all basically either decimal or quinary in nature. This is, undoubtedly, because of the fact that we were endowed with five fingers per hand, and two hands, so that our counting system grew in a manner which was determined by our own physical make-up. In contrast, high-speed machines do not have fingers or hands, so that there is no logical reason to assume that the counting system that these machines should use for best efficiency would in any way be dependent upon a decimal counting system. We therefore investigate the basis of our particular counting system to see if there are not some general rules which underlie the general counting process. Such rules might then be applied to the evolution of, not a new, but a different counting system more adaptable to machine use.

Any counting system must employ an ordered sequence of marks. This ordered sequence in the case of the decimal system is ten in number and consists of the ten decimal numerals: 0, 1, 2, 3, 4, 5, 6, 7, 8, and 9. This sequence of marks or *numerals* is *ordered* in that there is a first, 0; and a last, 9; and the rest have a definite sequence in proceeding from the first to the last. The name of the numbering system is dependent upon the number of allowable marks in this sequence. For a decimal system, it is ten; for a quinary system, it would be five; for an octal system, eight, and so on.

To count, we start with the first mark and proceed in order by writing down, as we count, the numerals until the last mark is reached. When the last mark has been used, the next count is indicated by replacing it by the first mark and increasing by one the mark in the column immediately to the left of this numeral. Thus, the next number after decimal 19 would be 20, in that the 9 was replaced by the 0 (the first mark), and the 1 was replaced by the 2 (the next mark in order). By following these simple rules, any counting sequence may be established. Counting sequences for a decimal system, an octal system, a quinary system, and a binary system are illustrated in Table 1.

Table 1. Number Representations in Various Counting Systems

Decimal	Octal	Quinary	Binary
000	000	000	00000
001	001	001	00001
002	002	002	00010
003	003	003	00011
004	004	004	00100
005	005	010	00101
006	006	011	00110
007	007	012	00111
008	010	013	01000
009	011	014	01001
010	012	020	01010
011	013	021	01011
012	014	022	01100
013	015	023	01101
014	016	024	01110
015	017	030	01111
016	020	031	10000
017	021	032	10001

We notice that all of these sequences can be generated by following the above outlined rule. One inquires now, of all the possible numbering sequences, which one might be the simplest for machine use? A first guess (at this point it would only be a guess) would be that the binary numbering system might offer advantages over any numbering system, primarily because it employs the fewest number of marks. In the binary system, only the marks 0 and 1 are employed, the smallest number that could be employed to generate a counting sequence. This in itself represents a kind of simplicity that may be easiest to incorporate into a machine.

Addition may always be accomplished by a process of counting. To add the number 11 to the number 3 in a decimal numbering system, one could always start with 11 and count three more numbers, i.e., starting with 11, one may count 12, 13, 14, which is, of course, the required answer. These simple steps, however, hide the real difficulty with this procedure; in adding two large numbers, the time consumed in counting would greatly slow down the arithmetic process. To overcome this difficulty, counting need only be done in each place. For example, to add the number 42 to the number 21, start with the right-hand digit of the second number, which is a 1, and count two more, giving 3. Then start with the next digit to the left of the first, which is a 2, and proceed to count four more to 6. The answer, then, is 63. To make this procedure general, it is only necessary to make an additional count in the column to the immediate left of any column that counts through 0. Although decimal machines have been built that perform addition using this simple counting procedure, we will now attempt to show that binary numbers are indeed best for machine use when several other features of this numbering system are noted. Before doing this, however, let us examine how one would perform arithmetic in the binary numbering system.

Most students in our elementary schools are taught to do addition by memorizing a set of rules that constitute what is called an addition table. They first learn the 0's table: that 0 and 0 are 0; 0 and 1 are 1; 0 and 2 are 2; 0 and 3 are 3, etc., up to 0 and 9 are 9. Then they learn the 1's table: 1 and 0 are 1; 1 and 1 are 2; 1 and 2 are 3; all the way up to 1 and 9 are 10; and, then, similarly, through the 2's, 3's, 4's, all the way up to the 9's table. They have now learned a complete addition table for the decimal system. Ignoring for the moment that addition is distributive and that some of the results in our table are actually duplicated, one could say that for decimal addition, a 10 by 10 table consisting of one hundred entries must be learned before all of the rules would be at hand.

To build a machine which would do addition using a decimal system, one must in effect "teach" it the decimal system. This "teaching" is accomplished by the particular wiring pattern of the machine. If this number of facts needed to do additions could be reduced, then the complexity of the wiring of the machine could likewise be reduced, resulting in a simpler machine. If the numbering system employed for our arithmetic used fewer than ten symbols, the number of entries in the addition table would be less than one hundred. Specifically, for a *quinary* system, based on five symbols, there would be twenty-five entries, and for a *binary* system, based on two symbols, there would be only four. The binary system, which uses the fewest number of marks or numerals, is thus the system having the fewest number of entries in its addition table. This addition table, which is so simple that one can easily learn it at a glance, is illustrated in Table 2. A

Table 2. Binary Addition

0	0	1	1
+0	+1	+0	+1
—	—	—	—
0	1	1	10

binary machine must be taught these four facts by so arranging the wiring that when signals representing the two numbers to be added are supplied to the machine, other signals representing the sum will be produced.

To see how this is accomplished, one must first examine how the numbers are represented electrically in the machine. It is conventional to represent a 0 by the absence of a voltage and a 1 by the presence of a voltage. Thus, one speaks of a circuit either as "having an input" (that is, having a 1 present) or "*not* having an input" (that is, having a 0 present), depending upon whether a voltage is present or absent.

COMPUTER CIRCUITS AND MACHINE LOGIC

Modern machines use many and complex circuits, but, despite this apparent complexity, there are basically only three types, these being illustrated by Figure 1. These circuits are named the "and" circuit, the "or" circuit, and the "inverter" or "not" circuit. The "and" circuit is so named because it supplies an output when it simultaneously has an input on the first lead, *and* the second lead, *and* the third lead, *and*

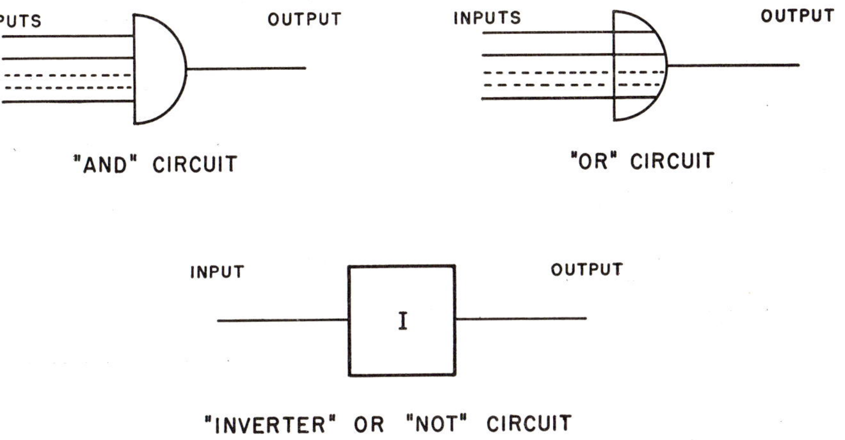

Figure 1. Representation of basic logic elements.

so on. The "or" circuit is so named because it produces an output when there is an input on the first lead, *or* the second lead, *or* the third lead, *or* and so on. The "inverter" circuit yields an output when there is *not* an input, and has no output when there is an input. These circuits may be realized in a number of ways. In the early machines they were realized with relays, in later machines by vacuum tubes, then by semiconductor diodes, and now in the new machines by transistors. The trend has been to increase the speed with which the circuits may respond and simultaneously to reduce their size. Several representations of these circuits are shown in Figure 2.

To construct a computer circuit that will do addition in accord with the table for binary numbering systems, it is first necessary to formulate the addition rules given by the addition table in terms of the connectives *and*, *or*, and *not*. Once these rules have been formulated in this manner, it is possible to introduce "and," "or," and "not" circuits between the wires representing the inputs and the outputs so as to simulate these same rules electrically. As an example, the binary addition table is first rewritten, as in Table 3, to illustrate the sum

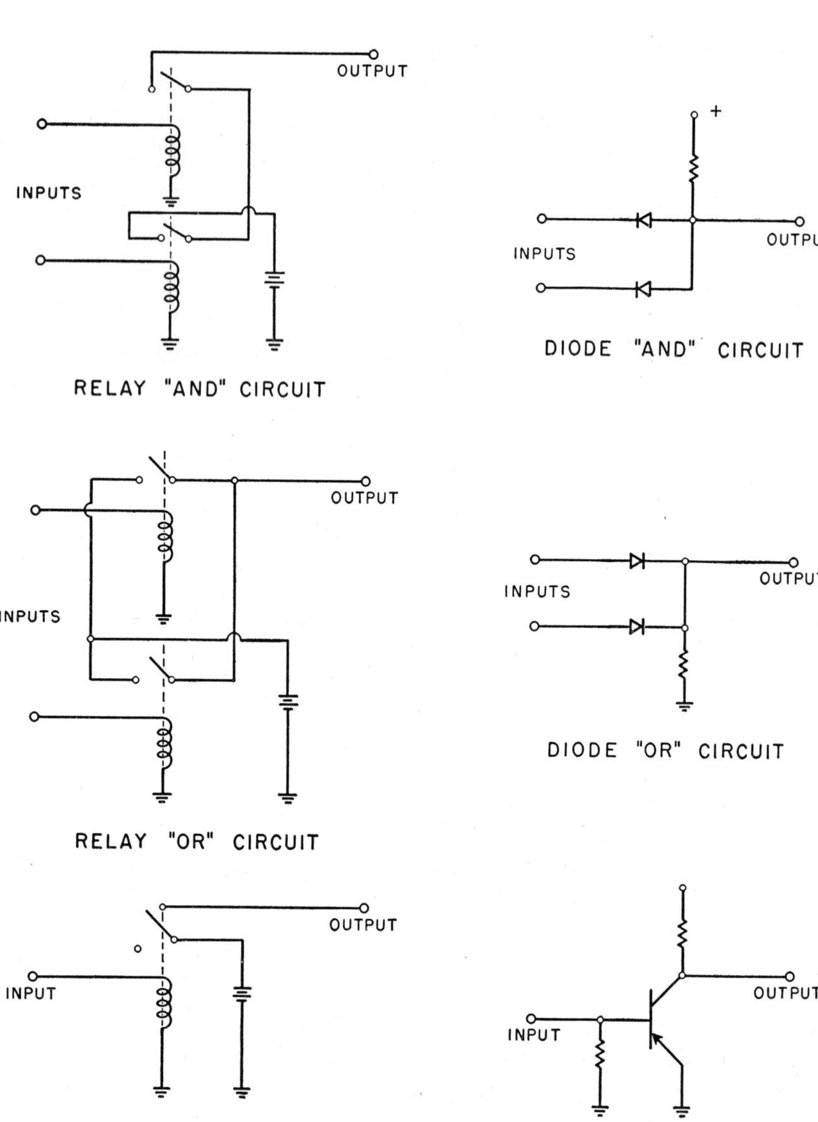

Figure 2. Circuit realizations of logic elements.

Table 3. Inputs and Outputs for Binary Adder

A	0	0	1	1
B	0	1	0	1
Sum	0	1	1	0
Carry	0	0	0	1

digit and the carry digit resulting from the addition process. The task is to construct a computer circuit as shown in Figure 3, which has an

Figure 3. A binary adder.

input for each of the two numbers to be added, and two outputs, one representing the sum and the other the carry.

One now formulates from Table 3 the logical rules that govern the two required outputs. These two rules are: *the sum is* A *and not* B, *or* B *and not* A; and, *the carry is* A *and* B. The completed circuit resulting from the mechanization of these two rules, using "and," "or," and "not" circuits, is shown in Figure 4. Here it is seen that the

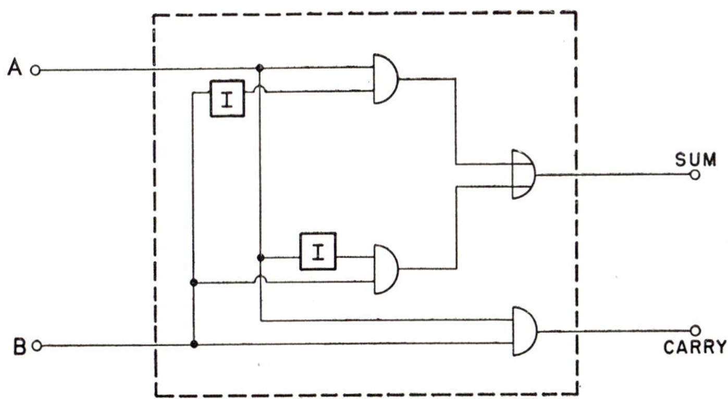

Figure 4. A realization of a single binary adder according to the rules: the sum is (A and not B) or (B and not A); and the carry is (A and B).

connectives between the input lines representing A and B and the output lines representing the sum and carry digits are the circuit equivalents of the connectives used in stating the two rules. It is easy to verify that the circuit gives the correct values of the sum and carry digits for any allowable combination of inputs.

The same procedure may be used to construct any kind of computational circuit, whether it be an adder for a binary, quinary, or a decimal system, or a multiplier, and so on. This procedure is valid because it can be shown that the logical connectives *and, or,* and *not* are sufficient to formulate any logical statement. An interesting feature of this procedure is that the rules as formulated are not unique, even for the simple binary adder. For instance, an alternative rule for the sum digit is: *the sum is* (A *or* B) *and* (*not* A *and* B.) This particular rule may be realized by the circuit, shown in Figure 5, which is an-

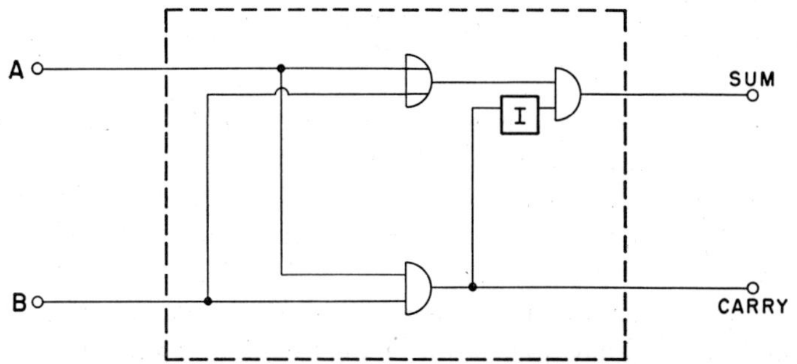

Figure 5. Another realization of a single binary adder according to the rule: the sum is (A or B) and (not A and B).

other, equivalent way of constructing our binary adder. Although these two rules for the "sum" digit which were previously mentioned are equivalent, their circuit realizations are different. In particular, one of them requires fewer parts.

An important step in the design of a computer is to investigate all possible circuit configurations which are logically equivalent and to make a choice for the final design which is optimum in some sense, such as requiring the fewest number of parts or being most reliable, etc. Although this is a major problem in systems engineering, it is not directly the concern of this chapter. Suffice it to say that there are procedures whereby a whole list of equivalent statements may be derived from any one of these statements, and their realizations may be examined to see which yields the optimum circuit.[1] Thus, it is not necessary to foresee all possible combinations which yield equivalent results; it is sufficient to determine only one of them, from which all the others may be derived in a formal manner.

We have now seen how to add two binary *digits,* but our problem is how to add two binary numbers, each of which may consist of a large number of binary digits. To do this, we use the digit adding rules over and over again for each digit of the binary number in a manner quite similar to the way in which the similar rules are used in a decimal system. Table 4 illustrates this. Here one sees two binary

Table 4. Example of Binary Addition

	1110110	Carry from previous digit
87 =	01010111	Number A
54 =	00110110	Number B
141 =	10001101	Sum

numbers, which are to be added together, and their decimal equivalents. The two least significant digits—those at the right—are added together, and the sum is recorded below and the carry entered above and to the left. One next proceeds to add the digits in the second column which, because it contains three digits and our rules only tell us how to add two, must be performed in two steps. First, the carry and the second least significant digit of the first number are added together according to our rules. Second, to this sum is added the second least significant digit of the second number, giving a sum digit which is recorded below the column and a new carry digit which is entered above to the left. This procedure is repeated until all of the digits in the number have been added.

In the addition of a column of three binary digits, at most only one carry digit can be generated since if a carry digit is generated upon the addition of the first two digits, none will be generated in the resulting addition with a third, or vice versa. This fact enables us to build an adder for handling binary numbers in a relatively simple and

[1] See, for example, Chapter 17 on Symbolic Logic.

straightforward way, using circuits capable of adding binary digits.

Before proceeding, let us first examine two ways in which binary numbers are commonly represented physically in machines. We have mentioned that the binary digits correspond to the presence or an

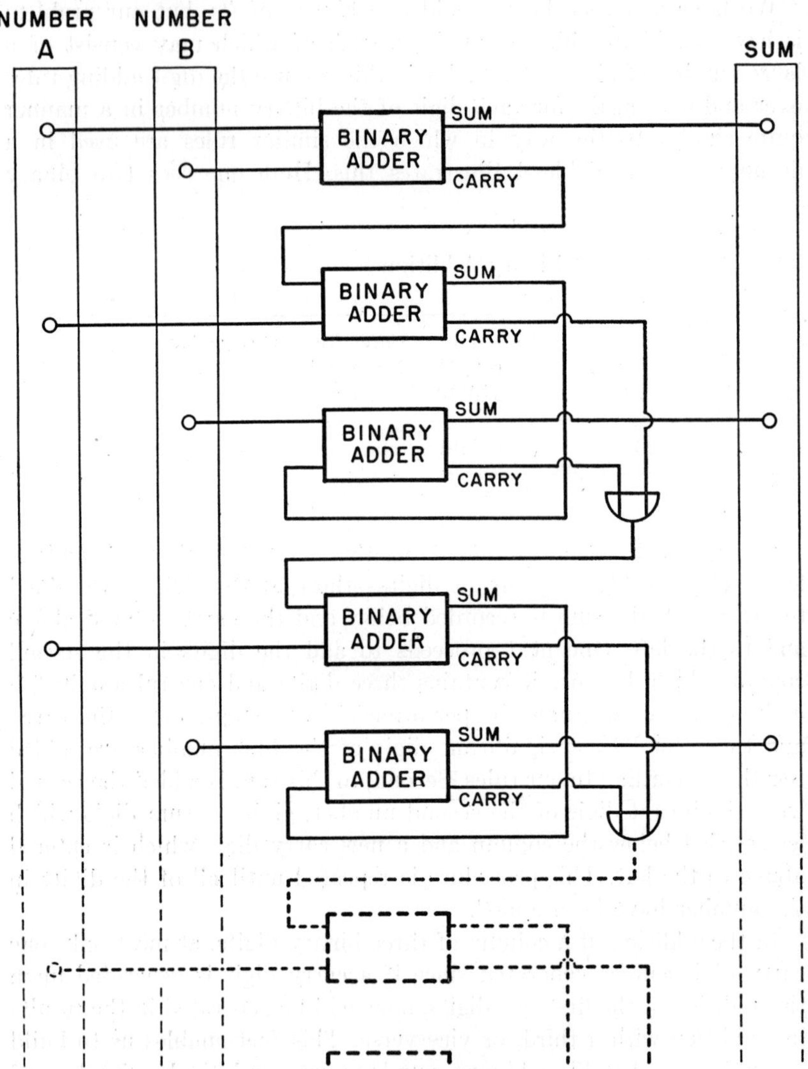

Figure 6. A parallel binary adder.

absence of voltage on a wire. To represent a multi-digit number at a given instant, we may provide as many wires as there are digits in the number, with the voltage of each wire corresponding to the value of associated digit at that instant. This is called a *parallel representation,* since all of the digits are represented concurrently, or in parallel.

Another way of representing multi-digit numbers requires only one wire upon which the digits are presented in sequence; that is, first the least significant, then the next least, and so on, until the most significant digit has been presented. This is called a *serial representation.* Adders for multi-digit numbers are shown in Figures 6 and 7. In Figure

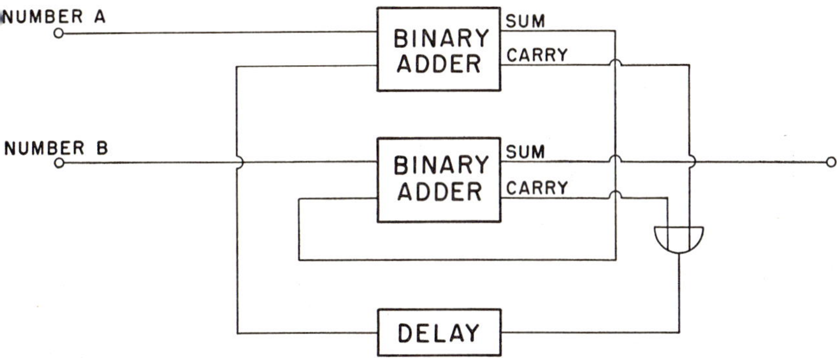

Figure 7. A serial binary adder.

6 the wires representing the digits of the two numbers are connected to the inputs of the appropriate adders so that addition is carried on in the manner previously outlined. Except for the time delay through the circuit, the sum is presented simultaneously with the application of the two numbers. In this way addition can be accomplished in very short intervals of time.

The serial adder performs addition in exactly the same manner. However, the answer is not obtained as rapidly as in the case of the parallel adder, since the addition is accomplished digit by digit. The output of the carry is sent through a delay circuit, whose delay is exactly the same time as that between digits of the input numbers, so that it arrives back at the input of the adder in coincidence with the arrival of the next most significant digits as the rules require. Thus, the parallel machine requires much equipment but gives an answer in a short time interval, while the serial machine requires a length of time equal to that required to present a number but requires very little equipment.

MACHINE MULTIPLICATION

Before proceeding to see how multiplication may be accomplished, it is essential to investigate another very important part of a computer —the register. A register is nothing more than a device capable of storing binary numbers. One such store may be a series of flip-flop

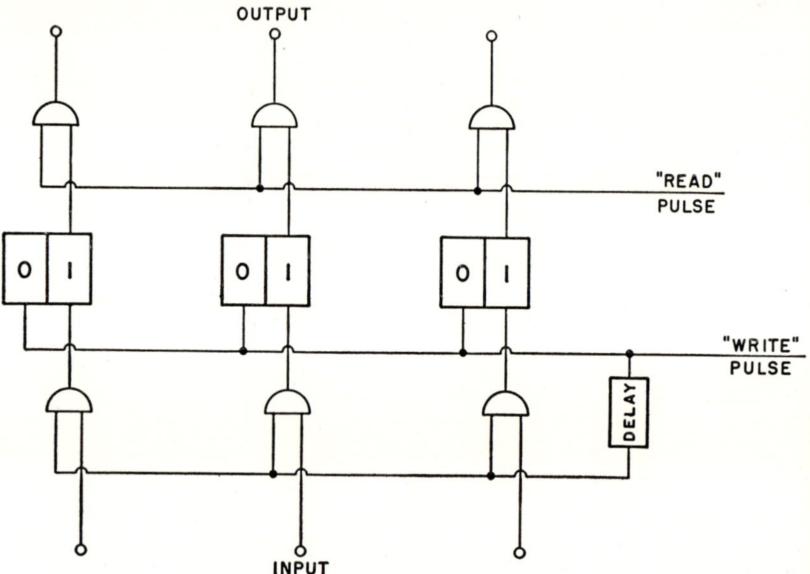

Figure 8. A parallel register.

circuits[2] where, for instance, a 1 may be recorded by the right-hand triode being in the conducting state, and a 0 recorded by the left-hand

[2] A flip-flop circuit is the electronic analogue of an ordinary toggle switch, such as is used to control the electric lights in a room. It has two states and will remain in either state until an external signal drives it into the opposite state. It differs from a toggle switch in that it may operate in a fraction of one-millionth of a second.

triode being in the conducting state. Thus, a single flip-flop is capable of storing or registering one binary digit. In a number of these flip-flops, equal to the number of digits in a binary number, one may store or register one complete binary number.

We may arrange circuits so that a number can be "written" *into* the register by application of a command signal, called a "write" pulse, and the same number may be later "read" *out* of the register by application of a command signal called a "read" pulse. As shown in Figure 8, the number is supplied in parallel form and is recovered in parallel form.

In order to be able to register or store numbers in serial form, a different kind of register, called a *shifting register,* is used. As shown in Figure 9, upon application of a command signal called a *shift pulse,*

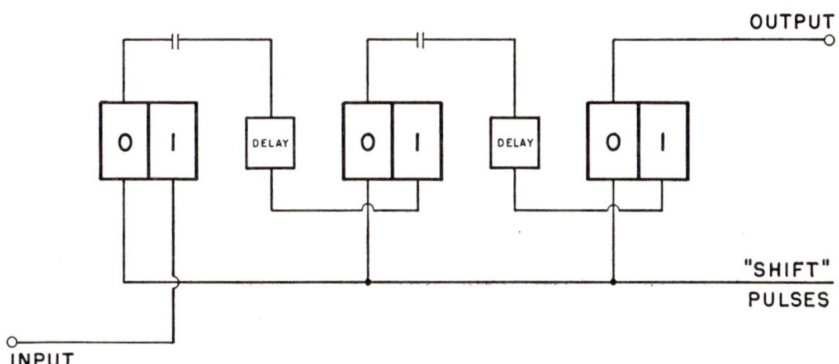

Figure 9. A shift register.

the binary digit stored in one flip-flop is transferred to the flip-flop immediately to the right. Thus, if a serial, binary number is fed into the register at the extreme left-hand end in synchrony with application of the command or shift-pulse digits, the successive digits will be successively shifted from the left toward the right until, after the application of as many shift pulses as there are digits in the number, the serial number will be completely stored in the register.

To read the number out of the register, it is only necessary to repeatedly shift the pulses one step to the right and observe the successive digits appearing at the extreme right-hand end. The digits of the number will then reappear in the same sequence in which they were stored.

It is possible to build a register combining the circuits of Figures 8 and 9 so that one has a shift register with parallel access and read-out. In this register, a number may be written in serially with parallel read-out or written in parallel with serial read-out. By using this device it is possible to convert from a parallel to a serial representation or vice versa.

In order to perform multiplication efficiently, every schoolboy must first learn a multiplication table. Here again the binary numbering system has an advantage in that the multiplication table consists of only four entries, as illustrated in Table 5. Upon examining this table,

Table 5. Binary Multiplication

0	0	1	1
×0	×1	×0	×1
—	—	—	—
0	0	0	1

one can write the simple rule for multiplication—the product is "A and B." Thus, the simple "and" gate in Figure 10 is a multiplying

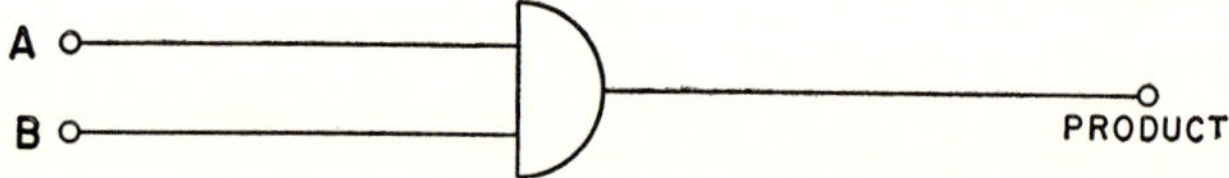

Figure 10. A single "and" gate element functions as a binary multiplier.

circuit for a pair of binary digits. Two binary numbers, each consisting of several digits, are multiplied in a manner similar to that of ordinary decimal multiplication. This is illustrated in Table 6 where two binary numbers, together with their decimal equivalents, are shown.

One first multiplies the multiplicand by the least significant digit of the multiplier. The multiplication is accomplished digit by digit of the multiplicand. One next multiplies the multiplicand by the next least significant digit of the multiplier and records this under the previously obtained partial product but shifted one place to the left. This procedure is repeated for each digit of the multiplier, and the resulting

Table 6. Example of Binary Multiplication

Binary	Decimal
101101	= 45
1011	= 11
———	—
101101	45
101101	45
000000	—
101101	495
———	
111101111	= 495

partial products are summed to give the total product of the two original numbers.

A circuit for accomplishing this is shown in Figure 11. Although this circuit illustrates serial addition and multiplication for reasons of

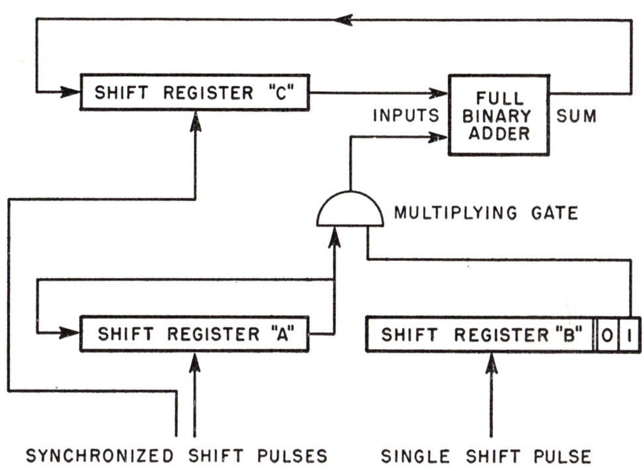

Figure 11. A serial multiplier.

simplicity, other circuits may be devised that will perform multiplication serially and addition in parallel, or multiplication in parallel and addition in parallel. In operation, the multiplicand and the multiplier are entered respectively into the shifting registers A and B. Numbers may be placed in these registers either in a serial or a parallel form as previously explained. The product of A and B will be recorded in shifting register C, which is initially set so that its contents at each digit place is a zero. Shifting pulses are applied to registers A and C simultaneously, and the output from register A is passed through the "and" gate, where it is multiplied by the least significant digit of the multiplier, and then to one input of a binary adder where it is added to the content of the register C. Since the contents of the register C were initially zero, this places the product of the multiplicand and the least significant digit of the multiplier in the register C.

A single shift pulse is now applied to register B so that the second least significant digit of the multiplier appears in the extreme right-hand flip-flop of register B. Shifting pulses are again applied to registers A and C so that the multiplicand is multiplied by the second least significant digit of the multiplier and added to the contents already stored in register C. These shifting pulses are so synchronized that the contents of register C arrive at the binary adder one digit time ahead of the new partial product. Thus, the product of the multiplicand and the second least significant digit of the multiplier is added to the previous partial product but shifted one place to the left. This process is repeated until each digit of the multiplier has been accounted for, at which time the product of A times B will be stored in register C. By developing equipment designed to function according to rules of arithmetic, the processes of addition and multiplication may be performed.

By an extension of these techniques, subtraction and division may also be performed in the same device, usually called the arithmetic unit, which, when presented with two numbers and a command signal, will operate on the two numbers in a manner determined by the command signal. For instance, it may add, subtract, multiply, or divide the two numbers, depending upon which command was given. The command is usually in the form of a voltage placed on various wires, which, when energized, cause the arithmetic unit to do addition, subtraction, multiplication, or division.

MACHINE CONTROL AND MEMORY

The foregoing section has described how the machine performs operations. However, even if a machine were capable of performing arithmetic operations at extremely rapid rates, it would be of little value if human intervention were required to supply the numbers and commands for the arithmetic operations being performed. Means must be provided by which the arithmetic unit may be made to perform the required sequence of arithmetic operations, and data storage is needed to supply the numbers to be operated upon and to record the results of the arithmetic operations. The first of these functions is fulfilled by the control unit and the second, the memory.

For the memory we need a device capable of storing or registering a great many binary numbers. The numbers in the memory must be capable of being read out rapidly or of being written in rapidly. The registers themselves may be used as a memory element so that the memory could consist of a collection of these registers. As a matter of fact, the first large-scale high-speed digital computer constructed, the ENIAC, used this form of the memory. The disadvantage of this register-type of memory is the excessive physical size, power consumption, and cost per digit stored. In most modern high-speed general-purpose digital computers, the binary digits are stored in small toroidal ferrite cores, each of which measures approximately one-tenth of an inch in diameter and can store one binary digit. Magnetic cores are of small size, low cost, and require no standby power.

To show how a memory could be constructed using registers, Figure 12 illustrates a memory which accepts numbers in parallel form and gives them back also in parallel form.

To write a number into a memory location such as that of the second register, one supplies the number in parallel form on the memory input wires, a signal to write on the write wire, and a signal on the control wire for the No. 2 memory location. This combination of control signals gives a command to the input of the No. 2 register such that the number coming along on the input wires will be recorded in the No. 2 register and the No. 2 register only. To read out a number, for instance the number which was just stored in the No. 2 register, one supplies a signal on the read-out wire and a signal on the No. 2 register wire. This combination of control signals enables the number stored in register No. 2, and only that number, to appear on the memory output wires. Although the system shown is capable of storing only four binary numbers, the principles outlined may be extended to cover any quantity of numbers. To activate the memory one needs to supply a signal which tells it either to read or write and a signal which identi-

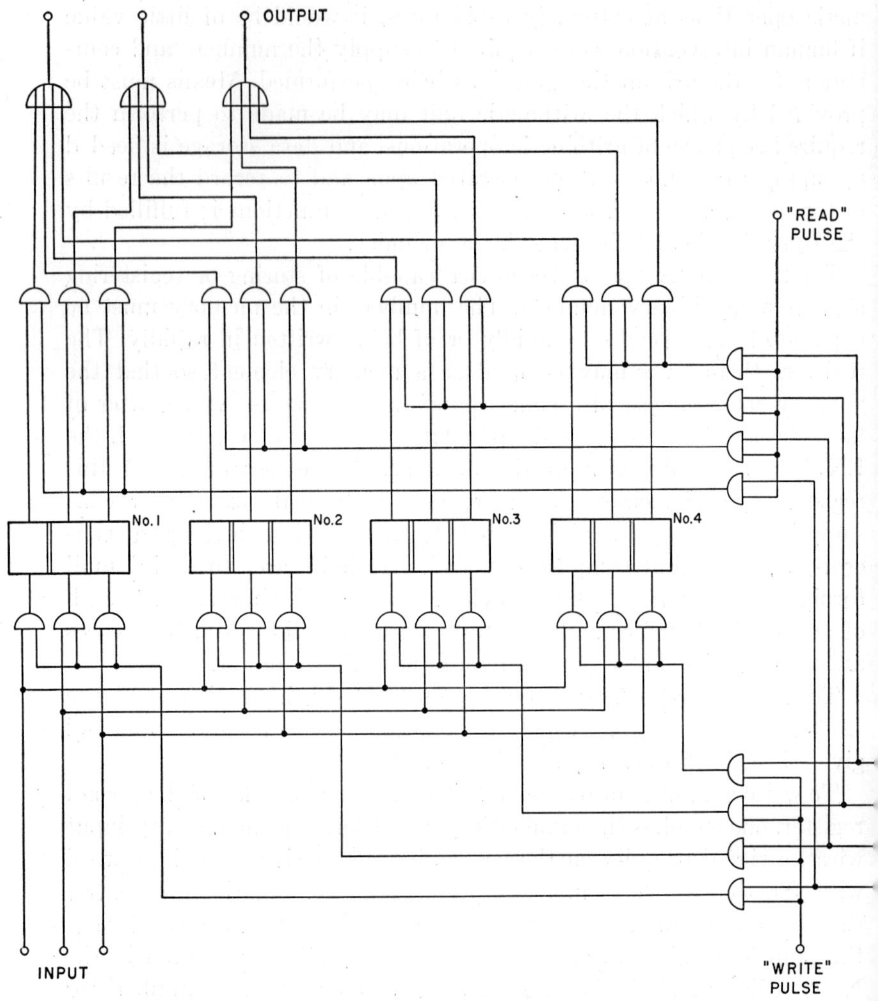

Figure 12. A simple memory unit capable of storing four binary digits.

fies which register is to be acted upon. These individual registers or memory locations are usually numbered consecutively, and these numbers provide the *address* of the memory locations.

At this point in our discussion another advantage of the binary numbering system may be recognized. Suppose, for purposes of illustration, that our computer is to operate with an accuracy of one part in a thousand. That is, no matter what numbering system is incorporated in the computer, it must employ sufficient digits in that numbering system to distinguish between any one of a thousand different counts. To do this with a decimal system would require three decimal digits, forming the numbers 000 through 999. Each memory cell of a decimal machine would then have to be able to record three decimal digits, each decimal digit being any one digit out of a list of ten different digits. Since our basic storage cells are binary in nature (that is, flip-flops, either one side or the other conducting; or the magnetic core magnetized in one direction or the other), in order to store the decimal digit one would have to employ ten of these basic binary units, nine of them sitting on 0 and one of them sitting on 1. The cell on 1 would indicate the required stored decimal digit. To store a three-digit decimal number would thus require a total of thirty binary storage blocks.

In contrast, our binary machine for the same accuracy must be capable of storing ten binary digits, since a count from 0000000000 to 1111111111 is a total of 1,024 counts. However, to store a ten-digit binary number requires only ten basic binary storage elements, one for each digit. Thus a decimal machine of the same accuracy as a binary machine requires three times as much storage media as the corresponding binary machine. When one realizes that most general purpose computers are not ten-digit binary numbers but thirty to forty digit numbers, typically providing storage spaces for up to 16,000 different numbers, or total storage for 480,000 or 640,000 bits, the three-to-one saving achieved by a binary machine is notable.

The control unit is a device in which the instructions given the machine are interpreted. An instruction must include specification of the desired operation, the locations (i.e., the addresses) of the numbers to be operated upon, the address where the result of the operation is to be placed, and an indication of the next instruction to be performed. The operations to be performed are usually numbered, or encoded, into a numerical form, so that the complete instruction appears as a number and may be handled by the machine like any other number. In particular, it may be stored in the memory. When this is done, the next instruction to be performed may be indicated by giving the address in the memory where it is stored.

Although four addresses are required for a complete instruction, not all four addresses may be stated explicitly; some may be implied. As noted previously in the multiplier example, there are registers of storage elements in the arithmetic unit, so that a complete instruction may be given by stating an operation and a single address, the other three addresses being implied as follows: The operation called for is performed on the number stored in the single address given, and upon the number stored in the arithmetic unit, the result is left stored in the arithmetic unit, and the next instruction which the machine performs is the one stored in the next memory address in numerical order. The machine starts at memory location 1, performs that instruction, goes to location 2, performs that instruction, then to locations 3, 4, and so on. Between a single-address machine (with three addresses implied) and the four-address machine, one has various combinations of stated and implied addresses, yielding two-address machines, three-address machines, which categories may be further broken down according to which locations are implied and which are explicitly stated.

The basic machine will work on a two-part cycle. During the *instruction part* of the cycle, the instruction is obtained from memory and the machine is set up to perform on the numbers specified the operation called for. During the *execution part* of the cycle, the actual operation is performed. In some machines these two parts go on sequentially; in others, concurrently.

During the first part of the cycle the proper instruction is obtained from memory and stored in the control unit. Each part of the instruction is stored in a separate register. The operation being performed is stored in a register called the *operation register*, and the addresses are stored in separate *address registers*. In a multi-address machine one of these address registers is usually the address of the next instruction to be performed, while in a single-address machine a separate register, usually called an *instruction counter*, is used.

After the execution part of the cycle, the next instruction is obtained from the memory as the number which is stored in the address location, given by the number in the address register or the instruction counter. Thus, in a single-address machine, after each execution cycle, the instruction counter automatically increases its count by one, so that the instruction sequence followed is given by the instructions stored in numerical order in a given part of the memory.

Since the instructions to be performed by the computer must be compiled by an operator and fed into the machine, one might ask why the person who is compiling and writing them down could not perform the operations and thereby compute the desired result? One answer to this question is that for many applications the same kinds of problems have to be run over and over again, with just a change in the input data. In such cases a single compilation of instructions, or *program,* as it is called, can be reused, with a saving of time for each repetition of the problem. Moreover, in most problems there is a large amount of repetition in the kinds of instructions that are given to the computer. Because of this and because of the fact that the instructions stored in the machine are numbers and may themselves be modified by the machine, the computer may be used to generate part of its own program.

By way of illustration, consider a decimal computer operating with a single-address instruction. Further assume that this machine operates on five-digit decimal numbers and that, for our instruction, the first two digits will be the operation to be performed by the machine, such operations including the ordinary arithmetic operations, while the remaining three digits shall be the address of one of the operands. Suppose further that the machine has 1,000 memory locations numbered or addressed by the numbers 000 through 999. Suppose one wishes to compute a table of factorials. One way in which the instructions may be compiled is shown in Table 7.

Here it is noted that to compute each factorial, two instructions are necessary, one to perform the required multiplication and the other to store the result in the memory. For each factorial, two additional memory locations are required, one to store the number whose factorial is desired and the other to store the factorial itself. Thus, only 250 factorials could be computed with the memory space available, and this would require the operator to compile a list of 500 instructions, 250 multiplications, and 250 store instructions. If the length of instructions is excessive, large amounts of time are required to prepare them. Furthermore, this long list of instructions takes valuable memory space and limits the useful capacity of the machine.

The machine itself, however, may be used to write part of its detailed instruction by following the program shown in Table 8. Note that in the program of Table 7, the only difference between the successive multiply and store instructions is a difference in the address by the number 1. The new instructions are produced by taking the old instructions and adding one to them. After the instructions have been modified, it is necessary for the machine to go back and perform the same series of operations over again. In a multiple-address machine this would present no difficulties because the last instruction would specify the memory instruction of the next instruction. However, in

Table 7

Memory Location	Contents	
	Instruction (Two Digits)	Address
000	Clear and add	500
001	Store	750
002	Multiply	501
003	Store	751
004	Multiply	502
005	Store	752
006	Multiply	503
007	Store	753
008	Multiply	504
'	'	'
'	'	'
'	'	'
500	00001	
501	00002	
502	00003	
503	00004	
504	00005	
'	'	
'	'	
'	'	
749	00250	
750	Blank	
751	"	
752	"	
'	'	
'	'	
'	'	
999	Blank	

our single address machine normally the next instruction is the one in the next memory location. In order to break this sequence, the machine is furnished with an additional instruction called a *branch* or *transfer instruction.* When this instruction is given, the instruction counter is not advanced by 1, but the address given, along with the transfer instruction, is stored in the instruction counter. Thus, the computer may be programmed to go through an iterative loop to calculate the factorial outlined by the program in Table 8.

Table 8

Memory Location	Contents	
	Instruction (Two Digits)	Address
000	Clear and add	506
001	Multiply	014
002	Store	507
003	Clear and add	000
004	Add	013
005	Store	000
006	Clear and add	001
007	Add	013
008	Store	001
009	Clear and add	002
010	Add	013
011	Store	002
012	Transfer	000
013	00001	
014	00002	
015	00003	
⋮	⋮	
505	00493	
506	00001	
507	Blank	
⋮	⋮	
999	Blank	

In this case it requires thirteen instructions for the machine to be able to compute one factorial and automatically program itself to compute the next one, while in the program of Table 7 it required only two instructions per factorial so that the machine operates at an overall slower rate. The actual list of instructions, however, that an operator would have to compile is only thirteen, while in the previous case it would be 500, two for each of the 250 numbers to be calculated. Furthermore, by requiring only nine memory locations to store the instructions, almost twice as many factorials (actually 493) could be computed with the limited memory space that the machine has.

There is another disadvantage to the program shown in Table 8: once this iterative procedure has been started, there is no means provided to stop it. If all one wanted were to calculate factorials, this would present no problem, for the machine would continue to calculate factorials. However, in some problems it is desirable to start an iterative procedure, repeat it a specified number of times, and then go on to something else. How this is accomplished is shown in the program

Table 9

Memory Location	Contents	
	Instruction (Two Digits)	Address
000	Clear and add	035
001	Multiply	016
002	Store	036
003	Clear and add	000
004	Add	015
005	Store	000
006	Clear and add	001
007	Add	015
008	Store	001
009	Clear and add	002
010	Add	015
011	Store	002
012	Subtract	056
013	Transfer if negative	000
014	Transfer	057
015	00001	
016	00002	
'	'	
'	'	
'	'	
034	00020	
035	00001	
036	Blank	
'	'	
055	Blank	
056	-- 056	
057	Next Instruction	

of Table 9. Here, by the addition of two more instructions, the machine is enabled to calculate the first twenty factorials and then proceed with some other kind of instruction routine. This is accomplished by a *subtraction instruction* and a *conditional transfer instruction* or a *"branch if negative" instruction.*

After both of the multiply and store instructions have been updated, the last number contained in the arithmetic unit is the updated store instruction. If one subtracts from this number a number representing an identical instruction, but with an address one larger than the largest address used to store the numbers whose factorials one wishes to compute, the result of this subtraction will be a negative number for all of those cases for which one desires to calculate an additional factorial, and will be zero when the last factorial has been computed. By the use of a conditional transfer instruction, which replaces the number in the instruction counter with the address given in that instruction if the contents of the arithmetic unit is negative, and which simply adds one to the instruction counter if the contents of the arithmetic unit is a zero or positive number, the computer will automatically continue in the iterative loop until the last factorial has been computed, and then proceed with the next desired instruction. In this way the skillful programmer may use the machine to help him write the detailed list of instructions that the machine is to follow.

It should also be noted that the conditional transfer instruction enables the computer to make elementary decisions. We have seen an example of this, in which the decision involved is a choice between continuing on one routine or switching to another. As noted previously in a discussion of basic computer building blocks, any complex decision or logical function can be decomposed into a series of elementary, logical functions or decisions. Thus, the computer can be made to perform seemingly complex decisions. Alternately, it may be programmed to determine the basis for making a correct decision by operating under different decision doctrine and then examining which set of results is the most favorable.

The computer itself may also be programmed to simulate computers not yet built, and in this way may aid in the design of new computers. The slow speed and limited capacity for quantitative detail of individual human beings handicap an engineer in completing a good design in the time he has available. Because of this, second and third generation computers, whose very designs are predicated upon the existence of an earlier generation of computers, are appearing or soon will be. It is interesting to speculate where these procedures will lead as we find our capacities for design chained to or at least dependent upon the characteristics of machines which are seemingly able to reproduce themselves in a manner determined by their own capabilities.

REFERENCES

(1) Canning, R. G. *Electronic Data Processing for Business and Industry.* New York: John Wiley and Sons, 1956.
(2) Culburtson, J. T. *Mathematics and Logic for Digital Computers.* New York: D. Van Nostrand, Inc., 1958.
(3) McCracken, D. D. *Digital Computer Programing.* New York: John Wiley and Sons, 1957.
(4) Phister, M. *Logical Design of Digital Computers.* New York: John Wiley and Sons, 1958.
(5) Richards, R. K. *Arithmetic Operations in Digital Computers.* New York: D. Van Nostrand, Inc., 1955.

Twelve

ELEMENTS OF INVENTORY SYSTEMS

ELIEZER NADDOR

INTRODUCTION

Terminology

Interest in the subject of inventory systems is constantly increasing, and its development in recent years parallels closely the development of operations research in general. Some authors even claim that, "More operations research has been directed toward inventory control than toward any other problem area in business and industry." (10, p. 195)

In the past, attempts have been made to define such terms as "the inventory problem," (4, p. 187), "inventory control," (16, p. 289), and "inventory process," (1, p. 271), but no generally accepted terminology yet exists in this area. Not only do we not have commonly accepted terms, but there also seem to be numerous viewpoints on what the inventory research field should encompass.

Some consider inventory systems those dealing with record-keeping of the amounts of commodities in stock. An inventory problem, for these people, is the problem of determining who should make the entries in the records, what details should be recorded, where and when entries should be made, etc. Others look at inventory systems from a broader point of view. For them, the inventory system usually does not involve specific commodities, but rather the totality of commodities and the investments in all the stocks in inventory. Their problems are problems of inventory turnover, problems of financing the investments tied up in stocks, etc. They are the people who are concerned more often than not with excessively high inventories and with how to reduce them.

Other groups consider inventory systems from still another point of view. Their problems are the problems of what items to stock, when to stock them, how many to stock, etc. They are also usually concerned with labor stability problems, utilization of equipment and facilities, customer relations, etc.

Because of the existence of this wide range of different points of view, it is necessary to devote some time to definitions of the inventory terms to be used. Most of these terms are proposed here for the first time, but they do not contradict or conflict with other known inventory definitions. It is hoped that the terminology given here may eventually be adopted for general usage in the inventory research field.

INVENTORY SYSTEMS

An inventory system is one in which only the following three types of costs are significant, and in which any two or all three are subject to control:

1. The cost of carrying inventories,
2. The cost of incurring shortages,
3. The cost of replenishing inventories.

An example in which these types of costs arise and in which they are subject to control is a production system in which future requirements are anticipated. In such a system costs may be controlled by making appropriate decisions with respect to ordering of raw materials, manufacturing of semi-finished and finished goods, and inventory of goods in readiness for shipment to consumers.

The first type of cost in this production system, the cost of carrying inventories, is the cost of the investment in inventories, the cost of storage, of handling, of obsolescence, etc.

The second type of cost, the cost of incurring shortages, is the cost of lost sales, loss of good will, overtime payments, the cost of special administrative efforts (telephone calls, memos, letters), etc.

The third type of cost, the cost of replenishing inventories, is the cost of machine setups for production runs, the cost of preparing orders, handling a shipment, etc.

These three types of costs will be referred to in general as the *surplus cost, shortage cost,* and *setup cost,* respectively. The sum of these costs will be referred to as the *total cost.*

The term *cost* is used in a broad and general sense, beyond the dollars and cents which are usually associated with the word. Cost is the measure of effectiveness used for the systems under study. We assume that surpluses, shortages, and setups can all be measured by a common unit of effectiveness, so that meaningful comparisons can be performed. It seems natural that effectiveness in this case should be referred to as cost.

By defining inventory systems as those in which the costs of surplus, shortage, and setup are significant and are subject to control, we imply that other costs are not pertinent. This does not mean that the methodology and results presented in this chapter do not apply to a wide range of systems in which there are other significant costs in addition to those associated with surplus, shortage, and setup. On the contrary, the methods and results do apply and examples can be given to show how. The above definition of inventory systems is necessary as a frame of reference and as an indication of the scope of the subject matter to be studied. Many systems encountered in practice are inventory systems in the sense defined above.

INVENTORY PROBLEMS

An *inventory problem* is a problem of making optimal decisions with respect to an inventory system. In other words, an inventory problem is concerned with the making of decisions which minimize the total cost of an inventory system. When one cost is decreased (or increased) the other costs may increase (or decrease). There is thus the problem of controlling the costs so that their sum will be the lowest. The inventory problem is thus defined in terms of making *optimal decisions* with respect to costs.

In reality the decisions which are made affect the costs, but they can rarely be made directly in terms of costs. The decisions which are usually made are in terms of time and quantity:

1. When should the inventory be replenished?
2. How much should be added to inventory?

The time element and the quantity element are the *variables* which are subject to control in an inventory system. (They are also referred to as the *controllable variables.*) They in turn affect the costs of surplus, shortage, and setup, and their sum, the total cost. The inventory problem is to find the specific values of the variables which minimize the total cost.

Although finding the variables which give the minimum total cost is the main part of the inventory problem, there are other problems to be answered. One of these is finding the minimum cost itself. The research worker developing decision rules to be used by operating personnel should not only say, "This is the rule." He should also be able to state, ". . . and this is how much it is going to cost."

Example 1 gives a simple illustration of an inventory problem:

Example 1: A customer orders from a manufacturing company shipments of a certain part at a uniform rate of 2,400 parts per year. The company makes the parts on a machine which can produce any number of parts at a time. The cost of setting up the machine for a production run is $42.00. The cost of carrying inventory is $0.56 per part per year. No shortages are allowed to occur.

The manufacturing company's inventory problem is to find how many parts should be produced for each production run. This variable is referred to as the lot size q.

If q is chosen to be 100 parts, then the inventory fluctuations during the first half of the year may be represented by the full lines in Figure 1.

There will be 50 parts, on the average, in inventory during the year. Hence the inventory carrying cost will be $50 \times 0.56 = \$28.00$ per year. Also, if q is 100, then there will be 24 set-ups during the year. These

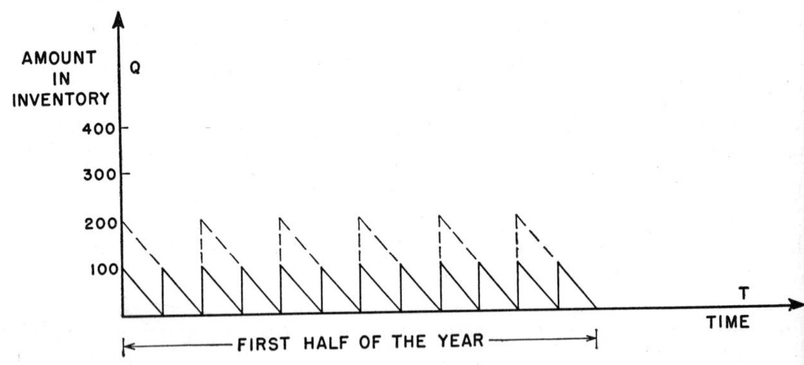

Figure 1. Inventory Fluctuations of Parts During the First Half of the Year.

will cost $24 \times 42.00 = \$1{,}008$ per year. Since no shortages occur in this case, the total cost of the inventory system, for a lot size of 100, will be $1,036.

Now suppose that the lot size is 200, i.e., q is chosen to be 200 parts. The inventory fluctuations during the first half of the year may be represented by the broken line in Figure 1 above. The total cost per year of the inventory system in that case will be:

$$100 \times 0.56 + 12 \times 42.00 = 56.00 + 504.00 = \$560.00 \text{ per year.}$$

Table 1 gives the total cost of the inventory system for various values of the variable q.

Table 1. Total Cost per Year of an Inventory System as a Function of Various Lot Sizes

Lot Size q	Average Inventory	Inventory Carrying Cost/Year	No. Setups/Year	Cost of Setups/Year	Total Cost/Year
100	50	$ 28.00	24	$1,008.00	$1,036.00
200	100	56.00	12	504.00	560.00
300	150	84.00	8	336.00	420.00
400	200	112.00	6	252.00	364.00
500	250	140.00	4.8	201.60	341.60
600	300	168.00	4	168.00	336.00
700	350	196.00	3.43	144.06	340.06
800	400	224.00	3	126.00	350.00
900	450	252.00	2.67	112.14	364.14
1000	500	280.00	2.4	108.00	388.00
1100	550	308.00	2.18	91.56	399.56
1200	600	336.00	2	84.00	420.00

Here, then, a lot size of 600 parts will minimize the total cost of the system and this minimum is $336 per year. (Later we will show how the answers can be obtained without computing the total cost for numerous values of the variable q.)

In this example the two decisions as to when to replenish and by how much are answered as follows: replenish every three months (whenever inventory reaches a zero level) and make a lot of 600 parts. These decisions will result in a minimum total yearly cost of $336.

An inventory system was defined as a system in which three types of costs were subject to control. An inventory problem can also be said to be a problem of *balancing* these costs. In other words, an inventory problem can be defined as the problem of balancing the three types of costs so that their sum will be minimized.

It is interesting to note that in Example 1, above, the minimum cost is obtained when the surplus cost and setup cost are equal. We will see later that this result is not a coincidence and that it holds for a certain class of inventory systems.

SOLUTION OF INVENTORY SYSTEMS

When we speak about the *solution of an inventory system* we have in mind the solution of the inventory problem. That is, the solution is a set of specific values of the controllable variables which minimize the total cost of the inventory system. In some cases specific values cannot be obtained because of the general nature of the problem studied or because of other reasons which are explained later. In these cases the solution is given by a set of *decision rules.* The rules specify how the specific values of the controllable variables can be obtained. We will refer to these rules as the *optimal decision rules.* The word optimal is added to remind us that the inventory problem and its solution are concerned with an optimization process. In our case this process is a minimization process, but rather than use the phrase "minimum decision rules," which may have several connotations, we prefer the phrase "optimal decision rules."

It will be recalled that the decisions pertaining to an inventory problem deal with two questions: "when?" and "how much?" The first question is usually answered in one of two ways:

a. Inventory should be replenished every t_o units of time.[1]
b. Inventory should be replenished when the amount in inventory reaches a level of s_o units of quantity.

The second question is also usually answered in one of two ways:

a. The quantity to be ordered, the lot size, is q_o units of quantity.
b. A quantity should be ordered so as to bring the amount in inventory to a level of S_o units of quantity.

In these terms, the solution of the inventory system described in Example 1 can also be stated as follows: $s_o = 0$ (i.e., inventory should

[1] The subscript $_o$ indicates that t_o is an optimal quantity.

be replenished when the amount in inventory reaches a level of 0 parts) and $q_o = 600$ parts (i.e., the quantity ordered, the lot size, is 600 units).

(In this example we have not shown formally why s_o should be zero, but we have assumed that the reader will agree intuitively with this result. Here, also, there exists a simple relation between the quantity to be ordered q_o and the optimal levels s_o and S_o, namely $q_o = S_o - s_o$, Since we assumed that $s_o = 0$, it follows that $q_o = S_o$.)

MODELS OF INVENTORY SYSTEMS

What method should one use to find the solution of an inventory system? In Example 1, the problem was to find the optimal lot size. In Table 1 we listed twelve alternative lot sizes and found the total cost of the system for each alternative. We noticed that when $q = 600$, then the total cost was minimum and concluded that the optimal lot size should be 600. This method of approach, though suitable in some instances, is generally not the best method, since (a) it does not insure that the best alternative is actually included in the list of alternatives and (b) it is lengthy and time-consuming.

Many inventory systems can be solved by another method which is generally more effective: a mathematical model describing the system is obtained and optimal decision rules are derived from the model.

A *model* is a representation of a system under study. A mathematical model is a model in which the system is represented by symbols which can be manipulated mathematically.

Consider, for example, the inventory system in Example 1. Let the lot size be q, and let the annual total cost of the system be C. Then the mathematical model of this system is given by:

$$(1) \ldots \qquad C = (q/2) \times 0.56 + (2400/q) \times 42.00 = 0.28q + 100800.00/q$$

The first term on the right side of equation (1) represents the surplus cost of the system per year. The second term represents the setup cost per year. The shortage cost is zero in this system. Thus equation (1) gives the three types of costs of the inventory system in symbolic terms. Here the controllable variable q, the lot size, and the total cost C per year can be manipulated mathematically. That is, it is possible to find in this case by the use of mathematical techniques what value of q will give the minimum total cost per year. The techniques of analytic geometry and calculus are useful for this purpose.

Using analytic geometry, one plots the graph of the curve $C = 0.28q + 100800.00/q$ as shown in Figure 2. From the graph we find by in-

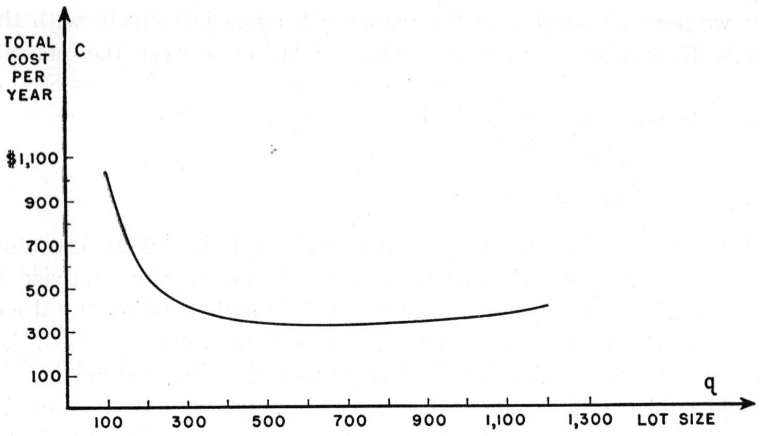

Figure 2. Total Cost Per Year as a Function of the Lot Size.

spection that the optimal lot size is 600 parts and that the minimum total cost is $336 per year. Actually there is almost no difference between using this method of solution and that employed in the derivation of Table 1.

Using the calculus ($dC/dq = 0$ etc.)[2] one obtains the following result:

$$q_o = \sqrt{100800.00/0.28} = 600 \text{ parts},$$

and

$$C_o = 2\sqrt{0.28 \times 108.00.00} = \$336/\text{year}$$

Most of the models to be discussed in this chapter will be of the form

$$(2) \ldots \qquad C = C_1 + C_2 + C_3$$

where C_1 will designate the surplus cost per unit time, C_2 the shortage cost per unit time, C_3 the set-up cost per unit time, and C the total cost per unit time. Each cost will generally be a function of one or more controllable variables (as q in Example 1) and of some parameters of the system (as 2400 parts, $0.56, $42.00 in the example).

[2] For the detailed derivation see page 338.

The primary purpose of the model will be to obtain solutions of inventory systems:

(i) Develop optimal decision rules,

(ii) Determine minimum total cost of systems.

It can also be used to:

(iii) Analyze the sensitivity of results,

(iv) Compare different inventory policies.

PROPERTIES OF INVENTORY SYSTEMS

In order to develop a model of an inventory system it is first necessary to know the properties of the system.

In Example 1 the customer orders shipments at a uniform rate of 2,400 parts per year. This "uniform rate" is a *property of the system* and so also is the fact that no shortages are allowed to occur, the cost of carrying inventories is $0.56 per part per year, and the setup cost is $42.00; these, too, are properties of the system. Had the example stated that the customer is *expected* to order 2,400 parts per year, this would have implied that he might order other quantities, or perhaps nothing at all. Such an expectation, as opposed to a known uniform rate, is yet another property that an inventory system may have. It can be shown that systems in which the rate of shipments is not uniform have substantially different optimal decision rules from systems with uniform rates.

The knowledge of the properties of a system is essential for obtaining a model which represents the system under study. Omitting properties, or making wrong assumptions about properties, will usually lead to incorrect decision rules. It should be mentioned, though, that correct (optimal) decision rules are sometimes obtained although wrong assumptions are made about a system. Similarly, the difference between the incorrect and correct decision rules may sometimes be rather insignificant. Only a knowledge of the properties of a system can enable us to say whether the decision rules are correct or whether the differences between correct and incorrect rules are insignificant.

Selection of Inventory Systems for Analysis

The problem of selecting inventory systems for analysis is not a simple one. When the properties of inventory systems in general are studied, one recognizes immediately that there exist almost an infinite number of different inventory systems. Which are the systems that deserve analysis?

Essentially, several approaches to the selection of inventory systems for analysis are possible. One may attempt to select the most general inventory system, analyze it in detail, and hope to be able to apply the results to any specific inventory systems. A second approach is the case study approach. The users of this approach elect and analyze specific inventory systems, and suggest that other inventory systems can be analyzed in a similar way. Still another approach is possible. First relatively simple inventory systems are analyzed and then more and more complex systems. The author has found the third approach the most suitable. It is certainly the most suitable for an introduction to the subject. But it is equally good as a basis of research work. It provides the "building blocks" on which the analysis of any system can be based.

BRIEF HISTORICAL REVIEW

The earliest known analysis of an inventory system is that of F. W. Harris in 1915 (15, p. 121). Harris is assumed to be the first person to have published the lot size formula:

$$q_o = \sqrt{2rc_3/c_1} \qquad (3)$$

where q_o is the economic lot size, r is the rate of requirements per unit time, c_1 is the cost of carrying one unit in inventory for a unit of time, and c_3 is the cost of replenishing inventory. This formula has had more applications than any other obtained from the analysis of inventory systems.

The system for which this formula holds is a very special one. The rate of requirements is assumed to be known with certainty, no shortages are allowed to occur, lead-time is assumed to be insignificant, etc. This means that the formula may be used only for systems in which the above assumptions hold, otherwise it may not give the economic lot size at all.

Researchers have recognized the shortcomings of the above formula for quite some time. Cooper in 1926 analyzed an inventory system in which the rate of production is considered. (In Harris' system the rate of production is assumed to be infinitely higher than the rate of requirements.) Fry in 1928 studied an inventory system in which requirements are not known with certainty. He showed how the theory of probability can be used to solve some inventory problems.

The first attempt to deal with a large variety of inventory systems and to present the beginning of a theory of inventory systems has been made by Raymond (15) in his book *Quantity and Economy in Manufacture,* published in 1931 and now out of print.

Interest in the study of inventory systems has increased since World War II. Numerous publications have been devoted to this subject. An excellent review of the systems which were studied until 1951 is given by Whitin (19) in his book *The Theory of Inventory Management.* The new edition of this book includes some more recent developments as well. Whitin's bibliography contains 180 entries and covers the period 1923 to 1951. The new edition has 43 additional entries which cover publications up to 1956.

The publication of the paper "Optimal Inventory Policy" by Arrow, Harris, and Marschak (2) in 1951 marks the beginning of what may be called the modern analysis of inventory systems. This paper constitutes a considerable advance in the study of inventory systems and has influenced numerous research workers in their approach to inventory systems. In the conclusion of their paper, the authors outline the general nature of other systems which should be studied.

An attempt to analyze the general systems proposed by Arrow, Harris, and Marschak was made by Dvoretsky, Kiefer, and Wolfowitz (4) in their paper "The Inventory Problem," published in 1952. In this paper very powerful mathematical and statistical tools have been employed and the systems studied have been of a very general nature.

Since 1952 more and more work has been devoted to inventory systems. Most of this work has been carried out by operations researchers. A considerable number of articles on the subject now appear in such journals as *Operations Research, Management Science, Naval Research Logistics Quarterly, Industrial Engineering,* and many others.

The references at the end of the chapter list reviews, articles, and books pertaining to inventory systems which have appeared in recent years.

APPROACHES TO SELECTION OF INVENTORY SYSTEMS FOR STUDY

When one examines the systems which have been studied these last forty years, one recognizes three approaches to the selection of inventory systems for study. The first is used by those who believe that it is possible to analyze some very general inventory systems and that the analysis could be used for the study of any inventory system. Such has been the approach of Dvoretzky, Kiefer, and Wolfowitz (4). They have actually succeeded in obtaining some general results mainly related to the existence and uniqueness of solutions. These results, however, are particularly of mathematical interest, and it is generally rather difficult to see how they could be used in the analysis of specific inventory systems.

The second approach is used by those who believe that, by analyzing some specific, hypothetical inventory systems encountered in practice, the ground can be prepared for analyzing any inventory system. This has been the approach of most research workers. This approach has been adopted by Arrow, Harris, and Marschak (2), Whitin (19), Welch (17), and numerous others. This approach usually leads to explicit decision rules for the specific systems studied, and the results can be applied to corresponding actual systems encountered in practice. In general, however, the actual and hypothetical systems do not correspond and adjustments are necessary in order to apply the results. Carrying out these adjustments is not always an easy task.

In this chapter the author has adopted the third approach. The users of this approach select relatively simple hypothetical inventory systems for analysis. They are primarily interested in developing *methods* of analysis rather than in obtaining general results or results for specific systems encountered in practice. Thus, we shall primarily be interested in setting up a methodology for analyzing inventory systems as *extensions* of *elementary systems*. Generally speaking, an elementary system is a system with simple properties. (The decision as to what are simple properties and hence what are more complex properties is arbitrary.) We will define simple inventory properties later and will analyze in detail the three inventory systems having these simple properties. These systems will be referred to as *elementary inventory systems*. All other systems will be assumed to be extensions of the elementary systems. Five extensions of the elementary inventory systems will be analyzed in detail.

PROPERTIES OF INVENTORY SYSTEMS

In order to be able to analyze an inventory system its properties must be established. This section deals with the identification of the various components of every inventory system and its properties. Three components are recognized for every system: *output, input,* and *costs.*

Briefly stated: "output" refers to what is taken *out* of inventory; "input" refers to what is put *in;* and "costs" refers to the pertinent *costs* associated with positive and negative inventories (the *surplus* and *shortage costs*); and with production (or ordering) operations (the *setup costs*).

Output

Output is discussed first because this is essentially the most important component. Inventories are kept to meet demands, to fill orders, to satisfy requirements. These demands, orders, and requirements we call *output.* Inventory problems exist only because there is output; otherwise we have no inventory problems.

Generally, output cannot be controlled directly and in many cases it cannot be controlled even indirectly, because output usually depends on decisions made by people outside the organization. Although output is generally not controllable its properties may be studied. When do customers place their orders? How much do they require? Is demand higher at the beginning of the month or at the end of the month? Do we have accurate information on future requirements, or do we have to estimate average demands and ranges of demand? All these properties are significant and affect the solutions of inventory systems.

OUTPUT SIZE

The quantity required for output is called *output size.* The size may be 20.3 pounds of some chemical, or 531 nails, or 37 automobiles, etc. We denote the size by x. The physical dimension of x is thus the *quantity dimension.* We denote this as follows:

$$[x] = [Q] \qquad (4)$$

where the square brackets refer to dimension and Q refers to quantity.[3]

When the output size is the same from period to period we say that it is *constant.* Otherwise we refer to it as *variable.* When we have precise advance information about the output size we say that it is *known with certainty.* When it is not known with certainty it is sometimes

[3] The other dimensions to be used in the chapter are those of time [T] and cost [$].

possible to ascertain its *probability distribution.* This distribution will be generally denoted by $P(x)$. (See page 346 for an example of a system with a probability distribution of output.)

OUTPUT RATE

Output rate is the output size per unit time. Output rate will be denoted by r. The dimension of the output rate is thus:

$$[r] = [Q]/[T] \quad (5)$$

where Q refers to quantity and T refers to time.

In Example 1 the customer required 2,400 parts per year. This is the output rate, and we can write: $r = 2{,}400$ parts/year. In this case the unit quantity is a part and the unit time is a year. The numerical value of the output rate depends on the units involved. If, say, the parts came in boxes of 100 parts each, the output rate can be given as:

$$r = 2{,}400 \text{ parts/year} = 24 \text{ boxes/year} = 200 \text{ parts/month} = 2 \text{ boxes/month, etc.}$$

If an output of x units occurs over a period of t time units, then the output rate is given by:

$$r = x/t \quad (6)$$

Just as the output size may be constant or variable, so can the output rate. Hence we may also speak about the probability distribution of the output rate, $P(r)$, in a manner analogous to that of the output size.

Numerous inventory systems, in which the rate of output is variable, may be solved without the knowledge of the distribution. It can be shown that it is sometimes sufficient to know only the *average* rate of output. This average will be denoted by $\bar{r}$. Formally the average is given by:

$$\bar{r} = \sum_{0}^{\infty} r\,P(r) \quad (7)$$

OUTPUT PATTERNS

If we consider a period of time over which an output of size x occurs, it is possible to recognize numerous ways by which quantities are taken out of inventory. All x units may be withdrawn at the beginning of the period; they may be all withdrawn at the end of the period; they may be withdrawn uniformly throughout the period, etc. These different ways by which output occurs during a period are referred to as *output patterns.* Figure 3 illustrates five such patterns.

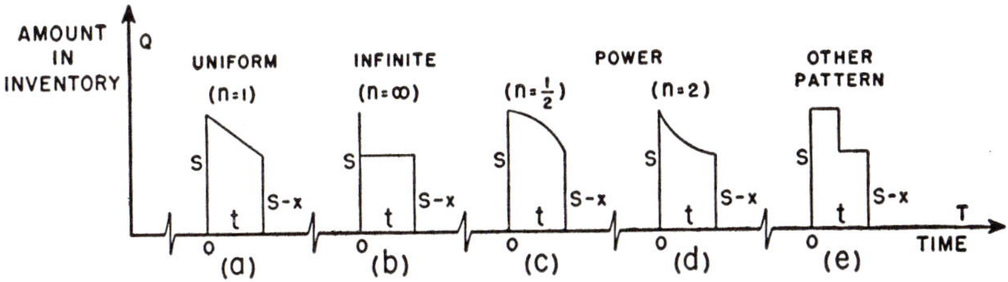

Figure 3. Output Patterns.

In all illustrations there are S quantity units in inventory at the beginning of the period, the length of the period is t time units and the output size is x quantity units. Illustrations (a), (b), (c), and (d) can be considered as a general class of patterns which are represented by:

$$Q(T) = S - x\sqrt[n]{T/t} \qquad (8)$$

where $Q(T)$ is the amount in inventory at time T,
S is the amount in inventory at the beginning of the period, i.e., when $T = 0$,
x is the output size during the period t,
and n is the output pattern index.

The patterns belonging to this class are referred to as *power patterns,* and their nature is entirely determined by n, the *pattern index.*

When $n = 1$ we have a uniform pattern (Figure 3a). When $n = \infty$ we have an *infinite pattern* (Figure 3b); this pattern occurs when output is only at the beginning of the period. Figures 3c and 3d illustrate power patterns with $n = \frac{1}{2}$ and $n = 2$, respectively. When n is larger than 1, this indicates that a larger portion of output occurs towards the beginning of the period. When n is smaller than 1, a larger portion of output occurs at the end of the period. When $n = 0$ all output occurs at the end of the period and the quantity S is carried in inventory throughout the period.

Input

The input component of inventory systems is generally that component which can be controlled by decision-makers within the organization. In general *input* refers to: the quantities which are scheduled to be put into inventories; the time when decisions are made with respect to these quantities; and the time when they actually are added to stock. It was pointed out previously that the inventory problem is the problem of finding when to order (or produce) and how much. Thus the inventory problem can also be considered as an input problem.[4]

In considering the input component we recognize the following elements. The *scheduling period* is the length of time between consecutive decisions with respect to inputs. The scheduling period is not always controllable. The *input size* is the quantity ordered. This quantity is, in general, controllable. The *lead-time* is the length of time between scheduling an input size and its actual addition to stock. Lead-time is generally not subject to control.

SCHEDULING PERIOD

The scheduling period is the length of time, measured in time units, between consecutive decisions with respect to input. The scheduling period will be denoted by t. Dimensionally, then, we have

(9) . . . $$[t] = [T]$$

Consider an inventory system in which decisions are made at times $T_1, T_2, T_3, \ldots T_i, T_{i+1}, \ldots$. The first scheduling period, t_1, is of length $T_2 - T_1$; the second, t_2, is of length $T_3 - T_2$, etc. In general the ith scheduling period, t_i, is given by:

(10) . . . $$t_i = T_{i+1} - T_i$$

These relations may also be illustrated as in Figure 4.

Scheduling periods may be prescribed or they may not be prescribed. When the scheduling periods are *prescribed,* the decision-maker cannot control them. In that case the t_i's are not variables subject to control, but they are given parameters. The only variables which can be

[4] In contrast, a queueing problem may be considered an output problem, since in that problem input is generally not controllable but output is.

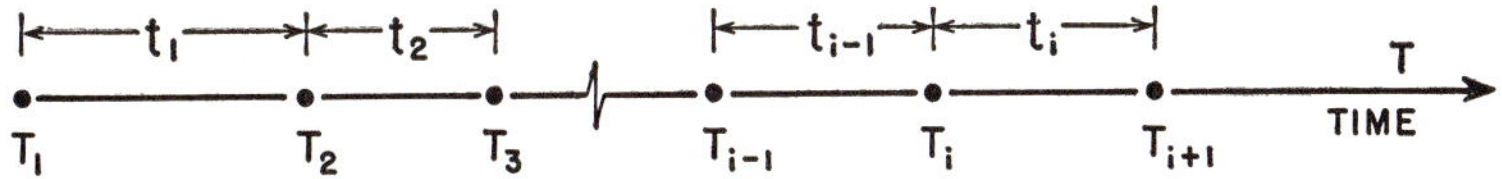

Figure 4. Scheduling Periods.

controlled in that case are the input quantities. When the t_i's are prescribed we usually find that they all have the *same* magnitude. This same prescribed scheduling period will be denoted by t_p, i.e.:

(11) . . . $$t_p = t_1 = t_2 = \cdots = t_i = \cdots$$

When the scheduling periods are *not prescribed* the t_i are *variables* subject to control. Here, again, we may have the same scheduling period. These will be denoted by t,

(12) . . . $$t = t_1 = t_2 = \cdots = t_i = \cdots$$

In that case one part of the inventory problem is always to find the *optimal scheduling period*. This period will be denoted by t_o.

LEAD-TIME

Lead-time is the length of time, measured in time units, between the making of a decision with respect to an input and its actual addition to stock. Lead-time will be denoted by LT. Its dimension is that of time, namely,

(13) . . . $$[LT] = [T]$$

In Figure 4 we illustrated a few scheduling periods. There, decisions with respect to input were made at times $T_1, T_2, \ldots T_i \ldots$. Consider the decision at time T_1 and assume that a quantity q_1 was ordered then and that this quantity was actually added to stock at time T_1'. In this case, the corresponding lead-time, LT_1, is $T_1' - T_1$. Similarly $LT_2 = T_2' - T_2$ and in general

(14) . . . $$LT_i = T_i' - T_i$$

Lead-time is ordinarily prescribed and is thus not subject to control. It is therefore one of the parameters of the inventory system. When lead-time is very small, we say that it is *insignificant* and we treat it as if $LT = 0$. In practice this can never occur, since some time always elapses between the making of a decision with respect to input and its

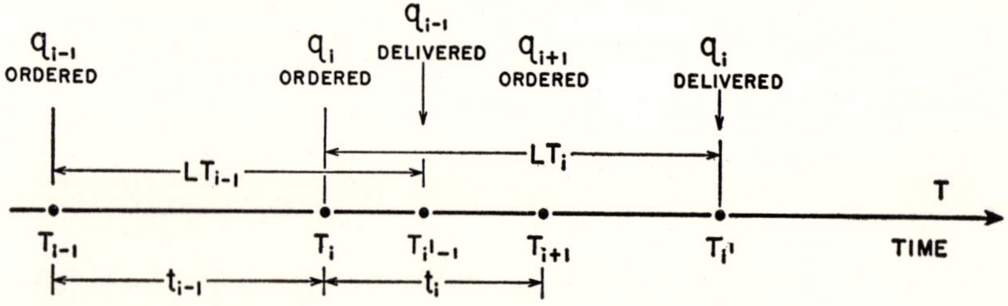

Figure 5. Lead-Time.

addition to stock. However, it is quite instructive to deal with systems in which $LT = 0$.

When lead-time is *significant* we write $LT \neq 0$. If lead-time is the same for each decision we say that it is *constant* and denote it by LT, namely,

$$(15) \ldots \qquad LT = LT_1 = LT_2 = \cdots = LT_i = \cdots$$

If lead-time is not the same for each decision we say that it *varies,* and we can speak about the probability distribution of lead-time in an analogous way to the probability distribution of output.

INPUT SIZE

In an analogy to the output size we define the *input size* as the quantity scheduled for input. This quantity will be denoted by q_i (i.e., it is the input size ordered at time T_i, the beginning of the ith scheduling period as in Figure 5). Dimensionally we have for the input size

$$(16) \ldots \qquad [q_i] = [Q]$$

Frequently the inventory problem is centered about this quantity q_i. How much to order (or produce)?

Whenever the quantity ordered is the *same* for every order we denote it by q and we refer to it as the *lot size.* We then have:

$$(17) \ldots \qquad q = q_1 = q_2 = \cdots = q_i = \cdots$$

In these cases the problem is to find the *optimal lot size,* q_o. A simple illustration of such a problem was given in Example 1.

In all systems to be analyzed in this chapter it is assumed that when an input q_i is ordered, then exactly this input is delivered and added to stock. However, there are also systems in which the quantity delivered is only subject to *statistical control.* In these systems the amount delivered, q_i', has a probability distribution in which one of the parameters is the input size q_i.

INPUT PERIOD, INPUT RATE, AND INPUT PATTERN

The *input period* is the length of time during which the input size q is being added to inventory. This period will be denoted by t'. The *average input rate, p,* is defined as the ratio of the input size and the input period. That is,

$$p = q/t' \qquad (18)$$

Dimensionally we have

$$[p] = [Q]/[T] \qquad (19)$$

When the input period is *insignificant* we have $t' \approx 0$ and we say that the corresponding rate is *infinite* (i.e., $p \approx \infty$). In this case we also say that input is *instantaneous:* all inputs in Figure 1, for example, are instantaneous.

Whenever input is not instantaneous we may be interested in the manner in which the input size is being put into inventory. This manner will be referred to as an *input pattern.*

Figure 6 gives a few illustrations of input patterns.

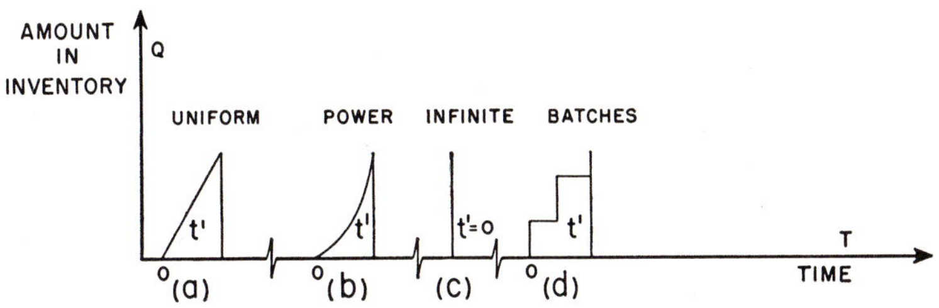

Figure 6. Input Patterns.

In all these illustrations an input of size q is being added to inventory at an average input rate p. The input pattern in illustrations (a), (b), and (c) can be considered as a general class of patterns which are represented by

(20) . . . $$Q(T) = q\sqrt[n]{T/t'} = q\sqrt[n]{Tp/q}$$

where $Q(T)$ is the amount in inventory at time T
q is the input size
t' is the input period
p is the average input rate, i.e., $p = q/t'$
n is the input pattern index

The input patterns belonging to this class are referred to as *power* patterns and their nature is determined by the *pattern index, n.*

When $n = 1$, we have a *uniform pattern* (Figure 6a). In this case

(20a) . . . $$Q(T) = qT/t' = Tp$$

When $n = \infty$, we have an insignificant input period ($t' \approx 0$), an infinite output rate ($p \approx \infty$) and we say that input is instantaneous.

The input pattern in illustration 6d is typical of cases where input is in batches.

Costs

An inventory system has been defined as a system in which only the following three types of costs are significant, and in which any two or all three are subject to control:

1. The surplus cost
2. The shortage cost
3. The setup cost

In order to use these costs as measures of effectiveness of the controllable variables of an inventory system they all have to be computed for the same period of time. Generally a year is chosen for this period of time, but any other unit of time is equally suitable. The corresponding measures of effectiveness are thus defined as follows:

C_1: The surplus cost per unit of time
C_2: The shortage cost per unit of time
C_3: The setup cost per unit of time
$C = C_1 + C_2 + C_3$: The total cost per unit of time

The dimension of each of these measures is:

(21) . . . $$[C] = [C_1] = [C_2] = [C_3] = [\$]/[T].$$

The properties of an inventory system should enable us to determine the relations between the controllable variables and the measures of effectiveness of the system. The output and input properties establish the quantities of surplus and shortage and the number of setups as a function of the controllable variables. Thus, in order to find C_1, C_2, and C_3 we have to know something about the unit costs of surplus, shortage, and setup. These *unit costs* will be denoted by c_1, c_2, c_3 respectively. The dimensions of these unit costs differ from system to system. For many systems, however, the dimensions of the unit cost are as follows:

(22) . . . $$[c_1] = [\$]/([Q][T])$$
(23) . . . $$[c_2] = [\$]/([Q][T])$$
(24) . . . $$[c_3] = [\$]$$

That is, the unit cost of carrying a surplus in inventory is $\$c_1$ per quantity unit per unit of time; the unit cost of incurring a shortage in inventory is $\$c_2$ per quantity unit per unit time; and the unit cost of setup is $\$c_3$ for each setup.

Thus, for these systems we have

(25) . . . $$C_1 = I_1c_1$$
(26) . . . $$C_2 = I_2c_2$$
(27) . . . $$C_3 = Nc_3$$

where I_1 and I_2 are respectively the average amounts of surplus and shortage in inventory during the unit of time and n is the number of setups during the unit of time.

The next three subsections cover the unit costs c_1, c_2, and c_3. But because detailed investigation of these costs is beyond the scope of this chapter, some general observations only will be given.

THE UNIT COST OF SURPLUS

The unit surplus cost covers the following cost elements associated with having surplus inventories:

1. The cost of money tied up in inventories,
2. The cost of storage,
3. The cost of taxes on inventories,
4. The cost of obsolescence,
5. The cost of insurance of inventories,
   and other costs.

It is customary to compute this unit cost as a fraction of the cost of items carried in inventory per unit time. Thus, in many systems, if the cost of a unit of quantity is $d and if the fraction is P per unit time we have

(28) . . . $$c_1 = dP$$

where the dimensions of d and P are:

(29) . . . $$[d] = [\$]/[Q]$$

(30) . . . $$[P] = 1/[T]$$

The fraction P, usually referred to as a percentage, depends on the nature of the cost elements given above. For many industrial applications its numerical value varies from about 5 per cent to about 25 per cent. Unfortunately, this value cannot be determined directly from present accounting records. A certain amount of research is always necessary for its determination. The accuracy required for P will be discussed in the section on sensitivity of results below.

THE UNIT COST OF SHORTAGE

This unit cost is usually the most difficult to determine. Some of the costs associated with the unit may include:

1. Overtime costs,
2. Special clerical and administrative costs,
3. Loss of specific sales,
4. Loss of good will,
5. Loss of customers,

and other costs.

Whenever the shortage cost is the cost of loss of specific sales the unit cost of shortage would equal the value of lost profit per unit quantity, say, $f per unit:

(31) . . . $$c_2 = f$$

where dimensionally

(32) . . . $$[c_2] = [f] = [\$]/[Q]$$

In many other cases the dimension of the unit cost of shortage is the same as the dimension of the unit cost of surplus,

(33) . . . $$[c_2] = [\$]/([Q][T])$$

In these cases the cost of shortage depends both on the quantities short and on the duration of time over which shortages occurred.

One encounters numerous managers who state that they do not allow any shortages to occur. Their statements imply that they assume that the unit cost of shortage is infinite. Such a unit cost does not seem realistic. What the managers are really trying to say is that the unit shortage cost is relatively high and that they therefore endeavor not to have any shortages. Systems A, D, and F below will show how the unit cost of shortage affects optimal solutions of these systems.

One also encounters numerous managers who categorically state that the unit cost of shortage cannot be measured. It is true that this unit it very difficult to measure. This, however, does not mean that the unit does not have specific value. Decisions affecting surpluses and shortages are made all the time by managers. Shortages do occur in most inventory systems. If the decisions are good ones (and the managers claim that they are) it is quite evident that in making the decisions the managers are actually placing a value on the unit cost of shortage. They may not be able to state numerically what this value is, but their decisions imply that such a value exists.

THE UNIT COST OF SETUP

Two classes of costs must be distinguished in discussing the unit cost of setup: (a) costs in which a setup refers to ordering an input of q units from an agency outside the organization studied, and (b) costs in which a setup refers to manufacturing of the input of q units within the organization itself.

In the first class the costs associated with the unit cost of setup may include:

1. Clerical and administrative costs,
2. Transportation costs,
3. Unloading costs,

and other costs.

In the second class the costs associated with the unit cost of setup may include:

1. Clerical and administrative costs,
2. Labor and machine setup costs,
3. Cost of materials used during setup testing,

4. Cost of time during which production cannot take place because of setups,

and other costs.

The dimensions of unit setup cost are almost always given by

$$(34) \ldots \qquad [c_3] = [\$]$$

An important consideration in determining the unit cost of setup is the number of different items pertaining to one setup. Whenever two or more different items have a common setup, the solutions *cannot* be obtained by considering each item individually and solving its inventory system. Only one inventory system exists in such a case in which several items affect the costs of the whole system. (See, for example, references 13, p. 35, and 14.)

SOLUTIONS OF SOME INVENTORY SYSTEMS

Eight inventory systems, systems A through H, are solved in this section. Systems A, B, and C constitute the elementary inventory systems; the others are extensions of the elementary systems. The five extensions given in this chapter should be considered only as a very small sample of possible extensions. They have been selected to illustrate the effect of some of the properties discussed in the previous section and to describe the general methodology of approach. It can also be shown that most other inventory systems described in the literature can be suitably formulated so that they, too, may be considered as extensions of the elementary systems.

Elementary Inventory Systems

An elementary inventory system is defined as a system involving the balancing of only two of the three types of costs, C_1, C_2, C_3, and having the following simple properties:

1. The output rate is known and is constant, i.e., r = known and constant
2. The output pattern is uniform, i.e., $n = 1$
3. The scheduling period is constant, i.e., t = constant
4. Lead-time is insignificant, i.e., $LT = 0$
5. The input size is constant, i.e., q = constant
6. The input rate is infinite, i.e., $p = \infty$
7. The costs are defined by equations (35), (36) and (37):

(35) . . . $C_1 = I_1 c_1$
(36) . . . $C_2 = I_2 c_2$
(37) . . . $C_3 = N c_3 = c_3/t$

where I_1 and I_2 are respectively the average surplus and shortage during a unit of time; where the dimensions of the unit costs are respectively $[\$]/([Q][T])$, $[\$]/([Q][T])$, and $[\$]$; and where N is the number of setups in a unit of time (i.e., $N = 1/t$).

8. The unit costs are given constants, i.e., c_1, c_2, c_3 are given constants.

There are thus only three elementary systems:

1. A system involving the balancing of surplus and shortage costs;
2. A system involving the balancing of surplus and setup costs; and
3. A system involving the balancing of shortage and setup costs.

In System A t_p is prescribed (and hence C_3 is a constant not subject to control). In System B, I_2 is zero (and hence C_2 is zero), i.e., no shortages are allowed to occur. In System C, I_1 is zero (and hence C_1 is zero), i.e., no surpluses are allowed to occur.

These are theoretical systems. They may sometimes be identical with actual inventory systems, but the main purposes of their presentation and analysis here are the following:

(a) To establish a starting point for further discussion.
(b) To illustrate the method of determining C, the total cost per unit time.
(c) To illustrate the methods for finding the optimal decision rules. These methods, too, will be used in the solution of other systems.
(d) To establish standards against which the solutions of other systems can be compared. (Specifically, the optimal decision rules and the corresponding minimum total costs per unit time.)

SYSTEM A. BALANCING OF SURPLUS AND SHORTAGE COSTS

The inventory system may be represented graphically as in Figure 7. In this system the following parameters are given:

t_p the scheduling period in time units; a prescribed constant (hence C_3 is constant)
r the output rate in quantity units per unit time; a constant
c_1 the cost of carrying one quantity unit in inventory per unit time; a constant
c_2 the cost of incurring a shortage of one unit per unit time; a constant

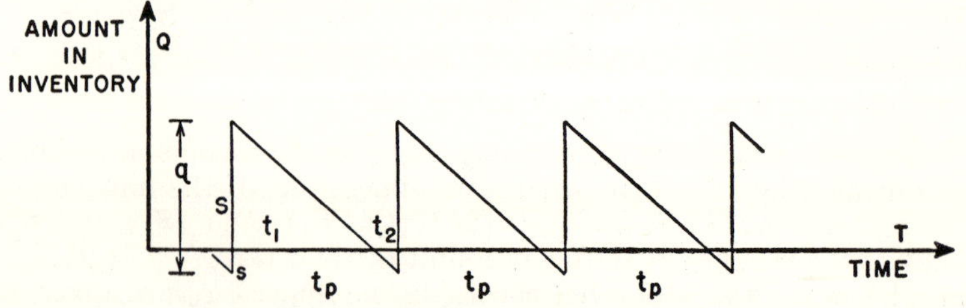

Figure 7. System A.

There is only one variable subject to control in this system, namely,

S the inventory level at the beginning of each scheduling period

The input q is actually prescribed, too, in this case, since we assume here that shortages are allowed to be carried over, i.e.,

(38) . . . $$q = rt_p$$

The other inventory level in this system, s, the level at the end of each scheduling period, depends on S and the other parameters of the system:

(39) . . . $$s = rt_p - S$$

(Note that s is a positive quantity by this definition, although it refers to a negative inventory.)

In order to find S_o, the optimal inventory level, and C_o, the minimum total cost per unit time, we proceed as follows:

From Figure 7 and equations (35), (36), and (37) we have:

(40) . . . $$C_1 = \tfrac{1}{2}S(t_1/t_p)c_1 = S^2c_1/(2rt_p)$$
(41) . . . $$C_2 = \tfrac{1}{2}s(t_2/t_p)c_2 = s^2c_2/(2rt_p) = (rt_p - S)^2c_2/(2rt_p)$$

The total cost per unit time subject to control in this case is $C = C_1 + C_2$, since C_3 is a constant. Hence

(42) . . . $$C = C_1 + C_2 = [S^2c_1 + (rt_p - S)^2c_2]/(2rt_p)$$
$$= [1/(2rt_p)](c_1 + c_2)S^2 - c_2S + c_2rt_p/2$$

Equations similar to (42), in which the total cost is given as a function of the variables subject to control, will be referred to as the *total cost equations*. This total cost equation generally enables the determination of the optimal decision rules.

Solving $dC/ds = 0$, leads to

(43) . . . $$S_o = rt_pc_2/(c_1 + c_2)$$

This value of S gives a minimum C, since d^2C/dS^2 is larger than zero. Substitution of this value in equation (42) gives the minimum total cost, C_o.

(44) . . . $$C_o = \tfrac{1}{2}rt_pc_1c_2/(c_1 + c_2)$$

Equation (43) represents an optimal decision rule. That is, if the inventory level at the beginning of each scheduling period is $rt_pc_2/(c_1 + c_2)$, then the total cost is minimized.

Example 2: A refinery uniformly uses 2,400 parts of a certain item x during the year. Item x is a component part of many pumps in the refinery and is subject to rapid deterioration. Whenever a replacement is not available leakages occur which cost $90 per month. Every month 200 parts of item x are manufactured. Manufacturing time is about one day. The cost of each part is $80. The percentage of inventory carrying cost is 15 per cent per year.

What should be the optimal inventory level of item x at the beginning of each month after all shortages have been made up?

This is an example of System A with the following given parameters:

$t_p = 1/12$ year; $r = 2{,}400$ parts per year; $c_1 = \$80 \times 0.15 = \$12$ per part per year; $c_2 = \$9$ per part per month $= \$108$ per part per year.

The model of the system is thus (by equation 42):

(45) . . . $$C = 0.3\,S^2 - 108S + 10{,}800$$

If no shortages were allowed, the inventory level S would have to be 200 parts. In this case the total cost of the inventory system would be $1,200 per year.

If, on the other hand, one would want to have equal surplus and shortage quantities, S would have to be 100. In that case the total cost of the inventory system would be $3,000 per year.

Using the decision rule in equation (43), we find that the optimal inventory level is $S_o = 180$ parts. This would lead to a minimum total cost of $1,080 per year.

SYSTEM B. BALANCING OF SURPLUS AND SETUP COSTS

The inventory system may be represented graphically as shown in Figure 8.

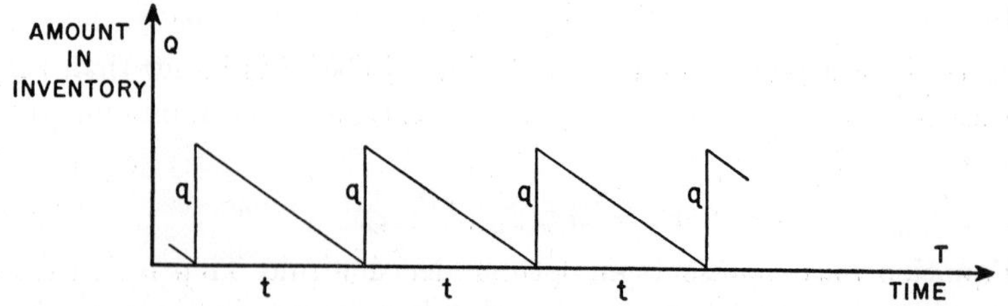

Figure 8. System B.

In this system the following parameters are given:

- r the output rate in quantity units per unit time; a constant
- c_1 the cost of carrying one quantity unit in inventory per unit time; a constant
- c_3 the cost of one setup
- I_2 is zero; i.e., no shortages are allowed.

Here the variables are q and t, but they are related by the equation

(46) . . . $$q = rt$$

We shall assume that q is the variable subject to control. (In this system $S = q$ and $s = 0$.)

The average inventory carried during the scheduling period t is $I_1 = \frac{1}{2}q$, hence, by (35)

(47) . . . $$C_1 = qc_1/2$$

And by (37) and (46) we have

(48) . . . $$C_3 = c_3/t = rc_3/q$$

Therefore, the total cost equation is

(49) . . . $$C = C_1 + C_3 = qc_1/2 + rc_3/q$$

Solving $dC/dq = 0$, leads to the classical formula

(50) . . . $$q_o = \sqrt{2rc_3/c_1}$$

This value of q gives a minimum C, since for that value d^2C/dq^2 is larger than zero. Substitution of this value of q in (46) and (49) gives:

(51) . . . $$t_o = \sqrt{2c_3/(rc_1)}$$

(52) . . . $$C_o = \sqrt{2rc_1c_3}$$

Equations (50) and (51) provide the optimal decision rules for System B. The minimum total cost per unit time of that system is given by (52).

A number of different inventory systems have the form of equation (49). This form may be stated in general as

(53) . . . $$C(z) = C_1 + C_3 = az + b/z$$

Solving for z we find that

(54) . . . $$z_o = \sqrt{b/a}$$

(55) . . . $$C_1(z_o) = \sqrt{b/a}$$

(56) . . . $$C_3(z_o) = \sqrt{b/a}$$

(57) . . . $$C(z_o) = 2\sqrt{ab}$$

These results apply for the System B (where $a = c_1/2$, $b = c_3$ and $z = q$) and will be used in subsequent work. It should be noted that for systems having a cost equation of the form (53) we have:

(58) . . . $$C_1(z_o) = C_3(z_o)$$

Example 3: In the refinery problem of Example 2, assume that the scheduling period is not prescribed; that no shortages are allowed; and that the setup cost of a production run is $350.

How often should the parts be manufactured?

This is an example of System B with the following given parameters:

r = 2,400 parts per year; c_1 = $12 per part per year;
c_3 = $350 per setup.

By equations (50), (51), and (52) we find the solution of the system.

t_o = 1.87 months; q_o = 374 parts; C_o = $4,488 per year.

That is, the items should be produced every 1.87 months (the optimum lot size being 374 parts). This would lead to a minimum total cost of $4,488 per year.

The following results correspond to some other scheduling periods:

t = 1 month	q = 200 parts	C = $5,400 per year
t = 1½ months	q = 300 parts	C = $4,600 per year
t = 2 months	q = 400 parts	C = $4,500 per year
t = 2½ months	q = 500 parts	C = $4,680 per year

SYSTEM C. BALANCING OF SHORTAGE AND SETUP COSTS

The inventory system may be represented graphically as shown in Figure 9.

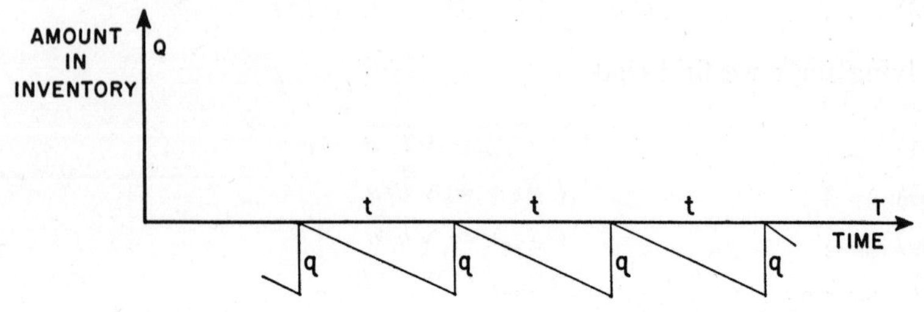

Figure 9. System C.

This system is similar to System B above, except that here surpluses are not allowed, but shortages are allowed. The analysis, therefore, is also similar. The solution of the system is:

(59) . . . $$q_o = \sqrt{2rc_3/c_2}$$

(60) . . . $$t_o = \sqrt{2c_3/(rc_2)}$$

(61) . . . $$C_o = \sqrt{2rc_2c_3}$$

Example 4: In the refinery problem of Example 2, assume that the scheduling period is not prescribed; that no surpluses are allowed (i.e., no inventory is carried); and that the setup cost of a production run is $350.

How often should parts be manufactured?

This system is analogous to System C. Here we have the following given parameters:

r = 2,400 parts per year; c_2 = $108 per part per year;
c_3 = $350 per setup.

By equations (59), (60), and (61) the solution of the system is:

t_o = 0.62 months; q_o = 125; C_o = $13,464 per year.

This means that for an optimal solution shortages should be accumulated until there are 125 parts short. Then a production run of 125 parts should be manufactured and all shortages made up.

For other scheduling periods we would get:

$t = \frac{1}{2}$ month	$q = 100$	$C = \$13{,}800$ per year
$t = 1$ month	$q = 200$	$C = \$15{,}000$ per year
$t = 1\frac{1}{2}$ months	$q = 300$	$C = \$19{,}000$ per year

System D: Balancing of Surplus, Shortage, and Setup Costs

The inventory system may be represented graphically as shown in Figure 10.

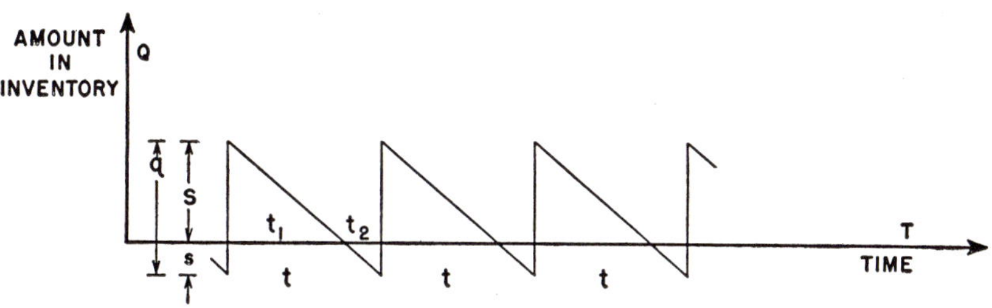

Figure 10. System D.

This system may be considered an extension of System A, of System B, or of System C. In System A, t is prescribed; here t is not prescribed but is a constant. In System B no shortages are allowed; here shortages are allowed. In System C no surpluses are allowed; here surpluses are allowed.

The parameters given for this system are therefore: r, c_1, c_2, c_3. The variables subject to control are now, say, S and t. The other variables related to them, q and s, are:

(62) . . . $$q = rt$$

(63) . . . $$s = rt - S$$

The total cost equation of this system is:

(64) . . . $$C = C_1 + C_2 + C_3 = [S^2c_1 + (rt - S)^2c_2]/(2rt) + c_3/t$$

(This is an immediate extension of System A.)

One way of solving this system is to treat C as a function of the two variables S, t. That is, solve $\partial C/\partial S = 0$, and $\partial C/\partial t = 0$, etc. A simpler way is to use the results obtained earlier.

For a fixed value t, this model is also a model of System A. Hence (for a fixed t) by System A above:

$$(65) \ldots \qquad S_o = rtc_2/(c_1 + c_2)$$
$$(66) \ldots \qquad C = \tfrac{1}{2}\, rtc_1c_2/(c_1 + c_2) + c_3/t$$

But equation (66) is of the form of equation (53) of System B, i.e., $C(t) = at + b/t$. Hence the solution:

$$(67) \ldots \qquad t_o = \sqrt{2c_3/(rc_1)}\,\sqrt{(c_1 + c_2)/c_2}$$
$$(68) \ldots \qquad C_o = \sqrt{2rc_1c_3}\,\sqrt{c_2/(c_1 + c_2)}$$

and by (62), (63), (65) and (67) we have

$$(69) \ldots \qquad q_o = \sqrt{2rc_3/c_1}\,\sqrt{(c_1 + c_2)/c_2}$$
$$(70) \ldots \qquad S_o = \sqrt{2rc_3/c_1}\,\sqrt{c_2/(c_1 + c_2)}$$
$$(71) \ldots \qquad s_o = \sqrt{2rc_3/c_2}\,\sqrt{c_1/(c_1 + c_2)}$$

Example 5: In the refinery problem of Examples 2, 3, and 4 assume that there are no restrictions. That is, the scheduling period is not prescribed and both surpluses and shortages are allowed to occur. The parameters are:

r = 2,400 parts per year; c_1 = \$12 per part per year;
c_2 = \$108 per part per year; and c_3 = \$350 per set-up.

By equations (67) through (71) we find the following solution:

t_o = 1.97 months; q_o = 394 parts; S_o = 355 parts;
s_o = 39 parts; and C_o = \$4,260 per year.

This solution should be interpreted as follows:

When there are 39 parts short, a production run of 394 parts should be made. After the shortages have been made up there will be 355 parts in stock. These will be used up and then shortages will again occur. After 1.97 months, when there are again 39 parts short another production run is made, and so on. This procedure is the best under the circumstances described and will lead to a minimum total cost of the system of \$4,260 per year.

System E: A System With A Finite Production Rate

When the input rate p is not infinite, and when the input pattern is uniform, the inventory system extended from System B may be represented graphically as in Figure 11.

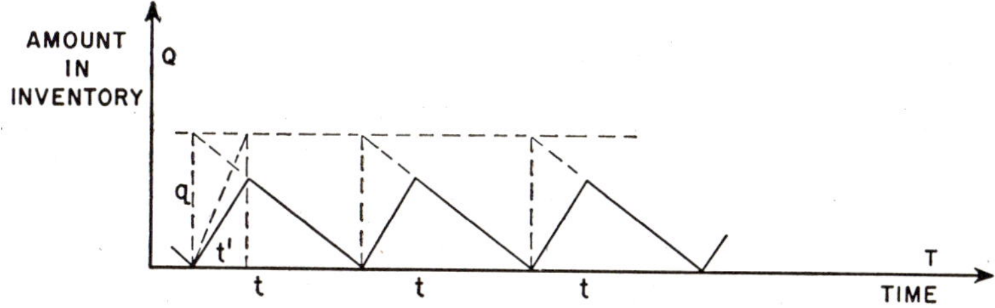

Figure 11. System E.

In System B the average amount in inventory is $q/2$. In System E it is $(1 - r/p)q/2$. This can be shown as follows:

It takes $t' = q/p$ units of time to produce q quantity units at a production rate of p quantity units per unit time. During this period the amounts in inventory increase at a rate of $p - r$ quantity units per unit of time. Hence at the end of the production run there will be $(p - r)q/p = (1 - r/p)q$ quantity units in stock. Therefore, the average amount in inventory is $(1 - r/p)q/2$.

The total cost equation of System E and its solution are therefore:

(72) . . . $$C = (1 - r/p)qc_1/2 + rc_3/q$$

(73) . . . $$q_o = \sqrt{2rc_3/c_1} \,/\, \sqrt{1 - r/p}$$

(74) . . . $$C_o = \sqrt{2rc_1c_3}\,\sqrt{1 - r/p}$$

Example 6: In the refinery problem assume that parts can be produced at a rate of 9,600 parts per year. Also assume that no shortages are allowed.

What is the optimal lot size and how frequently should production runs be scheduled?

Here, then: $r = 2{,}400$ parts per year; $p = 9{,}600$ parts per year; and $c_3 = \$350$ per setup.

The solution by (73) and (74) and (46) is therefore:

$$q_o = 432 \text{ parts; } t_o = 2.16 \text{ months; } C_o = \$3{,}888 \text{ per year.}$$

That is, every 2.16 months a production run of 432 should be scheduled. It would take $t' = 432/800 = 0.54$ months to produce the parts, and at the end of the production run there will be $0.54(800\text{-}200) = 324$ parts in stock. These 324 will be used up in $324/200 = 1.62$ months. At that time a new production run of 432 will be scheduled.

Systems with Variable Output

Most inventory systems encountered in practice have variable output. We will present two systems, F and G, which illustrate approaches to the solution of systems with variable output. System F is an extension of System B and for its solution we only need to know the average rate of output. System G is an extension of System A and for its solution we need to know the probability distribution of output.

SYSTEM F: CONSTANT RATIO OF MAXIMUM OUTPUT TO AVERAGE OUTPUT

The inventory system may be represented graphically as shown in Figure 12.

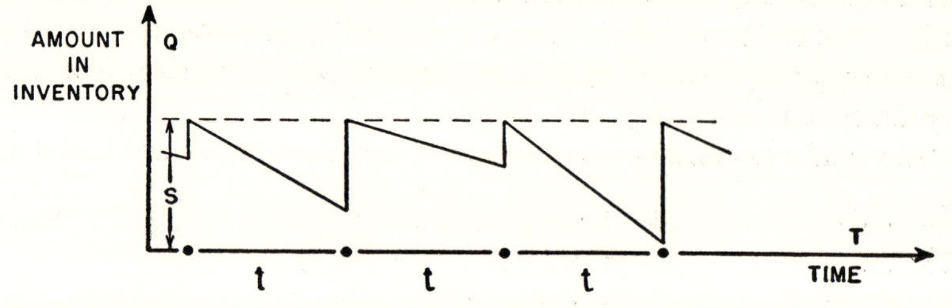

Figure 12. System F.

In this system the scheduling period is constant. A variable input q_i is scheduled at the beginning of each period to bring the inventory to a constant level S. Since no shortages are allowed (this is an extension of Model B), the level S must be large enough so that no shortages occur. (Such a system is particularly applicable in the case of several items having a common setup. We shall, however, limit the discussion to one item.)

Let $\bar{r}$ be the average rate of output. Let A be a factor such that

$$S = A\bar{r}t \qquad (75)$$

A is therefore the ratio of the highest possible output to the average output during any scheduling period t.

In this system, therefore, the total cost equation is

$$6) \ldots \quad C = (S - \bar{r}t/2)c_1 + c_3/t = (A - \tfrac{1}{2})\bar{r}tc_1 + c_3/t$$

If A is a constant (and it must be equal or larger than one by defi-
tion), then the solution is:

$$7) \ldots \quad t_o = \sqrt{2c_3/(\bar{r}c_1)} \,/\, \sqrt{2A - 1}$$

$$8) \ldots \quad S_o = A\sqrt{2\bar{r}c_3/c_1} \,/\, \sqrt{2A - 1}$$

$$9) \ldots \quad C_o = \sqrt{2\bar{r}c_1c_3}\,\sqrt{2A - 1}$$

This system is typical of the case when management requires that
ıe inventory at the beginning of each scheduling period be A times
ıe average output during the period, and where A is large enough so
ıat no shortages can occur.

Example 7: In the refinery problem assume that the usage of parts
aries from week to week and that the average usage is 2,400 parts per
ear. Assume that at the beginning of each scheduling period there is
lways 2½ times as much stock as would be used on the average
uring that period and that this insures that no shortages will occur.
ssume an infinite rate of production.

What is the optimal scheduling period and how much stock should
ıere be at the beginning of each scheduling period?

Here, then: $\bar{r}$ = 2,400 per year; $A = 2\tfrac{1}{2}$; c_1 = \$12 per part per year; and c_3 = \$350 per setup.

able 2. Inventory Records of System F in Example 7

Time	Amount in Inventory	Amount Produced	Amount Used During Scheduling Period	Amount in Inventory at End of Scheduling Period
0	0	470	210	260
0.94	260	210	160	310
1.88	310	160	400	70
2.82	70	400	80	390
3.76	390	80	120	350

By equations (77) (78) and (79) we have:

$$t_o = 0.94 \text{ months}; \; S_o = 470 \text{ parts; and } C_o = \$8{,}980 \text{ per year.}$$

If this solution is used, then, on the average $0.94 \cdot 200 = 188$ parts will be used per scheduling period. The highest possible usage would be $188 \cdot 2\frac{1}{2} = 470$ parts. Table 2 describes possible operation of the inventory system for the above solution.

SYSTEM G: A SYSTEM WITH PROBABILITY DISTRIBUTION OF OUTPUT

The inventory system may be represented graphically as shown in Figure 13.

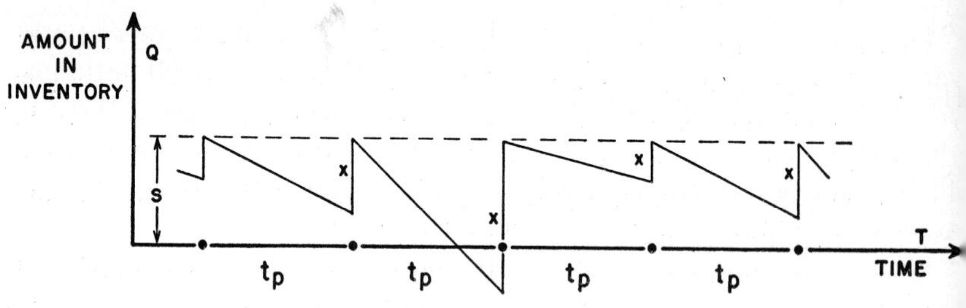

Figure 13. System G.

Thus during any prescribed scheduling period t_p the inventory system may also be represented graphically as shown in Figure 14.

Case (a) Case (b)

Figure 14. Typical Cases in System G.

Case a is a typical situation when the output size x is smaller or equal to the beginning level S. Case b is a typical situation when x is larger than S.

Let $P(x)$ be the probability that an output of size x will occur during the prescribed scheduling period t_p and let us assume that x can have the values $x_1, x_2, \ldots x_{\max}$.

Whenever $x \leqq S$ (i.e., Case a) the average surplus amount in inventory will be $S - x/2$ and hence the total cost per unit of time will be $(S - x/2)c_1$. The total *expected* cost C_a in Case a (for all $x \leqq S$) will be:

$$C_a = \sum_{x=x_1}^{x=S} (S - x/2)P(x) \tag{80}$$

Whenever $x > S$ (i.e., Case b) the total cost per unit time will be[5] $[S^2c_1 + (x - S)^2c_2]/(2x)$. The total expected cost C_b is Case b (for all $x > S$) will be

$$C_b = \sum_{x=S+1}^{x=x_{\max}} \{[S^2c_1 + (x - S)^2c_2]/(2x)\}P(x) \tag{81}$$

Hence the total cost equation is

$$C = c_1 \sum_{x=x}^{x=S} (S - x/2)P(x) + c_1 \sum_{x=S+1}^{x=x_{\max}} \tfrac{1}{2}S^2[P(x)/x] + c_2 \sum_{x=S+1}^{x=x_{\max}} \tfrac{1}{2}(x - S)^2[P(x)/x] \tag{82}$$

The solution of this equation will be omitted here, but how it can be used in a specific example will be shown. The reader who is interested in the analytical solution is referred to reference (11). Model IV in that reference represents System G.

Example 8: In the refinery problem assume again that the scheduling period is prescribed and is one month long. Assume that the total usage during the scheduling period has the probability distribution given in Table 3.

Table 3. Probability Distribution of Output in Example 8

Output Size during t_p	x	0	100	200	300	400	500
Probability of Occurrence	$P(x)$	.15	.20	.35	.15	.10	.05

What is the optimal inventory level?

[5] See equation (42) in System A for the detailed derivation. Also note that $x = rt_p$.

Table 4. Solution of System G

(1)	(2)	(3)	(4)	(5)	(6)	(7)	(8)	(9)	(10)	(11)
S	x	$P(x)$	I_{1_a}	I_{1_b}	I_{2_b}	C_1	C_2	$(C_1 + C_2)\,P(x)$	C_a, C_b	C
100	0	0.15	100			1,200		180		
	100	0.20	50			600		120	300	
	200	0.35		25	25	300	2,700	1,050		
	300	0.15		17	67	204	7,236	1,116		
	400	0.10		12	112	144	12,096	1,224		
	500	0.05		10	160	120	17,280	870	4,260	4,560
150	0	0.15	150			1,800		270		
	100	0.20	100			1,200		240	510	
	200	0.35		56	6	672	648	462		
	300	0.15		38	38	456	4,104	684		
	400	0.10		28	78	336	8,424	876		
	500	0.05		22	122	264	13,176	672	2,694	3,204
200	0	0.15	200			2,400		360		
	100	0.20	150			1,800		360		
	200	0.35	100			1,200		420	1,140	
	300	0.15		67	17	804	1,836	396		
	400	0.10		50	50	600	5,400	600		
	500	0.05		40	90	480	9,720	510	1,506	2,646
250	0	0.15	250			3,000		450		
	100	0.20	200			2,400		480		
	200	0.35	150			1,800		630	1,560	
	300	0.15		104	4	1,248	432	252		
	400	0.10		78	28	936	3,024	396		
	500	0.05		62	62	744	6,696	372	1,020	2,580
300	0	0.15	300			3,600		540		
	100	0.20	250			3,000		600		
	200	0.35	200			2,400		840		
	300	0.15	150			1,800		270	2,250	
	400	0.10		112	12	1,344	1,296	264		
	500	0.05		90	40	1,080	4,320	270	534	2,784

Note: See footnote p. 349.

To solve the problem use equation (82). It is best to organize the computation in tabular form as is done in Table 4.

SYSTEM H: A SYSTEM WITH SIGNIFICANT LEAD-TIME

When lead-time is significant, the decision rules for the elementary systems A, B, and C, as well as for their extensions D and E, are not affected. In these systems output is known and is constant; therefore, it is possible to plan inputs ahead of time, and the inventories will be the same whether lead-time is zero or not.

This, however, is not the case when output varies. Consider, for example, an extension of System F, the system represented by Figure 12. Let the lead-time be LT, the scheduling period t, the average output $\bar{r}$ per unit time, and the ratio of the highest output to the average output A. Consider a point of time T', LT before the next input becomes available. Let the inventory at that point be s; let the sum of all previous inputs which become available during the lead-time period be Σq. In order that no shortages occur during the scheduling period t', then,

$$(83) \ldots \qquad s + \Sigma q + q = A(LT + t)\bar{r}$$

We define the left-hand side of this equation as the inventory level Z, i.e.,

$$(84) \ldots \qquad Z = s + \Sigma q + q$$

A knowledge of Z also implies the knowledge of q, the input to be scheduled, since at the time a decision is made s and Σq are known. Z is a hypothetical inventory level. (It is a real level, equal to S, when $LT = 0$.) It may be described graphically as in Figure 15.

The average inventory in this system during any scheduling period

---

Columns (1), (2), and (3) are self-explanatory. Column (4) gives the average surplus inventory in Case a and Columns (5) and (6) give respectively the average surplus and shortage inventories in Case b. Columns (7) and (8) respectively give the surplus and shortage costs for a given output x. Column (9) gives the total expected cost for a given x. Column (10) gives C_a and C_b as per (80) and (81), and Column (11) gives C as per (82).

By repeating the computations again we can finally find that $S_o = 235$ parts and the corresponding minimum total cost is \$2,576 per year ($C_a = \$1,434$ and $C_b = \$1,142$).

Thus, the optimal inventory level for Example 8 is 235 parts, and this will lead to a minimum total cost of \$2,576 per year.

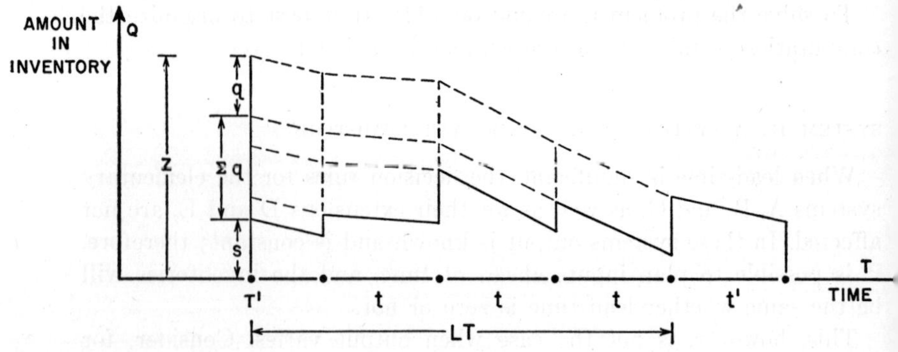

Figure 15. System H.

will be found to be $(A - 1)LT\bar{r} + (A - \frac{1}{2})t\bar{r}$. The total cost equation is therefore

$$(85) \ldots \quad C = (A - 1)LT\bar{r}c_1 + (A - \tfrac{1}{2})t\bar{r}c_1 + c_3/t.$$

For the case A = constant the solution of (85) is

$$(86) \ldots \quad t_o = \sqrt{2c_3/(\bar{r}c_1)} \; / \; \sqrt{2A - 1}$$

$$(87) \ldots \quad Z_o = ALT\bar{r} + A\sqrt{2\bar{r}c_3/c_1} \; / \; \sqrt{2A - 1}$$

$$(88) \ldots \quad C_o = \sqrt{2\bar{r}c_1c_3}\,\sqrt{2A - 1} + (A - 1)LT\bar{r}c_1$$

Example 9: This example is an extension of Example 7 which was given on page 345. In addition to the assumptions of Example 7 we now add the assumption that lead-time is two months.

The problem is to find the optimal scheduling period and the optimal hypothetical inventory level, Z_o.

Thus we have here: $\bar{r} = 2{,}400$ per year; $A = 2\frac{1}{2}$; $LT = 2$ months; $c_1 = \$12$ per part per year; $c_3 = \$350$ per setup.

By equations (86), (87), and (88) we have the solution $t_o = 0.94$ months; $Z_o = 1{,}470$ parts; $C_o = \$16{,}180$ per year.

Table 5 describes a possible operation of the inventory system for the above solution.

SENSITIVITY OF RESULTS

The preceding section emphasized the methodology in obtaining optimal decision rules for several inventory systems. We now attempt to study the following questions:

Table 5. Inventory Records of System H in Example 9

Time	Amount in Inventory	Outstanding Orders	Amount Ordered	Amount Used During t_o	Amount Received During t_o	Amount in Inventory at End of t_o
T	s	Σq	q	rt_o	q	s
0	500	600	370	210	400	690
0.94	690	570	210	160	200	730
1.88	730	580	160	400	370	700
2.82	700	370	400	80	210	830
3.76	830	560	80	120	160	870
4.70	870	480	120	450	400	820
5.64	820	200	450	430	80	470
6.58	470	570	430	400	120	190
7.52	190	880	400	50	450	590

(a) How do the various systems compare with one another?
(b) How important is it to use the optimal decision rules, rather than some other rules?
(c) How are the total costs affected if estimates, instead of the true values, are used?

These questions very often arise in practice for several reasons. For example, (1), the decision-makers may have some control on the properties of a system and may want to know whether it is worthwhile to change the properties; (2), the optimal decision rules cannot be employed, or their employment may be more costly than some alternative rules; and (3), the estimation of the parameters of the systems is costly.

Comparisons of Systems

Eight systems have been solved in the preceding section. If we consider them in the order in which they were solved, and if we ask which system was an extension of which, we find the following relations:

System	*Extended to Systems*
A	D, G
B	D, E, F, H
C	D
D	—
E	—
F	H
G	—
H	—

It therefore seems that one way of comparing the various systems is to study the following groups:

1. Systems A, B, C, D
2. Systems B, E
3. Systems A, G
4. Systems B, F, H

In addition, it also seems instructive to compare the numerical results of Examples 2-9 which followed Systems A-H.

COMPARISON OF SYSTEMS A, B, C, D

Systems A, B, and C were our elementary systems. In System A the scheduling period was prescribed; hence the setup cost per unit time was not subject to control. In System B no shortages were allowed; hence the shortage cost per unit time was zero. In System C no surpluses were allowed; hence the surplus cost per unit time was zero. In System D no restrictions were imposed; hence all three types of costs were subject to control.

System D can thus be considered as an extension of each of the Systems A, B, and C. In particular we can make the following observations:

(1) If we assume in System D that the unit shortage cost is infinite, i.e., $c_2 = \infty$, then System D becomes System B. And indeed, if we have $c_2 = \infty$ in the solution of System D, equations (67), (68), and (69) become equations (50), (51), and (52) and they constitute the solution of System B.

(2) Similarly, if we assume in System D that the unit surplus cost is infinite, i.e., $c_1 = \infty$, then System D becomes System C.

(3) If an inventory system is System D (i.e., $c_2 \neq \infty$) and we assume that no shortages are to be allowed (i.e., we use System B for finding the optimal decision rules), additional costs will be incurred by using a wrong assumption. The additional costs will be given by the difference of the cost of equations (52) and (68).

$$(89) \quad C_B - C_D = \sqrt{2rc_1c_3} - \sqrt{2rc_1c_3}\sqrt{c_2/(c_1 + c_2)} = \sqrt{2rc_1c_3}\sqrt{c_2/(c_1 + c_2)}\,(\sqrt{(c_1 + c_2)/c_2} - 1)$$

Thus, unnecessary expenses of $(\sqrt{(c_1 + c_2)/c_2} - 1)100\%$ are made if one assumes that the unit cost of shortage is infinite when in fact it is not.

(4) System A can be compared with System D if we add to the minimum total cost of System A as given by (44) the cost of the uncontrollable setup cost per unit time, c_3/t_p.

$$(90) \quad C_A = \tfrac{1}{2}rt_pc_1c_2/(c_1 + c_2) + c_3/t_p$$

We can now ascertain what is the effect of prescribing the scheduling period (as in System *A*) in comparison to not prescribing it (as in System D). If the period is prescribed the total cost is given by (90). If it is not, the total cost is given by (68). Hence the additional costs are:

$$(91) \quad C_A - C_D = \tfrac{1}{2}rt_pc_1c_2/(c_1 + c_2) + c_3/t_p - \sqrt{2rc_1c_3}\sqrt{c_2/(c_1 + c_2)} = \sqrt{2rc_1c_3}\sqrt{c_2/(c_1 + c_2)}\,(1 - t_p/t_o)^2/(2t_p/t_o)$$

where t_o is the optimal scheduling period of System D as per (67).

In this case, then, unnecessary expenses of $[(1 - t_p/t_o)^2/(2t_p/t_o)]100\%$ are made if a scheduling period is prescribed. The significance of this result is best illustrated by Table 6.

Table 6. Comparison of Systems B and D

t_p/t_o	1/5	1/4	1/3	1/2	1/1	2/1	3/1	4/1	5/1	6/1
$(C_A - C_D)/C_D]$ 100%	160%	112%	67%	25%	0	25%	67%	112%	160%	208%

From the table we note, for example, that if the prescribed scheduling period is twice as large as the optimal period, this results in total costs which are 25 per cent higher than the minimum costs.

COMPARISON OF SYSTEMS B AND E

The only difference between Systems B and E is with respect to the input rate, p. In System B it is infinite, i.e., $p = \infty$. In System D it is finite, i.e., $p \neq \infty$. Thus we may actually look on System B as a special case of System E. And indeed, if in System E we have $p = \infty$, then equations (72), (73), and (74) reduce to equations (49), (50), and (52) of System B.

An interesting question that can be raised in this connection follows. Assume that in an actual case the inventory system is System E, but the decision rule used is that given by (50). In other words, $p \neq \infty$, but a decision rule is based on the assumption that $p = \infty$. (This situation is known to occur frequently in practice.) The question is: What additional cost is incurred as a result of the use of the "wrong" decision rule?

Let the input used be $q' = \sqrt{2rc_3/c_1}$, and the corresponding cost C'. By (72), (73) and (74) we find that

$$C'/C_o = \tfrac{1}{2}(2 - r/p)/\sqrt{1 - r/p} \qquad (92)$$

Table 7 gives some values of C'/C_o as a function of r/p.

Table 7. Comparisons of Systems B and E

r/p	0.1	0.2	0.3	0.4	0.5	0.6	0.7	0.8	0.9	1.0
C'/C_o	1.001	1.006	1.016	1.033	1.061	1.107	1.187	1.342	1.739	∞

From the table we see, for example, that when the rate of output is half as large as the rate of input, the use of the "wrong" decision rule results in an additional cost of 6.1 per cent.

COMPARISON OF SYSTEMS A AND G

Systems A and G are similar in all respects except one. In System A the output size was constant and was known with certainty, i.e.,

$x = rt_p$ = known constant. In System G the output size was variable and its probability distribution was known, i.e., $P(x)$ was known.

System A is obviously a special case of System G. For if $x = rt_p$ is a known constant, then $P(x)$ is given by

$$(93) \ldots \qquad P(x) = \begin{cases} 0 & \text{for } x \neq rt_p \\ 1 & \text{for } x = rt_p \end{cases}$$

Substituting this value of $P(x)$ in equation (82) of System G we get:

$$(94) \ldots \qquad C = \begin{cases} \dfrac{c_1S^2}{2rt_p} + \dfrac{c_2(rt_p - S)^2}{2rt_p} & \text{for } 0 \leqq S \leqq rt_p = x \\ c_1(S - rt_p/2) & \text{for } S \geqq rt_p = x \end{cases}$$

Equation (94) is the total cost equation of System A. In developing equation (40) of System A we neglected to consider the case $S > rt_p$ since it seemed intuitively correct to assume that in that System $0 \leqq S_o \leqq rt_p$. Equation (94) represents more generally the total cost of System A.

We did not develop decision rules for System G. Hence, it is difficult to compare it with System A. Let us, therefore, only compare the numerical Examples 2 and 8 pertaining to these systems.

In both examples we have:

$t_p = 1$ month; $c_1 = \$12$ per part per year; $c_2 = \$108$ per part per year.

In Example 2 the output size x during t_p is 200 parts. In Example 8 the output size x during t_p has a probability distribution $P(x)$ given by Table 3 (p. 347); in that case the average output size x is 200 parts. The only difference between the two examples can also be stated as follows:

In both examples the average output size is 200 parts. In Example 2 the output size does not vary; in Example 8 the output size varies: there may be no output at all (with a probability of 0.15); there may be an output size of 100 (with a probability of 0.20), etc. The solution of Example 2 was $S_o = 180$ parts and $C_o = \$1{,}080$ per year. The solution of Example 8 was $S_o = 235$ parts and $C_o = \$2{,}576$ per year. Thus because of the fact that output varied during the scheduling period, we have different decision rules and the minimum costs differ. In the examples cited the variability of output leads to an increase of 139 per cent in minimum costs.

COMPARISON OF SYSTEMS B, F, AND H

System F is an extension of System B. In System B the output size was constant; in System F the output size varied but during any period the ratio of the maximum output size to the average output size was a constant A. System B is evidently a special case of System F. When A = 1 in System F then equations (76), (77), (78), and (79) reduce to equations (49), (50), (51), and (52) of System B.

The effect of the variability of output can be obtained by comparing (52) and (79). Let C_B and C_F designate respectively the minimum costs of Systems B and F; then from (52) and (79) we have:

$$(95) \ldots \qquad C_F/C_B = \sqrt{2A-1}$$

Thus, for example, if $A = 2\frac{1}{2}$, the minimum cost of System F with variable output is twice as large as the minimum cost of System B with constant output.

System H is an extension of System F. In System F lead-time was insignificant, i.e., $LT = 0$; in System H lead-time is significant, i.e., $LT \neq 0$. System F is evidently a special case of System H. When $LT = 0$ in System H, then equations (85), (87), and (88) reduce to equations (76), (78), and (79) of System F. (It is interesting to note that the optimal scheduling period, t_o, is the same for both systems no matter what the value of LT.)

Systems F and H can be used to find the effect of lead-time on the minimum total costs. Let C_F and C_H designate respectively the minimum total costs of System F and H. Then by equation (79) and (88) we have

$$\begin{aligned}(96) \ldots \qquad C_H/C_F &= 1 + LT\,(A-1)\bar{r}\,c_1/C_F \\ &= 1 + LT[(A-1)/\sqrt{2A-1}]\,\sqrt{\bar{r}c_1/(2c_3)} \\ &= 1 + (A-1)LT/t_o\end{aligned}$$

where t_o is the optimal scheduling period of both systems.

We can also compare Systems B and H directly. System H is an extension of System B. In turn, System B is a special case of System H, when in System H we have both $A = 1$ and $LT = 0$.

We can now also derive an expression for the simultaneous effect of variability of the output size and the significant lead-time. From equations (95) and (96) we get

$$(97) \ldots \qquad C_H/C_B = \sqrt{2A-1} + LT(A-1)\,\sqrt{\bar{r}c_1/(2c_3)}$$

SUMMARY OF RESULTS—EXAMPLES 2 TO 9

As a final comparison of Systems A to H, we give in Table 8 below a summary of the numerical results of Examples 2 to 9.

Table 8. Summary of Results—Examples 2 to 9

System		A	B	C	D	E	F	G	H
Example		2	3	4	5	6	7	8	9
Output	Given	r	r	r	r	r	$\bar{r}$	$P(x)$	$\bar{r}$
	A	1	1	1	1	1	$2\frac{1}{2}$	$2\frac{1}{2}$	$2\frac{1}{2}$
Input	t_p	1.00	—	—	—	—	—	1.00	—
	LT	0	0	0	0	0	0	0	2
	p	∞	∞	∞	∞	9,600	∞	∞	∞
Optimal Solution	t_o	—	1.87	0.62	1.97	2.16	0.94	—	0.94
	s_o	20	0	125	39	0	—	—	—
	q_o	—	374	125	394	432	—	—	—
	S_o (or Z_o)	180	374	0	355	—	470	235	1,470
Minimal Costs	C_{1o}	972	2,245	(0)	1,917	1,944	4,490	1,715	11,690
	C_{2o}	108	(0)	6,735	213	(0)	(0)	861	(0)
	C_{3o}	(4,200)	2,245	6,735	2,130	1,944	4,490	(4,200)	4,490
	C_o	1,080	4,490	13,470	4,260	3,888	8,980	2,576	16,180
	(C_o)	(5,280)	(4,490)	(13,470)	(4,260)	(3,888)	(8,980)	(6,776)	(16,180)
System		A	B	C	D	E	F	G	H

Similar data have been used in all examples. The output is either exactly 2,400 parts per year (Examples 2-6) or it is on the average

2,400 (Examples 7-9). Whenever output was variable, the ratio of maximum output to average output is $A = 2\frac{1}{2}$ (Examples 7-9). Whenever the scheduling period is prescribed we have $t_p = 1.00$ month (Examples 2, 8).

In addition, in all examples we have:

c_1, the unit cost of surplus is $12 per part per year;
c_2, the unit cost of shortage is $108 per part per year;
c_3, the unit cost of setup is $350 per setup

The yearly costs in brackets are the costs which were not subject to control in the corresponding examples. For instance, in Example 2, the setup costs were not subject to control; hence $4,200 is in brackets. Similarly (C_o) designates the minimum total costs plus the uncontrollable costs.

Deviating from Optimal Solutions

The effects of deviating from optimal solutions will be given only for Systems A and B. The methodology of approach, however, is general, and is equally applicable to any inventory system.

SENSITIVITY OF SYSTEM A

For System A the total cost equation is

$$(98)\ \ldots \qquad C = (S^2c_1 + s^2c_2)/(2rt_p)$$

and its solution is

$$(99)\ \ldots \qquad S_o = rt_pc_2/(c_1 + c_2)$$
$$(100)\ \ldots \qquad s_o = rt_pc_1/(c_1 + c_2)$$
$$(101)\ \ldots \qquad C_o = \tfrac{1}{2}\, rt_pc_1c_2/(c_1 + c_2)$$

Now consider a decision rule calling for $S = S'$, such that

$$(102)\ \ldots \qquad S' = w_SS_o$$

where w_S is some non-negative constant. This decision rule leads to a total cost C', which is obtained from (98) and (94), by substituting S' for S and $s' = rt - S'$ for s. By some algebraic calculations we find that

$$(103)\ \ldots \qquad C'/C_o = \begin{cases} = 1 + (c_2/c_1)(1 - w_S)^2 & \text{when } w_S \leqq (c_1 + c_2)/c_2 \\ = 2w_S - (c_1 + c_2)/c_2 & \text{when } w_S > (c_1 + c_2)/c_2 \end{cases}$$

The factor w_S represents the amount of deviation from the optimal decision rule. The ratio of C' to C_o represents the effect of this deviation. We note, from (103), that this ratio depends both on w_S and on the ratio of the unit costs of shortage and surplus.

Table 9 gives some numerical values for C'/C_o for various values of w_S and c_2/c_1.

Table 9. Sensitivity of System A

	c_2/c_1 \ w_S	0.5	0.8	0.9	1.0	1.1	1.2	1.5	2.0	3.0
	0.1	1.025	1.004	1.001	1.000	1.001	1.004	1.025	1.100	1.400
	0.5	1.125	1.020	1.005	1.000	1.005	1.020	1.125	1.500	3.000
	1.0	1.250	1.040	1.010	1.000	1.010	1.040	1.250	2.000	4.000
C'/C_o	2.0	1.500	1.080	1.020	1.000	1.020	1.080	1.500	2.500	4.500
	3.0	1.750	1.120	1.030	1.000	1.030	1.120	1.700	2.700	4.700
	5.0	2.250	1.200	1.030	1.000	1.050	1.200	1.800	2.800	4.800
	10.0	3.500	1.400	1.100	1.000	1.100	1.300	1.900	2.900	4.900
	20.0	6.000	1.800	1.200	1.000	1.150	1.350	1.950	2.950	4.950

According to this table, for example, if $S_o = 100$ lbs. but the level chosen is $S' = 120$ lbs., i.e., $w_S = 1.2$, and if $c_2/c_1 = \frac{1}{2}$ then the additional cost as a result of deviating from the optimal rule is 2 per cent of the minimum cost.

SENSITIVITY OF SYSTEM B

For System B the total cost equation is

(104) . . . $$C = qc_1/2 + rc_3/q$$

and its solution is

(104) . . . $$q_o = \sqrt{2rc_3/c_1}$$

(106) . . . $$C_o = \sqrt{2rc_1c_3}$$

Now consider a decision rule calling for $q = q'$, such that

(107) . . . $$q' = w_q q_o$$

where w_q is some non-negative constant. Let the total cost associated with q' be C'. Then from the above four equations we have:

$$(108) \ldots \qquad C'/C_o = (1 + w_q^2)/(2_{w}q)$$

Here the effect of deviating from the optimal decision rule is only a function of the factor of deviation w_q. Table 10 gives some numerical values of C'/C_o for various values of w_q. Similar relationships are illustrated in a graphical form in Figure 16.

Table 10. Sensitivity of System B

w_q	0.5	0.8	0.9	1.0	1.1	1.2	1.5	2.0	3.0
C'/C_o	1.250	1.025	1.006	1.000	1.005	1.017	1.083	1.250	1.667

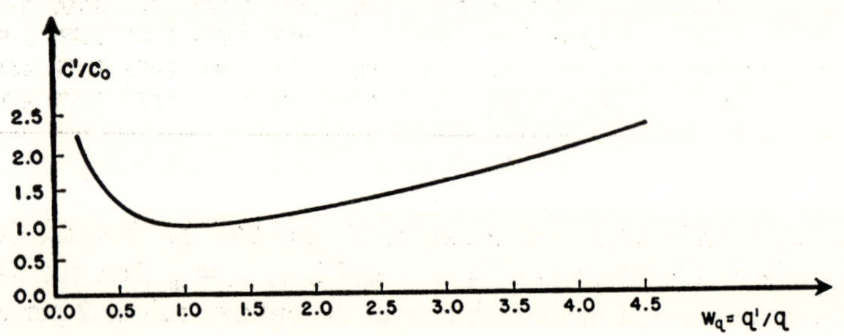

Figure 16. Sensitivity of System B.

In this case we see that small deviations from the optimal decision rule have relatively very small effects on the total cost.

Effects of Estimating Parameters

The effects of using estimates for the solution of inventory systems will be given again only for Systems A and B. The methods used can readily be extended to other systems.

The parameters used in Systems A and B are the rate of output r, and the unit cost parameters c_1, c_2, c_3. Let the true values of these parameters be r, c_1, c_2, c_3, and let their estimated values be r', c_1', c_2', and c_3'.

How does the use of estimated values affect the costs of the system? The effect is by the use of decision rules based on the estimated parameters rather than the true parameters. In System A, for example, the decision-maker, assuming that r', c_1', c_2' are the true parameters, chooses an inventory level $S' = r't_pc_2'/(c_1' + c_2')$, whereas the optimum level should be $S_o = rt_pc_2(c_1 + c_2)$. In this case, therefore,

$$(109)\ \ldots \qquad w_S = S'/S_o = (r'/r)(c_2'/c_2)/[(c_1' + c_2')/(c_1 + c_2)]$$

Let the total cost be C' when the estimated parameters are used, and let C_o be the minimum cost based on the true parameters. Then, by the results of a previous section

$$(110)\ \ldots \qquad C'/C_o \begin{cases} = 1 + (c_2/c_1)(1 - w_S)^2 & \text{when} \quad w_S \leqq (c_1 + c_2)/c_2 \\ = 2w_S - (c_1 + c_2)/c_2 & \text{where} \quad w_S > (c_1 + c_2)/c_2 \end{cases}$$

where now w_S is given by (109) above.

A similar analysis holds for System B. The input quantity chosen by the decision-maker is $q' = \sqrt{2r'c_3'/c_1'}$ whereas the optimal input is $q_o = \sqrt{2rc_3/c_1}$.

Hence

$$(111)\ \ldots \qquad w_q = q'/q_o = \sqrt{(r'/r)(c_3'/c_3)/(c_1'/c_1)}$$

and

$$(112)\ \ldots \qquad C'/C_o = (1 + w_q^2)/(2w_q)$$

where w_q is given by (111) above.

Equations (110) and (112) represent the effect of use of estimated parameters. It should be pointed out that there is no effect, i.e., $C'/C_o = 1$, not only when the estimated parameters are the true parameters, but also in those cases when the estimates are such that $w = 1$.

Tables 11 and 12 give some values of C'/C_o for various estimates of the parameters for Systems A and B respectively.

From the tables we find that, if, for example, all estimates are 20 per cent higher than the true values of the parameters, then in System A for $c_2/c_1 = 2$, this results in an inventory level 20 per cent higher than the optimal, but the additional total cost is only 8 per cent. In System B, for 20 per cent higher estimates, the input is 10 per cent higher than the optimal input, and the additional total cost is only 0.5 per cent.

Table 11. Effect of Estimates on System A

c_2/c_1	2.0	10.0	2.0	2.0	2.0	10.0	10.0	10.0
r'/r	1.20	1.20	0.80	1.20	1.20	1.20	0.80	1.20
c_1'/c_1	1.20	1.20	1.20	1.20	0.80	0.80	1.20	1.20
c_2'/c_2	1.20	1.20	1.20	0.80	1.20	1.20	1.20	0.80
S'/S_o	1.20	1.20	0.80	1.03	1.35	1.24	0.80	1.15
C'/C_o	1.080	1.300	1.080	1.002	1.245	1.576	1.400	1.225

Table 12. Effects of Estimates on System B

r'/r	1.20	0.80	1.20	1.20	0.80	0.80	1.20	0.80
c_1'/c_1	1.20	1.20	0.80	1.20	0.80	1.20	0.80	0.80
c_3'/c_3	1.20	1.20	1.20	0.80	1.20	0.80	0.80	0.80
q'/q_o	1.10	0.90	1.35	0.90	1.10	0.73	1.10	0.90
C'/C_o	1.005	1.006	1.042	1.006	1.005	1.050	1.005	1.006

SUMMARY AND CONCLUSIONS

This chapter presented some elements of inventory systems. It first defined such general terms as *inventory system, inventory problem,* and then proceeded to describe the *properties of inventory systems.* It pointed out the importance of knowing the properties of an inventory system, so that a *model* of the system can be constructed from which a *solution* of the system can be obtained.

It illustrated the methods for constructing models and solving them for eight different inventory systems. The first three were elementary systems with simple properties. The other five were extensions of the elementary systems. Each system was illustrated by a numerical example.

Finally, it discussed the sensitivity of the systems studied. It compared the optimal decision rules for the various systems and the numerical solution of their illustrations. For two elementary systems it also brought out the effects of deviating from optimal decision rules and the effects of using estimates of parameters on the total costs of the systems.

The results given above should be treated primarily as a contribution to the methodology in solution of inventory problems. Although some of the formulas which were developed may apply to specific inventory systems encountered in practice, the reader is advised to use them with caution.

It is strongly recommended that in practical situations the properties of the inventory system be first ascertained—then a model of the system constructed and solved—then the sensitivity of the system examined. Only after these steps have been taken should a detailed program for collection of data be launched.

REFERENCES

(1) Ackoff, Russell L. "The Development of Operations Research as a Science," *Operations Research,* Vol. 4, No. 3 (June, 1956), 271-73.

(2) Arrow, Kenneth J., Harris, Theodore, and Marschak, Jacob. "Optimal Inventory Policy," *Econometrica,* Vol. 19, No. 3 (July, 1951), 250-72.

(3) Arrow, Kenneth J., Karlin, Samuel, and Scarf, Herbert. *Studies in the Mathematical Theory of Inventory and Production.* Stanford: Stanford University Press, 1958.

(4) Dvoretsky, A., Kiefer, J., and Wolfowitz, J. "The Inventory Problem," *Econometrica,* Vol. 20, No. 2 (April, 1952), 187-222.

(5) ———, ———, and ———. "On the Optimal Character of the (s, S) Policy in Inventory Theory," *Econometrica,* Vol. 21, No. 4 (October, 1953), 586-96.

(6) Flagle, Charles D. "Queueing Theory and Cost Concepts Applied to a Problem in Inventory Control." *Operations Research for Management.* Vol. II, edited by Joseph F. McCloskey and John F. Coppinger. Baltimore: The Johns Hopkins Press, 1956.

(7) Laderman, J., Littauer, S. B., and Weiss, Lionel. "The Inventory Problem." *Journal of the American Statistical Association,* Vol. 48, No. 264 (December, 1953), 717-23.

(8) Magee, John P. "Guides to Inventory Policy," *Harvard Business Review.* Vol. 34, Nos. 1, 2, and 3 (January, March, and May, 1956).

(9) ———. *Production Planning and Inventory Control.* New York: McGraw-Hill, 1958.

(10) Naddor, Eliezer. "Elementary Inventory Models," *Introduction to Operations Research.* Edited by C. West Churchman, Russell L. Ackoff and E. Leonard Arnoff. New York: John Wiley and Sons, 1957.

(11) ———. "Some Models of Inventory and an Application," *Analyses of Industrial Operations.* Edited by Edward H. Bowman and Robert B.

Fetter. Homewood, Ill.: Richard D. Irwin, Inc., 1959.

(12) ———. "Economic Lot Sizes for Some Elementary Inventory Systems," *Bolletino del Centro per la Ricerca Operativa,* No. 4 (1957), 1-10.

(13) ———. *Properties, Models and Solutions of Elementary Inventory Systems and Their Extensions.* University Microfilms, Ann Arbor, Mich. (Publication No. 24,800, 1958; Library of Congress Card No. Mic 58—4126.)

(14) ———, and Saltzman, Sidney. "Optimal Reorder Periods for an Inventory System with Variable Cost of Ordering," *Operations Research,* Vol. 6, No. 5 (September-October, 1958), 676-85.

(15) Raymond, Fanfield E. *Quantity and Economy in Manufacturing.* New York: McGraw-Hill, 1931.

(16) Simon, Herbert A., and Holt, Charles C. "The Control of Inventories and Production Rates—A Survey," *Operations Research,* Vol. 2, No. 3 (August, 1954), 289-301.

(17) Welch, W. Evert. *Scientific Inventory Control.* Greenwich, Conn.: Management Publishing Corp., 1956.

(18) Whitin, Thomson M. "Inventory Control Research: A Survey," *Management Science,* Vol. 1, No. 1 (October, 1954), 32-40.

(19) ———. *The Theory of Inventory Management.* Princeton: Princeton University Press, 1957.

Thirteen

LINEAR PROGRAMMING

VINCENT V. McRAE

Any consideration of methodology for operations research and systems engineering, no matter how extensive, would be incomplete without a discussion of linear programming, for the techniques of linear programming have been applied so often and to such advantage that many persons unfamiliar with the new sciences tend to associate them exclusively with linear programming and its applications. Although often applied and extremely useful, linear programming techniques are merely a special class of methods applicable to some, but by no means to all, of the problems encountered in programming or in developing a method of operation designed to optimize returns from a given system (27). Yet there are so many systems in which the fundamental relationships of the quantities which can be varied by the entrepreneur are linear in form, that a systematic and scientific study of such systems and the way in which their operations could be improved has of necessity led to the body of theory to be briefly considered here, that of linear programming.

The development of linear programming is closely related to that of economic theory, and it is often stated that the former received its principal stimulus from the latter. For a long time that part of economic theory which was based purely upon mathematical concepts derived from various applications of calculus. In marginal analysis, for example, cost production and price functions were examined, and these functions were assumed to be continuous functions of independent variables and to possess continuous partial derivatives of the first and second orders. Moreover, the only types of functions admitted to this theory were those representable by exact equalities. For economic systems obeying these postulates the theory provided a basis for the determination of criteria for an optimal production program and for optimal utilization of resources. These criteria were expressed as equalities among the ratios of certain partial derivatives. Although the theory was applicable to some economic systems to a good degree of approximation, particularly agricultural systems, it was found to be practically useless for many industrial systems, for in these cases the assumptions were not applicable even to a first approximation.

Whereas the agricultural systems could absorb increments of input in almost any quantity and adjust methods of operations so that these inputs could be utilized, the industrial systems were found to require individual inputs in definite increments and groups of inputs in definite proportions if any of the inputs were to be utilized. That this is so may be easily seen by naively considering automobile production.

To produce an automobile for market one obviously requires a chassis, a motor, and four wheels, in addition to many other components. An increment of a motor will certainly not suffice, and, in fact, a motor without a chassis and without four wheels will fail to increase the production of the firm. This need for components in integral numbers and in definite proportions is reflected throughout the automobile industry and, indeed, throughout many modern industries. Consequently, industrial systems possess production functions which are not continuous functions of the inputs, and are not functions of independent inputs, and accordingly marginal analysis loses its utility to a great extent in considering systems of this kind. The problems faced by management today even in a small firm are essentially different from those for which marginal analysis was developed. Consequently, the mathematical structure well suited to the solution of these problems may be expected to differ significantly also. Linear programming is a small but very significant component of this new mathematical structure.

The precise areas from which linear programming developed are fairly difficult to specify. However, Samuelson attributes the determination of four sources of inspiration to Professor Koopmans (2). These are as follows:

1. Recognition in the early 1930's by such continental economists as Neisser, Von Stackelberg, Zeuthen, Schlesinger, Von Neumann, and Wald that the simple castle version of Walrasian general equilibrium

could not be adequately appraised by an uncritical counting of the number of its equations and the number of its unknowns.

2. The "new welfare economics" in the various versions of Lerner, Bergson, Kaldor, Hicks, Samuelson, Lange, and Arrow, which threw new light on the early writings of Smith, Walras, Von Wieser, Marshall, Pareto, Barone, Pigou, Von Mises, and Frederick Taylor.

3. The interindustry input-output theories and measurements of Leontief, Evans, Hoffenberg; and the related multisector analysis of Keynes, Harrod, William Salant, Machlup, Metzler, Solow, Goodwin, and Chipman.

4. The specific programming and optimizing problems raised by defense and military problems—and, more generally, the numerous optimizing problems that business firms have always had to solve in their quest for profits and survival.

Samuelson also adds several additional sources for the development of linear programming theory from economic theory; these generally are somewhat related to the ones cited by Koopmans. In any event it is sufficient to observe that there has been an interplay between economic theory and the development of linear programming, one having been spurred on by developments in the other.

This chapter deals with the general mathematical formulation of the linear programming problem, examines the assumptions implicit in this formulation, and explores methods of solution. Several special forms of the linear programming problem and the relationship of duality to these problems will also be examined.

THE GENERAL LINEAR PROGRAMMING PROBLEM

To formulate the general linear programming problem precisely, let a_{ij}, b_i, and c_j be sets of constants $(i = 1, \ldots, m;\ j = 1, \ldots, n)$ and x_j $(j = 1, \ldots, n)$ be a set of variables. We seek solutions $X = (x_1, x_2, \ldots, x_n)$ which satisfy the inequalities

$$\sum_{j=1}^{n} a_{ij}x_j \leqslant b_i, \quad i = 1, \ldots, m; \tag{1}$$

$$x_j \geqslant 0, \quad j = 1, \ldots, n; \tag{2}$$

and at the same time maximize the linear functional

$$z = \sum_{j=1}^{n} c_j x_j \tag{3}$$

The obtaining of such solution is the linear programming problem. Examine its structure.

Equation (3) defines the *objective function* z and this function is linear in each of the x_j. It is a function of the vector $X = (x_1, x_2, \ldots, x_n)$, and hence we may express (3) as

$$z = f(x). \tag{4}$$

The function is defined for all values of x with finite components; however, our attention is restricted to those values whose components satisfy restrictions (1) and (2).

We see that in equations (2) we require that all components be non-negative and that in equations (1) we require that the components satisfy *m linear* inequalities.[1] We commonly refer to equations (2) as the non-negativity restrictions and to the inequalities (1) as the *functional constraints.* In matrix terminology we may represent the functional constraints by the equation

$$AX \leqslant B \tag{5}$$

in which $A = (a_{ij})$ and $B = \text{col}\,(b_i)$.

Whereas the non-negativity restrictions limit the set of admissible vectors to vectors X with positive or zero components, the functional constraints limit this set to those vectors satisfying the matrix in equation (5). These restrictions and the functional z characterize that part of linear programming in which we attempt to maximize an objective function.

The linear programming problem may be stated in an equivalent form in which a linear functional is to be minimized and the functional constraints are '$\geqslant$' relations rather than '$\leqslant$' relations. Since this problem is obtained from the one stated by multiplying equations (1) by -1 and maximizing $-z$, this case can be covered adequately by discussions of the maximizing problem.[2]

An Economic Interpretation

An economic interpretation will serve as a convenient tool for fixing fundamental concepts and for a consideration of conditions necessary for application of the techniques to be discussed. Available economic

[1] Some of these relations may be equalities as well.

[2] The minimization problem is often more difficult to solve because first feasible solutions are not as readily obtained.

literature contains an abundant supply of such interpretations and the example to be presented is merely one of the many which would suffice (5, 7, 15, 30).

This example involves the case of a firm manufacturing several products. It may be assumed without too much loss of accuracy that the object of the firm is to maximize its profits, and hence if the linear programming theory is to apply, the objective function must represent the firm's profits. It is observed in equation (3) that if c_j is positive, a unit increase in the component x_j causes an increase of c_j in the objective function z. This must be interpreted as an increase in profit and hence c_j is the unit profit associated with the component x_j. Since the profits are associated with the sale of goods, it is natural to interpret the variable x_j as the quantity of the j-th type of product sold. With this interpretation, we associate a negative c_j with a product sold at a loss. The constraints (1) and (2) must be interpreted in a manner consistent with the interpretation of z, the c_i and x_i; however, we have already made several implicit assumptions about the economic process of the firm.

In the first place, since the objective function is linear, by interpreting c_j as unit profit and x_j as the number sold, we have assumed that each average or unit profit is independent of the number of products of that type sold. Thus our present interpretation is applicable to cases in which supply is roughly equivalent to demand, or more specifically to those in which unit profit is constant. If increased production of one of the firm's products beyond a certain point will tend to decrease price and thereby decrease profit, then our interpretation is exact only up to the point where such decreases occur and is approximate thereafter. But we have also assumed implicitly that the unit profit for each item of production is independent of that of all other items produced. For a unit increase in several values of the x's generates an increase in profit which is the sum of the unit profits. Thus at the levels of production permitted by the constraint equations, the unit profit structure is independent of the relative proportions of the products produced by the firm.

The non-negativity constraints of equations (2) are easily seen to be necessary, since they state the obvious condition that the firm is concerned only with the number of these products sold, or that it does not intend to buy any of the product which it produces.

In interpreting equations (3), several choices which are consistent with the previous interpretation are available, but the most common one associates these constraints with the resources of the firm. But regardless of choice, the inequalities place restrictions on the number of items of each type which can be made and/or sold, in addition to restricting the combinations in which they can be sold. Since they are inequalities, it is not required that these items be sold in given proportions, but only that their existence or sale in given proportions cannot exceed a given scale. Such restrictions on sales can arise from several sources, but the linear programming model is quite often applied to the production process since the proportions in which products can be made is often determined ultimately by the resources of the firm. Yet the model permits the constraints represented by these relations to be either production constraints, marketing constraints, or both. It is only essential that, for the case being considered, all extremes be represented by the numbers b_i and that the constraints imposed on the quantities sold by these extremes be linear functions of the numbers sold. The discussion here will be limited to the resources of the firm; however, the inclusion of linear marketing constraints can be made without difficulty.

In general, the firm has several resources which are limited in quantity and these limitations are expressed in the constants b_i. It is not necessary that different b_i be in the same units; therefore, the first constraint may represent the number of man-hours of a particular type available, while the second constraint represents the number of machine hours available on a particular type of machine, while the third constraint represents the supply of a particular type of material that is available for the production process, etc. By reasoning analogous to that used in the case of the profit function, since the a_{ij} are constants and the constraints linear, the a_{ij} must be interpreted as unit requirements. Thus for a production constraint, we interpret a_{ij} as the requirement of resource i for the production of a unit product of type j. Therefore, equations (1) state that the total use of resources of each type cannot exceed the constants b_i, the resources of that type available to the firm. But since inequalities are admitted, it is not required that all resources be used.

We may now state the production problem of the firm and the associated economic assumptions more concisely as follows: Given a firm with m fixed resources, manufacturing n products with production processes such that each unit of each product requires a constant quantity of each type of resource, selling each product at a fixed unit profit, and able to sell any combination of products produced within the limits of its resources, we are to determine the distribution of products to be produced so that the profit of the firm is maximized. To express this problem mathematically we let x_j be the number of products of type j to be produced, a_{ij} be the requirement of resource i for the production of a unit of product j, b_i be the quantity of resource i available, c_j be the unit profit associated with product j, and z be the total profit of the firm. The problem then becomes one of maximizing z as given by equation (3) subject to the constraints given in equations (1) and (2). A numerical example of a production problem of a firm will be considered after the general linear programming problem is interpreted geometrically.

The Geometric Interpretation

Several fundamental concepts are required for the geometric interpretation, as seen below:

1. With the ordered sequence of real numbers $x_1, x_2, \ldots, x_n$ we associate a *point* and a *vector* drawn to that point from the origin of the n-dimensional space containing all points. We designate either the vector or the point by $X = (x_1, x_2, \ldots, x_n)$; however, little confusion as to meaning arises since the context usually serves to remove ambiguity.

2. If X and Y are points in n-space, then if a is any non-negative real number less than unity, $Z = aX + (1 - a)Y$ is said to be a convex linear combination of the points X and Y. It is easily verified that as a varies from 0 to 1, Z traces the line segment joining X and Y if n-space is ordinary Euclidean space of three dimensions. More generally, we say that $Z = \sum_{i=1}^{r} a_i X_i$; $0 \leqslant a_i \leqslant 1$, is a convex linear combination of the points X_i if $\sum_{i=1}^{r} a_i = 1$.

3. A set of points S is said to be a *convex set* if for all X and Y in S, S contains every convex linear combination of X and Y.

4. A point X in a set S of points is an extreme point of S if it is impossible to express X as a convex linear combination of two other points of S.

5. The set of points $X = (x_1, \ldots, x_n)$ of n-space whose coordinates satisfy the equation $\sum_{i=1}^{n} a_i x_i = b$ is called a hyperplane, or simply a

plane in n-space. These notions are obvious extensions of concepts already familiar in two and three dimensional Euclidean space, and hence they will aid the intuition in interpreting equations (1), (2), and (3).

It is seen that if any of the relations (1) are strict equalities, the points X which satisfy the constraints lie on a hyperplane in n-space. In fact, if there are several equalities, then these points must lie on the intersections of the planes associated with these constraints. But if a constraint is an " $\leqslant$ " relation, such points constitute a "half-space" in the sense that apart from the plane representing the boundary or the strict equality, there is a one-to-one correspondence between points which satisfy the relation and those which do not. We shall use the term "half-space" for the set of points, including the bounding plane, which satisfy the inequality. Thus we say that the points X which satisfy all of the inequalities (1) must lie in the intersection of m half-spaces and/or planes associated with these relations.

Equations (2) further restrict the set of admissible points, for they require that the components of each vector be non-negative. That part of n-space in which all vectors have non-negative components is called the positive orthant of n-space and hence the set of admissible points is precisely those points in the intersection of the m half-spaces and the positive orthant of n-space.

For a fixed z, equation (3) denotes a hyperplane and the distance of this plane from the origin in n-space is proportional to z. Thus hyperplanes associated with larger values of the objective function lie at greater distances from the origin and the problem of maximizing the objective function is thus seen to be equivalent to finding the hyperplane which intersects the set of points defined in the preceding paragraph at the greatest distance from the origin. However, the family of hyperplanes designated by a variable z contains the set of points forming the positive orthant of n-space and hence successive planes in the family pick up each point in the solution set. Thus the geometric interpretation would lead to the conclusion that if the solution set were non-empty, then a maximum exists. This reasoning can be easily verified analytically and leads to the well-known fact that if the linear programming problem has feasible constraints, i.e., if a vector exists which satisfies these constraints, then the maximization problem has a solution. The maximum may in fact be infinite, but this can occur only if the solution set is unbounded.

A clearer idea of the nature of the solution space and of optimal programs can be arrived at if we consider the particular case in which $n = 3$. This designates the Euclidean space of three dimensions and is most familiar to us. In this space we see that the positive orthant is that octant of three-space determined by the positive extensions of the three axes. Each inequality of (1) determines a plane which divides the space into two sets and all points which satisfy this inequality lie on the plane or on one side of it with respect to the origin. Thus, if the solution set is bounded, it is a polyhedron in the positive orthant. The edges of the polyhedron are lines of intersection of the planes determined by the inequalities, its faces are some of the planes themselves, and its vertices are the extreme points of the solution set.[3]

Now since the maximization problem is interpreted geometrically as equivalent to considering members of the family of planes with successively larger distances from the origin, on bounded solution sets, this process must end and maxima be determined. If the polyhedron is properly oriented, then the plane which intersects the solution set at the maximum distance from the origin intersects it at a vertex, the maximum is unique, and only a single optimal program exists. But if this intersection is a line or a face of the polyhedron, then an infinite number of optimal programs exist. In each case we see that there is an optimal program associated with an extreme point of the solution set, for an edge contains two extreme points and a face at least three. This property of maxima and of optimal programs is very important, for we see that if the solution set is bounded and if we can examine the values of z at the extreme points of the solution set, we must necessarily determine an optimal program. Moreover, by examining the values of z at the extreme points we shall determine enough information to see whether alternate programs exist, the necessary and sufficient condition for this being that maxima exist at two or more extreme points.

The association of the solution set with a polyhedron in n-space has led to several fundamental properties of the solution set and to several theorems related to optimal programs. Although the proofs require more detail than that presented herein, the following properties are at least made plausible by the geometric arguments presented thus far:

a. The set of solutions to a system constrained by linear inequalities and/or equalities is a convex set.

b. A maximum (or minimum) of a linear functional constrained by

[3] Observe that not all planes are faces, as some of the constraints may be redundant.

a system of linear inequalities and/or equalities occurs at an extreme point of the solution set.

c. Alternate maxima (minima) or alternate optimal linear programs exist if and only if the linear functional takes on its maximum (minimum) value at two or more extreme points; if it does, then every maxima (minima) or every optimal program is a convex linear combination of those associated with the extreme points at which the maxima (minima) occur; conversely, any convex linear combination of those associated with the maxima (minima) which occur at the extreme points is a maximum (minimum) or optimal program.

d. If a linear program is feasible, then an optimum program exists and this program is associated with a finite value of the objective function if the solution set is bounded.

These facts will be illustrated in the section concerned with the geometric method of solution.

METHODS OF SOLUTION

The principal methods of solution of general linear programming problems are the geometric method, the relaxation method, the method of double description, and the simplex method. We shall consider the geometric method so as to illustrate basic concepts used in the geometric interpretation and discuss and illustrate the simplex method since it is the method used more commonly, both for hand calculation and for solutions made with the aid of digital computers.

The Geometric Method

The geometric method of solution is applicable only to problems involving spaces of relatively small dimension, and hence we shall limit its treatment to an illustrative example. We shall consider a very simple production problem, one associated with a firm that manufactures only two products, but products which require the use of common facilities and manpower. The objective of the firm is to maximize profits from the sale of the two commodities; it is assumed that all products made with the resources at hand can be sold at fixed prices. It is also assumed that there are six fixed resources to be allocated to the production of the two types of items, and we may think of these as being the number of hours of skilled and unskilled manpower available, the number of machine hours available on two machines (each machine being required for the production of either commodity), and two types of supplies or materials from which the products are manufactured. The resources are assumed to be fixed with respect to the period of time for which the program is to be developed; however, maximum utilization of any resource is not required if such utilization would tend to reduce profits.

Letting x_1 be the quantity of the first commodity to be produced and x_2 be that of the second commodity, we formulate the problem mathematically as follows: Assuming 5 and 8 to be respective unit profits on products x_1 and x_2, corresponding to the c_j in equation 3, page 367, we are to maximize

$$z = 5x_1 + 8x_2$$

subject to the functional constraints,

$$\begin{aligned} 10x_2 &\leqslant 30, \\ 10x_1 + 20x_2 &\leqslant 70, \\ 20x_1 + 10x_2 &\leqslant 80, \\ 30x_1 + 10x_2 &\leqslant 130, \\ 12x_1 + 9x_2 &\leqslant 120, \\ 15x_1 + 10x_2 &\leqslant 100; \end{aligned} \tag{6}$$

and the non-negativity conditions

$$\begin{aligned} x_1 &\geqslant 0, \\ x_2 &\geqslant 0, \end{aligned} \tag{7}$$

where the coefficients of x_1 and x_2 in equation (6) are the unit resource requirements, corresponding to the a_{ij} of equation 1, page 367. Since the system of inequalities (6) will be replaced by an equivalent system if we divide any of the inequalities by a positive number, we may express the relations (6) as follows:

$$\begin{aligned} x_2 &\leqslant 3, \\ x_1 + 2x_2 &\leqslant 7, \\ 2x_1 + x_2 &\leqslant 8, \\ 3x_1 + x_2 &\leqslant 13, \\ 4x_1 + 3x_2 &\leqslant 40, \\ 3x_1 + 2x_2 &\leqslant 20. \end{aligned} \tag{8}$$

Since we are concerned with 2-space, these relations are associated with hyperplanes which are lines. These lines are shown in Figure 1.

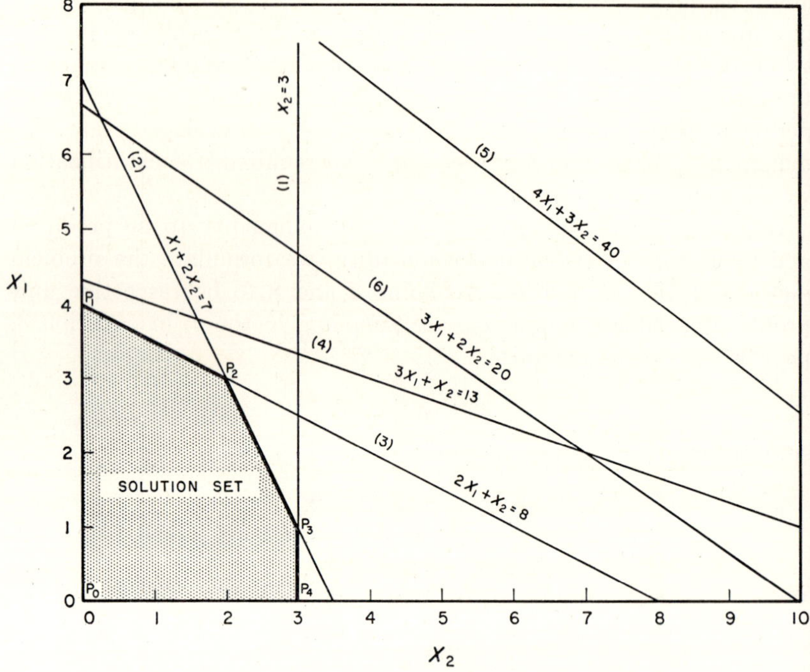

Figure 1. The set of solutions.

Figure 1 shows only the first quadrant, but since this is the positive orthant for the two-dimensional space, this is the only area of interest. The points which satisfy any of the given constraints are those in the area bounded by the line and the positive x_1 and x_2 axes. The solution space, being the intersection of all such areas, is represented by the polygon $P_0P_1P_2P_3P_4$. Thus we see that the solution set is bounded and as indicated previously is convex. In addition to assuring at least one optimal program for any objective function, the solution set is a polygon, and therefore there are alternate optimal programs only if the objective function is parallel to one of the sides of the polygon. But with the given price function there is a unique optimal program.

Figure 1 also gives a great deal of information about the balance of resources for the production of the firm; for we see immediately that the only constraints which affect the optimal program are those giving rise to the sides of the polygon. For all others, the resources are excessive for any feasible program and the restraints redundant. Full utilization of facilities and resources is limited by the availability of skilled manpower, unskilled manpower, and machine hours on the first machine. Yet even the availability of some of these may be excessive for optimal programs.

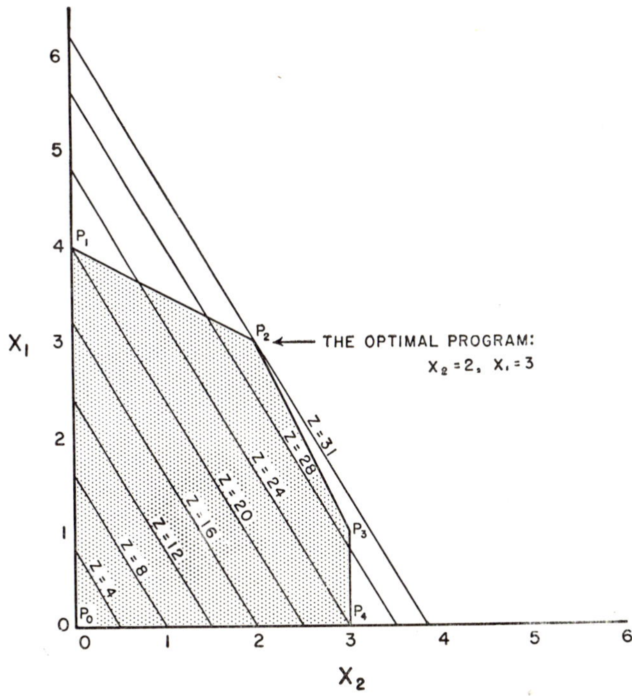

Figure 2. The optimal solution.

Figure 2 illustrates the geometric method of seeking optimal programs by considering the family of hyperplanes associated with the objective function. We see that the hyperplane in this family which intersects the solution set and is associated with the maximum z, because it is at the greatest distance from the origin, is a line through P_2 with slope equal to $-8/5$. Hence there is a unique optimal program in which 2 units of x_2 and 3 units of x_1 are produced. For this program, the supply of skilled manpower is not limiting, the only resources which are fully utilized being the supply of unskilled manpower and available time on machine 1.

Not only have we determined the optimal program for the production of the firm and found that with the resources available only thirty-one units of profit are attainable, but we are in position to adjust inventories and resources so that profits for future periods can be improved. It is obvious that if all resources are to be fully utilized in the future and if the quantities cited cannot be increased, then the constraint lines should be adjusted so that they all pass through P_2 or are as close to P_2 as is feasible. If production is to be increased and an optimal program maintained, then the supply of unskilled manpower must be increased and more hours be provided for machine 1. The extent to which these adjustments are possible, without altering other resources and the increased profit which could be obtained, can be easily determined from the graphs presented. Moreover, the costs associated with deviations from the optimal program are easily obtained. Further adjustments are possible and the results can also be found by use of these graphs; however, the examples cited are sufficient to illustrate the fact that the solutions to the optimization program usually contain information on which future programs can be based, and, accordingly, the technique is a powerful tool for management. This information is readily accessible for geometric solutions and often is as easily obtained from analytic ones, particularly those involving the simplex technique which will be considered next.

THE SIMPLEX METHOD

Perhaps the most commonly used method of solving systems of linear inequalities and/or systems composed of inequalities and equalities is the simplex method. This method was developed by Dantzig, Wood, and their associates at the Directorate of Management Analysis, DCS/Comptroller, USAF, and their work provided the first practicable and simple method for solving such systems when large numbers of variables were involved (9, 10, 11, 12, 13). In addition, it provided a stimulus for further theoretical research in linear programming. The method as originally proposed by Dantzig failed to provide adequately for some special problems associated with degenerate cases; however, work by Charnes resolved these difficulties and made the method applicable without the restriction of non-degeneracy (5).

The most complete mathematical treatment of the simplex method is contained in Charnes' lectures (5). This treatment, however, is not recommended initially for those merely interested in learning to apply the technique. A much more elementary exposition of the method by Wagner is contained in reference (33), where the method is presented in terms suitable for those unfamiliar with linear programming problems. In indicating the procedure for application of the method, the present discussion will draw freely from the exposition given by Wagner, and although no attempt at rigor will be made, some important theoretical concepts will be introduced in order to give an intuitive understanding of the logic leading to the development of the method and the principal mathematical relationships on which the technique is based. It is hoped that this will promote a fuller appreciation of the nature of the solution presented and a better understanding of why maxima are obtained by the procedure specified. Of course, those persons interested in the theoretical development or in extensive applications should consult the references given at the end of the chapter.

The theory involved in the simplex method can be developed by combining several results from the theory of sets with well-known facts established in matrix theory or in the theory of vector spaces. The functional constraints stated in equations (1) and (5) can be expressed in terms of the column vectors of the matrix A. If A_i denotes the i-th column vector of the matrix A, then we may express the relationship indicated by equations (1) and (5) as follows:

$$\sum_{i=1}^{n} x_i A_i = B. \tag{9}$$

In this form it is easily seen that the matrix B, a column matrix also, is a linear combination of the matrices A_i and that the x_i are the coefficients of these vectors.

A great deal is known about vectors and the systems generated by linear combination of them, vector spaces. One of the primary results in the theory involved is that any vector space has a basis, i.e., there exists a set of q linearly independent vectors which generate the vector space, but the space cannot be generated by less than q vectors. Furthermore, any linearly independent set of q vectors in the space form a basis.

A second fact of importance is that the expression of a single vector such as B in terms of a basis is unique; this establishes a one-to-one correspondence between some vectors X and bases of the vector space generated by the A_i. We are concerned only with the vectors A_i and not with bases which include other vectors, and there are only a finite number of bases which can be selected from among the A_i. These facts yielded results of significance when combined with several notions from set theory, or more particularly, the theory of convex sets.

In application of the set theoretic approach, it was established rigorously that a maximum of the functional z occurred at an extreme point of the solution set, and when it was found that each extreme point was associated with a linearly independent set of vectors among the A_i and that the coordinates of this extreme point were the coefficients of the A_i in equation (9), not only was the basis for the development of the simplex method partly established, but several facts of immediate interest were at once evident. First, since A was assumed to be an m by n matrix, a basis could contain at most $\min(m, n) = e$ of the vectors A_i, and, accordingly, if a maximum existed, at least one optimal program could be obtained in which at most e of the vectors appeared (or an X with at most e non-zero components if any optimal program were possible). But as had been indicated earlier, an optimal program was found to exist if a feasible solution existed. Thus it was possible to conclude that if the constraints were feasible, an optimal program involving no more products or processes than resources existed and that one such program could be determined by examining all groups of linearly independent vectors among the A_i of equation (9).

A procedure for determining a maximum of the functional was now evident. One could consider all linearly independent subsets of the vectors A_i, compute the vectors X associated with the expression of B in terms of the elements of each set when X could be so expressed, compute the value of z associated with each such X, and know that the maximum of the values determined was the maximum value of the functional. Geometrically, this amounts to examining the value of the functional at each extreme point and selecting the maximum from the set of values so obtained. The process was feasible, for there were a finite number of combinations of vectors to be examined and each calculation required could actually be carried out. But for large values of m and n, it was anything but practical, for it was extremely tedious if done by hand and very time-consuming if done either by hand or by computing machine. What was needed at this stage was a more systematic method of determining and examining the linearly independent sets, or, equivalently, the extreme points of the solution set.

If instead of continuing to examine linearly independent sets associated with undetermined values of z, we could examine them in such a way that the value of z always increased, then the maximum could be obtained in much shorter time and with much less effort. Geometrically, after having examined the function z at an extreme point or corner of the tetrahedron in space, we would like to examine another corner which could be found by traveling some distance away from the origin in the direction perpendicular to the hyperplane determined by the objective function. Then after sufficient repetitions we would ultimately reach a maximum and do so in less time than required for the inspection of all corners or extreme points. The simplex technique does precisely this.

In this technique, one introduces "slack vectors," or "idle factors," or, in essence, additional variables so that all inequalities become equalities and an easily obtained first feasible solution becomes available. This solution (in terms of slack variables) is associated with an extreme point of the solution space by virtue of the independence of the vectors A_i which appear in the expression for B. One then proceeds to examine other extreme points which increase the value of z. After several repetitions of the process involved, a maximum value for z is obtained. We shall explore this process through an economic example, drawing freely from Wagner's exposition as to the format for the solution (33).

Illustrative Example

In this example we consider the same firm as in the preceding example, but assume that the firm is contemplating a modification of its existing facilities and would like to evaluate the proposal in terms of future profits under an optimal production program. It is assumed that the supply of skilled and unskilled labor can be readily increased, as can the supply of material of type one, that an additional machine of type one can be obtained, and that the labor force can be doubled, if desirable, but not increased beyond that level. Since the materials of the first type are not considered limiting for the future time period, this constraint is dropped from consideration; however, its costs are included in determining unit profits. Along with the purchase of a new machine, the firm wishes to reconsider its selection of products for manufacture and finds four additional products which, it appears, can be easily marketed at fixed prices. The new production problem is formulated mathematically as before; but, since six possible products are involved, the vector space under consideration is of dimension six and the solution by the geometric method is no longer simple.[4] We shall solve this problem using the simplex method and attempt to do so in such a manner that the technique involved can be readily adapted to any linear programming problem.

It must be observed that the example chosen for illustrating the simplex technique avoids the problem of degeneracy and that of unbounded solution sets; however, methods for handling these special cases are available in reference (5).

We let x_3, x_4, x_5, and x_6 denote the quantities of the additional products to be produced and, proceeding as before, arrive at the set of functional constraints:

$$\begin{aligned} 5x_2 + 2x_3 + x_4 + 2x_5 + 3x_6 &\leqslant 60, \\ x_1 + 2x_2 + x_3 + 3x_4 + 2x_5 + x_6 &\leqslant 14, \\ 2x_1 + x_2 + 3x_3 + x_4 + 2x_5 + 3x_6 &\leqslant 8, \\ 3x_1 + x_2 + 2x_4 + 3x_5 + 6x_6 &\leqslant 13, \\ 3x_1 + 2x_2 + x_3 + 2x_5 + 2x_6 &\leqslant 20. \end{aligned} \tag{10}$$

The objective function or the expression for the profit of the firm is taken to be given by

$$z = 5x_1 + 8x_2 + 3x_3 + 4x_4 + 7x_5 + 8x_6. \tag{11}$$

Hence our problem is to maximize z as given in equation (11) subject to the non-negativity conditions imposed on the x_i and the functional constraints (10).

The first aim in the simplex technique is to obtain a feasible solution associated with an extreme point of the solution set. This will obviously be achieved if we obtain a solution in terms of vectors forming a basis of the solution space and we proceed so as to obtain such a solution even though we increase the number of variables to be considered. We introduce variables representing the quantities of the resources which are not used in a program and thereby convert the inequalities to equalities and provide a convenient basis for the solution space. This results in the functional constraints given in equations (12).

[4] This method can still be used to determine lower bounds to the profit functions, as indicated in the section on duality.

$$
\begin{array}{lr}
5x_2 + 2x_3 + x_4 + 2x_5 + 3x_6 + x_7 & = 60, \\
x_1 \boxed{+ 2x_2} + x_3 + 3x_4 + 2x_5 + x_6 \qquad + x_8 & = 14, \\
2x_1 + x_2 + 3x_3 + x_4 + 2x_5 + 3x_6 \qquad + x_9 & = 8, \\
3x_1 + x_2 \qquad + 2x_4 + 3x_5 + 6x_6 \qquad + x_{10} & = 13, \\
3x_1 + 2x_2 + x_3 \qquad + 2x_5 + 2x_6 \qquad + x_{11} & = 20, \\
z - 5x_1 - 8x_2 - 3x_3 - 4x_4 - 7x_5 - 8x_6 & = 0.
\end{array}
\tag{12}
$$

Observe that the vectors A_i, $i = 7, \ldots, 11$; form a basis of the vector space of dimension five and that if we set all x_i equal to zero for $i < 7$, we obtain our first feasible solution and the value of z associated with it from equations (12). This solution is $x_7 = 60$, $x_8 = 14$, $x_9 = 8$, $x_{10} = 13$, $x_{11} = 20$, and $z = 0$. Its interpretation is obvious, for it says that if we produce no products, then we obtain no profit. More important at this stage is the form of the equations given in (12). This form is such that if we set all x_2[1] but a certain set equal to zero, we obtain without further calculation both a program associated with an extreme point of the solution set and the profit to be derived from this solution. The set of x's are coefficients of vectors A_i which form a basis of the solution space, and in each row one x in the set has a coefficient of one, while all other x's in the set have zero coefficients in that row. We propose to maintain these properties at each stage of the iterative process and, in addition, to increase z at each stage. We do so by introducing a properly chosen vector into the basis for the space, replacing one of the vectors comprising the present basis. The critical questions are: Do we neglect an extreme point if we only consider bases of the solution space? Which vector do we introduce? Which vector does it replace, or, equivalently, which coefficient of a vector now in the basis do we set equal to zero?

Although we cannot establish it rigorously at this level of mathematical sophistication, the answer to the first question is no. So far as the second question is concerned, we observe in the last equation that the introduction into the program of x_i, $i \leqq 6$, at unit level will increase the profit attainable by the negative of the coefficient of that x_i. That is, if we introduce x_1 into the basis in such a way that the program calls for the production of a single unit, the profit will be increased by five units. Thus it appears desirable to introduce either x_2 or x_6 into the basis, for the profit associated with these commodities is eight units and is the maximum unit profit available. However, the introduction of any of the first six vectors will increase the profit function since their coefficients are negative. This will be true of the coefficients of the last equation at any stage in the iteration and serves to determine the possible profitable next stages. It can be shown that if the introduction of any new vector necessarily decreases the profit function, an optimal solution has been obtained. This leads to the first simplex criterion:

Criterion I. If in the present solution there are unknowns assigned values of zero which have negative coefficients in the last row, select the one having the coefficient of largest absolute value, say x_j, to test for entrance into the next solution. If there are no negative coefficients in the last row, an optimal solution has been obtained.

If we found an x_j, or if a next iteration is necessary, we must be certain that the introduction of this vector produces an extreme point, determine which vector to replace, and see how large we can make x_j without violating the functional constraints. To do so we must examine the coefficients x_j in the present set of equations. We observe that these coefficients cannot all be zero because of the manner in which the new sets of equations will be derived and state:

Criterion II. If *all* coefficients of x_j are negative, then x_j can be introduced with any positive value, the solution set is unbounded, and the profit function attains an infinite value as x_j is increased without limit.

If infinite values of z are impossible, then at least one of the coefficients of x_j will be found to be positive and we apply:

Criterion III. (a) Take the ratios of right members of the equations in the present set (These are the present solution and designate values of the variables in each equation with unit coefficients, which are associated with vectors in the present basis.) to the coefficients of x_j in the same rows if these coefficients are positive. (b) Select the minimum of the ratios so obtained and designate it by q. In the new program, the value of x_j shall be q and the vector to be replaced is A_k where x_k appears in the present solution and in the row from which the minimum ratio was obtained.

The selection of the minimum ratio q as the value of x_j and the determination of the vector A_k to be removed, or the x_k to be assigned a value of 0 in the next solution as indicated in Criterion III, were shown by Dantzig (11) to satisfy the functional constraints and to produce a new extreme point of the solution set. The calculation determining q and x_k is illustrated in Table 1.

We are now in position to form a new set of equations with the

Table 1

Row	Vector or Variable	Present Solution	÷	Positive Coef. of x_2	=	Ratios	Min	Next Solution
1	x_7	60		5		12		
2	x_8	14		2		7	7	$\begin{cases} x_8 = 0, \\ x_2 = 7 \end{cases}$
3	x_9	8		1		8		
4	x_{10}	13		1		13		
5	x_{11}	20		2		10		

property that the assignment of zero values to all variables not associated with vectors in the new basis [as in equation (9)] will yield the values of the remaining variables in the current program and the value of the functional z determined by this program. We merely divide the coefficients of both members of the equation in which the minimum ratio occurred by the coefficient of x_j in that row. This makes the coefficient of x_j in that row equal to 1. Multiplying both members of the new equation by appropriate constants and adding to the remaining equations, we eliminate x_j in the remaining equations, including the profit equation, and thereby obtain the next set of equations in the desired form, the new program, and the profit obtainable through the new program. Repeated application of the procedure already outlined must eventually yield a maximum. In the present example only two iterations are required, and these result in the sets of equations which follow.

$$
\begin{aligned}
-2.5x_1 \quad -0.5x_3-6.5x_4+3.0x_5+0.5x_6+x_7-2.5x_8 \quad &= 25, \\
0.5x_1+x_2+0.5x_3+1.5x_4+\ x_5+0.5x_6 \quad +0.5x_8 \quad &= \ 7, \\
1.5x_1 \quad +2.5x_3-0.5x_4+\ x_5+2.5x_6 \quad -0.5x_8+x_9 \quad &= \ 1, \\
2.5x_1 \quad -0.5x_3+0.5x_4+\ 2x_5+5.5x_6 \quad -0.5x_8 \quad +x_{10} \quad &= \ 6, \\
2.0x_1 \quad -3.0x_4 \quad +\ x_6 \quad -\ x_8 \quad +x_{11} \quad &= \ 6, \\
z-\ x_1 \quad +\ x_3+8.0x_4+\ x_5-\ 4x_6 \quad +\ 4x_8 \quad &= 56.
\end{aligned} \tag{13}
$$

We write the next set of equations in a modified matrix form in which the variables for elements in each column appear at the top of that column.

z	x_1	x_2	x_3	x_4	x_5	x_6	x_7	x_8	x_9	x_{10}	x_{11}	
	−2.8		−1	−6.4	+2.8		+1	−2.4	−0.2			=24.8,
	0.2	1		1.6	0.8			0.6	−0.2			= 6.8,
	0.6		1	−0.2	0.4	1		−0.2	0.4			= 0.4,
	−0.8		−6	−0.6	−0.2			0.6	−2.2	1		= 3.8,
	1.4		−1	−2.8	−0.4			−0.8	−0.4		1	= 5.6,
1	1.4		5	7.2	2.6			3.2	1.6			=57.6.

(14)

We see that in the last set of equations, all coefficients of the profit equation, or the last one, are positive, and hence by Criterion I no further improvement is possible. The profit associated with an optimal program is 57.6 units; this quantity and costs of the new equipment form the basis for the decision facing the firm. We observe that x_7, x_{10}, and x_{11} are slack variables with non-zero values in the final program. This is interpreted to mean that the resources involved are available in excess supply. x_2 and x_6 also enter at positive level, indicating that the firm would be well advised to restrict production to these commodities. The positive coefficients of x_1, x_3, x_4, x_5, in (14) indicate the net loss accruing to the firm by producing these commodities at unit level. Those associated with the variables x_8 and x_9 indicate the losses which would accrue for a unit utilization of slack or a deficit in resources 2 and 3. This enables the firm to estimate the risks involved in maintaining the supplies of these resources. There is no risk involved with the remaining resources as long as they attain the level required by the program, yet the losses which would accrue if these levels were not met can be easily obtained by adjusting the estimated resources and modifying the calculation presented. In sum, the simplex technique can provide a basis for many decisions relative to the proposed venture.

Further Details of the Simplex Method

Alternate Optimal Programs. It is frequently desirable to determine alternate programs yielding maximum values of the objective function. Alternate programs occurring at extreme points are indicated in the final profit equation by zero coefficients for variables not contained in the optimal basic program. If alternate optimal programs exist at extreme points, then any of the variables not in the final solution with zero coefficients in the last row can be assigned positive values in the solution without changing the value of the objective function. Moreover, any convex linear combination of optima is also an optimum, and conversely any optimum is a convex linear combination of the optima which occur at extreme points.

Criterion I. If there is a choice in the vectors to be introduced into the basis, or in variables to be assigned positive values in the next solution (as in the first step in the illustrative example), then either vector may be introduced without altering the final maximum.

Criterion III. If several vectors can be removed from the basis because the row in which the minimum ratio occurs is not uniquely determined, then degeneracy exists. Charnes has given a method for the selection of the vector to be removed from the basis in this case (5), as follows.

We call the variables associated with vectors in the basis a basic set of variables and observe that when degeneracy does not occur all variables in the basic set are assigned positive values in the program, but all variables not in the basic set are assigned zero values. If degeneracy occurs, two or more variables in the current basic set will be assigned values of zero in the new program; we must determine which one of these to delete from the basic set. If the candidates for deletion have coefficients of one in rows $r_1, r_2, \ldots$, etc., and, of necessity, zero coefficients elsewhere, and if x_j is the vector to be entered into the basic set, then we form ratios of the coefficients of x_1 to the coefficients of x_j in rows $r_1, r_2, \ldots$, etc., and select the minimum ratio if one exists. If a minimum ratio exists for row r_u, and only r_u, then the vector in the basic set with a coefficient of one in row r_u is the vector to be deleted from the basic set. If no minimum ratio occurs, then we examine the coefficients of $x_2, x_3, \ldots$, etc., in succession and in like manner until a minimum ratio is obtained; the first minimum ratio determines the variable to be deleted from the basic set. If at any stage a minimum ratio occurs, but is obtained for several rows, delete from the set of r's those rows not yielding the minimum ratio and begin the process again.

DUALITY IN LINEAR PROGRAMMING

As in the relationships between points and lines in geometry, there are certain dual relationships which appear in linear programming.

These are often useful in reducing the labor in solving programming problems, and hence duality will be considered here briefly.

The following pair of linear programming problems are known as duals of each other:

Determine $X = (x_1, \ldots, x_n)$ such that:	Determine $Y = (y_1, \ldots, y_m)$ such that:
$x_i \geqslant 0,$	$y_i \geqslant 0,$
$\sum_j a_{ij}x_j \leqslant b_i,$ and	$\sum_i a_{ij}y_i \geqslant c_j,$ and
$\sum_j c_jx_j = z$ is a maximum	$\sum_i b_iy_i = w$ is a minimum

where the a_{ij}, b_i, and c_j are constants with identical values in the two problems.

Since the constants in both problems are identical, it is to be expected that consistent interpretations for the same physical conditions would exist and that there would be a close relation between the solutions in the two cases. Both are true. If, as previously, we interpret the maximization problem in terms of the production of a firm, the functional constraints representing resource limitations and the vector X representing the production program of the firm, then the minimization problem can be interpreted in terms of unit accounting values to be assigned to each type of resource of the firm. Thus in the dual problem one wishes to minimize the total accounting value of the resources of the firm subject to the constraint that the accounting value of a unit produced can never exceed the total accounting values of the materials used in producing the unit and to the natural requirement that all accounting values be non-negative. The relationships between available stocks and optimal programs can be better understood by a study of the dual programs; many fundamental economic relationships have been established by this means.

Among the principal properties of dual programs, we find the following of most immediate interest (26):

a. Either both the minimization and maximization problem have optimal solutions or neither does.

b. A necessary and sufficient condition that one (and hence both) of the dual problems has an optimal solution is that both have feasible solutions.

c. A feasible vector X is optimal if, and only if, there is a feasible vector Y such that $z = w$ for these vectors. Dually, a feasible vector

Y is optimal if, and only if, there exists a feasible vector X such that $z = w$ for these vectors.

d. If both problems have optimal solutions, then $\max z = \min w$.

Properties a. and c. indicate quite clearly that we may find the maximum value of z by finding the minimum value of w (and conversely) and that if w has no minimum value, then we must infer that z has no maximum value. This suggests that for any linear programming problem in which we seek only $\max z$ or $\min w$, we may solve either the primal, the problem as stated, or its dual. This flexibility is very important because one of the two problems may be much easier to solve than the other. Furthermore, for many of the available techniques, in particular the simplex technique, the dual optimal program is readily obtainable from the tableau representing the optimal solution of the primal. Thus, the selection can usually be made without reservation.

Property b. indicates a necessary and sufficient condition for the existence of a solution to either problem, and since, when no solution exists, it is often easily demonstrable that one of the sets of constraints is infeasible, considerable effort may be conserved by examining both problems for feasibility.

Property c. gives a test for optimality, although its application is not always easy. Nevertheless, the combined set adds greatly to the flexibility of choice in methods of solutions.

The utility of this flexibility can be easily seen by considering the numerical examples presented thus far. In the first example given, the primal problem is one with a solution space of two dimensions, and hence it is readily solvable by the geometric method. The dual problem is stated as follows: we are to minimize

$$w = 70y_1 + 80y_2 + 130y_3 + 120y_4 + 100y_5$$

subject to constraints

$$y_i \geqslant 0;\ i = 1, \ldots, 5;$$
$$10y_1 + 20y_2 + 30y_3 + 12y_4 + 15y_5 \geqslant 5,$$
$$20y_1 + 10y_2 + 10y_3 + 9y_4 + 10y_5 \geqslant 8.$$

Thus the dual problem is in a space of five dimensions and cannot be conveniently represented or solved geometrically.[5] If the primal was in fact the minimization problem, then it is clear that the dual would provide a simpler solution.

[5] The geometric method can be used to determine approximate solutions for this case by considering the constraints two or three at a time.

In general, in order to facilitate the proper selection of the problem to be solved, the dual problems are written in the following form:

	x_1	x_2	$\leqslant$
y_1	10	20	70
y_2	20	10	80
y_3	30	10	180
y_4	12	9	120
y_5	15	10	100
$\geqslant$	5	8	

As in the present example, the problem involving the smallest number of variables is the easiest to solve or to approximate if geometric methods are used. However, the converse is true for the simplex method.

Churchman, et al., (8) states as a rule of thumb for the simplex method, that the number of iterations required is equal to from 1 to 1.5 times the number of rows, or functional constraints. Hence the above problem solved by this method would be expected to require about 5/2 as much time if the dual were solved instead of the primal. For this method, converting an optimal program for a dual into one for the primal is such an extremely simple matter that the conversion time required may be ignored and the choice made on the minimum number of constraints.

In making this conversion, we first see that min $w = \max z$, so the optimal value of the objective function is at hand. Further, if we have computed an optimal program, say for z, then we have the last equation (as in the sets illustrated) in which one obtains max z by substituting the values of the x_i in the optimal program. If we set y_i equal to the value of the coefficient of first slack variable in this equation, y_2 to the value of the coefficient of the second, etc., until all slack variables are utilized, and then begin with the first program variable, or x_1, and continue with sequential values of the subscripts of y, we obtain an optimal program for the minimization problem.[6] Thus the simplex technique is well suited to the solution of dual problems, and duality serves to reduce the labor involved. The dual theorems have also been used effectively in developing more efficient techniques of solution for special problems.

[6] In the illustrative example, an optimal program for the dual problem is $Y = (O, 3.2, 1.6, 0, 0, 1.4, 5, 7.2, 2.6)$.

SPECIAL LINEAR PROGRAMMING PROBLEMS

The linear programming problem occurs in many forms which permit the application of less general techniques for its solution, but techniques which are more efficient for these cases. Only two of this type will be discussed in detail. The frequency of their occurrence, however, and the comparative ease with which desired solutions can be obtained make them very important. Both are allocation problems and are commonly known as the assignment problem and the transportation problem.

The Assignment Problem

In its simplest form, the assignment problem is stated as follows: Given n resources, n uses to which these resources can be put, and a measure of efficiency or of lack of efficiency for the assignment of any resource to any use, we are to maximize the efficiency of assignment, or to minimize the inefficiency of assignment. The resources could be supplies, men, machines, etc., and the corresponding uses to which they are to be put could be indicated as demands, jobs, and production programs in the respective cases. The problems are abstractly identical in that we are to assign n items to perform n functions and wish to do so with maximum efficiency.

Assignment problems of this type are extremely common, and several programming techniques for their solution have been developed. Among the people who have made significant contributions are Flood (17), Kuhn (25), Dwyer (16) Votaw and Orden (32). Several others are indicated in reference 30. These techniques have been applied quite profitably in determining optimal assignments of personnel to jobs, man-hours to machines, plants to locations, tractors to trailers, and even electronic equipment to ships. References to such applications are contained in the recently published bibliography by Riley and Gass (30). Although these problems were solved using different techniques in many cases, the structures of the problems were identical in form.

The assignment problem is a special linear programming problem. Since the variables take on only values of 0 or 1, the i-th supply either is allocated to the j-th demand or is not. Here it is assumed that the supply exactly meets the demand so that the only problem is one of assignment. One might try all of the permutations which are possible and select from these the most profitable one; however, such a procedure would be ill-advised when n were sizable. If n is as large as 8, then 8! or 40,320 combinations would have to be examined, and the time required would be quite excessive. Using the simplex technique previously presented involves n^2 variables, although no adjustments must be made for the discreteness in this case. This approach is not in general recommended in this case, though it is often used, for it is very time-consuming relative to other known procedures. With currently available techniques, an average solution of an 8 by 8 matrix requires on the order of fifteen minutes and can be done quite easily without computing machines.

The technique to be considered here was developed by Flood (18). For its use the problem may be defined as follows: Given an n^2 cost matrix $A = (a_{ij})$ with $a_{ij} \geqslant 0$ for $i, j = 1, 2, \ldots, n$, we are to find an n^2 assignment matrix $X = (x_{ij})$ such that

$$\sum_{i=1}^{n} x_{ij} = \sum_{j=1}^{n} x_{ij} = 1, \quad x^2{}_{ij} = x_{ij} \tag{15}$$

and

$$C = \sum a_{ij} x_{ij} \tag{16}$$

is a minimum. In this form it is seen that the indexes a_{ij} designate or measure the inefficiency or cost of the assignment. They may be costs in terms of hours required to do a job, dollars spent in meeting a program, . . . etc. The conditions given in equations (15) are sufficient to describe a permutation matrix, or one in which there are n ones in the matrix, the remaining elements being zeros, and a one appears in each row and in each column. This insures that all resources are assigned and all functions performed.

The solution to be described is based upon two fundamental theorems, one by the Hungarian mathematician Koenig and the other by Kuhn (31). Koenig proved that: If the elements of a matrix are divided into two classes by a property R, then the minimum number of lines (rows and/or columns) that contain all of the elements with property R is equal to the maximum number of elements with property R, with no two on the same line. It was shown in Kuhn's article that: Given a cost matrix $A = (a_{ij})$, if we form another matrix $B = (b_{ij})$ where

$$b_{ij} = a_{ij} - u_i - v_j$$

and where the u_i and v_j are constants, the solution to the assignment problem with cost matrix A is the same as that for the same problem with cost matrix B.

The second result implies that we may make some of the elements in the cost matrix zero by the appropriate selection of values for u_i and v_j, and this procedure seems eminently desirable in view of our aim to minimize the total cost C. In fact, if we could obtain a set of n zeros such that one was in each row and one was in each column, then the problem would be solved for the matrix B, for the minimum total cost for the assignment represented by the set of zeros would obviously be zero and would be obtained by the assignments indicated by the set of zeros. Since the solution for A is the same as that for B, our problem would be solved.

The first result aids in determining such a set of zeros, for the minimum number of lines containing all of the zeros in a matrix is fairly easy to determine, and since this implies the existence of a set of n_i zeros, one of which is in each column and one in each row, the procedure for solution is easily determined. The procedure follows:

a. Select the minimum entry in each row of the cost matrix A and decrease the elements of each row by the minimum entry in that row, thus forming the matrix A_1.

b. Determine S_1 the minimum number of lines containing all zeros in A_1 and let n_1 be the number of lines in S_1. If $n_1 = n$, there is a set of n zeros no two of which are in the same line, and these positions determine an optimal solution. If $n_1 < n$, proceed to step c.

c. Repeat step a for columns instead of rows and for A_1 instead of A, thus forming A_2. Determine S_2 the minimum number of lines containing all zeros of A_2 and let n_2 be the number of lines in S_2. If $n_2 = n$, then A_2 contains a set of zeros of the desired type, and these specify an optimal assignment. But if $n_2 < n$, proceed to step d.

d. Select n_2 as the smallest element of A_2 not in any line of S_2. Subtract n_2 from all elements of A_2 not in lines of S_2 and add n_2 to all elements, if any, which lie at the intersections of lines of S_2; designate the resulting matrix by A_3. Determine S_3 and n_3 as S_2 and n_2 were determined and test to see if $n_3 = n$. If so, A_3 contains an appropriate set of zeros, and these specify an optimal solution. If $n_3 \nless n$, proceed to e.

e. Repeat d after increasing the subscripts used in the preceding application of d by one.

This procedure will ultimately produce a minimum cost program; it will be illustrated by a numerical example.

The assignment problem often occurs in a form in which the matrix is not square, for often the demand for personnel far exceeds the available supply. The problem is still one of assigning the available supply to activities which minimize total cost of assignment. We use the same technique for solving this problem after having made the matrix square by considering hypothetical personnel in the required number which can be assigned at no cost. This introduces lines of zeros in the cost matrix and the solution proceeds as before.

More complex assignment problems arise if we consider cases in which the demands are for multiples of the basic commodity, and

Let

$$A = \begin{Vmatrix} 22 & 30 & 26 & 16 & 25 \\ 27 & 29 & 28 & 20 & 32 \\ 33 & 25 & 21 & 29 & 23 \\ 24 & 24 & 30 & 19 & 26 \\ 30 & 33 & 32 & 37 & 31 \end{Vmatrix}. \text{ Then}$$

$$A_1 = \begin{Vmatrix} 6 & 14 & 10 & 0 & 9 \\ 7 & 9 & 8 & 0 & 12 \\ 12 & 4 & 0 & 8 & 2 \\ 5 & 5 & 11 & 0 & 7 \\ 0 & 3 & 2 & 7 & 1 \end{Vmatrix}, \; n_1 = 3,$$

$$A_2 = \begin{Vmatrix} 6 & 11 & 10 & 0 & 8 \\ 7 & 6 & 8 & 0 & 11 \\ 12 & 1 & 0 & 8 & 1 \\ 5 & 2 & 11 & 0 & 6 \\ 0 & 0 & 2 & 7 & 0 \end{Vmatrix}, \; n_2 = 3, \; h_2 = 2,$$

$$A_3 = \left\| \begin{array}{ccccc} 4 & 9 & 8 & 0 & 6 \\ 5 & 4 & 6 & 0 & 9 \\ 12 & 1 & 0 & 10 & 1 \\ 3 & 0 & 9 & 0 & 4 \\ 0 & 0 & 2 & 9 & 0 \end{array} \right\|, \; n_3 = 4, \; h_3 = 4$$

$$A_4 = \left\| \begin{array}{ccccc} 0 & 5 & 4 & 0 & 2 \\ 1 & 0 & 2 & 0 & 5 \\ 12 & 1 & 0 & 14 & 1 \\ 3 & 0 & 9 & 4 & 4 \\ 0 & 0 & 2 & 13 & 0 \end{array} \right\|, \; n_4 = 5.$$

Solution: $x_{11} = x_{42} = x_{33} = x_{24} = x_{55} = 1$
$C = 22 + 24 + 21 + 20 + 31 = 118$

assume that supplies of the basic commodity are concentrated in distinct groups and that the costs of providing a unit demand from distinct concentrations differ. The complexity arises because of the additional number of units of the basic commodity, for although we may treat each unit separately and proceed as before, the order of the matrix involved becomes increasingly unwieldy as n increases. Because problems of this type occur quite often in transportation activities, they are included here in a formulation of the classic transportation problem, along with indications of available methods of solution.

The Transportation Problem

We limit our formulation to the problem of selecting optimum programs for distributing a homogeneous commodity where s_i units are located at the origin O_i and d_j units are required at destination D_j; the cost associated with the transportation of a single unit from O_i to destination D_j is denoted by c_{ij} and we desire to minimize $C = \sum_{i,j} x_{ij} c_{ij}$, the total costs of distribution when supply equals demand. If supply is less than demand and no preferential destinations exist, then we may introduce enough fictitious supplies at imaginary locations to meet demands, assume that these can be distributed without cost, and proceed as if supply and demand were equal. Thus the restriction of supply equal to demand does not limit the applicability of the methods developed. The variable x_{ij} designates the quantity of supply moved from O_i to D_j, and consequently $\sum_j x_{ij} = s_i$, $\sum_i x_{ij} = d_j$. Since supply equals demand, $\sum_i s_i = \sum_j d_j$.

Several techniques are available for the solution of problems of this type, the so-called "transportation technique" perhaps being the best known (6). The problem can be solved by modification of the method of reduced matrices (16), by the simplex method (14), or by several others which reduce the size of the matrix to be handled (31). It is important to observe that so far as logical structure is concerned, the movement of resources and the concepts of origins and destinations are not essential elements in the problem. In fact, if there is a single resource in separate groups which must be allocated to meet multiple fixed requirements, and if the cost associated with an allocation of a unit of resource from a given group to any of the fixed requirements is constant as we increase the allocation from the group to that requirement, then the problem is abstractly identical to the transportation problem described and can be solved using the same technique.

Other Special Problems and Applications

Space does not permit more than a mention of the many special applications of linear programming. Accordingly, there was an attempt to include some of the more important ones in the examples given earlier. But there are a few that have been omitted thus far and which deserve particular mention. The problem of the traveling salesman is in this class. It has received extensive treatment and at first sight may appear to be a linear programming problem; yet much progress towards its solution has been made through the use of linear programming techniques, the objective function of the problem being linear.

The traveling salesman problem is that of finding an optimal route between a series of locations so that each point is visited once and only once and a return to the origin is made. The optimization takes place in terms of time involved, costs of travel, or some other linear function of the distances between the points to be served. The abstract problem appears in several forms, but all relate to the sequencing of operations so as to minimize or maximize a linear function.

When the cost associated with travel between two points is independent of the path taken, the problem is said to be symmetric; Fulkerson, Dantzig, and Johnson have suggested devices which are sometimes useful for solving this case (19). Flood has suggested a technique for this case also, and has developed a technique for testing the optimality of a given program (7, 9). Nevertheless, no rigorous analytical solution is known to exist at this writing.

The second topic which should be mentioned is the relationship between linear programming and the finite two-person zero sum game. This correspondence is one to one and is such that a solution to the game exists if, and only if, a solution to the corresponding linear programming problem exists. If solutions to both problems exist, they are related by simple transformation of variables (20). These properties are important, for because of them greater understanding of either problem can be gained by a study of both; moreover, numerical methods applicable to either can readily be applied to the corresponding problem.

Finally, further applications of linear programming are available in the very extensive literature now in existence, in spite of the comparative recentness of developments in this field. The most recent bibliography is an excellent one by Riley and Gass (30).

REFERENCES

(1) Allen, R. G. D. *Mathematical Economics.* New York: St. Martins, 1956.

(2) Antosiewicz, H. A. (ed.). *Second Symposium, Linear Programming.* Washington, D. C.: National Bureau of Standards, 1955.

(3) Arnoff, E. Leonard. "An Application of Linear Programming," *Proceedings of the Conference on Operations Research in Production and Inventory Control.* Cleveland: Case Institute of Technology, 1954.

(4) Charnes, A. "Optimality and Degeneracy in Linear Programming," *Econometrica,* Vol. 20 (1952), 160-70.

(5) Charnes, A., and Henderson, A. *An Introduction to Linear Programming.* New York: John Wiley and Sons, 1953.

(6) Cooper, W. W., and Charnes, A. "Transportation Scheduling by Linear Programming," *Proceedings of the Conference on Operations Research in Marketing.* Cleveland: Case Institute of Technology, 1953.

(7) Cooper, W. W., and Farr, D. "Linear Programming Models for Scheduling Manufactured Products." Pittsburgh: Carnegie Institute of Technology, September 1, 1952.

(8) Churchman, C. W., Ackoff, R. L., and Arnoff, E. L. *Introduction to Operations Research.* New York: John Wiley and Sons, 1957.

(9) Dantzig, G. B. Chapters I, II, XX, XXI, and XXIII in *Activity Analysis of Production and Allocation,* Cowles Commission Monograph No. 13. Edited by T. C. Koopmans. New York: John Wiley and Sons, 1951.

(10) ———. "Computational Algorithm of the Revised Simplex Method," *RAND Memorandum RM-1266,* 1953.

(11) ———. "Maximization of a Linear Function of Variables Subject to Linear Inequalities," *Activity Analysis of Production and Allocation,* Cowles Commission Monograph No. 13. Edited by T. C. Koopmans. New York: John Wiley and Sons, 1951.

(12) ———. "Upper Bounds, Secondary Constraints, and Block Triangularity in Linear Programming," *RAND Memorandum RM-1367,* 1954.

(13) ———, and Orchard-Hays, W. "Alternate Algorithm for the Revised Simplex Method," *RAND Memorandum RM-1268,* 1953.

(14) ———, Orden, A., and Wolfe, P. "The Generalized Simplex Method for Minimizing a Linear Form under Linear Inequality Restraints," *RAND Memorandum RM-1264,* 1954.

(15) Dorfman, R. *Application of Linear Programming to the Theory of the Firm.* Berkeley: University of California Press, 1951.

(16) Dwyer, Paul S. "The Solution of the Hitchcock Transportation Problem with a Method of Reduced Matrices." Ann Arbor: University of Michigan, December, 1955.

(17) Flood, M. M. "On the Hitchcock Distribution Problem," *Pacific Journal of Mathematics,* Vol. 3 (1953), 369-86.

(18) ———. "The Traveling-Salesman Problem," *Operations Research for Management.* Vol. II. Edited by Joseph F. McCloskey and John F. Coppinger. Baltimore: The Johns Hopkins Press, 1956.

(19) Fulkerson, D. R., and Dantzig, G. B. "Computation of Maximal Flows in Networks," *RAND Memorandum RM-1489,* 1955.

(20) Gale, D., Kuhn, H. W., and Tucker, A. W. "Linear Programming and the Theory of Games," *Activity Analysis of Production and Allocation,* Cowles Commission Monograph No. 13. Edited by T. C. Koopmans. New York: John Wiley and Sons, 1951.

(21) Harrison, Joseph O., Jr. "Linear Programming and Operations Research," Seminar Paper No. 2 in *Informal Seminar in Operations Research.* Baltimore: The Johns Hopkins Press, 1953.

(22) Henderson, A., and Schlaifer, R. "Mathematical Programming," *Harvard Business Review,* May-June, 1954.

(23) Koopmans, T. C. (ed.). *Activity Analysis of Production and Allocation,*

Cowles Commission Monograph No. 13. New York: John Wiley and Sons, 1951.

(24) Koopmans, T. C. (ed.). "Optimum Utilization of the Transportation System," *Proceedings of the International Statistical Conferences,* Cowles Commission Paper, New Series, No. 34. Washington, D. C., 1947.

(25) Kuhn, H. W. "The Hungarian Method for the Assignment Problem," *Naval Research Logistics Quarterly,* Vol. 2 (1955), 83-98.

(26) ———, and Tucker, A. W. *Contributions to the Theory of Games,* Annals of Mathematics Study No. 24. Princeton: Princeton University Press, 1950.

(27) Markowitz, H. M., and Manne, A. S. "On the Solution of Discrete Programming Problems," *Econometrica,* Vol. 25 (1957), 84-110.

(28) Orden, A. "Survey of Research on Mathematical Solutions of Programming Problems," *Management Science,* Vol. 1 (1955), 170-72.

(29) Project SCOOP. *Symposium of Linear Inequalities and Programming.* Washington, D. C.: U. S. Air Force, 1952.

(30) Riley, Vera, and Gass, Saul I. *Linear Programming and Associated Techniques: A Comprehensive Bibliography on Linear, Non-Linear, and Dynamic Programming.* Baltimore: The Johns Hopkins Press, 1958.

(31) Vidale, M. L. "A Graphical Solution of the Transportation Problem," *Journal of the Operations Research Society of America,* Vol. 4, No. 2 (April, 1956), 193-203.

(32) Votaw, D. F., and Orden, A. "The Personnel Assignment Problem," *Symposium on Linear Inequalities and Programming.* Washington, D. C.: Project SCOOP, U. S. Air Force, 1952.

(33) Wagner, Harvey M. "The Simplex Method for Beginners," *Journal of the Operations Research Society of America,* March-April, 1958, pp. 190-99.

Fourteen

QUEUEING THEORY

CHARLES D. FLAGLE

INTRODUCTION

Queueing theory—or waiting line problems, as the field is sometimes called—is the formal attempt to deal with the perplexing problem of organization and planning in the face of randomly fluctuating demand for some service to be performed. This fluctuation may be due to complete lack of control over the elements of demand, as in the placing of telephone calls on an exchange, or it may be the result of inability to maintain tight control of a scheduled process, as in arrivals of aircraft at an airport.

In either case, the occurrence of long gaps between arrivals of elements of demand may result in periods of idleness of service facilities. The time thus lost is lost forever, and subsequent short intervals between arrivals result in formation of a queue. Even though the service facility has excess potential capacity, some intermittent congestion will occur. (At the outset, it should be stated that we are considering here only systems in which the potential service capacity is greater than the demand. Congestion caused by basic inadequacy of the service facilities to handle demand is another problem.)

The organizational problem posed by randomly occurring, intermittent queues is one which has no categorical solution. An increase in service capacity to relieve congestion periods results in increased idleness. Conversely, a decrease in service facility reduces intermittent idleness with aggravation of congestion. Since there are costs associated with both idleness and congestion, the problem is to control demand and service rates to minimize total cost and to maintain congestion within tolerable bounds.

Queueing theory had a very practical and effective origin in the design of telephone communication systems, where it offered a much needed rational approach to control of congestion. The earliest work is attributed to A. K. Erlang (2) about half a century ago. The models which were developed by him and others for mass communications had one property in common—the elements of demand on the system (e.g., the placing of a call) were assumed to be independent of each other and outside the control of the system.

This assumption leads to an input which is a Poisson process—a phenomenon of great importance to be described later—one characterized by a negative exponential distribution of intervals between calls. The runs of short intervals between calls, characteristic of this process, promote congestion, while the significant number of long intervals, equally characteristic, results in wasteful idleness of facilities. For reasons which will be more apparent later, this model is adequately descriptive of the basic nature of the input to a mass communication system, being troublesome only in that it does not account for systematic changes in the underlying rate of placement of calls.

Some effective applications were also made to estimation of congestion of motor vehicle traffic where, again, the elements of demand (e.g., appearance of vehicles on a particular segment of highway) are relatively independent.

During World War II, as loads on communication and transportation systems grew, a new and broader emphasis was placed on extending the application of these theories and use of the models. The congestion characteristic of such systems is extreme sensitivity to load factor (the ratio of demand rate to potential service rate). Since World War II, with the founding of the professional societies devoted to operations research, queueing theory has at times all but pre-empted the pages of the journals. (Extensive bibliographies are available in references (9) and (12).) It is significant to note that the first two papers awarded the Lanchester Prize[1] were on the application of queueing theory. These were in the field of vehicular traffic by Edie (4) and production by Brigham (1).

Queueing theory has been found useful in attacking congestion diffi-

[1] The Lanchester Prize is awarded each year by The Johns Hopkins University to the author of the paper on operations research judged best by a committee of the Operations Research Society of America.

culties in some internal organizational problems concerned with flow of work, people, or communications. Here, however, a danger arises when the already developed models are used, for the essential character of independence of elements of demand on service points often does not exist (except in such cases as accidental breakdown of machines). What does exist most often is a scheduled process in which there occurs some random variation of actual demand time from scheduled time. If we would use queueing theory in the important area of organizational scheduling and control, it is important to learn when it is applicable and when it is not.

THE BASIC MODELS OF QUEUEING THEORY

The usefulness of queueing theory in attempting to solve problems of flow of people, products, or communications in organizations can be seen more clearly if some classification of operations is made. For the present discussion, a meaningful classification system for operations is a set of types similar to those developed by psychologists for study of personality. Each type is an abstraction having some pure extreme set of characteristics. All real systems represent to some degree a mixture of the types, but the isolation and rational analysis of the type permits insight into the character and behavior of the combinations.

For purposes of this discussion we need consider only two idealized types of operation, one of which may be called "determinate" and the other "indeterminate." The determinate operation is defined as one in which all aspects of the system are precisely known as functions of time. At any time we can predict precisely the state of the system. This is the idealized moving assembly line, in which every operation has been broken down into a sequence of phases all of equal and constant length. The desirable character of such a system is that each phase of the operation can be loaded to 100 per cent capacity, theoretically. There need be no standby or idle time, no in-process inventory; the efficiency of the operation is not affected by the number of phases; therefore, the ultimate in division of labor can be effected. Thus our first type, the determinate operation—if it were possible to achieve—would give the most effective conceivable utilization of productive resources.

Consider now the second type, the antithesis of the first, the indeterminate operation, in which any aspect of the system at any given time is known only in probabilistic terms, and the transition from one state to another during an interval can be expressed only as a set of probabilities. If at each phase of such a system the probability of occurrence of a demand for the service offered at that phase, or the probability of the completion of such service, during an interval remains constant, the behavior of that phase of the system is described by the theory of the single-channel queue.

THE SINGLE-CHANNEL QUEUE

A real example which approaches this type is the communication system in which messages arrive at a single station independently of each other and contain a preponderant number requiring short handling times. We can bridge the gap between this real example and its idealized mathematical model by considering some imaginary but physically possible system in which input and output are governed simply by chance, say by the throw of a pair of dice.

Let the first rule for this chance system be that an arrival (an input of one unit to the system) occurs when a 5 is rolled. We expect this to happen on the average 4 times in 36 throws. Let the second rule be that a departure of one unit (implying completion of service) occurs when a 6 is rolled, provided there is a unit in the system which can depart. For all other numbers rolled, the system remains unchanged. Since the 6 will appear on the average 5 times in 36, at each throw there is a greater chance of departure than an arrival—if the opportunity for departure exists at all. For the time being we need not be concerned with the order in which outputs take place. The important quantity is the highly variable number in the system, noted as n. In the illustration, $n = 4$. To relate this to the original problem, imagine that one of the four is in the process of some service which will be terminated upon the next roll of 6, at which point it leaves the system abruptly.

Were it not for the fact that some rolls of 6 occur when the system is empty, the rate of arrival would be only four-fifths the rate of departure—that is to say, our system is loaded only to 80 per cent of its potential capacity. It will be empty and its hypothetical facilities idle 20 per cent of the time.

The reader who has a pair of dice and nothing better to do can develop a feel for the behavior of such a model by keeping score of n,

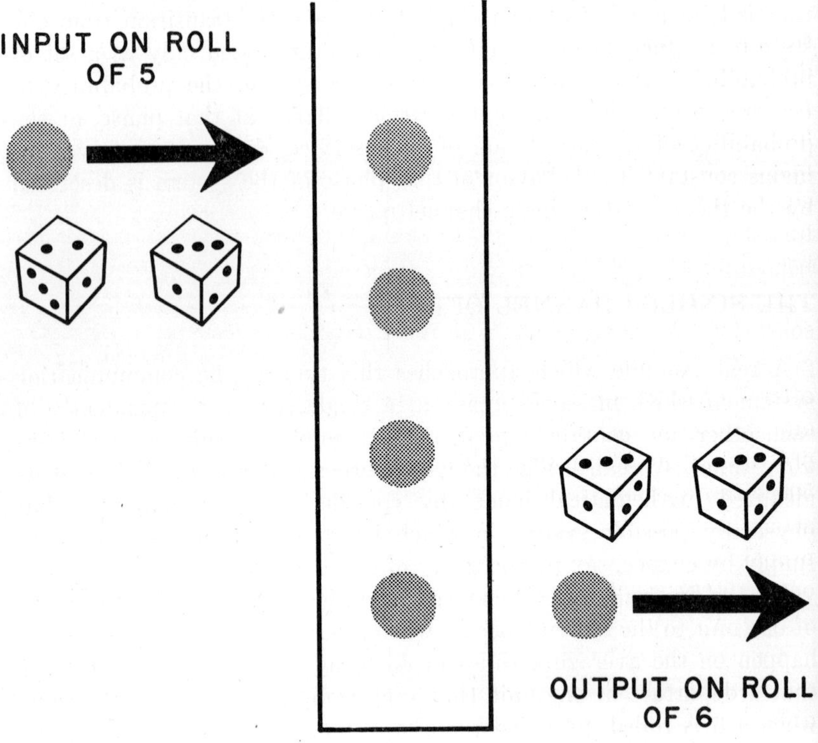

Figure 1. Single-channel queue controlled by throw of dice.

the number of units in the system (to be called the state of the system) for several hundred rolls of the dice. This is good preparation not only for reading the remainder of this chapter, but also the following on simulation techniques, for our game of dice yields a crude Monte Carlo solution of the mathematical model of the game.

A record of the "random walk" of the dice-controlled system will reveal first a surprising number of occurrences of large queues or high states of the system. In fact, over a long period of time, the average number in this system will be very nearly four. Another fact revealed by the record will be that there exists no systematic pattern of the state of the system. A plot of the system state (or queue length) against time shows no functional relationship between the two.

However, if we make repeated long runs of the dice model, even though the time history of the system state in each run will be quite different, another property will be consistently similar. This is the fraction of time the system spends in each state. In the example, in long runs (many throws of the dice) the system will be empty approximately 20 per cent of the time in each. This we can predict, and it is equivalent to saying that the probability is 20 per cent that we will find the system empty at any randomly chosen instant long after the beginning of system operation.

What can we say about the probability of finding the system in some state, n, after the 500th throw of the dice? Under the rules of the game, the probability is strongly related to the state of the system after the 499th throw, for on the 500th throw we can move at most one step up or down. Therefore, the statement can be made that the probability of finding the system in some state n (for $n \geqq 1$) on the 500th throw is equal to the probability that it is already in that state after the 499th and no output or input occurs on the 500th, plus the probability that it is in state $n - 1$ on the 499th and an input occurs on the 500th, plus the probability that it is in $n + 1$ on the 499th and an output occurs on the 500th.

This cumbersome but quite simple statement, when re-expressed in mathematical form, is the basis for the development of the differential difference equation approach in analytical queueing theory. Before proceeding to this re-expression, it is necessary to point out that the statement in both its verbal and mathematical form is based on the assumption of independence of individual inputs and outputs from each other and from the state of the system. Dice have this property, since whether or not a given number appears on the 500th throw is in no way influenced by what has happened on previous throws, nor does its occurrence influence future outcomes. It is this quality of independence that permits us to add the three phrases of the statement (or terms of the equation) and to express the probabilities stated in each phrase as the product of the probabilities of the events which must coincide to produce the desired state, n.

The choice of the 500th throw as our time of examination of the system is meant to imply that the system has been in operation for a sufficient time to obliterate the effects of its starting state—that the likelihood of finding it in any state would be the same if we chose the 1,000th throw. This in turn implies that the state probabilities ultimately stabilize (even though the state itself does not), and that there is a "statistical equilibrium" which characterizes these probabilities and renders them not a function of time.

We can broaden the understanding of the model by transforming it to mathematical terms. This necessitates some nomenclature, and the following is used since it is consistent with that most frequently encountered.

NOMENCLATURE

n	— the number of units in the system, including those waiting and those in service.
$E(n)$	— the expected value of n, $E(n) = \sum_{n=1}^{\infty} np_n$.
t	— an instant in time.
Δt	— a small increment of time.
$P_n(t)$	— the probability that n units are in the system at time t.
p_n	— the equilibrium probability that n units are in the system (the fraction of time the system is in state n).
λ	— the rate at which units arrive for service.
$1/\lambda$	— the average interval between arrival of units.
μ	— the potential service rate.
$1/\mu$	— the average service time.
$\lambda\Delta t$	— the probability that an arrival occurs during Δt.
$\mu\Delta t$	— the probability that a service is completed during Δt.
ρ	— the ratio of arrival rate to potential service rate, where $\rho = \lambda/\mu$ (called the load factor).
a	— number of parallel channels in a system.

In the transformation, "after the 499th throw" becomes a point in time, t, and the "outcome of the 500th throw" becomes the occurrence in some small interval of time, Δt, which immediately follows t. The chance of throwing a 5, signifying an input or arrival, is replaced by the quantity $\lambda\Delta t$, where λ is the underlying constant arrival rate. The implication is that in each increment of time, Δt, some small but constant probability exists of occurrence of an arrival no matter what has happened—or failed to happen—in previous intervals. This probability is proportional to the length of Δt and the arrival rate λ. Over a long period of time, t, the expected number of arrivals is λt. More will be said of this process later.

For the output process, we make a similar substitution. The chance of rolling a 6, signifying an output (or completion of service) is replaced by the product $\mu\Delta t$, where μ is defined as the potential output rate. The ratio of input to output rates, λ/μ, may be thought of as the load factor or system untilization factor. It should be borne in mind that we are considering cases in which this ratio is less than 1, i.e., where the system is loaded only to a fraction of its capacity.

In one respect, the mathematical process does not behave as well as the dice model. There is nothing to prevent both an arrival and departure, or several of each, from occurring in an interval Δt. If the events occur independently of each other, as specified, the probability of coincidence of events is equal to the product of the probabilities of each event. Thus the probability of two or more events will be a term containing $(\Delta t)^2$ or a higher order, and by assuming very small values of Δt, these coincidences can be neglected.

Having conveniently eliminated coincidences of events, we can consider that in each Δt, an arrival either does occur—with probability $\lambda\Delta t$, or it does not occur—with probability $(1 - \lambda\Delta t)$. Similarly for departures, if the system is empty the probability of a departure in Δt is 0 and of no departure is 1. If the system is occupied, the probability of a departure is $\mu\Delta t$ and of no departure is $(1 - \mu\Delta t)$.

This permits us to express the long verbal statement of state probability in a mathematical form as follows:

$$P_n(t + \Delta t) = P_n(t)[1 - \lambda\Delta t][1 - \mu\Delta t] + P_{n-1}(t)[1 - \mu\Delta t]\lambda\Delta t + P_{n+1}[1 - \lambda\Delta t]\mu\Delta t \qquad (1)$$

The manipulation of this expression into a set of differential difference equations and the solution of these equations is covered in several texts, e.g., references 3, 5, 6, 10 and 12. There is little point in repeating these derivations again, but a presentation and interpretation of the important result is in order. The first principal result is that the equilibrium state probability appears as a function of the system load factor, in the following form:

$$p_n = \rho^n(1 - \rho) \qquad (2)$$

where $(1 - \rho)$, as we had intuitively expected, is p_0, which is equivalent to the fraction of time the system is idle or empty.

From this expression of state probabilities it is possible to derive the average or expected state of the system,

$$E(n) = \sum_{n=1}^{\infty} np_n = \frac{\rho}{1 - \rho} \tag{3}$$

The properties of the single-channel queue expressed in equations (2) and (3) are illustrated in Figures 2, 3, and 4, in which state probabilities and average line length are plotted against the system load factor. They reveal the extreme sensitivity of such systems to the

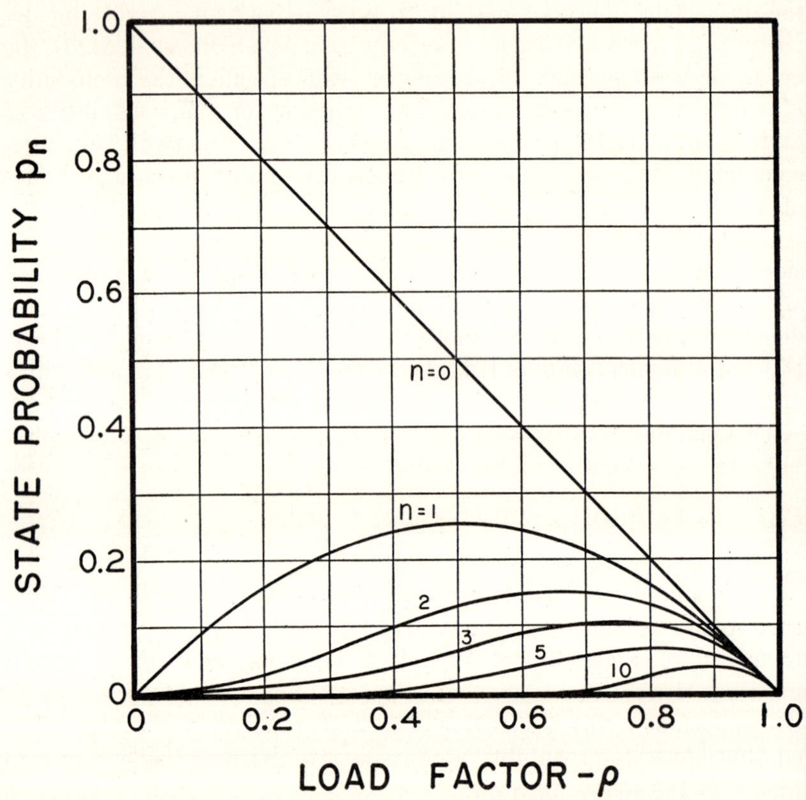

Figure 2. Probability of system being in state n.

load factor. Figure 2 gives the distribution of state probabilities, i.e., the probability that there will be 0, 1, 2 . . . n items in the system. Note that the probability of an ideal state of 1 in the system, implying that the facility is occupied but none is waiting, is a maximum of 25 per cent at 50 per cent load factor ($\rho = 0.5$). Under this condition, the system will, unfortunately, be idle 50 per cent of the time, while during the remaining 25 per cent of the time, some will be waiting.

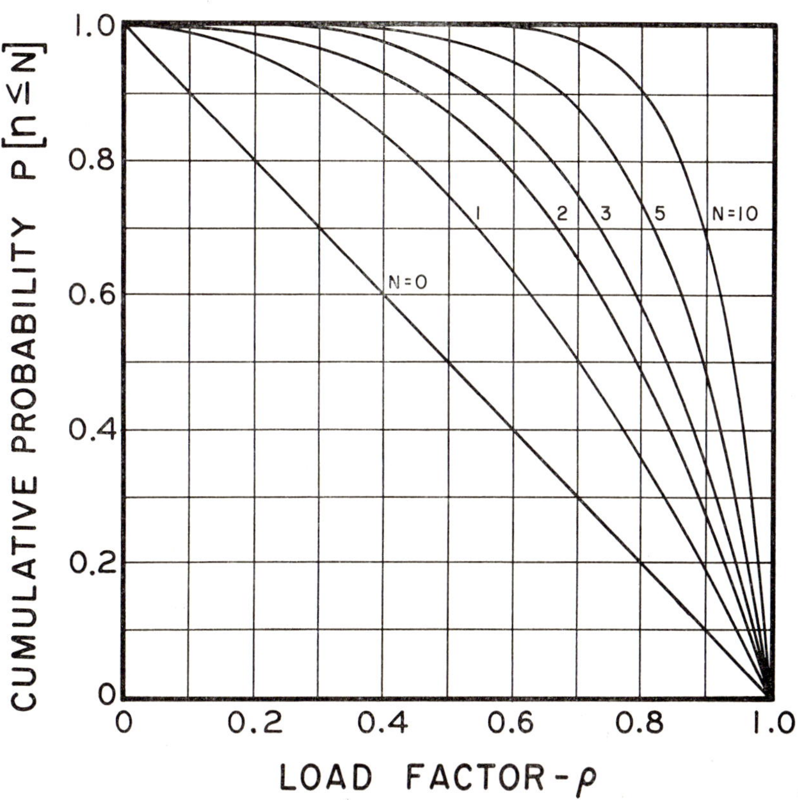

Figure 3. Probability of system being in state N or less.

Figure 3 gives the cumulative probability, i.e., the probability that there are n or less items in the system. Note that the probability of ten or less in the system is practically 100 per cent until the load factor reaches the vicinity of 60 per cent. By examining the system when loaded to 90 per cent of capacity ($\rho = 0.9$) we see that the system will be idle 10 per cent of the time and occupied by more than ten items for 30 per cent of the time. By reducing the load factor to 80 per cent ($\rho = 0.8$), although we increase idle time to 20 per cent, the system is still occupied by more than ten items for 10 per cent of the time.

A single-valued measure of the single-channel queue is given in Figure 4. This shows how the average number in the system varies

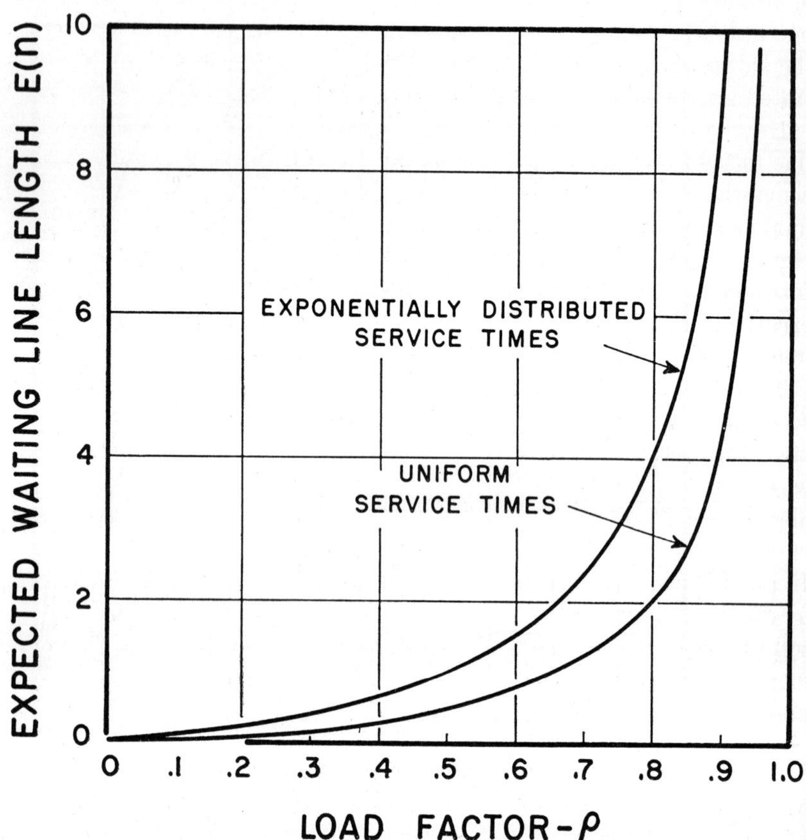

Figure 4. Expected state of the system, $E(n)$.

with the load factor, specifically that this average rises very rapidly as the load reaches two-thirds of the system capacity. A point of confusion may arise in the upper limiting load condition, for equation (3) states that as the load factor approaches 1, the expected number in the system approaches infinity. To explain this, we must recall that equation (2) gives "equilibrium" state probabilities, which strictly hold only when the system has been operating infinitely long. The infinitely long waiting line is a mathematical phenomenon, not a real one we are apt to encounter in our finite lives.

In general, the equilibrium state probabilities, for states other than 0, are higher than the probability at some finite time after the system operation has begun from an empty state. The higher the load factor, the slower the equilibrium probabilities are approached. This is illustrated graphically by Luchak (8) and Morse (10, p. 64).

Having shown in Figure 4 the sensitivity of the expected waiting line to load factor, the question arises, How sensitive is this matter of congestion to other properties such as the statistical character of the input and output processes? Kendall (7) shows that substitution of a regularly spaced output process, implying services of uniform length, for the chance process used in our model, cuts the expected line length exactly in half. Others have shown that giving priority always to the item which requires the shortest service time achieves the same effect.

There is not much encouragement in having to deal with a system so inherently inefficient as the idealized model we have been considering. However, we shall see later that there are strong natural forces which render the behavior of many real systems closely similar to this abstraction. The normal reaction in the real situation of this character is to find means of increasing the output rate when the queue is large. In human systems, this reaction is automatic to a degree as illustrated by Edie (4), who shows that decreasing service time is required by toll takers as traffic before a toll station mounts. A systematic way of increasing rate of flow is to provide alternative or parallel channels to handle congestion when it forms.

MULTIPLE-CHANNEL QUEUES

In the model previously treated, nothing specific was said about the inner workings, or "discipline," of the service facilities. The description of a single-channel queue was based on the constant rate of output, implying a single service facility. A step toward generalization can be taken by letting the output rate vary with the number in the system, up to some specified number of service facilities or channels. This fortunately does not significantly complicate the mathematical solution. In the basic equation (1), the service rate, instead of a constant μ, becomes $n\mu$ until the number in the system reaches the number of channels, a. Beyond this the output rate remains constant at $a\mu$. The resulting equilibrium state probabilities are:

$$p_n = \frac{1}{n!}\left(\frac{\lambda}{\mu}\right)^n p_0 \qquad n \leq a \tag{4a}$$

$$p_n = \frac{1}{a^{n-a}\, a!}\left(\frac{\lambda}{\mu}\right)^n p_0 \quad n \geq a \tag{4b}$$

$$p_n = \frac{1}{n!}\left(\frac{\lambda}{\mu}\right)^n e^{\frac{-\lambda}{\mu}} \tag{4c}$$

$$p_0 = \frac{1}{\displaystyle\sum_{n=0}^{a-1} \frac{1}{n!}\left(\frac{\lambda}{\mu}\right)^n + \frac{\left(\frac{\lambda}{\mu}\right)^a}{a!}\left[\frac{1}{1 - \frac{\lambda}{a\mu}}\right]} \tag{5}$$

An important special case of this model is that in which the number of channels, a, may be considered infinitely large. This implies that all arrivals into the system receive service immediately; there is no waiting line; and n is just the number simultaneously receiving service. State probabilities are given by Eqn. 4c, which is the expression of a Poisson distribution about an average value of $n = \lambda/\mu$. This model describes the variation of number of critically ill hospital patients, discussed in Chapter 25, p. 776.

The congestion reducing effect of parallel channels is apparent on comparison of equation (2) with equation (4), noting that the factor in the denominator, a^{n-a}, increases rapidly with n. This leads to some serious reconsideration of generally accepted philosophy of organizational design. This will be treated after discussion of one more generalization of the basic model.

INPUT FROM FINITE POPULATIONS

Another step toward generalization of the model is to let the input to the system be related to the state of the system in somewhat the same way the output was related to it in the multi-channel case. Here we assume that the input is from some finite population and that when an input occurs, the source population is reduced by one and the potential arrival rate from that source is reduced correspondingly. An example of this situation is the personnel turnover problem discussed in Chapter 25. Upon each resignation, the queue of positions to be filled increases and the number of remaining potential resignations is reduced by 1. The classical problem of this type deals with the maintenance of a finite group of machines, each of which has some known or estimated factors characteristic of rate of chance breakdown. If we designate the number of machines as m, the number of repair channels as r, and consider n the number of machines out of service, the state probabilities associated with the system are:

$$p_n = \frac{m!}{n!\,(m-n)!}\left(\frac{\lambda}{\mu}\right)^n p_0 \qquad n < r$$
$$p_n = \frac{m!}{r!\,(m-r)!}\left(\frac{\lambda}{\mu}\right)^r p_0 \qquad n \geq r \tag{6}$$

The major complications introduced in this model are not mathematical, but computational, since we have added new parameters. Dealing with these computational aspects is considerably simplified by the second of the Operations Research Society Monographs by Peck and Hazelwood (11).

Of all the possible real analogies to the models discussed, the machine breakdown problem leads most easily to a cost optimization model. Each machine that is down represents a loss of productive capacity, and each increase in the number of repair facilities or rate of repair represents a tangible increase in cost. With a very small but busy repair facility the downtime cost soars, while reduced downtime can be bought at a diminishing rate by increasing service facilities, where idleness increases simultaneously. Thus there is some repair facility size at which the sum of the cost of the facility plus loss due to machine downtime is minimum.

The machine breakdown problem is one example in which queueing theory lends itself to a simple cost-minimization model. There are other situations in which the principal value lies in its guide to the choice of form of organization. For an example of this we return to consideration of single- and multi-channel queues and the comparison of determinate and indeterminate systems.

APPLICATION

At this point, some significant conclusions can be drawn from the properties of single- and multi-channel queues, as they are summarized in equations (1) through (5). These equations tell us that, when input and output processes follow the prescribed rule of randomness, congestion can be reduced (that is, the probability of higher values of n can be lessened) if a single-channel operation is reorganized to form several parallel channels. This in turn says that under these conditions of random load, the form of organization which reduces congestion is quite different from the one which has been found best in the case of determinate input and output (or through-put) processes. In fact, the single-channel organization consisting of a number of independent

PARALLEL SYSTEM — 3 CHANNEL

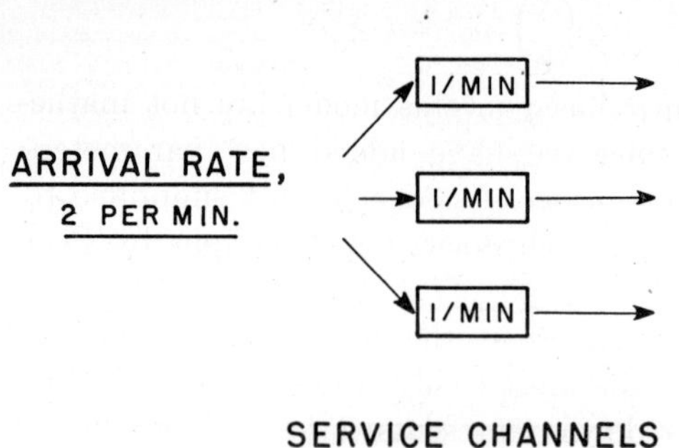

SERVICE CHANNELS
OPERATIONS 1 + 2 + 3

n	p(n)	F(n)
0	.1111	.1111
1	.2222	.3333
2	.2222	.5555
3	.1481	.7036
4	.0988	.8024
5	.0658	.8682
6	.0439	.9121
7	.0293	.9414

SERIES SYSTEM

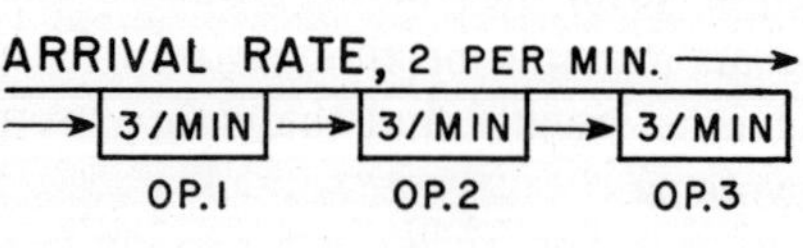

SERVICE CHANNELS

n	p(n)	F(n)
0	.0370	
1	.0728	.1098
2	.0974	.2072
3	.1093	.3165
4	.1099	.4264
5	.1007	.5271
6	.0904	.6175
7	.0783	.6958

Figure 5. Comparison of parallel and series systems.

steps in series—so effective in the machine-paced assembly line—turns out to be the worst possible form of organization from the point of view of delays and congestion for random loading. This argument is best illustrated by an example.

Feller (5, p. 379) gives an example of a three parallel-channel system randomly loaded to two-thirds of its capacity. It is interesting to see what happens if this system is reorganized, with division of labor, into three channels in series as shown schematically in Figure 5. Assume service to be divided into three equal parts; i.e., the service rate at each station is triple that in the parallel system. This is accompanied by economies in the form of less investment in facilities and less training of personnel. If the systems were loaded uniformly and with precise scheduling, there would be a net gain solely through specialization and division of labor.

However, with random loading, the reorganization in series form has devastating effects on the congestion characteristics of the system, as indicated in the tables of Figure 5. In the analysis, n is the state

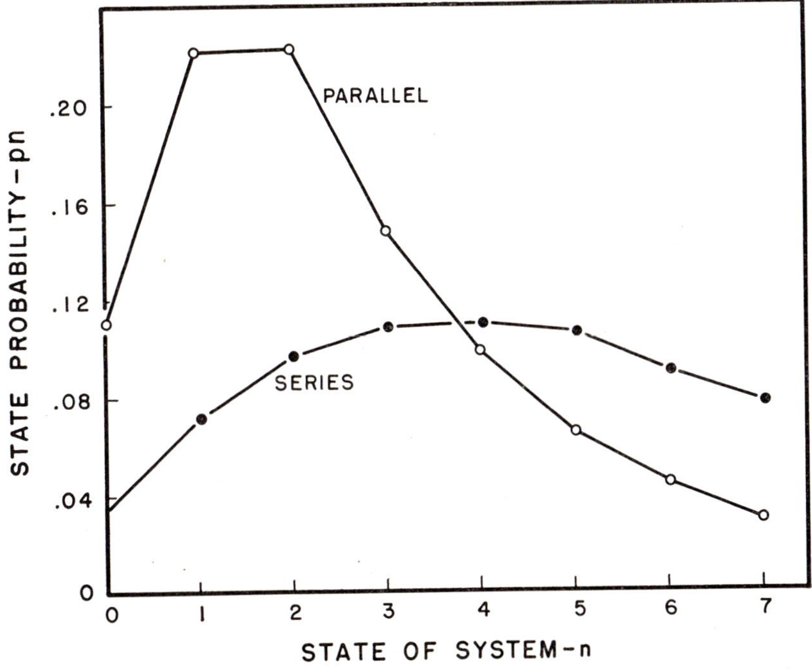

Figure 6. State probabilities for series and parallel system.

of the system, i.e., the number waiting and being served. The state probabilities are plotted in Figures 6 and 7.

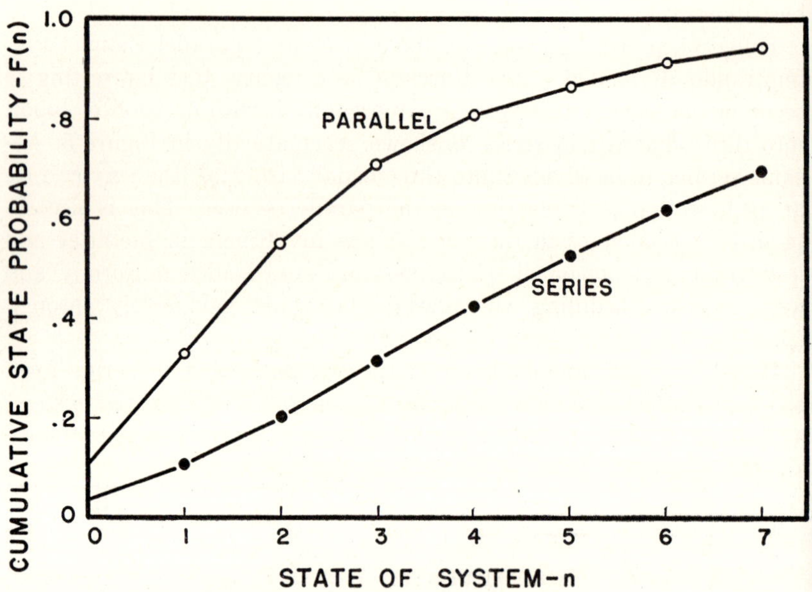

Figure 7. Cumulative state probability for series and parallel system.

The important thing to note is that the series system results in generally higher probabilities of congestion and necessitates greater waiting space. From the point of view of those who must wait for service, the probability of not having to wait at all is 0.555 in the parallel case versus 0.037 in the series system.

It must be kept in mind that the inputs and outputs of these models are a random process of a very special nature, as defined in the development—known technically as a Poisson process. As control is introduced to space arrivals and standardize service times, the characteristics approach that of the determinate system. Most real systems contain partial control of arrivals through scheduling, and the series system will not compare as unfavorably to the parallel system as in the models shown. Unfortunately, analytical means do not presently exist for handling these partially determinate systems. It is of interest to note, however, that the Poisson process can be considered as the limit of disintegration of the scheduled process, and that the effects of earliness and lateness in schedules is quickly felt. To describe this it is necessary to re-examine the model input process more carefully.

THE POISSON PROCESS

In all of the basic models discussed, the input process has been the same. It is known as the Poisson process because, in repeated periods of given duration, the number of events which occur will follow the Poisson distribution. The physical characteristic of the process is that a constant set of forces in the environment produces a constant probability that an event will occur in any small interval of time, no matter what has happened in the adjacent time intervals. Our dice example furnished an approximation of this in the input to the hypothetical single-channel queue. The mathematical model of that system, if we assume no outputs, $(\mu = 0)$, yields the properties of the Poisson process. In this case n, the state of the system, can be construed as the number of events which have occurred since starting time t_0. $P_n(t)$ is the probability that n events occur between t_0 and t. If we repeat this analysis for the input process alone, the following important results are obtained:

$$P_n(t + \Delta t) = P_n(t)[1 - \lambda \Delta t] + P_{n-1}(t)\lambda \Delta t \quad n > 0 \quad (7a)$$

$$P_0(t + \Delta t) = P_0(t)(1 - \lambda \Delta t) \qquad n = 0 \quad (7b)$$

from which, as $\Delta t \to 0$

$$\frac{d P_0(t)}{dt} = -\lambda P_0(t) \quad (8)$$

$$P_0(t) = e^{-\lambda t} \quad (9)$$

This says that chance of occurrence of no event by time t (which is equivalent to the probability that periods of time between events will exceed t) is negative exponential in character. If we let intervals between events be designated u, the cumulative distribution function of u, the probability that an interval is less than u, becomes

$$F(u) = 1 - e^{-\lambda t} \quad (10)$$

and the density function is

$$F'(u) = \lambda e^{-\lambda t} \quad (11)$$

This demonstrates the important property of the Poisson process of an exponential distribution of intervals between events, the characteristic of many small intervals yet a significant number of large ones. Here is the source of intermittent idleness and congestion. Viewed as the output process, the interval between events becomes the duration of the service time—thus the term "exponential service times" commonly used.

The other important result can be obtained from equation (7a) by recursion, knowing $P_0(t)$:

$$P_n(t) = \frac{(\lambda t)^n}{n!} e^{-\lambda t} \tag{12}$$

This says that the number of events which occur in time interval t is Poisson distributed about a mean of λt. This also tells us something of the characteristic variability of the process, for the Poisson distribution has a variance, λt, equal to the mean. If, as an example, we consider a period of time in which a Poisson process is operating at a rate which would produce an expected 100 events, the actual number of events, in 95 per cent of such periods, would range approximately between 80 and 120 events.

With such a highly variable and apparently erratic process it is difficult to develop an ability to estimate the rates at which events are basically being produced, or to sense changes in these rates. There are many real processes which follow very closely the Poisson process. The placing of telephone calls is the classic one, and the examples of receipt of military intelligence and receipt of orders, given in the chapter on simulation techniques, are others.

There are many more real input processes which are best described as scheduled with some loss of control. The arrival of aircraft over an airport, arrival of ships in port, arrival of appointment patients in a hospital—all are cases where a schedule exists but may be departed from by individuals who are either early or late. In the literature, the Poisson process is often used as the model for these cases.

There is some rational basis for using the Poisson process as a model for a real-system input, other than a wished-for ability to use some of the developed models. The Poisson process, which has as its basic postulate the independence of events, may be looked upon in another way—as the limit or disintegration of a scheduled process. It can be demonstrated that as the divergence of events from scheduled times becomes great relative to the interval between scheduled events, the statistical properties of the phenomenon rapidly converge upon those of the Poisson process. Stating that the Poisson process is the limit of disintegration of a scheduled process, the designer of partially determinate systems can use the assumption of Poisson input with somewhat the same attitude of a statistician assuming basic normality of a population from which he draws his samples.

It may be of interest to demonstrate this phenomenon quantitatively, and for this purpose Figure 8 represents a scheduled process in which actual events occur normally distributed about scheduled times. The scale of Figure 8 is such that the standard deviation σ is equal to the scheduled spacing $\overline{U}$. Events are numbered a_0, $a_{\pm 1}$, $a_{\pm 2}$, $a_{\pm n}$, indicating that we have chosen some event, a_0, as an arbitrary starting point of observation. Scheduled arrival or occurrence times are noted T_0, $T_{\pm 1}$, $T_{\pm 2}$, $T_{\pm n}$ and are spaced an interval, $\overline{U}$, apart.

Given a starting point in time, t_0, which is the occurrence time of event 0, followed by an interval of time u, there exists a finite probability that any other event may occur in this interval, although it is evident from Figure 8 that this probability is not significant for those events scheduled to occur more than 2 or 3 standard deviations away from the bounds of u.

The probability that the nth event will occur in interval u is given by the ordinate contained for that event between t_0 and $t_0 + u$, e.g., $P_{a_1}(u)$.

The probability that the nth event does not occur in u is one minus the probability that it does; and the probability that no event occurs in u at all is the product probability:

$$P_0(u) = \prod_{-\infty}^{\infty} [1 - Pa_n(u)]$$

$P_0(u)$, the probability that no event occurs in u, may be thought of as the probability or relative frequency of occurrence of intervals of length equal to or greater than u. That is, if an interval of u exists, it must exist in an interval of u or larger.

The probability that an interval u' is equal to or less than u, i.e., the cumulative distribution function of u', is

$$P(u' \leq u) = 1 - P_0(u)$$

By evaluating over a range of u for varying ratio of $\sigma/\overline{U}$ and initial starting time, t_0, it is possible to observe the characteristic of con-

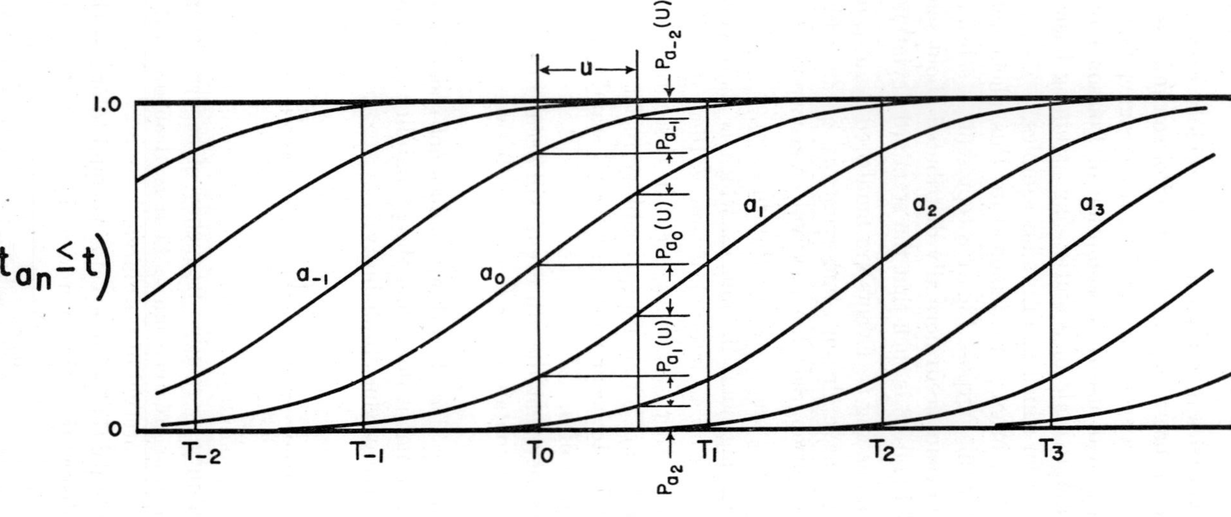

Figure 8. Probability distribution of arrivals about scheduled times.

vergence of the disintegration of the scheduled process onto the random process as plotted in Figure 9. The extremes of behavior are

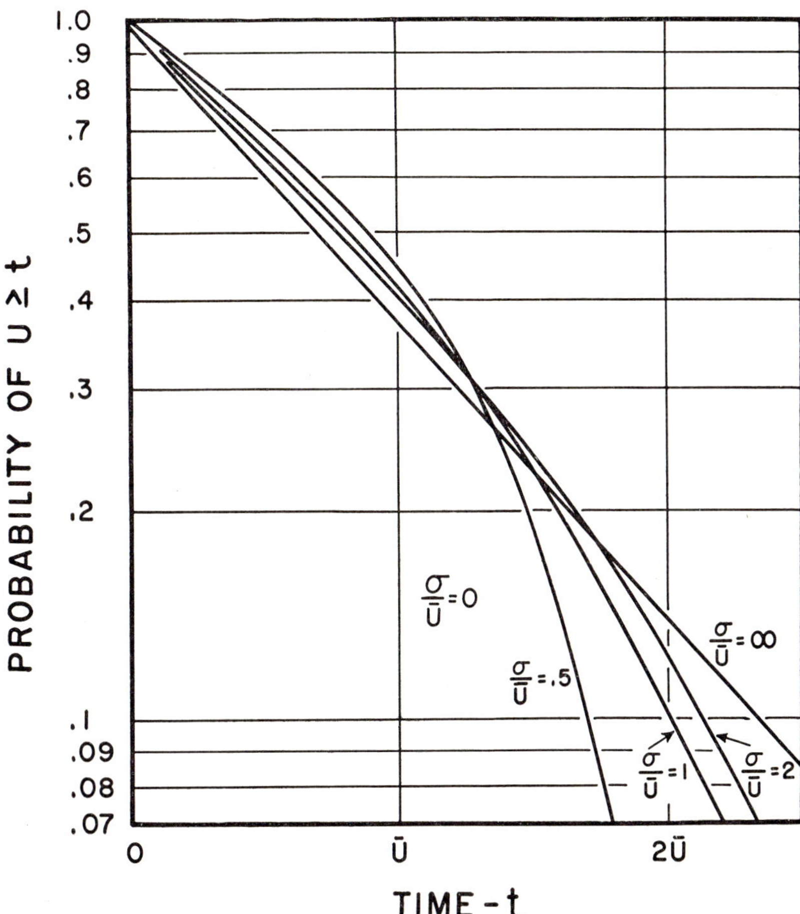

Figure 9. Distribution of intervals between arrivals in a disintegrating scheduled process.

represented by the perfect schedule, plotted as $\sigma/\overline{U} = 0$, a vertical line at $\overline{U} = 1$, indicating that all intervals are of the same duration, $\overline{U}$, on one hand and by the Poisson process, the straight sloped line, designated $\sigma/\overline{U} = \infty$.

Note how rapidly, as $\sigma/\overline{U}$ increases, the distribution approaches the Poisson case in which $\sigma/\overline{U} = \infty$. As would be intuitively expected, the greatest departure of the scheduled process from the purely random one is in the smaller probability of large intervals. The scheduled process, with fewer extremely long and extremely short intervals, will produce shorter waiting lines. (There are, reading from Figure 9, for example, 14 per cent of intervals in a purely random process which exceed twice the average, compared to 10 per cent when σ in a scheduled process is equal to the scheduled spacing.) Slightly over 8 per cent of the random intervals will exceed 2.5 times average vs. 5 per cent for the $\sigma/U = 1$ case. The Poisson approximation is a pessimistic one, and results based on this approximation are conservative estimates.

While the distribution of intervals between events in the imperfect scheduled process approaches the negative exponential of the Poisson process, it must be kept in mind that the ordering of the intervals is not purely random. When a short interval occurs because of earliness of one event, the probability of an adjacent long interval is enhanced and the probability of another short interval diminished. Because of this serial correlation, the convergence of the distribution of events in a period of time on a Poisson distribution occurs only when the deviations of individual events from schedule becomes large relative to the period of time considered. Thus, the choice of Poisson process as an input model will have the virtue of conservatism, but it may have the fault of costly overconservatism.

CONCLUSIONS

What can be done to make the theoretical developments in queueing theory useful in practical applications? What are the alternatives available for design of operations which fall between the types subject to precise analytical treatment? The following approach is suggested:

(1) Accept queueing models, not necessarily as precise predictors, but as a framework in which a problem can be approached to develop intuition concerning the basic functional relationship between variables. For example, whether or not the random loading of a service point is mathematically pure, the state of the system is constantly a randomly fluctuating quantity, and just the introduction of the con-

cept of state probability varying with load and other system parameters gives a foundation for rational consideration.

(2) Use the closest conservative analytical model of the real situation as a means of obtaining a first estimate of the solution to a problem. The theorem previously stated, that the Poisson process is the limit of disintegration of a scheduled process, provides a fortification for this. The acceptability of the solution can be tested by Monte Carlo methods or by the less general method of playback by feeding observed real input into the solution and computing resulting delays and idleness. The latter step should be applied as a check since any serial correlation or time dependence in the real input process can be revealed.

(3) Attempt wherever possible to merge randomly loaded systems. Free interchangeability takes advantage of the law of large numbers by permitting the load of a congested channel to be shifted to an idle one.

(4) Wherever possible, design services which are subject to randomly fluctuating demand so that their rate can be increased in congested periods, or so that some preparatory steps of the service can be performed in idle periods.

(5) Develop measures of cost associated with states of a system. For a given input rate, the solution of the queueing problem appears in the form of a set of state probabilities varying with potential service rate. Congestion cost can be expressed as the sum of the products of the cost of being in each state and the probability of being in the state. Since the state probabilities diminish as service rate is increased, we have a familiar situation in that, as service rates are increased, service costs rise while congestion costs fall. This leads to an equation of system cost as a function of a control variable, and optimization is possible.

A final and perhaps most pragmatic and desirable approach to queueing problems is to regard them as a form of social evil and attack them at the source, the random input. From the earlier discussion of classification of operation types, it will be recalled that in the indeterminate operation, the smoothing of service times provides some alleviation of congestion tendency; but the basic character is established by the input and still exists. Granted that some operational demands, such as telephone communication, are beyond control, others, such as flow of work in process or flow of patients through hospital clinics, can, through introduction of controls, be made to approach the determinate type. There will be a cost associated with such control, however, and when some mathematician can establish the functional relationship between congestion and degree of disintegration of the input schedule, queueing theory will extend even further our capabilities for the design of organizations and operations.

REFERENCES

(1) Brigham, G. "On a Congestion Problem in an Aircraft Factory," *Operations Research,* Vol. 3, No. 4 (1955).

(2) Brockmeyer, E., Halstrom, H. L., and Jensen, A. "The Life and Works of A. K. Erlang," *Transactions of the Danish Academy of Science* (Copenhagen: Akademiet for de Tekniske Videnskaber), Vol. 2 (1948).

(3) Churchman, C. W., Ackoff, R. G., and Arnoff, E. L. *Introduction to Operations Research.* New York: John Wiley and Sons, 1957.

(4) Edie, Leslie C. "Optimization of Traffic Delays at Toll Booths," *Operations Research,* Vol. 2, No. 2 (1954).

(5) Feller, W. *An Introduction to Probability Theory and Its Application.* Vol. 1. New York: John Wiley and Sons, 1950.

(6) Goode, H. H., and Machol, R. E. *System Engineering.* New York: McGraw-Hill, 1957.

(7) Kendall, W. G. "Some Problems in the Theory of Queues," *Journal of the Royal Statistical Society,* Series B, Vol. XIII, No. 2.

(8) Luchak, George. "The Solution of the Single-Channel Queueing Equations Characterized by a Time-Dependent Poisson-Distributed Arrival Rate and a General Class of Holding Times," *Operations Research,* Vol. 4, No. 6 (1956).

(9) McCloskey, Joseph F., and Coppinger, John L. (eds.). *Operations Research for Management.* Vol. II. Baltimore: The Johns Hopkins Press, 1956.

(10) Morse, P. M. *Queues, Inventories, and Maintenance.* New York: John Wiley and Sons, 1958.

(11) Peck, L. G., and Hazelwood, R. N. *Finite Queueing Tables.* New York: John Wiley and Sons, 1958.

(12) Saaty, T. L. "Résumé of Useful Formulas in Queueing Theory," *Operations Research,* Vol. 5, No. 2 (1957).

Fifteen

SIMULATION TECHNIQUES

CHARLES D. FLAGLE

INTRODUCTION

It must be evident that there is a serious gap between many of the analytical models discussed in this book and the real systems whose properties they are intended to portray. In many of the chapters, the subject matter is approached by choosing a simple abstraction of a real problem—an abstraction which can be represented mathematically—and approaching reality by building complexity into the model step by step. In all of the cases in which this approach is used, for example, in the treatment of elementary models in operations research, inventory systems, and queueing theory, it can be noted that as the simple model is modified toward realism, the mathematical complexities increase much more rapidly than the model itself approaches reality. In many cases, the mathematical model becomes unsolvable for the complex system, or must be subjected to such simplifying assumptions that the numerical results obtained from it are not applicable to the real problem. The mathematical abstraction may be valuable in giving insight into the functional relationship between the properties of a system and the variables which affect it; but some other analytical means must be found to bridge the gap between theory and useful design procedures.

To engineers, this is not an unusual situation, for in the design of physical systems it is often necessary to build and test a model—meaning now a physical model rather than a mathematical one—before building its prototype. In operations research, an activity we shall call "simulation techniques" is the counterpart of scale model testing in physical research and design. The parallel is worth exploring.

The aerodynamicist or mechanical designer, when unable to find mathematical expressions which predict completely or adequately the behavior of a component or system under design, falls back upon tests of a reduced scale model of that system under conditions which simulate those of the prototype in its environment. His position in doing this is rendered secure by a well-developed theory of similarity considerations and dimensional analysis. This theory in turn is based upon some inherent characteristics of physical systems: first, that all properties of both model and prototype can be described in the three basic dimensions of force, length, and time; and, second, that the substance from which the model and prototype are made and the environment in which they function are continuous and reproducible.

It is not sufficient in physical model testing simply to reduce the object under study by some linear scale factor, for this radically alters the relationship between the length, area, and volume of the object. The well-known "square-cube" law states that as a body is reduced in size, its area reduces with the square and its volume reduces with the cube of its linear dimension. This, of course, has serious effects on many properties of the object, the rate at which it cools or heats, its stresses and strains in acceleration, and its behavior in flowing through fluids.

In the case of the aerodynamicist estimating the properties of an airfoil section in a wind tunnel, the pressure distribution around the foil—producing its lift and drag—is similar in foils of the same shape but different size only if the velocity and viscosity of fluid around the two foils are adjusted so that the Reynolds number, the ratio of inertia to viscous forces, is the same.

These examples of the well-developed simulation techniques in the physical sciences have been mentioned to show not only the similarity but also the contrast of the situation to that of operations research. They are similar in the sense that the techniques stand between reality and complete mathematical abstractions of phenomena; but beyond this similarity of predicament, necessity, and motivation, the comparison breaks down.

Unfortunately, almost all the phenomena which gain the attention of operations research are not completely describable in simple dimensions of force, length, and time. In our operations, only the dimension of time stands as a consistently available quantitative measure. We may speak loosely of "trying things out on a smaller scale," but we can seldom describe size of operation or organization in units of length and force. Further, the environment in which operations take place is seldom a stable continuum. Operations are described most often in such terms as the rate of occurrence of events, the numbers of objects or people in or passing through a place or an activity at a given time. Most often these properties will be highly variable and subject to chance forces. Hence, the only meaningful expression of them is in terms of probability.

The problem of scale reduction of organization or operations, and its converse problem, the effect of growth, have received much attention in the social sciences. An interesting discussion and key to the literature is given by Haire (7), who observes the interrelationship of size, shape, and function in natural phenomena, noting,

> It seems clear from these examples that a certain shape only fits a particular size, and that the appropriate size may be rather narrowly limited and defined by rigorous physical laws. The same kind of thing is true in the relation between size and function. Particular ways of accomplishing certain functions are suited to one size, but are not at all appropriate as the size increases, and new solutions must be found.

Observing that in the phenomenon of human organization clear-cut relationships between size and form have not been established, Haire concludes, reviewing the evolution of some organisms,

> In these cases, a study of the history of growth and the relation between size and shape has given us a real lead to the primary forces in the environment in which the organism lives. One might hope, similarly, that an historical analysis of the growth of business, and a detailed understanding of the relation between size and shape and function in an industrial organization would give us a similar insight into the kinds of forces in its environment which shape its existence.

A combination of factors prevents us from simple reduction in scale, in addition to the lamentable fact that we do not have rigorous measures of size. When the process is a probabilistic or stochastic one, as so many operations are, its properties are a function of its absolute "size." If we want, for example, to estimate the congestion characteristics of a ten-channel communication system, we cannot do so by testing a five-channel system loaded to the same fraction of capacity.

This fact is revealed in equation (5), Chapter 14, in which system properties are a function of a, the number of channels. We must find some ten-channel analogy and model it in a reduced time scale.

In this respect at least—finding an analogy—the situation is fortunate, for it is not difficult to reproduce many of the stochastic processes of real life by some device, often a simple mechanical one like those used in games of chance. It is, in fact, the natural recourse to the accoutrement of gambling, both for purposes of sampling and simulation, that led to the name "Monte Carlo Method" for an analytical process closely related to our subject. Before proceeding with a technical discussion of the field of simulation techniques, which is evolving from Monte Carlo methods under the stimulus of war gaming in military operations research, it is necessary to say something by way of introduction about the relationship between real operational processes and the behavior of some mechanical random process generator such as a pair of dice.

PROBABILITY AND HUMAN AFFAIRS

The argument to be advanced is that many real and important phenomena in organized human life are affected and at times dominated by the same laws of chance which control dice, the roulette wheel, and the fall of cards.

In fact, this very similarity may account for the immemorial fascination with gambling. Our lives are filled with challenges whose nature and timing are not of our own making, but to which we nevertheless must respond. In many cases, we cannot wait to perceive the exact form of the challenge before preparing to take action. We must anticipate what will happen and our survival or success depends on making the right guess. So it is in games of chance. If simple games fascinate us because they "simulate" some real experience, is it possible that the similarity exists in many other less dramatic but equally important matters? The answer is affirmative.

Consider, for example, the appearance of the number 2 in a sequence of tosses of dice. Of the thirty-six possible combinations, a 2 can only appear in one; therefore, the probability of rolling 2 is one in thirty-six. We expect it to occur, on the average, once every thirty-six tosses, but in actuality it may occur in successive rolls or there may be over a hundred consecutive rolls without a 2; that is, there will be a wide range of gaps between occurrences. All we can say with certainty is that on each roll we have the same one chance in thirty-six of rolling a 2, no matter what has happened on previous rolls.

A very similar state of knowledge characterizes many situations. The chance of accident or sudden illness hangs over the heads of all of us and this is reflected in the emergency ward of our hospitals by the awareness that during any minute there is a probability of arrival of a patient—a probability which changes only slowly or remains just as great in the next minute and the next and the next. The same might be said for the probability that our phone may ring, or an order for a product may be received, or an employee may resign. In all these real cases we have the problem of organizing for the appropriate action to meet the demand created by the chance event.

Now the dice example under consideration has a property well known in the field of probability. The gaps between occurrences of independent events of low probability have a negative exponential distribution. There are many short gaps, and each time we increase the interval between events by some fixed amount, the frequency of occurrence of gaps diminishes by some fixed percentage.[1]

This behavior of dice becomes of more than academic interest if we note a few examples in which the same sort of behavior takes place. In these examples the act of throwing dice represents the passage of an interval of time and the roll of a 2 represents occurrence of some significant event. In the first case, Figure 1, the event is the recording of pieces of combat intelligence in a regimental journal during a World War II battle. Since the information recorded comes from a number of independent, unco-ordinated sources—much of it from enemy action —it is not surprising that the intervals between these events follow the same distribution as in our dice example. Similarly, for the receipt of orders for an industrial product sold nationally, the intervals between orders show the same pattern in Figure 2.

In both of these cases, the intervals between events are important, for in them some action must be taken. We are concerned with the design of these organizations—the number of people and the equipment and procedures needed to respond adequately to the demands. How long does it take to perform the services demanded? What skills are required? How many people are needed? Should the tasks be divided and specialized, sending each message or order through a num-

[1] The theory behind this is treated in the development of the Poisson process, Chapter 14, Queueing Theory, p. 417.

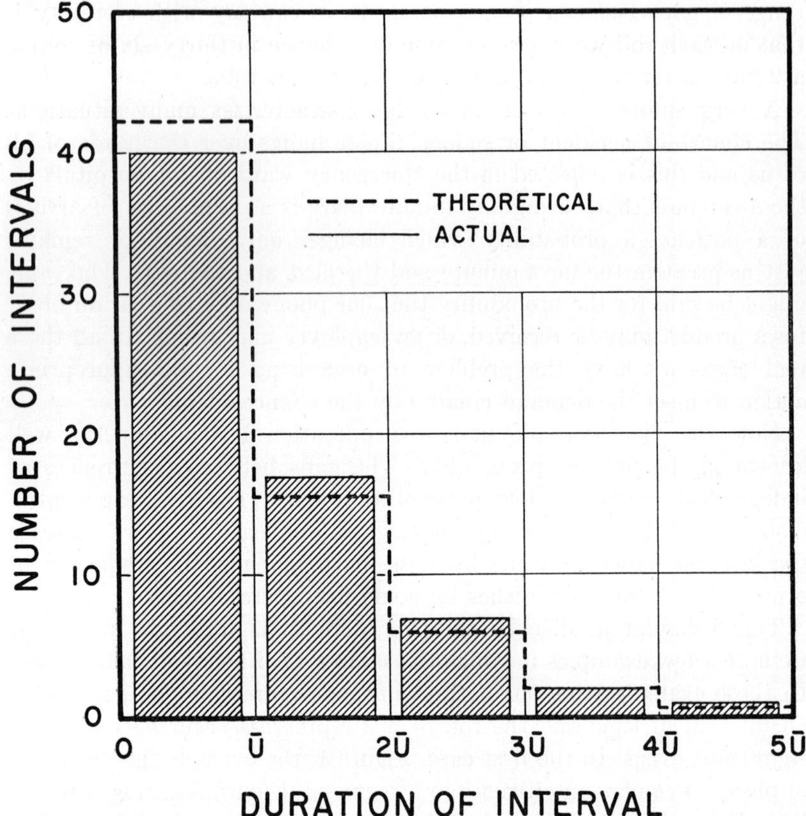

Figure 1. Distribution of intervals between arrivals of items of combat intelligence ($\bar{U}$ = average interval).

ber of hands in its processing, or should there be a number of parallel paths for complete processing to avoid intermediate waiting?

To answer these questions and determine the optimum solution, we must have means of estimating such system properties as time delays, congestion, etc., in terms of the input rates, potential service rates, and form of organization, the last two of which can be considered the variables which can be controlled.

To this point, the examples given are fortunate ones in that they may be simulated by dice, or other means still to be treated; or, since the processes are recognized by the independent nature of their events,

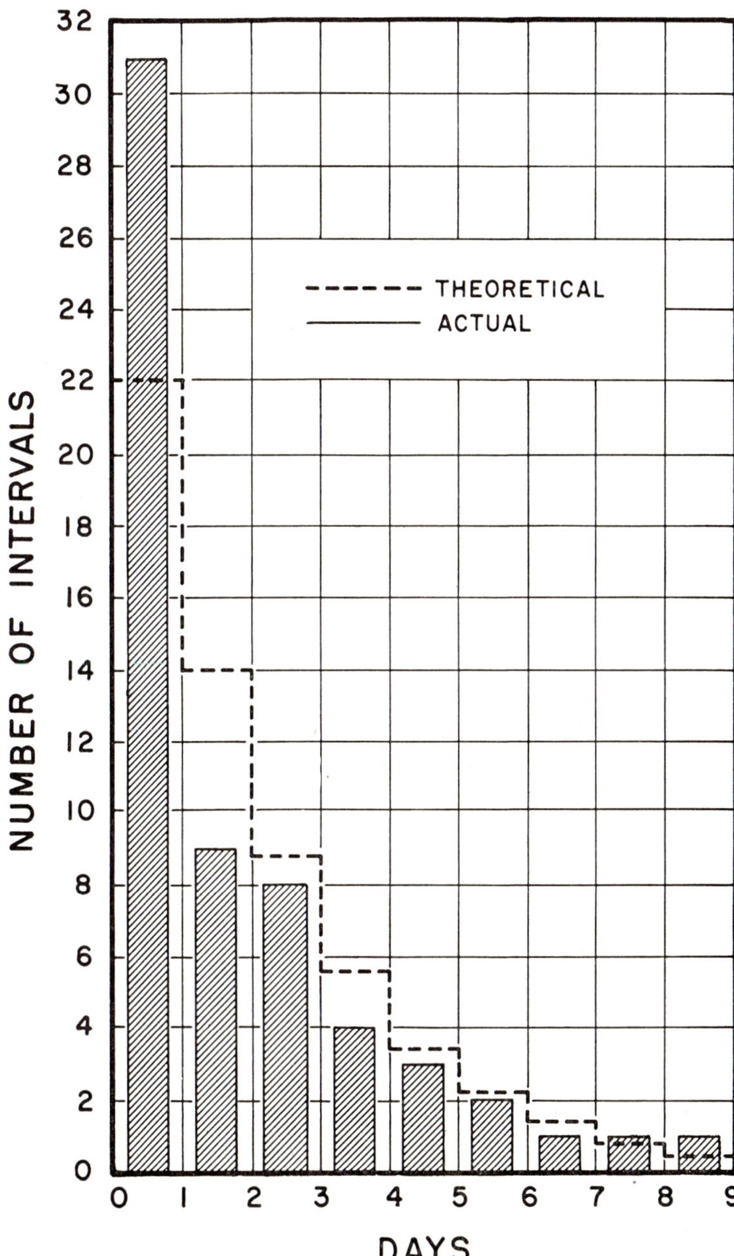

Figure 2. Distribution of intervals between receipt of orders.

as examples of the Poisson process, they may be expressed in mathematical terms. The next step requires determination of actual service times in the system and finding an analogy or simulation of their behavior, or if possible a mathematical expression of it.

If both the arrival pattern and service times are of the simple mathematical forms and the flow path is simple, i.e., either a single channel or several in parallel, all system properties can be estimated by use of queueing theory. If any of these simplifying conditions are not met, we can fall back on a simulation of the system and sample its behavior for estimates of system properties.

AN EXAMPLE—SIMULATION OF A SINGLE CHANNEL QUEUE

For example, assume that the relationship between the distribution of number in a single channel queue, with Poisson arrival process and exponentially distributed service times, could not be determined analytically. It has already been noted that the throw of dice for some number of low probability approximates the Poisson process. Consequently, this device could be used to simulate the system, as in fact it has been used to introduce the idea of the single-channel queue. By following this process as outlined on pp. 403 and 404, Chapter 14, we can construct a sample history of the simulation and use the observed fractions of time in various states as estimates of the equilibrium state probabilities for a system loaded to 80 per cent capacity. This tedious process can be avoided by recourse to another method of simulation involving use of a table of random numbers.

In effect, a table of random numbers is a roulette wheel with 100 slots, numbered from 00 to 99. The occurrence of any of the 100 numbers is equally likely. By using such a table, considerable time can be saved simply because each number represents the outcome of a trial with some time-consuming mechanical device. An effective means of simulating Poisson arrivals[2] or service times is to let several numbers represent arrivals and several others represent departures. If we let 00, 01, 02, and 03 signal arrivals and 95, 96, 97, 98, and 99 signal departures, it is possible to construct a history of the queue simply by scanning columns of random numbers. The arrival rate under these

[2] A more precise means of simulating the process is the rejection method of von Neumann described in reference (5).

conditions is four-fifths as great as the potential departure rate, just as in the dice problem, observing again the rule that no departure can take place if the system is empty. In this case, each random number represents the passage of a unit of time and the interval between arrivals and departures is proportional to the sum of unit intervals between the occurrences of arrival and departure signals. State probabilities can be estimated by making a set of sample observations of the state of the system at random intervals and computing the fraction of observations in each state. A sample of this means of simulation is illustrated in Table 1. Since arrivals occur only on 4 per cent of the numbers observed, and departures on 5 per cent, we again have a tedious process, although it will later develop as a practical approach when used with a computer.

Table 1. Simulation of Single Channel Queue by Scanning Random Number Table

Random Number	Arrival (A) or Departure (D)	State of System n
16		0
22		0
97		0
01	A	1
78		1
64		1
95	D	0
98		0
03	A	1
05	A	2
22		2

A point to note on the foregoing approach is that we have, in effect, divided time into a sequence of discrete intervals and examined each interval for the occurrence of events—of arrivals and departures. Since events occur in less than 10 per cent of these intervals, the process is one in which over 90 per cent of the observation adds nothing to the history of the process other than a recording of the passage of time. Let us designate this the "interval examination" technique to distinguish it from an alternative approach in which the system is examined only at the "instant of significant event." Later the feasibility of both approaches, using a digital computer, will be demonstrated; "interval examination" in this chapter, and "instant of significant event" as it has been used in the simulation of armored combat in Chapter 24.

An example of the alternative approach, applied to the single channel queue, using random number tables, is the following, in which every random number conveys some information about time of an arrival or departure. For each two digit random number from 00 to 99, we must find some corresponding interval length between arrivals which represent 1 per cent of the total possible intervals. This can most easily be done graphically from the cumulative probability distribution, as shown in Figure 3. As plotted, the rectangularly dis-

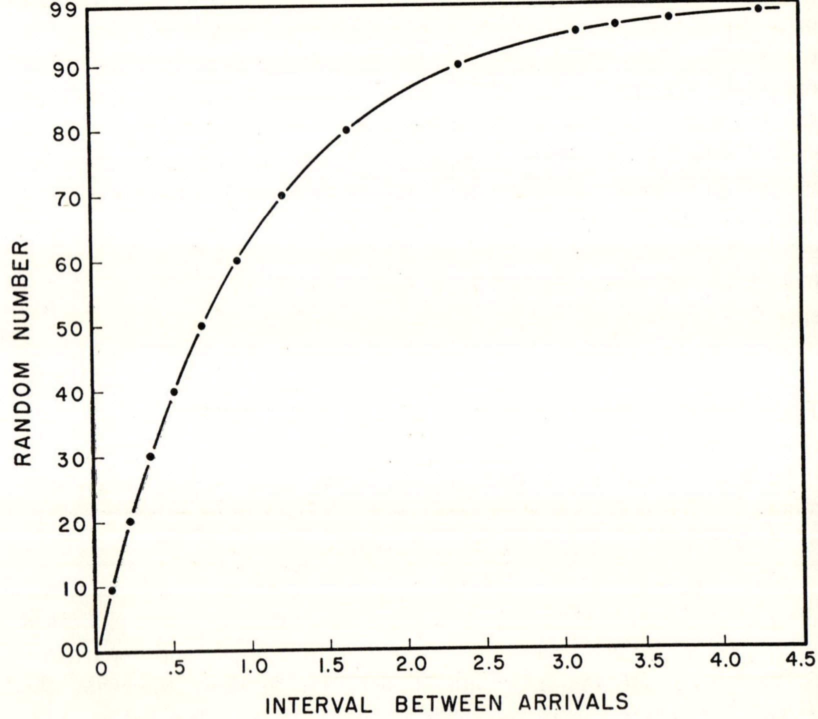

Figure 3. Graph for transformation of random numbers to a sample of exponentially distributed variates.

tributed random numbers are transformed to exponentially distributed ones.

Here, for unit arrival rate, the shortest 1 per cent of intervals lies in the range between 0 and .01 time units. We assign that shortest 1 per cent the random number 00 and characterize it by an interval of .005 time units, which is the midpoint of the range of the shortest 1 per cent. In the example at hand, the service rate is, on the average, 80 per cent less than the interval between arrivals. For a service time, then, random number 00 would yield $.8 \times .005$, or .004 time units.

It is important to note that any distribution, theoretical or empirical, could have been plotted and used in this way. In contrast to the analytical approach to the problem at hand, which is applicable only for a limited number of input and output distributions, the simulation techniques being described can accept any form of distribution.

With this means of transforming each of a sequence of random numbers[3] from 00 to 99 to a sequence of intervals between arrivals or a sequence of service times, it is simple to construct a schedule of the process, from which a sample history of system behavior may be obtained as shown in Table 2. Number in the system, delay in getting service, and total time in the system, may be easily computed from such a schedule. However, this approach, too, though economical in its use of random numbers, is a tedious business, requiring several hours to construct the history of a sample of 500 arrivals. But how good are the estimates of state probabilities obtained from the sample history of 500 observations? How large must a sample be to give a specified precision of results within acceptable confidence limits?

PRECISION, CONFIDENCE, AND SAMPLE SIZE

If, in a simulation, X observations of the state of the system are made, and X_n of the times, the system is found in state n, the fraction of observations in n is an estimate of state probability, p_n.

$$\hat{p}_n = \frac{X_n}{X} \tag{1}$$

If the observations of the system are random, the actual number

[3] The number 99 covers the range of intervals from 4.66 to ∞, which cannot be represented by a midpoint. The last percentile must be divided and treated in a manner similar to the first 99.

Table 2. Simulation by Random Scheduling

1 Arrival No.	2 Arrival Random No.	3 Arrival Interval	4 Time of Arrival	5 Service Random Number	6 Normalized Service Time	7 Scaled Service Time	8 Time Service Began	9 Time Ended	10 State of System Before Arrival
0			0						
1	10	.11	.11	66	1.09	.87	.11	.98	0
2	37	.47	.58	31	.378	.30	.98	1.28	1
3	08	.09	.67	85	1.93	1.54	1.28	2.82	2
4	99(7)	5.99	6.66	63	1.01	.81	6.66	7.47	0
5	12	.13	6.79	73	1.33	1.06	7.47	8.80	1
6	88	2.16	8.95	45	.61	.48	8.95	9.56	0
7	74	1.37	10.32	78	1.54	1.23	10.32	11.55	0
8	87	2.08	12.40	01	.015	.01	12.40	12.41	0
9	03	.04	12.44	22	.254	.20	12.44	12.64	0
10	95	3.10	15.54	94	2.90	.23	15.54	15.77	0

of times the system is found in any state, X_n, will be distributed binomially (see Chapter 9, p. 233) with a variance

$$\text{Var.}\ (X_n) = X \cdot p_n\,(1 - p_n) \tag{2}$$

and

$$\text{Var.}\ (\hat{p}_n) = \frac{p_n\,(1 - p_n)}{X} \tag{3}$$

When the expected value of X_n is large, say greater than 30, the distribution of X_n is very nearly normal and the usual confidence limits can be applied in terms of the standard deviation.

$$\sigma_{\hat{p}n} = \sqrt{\frac{p_n\,(1 - p_n)}{X}} \tag{4}$$

In many repetitions involving X observations, in 95 per cent of them, the observed value, $\hat{p}_n$, will fall in the range

$$\hat{p}_n = p_n \pm 1.96\sqrt{\frac{p_n\,(1 - p_n)}{X}} \tag{5}$$

Obviously, the precision of estimate increases as the number of observations, X, is increased. If some levels of precision and confidence are specified, the number of observations required is obtained by equating the permissible tolerance to the statistical variation within the confidence limits. For illustration, if it is desired to know p_n within 10 per cent of its true value, with 95 per cent confidence

$$p_n\,(1 \pm .1) = p_n \pm 1.96\sqrt{\frac{p_n\,(1 - p_n)}{X}} \tag{6}$$

and

$$.1\,p_n = 1.96\sqrt{\frac{p_n\,(1 - p_n)}{X}}$$

from which

$$X = \left(\frac{1.96}{.1}\right)^2 \cdot \frac{(1 - p_n)}{p_n} \tag{7}$$

We still must know something about the magnitude of quantity being estimated, p_n, for the required sample size, X, varies with it. This may require a guess, or some *a priori* knowledge, followed by trial runs, converging simultaneously on X and p_n.

In the example previously described, the single channel queue loaded to 80 per cent capacity ($\rho = .8$), we know the true value of p_n from queueing theory and may take advantage of it to illustrate the number of observations which must be made in a simulation to achieve some specified precision. Take the probability of an idle system, i.e., in state zero—a condition which we expect to prevail 20 per cent of the time.

$$p_0 = .2$$
$$1 - p_0 = .8$$

For precision within 10 per cent, at 95 per cent confidence, from equation (7)

$$X = \left(\frac{1.96}{.1}\right)^2 \cdot \left(\frac{.8}{.2}\right) \cong 1600$$

The estimation of other properties will require different sample sizes, perhaps smaller. Take, for example, the problem of estimating the average state of the system $E(n)$. From equations (3) through (6) of Chapter 14, we know for the single channel queue that

$$E(n) = \sum_{n=1}^{\infty} np_n = \rho/(1 - \rho) \tag{8}$$

$$p_n = \rho^n(1 - \rho) \tag{9}$$
$$q_n = (1 - p_n) = 1 - \rho^n(1 - \rho) \tag{10}$$

From X observations of the system, $E(n)$ is estimated from the observed frequencies of occurrence of each state.

$$\hat{E}(n) = \sum_{n=1}^{\infty} n\hat{p}_n$$

Substituting from equation (3)

$$\text{Var } \hat{E}(n) = \sum_{n=1}^{\infty} n^2 \frac{p_n(1 - p_n)}{X} \tag{12}$$

and from (9) and (10)

$$\text{Var } \hat{E}(n) = \sum_{n=1}^{\infty} n^2 \cdot \frac{\rho^n(1 - \rho)[1 - \rho^n(1 - \rho)]}{X} \tag{13}$$

which yields

$$\text{Var } \hat{E}(n) = \frac{1}{X}\left[\rho \frac{1 + \rho}{(1 - \rho^2)} - \frac{\rho^2(1 + \rho^2)}{(1 - \rho)(1 + \rho)^3}\right] \tag{14}$$

To estimate $E(n)$ with the same precision and confidence previously used for estimation of a single state probability, as in equation (6).

$$E(n)\,[1 \pm .1] = E(n) \pm 1.96 \sqrt{\frac{1}{X}\left[\frac{\rho(1+\rho)}{(1-\rho)^2} - \frac{\rho^2(1+\rho^2)}{(1-\rho)(1+\rho)^3}\right]} \quad (15)$$

rom which, using $E(n) = \rho/1 - \rho$

$$X = \left(\frac{1.96}{.1}\right)^2 \left(\frac{\rho}{1-\rho}\right)^{-2} \left[\frac{\rho(1+\rho)}{(1-\rho)^2} - \frac{\rho^2(1+\rho^2)}{(1-p)(1+\rho)^3}\right] \quad (16)$$

and in the case at hand, in which $\rho = .8$, $X = 842$.

It is apparent, then, comparing this to the 1600 observations needed to estimate p_0, that sample size is determined by the nature and magnitude of the quantity to be estimated.

Within the same confidence interval, we can increase precision only by increasing sample size. Since the standard deviation, and consequently the tolerance, diminishes as the square root of the sample size, we must expect to pay for increased precision with very large samples. To increase precision by a factor of ten, i.e., to ±1.0 per cent at the level of confidence chosen, the sample for estimation p_0 would have to be increased to 160,000 independent observations of the system. All this is for estimating the probability of the most frequently encountered state. If we wish to determine some property of lower probability, say 5 per cent, the sample would have to be quadrupled to 640,000 independent observations to achieve the same precision. Anyone who has followed the simple calculation required to produce Tables 1 and 2 will realize the undesirability of the task of carrying out such a volume of tedious calculations.

USE OF THE DIGITAL COMPUTER IN SIMULATION

Happily, we have at hand to aid us that grandest of all slaves, the digital computer. Its capabilities are admirably suited to the type of computations illustrated in Tables 1 and 2. It can perform these with great speed and can accept a considerable increase in complexity of the simulated system. Furthermore, the computer is capable of generating the necessary random or pseudo-random numbers[4] for carrying out simulation. In the case of arrival or service processes which can be expressed in mathematical form, the computer can gen-

[4] Pseudo-random numbers are produced by a process yielding a sequence which is ordered and reproducible, yet possesses the properties of a sequence randomly drawn from some specified population. See Chapter 15 of reference (11).

erate samples of intervals directly. For empirical distributions, tables of intervals corresponding to rectangularly distributed random numbers can be developed in the manner used in constructing Figure 3.

To illustrate the usefulness of the digital computer for simulation purposes, the simple problem of the single channel queue has been programmed for the Univac 1103A at the Operations Research Office.[5] The mechanics of the programming are such that the addition of parallel channels is quite simple and costs little in programming time. The computer process is basically the same as the preceding example of "interval examination" of manual calculation and is pictured in the flow chart, Figure 4.

For purposes of this simulation, time has been expressed as a numbered sequence of discrete intervals of duration Δt. Corresponding to each of 1,000 three-digit random numbers there is assigned a discrete number of intervals which yield the time elapsed between consecutive arrivals, or when modified by a factor, the number of intervals spent in service.

The process begins with several sub-routines which establish a sample of the history of the system. The first random number is selected, and its corresponding number of time intervals determines the interval of the first arrival. A second random number determines the number of time intervals for service of the first arrival, and this number is added to the arrival to determine the interval of departure of the first arrival. The average number of intervals between arrivals has been chosen as twenty, the average duration of service being sixteen intervals. For each interval, two quantities are stored; Δt_λ, which is the number of intervals until the next arrival, and Δt_μ, the number of intervals until the next completion of service.

The computation then begins at the circled *a* in Figure 4 by examining information stored for the first interval to answer the question, is there an arrival? If $\Delta t_\lambda = 0$, there is; if greater than zero, there is not. In the event there is not an arrival, one time interval is subtracted from the Δt_λ and Δt_μ of all subsequent intervals and the examination for the next time interval is made. The computation process, when no arrival or departure occurs, requires about 140 micro-seconds.

In the first interval in which $\Delta t_\lambda = 0$, signalling an arrival, the state

[5] The author is indebted to Mr. John Moss of the Operations Research Office, who programmed and carried out this computer simulation while participating in the first formal course in simulation techniques at The Johns Hopkins University.

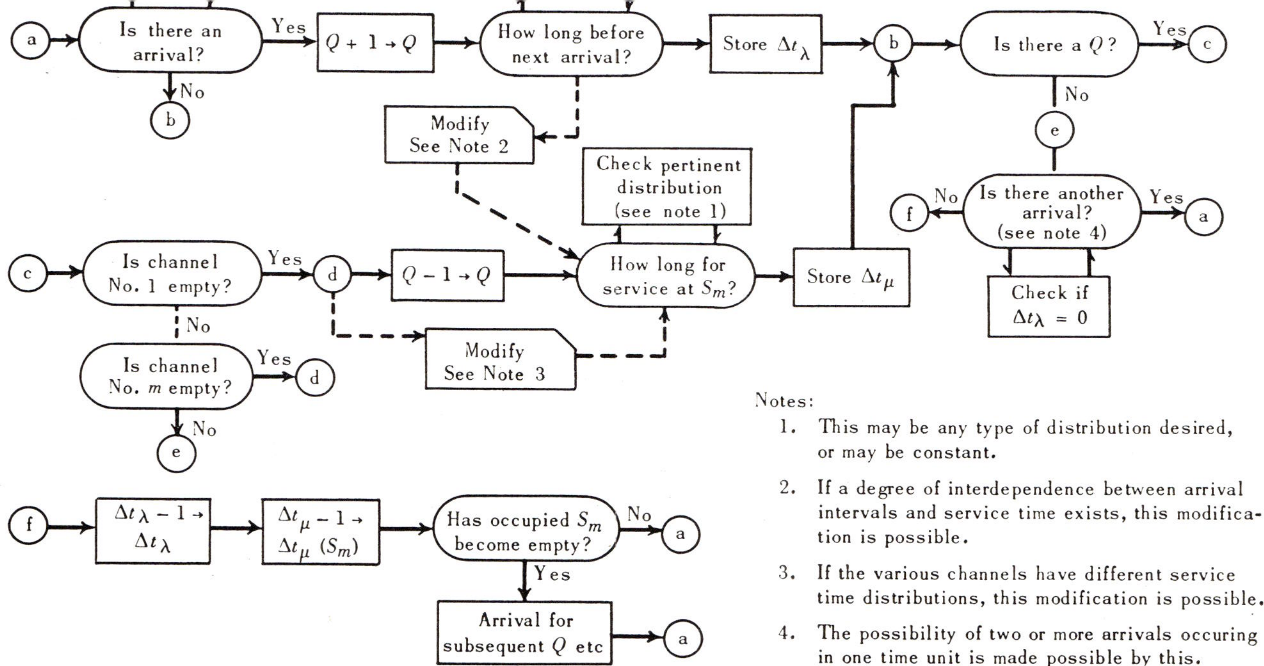

Δt_λ = Interval of time till next arrival
Δt_μ = Interval of time for service
Q = Number in queue
S_m = Service channel

Figure 4. Flow-chart for computer simulation of a queue.

of the system is increased by one, the sub-routines for establishing intervals of arrivals and departure of the next item are activated, one time interval is subtracted from all pertinent locations, and the examination is repeated. The process which accounts for an arrival and departure requires approximately five milli-seconds.

At this speed of computation a history of 100,000 arrivals can be recorded in ten minutes. The analysis of the system is made by printing out the desired properties at random or regular intervals.

By addition of sub-routines the following can be recorded:

1. Elapsed time.
2. Total number of arrivals at time t.
3. Number of units awaiting service (in the queue).
4. Number of units that have been serviced by time t.
5. State of occupancy of channels at time t.

Once programmed (an activity which takes about sixteen hours for the simple case illustrated), the time required to observe very large sample flows through the system is almost insignificant. Table 2 gives the results of a run of 100,000 arrivals into the system. To estimate probabilities, the state of the system was sampled before each hundreth arrival. The observed probability is the fraction of times the system was found in the corresponding state. Theoretical values are beside the observed ones. Also shown on Table 3 is the precision of estimate, which we are fortunate in knowing in this case by virtue of

Table 3. Results of Computer Simulation of Single-Channel Queue—100,000 Arrivals (1,000 Observations)

State of System	State of System, P_n		95 Per Cent Confidence Interval	
	Theoretical	Simulated	Lower Limit	Upper Limit
0	.200	.195	.187	.213
1	.160	.160	.148	.172
2	.128	.126	.117	.139
3	.103	.102	.093	.113
4	.082	.084	.073	.091
5	.066	.070	.058	.074
6	.052	.051	.044	.059
7	.042	.041	.036	.048

our knowledge of the theoretical probability. Had we not known the theoretical probability, a close approximation to the precision could have been derived from the observed probabilities or from multiple replication of the process.

EXAMPLES OF COMPUTER SIMULATION

Some excellent examples of the digital computer simulation are available. Notable among these is the simulation of armored combat reported by Zimmerman in Chapter 24 of this volume.

Jennings and Dickens (9) have simulated the peak hour traffic in a bus terminal for the Port of New York Authority. Other examples are those of Clapham (4) and Banberry and Taylor (1).

A variation of the theme is the multi-person business game (3) in which human decision plays a role in the course of events which are calculated and reported by a digital computer.

An interesting simulation is now under way in the intramural research program of the U.S. Public Health Service. This has developed in study of the progressive-care hospital system at Manchester, Connecticut. The author and John Moss, as consultants to the Public Health Service, are constructing a Birth and Death Process model of the community in order to forecast the fraction of the population needing various forms of hospital care. Both environmental and analytical considerations have led to recourse to simulation.

In the experimental hospital system under study, one area is equipped and staffed for critically ill patients, another for the less seriously ill, and still another for the self-sufficient patient who may require diagnostic services, or teaching, but otherwise makes little demand on medical and nursing resources. The term "progressive care" derives from the plan for progression of the patient through each of these areas as his needs vary through illness and convalescence.

As a test of feasibility of such a system, limited facilities for intensive and self-care patients have been created at Manchester. While these relatively small areas have been satisfactory for testing and demonstration of the general concept, they are not of sufficient capacity for testing the whole system. In particular, they do not yield estimates of the total number of beds required in each of the progressive areas to meet community needs. This determination of capacity required is a central problem of hospital management, one which is difficult to solve within a specific institution, since the population in each of the existing services is influenced by the physical limitations of the facilities and is in any event highly variable.

In this case a hypothetical hospital is being built on a digital computer. The first step of the process is measurement of input rate to and transition between the various services as the demands for these services are generated by the illness and accident phenomena of the community. This is done by daily classification of patients throughout the hospital to determine types of service needed. The process yields simultaneously sample hypothetical distributions of length of stay in the services.

In some cases the arrival rates and service rates are sufficiently well approximated by a Poisson process to permit estimation of the population distribution analytically, but in other cases the process is simulated to give a forecast of population distribution where there are no limitations on facilities. From this forecast the effect of various allocations of facilities on such vital factors as occupancy rates, waiting times for admission, and costs of operation can be assessed and the alternative which maximizes some criteria such as total days of patient care rendered can be chosen. Further simulation of the chosen alternative can be carried out for purposes of testing the effect of various policies or procedures or sensitivity of the system to changes in value of some of the parameters.

An important aspect of this simulation is its dynamics. The input to the hospital services is a function of the size of the population and the age distribution within it—and perhaps other factors. Both population and its age distributions are functions of time and birth and death rates. Economic factors affecting immigration and emigration must also be taken into consideration, and these are drawn from existing community forecasts for the next decade.

It has thus become possible to forecast through simulation the change in population of the community over a period of planning interest, with particular reference to demands for hospital facilities.

The speed with which the computer handles the many but simple calculations involved is great. Ten years of real life are simulated in less than an hour of computer time. Basic programming requires about four man-months, but once completed is easily adjusted for changes in such parameters as birth or immigration rate, or incidence and duration of a particular disease as affected by medical breakthrough.

THE ROLE AND STATUS OF SIMULATION IN OPERATIONS RESEARCH AND MANAGEMENT

We can conceive of the management process in four stages as shown schematically in Figure 5. We begin somewhat arbitrarily with meas-

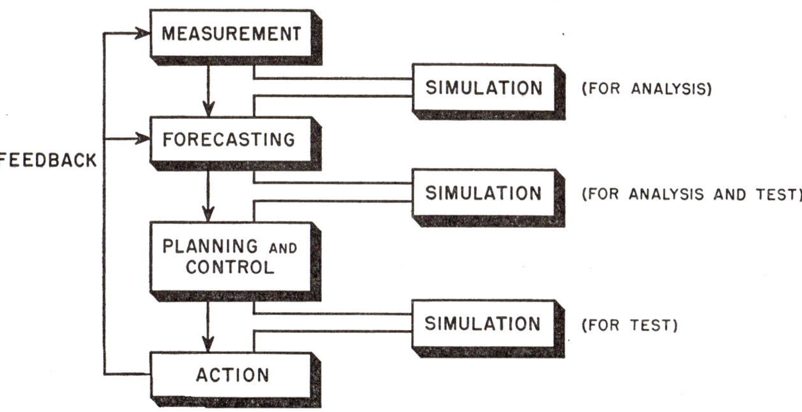

Figure 5. Simulation in the management process.

urement—that is, the recording of relevant facts and figures about an operation—and from these produce a forecast of the impact of these measured factors on the operation. This must be followed by the construction of a plan of action and the establishment of the administrative controls and the resources necessary to carry out the chosen course of action. If the action itself has an impact upon the environmental factors, measurements of this effect must be fed back into the process to adjust forecasts and plans. Between each of these stages of the management process some analysis is required. To the extent that intuition, mathematical models, or simple processes of observation and analysis are not sufficient to yield the necessary answers, simulation techniques are available both for purposes of analysis and test.

The hospital facilities forecast just cited is an example of the use of simulation in several planning and management stages.

At the time of this writing, the status of simulation techniques is perhaps best illustrated by the state of the literature in the field. Two of the recent past winners of the Lanchester Prize have been devoted to the topic, papers by Zimmerman (13) and by Thomas and Deemer (12). Yet no comprehensive text on simulation has been written. Some text material may be found in such works as Meyer (11), Goode and Machol (6) and Beckenbach (2). In addition to the examples previously cited, several interesting philosophical papers have appeared in the journals, notably those of Harling (8) and King (10).

From all of these experiments and writings, a fairly consistent definition of simulation techniques emerges, expressed well by Harling:

> By simulation is meant the technique of setting up a stochastic model of a real situation, and then performing sampling experiments upon the model. The feature which distinguishes a simulation from a mere sampling experiment in the classical sense is that of the stochastic model. Whereas a classical sampling experiment in statistics is most often performed directly upon raw data, a simulation entails first of all the construction of an abstract model of the system to be studied.

With this definition, simulation may be regarded as one of several forms of application of Monte Carlo techniques. Monte Carlo techniques then constitute a broader field which includes use of random numbers for approximating solutions of differential equations, an application which does not concern us here.

The field of simulation techniques is in a dynamic phase of development, paced by the evolution of the digital computer and the imaginations of those who apply the techniques. Through it all, it must be remembered that simulation yields only an empirical form of knowledge, fraught with the danger that the stochastic model is not truly representative. In this sense it is admittedly inferior to mathematical techniques, which yield functional relationships between variables. Ultimately, we may hope that the more slowly developing mathematical capabilities will make simulation as we now practice it unnecessary. Until that time comes, however, simulation techniques stand as powerful aids to the operations analyst who must produce useful results.

REFERENCES

(1) Banberry, J., and Taylor, R. J. "A Study of Congestion in a Melting Shop of a Steelworks," *Operations Research Quarterly,* Vol. 9, No. 2 (June, 1958).

(2) Beckenbach, E. F. (ed.). *Modern Mathematics for Engineers.* New York: McGraw-Hill, 1956.

(3) Bellman, R., *et al.* "On the Construction of Multi-Stage Multi-Person Business Game," *Operations Research,* Vol. 5, No. 4 (August, 1957).

(4) Clapham, J. C. R. "A Monte Carlo Problem in Underground Communications," *Operations Research Quarterly,* Vol. 9, No. 1 (March, 1958).

(5) Forsythe, G. E., Germand, H. H., and Householder, A. S. (eds.). *Monte Carlo Method.* NBS Applied Mathematics Series. Washington, D. C.: U. S. Government Printing Office, 1951.

(6) Goode, H. H., and Machol, R. E. *System Engineering.* New York: McGraw-Hill, 1957.

(7) Haire, Mason. "Size, Shape and Function in Industrial Organizations," *Human Organization,* Spring, 1955.

(8) Harling, J. "Simulation Techniques in Operations Research—A Review," *Operations Research,* Vol. 6, No. 3 (May-June, 1958).

(9) Jennings, N. H., and Dickens, J. H. "Computer Simulation of Peak Hour Operation in a Bus Terminal," *Management Science,* Vol. 5, No. 1 (October, 1958), 106-20.

(10) King, G. "The Monte Carlo Method as a Natural Mode of Expression in Operations Research," *Operations Research Society of America,* Vol. 1, No. 2 (February, 1953).

(11) Meyer, H. A. (ed.). *Symposium on Monte Carlo Methods.* New York: John Wiley and Sons, 1956. (In particular, "Use of Different Monte Carlo Sampling Techniques," Herman Kahn, The Rand Corporation; and "Generation and Testing of Pseudo-Random Numbers," Olga Taussky and John Todd.)

(12) Thomas, C. J., and Deemer, W. L., Jr. "The Role of Operational Gaming in Operations Research," *Operations Research,* Vol. 5, No. 1 (1957).

(13) Zimmerman, Richard E. "A Monte Carlo Model for Military Analysis," *Operations Research for Management,* Vol. II. Edited by Joseph F. McCloskey and John F. Coppinger. Baltimore: The Johns Hopkins Press, 1956.

Sixteen

INTRODUCTION TO THE THEORY OF GAMES

ELIEZER NADDOR

CONCEPTS AND TERMINOLOGY

Players in n-Person Games

The theory of games deals with systems in which two or more decision-makers are in competition. Such systems are called *games.* The decision-makers are called *players,* and they are referred to as P_1, P_2, ... etc. A game in which n players participate is called an *n-person game.*

A decision-maker need not necessarily be considered as a single individual. Thus, a baseball game is assumed to be a two-person game, as are bridge and chess. Three oil companies in competition may be considered as a three-person game. Wars are generally regarded as two-person games.

Strategies

The alternatives that each player has in playing are called *strategies.* Each alternative will generally have numerous factors. Some of the factors may be quantifiable and some may only be described qualitatively. If the number of factors is r, and if the ith factor can have m_i possible values or attributes, then the total number of alternatives m is $m = m_1 \times m_2 \times \ldots \times m_r$.

When m, the total number of alternatives for each player, is finite, the game is called a *finite game*. Otherwise it is called a *game with infinitely many strategies*.

Payoffs

The outcome of playing a game is described in terms of what each player gets at the end of the game. This is called a *payoff*. Although many games do not involve any payments at all, their outcomes can generally be described quantitatively and these quantities are called payoffs. In a bridge game the scores of each team are payoffs. In chess a win is generally scored as $+1$, a loss as 0, and a draw as $\frac{1}{2}$. If the outcomes cannot be described quantitatively, the methods of the theory of games generally cannot be used.

Zero-Sum and Non-Zero-Sum Games

When the total payoffs to all players at the end of each game is zero the game is called a *zero-sum game*. Bridge is a zero-sum game when, in scoring at the end of the game, the winning team gets a plus and the losing team a corresponding minus score.

When the total payoffs to all players at the end of each game is a constant, the game is called a *constant-sum game*. If the scoring method mentioned before is used for chess, then it can be considered as a constant-sum game (i.e., $1 + 0 = 1$ in the case of a win, $\frac{1}{2} + \frac{1}{2} = 1$ in the case of a draw).

Zero-sum games are thus special cases of constant-sum games, the constant being zero. It can be shown that the properties of zero-sum games and constant-sum games are the same. Therefore, it has become customary to refer to all of these games as zero-sum rather than constant-sum games.

All games which may end with a total payoff to all players which is not constant are called *non-constant-sum games*. These, too, are generally referred to as *non-zero-sum* games.

Payoff Matrices

In finite games the payoffs to the players are usually displayed in tabular form, as in Tables 1 and 2.

Such tables are referred to as payoff matrices. The payoff matrices in Tables 1 and 2 should be interpreted as follows:

Table 1. P_1's Payoff Matrix

	Alternative	P_2: B_1	B_2	B_3
P_1	A_1	1	−2	0
	A_2	−9	3	−3

Table 2. P_2's Payoff Matrix

	Alternative	P_2: B_1	B_2	B_3
P_1	A_1	−1	2	0
	A_2	9	−3	3

Player P_1 has only two alternatives, A_1 and A_2; Player P_2 has three alternatives, B_1, B_2, B_3. When Player P_1 chooses alternative A_1 and Player P_2 chooses alternative B_1, the outcome will be a gain of one for P_1 and a loss of one for P_2. When P_1 chooses A_1 but P_2 chooses B_2, then P_1 will lose two and P_2 will gain two. The four possible outcomes can be interpreted in a similar manner.

One notes here that the sum of the payoffs for any choice of alternatives is always zero. Hence the game whose payoff matrices are as in Tables 1 and 2 is a zero-sum game.

Rectangular Games

There is a certain class of games which have been studied extensively and for which many results have been obtained. These games have the following characteristics:

(a) Only two players participate;
(b) Each player has a finite number of alternatives;
(c) The sum of the payoffs is always zero.

These games are called *rectangular games*. Using the terms previously defined, rectangular games are finite two-person zero-sum games.

The payoff matrices of rectangular games can always be given in the form of Tables 1 and 2. Since they are zero-sum, it is not necessary to give both matrices. It is customary to give only P_1's payoff matrix.

The size of the matrix will depend on the number of alternatives that players have. In general, if P_1 has m alternatives and P_2 has n alternatives, the payoff matrix will have m rows and n columns.

The name given to these games is derived from the rectangular shape of the m by n payoff matrix.

ILLUSTRATIVE PAYOFF MATRICES

The River Tale[1]

"Steve is approached by a stranger who suggests they match coins. Steve says that it is too hot for violent exercise. The stranger says, 'Well, then, let's just lie here and speak the words "*heads*" or "*tails*"— and to make it more interesting I'll give you $30 when I call "*tails*" and you call "*heads*" and $10 when it's the other way around. And— just to make it fair—you give me $20 when we match.'. . ."

Steve, Player P_1, has two alternatives, A_1 and A_2. He can call "*heads*" (A_1) or he can call "*tails*" (A_2). Similarly, the stranger, Player P_2, has two alternatives: B_1 ("*heads*") and B_2 ("*tails*"). This is a rectangular game. The payoff matrix is given in Table 3.

Table 3. The River Tale Payoff Matrix

		P_2	
	Alternative	B_1 (*Heads*)	B_2 (*Tails*)
P_1	A_1 (*Heads*)	−$20	$30
	A_2 (*Tails*)	$10	−$20

Two-Finger Morra[2]

". . . 'Two-Finger Morra' has been played in Italy since classical antiquity. This game is played by two people, each of whom shows

[1] This game has been taken from J. D. Williams, *The Compleat Strategyst* (15, pp. 50-51). This book presents in witty and non-technical terms an excellent introduction to the theory of games.

[2] This game has been taken from J. C. C. McKinsey, *Introduction to the Theory of Games* (8, p. 7). This book is the first text in the field and is suitable for seniors and graduate students with a good mathematical background.

one or two fingers and simultaneously calls his guess as to the number of fingers his opponent will show. If just one player guesses correctly, he wins an amount equal to the sum of the fingers shown by himself and his opponent; otherwise the game is a draw."

Each alternative in the game has two factors: the number of fingers shown (one or two) and the number called (one or two). Hence each player has four alternatives.

The payoff matrix is given in Table 4.

Table 4. Two-Finger Morra Payoff Matrix

				P_2			
	Alternative			B_1	B_2	B_3	B_4
		Factor 1 (Show)		1	1	2	2
			Factor 2 (Call)	1	2	1	2
P_1	A_1	1	1	0	2	−3	0
	A_2	1	2	−2	0	0	3
	A_3	2	1	3	0	0	−4
	A_4	2	2	0	−3	4	0

A Game with Several Moves

There are three moves in the following game. Player P_1 makes the first move. Player P_2 makes the second move. Player P_1 makes the third and last move, ending the game.

In the first move P_1 places a coin on a table. He has the choice of placing a "*head*" or a "*tail.*" In the second move, P_2, having seen what P_1 did in the first move, places a second coin on the table, choosing between a "*head*" and a "*tail.*" But he keeps the coin covered so that P_1 cannot see it.

In the third move P_1 places a third coin on the table, again choosing "*heads*" or "*tails.*"

At the end of three moves, three coins are on the table.

If all three are "*heads*" player P_1 wins 2¢; if all three are "*tails*" player P_2 wins 2¢. If one is "*heads*" and two are "*tails*" P_1 wins 1¢; and if two are "*heads*" and one is a "*tail*" P_2 wins 1¢.

Games with several moves can best be described by the use of *trees* as in Figure 1.

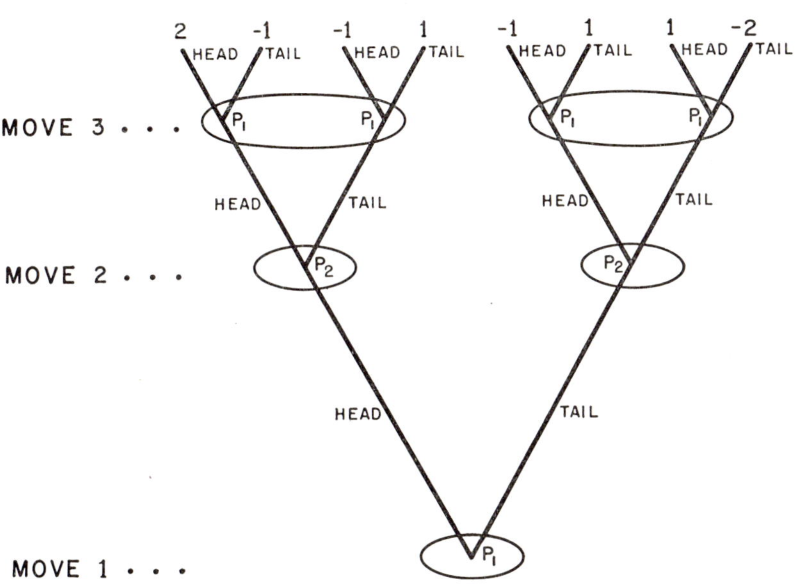

Figure 1. Tree of a game with several moves.

The circled *branches* of the trees indicate the state of knowledge of players. If several branches are circled together, this means that the player does not know which specific branch applies. The amount at the *top-points* are the payoffs to P_1.

Certain games with several moves, including the above, can be transformed into rectangular games. In our game player P_1 has four alternatives for placing two coins: "*heads*"-"*heads*" (A_1), "*heads*"-"*tails*" (A_2), "*tails*"-"*heads*" (A_3) and "*tails*"-"*tails*" (A_4).

Although player P_2 has only two choices, he has available four different strategies since he can see P_1's first coin. For each strategy he has to specify what he will do if P_1's first coin is either a "*head*" or a "*tail*." P_2's two different strategies are thus:

B_1: "*heads*" if first move was "*heads*" and "*heads*" if first move was "*tails*"

B_2: "*heads*" if first move was "*heads*" and "*tails*" if first move was "*tails*"

B_3: "*tails*" if first move was "*heads*" and "*heads*" if first move was "*tails*"

B_4: "*tails*" if first move was "*heads*" and "*tails*" if first move was "*tails*"

The payoff matrix is therefore as in Table 5.

Table 5. The Payoff Matrix for a Game with Several Moves

				P_2							
	Alternative			B_1		B_2		B_3		B_4	
				If		If		If		If	
		Move 1		*Head*	*Tail*	*Head*	*Tail*	*Head*	*Tail*	*Head*	*Tail*
			Move 2 / Move 3	*Head*	*Head*	*Head*	*Tail*	*Tail*	*Head*	*Tail*	*Tail*
P_1	A_1	*Head*	*Head*	2		2		−1		−1	
	A_2	*Head*	*Tail*	−1		−1		1		1	
	A_3	*Tail*	*Head*	−1		1		−1		1	
	A_4	*Tail*	*Tail*	1		−2		1		−2	

A Three-Person Game

The payoffs to the players in a three-person game in which P_1 has two alternatives (A_1, A_2), P_2 has two alternatives (B_1, B_2) and P_3

has three alternatives (C_1, C_2, C_3) may be displayed as in Table 6.

Using these payoffs one can determine payoff matrices for rec-

Table 6. Payoffs in a Three-Person Game

P_1's Alternative	A_1	A_1	A_1	A_1	A_1	A_1	A_2	A_2	A_2	A_2	A_2	A_2
P_2's Alternative	B_1	B_1	B_1	B_2	B_2	B_2	B_1	B_1	B_1	B_2	B_2	B_2
P_3's Alternative	C_1	C_2	C_3	C_1	C_2	C_3	C_1	C_2	C_3	C_1	C_2	C_3
Payoff to P_1	2	0	−2	1	3	2	3	−3	−4	5	−1	0
Payoff to P_2	1	−1	0	−1	−4	1	−5	−1	2	−2	0	4
Payoff to P_3	−3	1	2	0	1	−3	2	4	2	−3	1	−4

tangular games in which two players form a coalition and play against the third player.

If, for example, P_1 and P_2 form a coalition against P_3, we have a two-person game with players Q_1 and P_3 where Q_1 is the coalition of players P_1 and P_2. Our new player Q_1 has four alternatives (A_1, B_1; A_1, B_2; A_2, B_1; and A_2, B_2), which we designate by D_1, D_2, D_3, and D_4 respectively. The payoff matrix of Q_1 is as in Table 7.

Similarly, we can obtain Tables 8 and 9. Table 8 gives the payoff matrix for a coalition of P_1 and P_3, called Q_2, playing again P_2. The

Table 7. The Payoff Matrix for the Coalition Q_1 of Players P_1 and P_2

				P_3		
		P_1	P_2	C_1	C_2	C_3
Q_1 (P_1, P_2)	D_1	A_1	B_1	3	−1	−2
	D_2	A_1	B_2	0	−1	3
	D_3	A_2	B_1	−2	−4	−2
	D_4	A_2	B_2	3	−1	4

coalition Q_2 has six alternatives designated by $E_1, E_2, \ldots, E_6$ which correspond to $A_1, C_1; A_1, C_2, \ldots, A_2, C_3$.

Table 9 gives the payoff matrix for a coalition of P_2 and P_3, called Q_3, playing again P_1. The coalition Q_3 has six alternatives designated by $F_1, F_2, \ldots F_6$, which correspond to B_1, C_1; $B_2, C_2; \ldots, B_2, C_3$.

Table 8. The Payoff Matrix for the Coalition Q_2 of Players P_1 and P_3

				P_2	
		P_1	P_3	B_1	B_2
Q_2 (P_1, P_3)	E_1	A_1	C_1	-1	1
	E_2	A_1	C_2	1	4
	E_3	A_1	C_3	0	-1
	E_4	A_2	C_1	5	2
	E_5	A_2	C_2	1	0
	E_6	A_2	C_3	-2	-4

Table 9. The Payoff Matrix for the Coalition Q_3 of Players P_2 and P_3

				P_1	
		P_2	P_3	A_1	A_2
Q_3 (P_2, P_3)	F_1	B_1	C_1	-2	-3
	F_2	B_1	C_2	0	3
	F_3	B_1	C_3	2	4
	F_4	B_2	C_1	-1	-5
	F_5	B_2	C_2	-3	1
	F_6	B_2	C_3	-2	0

Closed Bidding

Two items worth \$5 and \$7 respectively are offered to two bidders, P_1 and P_2. Each bidder is invited to submit a closed bid for both items.

Bidder P_1 has a total of \$3 available. His alternatives A_1 and A_2 are to bid \$1 for the first item and \$2 for the second or \$2 for the first and \$1 for the second. Bidder P_2 has a total of \$4 available. His alternatives B_1, B_2, and B_3 are to bid \$1 for the first and \$3 for the second, \$2 for each, or \$3 for the first and \$1 for the second.

Each item goes to the man submitting the highest bid for it. In the case that both submit equal bids, a toss of a coin determines the man who will get the item for the amount bid.

The payoffs to the bidders are given in Table 10, where the first of each pair of members is P_1's payoff and the second is P_2's payoff.

Table 10. A Closed Bidding Payoff Matrix

				P_2		
	Alter-native			B_1	B_2	B_3
		Bid for Item 1		1	2	3
			Bid for Item 2	3	2	1
P_1	A_1	1	2	2, 6	$2\frac{1}{2}$, $5\frac{1}{2}$	5, 2
	A_2	2	1	3, 4	$1\frac{1}{2}$, $6\frac{1}{2}$	3, 5

When P_1 uses alternative A_1 and P_2 uses B_1, both bid \$1 for the first item and P_1 bids \$2 for the second item, whereas P_2 bids \$3 for it. P_2 is thus sure to get the second item for an expected gain of \$7 − \$3 = \$4. They toss a coin for the first item. Its worth is \$5 and the winner gets it for \$1 or a gain of \$4. Since it is equally likely that P_1 or P_2 will get it, the expected gain for each is $\frac{1}{2}$ of \$4 or \$2. Hence for alternatives A_1 and B_1, P_1 can expect to gain \$2 and P_2 can expect to gain \$2 + \$4 = \$6.

In a similar manner the other entries in Table 10 can be determined. The game represented by the payoff matrix of Table 10 is an example of a non-zero-sum game.

DEFINITION OF SOLUTION OF RECTANGULAR GAMES

Matrix Notation

The general payoff matrix of a rectangular game is represented in Table 11.

In this game P_1 has m alternatives and P_2 has n alternatives. When P_1 chooses A_1 and P_2 chooses B_1 then P_1 will get a_{11} and P_2 will get $-a_{11}$. And in general, when P_1 chooses A_i and P_2 chooses B_j then P_1 will get a_{ij} and P_2 will get $-a_{ij}$.

The element a_{ij} may be positive, zero, or negative. When it is positive this indicates a gain for P_1 and a loss for P_2. When it is zero neither player gains or loses. When it is negative a_{ij} indicates a gain for P_2 and a loss for P_1.

Table 11. General Payoff Matrix

		P_2					
		B_1	B_2		B_j		B_n
P_1	A_1	a_{11}	a_{12}	$\cdots$	a_{1j}	$\cdots$	a_{1n}
	A_2	a_{21}	a_{22}	$\cdots$	a_{2j}	$\cdots$	a_{2n}
		$\vdots$	$\vdots$		$\vdots$		$\vdots$
	A_i	a_{i1}	a_{i2}	$\cdots$	a_{ij}	$\cdots$	a_{in}
		$\vdots$	$\vdots$		$\vdots$		$\vdots$
	A_m	a_{m1}	a_{m2}	$\cdots$	a_{mj}	$\cdots$	a_{mn}

The payoff matrix of a rectangular game is thus completely specified if one knows a_{ij} for $i = 1, 2, \ldots, m$ and for $j = 1, 2, \ldots, n$.

Pure and Mixed Strategies

If a player chooses the same alternative in each play of the game we say that he uses a *pure strategy*. Otherwise we say that he uses a *mixed strategy*.

Consider a player P_1 who plays a rectangular game N times. Of these, N_1 times he chooses A_1, N_2 times he chooses A_2, and so on. The fraction of the time that A_1 is chosen is N_1/N, the fraction that A_2 is chosen is N_2/N, etc. We denote these fractions by $x_1, x_2, \ldots$ etc. They are sometimes called *relative frequencies*.

The following relations must hold for the relative frequencies $x_1, x_2, \ldots, x_m$:

(1) . . . $$x_i \geqq 0 \quad \text{for} \quad i = 1, 2, \ldots, m$$

(2) . . . $$\sum_{i=1}^{m} x_i = 1$$

Similarly, if P_2 chooses relative frequencies $y_1, y_2, \ldots, y_r$, we have:

(3) . . . $$y_j \geqq 0 \quad \text{for} \quad j = 1, 2, \ldots, n$$

(4) . . . $$\sum_{j=1}^{n} y_j = 1$$

Using the notion of relative frequencies we can describe a mixed strategy of player P_1 as a vector X, thus:

(5) . . . $$X = (x_1, x_2, \ldots, x_m)$$

Similarly, the vector Y would describe the mixed strategy of player P_2:

(6) . . . $$Y = (y_1, y_2, \ldots, y_n)$$

Whenever P_1 has a pure strategy, i.e., whenever he chooses some alternative $A_i{}^*$ all the time, then the pure strategy vector X_p would be of the form

(7) . . . $$X_p = (0, 0, \ldots, 0, 1, 0, \ldots, 0)$$

where all the x's are zero except $x_i{}^*$ in the j^*th position which is 1. Similarly, the pure strategy vector of P_2 having a pure strategy in which he chooses $B_j{}^*$ all the time is

(8) . . . $$Y_p = (0, 0, \ldots, 0, 1, 0, \ldots, 0)$$

where the 1 occurs in the j^*th position.

Expected Payoff

Let P_1 use a mixed strategy X and let P_2 use a mixed strategy Y. Payoff a_{11} would occur with a relative frequency of x_1y_1. Payoff a_{12} would occur with a relative frequency x_1y_2, and so on. Player P_1 can therefore expect to get

(9) . . . $$\begin{aligned} E(X, Y) &= a_{11}x_1y_1 + a_{12}x_1y_2 + \cdots + a_{mn}x_my_n \\ &= \sum_{i=1}^{m} a_{i1}x_iy_1 + a_{i2}x_iy_2 + \cdots + a_{in}x_iy_n \\ &= \sum_{i=1}^{m}\sum_{j=1}^{n} a_{ij}x_iy_j \end{aligned}$$

The expression for $E(X, Y)$ in (9) is called the *expected payoff*. It is the amount that P_1 can expect to get if he uses a mixed strategy X and P_2 uses a mixed strategy Y. The amount that P_2 will get in this case is $-E(X, Y)$.

Two numerical examples will illustrate how $E(X, Y)$ may be computed in specific cases.

(a) In the river-tale problem assume that Steve chooses "*heads*" 50 per cent of the time and "*tails*" 50 per cent of the time. That is, $X = (0.5, 0.5)$. Assume that the stranger chooses "*heads*" 60 per cent of the time and "*tails*" 40 per cent of the time. That is, $Y = (0.6, 0.4)$. The expected payoff by (9) is

$$(10) \ldots \quad E((0.5, 0.5), (0.6, 0.4)) = -20(0.5)(0.6) + 30(0.5)(0.4) + 10(0.5)(0.6) - 10(0.5)(0.4) = -\$2$$

That is, in this case Steve would be expected to lose, on the average, \$2 per game.

(b) In the two-finger Morra problem assume that P_1 shows one finger and calls "one" $\frac{1}{3}$ of the time, and shows two fingers and calls "two" $\frac{2}{3}$ of the time. That is, $X = (\frac{1}{3}, 0, 0, \frac{2}{3})$. Assume that P_2 has a pure strategy and continuously shows two fingers and calls "one." That is, $Y_p = (0, 0, 1, 0)$. The expected payoff, by (9), is

$$(11) \ldots \quad E(X, Y_p) = 0(\tfrac{1}{3})(0) + 2(\tfrac{1}{3})(0) - 3(\tfrac{1}{3})(1) + \cdots - 3(\tfrac{2}{3})(0) + 4(\tfrac{2}{3})(\tfrac{1}{3}) + 0(\tfrac{2}{3})(0) = \tfrac{5}{3}$$

That is, player P_1 would be expected to win, on the average, $\frac{5}{3}$ per game.

Definition of a Solution

We can now define a solution of a rectangular game.

From P_1's point of view a solution is an optimal strategy $X_o = (x_{1_o}, x_{2_o}, \ldots, x_{m_o})$ which makes $E(X_o, Y)$ of (9) as large as possible. From P_2's point of view a solution is an optimal strategy $Y_o = (y_{1_o}, y_{2_o}, \ldots, y_{n_o})$ which makes $E(X, Y_o)$ as small as possible.

These two points of view can be combined to give a general definition of the solution of rectangular games:

X_o is an optimal strategy for player P_1 and Y_o is an optimal strategy for player P_2 if for any other strategies X and Y relations (12) hold:

$$(12) \ldots \qquad E(X, Y_o) \leqq E(X_o, Y_o) \leqq E(X_o, Y)$$

The expected payoff $E(X_o, Y_o)$ is called the *value* of the *game* and is denoted by v_o:

$$(13) \ldots \qquad v_o = E(X_o, Y_o) = a_{11}\, x_{1o}\, y_{1o} + a_{12}\, x_{1o}\, y_{2o} + \cdots + a_{mn}\, x_{m_o}\, y_{n_o}$$

It can be shown, for example, that (14) and (15) represent a solution of two-finger Morra.

$$(14) \ldots \qquad X_o = (0, \tfrac{4}{7}, \tfrac{3}{7}, 0)$$

$$(15) \ldots \qquad Y_o = (0, \tfrac{3}{5}, \tfrac{2}{5}, 0)$$

The value of two-finger Morra is 0, as would be expected from the symmetric nature of the game:

$$(16) \ldots \qquad v_o = E(0, \tfrac{4}{7}, \tfrac{3}{7}, 0, 0, \tfrac{3}{5}, \tfrac{2}{5}, 0 = 0(0)(0) + 2(0)(\tfrac{3}{5}) + \cdots + 0(\tfrac{4}{7})(\tfrac{3}{5}) + 0(\tfrac{4}{7})(\tfrac{2}{5}) + \cdots + 0(0)(0) = 0$$

Here $E(X, Y_o)$ and $E(X_o, Y)$ are as follows:

$$(17) \ldots \qquad E(X, Y_o) = 2x_1(\tfrac{3}{5}) - 3x_1(\tfrac{2}{5}) - 3x_4(\tfrac{3}{5}) + 4x_4(\tfrac{2}{5}) = -x_4/5$$

$$(18) \ldots \qquad E(X_o, Y) = -2(\tfrac{4}{7})y_1 + 3(\tfrac{4}{7})y_4 + 3(\tfrac{3}{7})y_1 - 4(\tfrac{3}{7})y_4 = y_1/7$$

Thus, no matter what X is, $E(X, Y_o)$ by (17) will always be smaller than or at most equal to zero. Similarly $E(X_o, Y)$ by (18) will always be larger than or at most equal to zero. Hence (16), (17), and (18) indicate that X_o and Y_o of (14) and (15) satisfy (12). That is, $(0, \tfrac{4}{7}, \tfrac{3}{7}, 0)$ and $(0, \tfrac{3}{5}, \tfrac{2}{5}, 0)$ are indeed optimal strategies for the players and constitute a solution of two-finger Morra.[3]

DETERMINATION OF SOLUTION OF RECTANGULAR GAMES

Existence of Solutions

The following two theorems can be shown to be true for rectangular games.

[3] There are other solutions, too. The general solution is $X_o = (0, \tfrac{4}{7} + a/35, \tfrac{3}{7} - a/35, 0)$ where $0 \leqq a \leqq 1$ and $Y_o = (0, \tfrac{4}{7} + b/35, \tfrac{3}{7} - b/35, 0)$ where $0 \leqq b \leqq 1$.

Theorem 1. Every rectangular game has a value; a player of a rectangular game always has an optimal strategy.

Theorem 2. v^*, X^*, and Y^* are respectively the value of a rectangular game and optimal strategies of players P_1 and P_2, only if for every pure strategy X_p of P_1 and for every pure strategy Y_p of P_2 the following relations hold:

(19) . . . $E(X_p, Y^*) \leqq v^*$ for every pure strategy X_p of P_1

(20) . . . $E(X^*, Y_p) \geqq v^*$ for every pure strategy Y_p of P_2

Theorem 1 implies that every rectangular game has a solution and that its value is some finite number. The theorem does *not* say that each game has a unique solution. Two-finger Morra, for example, can be shown to have infinitely many solutions (see footnote 2 on page 451). That is, a player of a rectangular game may have more than one optimal strategy. The value of each game, on the other hand, is unique. Each game has one, and only one, value.

Theorem 2 helps us to check proposed solutions of rectangular games. If relations (19) and (20) hold for some proposed solution v^*, X^*, Y^*, then it is indeed a solution. If they do not hold then the proposed solution is not valid. In the latter case, v^* is not the value of the game, and/or X^* is not an optimal strategy of P_1 and/or Y^* is not an optimal strategy of P_2.

Maxmin and Minmax

The following functions play important roles in the theory of games:

$$\min_j a_{ij},\ \max_i a_{ij},\ \max_i \min_j a_{ij},\ \min_j \max_i a_{ij}$$

$\min_j a_{ij}$ is the smallest element of row i in matrix a_{ij}

$\max_i a_{ij}$ is the largest element of column j in matrix a_{ij}

$\max_i \min_j a_{ij}$ is the largest element of the elements $\min_j a_{ij}$

$\min_j \max_i a_{ij}$ is the smallest element of the elements $\max_i a_{ij}$

In the matrix of Table 4, for example, we have:

$$\min_j a_{1j} = -3,\ \min_j a_{2j} = -2,\ \min_j a_{3j} = -4,\ \min_j a_{4j} = -3$$

and hence

$$\max_i \min_j a_{ij} = \max_i (-3, -2, -4, -3) = -2$$

Similarly,

$$\min_j \max_i a_{ij} = \min_j (\max_i a_{i1}, \max a_{i2}, \max a_{i3}, \max a_{i4})$$
$$= \min_j (3, 2, 4, 3) = 2$$

In an analogous way one can define the functions $\min_Y E(X, Y)$, $\max_X E(X, Y)$, $\max_X \min_Y E(X, Y)$, $\min_Y \max_X E(X, Y)$ where $E(X, Y)$ is given by equation (9).

It can be shown that maxmin a_{ij} and minmax a_{ij} are not always equal. This is true for the example just given pertaining to the matrix of Table 4.

It can also be shown that for any rectangular game

(21) . . . $$\max_X \min_Y E(X, Y) = \min_Y \max_X E(X, Y) = v_o$$

Equation (21) is the basis of Theorem 1 stated above.

How do all these functions relate to the payoffs to players P_1 and P_2?

The least amount that P_1 can be sure of getting is $\min_j a_{ij}$, if he chooses alternative A_i. Therefore, $\max_i \min_j a_{ij}$ is the least amount he can ensure getting in the game. Similarly, $\min_j \max_i a_{ij}$ is the most P_2 can be sure of losing in the game. P_1 could possibly get more than $\max_i \min_j a_{ij}$; P_2 could possibly lose less than $\min_j \max_i a_{ij}$. These results pertain to pure strategies only. Evidently, then, the value of the game v_o is larger than the maxmin that player P_1 can ensure winning, and it is smaller than the minmax that player P_2 can ensure losing.

(22) . . . $$\max_i \min_j a_{ij} \leqq v_o \leqq \min_j \max_i a_{ij}$$

Games with Saddle Points

A study of inequation (22) leads to the following theorem:

Theorem 3. Every rectangular game for which

(23) . . . $$\max_i \min_j a_{ij} = \min_j \max_i a_{ij}$$

has a solution in which the optimal strategies of the players are pure strategies, and the value of the game is given by (23).

Theorem 3 applies to the payoff matrix of Table 12.

Table 12. Payoff Matrix for a Game with a Saddle Point

		P_2					
	Alternative	B_1	B_2	B_3	B_4	B_5	$\min_j a_{ij}$
P_1	A_1	−3	4	−2	0	−1	−3
	A_2	8	5	3	4	4	3
	A_3	3	1	2	1	2	1
	A_4	1	0	−1	2	0	−1
	$\max_i a_{ij}$	8	5	3	4	4	3

$= \max_i \min_j a_{ij}$

$= \min_j \max_i a_{ij}$

The value of the game is therefore $v_o = 3$. The optimal strategy of player P_1 is a pure strategy $X_o = (0, 1, 0, 0)$ and the optimal strategy of player P_2 is a pure strategy $Y_o = (0, 0, 1, 0, 0)$. Every time the game is played, P_1 will choose A_2 and P_2 will choose B_3. P_1 will win three per game and P_2 will lose three per game.

The element a_{23} in the above game has an interesting property. It is, at the same time, the lowest in its row and the highest in its column. Such an element is said to be at a *saddle point* of the payoff matrix a_{ij}.

It can be shown that the following theorem is always true:

Theorem 4. Any rectangular game in which $\max_i \min_j a_{ij} = \min_j \max_i a_{ij} = a_{i_o j_o}$ has a saddle point at (i_o, j_o). And conversely, if (i_o, j_o) is a saddle point, then $\max_i \min_j a_{ij} = \min_j \max_i a_{ij} = a_{i_o j_o}$.

Theorems 3 and 4 thus enable us to find solutions of games with saddle points. We inspect each matrix and check whether it has a saddle point (i.e., if it has an element which is least in its row and highest in its column). If it has such an element, then the value of the game is that element and the corresponding alternatives constitute the pure optimal strategies of the players.

Solution of 2 x 2 Games

We now proceed to discuss solutions of games which do *not* have saddle points. We first discuss 2 x 2 games, i.e., games in which each player has only two alternatives. Then we discuss 2 x n and m x 2 games, i.e., games in which at least one player has two alternatives. Finally we discuss m x n games, i.e., games in which P_1 has m alternatives and P_2 has n alternatives and where m and n are any finite integers.

The general payoff matrix of a 2 x 2 game is given in Table 13.

Table 13. A Payoff Matrix of a 2 x 2 Game

		P_2	
	Alternative	B_1	B_2
P_1	A_1	a_{11}	a_{12}
	A_2	a_{21}	a_{22}

Since the game has no saddle point some specific relations must exist among the four elements a_{ij}. By suitable identification of alternatives and without loss of generality, it can be shown that the relations are:

$$(24) \ldots \qquad \begin{array}{ll} a_{11} > a_{12} & a_{11} > a_{21} \\ a_{22} > a_{12} & a_{22} > a_{21} \end{array}$$

Otherwise a saddle point will exist.

Let the general solution of this game be v_o, $X_o = (x_o, 1 - x_o)$, $Y_o = (y_o, 1 - y_o)$ where $0 \leqq x_o \leqq 1$ and $0 \leqq y_o \leqq 1$. By Theorem 2 we have:

$$(25) \ldots \quad a_{11}y_o + a_{12}(1 - y_o) \leqq v_o \qquad \text{(by (19) for } A_1\text{)}$$
$$(26) \ldots \quad a_{21}y_o + a_{22}(1 - y_o) \leqq v_o \qquad \text{(by (19) for } A_2\text{)}$$
$$(27) \ldots \quad a_{11}x_o + a_{21}(1 - x_o) \geqq v_o \qquad \text{(by (20) for } B_1\text{)}$$
$$(28) \ldots \quad a_{12}x_o + a_{22}(1 - x_o) \geqq v_o \qquad \text{(by (20) for } B_2\text{)}$$

It can be shown that these four inequations have a solution only when the equalities hold. The complete solution is then:

$$(29)\ \ldots\quad v_o = [(a_{11}a_{22}) - (a_{12}a_{21})]/[(a_{11} + a_{22}) - (a_{12} + a_{21})]$$

$$(30)\ \ldots\quad x_o = (a_{22} - a_{21})/[(a_{11} + a_{22}) - (a_{12} + a_{21})]$$

$$(31)\ \ldots\quad y_o = (a_{22} - a_{12})/[(a_{11} + a_{22}) - (a_{12} + a_{21})]$$

This means that player P_1 should use an optimal mixed strategy

$$(32)\ \ldots\quad X_o = \left[\frac{a_{22} - a_{21}}{(a_{11} + a_{22}) - (a_{12} + a_{21})}, \frac{a_{11} - a_{12}}{(a_{11} + a_{22}) - (a_{12} + a_{21})}\right]$$

and that player P_2 should use an optimal mixed strategy

$$(33)\ \ldots\quad Y_o = \left[\frac{a_{22} - a_{12}}{(a_{11} + a_{22}) - (a_{12} + a_{21})}, \frac{a_{11} - a_{21}}{(a_{11} + a_{22}) - (a_{12} + a_{21})}\right]$$

J. D. Williams (15) has a simple method for interpreting equations (32) and (33), the *method of oddments*. He calls the difference between two numbers in a row (or a column) an *oddment*. He calls the ratio of the two relative frequencies of any player *odds*. In these terms equations (32) and (33) can be read as follows:

Player P_1's optimal mixed strategy is to play A_1 and A_2 with the odds of oddment $a_{22} - a_{21}$ to oddment $a_{11} - a_{12}$. Player P_2's optimal mixed strategy is to play B_1 and B_2 with the odds of oddment $a_{22} - a_{12}$ to oddment $a_{11} - a_{21}$.

Applying the method of oddments to the river-tale problem (Table 3) leads to the following results:

The optimal odds for Steve are (−20−10) to (−20−30) or 3 to 5. That is, A_1 should be played $\frac{3}{8}$ of the time and A_2 should be played $\frac{5}{8}$ of the time. Similarly, the optimal odds for the Stranger are 5 to 3 for B_1 and B_2 and the corresponding frequencies are $\frac{5}{8}$ and $\frac{3}{8}$. The value of that game is:

$$(34)\ \ldots\quad v_o = -20(\tfrac{3}{8})(\tfrac{5}{8}) + 30(\tfrac{3}{8})(\tfrac{3}{8}) + 10(\tfrac{5}{8})(\tfrac{5}{8}) - 20(\tfrac{5}{8})(\tfrac{3}{8}) = -\$1.25$$

On the average, then, Steve will lose $1.25 per game even if he uses an optimal strategy.

In concluding the discussion of 2 x 2 games we must add an observation. It can be easily verified in these games that if one player uses an optimal strategy then the value of the game is not changed if the other player does not use an optimal strategy. This is not always true for other games.

Solution of 2 x n *and* m *x 2 Games*

Games in which one of the players has only two alternatives may be solved graphically. We first consider 2 x n games and we use the pay-off matrix in Table 14 as an illustration.

Table 14. A Payoff Matrix of a 2 x 4 Game

		P_2			
	Alternative	B_1	B_2	B_3	B_4
P_1	A_1	11	−2	−4	4
	A_2	−4	3	6	−1

Let x^* be a relative frequency for player P_1 for playing alternative A_1 and let v^* be some value. Then, by inequation (20) of Theorem 2, x^* and v^* are respectively an optimal relative frequency and value of the game if the following four inequations are satisfied.

(35) . . . $$11x^* - 4(1 - x^*) \geqq v^*$$
(36) . . . $$-2x^* + 3(1 - x^*) \geqq v^*$$
(37) . . . $$-4x^* + 6(1 - x^*) \geqq v^*$$
(38) . . . $$4x^* - 1(1 - x^*) \geqq v^*$$

The relative frequency x^* must be larger than or equal to zero and it must be smaller than or equal to one. Inequation (35) thus represents all the points (x^*, v^*) in a plane which are below the line $11x^* - 4(1 - x^*) = v^*$ and which are between the lines $x^* = 0$ and $x^* = 1$ inclusive. All four inequations (35) through (38) together represent the points as in Figure 2.

Figure 2 can be used to find the expected payoffs for various strategies of P_1 and P_2.

Suppose for example that P_1 uses the mixed strategy $X^* = (0.8, 0.2)$. This corresponds to the line $x^* = 0.8$ in Figure 2. If P_2 uses a pure strategy involving only B_1, the expected payoff is 8. Similarly, for pure strategies B_2, B_3 and B_4 the corresponding expected payoffs are −1, −2, and 3 respectively. If P_2 uses a mixed strategy involving

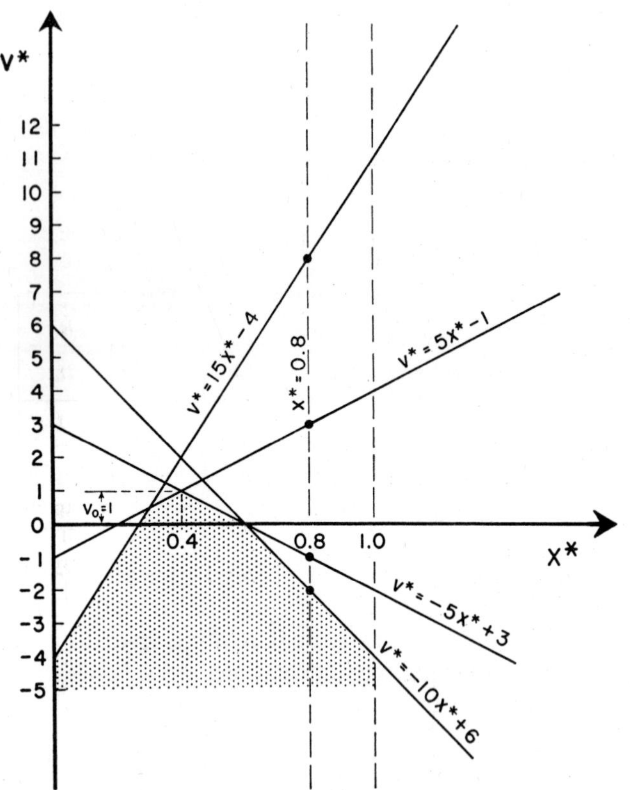

Figure 2. Graphical respresentation of a 2 x 4 game.

B_1 and B_2, say, $Y^* = (0.6, 0.4, 0, 0)$, then the expected payoff would be between 8 and -1: $0.6 \times 8 + 0.4 \times (-1) = 4.8$. It is obvious, then, that if P_1 chooses $X^* = (0.8, 0.2)$, P_2's optimal strategy is the pure strategy involving B_3: $Y^* = (0, 0, 1, 0)$ and P_1's expectation would be an average loss of 2 per game.

Similar analysis indicates that P_1's optimal relative frequency is $x_o = 0.4$ which ensures a gain of at least 1. P_2's optimal strategy should not include alternatives B_1 and B_3; otherwise P_2 will lose more than 1. The solution of the game can therefore be obtained by solving the 2 x 2 game of Table 15 in which only alternatives B_2 and B_4 appear.

Using the oddment method we find the odds for player P_1 are 4 : 6, and for player P_2 they are 5 : 5. The value of the game is 1.

Table 15. The Reduced Payoff Matrix of a 2 x 4 Game

		P_2	
	Alternative	B_2	B_4
P_1	A_1	−2	4
	A_2	3	−1

The complete solution is thus:

$$X_o = (0.4, 0.6)$$
$$Y_o = (0, 0.5, 0, 0.5)$$
$$v_o = 1$$

An analogous approach can be used in solving m x 2 games. We illustrate this using the 5 x 2 game of Table 16.

Table 16. A Payoff Matrix of a 5 x 2 Game

		P_2	
	Alternative	B_1	B_2
P_1	A_1	−15	5
	A_2	−6	4
	A_3	6	1
	A_4	9	−1
	A_5	13	−7

The inequations corresponding to (19) of Theorem 2 are:

(39) . . . $$-15y^* + 5(1 - y^*) \leqq v^*$$
(40) . . . $$-6y^* + 4(1 - y^*) \leqq v^*$$
(41) . . . $$6y^* + 1(1 - y^*) \leqq v^*$$
(42) . . . $$9y^* - 1(1 - y^*) \leqq v^*$$
(43) . . . $$13y^* - 7(1 - y^*) \leqq v^*$$

And the graphical representation of the game is as in Figure 3.

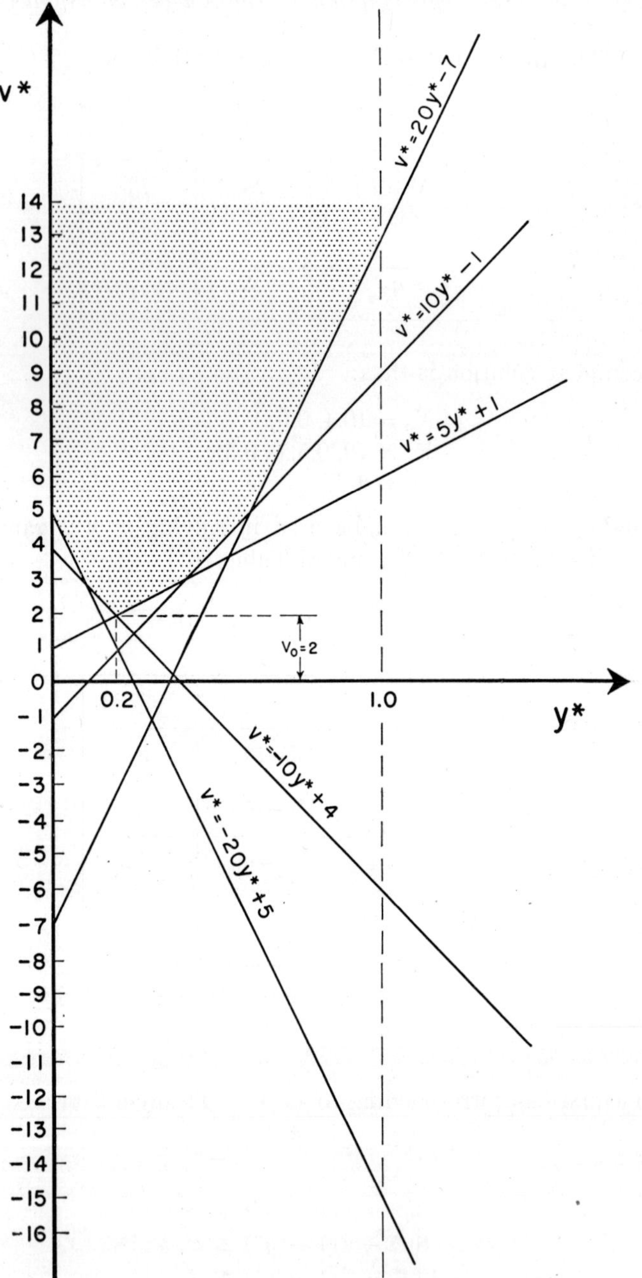

Figure 3. Graphical representation of a 5 x 2 game.

In this case the points satisfying inequations (39) to (43) lie *above* the corresponding lines.

The complete solution is:

$$Y_o = (0.2, 0.8)$$
$$X_o = (0, \tfrac{1}{3}, \tfrac{2}{3}, 0, 0)$$
$$v_o = 2$$

Solution of m x n *Games*

Several methods exist for finding solutions of m x n games. All are rather lengthy and will be described very briefly.

Before attempting to solve an m x n game it is useful to take two steps. First, a check should be made for existence of a saddle point. Second, certain alternatives may be eliminated by checking for *dominance.*

An alternative A_i of P_1 is said to *dominate* an alternative A_j if the payoffs for A_i are larger than or equal to the corresponding payoffs for A_j. In Table 12, for example, alternative A_2 dominates A_1 since the payoffs 8, 5, 3, 3, 4 are respectively greater than the payoffs -3, 4, -2, 0, -1. Similarly, A_2 dominates A_3 and it also dominates A_4.

When A_i dominates A_j, one may disregard A_j when searching for a solution. In the game of Table 12, A_2 dominates all other alternatives; hence a solution for P_1 is $X_o = (0, 1, 0, 0)$.

Similar reasoning may be used for player P_2. But in his case if B_i dominates B_j, then alternative B_i should be disregarded when searching for a solution, since player P_2 is interested in minimizing the payoffs to P_1.

The notion of dominance may be extended to considerations of more than two alternatives at a time. However, a discussion of this aspect is beyond the scope of this chapter. See, for example, references (8) and (10).

ALGEBRAIC SOLUTIONS

The algebraic method for finding solutions of m x n games is based on Theorem 2. General mixed strategy of P_1 and P_2 are $X = (x_1, x_2, \ldots, x_m)$ $Y = (y_1, y_2, \ldots, y_n)$, where $x_i \geqq 0$, $\overset{m}{\Sigma} x_i = 1$, $y_j \geqq 0$, $\overset{n}{\Sigma} y_j = 1$. The problem is to find specific x_i^*, y_j^* and v^* which satisfy these requirements as well as inequations (19) and (20). A process of trial and error eventually would lead to the solution. One tries successively various combinations of equalities and inequalities in (19) and (20) until x_i^*, y_j^* and v^* are found. This process, though very lengthy, is finite and will always lead to a solution by Theorem 1.

MATRIX SOLUTIONS

The matrix method involves a systematic examination of *all* square submatrices of the m x n payoff matrix. By the use of certain elementary matrix operations it is possible to get from each submatrix mixed strategies X^*, Y^* and a value v^*. These are then checked by Theorem 2, and if inequations (19) and (20) hold, X^*, Y^* and v^* constitute a solution.

This method, too, has only a finite number of steps. It ensures that a solution will always be found, and if more than one solution exists, it enables the finding of all solutions.

However, since the number of all square submatrices is rather large even for relatively small m x n games, a great amount of time is required in the use of this method.

FICTITIOUS PLAY SOLUTION

George W. Brown's method of finding solutions by fictitious play (3) is based on a hypothetical series of consecutive plays of the game. In the first play player P_1 chooses A_1 and player P_2 chooses B_1. Thereafter each player chooses at each play the optimum pure strategy against the mixture represented by all the opponent's past plays.

Table 17 represents fictitious play of the river-tale problem.

It can be shown that the value of the game always lies between the highest average payoff to P_2 and the lowest payoff to P_1. In the above example, the highest average payoff to player P_2 occurs in fictitious play 7 and is $-2\frac{6}{7}$. The lowest payoff to player P_1 occurs in plays 5 and 10 and is 0. Hence the value of the game v_o is

$$-2\tfrac{6}{7} \leqq v_o \leqq 0 \qquad (44)$$

By increasing the number of plays the range within which v_o lies can be decreased to any desired accuracy.

It can also be shown that the starred alternatives may *sometimes* give estimates of the optimal strategies. In the above example A_1 is

Table 17. Fictitious Play

		P_2: B_1	P_2: B_2	
P_1	A_1	-20	30	
	A_2	10	-20	
	-20	-20^*	30	1
	-5	-10^*	10	2
	$-3\frac{1}{3}$	0	-10^*	3
	$-7\frac{1}{2}$	10	-30^*	4
	-10	20	-50^*	5
	$-3\frac{1}{3}$	0	-20^*	6
	$-2\frac{6}{7}$	-20^*	10	7
	-5	-40^*	40	8
	$-6\frac{2}{3}$	-60^*	70	9
	-8	-80^*	100	10
	⋮	⋮	⋮	⋮

Average Payoff Per Game	10	10	10	$2\frac{1}{2}$	0	5	$8\frac{4}{7}$	5	$2\frac{2}{9}$	0	...
Total Payoffs Based on Past Plays (A_1)	-20	-40	-60	-30	0^*	30^*	60^*	40^*	20^*	0^*	...
Total Payoffs Based on Past Plays (A_2)	10^*	20^*	30^*	10^*	-10	-30	-50	-40	-30	-20	...
Number of Play	1	2	3	4	5	6	7	8	9	10	...

The first play of A_1 and B_1 results in totals -20, 10 for P_1 and -20, 30 for P_2. In fictitious play number 1, P_1's optimal strategy (marked with *) is A_2 (since $10 > -20$) and P_2's optimal strategy is B_1 (since $-20 < 30$). Hence for the second play, for P_1 the total payoffs based on past plays are $-20 + (-20) = -40$ and $10 + 10 = 20$ and for P_2 they are $-20 + 10 = -10$, $30 - 20 = 10$.

chosen 6 times out of 10 and A_2 is chosen 4 times out of 10. Hence an estimate for X_o is

(45) . . . $$X_o \approx (\tfrac{6}{10}, \tfrac{4}{10})$$

Similarly the estimate for Y_o is

(46) . . . $$Y_o \approx (\tfrac{6}{10}, \tfrac{4}{10})$$

The reader will recall that the exact solution of the river-tale problem is:

$$X_o = (\tfrac{3}{8}, \tfrac{5}{8}), \quad Y_o = (\tfrac{5}{8}, \tfrac{3}{8}), \quad v_o = -1.25.$$

LINEAR PROGRAMMING SOLUTION

It can be shown that every game problem can be transformed into a linear programming problem (4).

We shall illustrate how this can be done for the game of Table 14. The reader should be able to use the illustration as a model for solving *any* other game problem by converting it into a linear programming problem.

We first observe in Table 14 that some of the payoffs are negative. Hence the value of that game could be negative or zero. To use the method of conversion into linear programming, the value of the game should be a positive number. We therefore add a constant 5 to all payoffs of Table 14, giving us Table 18.

Table 18. A Payoff Matrix with a Positive Value of Game

		P_2			
		B_1	B_2	B_3	B_4
P_1	A_1	16	3	1	9
	A_2	1	8	11	4

If the value of the game of Table 18 is g_o, and of the value of the game of Table 14 is v_o, then it can be shown that

(47) . . . $$g_o = v_o + 5 \quad \text{where} \quad g_o > 0$$

It can also be shown that the optimal strategies of the players are not affected when a constant is added to *all* elements of the payoff matrix.

Let $X^* = (x_1^*, x_2^*)$ be a mixed strategy of player P_1 and let g^* be some value. Then, by inequation (19) of Theorem 2, X^* and v^* are respectively the optimal frequency and the value of the game if the following inequations are satisfied.

(48) . . . $$16x_1^* + x_2^* \geqq g^*$$
(49) . . . $$3x_1^* + 8x_2^* \geqq g^*$$
(50) . . . $$x_1^* + 11x_2^* \geqq g^*$$
(51) . . . $$9x_1^* + 4x_2^* \geqq g^*$$

Since x_1^* and x_2^* are relative frequencies we also have

(52) . . . $$x_1^* + x_2^* = 1$$

Since g_o is positive, we can divide each of the inequations (48) to (52) by g^* without affecting the corresponding inequations for the solution sought. This leads to

(53) . . . $$16(x_1^*/g^*) + (x_2^*/g^*) \geqq 1$$
(54) . . . $$3(x_1^*/g^*) + 8(x_2^*/g^*) \geqq 1$$
(55) . . . $$(x_1^*/g^*) + 11(x_2^*/g^*) \geqq 1$$
(56) . . . $$9(x_1^*/g^*) + 4(x_2^*/g^*) \geqq 1$$
(57) . . . $$(x_1^*/g^*) + (x_2^*/g^*) = 1/g^*$$

We now define u_1, u_2 and z, such that

(58) . . . $$u_1 = x_1^*/g^*$$
(59) . . . $$u_2 = u_2^*/g^*$$
(60) . . . $$z = 1/g^*$$

Substitution of (58), (59) and (60) into (53) to (57) leads to

(61) . . . $$16u_1 + u_2 \geqq 1$$
(62) . . . $$3u_1 + 8u_2 \geqq 1$$
(63) . . . $$u_1 + 11u_2 \geqq 1$$
(64) . . . $$9u_1 + 4u_2 \geqq 1$$
(65) . . . $$z = u_1 + u_2$$

Player P_1 is interested in as high a value of game, g_o, as possible. In other words, he is interested in maximizing g^* or, by (60), he is interested in minimizing z. Minimum z, to be denoted by z_o, will thus correspond to g_o.

Inequations (61) to (65) represent a linear programming problem. One has to find u_1 and u_2 which satisfy (61) to (64) which minimize z

in (65). Using linear programming techniques (in general the simplex method, here a graphical method can be used), one obtains the following solution:

(66) . . . $u_{1_o} = \frac{1}{15} \qquad u_{2_o} = \frac{1}{10} \qquad z_o = \frac{1}{6}$

Therefore, by (58), (59) and (60)

(67) . . . $g_o = 6, \qquad x_{1_o} = \frac{2}{5} \qquad x_{2_o} = \frac{3}{5}$

Thus the optimal strategy of P_1 and the value of the game by (67) and (47) are

(68) . . . $X_o = (\frac{2}{5}, \frac{3}{5}) \qquad v_o = 1$

This result checks, of course, with the graphical solution given earlier.

The optimal strategy for P_2 can be found in a similar manner. The resulting linear programming would be the dual of the problem represented by inequations (61) to (65).

TRIAL AND ERROR SOLUTIONS

In concluding this section on methods of solution of m x n games we also have to mention that it is quite all right to try to guess at a solution. Theorem 2 can always be used as a check. If the guessed at solution satisfied inequation (19) and (20), then it is a solution. This is referred to as a trial and error method.

Certain characteristics of the payoff matrix, or some additional information on the problem as a whole, may give clues for guessing. Consider, for example, the payoff matrix of Table 19.

Table 19. A Payoff Matrix of a Game Solved by Trial and Error

		P_2			
		B_1	B_2	B_3	B_4
P_1	A_1	2	1	0	−1
	A_2	−1	2	1	0
	A_3	0	−1	2	1
	A_4	1	0	−1	2

The payoff matrix has certain symmetry characteristics which indicate that both players should adopt mixed strategies $X^* = Y^* = (\frac{1}{4}, \frac{1}{4}, \frac{1}{4}, \frac{1}{4})$. This would imply that the value of the game is $v^* = \frac{1}{2}$. We check this trial solution with (19) and (20) and find that it satisfies the requirements of Theorem 2. Hence it is indeed a solution of the game.

OTHER GAMES

A detailed discussion of other than rectangular games cannot be undertaken within the bounds of one chapter. We therefore present a few illustrations of some non-rectangular games in order to give the reader an indication of the scope of these other games. For further information the reader is referred to references (7), (8) and (10).

Games in Extensive Form

The game with several moves illustrated before (pp. 452-453) is an example of a game in *extensive form*. This game was not stated originally as a rectangular game. Only by suitably defining all the alternatives of the players were we able to transform it into a rectangular game the payoff matrix of which was given in Table 5.

Games in extension form may always be represented graphically as in the tree of Figure 1. They can incorporate chance moves (not controllable directly by the players) and the players may be assigned various degrees of information on preceding moves. Most parlor games (tic-tac-toe, bridge, poker, etc.) are games in extensive form.

It can be shown that every two-person zero-sum game in extensive form can be transformed into an equivalent rectangular game by suitable definition of alternatives. The transformation is generally impractical (even for such a simple game as tic-tac-toe) since the number of alternatives is extremely large. But it is useful to know that transformation can be accomplished. For example, one can show that every two-person zero-sum game in extensive form with complete information has a saddle point. That is, in any such game each player has an optimal pure strategy.

Continuous Games

A game in which the players P_1 and P_2 can choose any number in the interval 0 to 1 and in which the payoff is, say,

$$(69) \ldots \qquad M(x, y) = (x - y)^2$$

where P_1 chooses x and P_2 chooses y, is an example of a *continuous game*. Here the number of alternatives for each player is infinite.

The payoff represented by $M(x, y)$ in (69) can be described graphically as in Figure 4.

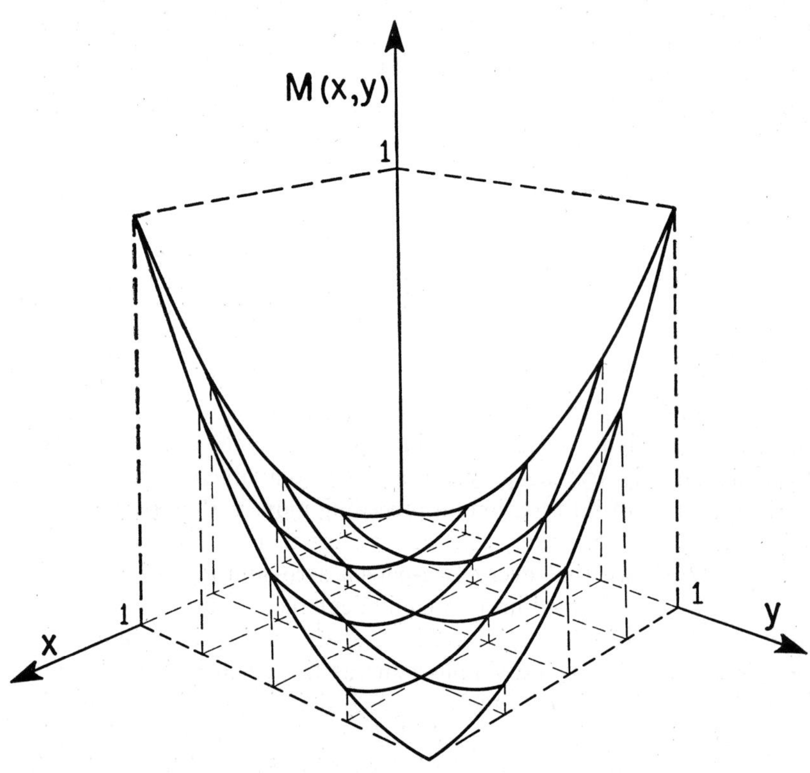

Figure 4. Payoff of a continuous game.

If the players were restricted in their choices of x and y, for example, if they could each only choose the values 0, $\frac{1}{4}$, $\frac{1}{2}$, $\frac{3}{4}$, 1, then the game would be a rectangular game. In that case the payoff matrix would be as in Table 20.

Table 20. The Payoff Matrix of a Restricted Continuous Game

	Alternative	$x \backslash y$	P_2: B_1	B_2	B_3	B_4	B_5
			0	$\frac{1}{4}$	$\frac{1}{2}$	$\frac{3}{4}$	1
P_1	A_1	0	0	$\frac{1}{16}$	$\frac{4}{16}$	$\frac{9}{16}$	$\frac{16}{16}$
	A_2	$\frac{1}{4}$	$\frac{1}{16}$	0	$\frac{1}{16}$	$\frac{4}{16}$	$\frac{9}{16}$
	A_3	$\frac{1}{2}$	$\frac{4}{16}$	$\frac{1}{16}$	0	$\frac{1}{16}$	$\frac{4}{16}$
	A_4	$\frac{3}{4}$	$\frac{9}{16}$	$\frac{4}{16}$	$\frac{1}{16}$	0	$\frac{1}{16}$
	A_5	1	$\frac{16}{16}$	$\frac{9}{16}$	$\frac{4}{16}$	$\frac{1}{16}$	0

For the continuous game without restrictions, it can be shown that P_1's optimal strategy is a mixed strategy and that P_2's optimal strategy is a pure strategy. Namely: P_1 should choose $x = 0$ with a relative frequency of $\frac{1}{2}$ and he should choose $x = 1$ with a relative frequency of $\frac{1}{2}$; P_2 should choose $y = \frac{1}{2}$ all the time. The value of this game is $\frac{1}{4}$. A rough check of this solution can be made by reference to Table 20.

It can be shown that all continuous games have solutions. For certain continuous games with special payoff functions, $M(x, y)$ we have methods for finding the solutions. But no general method yet exists for finding solutions of all continuous games, although we know that they have solutions.

n-Person Zero-Sum Games

The three-person game mentioned earlier (pp. 454-456) is an example of an n-person zero-sum game. No general acceptable definition of a solution of such a game yet exists. That is, we do not yet have an intuitively acceptable definition as to what constitutes a solution of such a game.

One reason for the lack of an acceptable definition is the fact that coalitions can form whenever more than two players participate in a game.

Von Neumann and Morgenstern (10) have suggested an approach which rests heavily on possible formation of coalitions. They suggest the following procedure:

1. First determine the value of all possible two-person games from the three-person game which results through formations of coalitions.

In the three-person game just mentioned there exist three such games, the payoff functions of which are given in Tables 7, 8, and 9. The value of the first game is -1, the value of the second game is 3, and the value of the third game is 2. Let $v(T)$ denote the value of the game when the coalition T plays against all other players in coalition. We then have the following values for $v(T)$:

$$(70) \ldots \quad \begin{array}{ll} v(P_1, P_2) = -1 & v(P_3) = 1 \\ v(P_1, P_3) = 3 & v(P_2) = -3 \\ v(P_2, P_3) = 2 & v(P_1) = -2 \end{array}$$

The function $v(T)$ is called the *characteristic function* of the zero-sum n-person game.

2. Next, observe that player P_1 can assure that he would never lose more than 2, or, stated differently, that he would gain at least -2. Similarly players P_2 and P_3 can ensure that they would gain at least -3 and 1 respectively.

What could in that case be the payoffs to the players as a result of bargaining? Say that P_1 would get x_1, P_2 would get x_2 and P_3 would get x_3. What can we say about these x_1, x_2, x_3?

By previous reasoning, it is obvious that P_1 would never agree that x_1 would be smaller than -2, since by playing by himself he can ensure that he will not lose more than -2. Similar reasoning pertains to P_2 and P_3, thus leading to the first observation about the x_1, x_2, x_3:

$$(71) \ldots \quad x_1 \geqq -2, \qquad x_2 \geqq -3, \qquad x_3 \geqq 1$$

We also recall that this is a zero-sum game. Hence

$$(72) \ldots \quad x_1 + x_2 + x_3 = 1$$

Any set of x_1, x_2, x_3 satisfying (71) and (72) is called an *imputation* and is denoted by $X = (x_1, x_2, x_3)$. We list below a few imputations:

$$X_1 = (-1, -1, 2), \quad X_2 = (0, -2, 2), \quad X_3 = (-1, 2, 3),$$
$$X_4 = (2, -3, 1), \quad X_5 = (0, -1, 1), \quad X_6 = (-2, 1, 1)$$

3. Next note that two players may sometimes prefer one imputation to another. For example, players P_1 and P_3 would prefer X_1 to X_6. We denote this as follows:

$$(73) \ldots \qquad X_1 \underset{P_1, P_3}{\succ} X_6$$

and we read (73) as follows:

"Imputation X_1 *dominates* imputation X_6 with respect to the players P_1 and P_3."

Similarly X_5 dominates X_3 with respect to players P_1 and P_2, and X_2 dominates X_4 with respect to players P_2 and P_3. No dominance relation exists between X_1 and X_2, for example, since there are no two players which prefer one to the other.

4. Finally observe that the following three imputations[4] have rather special properties.

$$(74) \ldots \quad S_1 = (0, -1, 1), \quad S_2 = (0, -3, 3), \quad S_3 = (-2, -1, 3)$$

(i) These imputations do not dominate one another.

(ii) Any *other* imputation is dominated by one of these imputations

(E.g. $S_2 \underset{P_1,P_3}{\succ} X_1$, $S_3 \underset{P_2,P_3}{\succ} X_2$, $S_1 \underset{P_1,P_2}{\succ} X_3$, $S_3 \underset{P_2,P_3}{\succ} X_4$, X_5 is S_1 and $S_2 \underset{P_1,P_3}{\succ} X_6$).

Von Neumann and Morgenstern call the set of imputations S_1, S_2, S_3 a *solution* of the three-person game. According to them, then, a solution is comprised of possible specific payoffs to the players. No two players prefer one of these specific payoffs to another. But one of these specific payoffs would always be preferred by two players to some other payoff not in the set.

The definition of a solution does not tell the players how to find optimal strategies, nor how to bargain with one another. It also does not indicate what the final outcome of the game would be. These are some of the reasons for no universal acceptance of Von Neumann and Morgenstern's definitions of n-person games.

[4] No attempt will be made to show here how these imputations were found. The interested reader is referred to reference (8) for details.

Non-Zero-Sum Games

The closed bidding illustration (pp. 456-457) is an illustration of a non-zero-sum game. For that game, too, there is no general acceptable definition of a solution.

Von Neumann and Morgenstern suggest the following approach:

1. First, introduce a fictitious player, P_3, and assign to him such payoffs that will transform the game into a zero-sum game. For the closed bidding illustration (Table 6) this leads to the payoff of Table 21.

Table 21. Payoffs of a Non-Zero-Sum Game with a Fictitious Player

P_1's Alternatives	A_1	A_1	A_1	A_2	A_2	A_2
P_2's Alternatives	B_1	B_2	B_3	B_1	B_2	B_3
P_3's Alternative	C_1	C_1	C_1	C_1	C_1	C_1
P_1's Payoff	2	$2\frac{1}{2}$	5	3	$1\frac{1}{2}$	3
P_2's Payoff	6	$5\frac{1}{2}$	2	4	$6\frac{1}{2}$	5
P_3's Payoff	-8	-8	-7	-7	-8	-8

The fictitious player P_3 obviously has no choices in this game. We indicated this by assigning to him only one course of action, C_1.

2. The methods of zero-sum games can now be applied to the non-zero-sum game which was extended into a zero-sum game. The only difference is that player P_3 is not considered as an active participant in coalitions. The characteristic function of this game is:

(75) . . . $v(P_1) = 2\frac{1}{4} \quad v(P_2) = 5\frac{2}{3} \quad v(P_3) = -8 \quad v(P_1, P_2) = 8$

Some imputations of this game are:

$$X_1 = (2\tfrac{1}{3}, 5\tfrac{2}{3}, -8) \qquad X_2 = (2\tfrac{7}{24}, 5\tfrac{17}{24}, -8) \qquad X_3 = (2\tfrac{1}{4}, 5\tfrac{3}{4}, -8)$$

The solution of the game (by Von Neumann's and Morgenstern's definition) is any imputation of the form

(76) . . . $S = (2\frac{1}{4} + a, 5\frac{2}{3} + b, -8)$ where $a \geqq 0, b \geqq 0, a + b = \frac{1}{12}$

There are thus infinitely many solutions of the game. As a matter of fact, any imputation (including X_1, X_2, X_3 above) is a solution if we use Von Neumann's and Morgenstern's definition. Result (76) can also be interpreted thus: Player P_1 can ensure a gain of at least $2\frac{1}{4}$; player P_2 can ensure a gain of at least $5\frac{2}{3}$. Both, in coalition, can ensure a gain of 8. Thus $\frac{1}{12}$ [i.e., $8 - (2\frac{1}{4} + 5\frac{2}{3})$] is open for bargaining between the players. P_1 gets the $\frac{1}{12}$ in X_1, P_2 gets it in X_3, and they share it equally in X_2.

CONCLUSION

The theory of games has applications in various branches of science: social science, operations research, statistics, and mathematics.

James R. Newman, in his commentary on the social application of mathematics in *The World of Mathematics* (11, p. 1264) has this to say: ". . . The Theory of Games is today the most promising mathematical tool yet devised for the analysis of man's social relations. . . ."

About Von Neumann's and Morgenstern's book, *Theory of Games and Economic Behaviour* (10), which was first published in 1944, his comments are:

> . . . The authors do not claim to have formulated a full-fledged mathematical theory of society; their book treats only of economic problems, and even in this area both theory and applications are in infancy. But it is a vigorous and promising infancy; the theory of games can fairly be said to have laid the foundation for a systematic and penetrating mathematical treatment of a vast range of problems in social science. (11, p. 1265)

A more recent survey of the theory of games is given by Luce and Raiffa in their book *Games and Decisions* (7), published in 1957. Although the authors indicate that thus far there have been relatively few applications of the theory of games in the social sciences, they make the following interesting observations:

> Judging by physics, the time scale for the impact of theoretical developments is often measured in decades. Second, although the present form of the theory may not be totally satisfactory—in part . . . —this does not necessarily mean that abandoning it is the only possible course for a social scientist. . . . Third, game theory is one of the first examples of an elaborate mathematical development centered solely in the social sciences. . . . (7, pp. 10-11)

Whereas applications of the theory of games in the social sciences are relatively few, this is not the case in application in operations research. The theory is known to have been applied in numerous cases in military operations research. Many of these applications have not been published because of the classified nature of the operations research studies. In industrial operations research the theory has been used in allocation of advertising expenditures (6), in competitive-bidding strategies (5), in solutions of refinery problems (12), in complex managerial decisions (14), to cite just a few examples.

In applying the theory of games to statistics one considers the following two-person game. There is the statistician, player P_1, playing against nature, player P_2. Thus, it is possible to regard statistical theory from a viewpoint of the theory of games. Such an approach has been adopted by Blackwell and Girshick in their book *Theory of Games and Statistical Decisions* (2).

Finally, mention may be made of applications of the theory of games in mathematics. Most of the research work to date which is associated with the theory of games is mathematical in nature. Various branches of mathematics have been used: set theory, topology, theory of matrices, numerical analysis, theory of real variables, etc. It is therefore not surprising that ". . . When known tools were not applicable new mathematics was created. . . ." (7, p. 11)

The interesting relation between the theory of games and linear programming also may be mentioned. We have already pointed out earlier (pp. 474-476) that every theory of games problem can be solved by converting it into a linear programming problem. It can also be shown that every linear programming problem can be solved by converting it into a theory of games problem (4).

From the foregoing it is clear that further applications of the theory of games to the social sciences, to operations research, to statistics, to mathematics, and to other fields may be anticipated in the future.

REFERENCES

(1) Blackwell, David. "Game Theory," *Operations Research for Management*. Vol. I. Edited by Joseph F. McCloskey and Florence N. Trefethen. Baltimore: The Johns Hopkins Press, 1954.

(2) ———, and Girshick, M. A. *Theory of Games and Statistical Decisions*. New York: John Wiley and Sons, 1954.

(3) Brown, George W. "Iterative Solution of Games by Fictitious Play," *Activity Analysis of Production and Allocation,* Cowles Commission Monograph No. 13. Edited by T. C. Koopmans. New York: John Wiley and Sons, 1951.

(4) Dantzig, George B. "A Proof of the Equivalence of the Programming Problem and the Game Problem," *Activity Analysis of Production and Allocation,* Cowles Commission Monograph No. 13. Edited by T. C. Koopmans. New York: John Wiley and Sons, 1951.

(5) Friedman, Lawrence. "A Competitive Bidding Strategy," *Operations Research,* Vol. 4, No. 1 (February, 1956), 104-12.

(6) ———. "Game Theory Models in the Allocation of Advertising Expenditures," *Operations Research,* Vol. 6, No. 5 (September-October, 1958), 699-709.

(7) Luce, R. Duncan, and Raiffa, Howard. *Games and Decisions.* New York: John Wiley and Sons, 1957.

(8) McKinsey, J. C. C. *Introduction to the Theory of Games.* New York: McGraw-Hill, 1952.

(9) Naddor, Eliezer. "The Theory of Games," *Introduction to Operations Research.* Edited by C. W. Churchman, R. L. Ackoff, and E. L. Arnoff. New York: John Wiley and Sons, 1957.

(10) von Neumann, J., and Morgenstern, O. *Theory of Games and Economic Behavior.* Princeton: Princeton University Press, 1947.

(11) Newman, J. R. *The World of Mathematics.* Vol. II. New York: Simon and Schuster, 1956.

(12) Symonds, Gifford H. *Linear Programming: The Solution of Refinery Problems.* New York: Esso Standard Oil Co., 1955.

(13) Vajda, S. *The Theory of Games and Linear Programming.* London: Methuen, Ltd., and New York: John Wiley and Sons, 1956.

(14) Weart, Spencer A. "Practical Application of the Theory of Games to Complex Managerial Decisions," *The Journal of Industrial Engineering,* Vol. VIII, No. 4 (July-August, 1957), 203-9.

(15) Williams, J. D. *The Compleat Strategyst.* New York: McGraw-Hill, 1954.

Seventeen

SYMBOLIC LOGIC

WALTER E. CUSHEN

Traditionally, the methods of operations research that have been applied with some measure of success in the "real" world have relied primarily on the areas of mathematics in which the techniques of solution are relatively clear. However, there is one persistent attribute of larger operational problems: they are a complex of poorly defined variables, related in ways that can be only partially described, and the preferred mode of solution for the whole problem can be described only in the vaguest of terms.

For illustration, consider the celebrated case of the Seabrook Farms Experiment.[1] The problem of getting the best harvest of peas can be asserted to be the criterion against which any alteration in modes of behavior should be measured. However, the variables at play in any given harvest season were multitudinous and non-linear—the weather and associated rate of crop maturity being only two. The essence of the problem was to find the significant variables affecting the harvesting process and to develop means for controlling them or adapting to them. The improved answer was the development of a climatic calendar.

Some years ago it was suggested that the techniques of symbolic logic might be gainfully employed in sorting out the variables that complicate a large problem, and might lead to the means of discover-

[1] Charles Warren Thornthwaite, "Operations Research in Agriculture," *Operations Research for Management,* eds. Joseph F. McCloskey and Florence N. Trefethen (Baltimore: The Johns Hopkins Press, 1954), pp. 368-80.

ing a manageable solution space.[2] There has as yet been no technological breakthrough known to the author that would make this technique manageable, although there have been a number of suggestions for methods of computer solutions of logical problems for large numbers of variables.[3] We are, therefore, still talking about the future of operations research, and not about distinguished accomplishments when we broach the subject of symbolic logic. The persistent problem of sketching in the "logic" of a problem solution remains, with experience and artistry, obviously a better practical bet than rational technique for arriving at a set of guidelines for an operations research group.

To some extent, this must always remain so. The world must always be regarded as dynamic and at least partially irrational. Reason, for at least a large number of metaphysical schools, is descriptive, rather than a reflection of an ontological principle. The saving characteristic of the physical world is that there are enough similarities among events so that repetitions of experiments can be made and physical laws (cast in the form of rigid mathematics) "justified," regardless of one's metaphysical persuasions.

Herein lies both the strength and weakness of logic. Given the assumption that the facts to be described in any given problem behave in a manner that can be rationally described, the most natural procedure is to adopt a set of axioms defining the logic to be applied, and set to work to calculate the natural consequences of the set of assumptions. Unswerving faithfulness to the laws of the selected logic (for there are many possible logics) will produce the last jot of information that can be inferred from the set of assumptions. The complications arise, not so much with the theory of the logical pattern, so far as practical problems are concerned, but with an inability to handle more than a trivial number of variables in the process of calculating the

[2] Walter E. Cushen, "Symbolic Logic in Operations Research," *Operations Research for Management,* eds. Joseph F. McCloskey and Florence N. Trefethen (Baltimore: The Johns Hopkins Press, 1954), pp. 187-202.

[3] Among the various approaches to computer solution of logical problems should be listed Wilton R. Abbot, "Computing Logical Truth with the California Digital Computer," *Mathematical Tables and Other Aids to Computation,* Vol. 5, (July, 1951), 115-84; and Robert S. Ledley, "A Digitalization, Systematization and Formulation of the Theory and Methods of the Propositional Calculus," *National Bureau of Standards Report No. 3363* (Washington, D.C.: U.S. Government Printing Office, 1954).

"answers" to the stated problem. The most customary procedure to adopt in a logical exercise is to assume that the rules of the analysis require that the logic be two-valued—that is, something is either true or false, black or white. A great deal of thought has gone into documenting a demonstration that a two-valued logic is an acceptable assumption. This can be quickly illustrated by showing that points on a value-scale can be defined in terms of greater-than x or not-greater-than x.

Symbolic logic grew and flourished with the substitution of symbols for words. During the early part of the twentieth century there was a growing mathematical school intent upon examining the basis upon which mathematical systems were founded, and an international conference of mathematicians at Paris in 1903 served as a crystallizing influence to bring together a number of persons who later figured largely in the development of symbolic logic. Among them were Bertrand Russell and Alfred North Whitehead, who together produced the three-volume classic in the field: *Principia Mathematica.*[4] These three volumes were predominantly the work of Russell, and served to show that the foundations of arithmetic rested squarely on presuppositions that must be regarded as logical in nature. Whitehead intended to pen a fourth volume that was to have served a similar purpose for geometry,[5] but the rapid growth of interest in non-Euclidean geometry diverted Whitehead from his intended task, and instead he turned to some exploratory systematics in the overlapping regions of geometry and physical theory.[6] Since the appearance of the classic, tremendous progress has been made by a large number of logicians in the various phases of logical theory.

However, the applications of this new field of work to the problem areas of the action world have not been as widespread as the pure theory merits. Symbolic logic is a useful tool in the design of computers, where the elementary elements are the presence or absence of

[4] Alfred North Whitehead and Bertrand Russell, *Principia Mathematica* (Cambridge: Cambridge University Press), Vol. 1 (1910), Vol. 2 (1912), and Vol. 3 (1913).

[5] Bertrand Russell, "Whitehead and *Principia Mathematica,*" *Mind,* Vol. 57 (1948), 138.

[6] Alfred North Whitehead, *An Enquiry Concerning the Principles of Natural Knowledge* (Cambridge: Cambridge University Press, 1919); and Whitehead, *The Principle of Relativity with Applications to Physical Science* (Cambridge, Cambridge University Press, 1922).

an impulse at a given point. The requisite assumptions for a direct application are therefore satisfied. Calculation of the effects of energizing certain circuits and inhibiting others is straightforward. The use of simplifying formulas enables a reduction of the resultant expression to its simplest form. The payoff comes from an easy translation of the results of the calculations—the simplest form describes the circuit with the fewest electronic redundancies, hence the lowest cost in components.

A somewhat more notorious application was made by Kalin and Burkhardt during their student days at the Harvard Law School. The application was a logical machine that would compute the "truth value" of a small set of variables representing the terms of a contract. The implications are clear: if the terms of a contract are mutually consistent, the resultant truth value is "true." If there are inconsistencies, the truth value is "false." Adjustment of the terms of a contract so as to yield a fully consistent pattern is therefore, in theory, possible.

This experiment is reminiscent of an earlier attempt by Leibniz to develop a universal logical calculus. Leibniz' intentions were apparently to facilitate common consent in the ratification of international agreements and treaties. He is reported to have developed a logical calculator, which he demonstrated to the Royal Society of London. But the basic difficulties remained: the systematization of statements about the world of action into a form usable in a logical calculus, and the complications in calculating methods arising from the use of a large number of variables.

Thus, one difficulty is interpretational in nature; the second procedural. The first determines the denotations of statements, and hence the substance of the argument; the second is purely methodological. It is with the second of these difficulties that this chapter is concerned. The interpretational problem is at least as severe, partly due to its qualitative nature. Recent thinking in British philosophical circles has tended to concentrate on this problem, and various facets of the problem can be found in almost any recent issue of *Mind* or *Philosophy*, and in the *Proceedings of the Aristotelian Society*. One of the better books treating of the problem in grand scope is Strawson's *Introduction to Logical Theory*.[7]

[7] P. F. Strawson, *Introduction to Logical Theory* (London: Methuen & Co., 1952).

BASIC NOTATIONS

The rapid recent development of the field of symbolic logic has led to the generation of various competing forms of symbolism for the basic connectives such as "and," "or," and "not." Each author has expressed his own preferences for symbols, and each area of interest in symbolic logic has generally adopted by common consent a form of symbolism for treating the arithmetic of that area of logic. Although there is by no means a universally accepted set of notations, the trend is in that direction, and it may not be many years before all writers on the subject are using the same basic symbol to mean precisely the same expression.

In this chapter there is need for using four basic symbols. The notations used will be those from one of the more common systems.

The considerations in this chapter will be specialized to one of the areas of symbolic logic, that of the calculus of propositions. Although there has been much academic discussion regarding a precise definition of what one means by a "proposition," this chapter will adopt the definition that a proposition is a sentence which says something about a subject. Thus, the sentence, "The musician went fishing with John," will be regarded as being a proposition that says something about the musician. In order to express this proposition in shorthand form, one can simply agree that the letter p will be defined to be equal to the sentence, "The musician went fishing with John," and in all subsequent discourse the appearance of the letter p will mean precisely that.

The second basic definition is that of the negative. Thus if one wishes to say, "It is false that the musician went fishing with John," one simply modified the expression p by putting a negative mark with the symbol. In this development the expression $\bar{p}$ will be used.

With these two definitions established, it can then be stated that the two-valued logic is one in which propositions that are made are either barred or unbarred, can take on no intermediate value between the two, and cannot be both barred and unbarred at the same time and in the same sense.

It is possible in pursuing connected discourse to have a series of connected propositions. It has been shown that all series of connected discourses can eventually be reduced to one of two connectives—"and" or "or." If, therefore, one is talking about two propositions together, p and q, one can symbolize this by simply stating $p \cdot q$.

One can also connect true and false statements in this manner. For instance, that p is true and q is false can be symbolized directly by the expression $p \cdot \bar{q}$.

The final connective that will be used in this chapter is the symbolism for "or." The statement p is true or q is false can be symbolized by the expression $p \vee \bar{q}$. Now the connective "or" can take on one of two forms. It may be either inclusive or exclusive. The "or" symbol that has been explained above is the inclusive "or." This can be illustrated simply by saying that the phrase $p \vee \bar{q}$ means either p is true or q is false, or both. The exclusive "or," on the other hand, would mean either p is true or q is false but not both. The exclusive "or" is symbolized by $\veebar$. It, therefore, becomes possible to express the necessary and sufficient conditions for a two-valued logic by saying $d \veebar \bar{p}$, that is, a proposition is either true or false, but not both at the same time.

Various symbols may be combined in any sequence so that the expression of complicated thoughts becomes symbolically clear. Frequently, parentheses are used to mark off complicated thoughts when they are themselves connected by a simple connective symbol with other complicated thoughts. Accordingly, one can symbolize the following expression by the use of parentheses. "*a* is true and *b* is false or else *c* is true and *d* is false" might take on any one of a number of meanings depending upon the degree to which some of the ideas are compounded. The idea might be rendered $(\mathrm{a} \cdot \bar{\mathrm{b}}) \vee (\mathrm{c} \cdot \bar{\mathrm{d}})$, or it might be rendered $(\mathrm{a}) \cdot (\bar{\mathrm{b}} \vee \mathrm{c}) \cdot \bar{\mathrm{d}}$. The meanings of the two different renditions are quite in contrast with each other. The advantages of a symbolic formulation are therefore immediately apparent, in contrast with the verbalisms of ordinary discourse.

COMPUTATIONAL PROCEDURES

There are a number of ways in which the procedures for carrying out the arithmetic of symbolic logic may be done. Just as in mathematics, there are algebraic transformations and there are matrix representations. One of the interesting methods of computation employs what is known as a "truth table." A truth table is nothing more than a matrix representation of the truth or falsehood value of propositions when they are taken together. Truth tables normally make use of a

1 whenever a proposition is true and a *0* whenever a proposition is false. If one is talking about two propositions, *a* and *b*, it is possible to symbolize the different ways in which *a* and *b* can be combined through the use of this truth table, and furthermore to show under what circumstances those combinations will be true or false when compared with other symbolic expressions. The basic reference table of truth values corresponding to the notations that have been described above are presented in Table 1.

Table 1. Reference Table of Truth Values

a	b	a	$\bar{a}$	a ∨ b	a ⊻ b	a · b
1	1	1	0	1	0	1
1	0	1	0	1	1	0
0	1	0	1	1	1	0
0	0	0	1	0	0	0

First, there are two propositions, *a* and *b*, and each may be either true or false. It follows that there are four possible ways in which *a* and *b* can be combined. These are illustrated to the left-hand side of the double bar. The first entry symbolizes the case in which both *a* is true and *b* is true. The second illustrates the combination *a* is true and *b* is false, and so on. To the right-hand side of the double bar are illustrated the five primitive expressions that have been defined above. The first of them is to show the value of a true proposition. In this case the proposition *a* takes on the truth value 1 whenever *a* to the left of the double bar has the truth value 1, regardless of the truth value of *b*. Thus, it is under the first two combinations, and only the first two combinations that the proposition *a* is true can be stated to have the truth value 1. Under the last two combinations the truth value of the proposition is 0.

The second of the propositions, $\bar{a}$, simply reverses the situation for the first proposition. Thus, the truth value is 0 for the first two cases and 1 for the last two. In the case $a \vee b$, the expression is true whenever either or both *a* and *b* to the left of the double line has a 1 as the truth value. Thus, there are three 1's corresponding to the first three cases under the proposition $a \vee b$. To illustrate the difference between the inclusive and the exclusive "or," the fourth proposition excludes the first case as having a truth value of 1. The first case says that both a and b are true, and this is not consistent with the use of the exclusive "or." The final proposition $a \cdot b$ takes on a truth value 1 only when both a and b are true. Hence, it has a 1 corresponding to the combination illustrated by the first case, and a truth value of 0 for all other cases.

It is apparent that the truth table can be extended to include larger numbers of variables within the limits of the physical capability of listing and computing the truth values of the propositions that would appear in a connected bit of discourse. Normally, calculations by hand bog down long before the number of variables has reached 10. With high speed digital computers the calculations probably become impractical to compute if the number of variables exceeds the number of binary digits in a word of standard computer length. For some computers this is 24; for others, 36, and so on.

This chapter introduces a technique believed by the author to be the first time that this short-cut computational method has been proposed, whereby it becomes more possible for hand computation of truth table values to be made for numbers of variables that are normally at the upper limit of what is practical to compute by hand, and certainly beyond the number of variables used in complicated textbook examples. For purposes of illustrating the method, one further definition will be made. In this case a proposition or set of propositions will take on one of three possible truth values. It may be 1, in which case the proposition is definitely true; it may be 0, in which case the proposition is definitely shown to be false. The third option is the case in which the proposition might conceivably be either true or false but it has not been finally shown which of the two truth values will apply. The symbolism used will be a *10*. In effect, it means that this particular variable is indifferent.

The method being proposed here is used in cases where it is necessary to identify precisely the association of one set of subjects with a second or third set of possible predicates for that subject. It is useful in examining the possible permutations of variables that might serve to make a set of conditions true or false. Further extensions of the matrix representation have not been explored, and it is therefore not proposed that the techniques might have a more general usefulness. It is apparent that the limitations on hand computations using this technique are probably placed at something like double the number of variables that currently limit the practicability of hand solutions.

The method may be described and illustrated by the use of two examples. These two examples are not original with the author, but are of the nature of the puzzle problems that get wide informal circulation in college undergraduate days. The authorship of the examples here used is unknown to this writer.

EXAMPLE NO. 1

There are three men—Jack, John, and Joe—each of whom is engaged in two occupations. Their occupations classify them as two of the following: bootlegger, chauffeur, musician, painter, gardener, and barber.

1. The chauffeur offended the musician by laughing at his long hair.
2. Both the musician and the gardener used to fish with John.
3. The painter bought a quart of gin from the bootlegger.
4. The chauffeur courted the painter's sister.
5. Jack owed the gardener $5.00.
6. Joe beat both Jack and the painter at quoits.

The problem is obviously to associate the names of the men with the two occupations in which they are engaged. By way of illustration of the cautious nature of logical thinking, it is apparent that one further presumption is necessary before a solution can be attempted. The presumption is that each of the three men are engaged in two and only two occupations and that no two men are engaged in the same occupation. Although the presumption is quite natural, it is not explicitly stated. In many practical problems these unstated assumptions are the very ones that attract the analyst to a solution in one direction or another. Frequently, it is only necessary to formulate the problem in order to show the possibilities of exploring a solution space which would not normally occur to a person if he had jumped in the problem without first laying out the basic features, looking for *all* the assumptions, and giving particular attention to those that were not stated.

The possible combinations of Jack, Joe, and John with respect to the occupations listed can be expressed in matrix form. Table 2 shows the propositions that would be associated with any combination. Thus, the proposition a is defined to mean Jack is a bootlegger; h means John is a chauffeur, etc.

Table 2. Propositions

	Bootlegger	Chauffeur	Musician	Painter	Gardener	Barber
Jack	a	b	c	d	e	f
John	g	h	i	j	k	l
Joe	m	n	o	p	q	r

With the initiation of the problem, and prior to the examination of any of the propositions, the truth value of the matrix is one of uncertainty; that is, all the propositions will be assigned value 10. This is illustrated in Table 3.

Table 3. Truth Value Table

	Bootlegger	Chauffeur	Musician	Painter	Gardener	Barber
Jack	10	10	10	10	10	10
John	10	10	10	10	10	10
Joe	10	10	10	10	10	10

The object of the exercise is to change as many of the entries as possible from their present indeterminate form to the more determinate 1 or 0. If there is a unique solution to the problem as stated, all the 10's will disappear and give way to entries that take on the truth value 1 or 0, but not both. If the propositions are inconsistent, one method of reasoning will give a 0, and another will give a 1.

With this kind of representation, the constraints of the problem are easier to apply. The fact that only one person may be involved in any of the occupations means that each of the columns should end up with a single 1 and two 0's in it. Therefore, as soon as it is determined with definiteness that a given occupation should be associated with one person, the natural consequence is that all other entries in that column will take on the value 0. Similarly, the constraint that a given person have two and only two occupations leads to the consequence that as soon as two 1's appear in any row, the remainder of the entries automatically take on the form 0. The reverse is also true. If in any column two 0's are established, the remaining entry must by the nature of the assumptions become 1.

The method of solution will then be to examine each of the propositions from 1 through 6 in an attempt to transform the uncertainties into certainties.

By proposition 6 it is known that j is true, d is false, and p is false. This permits definiteness to appear in the entire fourth column of the matrix as shown in Table 4.

Table 4. Effect of Proposition 6.

	Bootlegger	Chauffeur	Musician	Painter	Gardener	Barber
Jack	10	10	10	0	10	10
John	10	10	10	1	10	10
Joe	10	10	10	0	10	10

Proposition 5 establishes that e is false. The resultant matrix is Table 5.

Table 5.

	Bootlegger	Chauffeur	Musician	Painter	Gardener	Barber
Jack	10	10	10	0	0	10
John	10	10	10	1	10	10
Joe	10	10	10	0	10	10

Note that the identifications to the rows and columns in the matrix are now dropped since they are not essential to the matrix arithmetic, but only to the interpretation of the elements in the matrix.

It has already been established that j is true. By proposition 3 it is known that g cannot be true at the same time as j, and by proposition 4 it is known that h cannot be true at the same time as j. Furthermore, by proposition 2 it is known that i is false and k is false. Application of these four facts results in the matrix of Table 6.

Table 6.

	Bootlegger	Chauffeur	Musician	Painter	Gardener	Barber
Jack	10	10	10	0	0	10
John	0	0	0	1	0	10
Joe	10	10	10	0	10	10

The effects of boundary conditions now come into play. Each row must have two 1's and each column must have a 1. It, therefore, becomes possible to infer that proposition 1 is true, and that proposition q is true, and to insert a 1 in the truth values of those entries. Those facts having been established, by presumption from the boundary conditions, proposition f must be false and r must be false. The results are shown in Table 7.

Table 7.

	Bootlegger	Chauffeur	Musician	Painter	Gardener	Barber
Jack	10	10	10	0	0	0
John	0	0	0	1	0	1
Joe	10	10	10	0	1	0

By proposition 2 it is known that o and q cannot be true at the same time, but q has been established to be true, and therefore o must be false. This in turn makes proposition c true, resulting in the matrix of Table 8.

By proposition 1 it is known that b and c cannot be simultaneously true. Inasmuch as c has already been established to be true, b must

Table 8.

	Bootlegger	Chauffeur	Musician	Painter	Gardener	Barber
Jack	10	10	1	0	0	0
John	0	0	0	1	0	1
Joe	10	10	0	0	1	0

become false. This fact then enables the unique determination of a solution to the set of statements as illustrated in Table 9.

Table 9. Solution

	Bootlegger	Chauffeur	Musician	Painter	Gardener	Barber
Jack	1	0	1	0	0	0
John	0	0	0	1	0	1
Joe	0	1	0	0	1	0

For ease of the interpretation of the results, the identifications for the rows and columns have once again been included. Because of the small number of combinations in this particular problem, the relative savings in effort and time required for solution are marginal.

A second, somewhat more complicated, example will illustrate an occasion on which the matrix method is expanded to a more complicated situation.

EXAMPLE NO. 2

Five men were engaged in a poker game: Brown, Perkins, Turner, Jones, and Riley. Their brands of cigarettes were Luckies, Camels, Kools, Old Golds, and Chesterfields, but not necessarily respectively. At the beginning of the game the number of cigarettes possessed by each of the players was twenty, fifteen, eight, six, and three, but not necessarily respectively. Later in the evening, at a given instant of time when no one was smoking, the following sequence of events transpired:

1. Perkins asked for three cards.
2. Riley had smoked half of his original supply, or one less than Turner smoked.
3. The Chesterfield man originally had as many more, plus half as many more, plus two and a half cigarettes, than he has now.
4. The man who draws to an inside straight absentmindedly lights the tip end of his fifth cigarette, then softly mutters that he would light another, except he has only one left to last the night.
5. The man who smoked Luckies had smoked at least two more than anyone else, including Perkins.
6. Brown drew as many aces as he originally had cigarettes.
7. No one has smoked all his cigarettes, and no one has shared his cigarettes.
8. The Camel man asks Jones to pass Brown's matches.

PROBLEM: How many cigarettes did each man have to begin with, and of what brand?

Once again the problem solution is phrased in terms of the truth table matrix. In this case the statements that are made about each of the men involve possible permutations among two sets of predicates. The two sets of predicates involve the number and brand of cigarettes that each had. There are several unwritten assumptions. This is that no two men smoked the same brand of cigarettes, and that no two men possessed the same number of cigarettes to begin with. The goal once again is to reduce the uncertainties associated with the various predicates and to attempt to determine a unique solution to the problem that will satisfy each of the eight propositions made regarding the five players.

The format for the solution of the problem will be laid out in terms of five matrices, one for each of the five players. In this case the interpretation of the rows and columns will be that shown in Table 10.

At the initiation of the problem, all elements of each of the five matrices are indeterminate, that is, have a 10 as the statement of the undefined truth value of the element of the matrix. If it is determined that Brown has 20 Luckies at the end of the problem, then the solution matrix for Brown will show a 1 in the element corresponding to the first row and first column and 0's throughout the remainder of this matrix. Solutions that are not unique will show 10's in the appropriate matrix elements, and if the propositions are inconsistent, it will be possible to deduce both 1 and 0 for an element.

Table 10.

	20	15	8	6	3
Luckies					
Camels					
Kools					
Old Golds					
Chesterfields					

Proposition 6 leads to the conclusion that Brown originally had three cigarettes. The conclusion is reached by the use of the implicit assumption that there were four aces in the deck of cards to begin with. This conclusion determines the first matrix modification and results in the insertion of 0's throughout the first four columns of Brown's matrix. Inasmuch as it has not been determined what brand of cigarettes Brown has, the last column remains indefinite as in Table 11.

Table 11. Brown's Matrix

	20	15	8	6	3
Luckies	0	0	0	0	10
Camels	0	0	0	0	10
Kools	0	0	0	0	10
Old Golds	0	0	0	0	10
Chesterfields	0	0	0	0	10

By proposition 2 it is known that Riley must have had an even number of cigarettes, hence either 6, 8, or 20. This puts the first definiteness into Riley's matrix, as shown in Table 12.

By proposition 5 it is known that Perkins did not have Luckies. By proposition 8 it is known that Brown did not have Camels, and that Jones did not have Camels. By proposition 3 it is known that the man who had Chesterfields must have had an odd number. By the algebra implicit in proposition 3 it would have also been impossible for the Chesterfield man to have had just three cigarettes and still have an integral number remaining.

Table 12. Riley's Matrix

	20	15	8	6	3
Luckies	10	0	10	10	0
Camels	10	0	10	10	0
Kools	10	0	10	10	0
Old Golds	10	0	10	10	0
Chesterfields	10	0	10	10	0

It has also been determined that Brown has three cigarettes. By virtue of the assumptions at the beginning of the problem this means that no one else could have had cigarettes numbering 3. These various considerations give an intermediate matrix for each of the five people, as shown in Table 13.

Table 13. Intermediate Matrix—I

Brown				
0	0	0	0	10
0	0	0	0	0
0	0	0	0	10
0	0	0	0	10
0	0	0	0	0

Perkins			
0	0	0	0
10	0	10	10
10	0	10	10
10	0	10	10
0	10	0	0

Turner				
10	10	10	10	0
10	10	10	10	0
10	10	10	10	0
10	10	10	10	0
0	10	0	0	0

Jones				
10	0	10	10	0
0	0	0	0	0
10	0	10	10	0
10	0	10	10	0
0	10	0	0	0

Riley				
10	0	10	10	0
10	0	10	10	0
10	0	10	10	0
10	0	10	10	0
0	0	0	0	0

The establishment of the fact that the Chesterfields originally numbered 15, with the result that 10 were smoked, needs to be integrated with the implications of proposition 5. The net result is that the number of Luckies must have been 20 at the start of the evening. With the establishment of the initial Luckies supply at 20 and the requirement of proposition 5, it is deduced that the man who smoked Luckies had smoked at least 12. A comparison of Riley's performance and Turner's performance in terms of the algebra of proposition 2 shows that under no circumstances would Turner have smoked as many as 12 cigarettes. Therefore, Turner did not have Luckies, and, by association, did not have 20 cigarettes. At this stage of development the matrices appear as in Table 14.

Table 14. Intermediate Matrix—II

Brown

0	0	0	0	0
0	0	0	0	0
0	0	0	0	10
0	0	0	0	10
0	0	0	0	0

Perkins

0	0	0	0	0
0	0	10	10	0
0	0	10	10	0
0	0	10	10	0
0	10	0	0	0

Turner

0	0	0	0	0
0	10	10	10	0
0	10	10	10	0
0	10	10	10	0
0	10	10	0	0

Jones

10	0	0	0	0
0	0	0	0	0
0	0	10	10	0
0	0	10	10	0
0	10	0	0	0

Riley

10	0	0	0	0
0	0	10	10	0
0	0	10	10	0
0	0	10	10	0
0	0	0	0	0

A further example of a hidden assumption can be discovered by careful reference to proposition 4. This hidden assumption also dates the problem as having its origin at a time earlier than the recent broadening of choices in the cigarette field. The presumption is that Kools is the only *filter*-tip or cork-tip cigarette. Proposition 4 requires that the number of Kools be at least five, and since Brown had only three cigarettes to begin with it must be presumed that Brown, therefore, did not have Kools. This assumption determines Brown's matrix uniquely as being three Old Golds. Furthermore, since no one else could have Old Golds, it establishes the 0 truth value nature of the fourth row of every one else's matrix.

The ambiguity with respect to who had 20 Luckies resolves itself now into a competition between two people, Jones and Riley. If Riley had had the 20 Luckies, then by proposition 2 he would have had to smoke 10 since Turner had the 15 Chesterfields. Proposition 2 would have required that Turner smoke 11 of them. By proposition 3, however, it has been established that the Chesterfield man had smoked 10.

Therefore, to adopt the solution that Riley had the 20 Luckies would be a solution inconsistent with the premises of the problem, and therefore, Riley must be presumed not to have had the 20 Luckies.

Inasmuch as someone had to have the 20 Luckies and there is only one other candidate—Jones—it must be concluded that Jones had the 20 Luckies. At this stage of the argument the matrices for the various participants in the poker game is that of Table 15.

Table 15. Intermediate Matrix—III

Brown						Perkins						Turner				
0	0	0	0	0		0	0	0	0	0		0	0	0	0	0
0	0	0	0	0		0	0	10	10	0		0	10	10	10	0
0	0	0	0	0		0	0	10	10	0		0	10	10	10	0
0	0	0	0	1		0	0	0	0	0		0	0	0	0	0
0	0	0	0	0		0	10	0	0	0		0	10	10	0	0

Jones						Riley				
1	0	0	0	0		0	0	0	0	0
0	0	0	0	0		0	0	10	10	0
0	0	0	0	0		0	0	10	10	0
0	0	0	0	0		0	0	0	0	0
0	0	0	0	0		0	0	0	0	0

The next elimination involves the determination of who had the 15 Chesterfields. Reference to the preceding matrix shows that this competition exists only between Perkins and Turner. The reasoning will be as before, that is, the determination of the consistency of a hypothesis with the remaining propositions. If Turner had had the 15, it would have meant that he must have smoked 10, and that this must be one more than Riley. Therefore, if Turner had smoked 10, Riley must have smoked 9. This, however, is inconsistent with the options open to Riley: Riley had either 6 or 8 cigarettes. Therefore, it cannot be that Turner had the 15. A reference to the algebra of the other situations shows that the hypothesis that Perkins had the 15 Chesterfields is consistent. Therefore, the matrix for Perkins is determined to be that he had the 15 Chesterfields.

At this juncture of the argument there are 4 undetermined elements in the matrices of Turner and Riley. The argument centers around the permutations possible between 6 and 8 cigarettes and the Camels and Kools brands.

Now proposition 2 requires that Turner had smoked one cigarette more than Riley. Proposition 4 requires that someone had to smoke 5 cigarettes. If Riley had had 8 cigarettes to begin with, proposition 2 would require that Riley had smoked 4. If Riley had had 6 cigarettes proposition 2 would have required that he smoke 3. Since someone had smoked 5, it must therefore have been Turner. One is therefore forced to associate Turner with the Kools. If Turner then smoked 5 Kools and by proposition 2 was required to smoke one more than Riley then Riley must have smoked 4. Four cigarettes indeed would have been half of an original allocation of 8 cigarettes. Therefore, the net result is that Riley must have had 8 Camels to begin with and Turner 6 Kools.

The problem therefore does have a unique solution. Brown had 3 Old Golds. Perkins had 15 Chesterfields. Turner had 6 Kools. Jones had 20 Luckies, and Riley had 8 Camels.

GENERALIZATION

A number of general principles can now be demonstrated by reference to the example of the poker game. If it had not been possible to resolve the decision with respect to the ambiguity between the original assets of Riley and Perkins, it would have meant that there were four possible solutions to this particular set of assumptions. Imagine now a situation which centers not around the poker game but around a contract. Any of the four solutions would have been legally possible. A person who is preparing the contract therefore would be interested in knowing if another term should be added to the contract which would rule out a possibility that was believed to be undesirable. The methods shown here would have isolated the ambiguities in the terms of the proposed contract, providing, of course, that it could be reduced to the small number of variables handled in this way.

Presume on the other hand that the contract had already been agreed upon and that a company was following course-of-action A all within the terms of the contract. If course-of-action B was also possible and preferable it could be initiated within the terms of the same contract, providing the matrix showed the choice between A and B to be indifferent with respect to the truth value of the contract.

As a third general principle, it can be stated that it is possible to loosen, rather than tighten, the terms implicit in these propositions. Thus, if the unique solution achieved as a result of the propositions that have been stated prove to be unacceptable, then by successive relaxation of each of the propositions it is possible to determine where the uniqueness can be lost. In the case of a contract this is the phrase that should be eliminated, and a second clause inserted which would block the solution that has been believed to be undesirable.

The discussion of this contribution to the volume at hand was initiated with reference to the complicated problems that are frequently phrased in ambiguous terms and often have indeterminate or probabilistic parameters at work. Referring once again to the solution to the poker game problem at the juncture just before the final solution, in which there remained some ambiguity between the possibilities open to Riley and Turner, imagine that the statement of the propositions were such that this final determination could not be made on logical grounds. This would have meant that there were four possible solutions to the problem. If, again, instead of a poker game the problem had been one of determining an optimum cost effectiveness relationship, the merit of this approach would have been to reduce the problem from one of a large number of possible permutations to one that involves a concentration on two people, two brands of cigarettes, and two initial numbers to start the game. In that event the methods of symbolic logic employed here would have served a highly useful purpose. It would have isolated the critical parts of the problem that were in doubt and would have permitted attention to be focused on the solution space about which there was some ambiguity. Ideally, then, the problem should have been handed to an operations analysis group, which would have compared the respective merits and demerits of the four solutions that were available within the constraints specified by the propositions that acted as boundary conditions.

APPLICATIONS

The search for case studies epitomizing the unique values of symbolic logic is a barren one. Reference has already been made to its pertinence in problems of circuit design and legal inference. What is generally lacking is a typical operations research problem, identifiable as such, in which the play of logical reasoning suffices to solve any but the simplest problems.

Certain problem areas appear to possess characteristics that might lend themselves to logical analysis. For example, the action matrix that is implied in the pronouncements of the Kremlin could hardly fail to yield interesting, albeit expected, inconsistencies. It might also lead to instructive ambiguity-situations: are the poorly defined situations the most likely locales for future exploitation, and does the remainder of the action matrix offer information on what the action is likely to be? Does it also suggest to the user some courses of action that would give greater chance of counter-success?

Basically, however, logicians have been preoccupied with form, and only passingly with the content of the arguments. Logic in itself needs no apology; it is worshipped in every decision-making process. What is needed is some assurance that the demand for logic bears some real-world meaning. The important search is therefore one of attempting to identify and isolate those situations in which a rational pattern is traceable.

The search is complicated by the fact that actions are not always reasonable, when measured by some given pattern. People make errors in applying a well-known system of thought. Further, other people apply systems of thought that are not like our own; this does not mean that their actions cannot be rationally described. Finally, it is not clear that logic itself is unchangeable. There is a considerable volume of metaphysical discourse documenting the argument that an unchanging logical pattern underlies all of nature. Recent philosophical trends have pointed to the possibility that logical relations exist as a result of events, rather than prior to them. Evolution appears to be a fact, rather than a recombination of terms. Society discovers the urgency of the value of the individual, and in this discovery creates situations for the achievement of previously unthinkable goals.

The pertinence of symbolic logic to operations research? Every time the solution to an operational problem is discarded because the constraints of the situation prevent its implementation, we have a reassertation of the thesis that something is violating a logical requirement. The obvious and needed problem solutions in effectiveness, and the urgent cost comparisons have made their mark. The time is now here when the operations analyst must begin to examine the full context of the operations he is studying in an attempt to discover what kinds of problems need to be solved. He must devote attention to the question of the interplay of factors that narrow his solution space, with a view to the possible variation of those assumptions. Although problem solutions are required, identification of the lucrative problems may be more important.A cost-effectiveness study of a mode of operation is a useful pursuit. But it may be secondary to answering the question of whether that mode of operations is still competitive with other modes in the face of newer assumptions. In such discourses, the reliance on logic is apparent. Symbolic logic should be given its chance to assist, because these are the conditions most favorable to the growth of this high-risk, high-gain discipline.

Eighteen

THE DESIGN OF EXPERIMENTS

WILLIAM G. COCHRAN

INTRODUCTION

In ordinary speech, the word "experiment" has a broad meaning, covering anything that involves trying out something new. In the subject that has come to be known as the design of experiments, the word is used in a narrower sense. The essence of an experiment is that we deliberately apply two or more procedures, called the *treatments,* to some process or operation for the purpose of measuring and comparing their effects. In the preparation of alloys, for instance, the objective of an experiment might be to compare the effects of different concentrations of carbon, or different firing temperatures, or different types of furnaces, or different types of molds into which the molten effluent is poured, on the important qualities of the resulting alloy. The comparative experiment is one of the most powerful weapons available to the scientist. If successful, it provides trustworthy factual information which advances our understanding and contributes to sound practical decisions.

Systematic study of the principles involved in the planning and execution of experiments was initiated over thirty years ago by R. A. (now Sir Ronald) Fisher, with particular reference to agricultural experimentation. There is now a large literature, most of it concerned with the design of a *single* experiment. More recently, attention has shifted to the problem of planning an experimental program in exploratory or developmental research. In this, a series of experiments is to be conducted to discover which of a number of factors or variables have important effects on some response, and to learn something about the nature of these effects.

This chapter contains two main parts. The first presents a brief summary of some principles that should guide the conduct of a single experiment in which the treatments to be compared have already been chosen. In the second, several strategies that have been proposed for carrying out an experimental program will be discussed.

PART I. THE PLANNING OF A SINGLE EXPERIMENT

Variability and Its Effects

An old recipe for an ideal experiment is to keep everything constant except the imposed differences in treatments. If this could be done, any differences in results could be ascribed unequivocally to the differences in treatment. In most lines of experimentation, however, it is either impossible or prohibitively expensive to keep all other factors constant that might influence the results. Variability will be present in the raw materials used, in the state of the environment, in the operations required for the application of the treatments, and in the process of measurement. Thus the *observed* effect of a treatment is subject to an experimental error which represents the joint effect of all these other sources of variability.

Experimental errors have two undesirable consequences. They may produce biases or systematic errors in the estimates of the treatment effects. With two treatments, for example, suppose that each treatment is to be tried six times: in other words, to use a common technical term, the experiment is to have six *replications.* If six applications of a treatment can be processed and the results measured in a single day, one way of doing the experiment is to complete the six replications of treatment 1 on Monday, and to carry out those for treatment 2 on Tuesday. There might, however, be an unanticipated shift in the level of the responses from Monday to Tuesday, due to the use of a new batch of raw material, to a change in humidity, or to a zero error in a reading instrument. If this occurred, all six results from treatment 2 might be found to lie above those for treatment 1, suggesting a consistent superiority of treatment 2, although in fact the differences were caused by the diurnal shift in level.

When a bias of this type is present, the standard methods for the statistical analysis of the results of experiments are incapable of detecting or of correcting for it. Tests of significance and confidence limits become misleading. Biases confuse experimenter and statistician alike.

Experimental errors also produce fluctuations of a more erratic or "random" nature in the results. If several replications are made, these fluctuating errors cause the observed differences in effect between two treatments to vary from one replication to another. Looking over the experiment as a whole, we may not be clear as to what the true size of difference is, nor even whether the superiority of one treatment has been established.

Fluctuating errors are less serious than systematic errors in two respects. Since they do not consistently favor any specific treatment, their effects tend to cancel out over a series of replications. Further, the standard methods of statistical analysis measure the extent to which the estimated treatment effects are influenced by fluctuating errors, so that the experimenter is made aware of the degree of uncertainty in the results. This is small consolation, however, if fluctuating errors in an experiment are so large that no definite conclusions can be drawn.

Some General Principles

Three principles which flow obviously from this analysis of the effects of variability may be stated as follows:

1. To take any precautions necessary to reduce systematic errors to a negligible size.
2. To ensure that, despite the presence of fluctuating errors, the treatment effects are estimated with precision adequate for the purpose at hand.
3. To measure the amount of uncertainty that exists in the estimated treatment effects because of experimental errors, and to take account of this uncertainty when drawing conclusions from the results.

In biological research, in which experimental errors are often large, these principles have long been regarded as essential to sound experimentation. In the industrial field, experimental errors tend to be lower, and it has sometimes been thought that we need not worry about them, because any treatment effects that are large enough to be of practical interest will show up clearly despite the errors. While this may be true in certain experiments, there are important segments of industrial research in which experimental errors are just as high as in biological research.

In the next two sections, these principles will be discussed in more detail.

Precautions Against Systematic Errors

As we have seen, a systematic error is created when some source of variation tends to favor or handicap a specific treatment persistently in all replications. The first step in guarding against such errors is to check over the operations involved in the practical conduct of the experiment, in order to detect whether bias is likely to enter at any stage. While the opportunities for bias vary with the type of experiment, there are several vulnerable stages.

1. In the assignment of raw material to the treatments. For instance, if the raw material required for the replications of treatment 1 is allotted first, then that for treatment 2, and so on, it may happen that treatment 1 receives the most responsive batch of raw material, if the experimenter, without realizing it, picks out responsive material first.
2. In the order in which operations are carried out on the treatments. Unsuspected time trends are sometimes present in laboratory and factory procedures.
3. In the spatial allotment of treatments. There may be systematic differences between different shelves in an oven or different parts of a water bath.
4. In the process of measurement, particularly if this is wholly or partly subjective.

With regard to the first three sources of bias, a useful precaution is to make the allotment in question by the use of a table of random numbers. The unsystematic nature of random numbers guards against the persistent repetition of favorable conditions for any one treatment. In other words, randomization helps to ensure that the experimental errors will be fluctuating rather than systematic. A second useful device is that known as blocking or grouping (p. 513). This may be more effective than randomization, in that it may cause certain systematic sources of error to cancel out entirely. In references (2) and (3) numerous combinations of randomization and blocking, appropriate for different experimental situations, are described.

Where measurements are subjective, a standard precaution is to prevent the person making the measurements from knowing which treatment is being measured at any given time. Presentation of treatments in a randomized order is also advisable, in case the measurer has a drift or trend in his measurement errors.

Methods for Increasing Precision

The standard deviation of the experimental results, computed by the usual methods, provides a measure of the size of the fluctuating errors to which the results are subject. If the standard deviation is undesirably large, we should try to discover at what stages in the experiment the major sources of variability arise. Sometimes this requires a separate and time-consuming investigation of the variability of the raw material or of the precision of measuring devices.

The principal devices for increasing the precision of the final results are outlined below.

Refinements of technique. This covers such actions as the procurement of more uniform raw material or more accurate instruments of measurement, standardization of the environment, e.g., by humidity or temperature controls, and the development of small-scale methods of experimentation that can be done with high precision. Sometimes experimentation cannot proceed with any hope of getting results unless such refinements are made. Their disadvantages are that the refinements may be costly, or that, as in the case of small-scale research, the conditions of experimentation come to differ markedly from those to which it is hoped to apply the final results. For this reason, as is well known, small-scale research is often used primarily to discard bad ideas and screen out a few promising treatments that will be tested later under more realistic conditions.

Replication. Even with a high standard deviation, satisfactory precision in the treatment means can be obtained if the treatments are replicated often enough, with precautions such as randomization to avoid systematic errors. When experiments must be done under "practical" conditions, so that refinements of technique cannot be introduced, replication is sometimes the most useful weapon. Its disadvantage is the cost involved: since the standard error of the average result for a treatment is $\sigma/\sqrt{r}$, where r is the number of replications and σ the standard deviation, the standard error decreases rather slowly as r is increased.

Auxiliary measurements. In comparing different commercial feeds for chickens or hogs, the amount by which an animal increases in weight during a given period is usually correlated with its initial weight. The initial weight is an auxiliary measurement that predicts to some extent the final response (i.e., weight) of an animal. By a statistical technique known as the analysis of covariance, these auxiliary measurements can increase the precision of the final responses, the idea being to adjust the final weights so as to remove the effects of variations from animal to animal in initial weights. In general terms, covariance removes the error arising from some source of environmental variation that can be measured by the auxiliary variate but not conveniently controlled.

Blocking. By a careful arrangement of the way in which the experiment is done, potential sources of error can often be balanced out so that they affect all treatments equally. Two illustrations will be given. McGehee and Gardner (1) conducted an experiment to measure the effect of factory music programs on the productivity of women engaged in a repetitive operation in rug manufacturing. Four types of musical program (*A, B, C, D*) were compared with no music (*E*), a different program being tested each day. The schedule for the experiment was as follows:

Week	Mon.	Tues.	Wed.	Thurs.	Fri.
1	*A*	*B*	*C*	*D*	*E*
2	*B*	*C*	*D*	*E*	*A*
3	*C*	*A*	*E*	*B*	*D*
4	*D*	*E*	*A*	*C*	*B*
5	*E*	*D*	*B*	*A*	*C*

The features of interest are that each program appears once on each day of the week, and once during any given week, the arrangement of letters being known as a 5 x 5 latin square. Thus, if productivity is consistently higher on certain days of the week than on others, these differences will cancel when two programs are compared. The same remarks apply to consistent differences in productivity from week to week. A grouping of this kind, such that each treatment occurs equally often in the group, is often called a *block,* the term being borrowed from agricultural experimentation. This latin square employs double blocking, the blocks being respectively days of the week and weeks.

The following slightly more complex arrangement (Table 1) might be used to compare the durabilities of, say, types of leather or rubber on a machine that produces artificial wearing. The machine takes only two specimens in a single run.

Table 1. Incomplete Block Design Used to Compare Five Types of Material on a Wear Machine

	Run number									
	1	2	3	4	5	6	7	8	9	10
Right	*A*	*B*	*C*	*D*	*E*	*A*	*B*	*C*	*D*	*E*
Left	*B*	*C*	*D*	*E*	*A*	*C*	*D*	*E*	*A*	*B*

Two possible sources of error are that there may be a consistent difference between the amounts of wear on the right and left sides of the machine, and that there may be consistent differences in wear from run to run. The arrangement balances out the "right-left" difference by having each type (*A*, *B*, *C*, *D*, and *E*) run twice on the right side and twice on the left. The differences between runs cannot be balanced in this way, because each run accommodates only two specimens. It may be verified, however, that every pair of letters occurs just once in the same run. It follows that the difference between the average results for type *A* and the average results of its partners in the same runs (i.e., runs 1, 5, 6, and 9) is an estimate of the average performance of *A* relative to that of the four other types. From these figures it is easy to obtain comparable estimates of the average performances of the five types, from which any consistent run to run differences have been eliminated. The arrangement is one of a class known as incomplete block designs, the "run" being the incomplete block.[1]

The chief advantage of blocking is that frequently it involves little or no extra cost in the conduct of the experiment, so that even a modest increase in precision from blocking is all to the good. A considerable variety of designs utilizing blocking will be found in references (2) and (3).

Finally, none of the techniques described above has any inherent superiority over the others. In a specific situation the experimenter should utilize those that promise the best return for a given expendi-

[1] For this example the author is indebted to Dr. W. J. Youden.

ture of his resources. His knowledge about the primary sources of variability will, of course, be helpful in making this choice.

With regard to the third principle, i.e., that account must be taken of the experimental errors when drawing conclusions from the results, the standard methods of statistical analysis are now well developed to cope with this problem, provided that systematic errors are absent.

PART II. THE DESIGN OF AN EXPERIMENTAL PROGRAM

Under discussion next are some of the strategies that have been proposed for investigating the effects of a number of different factors or variables on one or more responses. The number of factors may vary from two or three to as high as twenty or more. The factors may be quantitative, like temperature or time, or qualitative, like type of leather or shape of mold. For each quantitative factor there will be a known range within which the factor may be varied in the experiments.

In this type of problem the demands of fundamental research are usually more taxing than those of applied research. In fundamental research the scientist is likely to wish to learn about the interrelationships among the effects of the factors throughout the whole of the range of each quantitative factor and for all variants of each qualitative factor. In applied research the objective is often the more restricted one of finding the levels at which the factors must be set in order to obtain a high response. With this goal we do not need to spend time investigating regions in which the response is low, and we can discard quickly variables that have only minor effects.

Factorial Experimentation

Perhaps the simplest method of attack is to run a separate experiment for each factor, having as many experiments as there are factors. As an alternative, Fisher pointed out that if all factors can be studied simultaneously in a single factorial experiment, this makes an economical use of resources and provides appropriate data for investigating the interrelationships among the effects of the factors. The point can be illustrated by a program which contains three factors, A, B and C, each to be investigated at only two levels, say low and high.

To give physical meaning to the explanation, we may consider as an example the three factors, A, B, and C, as the nitrogen, phosphorus, and potassium content of a fertilizer, each comprising 5 per cent of the fertilizer when at the low level and 10 per cent at the high. Each observation for a particular combination of factors would be a measure of the yield of a plot fertilized with a specified weight of fertilizer and quantity of seed.

In the experimental plans (Tables 2 and 3), the presence of a letter signifies that the corresponding factor is held at its high level; the absence of the letter denotes the low level. Thus, the symbol ac implies that factors A and C are being held at their high levels, while B is held at its low level. The treatment combination in which all three factors are at their low levels is usually denoted by the symbol (1). In the example, then, combination (1) is a 5-5-5 fertilizer, while at the other extreme, abc denotes the 10-10-10 mix.

Assuming that four replications are considered advisable, Table 2 shows how the three experiments might be conducted by the "one factor at a time" method. Note that in the first experiment, which is devoted to factor A, some decision must be made about the levels at which B and C are fixed during this experiment. We have supposed that B is held at its high level and C at its low level, so that Experiment I compares the treatment combination b with ab. Similar choices must be made in the second and third experiments. It is assumed that the high level of A was superior in Experiment I and that this level was used in Experiment II, ab being used in Experiment III for the same reasons. In Table 2, each letter is repeated four times to signify fourfold replication.

Table 2. Illustration of Single-Factor Experiments

Exp. I	(Factor A)	Exp. II	(Factor B)	Exp. III	(Factor C)
b	ab	a	ab	ab	abc
b	ab	a	ab	ab	abc
b	ab	a	ab	ab	abc
b	ab	a	ab	ab	abc

This series of experiments requires twenty-four treatment combinations to be tested. Its information about relations between the effects of the factors is weak. If someone asks, "Is the effect of A the same at low and high levels of C?", the series provides no data to answer the question, since A was tested only at the low level of C in Experiment I.

In factoral experimentation, all treatment combinations that can be made up from the low and high levels of each factor are compared simultaneously in a single experiment, which contains 2^3 or 8 combinations. Table 3 shows these combinations and illustrates a standard method of analyzing the results.

Table 3. A 2^3 Factorial Experiment

Factorial effect	Treatment combination								Divisor
	(1)	a	b	ab	c	ac	bc	abc	
A	−	+	−	+	−	+	−	+	4
B	−	−	+	+	−	−	+	+	4
C	−	−	−	−	+	+	+	+	4
AB	+	−	−	+	+	−	−	+	4
AC	+	−	+	−	−	+	−	+	4
BC	+	+	−	−	−	−	+	+	4
ABC	−	+	+	−	+	−	−	+	4

The eight treatment combinations provide four little experiments on the effect of factor A, namely (1) vs. a; b vs. ab; c vs. ac; and bc vs. abc. The average of the results of these four comparisons gives an estimate of the effect of A (high level vs. low level) based on four replications. The succession of − and + signs required to compute this average is shown in row A of Table 3.

Similarly, the factorial experiment provides four separate tests of the effect of B; (1) vs. b; a vs. ab; c vs. bc; and ac vs. abc. The succession of signs needed to compute this effect, with divisor 4 to obtain the average, appears in row B. In the same way the experiment gives four replicates for the average effect of C. These average effects are called the *main effects* of the factors.

Thus, by testing eight combinations in a single factorial experiment, we obtain fourfold replication for the average effects of each factor, whereas twenty-four combinations had to be tested by the single-factor approach. The secret of the high efficiency of the factorial method is that every treatment combination in the experiment supplies information about the effects of all three factors.

To revert to the question "Is the effect of A the same at low and high levels of C?", the factorial experiment throws some light on the answer. The comparisons

$$abc - bc \qquad \text{and} \qquad ac - c$$

are estimates of the effect of A at the high level of C, while the comparisons

$$ab - b \qquad \text{and} \qquad a - (1)$$

are estimates of the effect of A at the low level of C. Consequently the quantity

$$\left\{\frac{abc - bc + ac - c}{2}\right\} - \left\{\frac{ab - b + a - (1)}{2}\right\}$$

estimates the difference between the average effect of A for high C and the average effect of A for low C. This expression is called the AC interaction. (By a convention which will not be explained here, a divisor 4 is generally used instead of 2.) The series of $-$ and $+$ signs for computing the AC interaction is shown in the row labeled AC in Table 3. By the same process, AB and BC interactions may be computed. We can go further and examine whether the AC interaction is the same at low and high levels of B, since the experiment furnishes separate estimates of the AC interaction at these two levels. The difference between these two estimates, with a suitable divisor, is called the ABC three-factor interaction, which is computed as shown in the last line of Table 3.

Briefly, then, the principal advantages of factorial experimentation are that it provides (for a much smaller total amount of experimentation) equally precise estimates of the average effects of the factors as the single-factor method, and it also supplies a convenient body of data for examining relationships between the effects of different factors.

Additivity of Factorial Effects

The meaning of the various kinds of interactions and the rules for calculating them are explained in the standard textbooks. In many types of work, most of the interactions turn out to be small, particularly those among three or more factors. This is the reason why the analysis of the results of factorial experiments in terms of main effects and interactions has been found useful: frequently it is necessary to report and study only the main effects plus a few interactions between pairs of factors.

If all interactions are negligible, we have the simple situation in which any factor produces the same effect for all combinations of levels of the other factors. The factors are then said to be *additive* in their effects. The term is appropriate because if there are no interactions, the difference between the response to *abc* and the response to the combination (1) in which all three factors are at their low levels will be given by the sum of the main effects of A, B, and C. When effects are additive, the summary of results is particularly simple, since the main effects tell the whole story.

Whether effects are additive or not depends primarily on the way in which nature behaves. The experimenter can, however, exert an influence through the choice of factors, the choice of scale in which results are analyzed, and the choice of the range through which each factor is varied. For instance, if two factors are limiting, in the sense that the response will be poor unless both are in ample supply, we may expect to find an interaction. If the factors produce multiplicative rather than additive effects, analysis of the logs of the responses will restore additivity. As regards the widths of the ranges through which factors are varied, the assumption of additivity is likely to be a closer approximation to the true situation when the ranges are small, so that the effects of the factors are modest rather than large.

Additivity is stressed here because without it, especially if numerous interactions are large, the results of a factorial experiment are apt to be confusing both to interpret and to use. The search for types of factors and scales of analysis which produce approximately additive effects is therefore an important task in multifactorial research.

Limitations of Factorial Experimentation

Although the factorial experiment is a major contribution to the experimental sciences, there are limitations on its utility. The most obvious is that factorial experiments become prohibitively large and complex if the factors are numerous. Even if only two levels are included for each factor, an experiment with seven factors contains 2^7

or 128 combinations to be tested, and this number doubles for each additional factor. With quantitative factors whose effects are expected to be curvilinear, more than two levels are desirable, and the size of the experiment mounts still more rapidly. To test five different levels of each of six factors, the number of treatment combinations is 5^6 or 15,625.

In applied research in which high-yielding combinations of levels of the factors are the primary objective, a factorial experiment may devote too much experimentation to low-yielding combinations. In a 2^7 experiment, 64 of the tests are made at the low level of A. If this level gives uniformly poor responses, an alternative approach that revealed this fact quickly would save unproductive work.

As illustrated previously, an attractive feature of the factorial experiment is that it provides internal replication for estimating the main effects. This replication is extremely useful in lines of work in which experimental errors are high. But if experimental errors are low relative to the amount of change in the response as the factor levels are changed, a high degree of replication may not be needed for finding a good combination of factor levels. In a 2^7 factorial, any main effect is based on 64 replications: perhaps two or three would suffice.

The length of time required to obtain the results of an experiment will influence research strategy. When there is a long delay in obtaining results, as in experiments on farm crops or in tests of weathering in which treated surfaces are exposed for several months, it seems wise to obtain a large amount of information in a single complex experiment. When results become known quickly, on the other hand, the tendency will be to use smaller experiments of restricted scope, each planned so as to capitalize on the results of the preceding experiments.

Factorial Experiments in Fractional Replication

If a complete factorial experiment is too large, and the experimenter is unwilling to decrease the number of factors or the number of levels of each factor, he must accept some sacrifice in the quality of information obtained in order to reduce the size of the experiment.

By a selection of part of the treatment combinations in a complete factorial, it is possible to obtain useful information, though with some risk of being misled by the results.

Suppose, for example, that in a 2^3 factorial, we test only the four combinations

$$a, b, c, abc.$$

The main effect of A is estimated as

$$\frac{(abc) + (a) - (b) - (c)}{2}.$$

Note that this comparison is balanced for b and c: i.e., each level of b is represented both on the + and the − sides of the comparison. The same holds for c. Thus, this estimate of A is not affected by the size of the B and C effects.

However, if we attempt to compute the BC interaction from the four observations, Table 3 shows that it is given as

$$\frac{(abc) + (a) - (b) - (c)}{2}.$$

That is, the BC interaction and A are identical. A and BC are called *aliases*. It follows that if B and C have a positive interaction, this shows up as an effect of A. Consequently, in an experiment in which A has actually no effect, but B and C have effects and a positive interaction, we would erroneously report an effect of A. This is the type of hazard present in the use of fractional factorials.

With a 2^4 factorial, a possible half-replicate is the set

$$(1), ab, ac, ad, bc, bd, cd, abcd.$$

In this case it will be found that all four main effects are independent of two-factor interactions, although the two-factor interactions fall into alias pairs; $AB = CD$; $AC = BD$; $AD = BC$.

With a 2^5 factorial, 16 of the 32 treatment combinations can be picked to give independent estimates of all five main effects and all ten two-factor interactions. The hazard in this plan is that if some three-factor interactions are not negligible, the estimates of certain of the two-factor interactions are biased. The presence of four-factor interactions produces biased estimates of the main effects.

If all interactions can be assumed negligible, a greater reduction in the size of the experiment can be achieved. Table 4 shows the plan for an experiment with seven factors each at two levels, in which all the main effects are estimated from only eight tests ($\frac{1}{16}$th replicate). The eight treatment combinations appear in the second column. The remaining columns give the series of − and + signs used to estimate the seven main effects.

Table 4. A Fractional Factorial Experiment to Screen Seven Factors in Eight Tests

Test No.	Treatment	Factorial Effect						
		A	B	C	D	E	F	G
1	(1)	−	−	−	−	−	−	−
2	*abcd*	+	+	+	+	−	−	−
3	*abef*	+	+	−	−	+	+	−
4	*acfg*	+	−	+	−	−	+	+
5	*adeg*	+	−	−	+	+	−	+
6	*bceg*	−	+	+	−	+	−	+
7	*bdfg*	−	+	−	+	−	+	+
8	*cdef*	−	−	+	+	+	+	−

To illustrate the basic feature of the plan, consider the four combinations, *abcd, abef, acfg* and *adeg*, in which A is at the high level. Note that each of the other letters, b, c, d, e, f and g, appears twice in the four combinations. The same property holds for the four combinations in which a is absent. It follows that in the average effect of A, as estimated from

$$\left\{\frac{abcd + abef + acfg + adeg}{4}\right\} - \left\{\frac{bceg + bdfg + cdef + (1)}{4}\right\},$$

the effects of the six other factors cancel out. Thus an unbiased estimate of the average effect of A, and similarly of all other factors, is obtained with fourfold replication. If any interactions are present, some of these estimates are biased, this being the risk taken in order to obtain a drastic reduction in the size of the experiment.

We could, alternatively, obtain some information about the effects of all seven factors by testing the eight combinations (1), a, b, c, d, e, f and g. With this experiment, however, the estimates of the factorial effects are based on single replications and are positively correlated, since the response to the combination (1) appears with a negative sign in all the estimates $\{a - (1)\}$, $\{b - (1)\}$, etc. The advantages of the fractional replication plan in Table 4 are that we obtain estimates based on four replicates and free from this algebraic correlation.

A number of useful plans with fractional replication have been worked out for the 2^n series of factorials, (2), (3), (4), and have found numerous applications in industrial research. If interactions are negligible, the main effects of up to fifteen factors can be estimated in only sixteen tests. In general, the plans are appropriate for experimental situations with the following features.

(i) Experimental errors are high enough so that the internal replication provided by fractional factorials is needed.

(ii) Only two levels of each factor are necessary. This implies either that the factors are qualitative, or, if quantitative, that their main effects are expected to be approximately linear within the range covered.

(iii) Enough is known about the process so that most of the interactions can be assumed negligible or at least small relative to main effects.

These conditions may apply in a screening test in which only a few of the factors are expected to have important effects, but it is desired to include a substantial number of factors in case some factors thought to be minor should turn out to have major effects.

A good account of the uses of fractional factorials in a food research laboratory has been given by Carroll and Dykstra in their recent book, *Experimental Designs in Industry* (13).

With factors at more than two levels, the possibilities are limited. Practical designs have been developed for the 3^4 factorial in $\frac{1}{3}$ replicate and for combinations like the 4×2^4, $4^2 \times 2^3$. Horton (14) describes uses of the 4×2^4, the $4 \times 4 \times 8 \times 8$ and the $3^6 \times 2 \times 2$, with a careful account of the initial planning that led to the adoption of these designs.

When experiments can be completed quickly, a sequence of fractional factorials may form an efficient strategy. The first experiment is small, intended to estimate main effects and perhaps a few two-factor interactions. It may be clear from the results that several factors can be dropped, having only minor effects. The second experiment is then designed to estimate the effects of the principal factors more precisely and to delve somewhat deeper into the interrelations among these effects. A third experiment might be a complete factorial involving the two or three most important factors. Alternatively, results of the first experiment may suggest that all factors should be retained, but that their interactions require more elucidation. This can be done by making the second experiment a further fraction of the same replication. Care is required in constructing the second experiment if the desired information is to be obtained. The problem is discussed by Davies and Hay (5), who first suggested this use of fractional factorials, and in references (2) and (3).

In summary, fractional factorials have become one of the principal research weapons in numerous industrial problems.

Random Balance Designs

These designs, proposed by Satterthwaite (16), have the same general objective as fractional factorials. The difference is that the treatment combinations are constructed by taking for each factor an independent random selection of the levels of that factor which it has been decided to investigate. If three levels of the first factor are to be included, a series of random selections from the numbers 1,2,3 is made, for example 3,1,1,2,2,1. In the first treatment combination this factor appears at level 3, in the second at level 1, and so on. The restriction is usually made that over the whole experiment, levels 1, 2 and 3 will appear equally often. If the second factor has two levels, repeated random selections of the numbers 1, 2 are made to determine its levels in the successive treatment combinations.

Several advantages are claimed for this type of arrangement. The treatment combinations are easy to set up. There is complete flexibility as to numbers of factors, numbers of levels of each factor and number of combinations tested. In particular, designs can be written down for problems (12 factors at 2 levels and 5 factors at 3 levels in 24 tests) for which no fractional factorial design has been constructed. The experiment can be stopped and analyzed after any reasonable number of tests has been completed.

Random balance designs are, of course, another type of fractional factorial design, since they test only a fraction of the treatment combinations that constitute a complete factorial. The contrast between random balance designs and the older fractional factorials is that in the latter the treatment combinations are carefully selected so that those effects considered most important (usually the main effects of the factors) will be estimated independently of each other. With random balance, in its simplest form as described here, the tested combinations are a random selection from all combinations. For this reason, random balance designs are less efficient, in the sense that for a given size of experiment the principal effects are estimated with higher standard errors, and are harder to analyze and interpret clearly. Satterthwaite has stated (16) that he does not recommend random balance, except where its simplicity of construction is a preponderating consideration, for situations in which a balanced fractional factorial is already available. Critics have also objected that the inclusion of large numbers of factors in small experiments leads to biased and confusing results. An account of the pros and cons of random balance designs will be found in the papers and discussion following reference (16).

It seems likely that in practical applications the distinction between random balance and the older fractional factorial designs, and the controversy about their respective roles, will diminish. Proponents of random balance incorporate fractional factorials into their designs. For instance, if eight of the factors are to appear at two levels, this part of the design will be a balanced fraction of a 2^8 factorial rather than a random selection of levels. The development of random balance designs has stimulated research on extending the balanced type of fractional factorial so as to produce new designs for problems that are beyond the present scope of these designs.

Second-Order Designs

When the factors are quantitative it is natural to think of the response y as a mathematical function of the levels $x_1, x_2, \ldots, x_k$ of the factors. The purpose of the experiment is to investigate the nature of the response surface

$$y = \phi(x_1, x_2, \ldots, x_k).$$

Published research on this problem is confined thus far mainly to situations in which the function ϕ can be represented by a plane or by a quadratic function of the x's, although work is in progress on other functions.

The 2^k factorial, or fractional factorial, is a convenient design for fitting a plane surface. For instance, with a 2^3 factorial, the equation

$$y = b_1x_1 + b_2x_2 + b_3x_3$$

may be fitted easily. If the experiment is replicated, or the experimenter has an external estimate of his experimental error, the remaining four degrees of freedom for treatments may be used to judge whether the planar fit is satisfactory within the limits of experimental error.

For fitting a quadratic surface the 3^k factorial is convenient, but requires a large experiment if k exceeds 2. Box and Hunter (9) have developed a series of designs, called rotatable second-order designs, specifically for the fitting of quadratic response surfaces. Denote the lower level of any x by -1 and the upper level by $+1$, so that the "center" of the design is the point with coordinates $(0, 0, 0, \ldots, 0)$. These designs consist of three parts:

(i) A 2^k factorial or fractional factorial.

(ii) A "star," with tests made at the combinations $(\alpha, 0, 0, \ldots, 0)$; $(-\alpha, 0, 0, \ldots, 0)$; $(0, \alpha, 0, 0, \ldots, 0)$; $(0, -\alpha, 0, 0, \ldots, 0)$; etc.

(iii) One or more tests at the center $(0, 0, \ldots, 0)$.

With suitable choices of α and of the number of points at the center, this design will estimate a quadratic response surface with a standard error that is practically constant at all points within a distance 1 from the center. De Baun and Schneider (15) discuss chemical applications of these designs.

The strategies to be described in subsequent sections were developed for the more restricted problem of finding the combination of factor levels that gives a maximum response, and of mapping the general nature of the response in the neighborhood of this optimum set of levels. Since all the strategies leave much scope for variations in the way in which they are applied, they cannot be presented as a series of automatic steps. Only the central idea will be outlined. These methods are probably primarily of use when there is a single response, or only one response of predominating interest. With several responses of equal interest, the fractional factorials and second-order designs mentioned previously may prove more suitable.

The Single-Factor Method

This is the method already illustrated at the beginning of Part II. A good description is given by Friedman and Savage (7). We shall discuss the method for quantitative factors, although it applies also to qualitative factors. The experimenter first makes a guess at the optimum factor levels, say $x_{11}, x_{21}, \ldots, x_{k1}$. In the first experiment, several levels of the first factor (say 3 to 5) are compared, the levels of all other factors being kept fixed at the initial guesses x_{i1}. From the results of the first experiment, the level x_{12} of the first factor that gives the highest response is estimated. The second experiment tests several levels of the second factor, the other factors being fixed at the levels $x_{12}, x_{31}, \ldots, x_{k1}$, and so on. At the end of the first round we have reached a new approximation $x_{12}, x_{22}, \ldots, x_{k2}$ to the optimum combination.

We proceed with a second round, and so on. Various modifications may be introduced. The ranges may be narrowed in subsequent rounds, or some factors may be dropped. More than one factor may be varied simultaneously, as in the steepest ascent method to be described later. The process terminates when the response seems to be capable of little or no further improvement.

The amount of experimentation will depend on (i) the number of factors, (ii) how close the initial guess is to the optimum, (iii) the amount of replication needed in each single-factor experiment, and (iv) the magnitudes of the interactions between the effects of the factors. The method works best when experimental errors are small, so that a single replication suffices in each experiment, and when interactions are absent, i.e., the effects of the factors are additive. If these two conditions hold, a point close to the optimum should be reached at the end of the first round. Interactions slow up progress, because the best value of x_i when the other factors are held at their initial levels may be quite different from its best value when the other factors are held at new levels, so that several rounds are needed for the process to come near to the optimum.

At the beginning of Part II the single-factor method was criticized as being inefficient relative to a complete factorial. Some readers may wonder why it is now presented as a promising method. The reason is that the conditions of the comparison have changed. Suppose that there are three factors, and that five levels of each are considered necessary to locate the optimum level for a factor. A round of the single-factor method, using only one replication, takes 15 tests, whereas a complete replication of the 5^3 factorial requires 125 tests. The complete factorial supplies much internal replication and enables us to investigate interactions if we want to. But if experimental errors are small the replication is not needed, and if interactions are negligible they will not weaken the performance of the single-factor method in locating the optimum quickly.

The Method of Steepest Ascent

This method, introduced by Box and Wilson (8), starts from an initial guess about the position of the optimum. In the subsequent experiments, all factors are varied simultaneously in the hope of proceeding more directly towards the optimum than with the single-factor method. The first experiment is conducted in the neighborhood of the initial guess, and may be a fraction of a 2^n factorial. Its purposes are (i) to fit a linear approximation to the response surface, i.e.,

$$\hat{Y} = Y_0 + b_1x_1 + b_2x_2 \cdots + b_kx_k, \tag{1}$$

where $\hat{Y}$ is the estimated response, and x_i the level of the ith factor, and (ii) to provide data for a test of significance of the adequacy of the linear fit.

If the linear equation appears to fit adequately, a direction of steepest ascent is estimated from its results. Suppose that the origin of the x-space is taken as the initially guessed set of levels. Consider all points equidistant from this origin, i.e., lying on the sphere

$$x_1^2 + x_2^2 + \cdots + x_k^2 = r^2.$$

From equation (1) it is easy to show by calculus that the point on the sphere having the highest expected response has coordinates

$$x_i = rb_i/\sqrt{b_1^2 + b_2^2 + \cdots + b_k^2}.$$

Thus, the direction of steepest response is such that x_i must be changed by an amount proportional to b_i.

Having chosen a value of r, the experimenter makes a new test at this point on the path of steepest ascent. He continues tests on this path so long as the observed improvement in response agrees with the improvement predicted from the linear equation. When the observed improvement begins to fall short, a new linear approximation is fitted, a new path of steepest ascent is computed, and further tests are made on this path.

This process continues until either (i) the linear approximation still holds but the b_i are all small, suggesting that a plateau has been reached, or (ii) the linear equation no longer fits. In the latter event a design to which a second-degree equation can be fitted is used. For this purpose the rotatable second-order designs may be used. The second-degree equation may succeed in locating the position of the optimum, or, if the situation is more complex, analysis of its results suggests the direction of further experimentation.

The method is more sophisticated and complex in operation than the single-factor method. The experimenter's judgment must be exercised in choosing the range through which each factor is varied in the initial 2^k experiments. Although this may not be obvious at first sight, these choices have a marked effect on the direction which the calculations will indicate to be that of steepest ascent, and in which further tests are made. The sizes r of subsequent jumps must also be chosen by judgment. Mistakes in these choices may, however, be corrected later in the process.

The method has a high degree of flexibility for coping with different shapes of response surface, and in making advances in response in the face of complex interactions which impede progress by the single-factor method. Advantage is taken also of the internal replication associated with factorial experiments: this should help when experimental errors are sizable.

Use of Random Balance Designs

When the number of factors is large and little is known about the shape of the response surface, a random balance design may be a useful first step (6).

What can be done with the results of a series of such tests? The simplest procedure is to pick the winner, i.e., the test that has given the highest observed response, in the hope that the combination of levels used in this test will be somewhere near the optimum. If experimental errors are negligible, a simple and interesting result holds for the winner. Consider a subregion in the factor space made up of all points at which the true response exceeds some specified value. Then the probability that the winner lies in this subregion of high response is

$$P = 1 - (1 - f)^p$$

where p is the number of random tests made and f is the ratio of the volume of the subregion to the total volume of the factor space.

Table 5. Probability P that the Winner Falls in a Region of High Response

	Number of Tests		
f	10	20	30
.05	.40	.64	.87
.10	.65	.88	.99
.20	.89	.99	.99+

Table 5 shows that with 20 tests, the probability is 0.88 that the winner falls in the subregion containing the highest 10 per cent of all responses. With 30 tests, this probability jumps to 0.99, and the probability is 0.87 that the winner is in the subregion containing the highest 5 per cent of responses.

This result holds irrespective of the number of factors or of the shape of the response surface. It would apply, for instance, to 20 tests made on 28 or 34 factors. When there are many factors and experimental errors are negligible, the random method appears an inexpensive way of bringing us quickly into a subregion of high response. If the optimum is flat, any point in the best 5 per cent or 10 per cent subregion may give a response satisfactorily near the optimum.

At the end of the first set of tests, something more may be learned by studying the other combinations of levels that give responses close to that of the winner and by plotting the response against the levels of individual factors or small groups of factors. It may be possible to detect and eliminate factors whose levels have a minor influence on the response. A second set of tests can then be started on a smaller number of factors, with levels closer to that of the estimated optimum.

When experimental errors are present, the probabilities in Table 5 are decreased to an extent which will, of course, depend on the sizes of the errors but has not yet been investigated. Errors will also make it difficult to obtain clear-cut conclusions about the importance and nature of the effects of individual factors. Much further research on the utility of the method is needed.

Comparisons Between the Methods

It seems certain that no single method is best under all conditions (10). The relative advantages will depend on the number of factors, the shape of the response surface, and the sizes of the experimental errors. When the errors are small, the random method looks promising as a starter if many factors (say more than 10) must be included, or if multiple peaks are likely, because the other methods may converge on a peak that is not the highest one. The single-factor method should do well when experimental errors are small and the effects of the factors are approximately additive. The steepest ascent method provides more internal replication and has been claimed to be particularly effective with the type of "rising ridge" surface encountered in chemical experiments, but becomes complex if the factors are numerous.

Comparisons under actual operating conditions will seldom be feasible. In fact, since each strategy allows flexibility and has numerous variants, it is difficult to define the strategies specifically enough to permit comparisons. What can be done is to construct different types of response surfaces mathematically and test the performance of some variants of each method on each surface. Brooks (11) compared several variants of the random, the single-factor, the steepest ascent, and the complete factorial methods on four different surfaces in two variables (i.e., with two factors). Setting each strategy the same task, and taking the most successful variant of each, he found the steepest ascent method best on all four surfaces. The single-factor was practically as good on three of the surfaces but somewhat poorer on a fourth. The complete factorial usually placed third and the random method last. The methods fell in the same order whether experimental error was introduced or not.

Analogues of response surfaces in five variables have been constructed by means of electrical resistance networks (12). These are housed in little black boxes. One contains five dials on which the level of each factor is set. A second box, connected to the first, has a dial by which an experimental error may be added to the true response, and a voltmeter on which the observed response is read.

In the report given by McArthur and Heigl (12), the boxes were used to discover how research teams attack such problems rather than to compare specifically defined strategies. (The authors point out the difficulty already mentioned of defining strategies in sufficient detail for a direct comparison.) The element of cost was introduced into the problem by charging a certain sum for each test and crediting the method with another amount for each 1 per cent gain in the true response.

The two methods used most frequently by the teams were the single-factor and the $\frac{1}{4}$ replicate of a 2^5 factorial, followed perhaps by further tests. The fractional factorial did better, on the whole, on two of the three surfaces tried, the two strategies being about equally effective in the third, although more tests would be needed for definitive conclusions. Not enough use was made of the steepest ascent or the random methods to permit comparisons.

The results also furnish interesting suggestions about common mistakes in research strategy. These include lack of boldness, staying in a rut, being deceived by experimental errors, failing to eliminate unimportant variables, and not knowing when to stop. As the authors point out, however, no good positive rule about when to stop has emerged as yet from experience with the boxes.

In conclusion, it must be emphasized that knowledge in this area is rudimentary. Statements made in this chapter about the strengths and weaknesses of different strategies should be regarded as highly tentative. In view of the interest which the problem has aroused, there is reason to anticipate a useful volume of soundly based research in the near future.

REFERENCES

(1) McGehee, W., and Gardner, J. E. "Music in a Complex Industrial Job," *Personnel Psychology,* Vol. 2 (1949), 405-17.

(2) Davies, O. L. (ed.). *Design and Analysis of Industrial Experiments.* New York: Stechert-Hafner Co., 1956, 2nd edition.

(3) Cochran, W. G., and Cox, G. M. *Experimental Designs.* New York: John Wiley and Sons, 1957, 2nd edition.

(4) "Fractional Factorial Experiment Designs for Factors at Two Levels," *Applied Mathematics Series, 48,* National Bureau of Standards, 1957.

(5) Davies, O. L., and Hay, W. A. "The Construction and Uses of Fractional Factorial Designs in Industrial Research," *Biometrics,* Vol. 6 (1950), 233-49.

(6) Anderson, R. L. "Recent Advances in Finding Best Operating Conditions," *Journal of the American Statistical Association,* Vol. 48 (1953), 789-98.

(7) Friedman, M., and Savage, L. J. "Planning Experiments Seeking Maxima," *Techniques of Statistical Analysis.* New York: McGraw-Hill, 1947.

(8) Box, G. E. P., and Wilson, K. W. "On the Experimental Attainment of Optimum Conditions," *Journal of the Royal Statistical Society,* Series B, Vol. 13 (1951), 1-45.

(9) Box, G. E. P., and Hunter, J. S. "Multifactor Experimental Designs for Exploring Response Surfaces," *Annals of Mathematical Statistics,* Vol. 28 (1957), 198-240.

(10) Box, G. E. P. "Integration of Techniques in Process Development," *Transactions of the 11th Convention of the American Society for Quality Control* (1957), pp. 687-702.

(11) Brooks, S. "Comparisons of Methods for Estimating the Optimal Factor Combination." Unpublished Sc.D. thesis, The Johns Hopkins University, 1955.

(12) McArthur, D. S., and Heigl, J. J. "Strategy in Research," *Transactions of the 11th Convention of the American Society for Quality Control* (1957), pp. 1-18.

(13) Carroll, M. B., and Dykstra, C., Jr. "The Application of Fractional Factorials in a Food Research Laboratory," *Experimental Designs in Industry.* New York: John Wiley and Sons, 1958.

(14) Horton, W. H. "Experiences with Fractional Factorials," *Experimental Designs in Industry*. New York: John Wiley and Sons, 1958.
(15) De Baun, R. M., and Schneider, A. M. "Experiences with Response Surface Designs," *Experimental Designs in Industry*. New York: John Wiley and Sons, 1958.
(16) Satterthwaite, F. E. "Random Balance Experimentation." *Technometrics,* Vol. I (1959), 111-193.

Nineteen

HUMAN ENGINEERING [1]

ALPHONSE CHAPANIS

A DEFINITION OF HUMAN ENGINEERING

Human engineering is generally concerned with ways of designing machines, operations, and work environments so that they match human capacities and limitations. Another way of saying this is that the business of the human engineer is the engineering of machines for human use and the engineering of human tasks for operating machines.

The Outlook of the Human Engineer

The basic philosophy of the human engineer can be illustrated by considering the thousands of accidents which happen daily. "To err is human" is so ingrained in our everyday speech and thinking that it often diverts our attention from positive, remedial courses of action. For years it has been customary to write off most aircraft accidents as the result of "pilot error." Statistics compiled by insurance companies on home, street, railway, and industrial accidents are full of "causes" such as "carelessness," "faulty attitude," "inattention," and the like. Although these labels appear to be helpful, they really are not, primarily because they do not tell us how to reduce accidents.

[1] Preparation of this chapter was supported under Contract N5-ori-166, Task Order I, between the Office of Naval Research and The Johns Hopkins University. This is Report No. 166-I-215, Project Designation No. NR 145-089, under that contract. Reproduction in whole or in part is permitted for any purpose of the United States Government.

This should not be misunderstood. People do make mistakes. But the human enigineer raises these important questions: Is some of the blame to be found in the design of the equipment which people use? Do people make more mistakes on some kinds of vehicles than others? Is it possible to redesign machines so that human errors are reduced? Research over the past two decades provides us with answers which add up to a resounding "Yes!" to all these questions. This, then, is the rationale for the human engineer's approach: He does not heap more abuse and blame on the fallible human operator. On the contrary, he accepts as a basic premise the fact that people do make mistakes. But he also starts with the assumption that equipment, like people, can be "accident-provocative." Then he turns his attention to the machine and the job to see if he can make them more nearly error-free.

Other Goals of Human Engineering

These remarks about accidents do not mean that human engineering is synonymous with safety engineering. Increasing safety and decreasing accidents are major goals of human engineering, to be sure, but there are other important goals as well. Increasing the efficiency with which machines can be operated, increasing productivity in industrial operations, decreasing the amount of human effort required to operate machines, and increasing human comfort in man-machine systems—all these, singly or in combination, are the major objectives of human engineering.

Human Engineering Versus Industrial Psychology

Our definition of human engineering can be sharpened by comparing it with some other closely related areas of science which have been applied to industry, such as industrial psychology. The major emphasis in present day industrial psychology seems to be on personnel selection and training. Indeed, if the content on personnel were subtracted from many textbooks in industrial psychology the remainder would be very close to zero. Stated baldly, personnel selection aims to find the best man for the job. Personnel training aims to make the man fit the job once he has been selected. Human engineering, on the other hand, aims to make the job fit the man—as many men as possible.

Human Engineering Versus Time-and-Motion Study

One important antecedent of human engineering is time-and-motion study. Most time-and-motion engineers are used in industry to set fair rates of pay for diverse kinds of jobs, and it is in this connection that one most frequently thinks of time-and-motion study. But in the course of their work, time-and-motion engineers have come to realize that many machines and jobs are poorly designed for the average worker. As a result, these men have frequently been able to change the machine or the job, instead of trying to make workers do things that really do not match their abilities. When time-and-motion engineers have engaged in this kind of activity they have been dealing with some of the same sorts of problems discussed in this chapter.

Why then do we delimit a field of applied technology separate from time-and-motion study? The answer here is clear. Time-and-motion engineers have been concerned mainly with elementary, rule of thumb principles of movement. In the main, these have worked well. But human engineering draws on many scientific specialties to answer questions which the time-and-motion engineer has typically left untouched. From the experimental psychologist, the physiologist, the anthropologist, the toxicologist, and the physician have been assembled a whole host of findings dealing with man's behavior in a complex machine environment. There have been dozens of experiments on the design of dials, on the arrangement of dials, on the coding of controls, on the natural relationships between dials and controls, on the forces which man can apply with various controls, on the best diameters for cranks, on the intelligibility of speech in noise, on the information-handling capacity of the human, on human tolerance to heat, humidity, and noxious gases, and so on.

These and many other data related to them are so varied, so complex, so far reaching and so different from those of conventional motion study that most people have agreed that a new name, a new field, and a new specialist should be created. That field is human engineering; the specialist is the human engineer.

A Word about the Name "Human Engineering"

The field of technology with which this chapter deals is vigorous, lusty, and growing rapidly, although the members of the family are not yet sure what it ought to be called. Various writers in this country have, at one time or another, referred to it as applied experimental psychology, engineering psychology, biomechanics, human engineering, applied psychophysics, and psychotechnology. In England and on the continent the term "ergonomics" seems to be generally accepted for this purpose. Any of these terms could, with various degrees of precision, be used to describe the subject matter of this chapter.

The term "human engineering" has been chosen primarily because it seems to be the one which has the widest general acceptance in the United States today. Despite this common usage, the term "human engineering" has some different connotations. In the 1920's and 1930's, for example, it was used to refer to human relations problems in industry. Those people who remember this older use of the term are sometimes perplexed when they see the way it is used nowadays.

At any rate, it should be clearly understood that "human engineering" does not mean "engineering humans." This is not a chapter about how to get along with the boss, make employees happy, or improve relations with the company union. This is a chapter about the engineering of machines for human use.

A DEFINITION OF SYSTEMS ENGINEERING

In their textbook on systems engineering, Goode and Machol (18) abandoned any attempt at providing a formal definition of exactly what constitutes a large-scale system. Despite the difficulties of finding a universally acceptable definition of systems engineering, it is fair to say that the systems engineer is the man who is generally responsible for the over-all planning, design, testing, and production of today's automatic and semi-automatic systems. The first need for this kind of engineering appeared when it was discovered that satisfactory machine components do not necessarily combine to produce a satisfactory system. Even though the machine elements may satisfy individual specifications, very often the system as a whole will not work as planned when these components are merely joined together. Complexity of our modern machine assemblages has created the need for systems design and systems engineering.

The telephone system is a good illustration of this problem. It is not sufficient that a switching circuit or new piece of equipment be designed as a single independent development. Each new circuit must fit into and be compatible with the entire telephone network. For this reason each new circuit or item of equipment has to be tested, not only as a unit, but in conjunction with many other circuits which already exist, or—and this is a point which must not be overlooked—*with circuits which must be planned for the future.*

Some Characteristics of a System

Although they were not able to give a formal definition of a system, Goode and Machol (18) were able to identify some of the important characteristics of modern systems. They point out in the first place that the systems with which they are concerned are *equipment* systems. The primary function of the engineer is to make decisions involving choices and design of equipment. This excludes purely biological, social, or economic systems.

Another distinguishing characteristic is that each system has a certain integrity or unity. It may or may not be controlled from some central point, but in every case the parts of the system have some common purpose. An additional feature of systems is that they are large—large in terms of the numbers of parts which go into them, in the number of different types of parts, in the number of functions they perform, in the number of inputs they receive, and in terms of their cost. Systems are usually complex. Automatic machines today can accept many different kinds of inputs, perform alternative kinds of operations on these inputs, and present the results in different ways.

The last, and most important, characteristic of systems is that they are automatic. Although the degree of automaticity may vary over a wide spectrum, there are no systems in which human beings perform all control functions; conversely, there will probably never be systems in which *no* human beings are involved.

AUTOMATIC SYSTEMS STILL USE HUMAN OPERATORS

This last characteristic of automatic systems is so important for the whole thesis of this chapter that it deserves elaboration. Although many machine systems in use today are described as fully automatic, they all still use human operators in one way or another. Moreover, it is easy to overlook or to be misled about the amount of work human operators may still do in automatic systems.

For example, an article in a recent engineering journal (3) described

". . . a new automatic control system . . . designed to give the ultimate user increased machine productivity, higher machine speed and greater accuracy since it eliminates the human factor." This sounded like a completely automatic system. Yet a number of times the author, perhaps without realizing it, had to refer to human operators doing something in this "automatic system." Certain phrases from this article are quoted below to illustrate this point. Although these are quoted out of context, care has been taken to avoid distortion. For emphasis, those activities performed by human operators have been italicized.

> While little *machine modification* is necessary. . . . It is particularly suited for control of production machines on which a moderate number of pieces of one type are to be made, since it is simple to change from one variation of product to another simply *by providing another deck of director cards*. . . . Machine directions that would normally *be programmed*. . . . Standard card processing equipment . . . is used for *punching, sorting and stocking the cards. Positioning information is punched into the cards*. . . . This permits the *operator to read directions directly from the card*. . . . The cards also have space for . . . *operator's instructions*. . . .
>
> . . . *the operator initiates the control* and the first card directs the table motion. . . . *The operator then clamps the steel sheet on the table and initiates the control again.*
>
> *The positioning information can be read off a blueprint or table of coordinates. The remaining information punched into the cards includes drawing and part numbers, card sequence checking, machine instructions* . . . *operator's instructions* . . . etc.
>
> If power is removed during any part of a cycle it is only necessary *to push the 'Control On' button*. . . . *It is also possible to place the control on 'Hand' operation while running through a deck, move the various motions* (sic) *to any other point; and return the control to 'Auto'*. . . . By reading the cards in a definite order the machine operation may be halted at specified points, and through the illumination of an indicating light *the operator is told to check his planning card*. . . . If for any reason the sequence becomes disrupted, *the operator can reset the sequence to any desired number by means of manual switches provided on the control console.*

These selections do not exhaust all references to human activities in this system. But a definite impression remains that there is still a very substantial amount of human control required here. To be sure, the operators in this system are doing different things than the conventional skilled machine-tool operator does, but they are nonetheless an important and integral part of the system. Furthermore, this system cannot be operated by unskilled help. Reading blueprints, programming the operations, recognizing where and when a sequence should be reset—all these activities assume a high level of skill, perhaps even a higher level than that required for conventional machine shop work.

On a somewhat different topic, care must be taken not to confuse remote control with full automaticity. In some guided missile and drone aircraft systems, for example, the human operator is still in the control loop. He is merely sitting on the ground instead of in a cockpit. Substantially the same is true of robot tractors, ships, trains (2), and other vehicles where the operator may be at some fixed control location (as in the case of remotely controlled tractors and trains) or in another vehicle (as in the case of remotely controlled aircraft and ships).

The proportion of men to machine units is steadily decreasing in modern systems, to be sure, and operators are required to perform different kinds of functions than they do in non-automatic machine systems. Nevertheless, operators are there. No engineer today seriously considers dispensing with all human operators in any system which is planned for the foreseeable future. It is this fact—the fact that operators will still continue to be used in automatic systems—which makes this field an interesting and important one for behavioral scientists.

HUMAN FACTORS IN SYSTEMS ENGINEERING

Now that the fields of human engineering and systems engineering have been defined, let us examine how the one contributes to the other. The important human factors in systems engineering can be grouped under four major headings: (a) the allocation of responsibility between men and machines, (b) the synthesis of men and machines into systems, (c) the human engineering of systems components, and (d) the evaluation of systems. Although these factors are not always explicitly recognized and identified, they all enter into every system problem. However, the relative emphasis attached to each of the four ingredients varies with the particular system. In one, considerations about the allocation of responsibility between men and machines may be an especially important part of the initial planning. In another, the major psychological contributions can be made in the human engineering of components.

Allocation of Responsibility Between Men and Machines

In planning the design of a large system, a basic question which has to be answered by designers is not man *or* automatic machinery, but rather: What functions should the human be assigned in the best system for a particular job?

Several years ago, a report on human engineering for air navigation and traffic control (13) pointed out that it was possible to design several kinds of control systems which could be distinguished in terms of the *degree* of human participation in the control process. At one extreme, systems could be designed with primary control in the hands of human operators who would be assisted by data analysis, transmission, and display equipment. At the other extreme were those systems which would be almost "fully automatic" and in which almost every function would be performed by machinery. In these, human operators would be used for monitoring, checking, and maintaining the machines. Between these two extremes lay a spectrum of systems with still other degrees of human participation. Thus, a system could be designed so that its major work would be performed by semi-automatic machinery but in which the human operator would routinely perform certain critical functions; for example, planning and decision making.

Although the engineer may not always consider the problem in precisely these terms, he must, implicitly or explicitly, make some decision about the role of the human in every system he designs. When he says that there will be no people in the system, what he usually means is that human operators will be used solely for monitoring and maintenance. But this does not dispense with people; it merely means that the human tasks are different from those in conventional machine systems.

SYSTEMS SPECIFICATION

Rational decisions about the role of human operators in automatic systems can be made only after a careful specification has been prepared for the system. This specification must cover such points as these:

A. The over-all requirements of the system:
   1. The mission of the system and what it is supposed to accom-

plish stated in terms of minimum performance requirements,
2. The conditions under which it will be used and the general environment in which it will operate, including weather, ambient illumination, and temperature,
3. The total time line of its existence, including delivery and uncrating of equipment, training operators, training maintenance technicians, obtaining and storing replacement parts, and even abandonment, disposal, or conversion of the system when necessary,
4. Any constraints on design, such as a maximum allowable size or weight, or a requirement for mobility,
5. Emergency situations and other infrequent but possible uses of the system.

B. The external environment of the system in terms of:
1. The various signals and data which the sensing devices of the system must detect and display,
2. The unwanted noises, hazards, and obstructions in the environment that will limit performance of the system.

C. The internal environment of the system specified in terms of:
1. The lighting conditions, noises, and atmospheric conditions to which the internal parts of the system will be exposed,
2. The amount of space available within the system,
3. Special conditions such as movement, shaking, buffeting, and acceleration which might affect the performance of components of the system.

D. The data conversions which must occur in the system in terms of:
1. Computations,
2. Data-transformations,
3. Decisions,
4. Other linkages between the inputs and the outputs of the system.

E. The outputs of the system specified:
1. As to type, that is, whether the output is information, a finished product, or an action of some sort,
2. As to amount, that is, the total quantity of output the system is expected to produce.

FUNCTIONAL ANALYSIS

When the systems designer has completed a full specification of the system, he has prepared the groundwork for a listing of the functions it is supposed to perform. Responsibility for these functions should then be assigned after comparing what men can do better than machines and vice versa. For example, suppose that the inputs to the system consist of checks, drafts, and money orders submitted in payment of bills. The payments may be written or typewritten and will vary in size, form, and arrangement. If the first function of the system is to read these inputs and to transform them into coded form for computation, there is little question about who should have this responsibility. There is no known machine which can perform this kind of function.

To take another example, suppose that one of the functions of an air-traffic control system is to make space assignments for all routine flights and also to be prepared to issue special assignments and orders in cases of emergency. Here again, there is little question about who should have this kind of responsibility. At the present time there is no known way of programming a computer for all of the thousands of possible kinds of emergencies which could happen in the air.

Table 1. A Highly Abbreviated List of Some of the Relative Advantages and Disadvantages of Men and Machines (Adapted from Fitts, *et al.* (13) and Williams, *et al.* (33))

Men	Machines
Able to handle low probability alternatives, i.e., unexpected events.	Difficult to program. Difficult to anticipate all possible events and so virtually impossible to program for all such contingencies.
Able to perceive, i.e., to make use of spatial and temporal redundancies and so to organize many small bits of information into meaningful and related "wholes."	Zero, or very limited, ability to perceive. "Organization" has to be elaborately programmed, which is difficult to do because of the many alternative ways organization can be formed from elements.
Possess alternative modes of operation. Can accomplish same or similar results by alternative means if primary means fail, or are damaged.	Alternative modes of operation limited. May break down completely when partial injury or damage occurs. Not able to regenerate or heal.
Limited channel capacity, i.e., there is a maximum amount of information that can be handled per unit time, and this is small.	Channel capacity can be made almost as large as desired.
Performance subject to decrement over fairly short time periods because of fatigue, boredom, and distraction.	Behavior decrements only over relatively long periods of time.
Comparatively slow and poor computers.	Excellent and very rapid computers.
Flexible; can change programming easily and frequently. Very large number of programs possible.	Relatively inflexible. Flexibility in kind and number of programs can be achieved only at a great price.

To illustrate the considerations which enter into decisions of this kind, Table 1 contains a highly abbreviated list of some of the comparative advantages and disadvantages of men and machines as systems components. Two things must be stressed: (a) this is a highly abbreviated list, and (b) Table 1 lists these comparisons in very general terms. In most instances there is much more precise data about human and machine performances than these statements would suggest. For example, it is known that the maximum amount of information which has ever been transmitted through the human channel, under any circumstances, is only about 40 bits per second. Machine systems are capable of enormously greater rates of handling information.

SOME RESPECTS IN WHICH HUMANS SURPASS MACHINES

In these days when newspaper headlines daily extol the achievements of modern automatic machines, it is easy to get the impression that machines can do anything people can do. That this is far from the truth of the matter can be demonstrated by elaborating on a few of the items in Table 1.

Think, for example, about the extraordinary capacity man has for perception. For years, engineers and scientists have been trying to construct machines that will "read" printed text. Their best efforts so far have had extremely limited success because of the difficulties of building "the perception of wholes" into a machine. In the light of this experience, human ability to read material like that in Figure 1 is no mean achievement.

One of the best illustrations of the advantages of a living organism can be found in the fact that adjacent to one of America's most outstanding electronic laboratories is a farm on which seeing-eye dogs are trained. Despite years of research and development on radar, the simple fact of the matter is that electronic guidance devices for the blind cannot even begin to replace the perceptual capacities of a dog!

As an illustration of man's flexibility consider the extraordinary adaptability he exhibits when he (a) walks, (b) roller skates, (c) rides a bicycle, (d) drives an automobile, (e) flies an airplane, and (f) flies a helicopter. These are all radically different control systems, each with completely different dynamic characteristics. Constructing a servo system to do even one of these jobs efficiently is an enormously diffi-

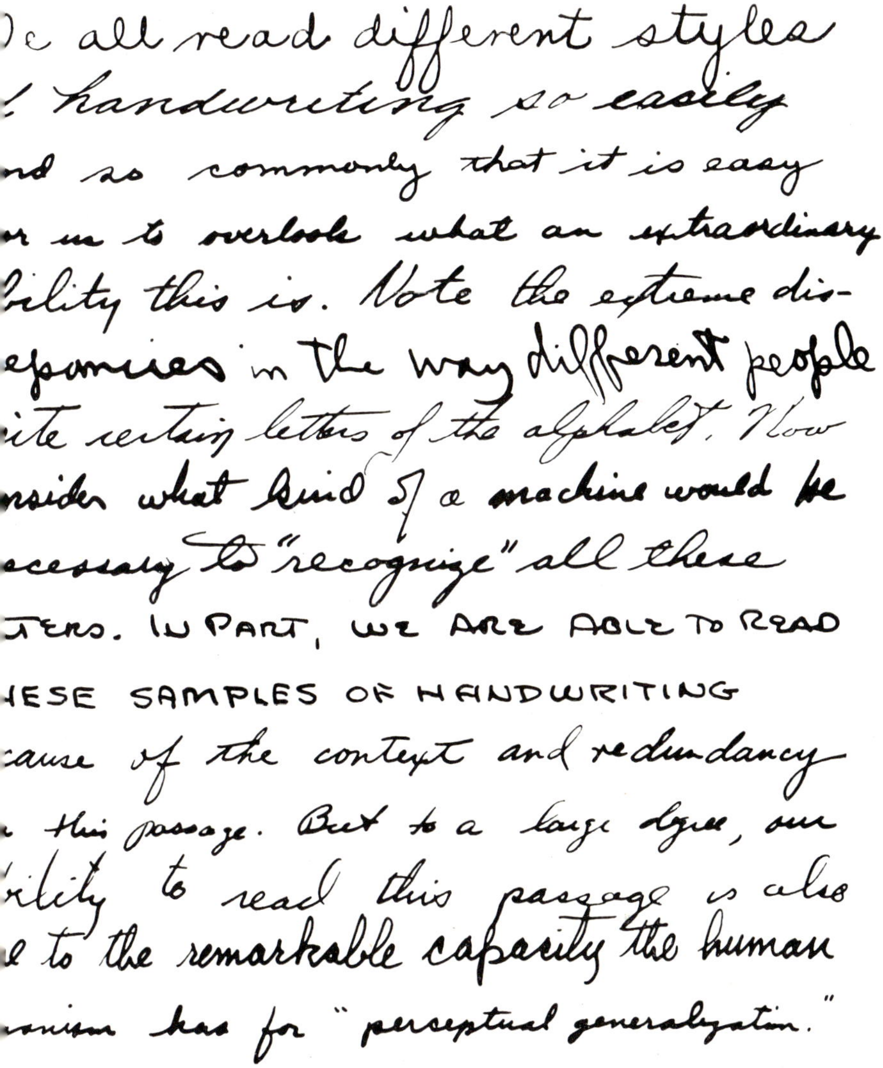

e all read different styles
f handwriting so easily
nd so commonly that it is easy
r us to overlook what an extraordinary
bility this is. Note the extreme dis-
epancies in the way different people
ite certain letters of the alphabet. Now
nsider what kind of a machine would be
ecessary to "recognize" all these
ters. IN PART, WE ARE ABLE TO READ
HESE SAMPLES OF HANDWRITING
cause of the context and redundancy
n this passage. But to a large degree, our
bility to read this passage is also
e to the remarkable capacity the human
anism has for "perceptual generalization."

Figure 1. Humans greatly surpass machines in perceptual capacity.

cult task for the designer. Yet the human controller has solved all these control problems and can handle them all seriatim.

THE RCA BIZMAC ELECTRONIC ACCOUNTING SYSTEM

Several of the considerations mentioned above are well illustrated in the design of the RCA Bizmac Electronic Accounting System—said to be the largest commercial data-processing system so far constructed (19, 21, 26). The installation covers about 20,000 square feet and consists of 357 separate equipment units, among them 12 major electronic machines and about 200 other fully automatic or semi-automatic equipments. This particular development was designed for the job of maintaining continuous and rapid stock and inventory control of over 250,000 catalog items. It also handles all the files, orders, receipts, payments, bills, and associated business paper work.

One of the first and most important series of psychological questions occurred at the planning stage. It was at this point that the systems engineers, in combination with psychologists, carefully sorted out and separated the functions of planning, decision-making, execution, and monitoring. The results of these early deliberations led to the decision that operators would be used in three ways: (a) for planning and preparing the work (that is, for programming the work); (b) for operating the system; and (c) for maintaining it (See Figure 2). The task of the first group of personnel is to specify the data-processing job and to prepare a detailed and specific set of operating instructions for the job. The task of the second group of personnel is to receive the operating instructions, direct, control, and monitor the processing of data through the machines. The task of the maintenance group is to inspect, test, and repair machines as required.

AN AUTOMATIC INTERCEPTOR AIRCRAFT

Another illustration of this kind of systems planning is contained in an article by Seamans and Pickford (29) which is concerned with the development of an automatic interceptor aircraft. The flight path of the vehicle may be controlled (a) in earth coordinates by a ground environment and data link, (b) in relation to the target as required under fire control operation, (c) in geographic coordinates of latitude and longitude in the navigation mode, and (d) in relation to the landing strip in the ILS mode. The decision of the engineers in this case

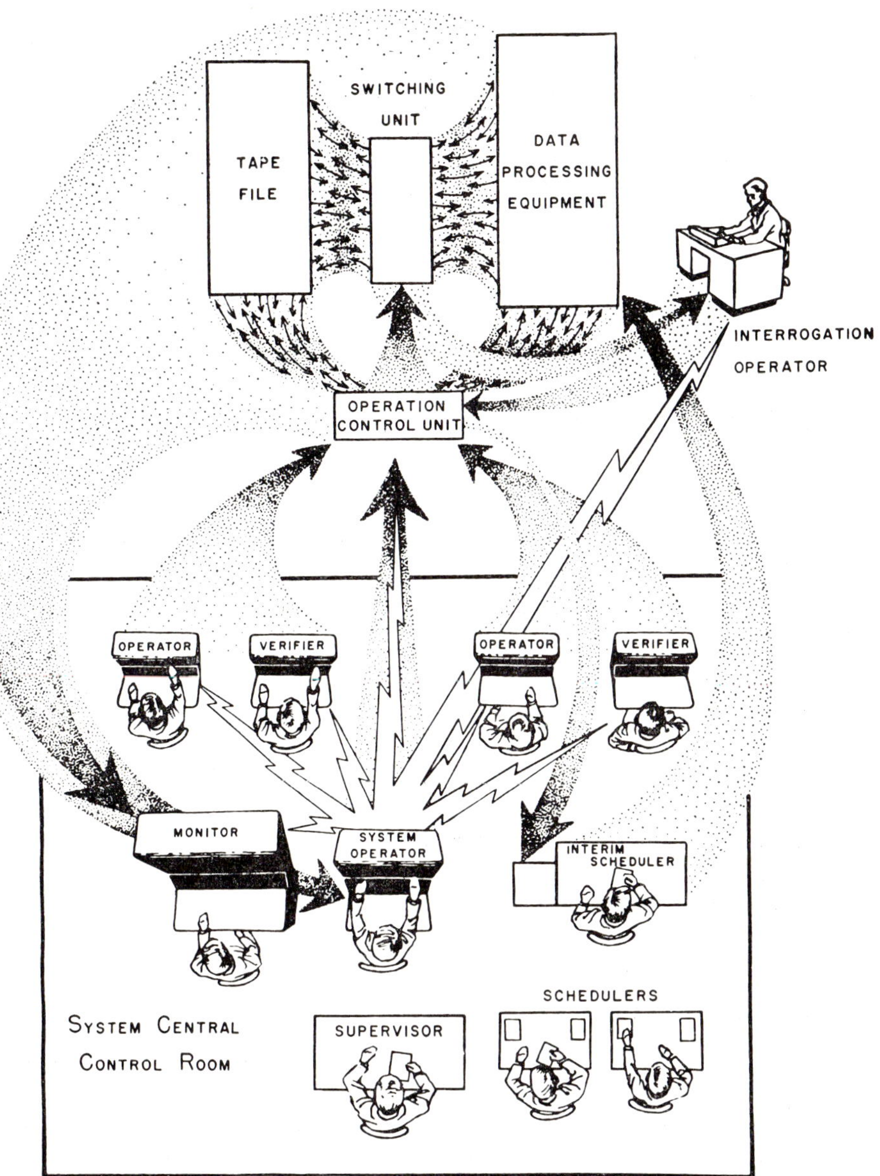

Figure 2. This diagram shows the chains of command issuing from the central control station in the Bizmac System. The lightning strokes show the system operator's control over the other operators and the Bizmac machines. The broad arrows show the flow of information from the operating team to the machines and the feedback from the machines to the operators. (Courtesy of the Radio Corporation of America.)

was that the pilot would be carried with the aircraft primarily for his superior powers of judgment and decision-making. Thus, one of the pilot's functions is that of exercising judgment as to the selection of the proper mode and the proper operation of the automatic elements. In addition, provision is included to allow the pilot to override the system in the event of failure or other emergency. Finally, we should note that some of the most important functions the pilot frequently serves in this kind of aircraft stem from his remarkable perceptual capacities. The pilot can recognize targets on the radar scope, distinguish friend from foe, discover targets of opportunity—targets which might have eluded or been undetected by radar—and bring back supplementary intelligence of great value.

THE BELL TELEPHONE TV MICRO-WAVE SYSTEM

An almost nearly fully automatic machine system is the Bell Telephone TV Micro-wave System, the series of 107 towers which span the United States. Each of these towers supports four giant "horns," which serve as receivers and transmitters of trains of ultra short-wave frequencies. Associated with each tower is a compact building containing automatic electronic receiving, amplifying, and transmitting devices. The buildings are heated automatically, and the air is cleansed by machinery. All the apparatus is in duplicate. Thus, if something happens to one piece, its alternate is cut in automatically or by pulsed signals from the nearest station which has human attendants. In this system, then, human operators serve as remote monitors and as maintenance men. Monitoring is done regularly; service and maintenance crews make occasional periodic check-ups.

RESEARCH NEEDED

It is important to point out that mere speculation and armchair philosophizing may not be enough to provide valid answers to this question of the allocation of duties. Consider, for example, a research program assigned by the Air Force to the Ryan Aeronautical Company (4). One of the basic questions to be answered is this one: Exactly what balance of automatic and manual control should be provided in an aircraft cockpit? The solution in this case is being sought by an experimental program using pilots who will "fly" aircraft simulators with various amounts of automaticity. In many instances a research solution may be the only dependable answer.

The Synthesis of Men and Machines into Systems

Having decided on the functions which operating personnel will have in a system and on the division of responsibility between men and machines, the engineer then faces the problem of integrating these personnel into the system; that is, of organizing the two kinds of components into an efficient working combination. This involves something much more general than the human engineering design of components for efficient use. Rather, it concerns some general principles about ways in which men and machines should complement each other in automatic or semiautomatic systems. This is an area of human engineering which has not received very much attention. Perhaps the best thing to do, then, is to illustrate this with some examples of what are considered to be general principles of man-machine systems design.

THE PRINCIPLE OF VERIFICATION THROUGH INDEPENDENT DUPLICATION

As mentioned earlier, during the planning of the RCA Bizmac system the decision was made to program the data into the system by human operators because of their flexibility and because of the nature of the input data. Yet it is generally known that even when equipment has been well engineered for maximum efficiency, human operators still make mistakes. These mistakes may amount to something on the order of 1 or 2 per cent, depending on the particular task. The accuracy requirements of this electronic accounting system were very much more stringent than this. How then was it possible to design the system to make use of the flexibility which characterizes human operators and yet to provide the extreme accuracy required?

The solution to this problem was to use two operator-verifier teams. The members of each team make identical set-ups on different consoles (see Figure 2). Special monitoring equipment has been designed to compare the output of the two human operators and to accept for processing only those items which agree exactly. Discrepancies are rejected for correction. Under certain conditions this technique will reduce the error rate to a few per million, even though the error rate for a single operator may be one in a hundred. What is the secret of this system? Basically, it is this: An error can get through the system only if both operators make the *same* error at the *same* time. Note the emphasis on the word *same* because this is the crux of the matter. Although it may be quite likely that both operators will make an error on the same setting, it is very unlikely that they will make the same error.

This state of affairs is quantified in the following equation:

$$P_{e_T} = \sum_{s=1}^{s=N} \left[P_s \sum_{i=1}^{i=n} \left(P_{e_{si}} P'_{e_{si}} \right) \right]$$

where P_{e_T} = the probability that an error will get into the system,
$P_{e_{si}}$ = the probability that operator 1 will make an error of the type i in making setting s
$P'_{e_{si}}$ = the probability that operator 2 will make an error of the type i in making setting s, and
P_s = the proportion of the time that setting s is required.

To see how this works in practice, consider the set of hypothetical data in Table 2, where it is assumed that only 10 settings may be called for. These are the settings 0, 1, 2, 3 . . . 9. The entries in the table show the probabilities that each of the values at the top of the table will actually be set. For example, when the operator is supposed to make a setting of 8, the probability is 0.994 that he will actually set his control to this value and 0.006 that he will make an error in setting his control. The errors, as are seen from Table 2, are distributed among three values: the probability is 0.001 that he will set the control to 6; 0.003 that he will set it to 7; and 0.002 that he will set it to 9. The probability that two operators, working independently, will make the same error at the same time is the sum of $(0.001)^2$, $(0.003)^2$, and $(0.002)^2$, or 0.000014.

If the ten settings in Table 2 are required equally often the value of P_s is 0.10 throughout. Thus, the probability that one operator will make an error in setting the equipment is $P_s \left[\sum_{i=1}^{i=n} P_{ei} \right]$ or $(0.1)(0.1) = 0.01$. That is, a single operator can be expected to make about 1 error per 100 settings. If two operators work independently the probability that both of them will make the same error is (0.1)(0.000308) or 0.0000308. That is, about 31 errors per million settings can be expected to get through the double operator system. This is a very much smaller probability than the 1 in 100 value obtained with a single operator.

Table 2. A Hypothetical Set of Data Showing the Gain To Be Expected in System Accuracy by Having Two Operators Set Values Independently into the System

Setting S	Values actually set										Probability that an operator will make an incorrect setting $\left(\sum_{i=1}^{i=n} P_{ei} \text{ or } \sum_{i=1}^{i=n} P'_{ei}\right)$	Probability that both operators will make the same error $\left(\sum_{i=1}^{i=n} P_{ei} P'_{ei}\right)$	Proportion of the time that each setting is required (P_s)	$P_s\left(\sum_{i=1}^{i=n} P_{ei}\right)$ or $P_s\left(\sum_{i=1}^{i=n} P'_{ei}\right)$	$P_s\left[\sum_{i=1}^{i=n} P_{ei} P'_{ei}\right]$
	0	1	2	3	4	5	6	7	8	9					
9									*0.002*	**0.998**	0.002	0.000004	0.06	0.00012	0.00000024
8							*0.001*	*0.003*	**0.994**	*0.002*	0.006	0.000014	0.08	0.00048	0.00000112
7					*0.001*	*0.002*	*0.003*	**0.990**	*0.003*	*0.001*	0.010	0.000024	0.10	0.00100	0.00000240
6				*0.001*	*0.003*	*0.005*	**0.986**	*0.003*	*0.002*		0.014	0.000048	0.12	0.00168	0.00000576
5			*0.002*	*0.003*	*0.005*	**0.982**	*0.004*	*0.003*	*0.001*		0.018	0.000064	0.14	0.00252	0.00000896
4		*0.001*	*0.003*	*0.004*	**0.982**	*0.005*	*0.003*	*0.002*			0.018	0.000064	0.14	0.00252	0.00000896
3		*0.002*	*0.003*	**0.986**	*0.005*	*0.003*	*0.001*				0.014	0.000048	0.12	0.00168	0.00000576
2	*0.001*	*0.003*	**0.990**	*0.003*	*0.002*	*0.001*					0.010	0.000024	0.10	0.00100	0.00000240
1	*0.002*	**0.994**	*0.003*	*0.001*							0.006	0.000014	0.08	0.00048	0.00000112
0	**0.998**	*0.002*									0.002	0.000004	0.06	0.00012	0.00000024
Sums											0.100	0.000308	1.00	0.01160	0.00003696

In most practical situations the various settings on an instrument are not used equally often. When this is the case, the probability values must be modified to take this into account. In Table 2 the additional assumption has been made that the settings near the middle of the range are used more frequently than those at the extremes. Thus, settings of 4 and 5 are needed 14 per cent of the time, whereas settings of 0 and 9 are called for only 6 per cent of the time. It is easy to see that this will increase the error rate because the operators are required to make more settings in the middle of the range—precisely where more errors are normally made. If a single operator makes the settings, slightly more than 1 in 100 errors may be expected when the errors are weighted for their frequency of use. The computed probability is 0.0116 or an error rate of 11,600 per million settings. When two operators work independently, the error rate is 37 per million settings.

Look more closely at the assumptions underlying this method. The first concerns the particular example in Table 2, but not the equation given above. In this example, it is assumed that the distributions of the errors made by the two operators are the same; that is, that each of the two operators makes errors with the frequencies shown at the left in Table 2. In general, this will rarely be the case. Practically, it means that the $\Sigma P_{esi} P'_{esi}$ values must be computed from two separate probability tables, one for the first operator, the second for the other operator. The net result of this situation is an even greater reduction in P_{e_T}. The extreme case, of course, is one in which the errors made by one operator are all different from those made by the other. If this were the case, no errors should be expected to get past our checking system.

Another important assumption underlying both the example in Table 2 and the equation given earlier is that the errors made by the two operators are independent; that is, uncorrelated. In general, if the two operators do not make errors independently, the net result is to increase the system error above that predicted from the equation. The extreme case occurs if both operators were to make exactly the same error at the same time. If this happened, the result would be equivalent to having only a single operator at work, so that nothing would be gained from using this principle.

One way of evaluating such a system is to try it. This is not as easy as it sounds for two reasons: (a) It takes an enormous number of observations to collect reliable data on something which occurs only a few times per million trials, and (b) it is extremely difficult to set up a test of the principle because the method of checking errors in the experiment must be even better than the procedure being evaluated.

There are two kinds of partial empirical answers to this problem. The first comes from an experiment by Weldon and Peterson (31). As part of their study these investigators had pairs of operators make independent but duplicate settings on three kinds of dials. Out of 2,850 such double settings there were no instances in which both operators made the same error on the same setting. This was the case even though the probability of a single operator making an error with one of these dials ranged from 0.013 to 0.029. The results of these double-setting observations suggest that the probability of error with such a system in this specific situation is less than 1 in 2,850, or less than 0.00035.

The second answer comes from practical experience in the punching and verifying of IBM cards. In this procedure coded numerical data are transferred to stiff cards for machine calculations. The operator uses a simple keyboard which punches holes in the card—the positions of the holes corresponding to numbers. IBM operators become very good at this, but even an experienced operator may make as many as two errors in every 100 numbers punched.[2] Although this is a very low error rate, it is much too great for most machine calculations. To guard against such punching errors, all of the IBM cards are run through another machine by a second operator. The second operator goes through all the motions of punching the data except that her machine does not actually punch holes. The verifier is essentially a comparator—corresponding to the monitoring circuit mentioned earlier. So long as there is a hole in the IBM card corresponding to every position she punches, everything goes along well. If, however, she punches a key for which there is no hole in the IBM card, the verifying machine signals the operator that there is a discrepancy between what is on the card and what the verifying operator punched. Although there appear to be no published figures on this point, IBM supervisors seem to agree that the punching and verifying operations may reduce errors to a few per 100,000 numbers punched.

[2] This is only a rough approximation, of course, because there are many factors which affect this error rate—the way in which the numerical data are originally encoded, the legibility of the copy from which the operator works, and so on.

THE PRINCIPLE THAT MACHINES SHOULD MONITOR MEN

In the report on human engineering for air-traffic control (13) it was suggested that "checking, verifying, and monitoring equipment be devised that will make it impossible for any human in an aircraft or on the ground to violate basic safety rules. . . ." This is another illustration of what may be considered a general principle of man-machine systems design. A simple illustration of the application of this principle is contained in an article by Collins and Hofman (11). This article describes a control panel used in television operating centers. This panel—basically a master distribution system—permits programs originating from any television network or city to be switched to any other network. Some special requirements are that the switching should be done at precisely the correct time and without error. The control panel is an elaborate series of buttons and lights, and has associated with it all of the necessary switching, holding, and delaying circuits.

This is another system which was designed to include human operators because of their flexibility and because of their monitoring function. The task of an operator is to connect incoming lines to the correct outgoing lines. The former are controlled by depressing black buttons; the latter, by white ones. To guard against the possibility that two incoming lines might be connected to a single outgoing line, appropriate circuits are arranged so that if two black buttons should be depressed at the same time as a white one, no pre-selection occurs and the operator is warned by the equipment that an error has been made. In short, the machine monitors the man and prevents him from making this important error.

AIDING THE OPERATOR IN CLOSED-LOOP TRACKING SYSTEMS

An important principle to observe in the design of closed-loop tracking systems is that the system should provide the operator with appropriate types of aiding. The primary purposes of aiding are (a) to relieve the operator of the necessity for making continuous control adjustments, and (b) to improve the performance of the system. To understand what these statements mean it is necessary to define the terms: closed-loop system and aiding.

Many man-machine systems can be diagrammed as in Figure 3.

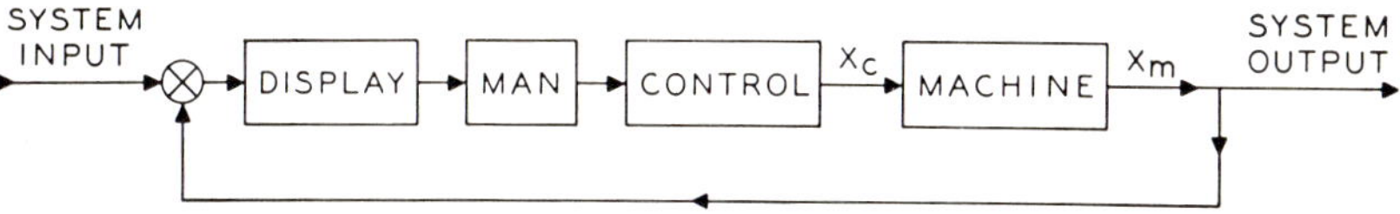

Figure 3. Schematic block diagram of a closed-loop system.

There is a display which usually portrays continuously changing information. The operator senses this information, ordinarily through his eyes or ears, and moves his control, or controls, accordingly. The output of the operator's control (X_c) produces some action in the machine. The output of the machine (X_m) constitutes the system output. As used here, *machine* refers to all the components in the system other than the display, the man, and the controls which the man operates. The thing that makes this a *closed-loop* system is that information about the performance of the system is fed back to the display. The display, therefore, combines information about the system input and the system output.

Driving an automobile is a familiar illustration of a closed-loop system. The system input is the panorama of the winding road, the stream of vehicular traffic, and the outline of the front end of the driver's own automobile framed in his windshield. What the driver sees (the display) is partly a function of objects not under his control (the road) and partly under his control (the position of his car on the road). The controls, of course, are the steering wheel, accelerator, and brake. The outputs of these controls act upon the machine (the automobile) to yield the system output (the movements of the automobile).

It is easy to find other examples of closed-loop systems which fit this model. Here are a few:

(a) Flying an airplane,
(b) Diving a submarine,
(c) Steering a ship,
(d) Controlling the water level above and below a dam,
(e) Operating a power shovel, and
(f) Controlling the flow of fluids in a continuous-process chemical plant.

Good human engineering practice is required in the design of every part of such closed-loop systems—on the displays and controls, for example—but of primary concern at the moment are the dynamics of the machine itself; that is, the way in which the control output is transformed into a system output.

The simplest kind of transmission (Figure 4) is a zero-order con-

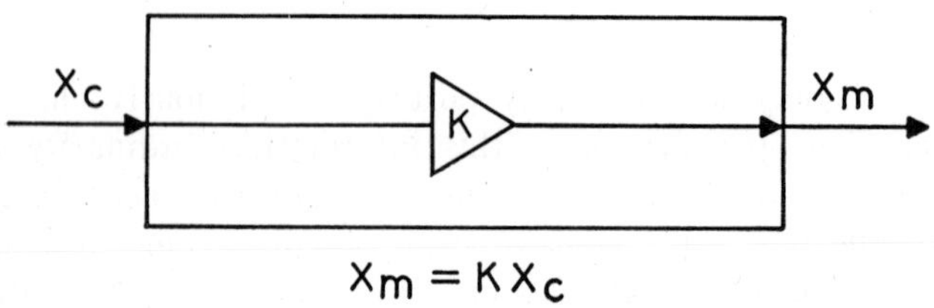

Figure 4. Zero-order control.

trol linkage in which the operator's control movements (X_c) directly determine the system output (X_m). K is a constant representing the *gain, amplification,* or *gearing ratio.*

In the top line of Figure 5 are two sets of hypothetical control movements. Both show a sequence of movements starting with the control at a neutral (or zero) position. The control is displaced in one direction (as, for example, to a or c), held at that position for a short period of time, and then returned to the neutral (or zero) position. After another short period of time the control is then displaced in the opposite direction (as, for example, to b or d), held there for a short period of time, and returned once more to the neutral position. The sequence of control movements at the left of the diagram shows mathematically exact *step inputs.* Although no control operates with such instantaneous precision, this artificial example will help to illustrate the nature of the control dynamics in each of the systems discussed. The sequence of control movements on the right is somewhat more realistic because it shows the control moving gradually from one position to another. Notice that the total displacement for the control movements on the right, that is, from 0 to c or from 0 to d, is only half those on the left, that is, from 0 to a or from 0 to b.

The curves in the second line of Figure 5 show the system output for each of these control sequences when a zero-order control system links the two. The curves representing the system output have exactly the same shapes as those representing the control output, being merely re-

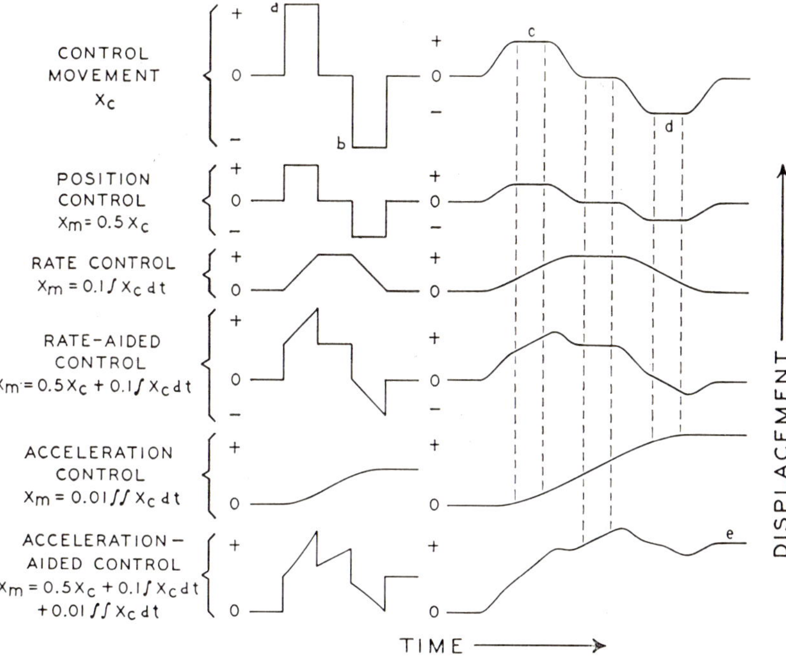

Figure 5. The top line shows two sets of hypothetical control movements. Underneath are the corresponding movements of the system (system output) for each of several kinds of control linkages.

duced by the factor K (0.5 in this example). K could, of course, be greater than 1, in which case the system output would be greater than the control output.

The next level of complexity, a first-order control (upper half of Figure 6), is one in which the operator's control output directly determines the *rate of change* of the machine output. This is commonly called a rate control, or velocity control. The third line of Figure 5 shows the machine output of a rate control system for each of the two control sequences in the top line. Note that a displacement of the control produces a rate of change of position in the machine—the rate of change in the machine being proportional to the displacement of the control. Except for time lags and certain other minor complicating factors, an automobile approximates a first-order control system. A fixed displacement of one of the controls (the accelerator) produces a

FIRST-ORDER CONTROL

A. RATE (OR VELOCITY) CONTROL

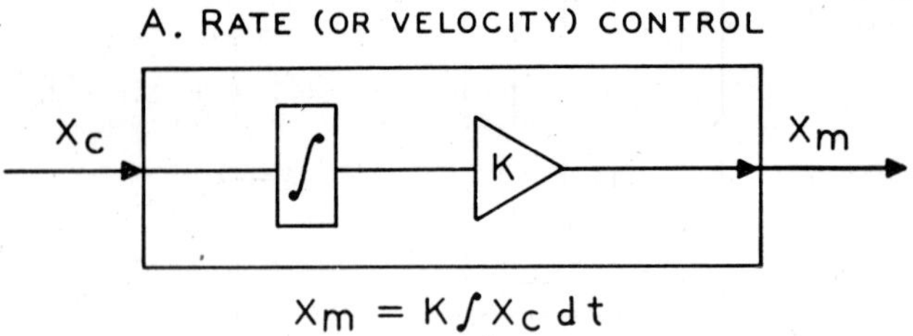

B. RATE-AIDED (OR VELOCITY-AIDED) CONTROL

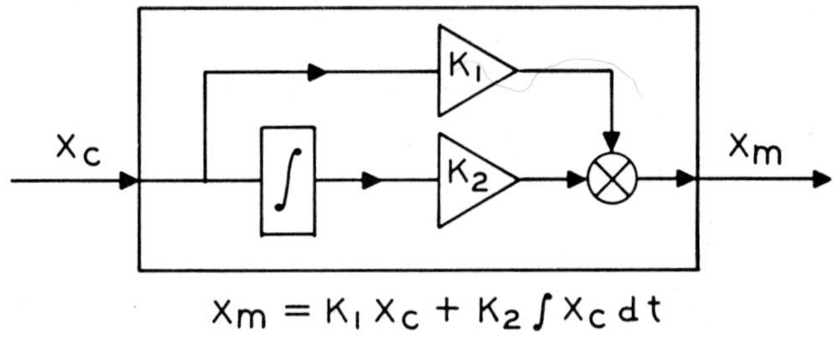

Figure 6. Two types of first-order control.

constant velocity in the system output (the forward movement of the automobile).

First-order control systems very often include the lower-order position term as well, in which case they are called rate-aided, or velocity-aided control systems (see the lower half of Figure 6). The output of such a control system, illustrated by the curves in the fourth line of Figure 5, is the algebraic sum of both the position and the velocity components of the system; i.e., the two curves on line four are the sums of the curves on the two lines immediately above. In the design of rate-aided systems, an extremely important consideration is the proper selection of values for the constants K_1 and K_2, or, to be more exact, the ratio K_1/K_2, called the rate-aiding constant. A considerable amount of research suggests that the best all-around value for this ratio is 0.5, with values between 0.2 and 0.8 constituting an acceptable range. The values shown in Figure 5 were selected because they seemed to be appropriate for this graphical illustration; they are clearly inappropriate for any real control system.

Still more complexity is shown in the second-order control system in the upper half of Figure 7. Here the output of the operator's control

SECOND-ORDER CONTROL

A. ACCELERATION CONTROL

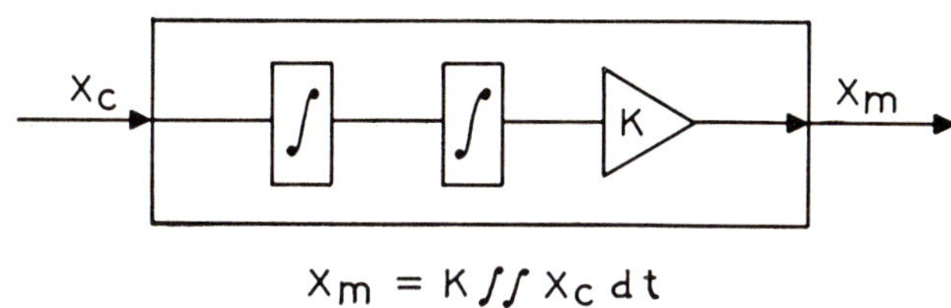

$$X_m = K \iint X_c \, dt$$

B. ACCELERATION-AIDED CONTROL

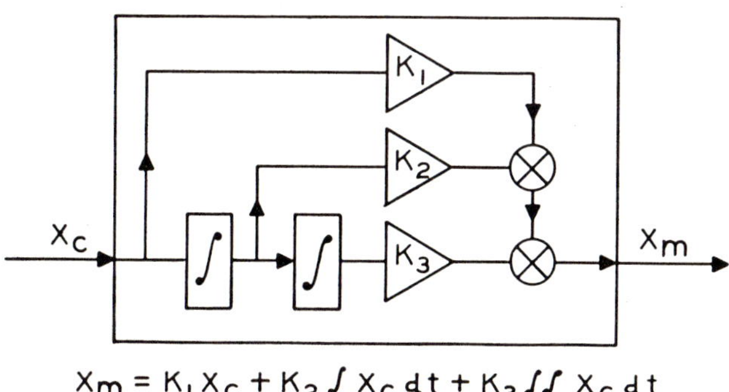

$$X_m = K_1 X_c + K_2 \int X_c \, dt + K_3 \iint X_c \, dt$$

Figure 7. Two types of second-order control.

determines the acceleration of the machine output. The schematic representation of such a system output is shown in the fifth line of Figure 5. Note that when the operator's control is displaced and held at a fixed position the machine movement keeps increasing at a faster and faster rate. When the operator's control is returned to its neutral position after an initial displacement, the machine continues to change its position at a constant rate. To stop the machine movement the operator must make another displacement in the opposite direction from the first.

Second-order control systems frequently include both the lower-order terms, one for position and one for rate. Such systems (illustrated in the lower half of Figure 7) are commonly called acceleration-aided systems. The performance of such a system is shown in the bottom line of Figure 5. As in the case of the rate-aided system, the selection of the appropriate constants is a critical matter if a man is to use the system effectively. Research suggests that these constants should be in the ratio 1:2:8 or 1:4:8. The constants illustrated in Figure 5, being selected for their graphical convenience, are clearly unsuited for any real control system.

Although control systems of still higher order occur commonly (See Figure 8), they follow the general pattern illustrated in the preceding discussion.

Now it is possible to go back to the beginning of this discussion and restate the general principle involved:

> Aiding should be used in closed-loop tracking systems when the input (or desired output) is characterized by a preponderance of movements which occur at a constant rate, a constant acceleration, or some constant higher derivative. The number of terms used in aiding should exceed by one the derivative of the input which is constant. Thus, for a constant input rate there should be three terms in the aiding (viz., position, rate and acceleration); for a constant input acceleration there should be four terms in the aiding, and so on. The outcome of applying this principle is that it relieves the operator of the necessity for solving higher-order differential equations and allows him to behave as a simple amplifier.

Since general synthetic principles are of concern here, and not specific mechanisms, the problem of how one designs integrations into control systems will be ignored, except for pointing out that a single integration can be simply achieved by using a viscously damped joystick. This type of control responds with an angular velocity proportional to the magnitude of the force applied by an operator. A double integration can be closely approximated by a control having high inertia. Such a control responds with an angular acceleration which is proportional to the force applied by the operator.

OPERATOR NEEDS IMMEDIATE INFORMATION ABOUT THE EFFECTS OF HIS CONTROL MOVEMENTS

Another important principle relating to the synthesis of men into closed-loop control systems concerns the information on the display which is provided the human operator. The preceding discussion showed that aiding, in the form of successive integrations of control outputs, can be very helpful in assisting the human operator to track *complex* inputs. The other side of the coin, however, is that aiding makes it very difficult to track *simple* inputs. For example, consider the kind of control movement an operator must make in producing a simple change in the position of a system from one point to another. If the operator has an acceleration-aided control he must make a complicated series of control movements of the general form shown in the upper right hand corner of Figure 5. The system output, shown in the lower right hand corner of the same figure, will be a simple change in position from 0 to e. Moreover, the control movements required to move the system from one place to another cannot be much simpler than those shown in the figure and will ordinarily be much more complicated. Control movements of this degree of complexity are difficult to learn and to execute with precision. As the control order increases (the integration terms increase) the control problem becomes almost impossibly difficult for a man to learn.

The situation may be still further complicated by the fact that the higher-order components may not have been deliberately designed into the system, but may instead be a function of the system itself and of the medium through which it moves. An example is provided by fleet-type submarines in which a single planesman controls the depth of the submarine. He has a control (a wheel), the position of which directly affects the rate of change (first derivative) of the bow plane angle (the control surface). The angle of the bow plane directly affects the acceleration (second derivative) of pitch, and the pitch of the submarine directly affects the rate of change of depth. As a result, the planesman controls the fourth derivative of depth, or, to turn this statement around, there are four integrations between the output of the operator's control and the system output (See Figure 8). Unfor-

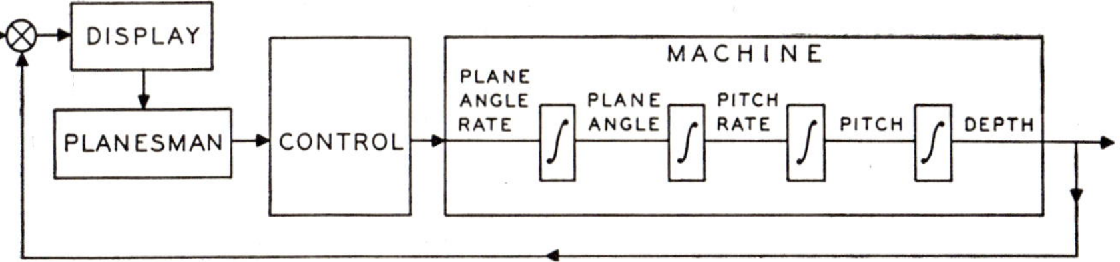

Figure 8. Block diagram of the control system for a fleet-type submarine. (After Ely, *et al.* (12).)

tunately, there is not very much the designer can do about this situation so long as submarines move through water and are controlled primarily by bow planes. Since the movements the planesman wants to produce are simple changes in position from one depth to another, he is provided an excessively complex system. Note, incidentally, that a similar kind of dynamic control situation occurs in aircraft, although the movements occur much more rapidly than in submarines.

It is virtually impossible for a human operator to learn to control a system of this sort if he is provided only with a single display showing the output of the system (depth in the case of the submarine). For this reason, higher-order systems of this type normally provide the operator with an array of displays. One provides output information and the rest provide derivative information. In the case of the submarine, the operator is usually provided with displays showing depth (system output), depth rate or pitch (the first derivative of depth), pitch rate (the second derivative of output), and bow plane angle (the third derivative of output). Even with such an array of displays, the operator's task is still quite difficult. He needs to know how each indicator must move in order to achieve a desired output, and he can function well only if he (a) reaches a high level of skill, and (b) pays close attention to all his displays.

The design principle which this discussion has been leading up to is that the system should contain a single display which gives the operator immediate knowledge of the results of his own control actions before they become available from sensing of the system's actual outcome. This is done by a process called "quickening." A quickened display is actually a composite of the system output (X_m) and all its derivatives (See Figure 9). The effect of a perfectly quickened display

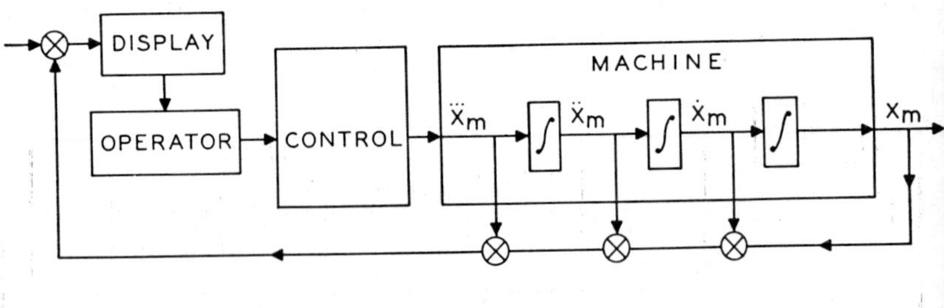

$$X_{DISPLAY} = K_1 X_m + K_2 \dot{X}_m + K_3 \ddot{X}_m + K_4 \dddot{X}_m$$

Figure 9. Quickening applied to the control system illustrated in Figure 8.

is to tell the operator exactly where to position his control at all times. Actual tests of such displays show impressive gains in operator and systems performance over conventional, unquickened displays (See (5, 27)). Two examples of quickened aircraft displays are the Sperry Zero Reader (20) and the Instrument Approach System Steering Computer (1).

DESIGNING THE SYSTEM FOR EFFICIENT FAULT LOCATION

Another general principle of systems design concerns the design of the entire assemblage for the efficient diagnosis of faults. This has become especially important in automatic machine systems, not only because of the scarcity of trained technicians, but also because of the heavy financial investment which automatic machinery represents. The more expensive the plant, the less the time it can afford to remain idle. With automation, breakdown of a single control unit may stop production in the whole factory until the fault is located and corrected.

Maintenance problems are even more pressing when the complex equipments which are now used by our military forces are considered. According to accurate figures compiled by electronics maintenance personnel, one radar gunsight required eighteen hours of maintenance for each hour of operation in the aircraft. The B-36 aircraft uses 2,700 vacuum tubes in its electronic equipment. Twenty of these tubes may be expected to fail during a routine twelve-hour mission. An Air Force spokesman has reported that if only one in 5,000 cold-solder joints fails, approximately 63 per cent of certain guided missiles will fail. All of these facts can be summarized in this way: To keep electronic equipment working throughout its useful life the military spends ten times the original cost of the equipment. The human part of this story becomes apparent when it is noted that, of the total time a typical piece of electronic gear is down, 60 to 80 per cent of that time is spent just *locating the trouble.*

The principles of information theory have been put to good use to simplify the task of fault-finding in complex systems. By means of a stepwise, successive halving of probabilities, it is possible to minimize the number of steps required to make diagnostic decisions (7, 17). Note, however, that such simplified diagnostic procedures cannot be applied to any system of machinery. The system has to be specifically designed with the maintenance man in mind. An illustration is provided by the design of the television micro-wave system already described. When faults occur in one of the television towers or its associated control equipment, remotely located engineers can interrogate the distant station to discover what the exact trouble is before they send out a repair crew.

The Human Engineering of Systems Components

By this time it should be abundantly clear that human operators are still used in one way or another in all automatic and semiautomatic machine systems. This being the case, there is usually a lot of human engineering work to be done on the components of these systems. So much research[3] has been done on this aspect of systems design that it would be impossible to give an adequate summary of it in only a few pages. Those interested in a more comprehensive review of this literature should consult appropriate textbooks (10, 25). Only a few of the hundreds of human engineering findings which can be applied to the design of systems components will be cited here.

THE SELECTION OF VISUAL INDICATORS

There are three basic types of symbolic indicators which are commonly used for conveying information from a machine to an operator or vice versa. These are: the moving pointer with a fixed scale, the moving scale with a fixed pointer, and the direct reading counter (See Figure 10). The most important factor which should govern the choice of one or another of these three indicators is the use to which the instrument is to be put. The results of a large number of experiments on such indicators are summarized in Figure 10 in the form of positive recommendations. Along the left side of the table are various usages of these indicators considered from the standpoint of the operator who must do something with the information. The signs inside the table contain the recommendations. Three plus signs indicate satisfactory usages; a single plus indicates an acceptable, but not completely satisfactory, usage; a minus sign indicates a usage which should be avoided.

Indicators are most commonly used to provide precise quantitative

[3] The bibliography by McCollom and Chapanis (24) contains over 5,000 titles; the one prepared at Tufts University (28), over 2,000 titles.

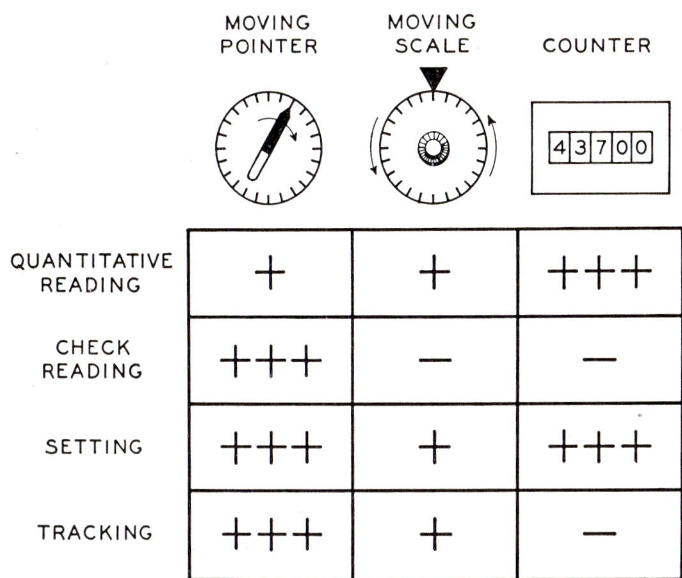

Figure 10. Recommended uses for three basic types of symbolic indicators.

information about a machine or instrument setting. In general, the counter is the best indicator for this purpose. Research studies show that operators can get exact numerical values fastest and with fewest errors when the counter is used for displaying quantitative data.

The second major use of visual indicators is for check reading. Much more often than many people suppose, the operator of a piece of equipment is not concerned with precise quantitative readings but only with a rough check on the functioning of his equipment. The pilot, for example, usually does not care whether the manifold pressure is 47.2 or 48.9 inches. He merely wants to know if the pressure is all right, or not all right. The typical driver of an automobile is not concerned about whether the temperature of his engine is 140, 150, or 160 degrees. He glances at the temperature gage occasionally to see if the engine is in the normal range of operating temperatures and becomes concerned only if it is too hot. Whenever an operator is interested merely in discovering whether a machine is functioning within or outside of a normal range, he is obtaining a check reading. For this purpose the moving pointer instrument is excellent. The location of the pointer is easily detected, and it provides a useful cue in making check readings. The operator soon learns that when the pointer is oriented in a particular direction the system is functioning normally. In fact, after only a little familiarization with such an indicator the operator usually does not read the numbers or the scale at all. In addition to these advantages, changes in setting are easily detected so that an operator can readily see if the quantity being measured is increasing or decreasing. The moving scale and counter, on the other hand, are poor for check reading purposes. Neither of these instruments permits an operator to judge the direction and magnitude of a deviation without first reading numbers or a scale.

The third major use to which indicators are often put is to set information into a machine. For this purpose the moving pointer and counter are both excellent, although sometimes the designer might want to use one rather than the other. With the moving pointer instrument it is possible to design simple and direct relations between the motion of the pointer and the motion of the control. In addition, the position of the pointer helps the operator to monitor the changes which he makes with his control. The counter also allows the operator to monitor numerical settings with great precision. However, the relationships between the motion of a control and the movement of the counter dials are always somewhat ambiguous. An additional disadvantage of the counter is that it is ordinarily not readable during rapid movements. The disadvantages of the moving scale are amplified below.

The fourth major use to which indicators are put is tracking. We have already mentioned several kinds of tracking systems earlier in this chapter. The display for a tracking system might be, for example, a compass or a speedometer. For tracking purposes, the moving pointer instrument is clearly the best. The position of the pointer is readily monitored and controlled. In addition, this instrument provides for the most simple and direct relationships between the movements of the pointer and the control which activates it. The moving scale indicator is only fair for tracking. There is no pointer position to aid in monitoring, and there is generally an ambiguous relationship between the movements of the scale and the control. The counter is poor for tracking purposes because changes on the counter are difficult to monitor. The relationships between the movements of the counter and the control which activate it are ambiguous, and the counter is not generally readable during rapid changes.

THE MOVING SCALE INDICATOR VERSUS COUNTER

An application of the research conclusions summarized in Figure 10 is illustrated in Figure 11. The control on the left is a moving scale

Figure 11. On the left is the Model 746 Counting Dial; on the right, the Model 1301 Microdial manufactured by the Borg Equipment Division of the George W. Borg Corporation.

indicator commonly found on many types of panels and consoles. In the center of the scale is a knurled knob which can be turned by hand. The scale shown above the knob moves in the same direction as the knob. Actually, there are two scales—one division on the inner one equals one complete revolution of the outer one. The moving scale indicator has some interesting psychological characteristics. It is impossible to design a conventional indicator of this type without violating one of three important design principles (8):

1. A clockwise rotation of the knob should increase the value of the thing being controlled. For example, if this control were used to set frequency on an oscillator, a clockwise rotation of the knob should increase frequency.
2. The knob and the scale should move in the same direction.
3. The numerals on the scale should increase in a clockwise direction around the scale in order to reduce reading errors.

One sees from Figure 11 that this moving scale indicator violates the third principle. After trying to read the setting on that indicator, perhaps one will appreciate why operators make errors in reading such indicators.

On the right of Figure 11 is another indicator making use of a direct reading counter. In this case the operator rotates the knurled flange around the outside of the control. Clockwise rotation of the flange increases the numbers in the windows; counterclockwise rotation decreases the values. Both of these control devices are available commercially from the same company. It should not be surprising to learn that experimental studies of these types of controls show that the one on the right leads to far fewer setting and checking errors (32). Incidentally, note that both indicators in Figure 11 are set to the same value.

THE DESIGN OF SCALES

A considerable amount of research effort has gone into the study of scales and the way in which they should be designed. Figure 12 shows in tabular form the recommendations which have come out of these studies.

The left column of this figure shows the graduation interval value. This is the value of the smallest division on the scale. Notice that there are only three values recommended—the 1, 2, and 5, or decimal multiples of these digits. No other values are acceptable because research shows that people cannot readily interpret or use scales which are graduated in other ways. The numbered interval values are the values between the major graduation marks to which numerals are attached. Here again the only recommended numerals are the 1, 2, and 5, or decimal multiples of them. Fortunately, research shows that one can multiply or divide the basic numerical values on scales by factors of ten or a hundred without appreciably affecting the speed or accuracy with which scales can be read. This means that a scale which is numbered 0, 1, 2, 3, etc., is just about as easily read as one which is numbered 0, 10, 20, 30, etc.

At the extreme right of Figure 11 are shown recommendations about the graduation marks to be used on the scales. Notice that there are three kinds of graduation marks: major, minor, and intermediate marks. Some scales make use of all three types of markings; some use only the major and minor marks; some use the major and intermediate

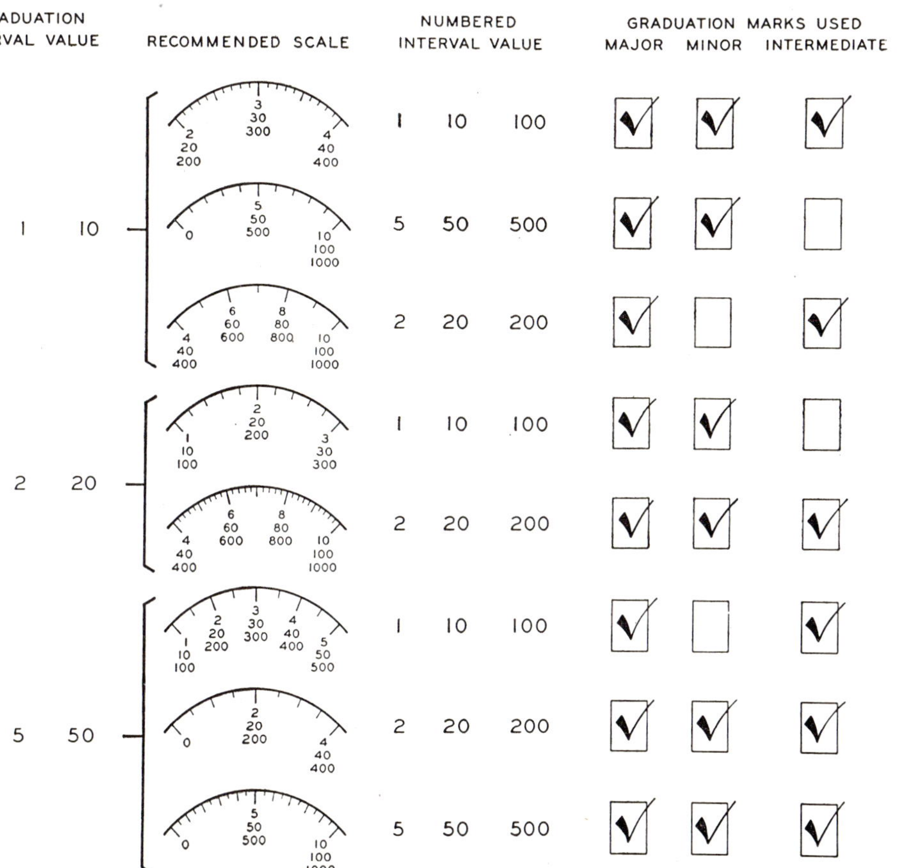

Figure 12. Recommendations concerning the design of scales. (Adapted from Baker and Grether (6).)

marks. With the samples shown in the figure it should be possible to redesign any scale to conform to these recommendations.

AN ILLUSTRATION OF THESE PRINCIPLES

It is important to point out that the problems discussed here are not abstract laboratory problems. Illustrations of poor dial design are easy to find. Figure 13 is only one of many such illustrations. This

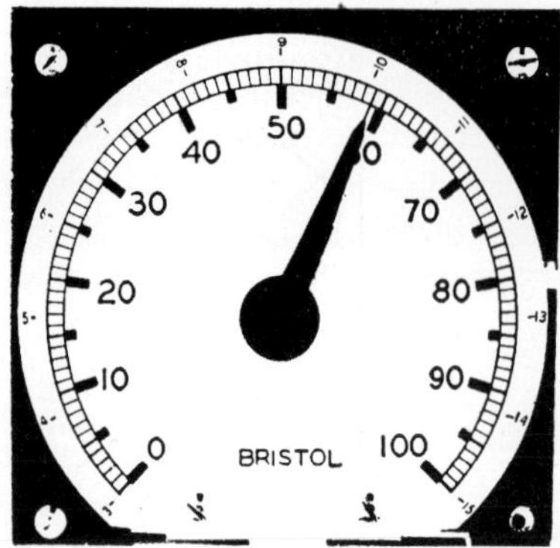

Figure 13. On the left is a pneumatic indicator dial as originally manufactured; on the right, the same dial after redesign. (After Fleming (14).)

is an interesting one because it is a "before-and-after" illustration. The dial on the left shows a type of scale which had been manufactured by the Bristol Company, the manufacturer of a wide variety of meters and recording instruments. The scale on the right is one which was redesigned on the basis of good human engineering practice. The comparison between the two is so striking that it hardly requires any discussion. In addition to being more legible, the scale on the right employs a much more satisfactory numbering system.

The experience of this company is instructive in this connection. Purchasers of one of their instruments frequently misread critical settings and so thought the instrument was out of calibration. After the dial had been redesigned, these misreading errors were greatly reduced and customers thought the instrument was much more stable. As reported by J. G. Fleming: "We are just amazed at the consequence of simply changing scale-markings." Other examples of "before-and-after" designs can also be found in Fleming's article (14).

Peak Clipping and the Intelligibility of Speech in Noise

Quite a different human engineering application concerns a technique for improving the intelligibility of speech in noise. This principle is primarily applied to the design of components used in communication systems. The technique involves doing something which sounds unbelievable. To make speech more intelligible in noise one starts by chopping some of it off and throwing it away. Since this kind of mutilation makes speech sound a little unnatural, it is referred to reasonably enough as a kind of distortion, and since the chopping is applied to the intensity, or amplitude, dimension of speech it is called *amplitude distortion.*

Figure 14 illustrates the kind of amplitude distortion to be discussed

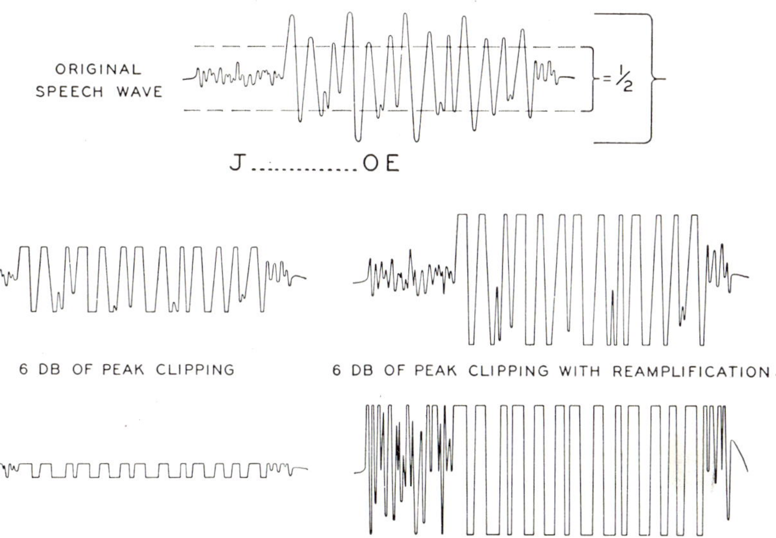

Figure 14. Peak clipping applied to a speech wave. (After Licklider, *et al.* (23).)

here. It is called peak clipping. At the top of the illustration is a schematic representation of the wave form resulting when a male voice says the word "Joe." Notice that there is first a series of low-amplitude, high-frequency vibrations corresponding to the "J" sound. These are followed by some high-amplitude, low-frequency vibrations corresponding to the "O" sound. The former does not refer to the sound of "J" when pronounced as a letter of the alphabet (as "Jay"), but rather to the sound of "J" in the word "Joe."

Start to say the word "Joe" but do not actually come out with the "O." Do this a couple of times. Now make the "J" sound and follow it with the "O" sound. Do this carefully several times and notice that the "O" sound is very much louder than the "J" sound—that is, the former has much more intensity or amplitude. In addition notice that the "O" sound is lower or deeper than the "J" sound—that is, it has a lower pitch and lower base frequency. The uppermost illustration in Figure 14 shows these two facts in terms of the sound waves involved.

Now imagine that a mechanism is available which could chop off the highest peaks of the sound wave, as is illustrated by the dotted lines at the top of Figure 14. Any part of the wave falling outside the dotted lines is clipped off and thrown away. In the illustration the distance between the dashed lines is approximately one-half of the vertical distance from the highest to the lowest points of the original wave. In the decibel system of notation this corresponds to almost exactly 6 decibels (db) of peak clipping. The next part of the figure shows 20 db of peak clipping. With this much peak clipping, about nine-tenths of the speech wave is cut off and discarded.

Figure 14 shows also that when one peak clips speech he also reduces the average amount of power: clipped speech is not as loud as unclipped speech. Suppose, however, that one were to reamplify or expand the clipped speech until it had the same peak intensity as the unclipped speech. The illustrations on the right of Figure 14 show what these waves would now look like. This procedure is a sensible one from an engineering point of view since there are many communication systems which have a limited amplitude-handling capacity. By peak clipping and reamplifying the wave better use is being made of this capacity; that is, speech is being packaged more efficiently.

How much is gained by this operation? Figure 15 shows only one out of many sets of data which have been collected on this problem. These data were obtained by putting the talker in a quiet room and the listener in a noisy one. The listener then tried to understand (a) unclipped speech at various peak amplitudes, and (b) speech which had been clipped by various amounts and then reamplified to the same peak amplitude. Take, for example, the data for speech at a peak amplitude of 80 db. Under the particular noise conditions in which these tests were made, the average listener could get scarcely 1 per cent of the unclipped words spoken. With 12 db of peak clipping, the average listener could hear and understand about 40 per cent of the words; and with 24 db of peak clipping, approximately 75 per cent of the words. This is an impressive gain.

The explanation for this rather unusual state of affairs is not hard to find. The low intensity components of the speech waves are the

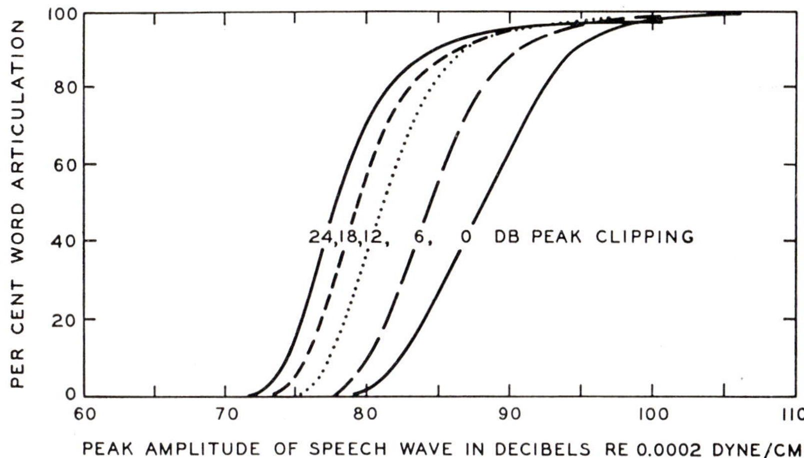

Figure 15. The effects of peak clipping on the intelligibility of speech. Each curve shows word articulation as a function of speech intensity for different amounts of peak clipping. (After Licklider (22).)

semivowels and consonants. The high intensity of the speech wave consists of the vowels and diphthongs. In general, the vowels may have as much as 1,000 times the power of some of the consonant sounds. It is very difficult to hiss in a loud voice, for example, but it is easy to shout "O." When one peak clips speech waves he cuts off some of the power of the vowel sounds. This is not very disastrous because the vowel sounds do not contribute very much to the understandability of speech. It is the consonant sounds which are more important for producing understandable speech. By clipping the speech and then reamplifying it, one chops off some of the intensity of the vowel sounds and expands the intensity of the consonant sounds. Another way of saying it is that one discriminates against the vowels in favor of the consonants.

This principle is a general one having wide applicability. One case in which it has been put to good use is in the design of radio transmitters where the amplitude-handling capacity of the system is limited by the amount of modulation possible on the carrier signal. Another application is in the design of hearing aids where the peak amplitude is limited by a psychological consideration—if the amplitude is too great it tickles the ear, or may even become uncomfortable or painful.

CONTROL FORCES AS A FUNCTION OF THE DESIGN OF CONTROLS

Now that there are all sorts of power-aiding mechanisms to assist man in operating controls, the problem of control forces is not quite as important as it used to be. It has not entirely disappeared, however, because there are still some situations in which man has to furnish all of the force needed to operate certain controls. Take, for example, handles for emergency doors and escape hatches in tanks, ships, and aircraft. If the vehicle is disabled, one cannot rely on power-aiding devices to help turn the handles. There may not be any engines running to provide the power needed. A further complication arises from the fact that if the vehicle has been disabled because of a collision the entire frame may have been twisted so that normal amounts of

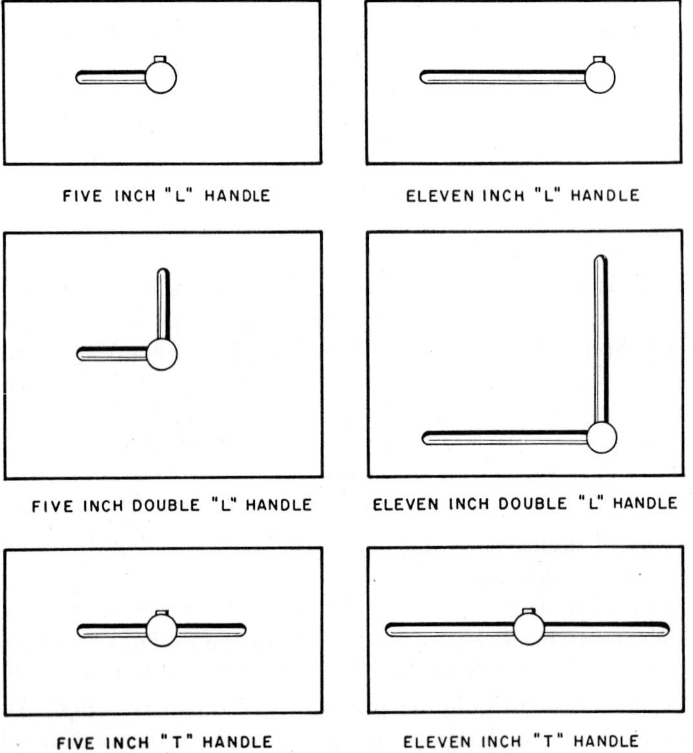

Figure 16. Six handles used in a study of control forces by King, Bruyere, and Swearingen. (Unpublished research)

force may not be enough to turn the handle. Under these circumstances, the designer will do well to provide a large margin of safety.

Some typical findings which have come out of research in this area are illustrated in Figure 16. Shown are six different door handles studied by King and his co-workers at the Civil Aeronautics Medical Research Laboratory. Although the study was concerned with door handles for aircraft, the findings are general and should be useful in many other situations as well.

Three types of handles, each in two different sizes, were studied. They were mounted on a mock-up of a main cabin door for an aircraft and were oriented in any one of eight positions 45 degrees apart. Subjects were required to turn the handles either clockwise or counterclockwise and with one or two hands. Figure 17 shows the average

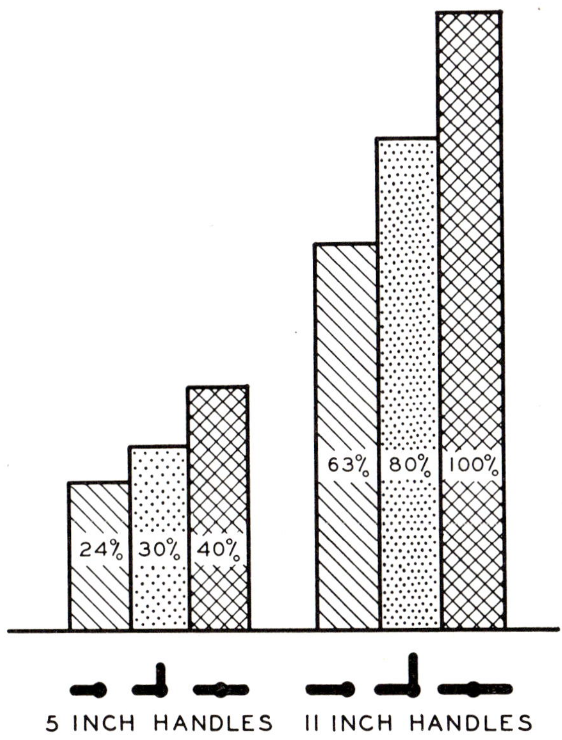

Figure 17. Average maximum static forces which could be exerted on the handles shown in Figure 16. The averages are expressed as percentages of the force which could be exerted on the 11-inch T-handle.

maximum forces which could be exerted with each of these controls. The 11-inch T-handle was by far the best. The average maximum force which could be exerted with it was roughly 4,000 inch pounds. This value has been assigned an arbitrary value of 100 per cent and all other handles have been compared with it. Two important general rules show up clearly:

1. The 11-inch handles are uniformly better than the 5-inch handles.
2. For a given size the T-handle is superior to the double L-handle, and the latter, in turn, is better than the single L-handle.

THE HUMAN ENGINEERING OF COMPONENTS FOR MAINTENANCE

Maintenance problems created by the rapid growth of automatic and semiautomatic equipment occur not only in large systems of machines, but throughout the full gamut of modern instruments and devices. The Consumer Service Division of the Detroit Edison Company, to take one example, has reported on the excessive amount of servicing and repairs required on both small and major appliances. In a typical year, this company had to repair 600,000 small appliances. One in every 7.5 toasters in the Detroit area, one in every 9 clocks, and one in every 5 coffeemakers required some type of repair. A major criticism which arose concerned the design of the equipment from the standpoint of its maintenance. In some instances appliances had to be taken apart completely to replace something as trivial as a pilot light.

The picture, however, is not completely negative. A company which has a good record in this respect is the telephone company. Because it is responsible for its own maintenance, the telephone company in America seems to have been more alert to the problems of maintenance than many other industries. In new developments for the telephone system, maintenance considerations are often given prominence equal to that given other factors. A typical illustration is contained in an article by Steward which describes the development of a special purpose switchboard for telephone answering services (30). The article describes the switchboard, considers carefully how it is integrated into the main telephone system, how subscribers' phones can be connected to the switchboard, and how subscribers' privacy can be insured. But included in the same article is a discussion of how the equipment can be serviced. Relay equipment and answering-service line units are in the rear of the board on vertical mounting plates, each of which is hinged at the base so that it can be lowered for easy maintenance. In addition, all units and line cables can be readily installed, removed, or replaced through the use of jacks and plugs. The important point is that this kind of convenience and accessibility did not just happen—it was designed into the equipment.

The human engineering recommendations which can be used in designing equipment for efficient maintenance are so varied and so numerous that it would be impossible to do them justice here. They are sufficiently important, however, that they should not be ignored completely. The figures below illustrate three of them graphically. Figure 18 shows one way in which test points designed into cable connections can help the maintenance man more readily localize sources of malfunction. Figure 19 shows how color coding can help the mainte-

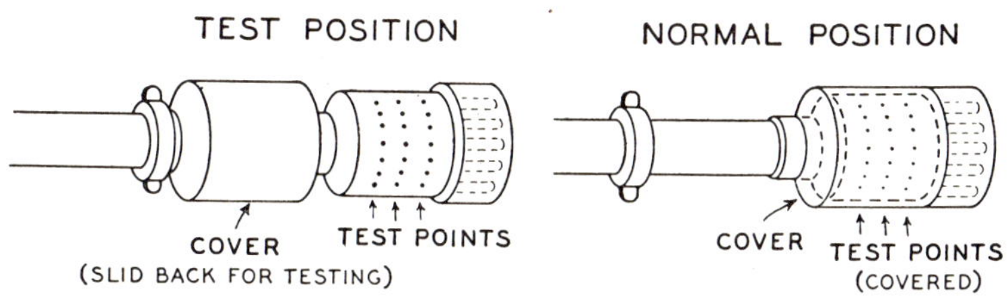

Figure 18. Test points on cables are one way of helping the maintenance man locate faults efficiently. (From Folley and Altman (15).)

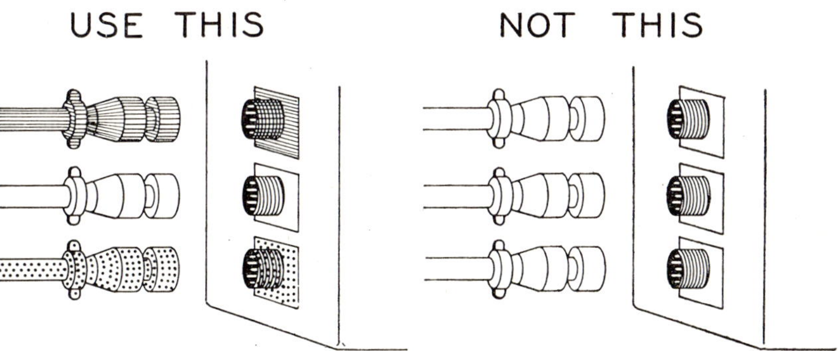

Figure 19. Color coding of plugs and receptacles is a convenient means of helping the maintenance man attach cables to the correct receptacles. (From Folley and Altman (15).)

nance man to untangle the numerous cables he is likely to find in complex equipment and to connect the correct ones together. Finally, Figure 20 illustrates only two of many sizing considerations which are so often overlooked in the design of modern equipment.

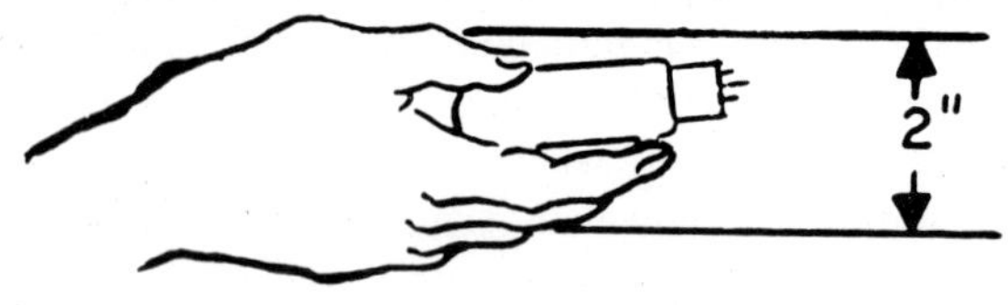

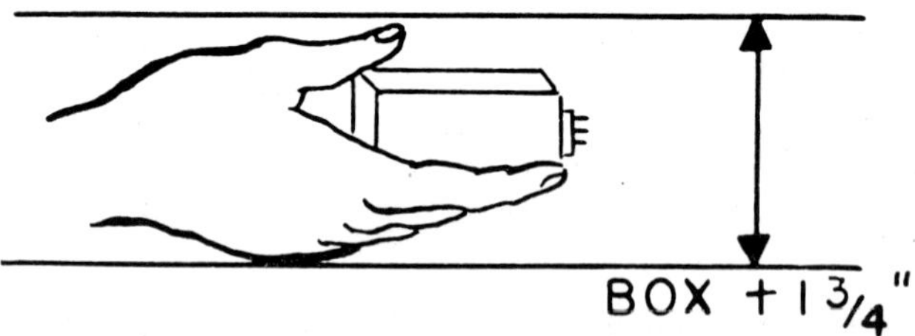

Figure 20. Minimum allowable access spaces for replacing a tube or miniature component. (From Folley and Altman (16).)

SUMMARY

It would be impossible to summarize adequately the major human engineering design practices which can be applied to the design of systems components. The few illustrations discussed above merely serve to illustrate the range of problems on which the human engineering specialist can give advice.

Systems Evaluation

The fourth and final area in which human factors become involved in systems engineering is in the evaluation of systems. Most current definitions of systems engineering emphasize the importance of evaluating the entire system before it is actually put into production. As noted earlier, many systems include one or more human operators as a vital part of the system. Thus, any reasonable attempt to evaluate the performance of the system must include an evaluation of the performance of the human operator working with the equipment provided him. The automatic interceptor aircraft referred to earlier carries a pilot. For this reason the systems engineer responsible for the design of this system points out that it is necessary to evaluate the capability of personnel to accomplish the functions assigned them. Tests have to be made of the interrelationships between the operator and the automatic equipment to find out how well the combination will work.

Another illustration is provided by the Bizmac system already referred to on several occasions. When the system was fairly well along in its design, tests were set up and run under simulated operating conditions: (a) to test the operating conditions; (b) to get some measure of operator load at the various operating positions; (c) to see what the effects of equipment breakdown were on over-all system performance, and (d) to get some indication of the adequacy of the proposed operator and machine combination.

There is little need to multiply examples of this sort. Systems engineers are frequently responsible for designing and running tests on complicated man-machine combinations. Those who have actually run some appreciate that such tests are much more complicated than doing ordinary engineering evaluations. People differ, and tests run on one subject may not be at all typical of the performance of the average person. People learn during the course of an experiment; often they become bored or fatigued; they are sensitive to the kinds of instructions which precede the experiment and to words of praise or reproof given during it; they interact with the experimenter in strange and sometimes unexpected ways—they may try to outguess the experimenter, or, on occasion, may deliberately sabotage the outcome of the tests. These and still other factors must be anticipated and controlled if the results of man-machine experiments are to be trusted.

Psychologists, of course, have been face to face with these problems for years and, in the course of their work, have evolved techniques for handling many of them. For this reason it is perhaps not surprising to find that psychologists are frequently consulted for their advice on problems of technique and methodology. A monograph (9) prepared recently is an attempt to pull together some of this information in a single source.

SUMMARY

In summary, this chapter has covered some of the human factors involved in the design of automatic and semiautomatic machine systems. It has demonstrated that there are still quite a few human problems in most such systems, and that a successful automatic system requires the engineer to consider carefully the role of the human operator and how he is designed into the system. Finally, it has suggested why the human factors specialist is often considered a member of the systems design team, and in what ways he can contribute to the important and challenging task of designing new systems for our automatic world of tomorrow.

REFERENCES

(1) Anderson, W. G., and Fritze, E. H. "Instrument Approach System Steering Computer," *Proceedings of the Institute of Radio Engineers,* Vol. 41 (1953), 219-28.

(2) Anon. "Look, No Hands on the Throttle—No Throttle!" *Industrial Science and Engineering,* Vol. 3, No. 1 (January, 1956), 27.

(3) Anon. "Numerical Positioning Control," *Electrical Manufacturing,* Vol. 56, No. 3 (September, 1955), 122-27.

(4) Anon. "Pilots Are Only Human!" *Ryan Reporter,* Vol. 17, No. 4 (August, 1956), 1, 39.

(5) Bailey, A. W. "Simplifying the Operator's Task as a Controller," *Ergonomics,* Vol. 1 (1958), 177-81.

(6) Baker, C. A., and Grether, W. F. "Visual Presentation of Information," *WADC Technical Report 54-160,* Wright Air Development Center, Wright-Patterson Air Force Base, Ohio, 1954.

(7) Bean, E. S., and Woodson, W. E. "Human-engineered Design for Reliability and Maintenance," *Electrical Manufacturing,* Vol. 56, No. 1 (July, 1955), 100-5.

(8) Bradley, J. V. "Desirable Control-display Relationships for Moving-scale Instruments," *WADC Technical Report 54-423,* Wright Air Development Center, Wright-Patterson Air Force Base, Ohio, 1954.

(9) Chapanis, A. *The Design and Conduct of Human Engineering Studies.* San Diego: San Diego State College Foundation, 1956.

(10) Chapanis, A., Garner, W. R., and Morgan, C. T. *Applied Experimental*

Psychology: Human Factors in Engineering Design. New York: John Wiley and Sons, 1949.
(11) Collins, C. A., and Hofman, L. H. "Switching Control at Television Operating Centers," *Bell Laboratories Record,* Vol. 35, No. 1 (January, 1957), 10-14.
(12) Ely, J. H., Bowen, H. M., and Orlansky, J. "Man-Machine Dynamics," *WADC Technical Report 57-582,* Wright Air Development Center, Wright-Patterson Air Force Base, Ohio, 1957.
(13) Fitts, P. M. (ed.). *Human Engineering for an Effective Air-Navigation and Traffic-Control System.* Washington, D.C.: National Research Council, 1951.
(14) Fleming, J. G. "Improve Power-Plant Instrumentation by Applying Human-Engineering Data," *Power,* Vol. 98, No. 1 (1954), 86-9 and *passim.*
(15) Folley, J. D., and Altman, J. W. "Selecting and Applying Wiring, Cables, and Connectors," *Machine Design,* Vol. 28, No. 15 (1956), 92-4.
(16) Folley, J. D., and Altman, J. W. "Basic Recommendations for Designing Maintenance Accesses in Electronic Equipment," *Machine Design,* Vol. 28, No. 16 (1956), 99-102.
(17) Folley, J. D., and Altman, J. W. "A Systematic Approach to Preparing Maintenance Procedures," *Machine Design,* Vol. 28, No. 25 (1956), 124-27.
(18) Goode, H. H., and Machol, R. E. *System Engineering.* New York: McGraw-Hill, 1957.
(19) Halstead, W. K. "The RCA Bizmac Electronic Accounting System," *RCA Engineer,* Vol. 1, No. 4 (1955-56), 12-21.
(20) Kellogg, S., and Fragola, C. F. "The Sperry Zero Reader," *Aeronautical Engineering Review,* Vol. 8, No. 11 (1949), 22-31.
(21) Leas, J. W. "Engineering the RCA Bizmac System," *RCA Engineer,* Vol. 1, No. 4 (1955-56), 10-11.
(22) Licklider, J. C. R. "Effects of Amplitude Distortion upon the Intelligibility of Speech," *Journal of the Acoustical Society of America,* Vol. 18 (1946), 429-34.
(23) Licklider, J. C. R., Bindra, D., and Pollack, I. "The Intelligibility of Rectangular Speech-waves," *American Journal of Psychology,* Vol. 61 (1948), 1-20.
(24) McCollom, I. N., and Chapanis, A. *A Human Engineering Bibliography.* San Diego: San Diego State College Foundation, 1956.
(25) McCormick, E. J. *Human Engineering.* New York: McGraw-Hill, 1957.
(26) Owings, J. L. "Human Engineering the Bizmac System," *RCA Engineering,* Vol. 2, No. 3 (1956), 16-22.
(27) Ritchie, M. L., and Bamford, H. E., Jr. "Quickening and Damping a Feedback Display," *Journal of Applied Psychology,* Vol. 41, No. 6 (1957), 395-402.
(28) Saul, E. V. (ed.). *Human Engineering Bibliography, 1955-1956, ONR Report ACR-24.* Medford, Mass.: Tufts University (Institute for Applied Experimental Psychology), 1957.
(29) Seamans, R. C., Jr., and Pickford, H. W. "Development of Airborne Systems," *RCA Engineer,* Vol. 2, No. 4 (1956-57), 37-41.

(30) Steward, G. D. "Switchboards for Telephone Answering Service," *Bell Laboratories Record,* Vol. 35, No. 1 (January, 1957), 6-9.

(31) Weldon, R. J., and Peterson, G. M. "Factors Influencing Dial Operation: Three-Digit Multiple-turn Dials," *Engineering Research Report SC-3659 (TR)*, Albuquerque: Sandia Corporation, 1955.

(32) Weldon, R. J., and Peterson, G. M. "Effect of Design on Accuracy and Speed of Operating Dials," *Journal of Applied Psychology,* Vol. 41 (1957), 153-157.

(33) Williams, A. C. Jr., Adelson, M., and Ritchie, M. L. "A Program of Human Engineering Research on the Design of Aircraft Instrument Displays and Controls," *WADC Technical Report 56-526,* Wright Air Development Center, Wright-Patterson Air Force Base, Ohio, 1956.

Twenty

INFORMATION THEORY

WILLIS C. GORE

There are two major aspects of the subject called information theory. The first is concerned with the quantitative definitions of the amount of information conveyed in a message and of the capacity of the communication channel to transmit information. With such measures defined, theorems involving these quantities may be developed and proved, and one may test the reasonableness of this theory against one's experiences. Fortunately, a surprisingly consistent agreement does exist, and the theory may be used to optimize the design of a communications system. The general implementation involves the translation of the messages to be transmitted into a form which is better matched to the communication channel. This aspect is usually referred to as the *coding problem.*

First attempts to formulate such a theory date to the middle 1920's when Nyquist and Hartley, both of the Bell Telephone Laboratories, attempted to formulate the capacity of a communication channel. However, it was not until the late 1940's that Shannon, of the Bell Telephone Laboratories, was able to formulate the ideas of channel capacity and information contained in a message into a theory which is remarkably consistent with our experiences.

The second major aspect of information theory had its origins after Hartley and Nyquist but before Shannon, and in this country resulted mainly from the work of Norbert Wiener of Massachusetts Institute of Technology. His ideas greatly influenced Shannon and are embodied in the latter's theory. The essential concept is that communication problems are related to games involving the laws of chance or probability theory, and, hence, a study of the stochastic (i.e. random) processes, can serve as a background to the studying of communication systems. Since the implications of this concept will be made evident in studying the coding problem, the major emphasis of this chapter will be upon Shannon's theory of communication. The attempt of this chapter will be, then, to correlate the ideas of this theory with our intuition, although the theory has a basis in mathematics which is independent of how useful or how consistent it is with our everyday experiences. None of the complex aspects of this mathematical background or proofs will be given here, as the intent is rather to provide a heuristic presentation of the major ideas of this theory.

THE INFORMATION MEASURE OF DISCRETE NOISELESS MESSAGES

It is first necessary to formulate a quantitative measure of the amount of information conveyed by a message. Suppose, for example, that we are confronted by a number of boxes arranged in a row, one of which contains a ten dollar bill. Further, we may select one of the boxes, and if it contains the money we may keep it. If someone should give us the correct message that the money is in the third box from the left, we might wonder how much information was conveyed by that message. To determine a quantitative measure of information is therefore an objective of this section.

It is important to differentiate between the *amount* of information conveyed by the message and the *value* of the message to the person receiving it. The concern here is with the former and not the latter. For our purposes the *amount* of information required to select the box containing the money is the same whether the box contains a ten dollar bill or a thousand dollar bill, although the *value* of the information would be one hundred times greater in the latter event.

As a guide to establishing a meaningful quantitative measure of the amount of information conveyed by the message, we may examine some qualitative properties that such a measure should have. Obviously, if there were twenty boxes, the amount of information conveyed by the message that specifies which box contains the money should be greater than if there were only two boxes. If there were only one box, the amount of information conveyed should be zero, since receipt of the message tells us nothing that we did not already know.

This illustration suggests the idea that the generation of a message involves the selection of one message out of a list or alphabet of possible messages, and that the amount of information conveyed by the message should increase as the number of possible messages increases. An alternate expression of the same idea would be to note that the larger the number of messages which might be selected, the greater the uncertainty as to which message will be received, the uncertainty becoming a certainty when there is only one possible message.

Thus, we desire that the amount of information conveyed by a particular message should increase the greater the uncertainty as to which message will be received. This suggests that the number of possible messages would serve as a quantitative measure for the amount of information conveyed by a message. However, any monotonic increasing function of the number of messages would serve equally well, such as the square or cube-root of the number of messages.

From this infinitely long list of possible measures for the amount of information in a message, do we have any reasonable basis for selecting one which is better than all of the others? The answer is,

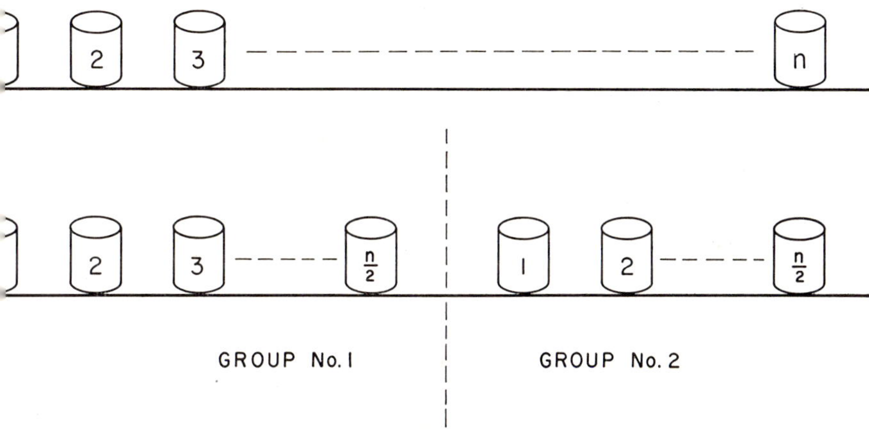

Figure 1. A set of n objects may be labeled by n digits assigned to the entire set, or, after dividing the original collection into two equal groups, by assigning $n/2$ digits to each group and designating the group.

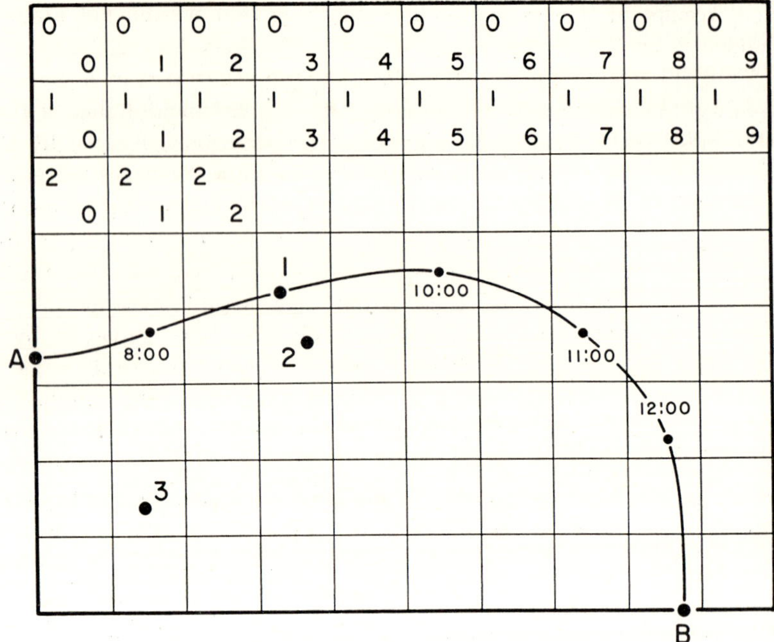

Figure 2. A map showing the movement of a hypothetical boat across a lake.

yes, there is a "best" one, and this may be selected by requiring an *additive property* for the amounts of information conveyed by several messages. That is, if the information conveyed by a single message may also be conveyed by two simpler messages, it would be desirable that the amount of information conveyed by this single message should be given by the sum of the amounts of information carried by the two simpler messages.

As an illustration, consider a group of boxes, numbered from 1 to n shown in Figure 1. Suppose that the boxes are further subdivided into two groups, both numbered 1 through $n/2$. Then, a message indicating which box contained the money would be equivalent to two messages one indicating which group the box was in, and the other which box of that group the money was in. As another example, consider the rectangularly shaped lake shown in Figure 2, which is divided by the parallels of latitude and meridians of longitude into equal-sized square areas numbered in consecutive order. A single message indicating in which square a boat was located would be equivalent to two messages, one specifying the column and the other message specifying the row of the square in which the boat was located.

It can be shown that of all monotonic increasing functions of n, only the logarithmic function has this additive property. The amount of information, I, may thus be defined as some constant times the logarithm of the number, n, of independent messages allowed.

$$I = K \log n \tag{1}$$

The fact that the information conveyed by the single message is equivalent to the sum of the information conveyed by the two separate messages may be demonstrated with reference to the situation shown in Figure 2. Let n_1 = number of rows, n_2 = number of columns, then $n_1 n_2$ = number of squares. The amount of information contained in a message that specifies which square contained the boat would then be

$$I = K \log n_1 n_2.$$

The information contained in a message specifying the row would be

$$I_1 = K \log n_1.$$

The information contained in a message specifying the column would be

$$I_2 = K \log n_2.$$

And, therefore, we have that

$$I = I_1 + I_2$$

since

$$\log n_1 n_2 = \log n_1 + \log n_2.$$

The constant K may be regarded as a factor which changes the units in which our information is expressed. Multiplying a logarithm by a constant is equivalent to changing the base of the logarithms in accord with equation (2).

$$(\log_c b)(\log_b a) = \log_c a \tag{2}$$

Therefore, the equation for the amount of information conveyed by a message is usually written as the logarithm of the number of allowable messages, where the base of the logarithm determines the information unit. If logarithms to the base 10 are used, the unit of information is called a *decit*. This is a contraction of *decimal digit*—a logical name since, if the number of allowable messages is 10, the amount of information conveyed by a single message ($\log_{10} 10$) is 1 and a single decimal digit could be used to indicate which of the 10 messages was transmitted by encoding the first message as *0*, the second as *1*, the third as *2*, and so on until the tenth message is encoded as *9*. If there were 1000 allowable messages, the information conveyed by any one message would be $\log_{10} 1000$, or 3 decits. That is to say, three decimal digits would be sufficient to indicate which of the 1000 messages was sent by encoding the first message as 000, the second as 001, and so on with the 1000th message encoded as 999.

The most important unit of information is obtained when the base of the logarithms is two. It is called the *bit*, which is a contraction of *binary digit*. If the messages are encoded into binary numbers, the amount of information is given by the number of binary digits necessary to indicate which message was transmitted.

There is, however, another factor to be taken into consideration which in our illustrations thus far has been ignored. This effect is the possibility that we might have some prior knowledge as to which message would be selected. This prior information could be expressed in the form of probabilities that the various messages would be transmitted. Thus, certain messages are often more likely to be transmitted than the others.

These considerations may be illustrated by the messages that might describe the position of an excursion steamer in the hypothetical lake of Figure 2. Let us suppose that the excursion steamer leaves the port at point A and follows a planned route that goes from point A to point B. We have marked on this path the times at which it is expected that the steamer will be at certain locations. Suppose one were to inquire as to the location of the steamer at nine o'clock. If everything goes according to plan, and there is no reason to suspect otherwise, then the location of the steamer should be at point 1, as shown in Figure 2. If, in fact, the steamer reported that it was at point 1, the message would convey little information, since this is where we expected the ship to be.

If, on the other hand, the steamer is reported to be at point 2, slightly off course, then we would have received some information in the message; while if it reported that it was at point 3, far off course, far from where we expected it, then we would have received even more information from this particular message.

Our expectations as to which messages will be received may be expressed in the form of probabilities. We assign unit probability, or a probability equal to 1, to the event which is certain to happen, while we assign probability 0 to the event which never happens (well, almost never!). To all other events, ranging between these two extremes, we give values of probability between 0 and 1. Thus, a probability of $\frac{1}{2}$ means that *on the average* a particular event happens one-half of the time. Thus, with reference to the steamer on the lake, we could, in principle, assign definite probabilities to the occurrences of the messages which indicate that the steamer is at point 1, point 2, or point 3. These probabilities would be assigned in accordance with the ability of the steamer to keep its schedule. If the steamer is known to be always on schedule, we would assign probability of unity to reception of the message that it is at point 1, while if it is on schedule one-half of the time, we would assign a probability of $\frac{1}{2}$, and so on.

To simplify the foregoing discussions, it has been assumed that all of the messages were equally likely. Since there are n possible messages of equal likelihood, the probability for each message is $1/n$. However, in general, the probabilities of occurrence of different messages will not be alike, and, hence, different messages will convey different amounts of information. We therefore define a new quantity which is the *average amount of information* associated with the selection of a single message from a given list. This average is equal to the sum over all possible messages of the amount of information conveyed by each particular message weighted by the probability that it was transmitted:

$$I = -\sum_{i=1}^{n} P_i \log P_i \tag{3}$$

It can be proved that the average amount of information associated with the selection of messages out of a given list is a maximum when all the probabilities are equal and is less than this maximum for any other distribution of probabilities.

Next, we define the *capacity* of a given channel for handling information. The *channel capacity* is commonly expressed in "information units per unit time" and may be measured by forming the ratio of the amount of information conveyed in the channel to the time interval used to transmit this information, as the time interval becomes infinitely long. Obviously, this quantity will depend upon the kind of information that is being sent over the channel, since some message or information sources might produce information at a faster rate than others. To obtain a unique measure, channel capacity will therefore be defined as the value of this ratio for that particular information source which makes this ratio a maximum. Having defined the average amount of information in a message source and the channel capacity, we are in a position to state the first fundamental theorem of information theory: *if the average amount of information per message from a source is* I *and if the channel has a capacity of* C, *then it is possible to encode the messages so that they may be transmitted over the communication channel at a rate* R *which has a maximum value equal to* C/I. *Furthermore, it is impossible to encode the messages so that they may be transmitted without error at a greater rate.*

To illustrate this particular theorem, consider the case in which there are four possible messages, A, B, C, and D, with probabilities of occurrence respectively given by $P_A = \frac{1}{2}$, $P_B = \frac{1}{4}$, $P_C = \frac{1}{8}$, and $P_D = \frac{1}{8}$. The average amount of information per message, when evaluated by using equation (3), is found to be

$$\begin{aligned} I &= -[(\tfrac{1}{2}) \log (\tfrac{1}{2}) + (\tfrac{1}{4}) \log (\tfrac{1}{4}) + (\tfrac{1}{8}) \log (\tfrac{1}{8}) + (\tfrac{1}{8}) \log (\tfrac{1}{8})] \\ &= 1\tfrac{3}{4} \text{ bits/message.} \end{aligned}$$

Thus, on the average there are $1\frac{3}{4}$ bits of information per message transmitted.

Carrying this further, suppose that our communication channel includes a typist who is operating a typewriter which has only four keys, labelled A', B', C', D'. If the typist strikes one key every second, what would be the limitation of capacity as imposed by the typist? In this example, it is possible to send one symbol (or message) per second, but we must evaluate the maximum amount of information that those messages convey. We have already seen that the amount of information associated with a set of four messages is a maximum when all four messages have equal probabilities. Therefore the maximum amount of information associated with each message of the optimum set having equal probabilities would be

$$I = -\log \tfrac{1}{4} = 2 \text{ bits/message.}$$

As there are two bits of information associated with each symbol, the channel capacity as limited by the typist is two bits per second. According to our theorem, then, it should be possible to encode the original messages, A, B, C, D, into such a form that it is possible to transmit these messages at a rate of

$$R = \frac{2}{1\frac{3}{4}} = \tfrac{8}{7} \text{ messages/second}$$

or $1\frac{1}{7}$ symbols per second.

It is to be noted that if the typist associated the A' key with the A message, the B' key with the B message, the C' key with the C message, and the D' key with the D message, she would be able to send only one message per second. The theory predicts that better encoding of the message would enable the typist to process messages at an average R, which is over 14 per cent faster than one message per second. The next question to be asked is, how does one arrive at the proper coding to be used?

In his proofs of this particular theorem, Shannon describes a scheme by which the code may be obtained (1). However, for this particular example, a much simpler, although equivalent, scheme devised by Fano may be used to arrive at the correct code. As illustrated in Figure 3, this procedure is to divide the list of messages A, B, C, and D into two parts so that the probability of each part is a half. We next assign the digit 0 to each one of the letters on the left half of the dividing line and a 1 to each one of the messages on the right half of the dividing line. This procedure is then repeated for each half of the original probabilities. Thus, as shown in Figure 3, since the left

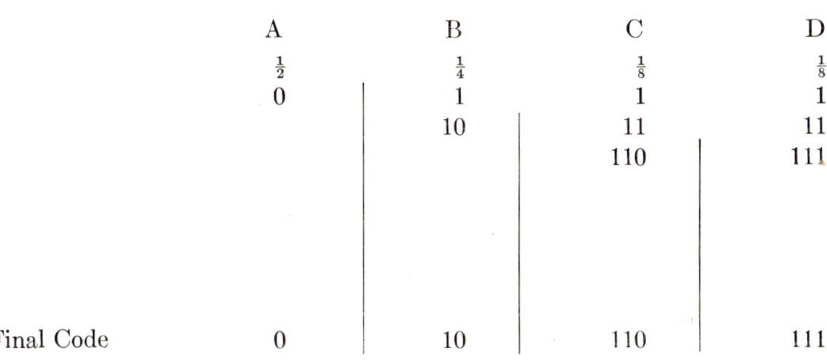

Figure 3.

half cannot be further subdivided, the right half alone is divided again in half, and each message on the right of this new dividing line is given a code 1 and each message on the left of this dividing line, a code 0. This process is repeated until each subdivision contains only one message; the code for each of our original messages is now obtained in binary form. We note that the particular messages which have a high probability of occurrence (i.e., those which occur most frequently) are encoded with the smallest number of binary digits, while those which occur less frequently are encoded into longer sequences of binary digits.

Next consider a typical sequence of messages from this information source, shown in Figure 4. We note that $\frac{1}{2}$ of the messages are A,

	A	B	A	A	C	B	A	D
	0	10	0	0	110	10	0	111

or 01000110100111

or 01 00 01 10 10 01 11

or B′ A′ B′ C′ C′ B′ D′

Figure 4.

$\frac{1}{4}$ of the messages are B, $\frac{1}{8}$ of the messages are C, and $\frac{1}{8}$ of the messages are D. (In general, such a short sequence of messages would not contain percentages of occurrences of the individual messages which exactly duplicate the probabilities of the assumed message source. However, as the length of sequence is increased, one is assured that the fractional occurrences approach the specified probabilities of occurrences.) We next encode each of the original messages into binary form, which is also shown in Figure 4. Here we note that the encoding is unique; that is, a particular sequence of original messages is encoded into one and only one sequence of 0's and 1's. It is also further noted that this sequence of 0's and 1's can be decoded into the original sequence of A, B, C, and D in but a single way. It is also further to be noted that the original message was encoded to a sequence of fourteen 0's and 1's (i.e., fourteen bits of information). This, by the way, is exactly what the theory predicts, since our calculation showed that there were $1\frac{3}{4}$ bits of information per message for a total of eight messages, giving a total of fourteen bits of information.

This, however, does not help our typist, who can only type an A', B', C', and D'. So we next devise a new binary code to relate the A', B', C', and D' to binary digits, and this we do for the particular case in which all of the messages are equally likely. Following the same procedure used before, we have in Figure 5 the encoding process for binary to our primed notations, by which it will be noted that

A′	B′	C′	D′
$\frac{1}{4}$	$\frac{1}{4}$	$\frac{1}{4}$	$\frac{1}{4}$
0	0	1	1
00	01	10	11

Figure 5.

each of our original messages is encoded into a sequence of two binary digits. This is again what the theory would predict, since our previous calculations indicated that it is possible to send two bits of information for each of these messages.

Returning to our original message which was encoded into a binary code (as shown in Figure 4), let us further encode this binary code using the symbols A', B', C', and D', according to the relationships in Figure 5. It is noted that the encoding process results in only *seven* of our new primed messages. Thus, in a seven-second interval (the time required to send seven of our primed messages), we may send the equivalent of *eight* of our unprimed messages. Stated another way, we are sending our original messages at an average rate of $1\frac{1}{7}$ messages per second, as the theory predicts.

Some difficulty may be associated with this procedure, if there is a particularly unlikely sequence such as the repeated occurrences of an improbable message, since the binary digits are then produced faster than they may immediately be transmitted by the primed messages. For this reason, some kind of storage mechanism or medium is necessary to preserve temporarily the incoming information until such time as the more frequent messages occur which are encoded into a small number of binary digits.

Difficulty of another kind may also occur. In the example previously considered, the probabilities were all expressible as integral powers of one-half, and, as a consequence, it was possible to divide the groups of messages successively into exact halves. In general, for any arbitrary distribution of probabilities this partitioning into successive halves cannot be done exactly, with the result that the codes become excessively long. This is illustrated in the next example.

In this example, shown in Figure 6, $P_A = \frac{1}{2}$, $P_B = \frac{1}{3}$, $P_C = \frac{1}{6}$, $P_D = 0$. From equation (3).

$$I = -[(\tfrac{1}{2}) \log (\tfrac{1}{2}) + (\tfrac{1}{3}) \log (\tfrac{1}{3}) + (\tfrac{1}{6}) \log (\tfrac{1}{6})] = 1.46 \text{ bits/message.}$$

A	B	C	D
$\frac{1}{2}$	$\frac{1}{3}$	$\frac{1}{6}$	0
0	1	1	1
	10	11	11
		110	111

	A	B	A	A	C	B	— 6 messages
	0	10	0	0	110	10	— 10 bits
or	01	00		01	10	10	
or	B′	A′		B′	C′	C′	— 5 primed messages
or	$I = \frac{10}{6} = \frac{5}{3}$ Bit per message						

Figure 6.

Also shown in Figure 6 is a typical sequence of six messages that is encoded into ten bits, which is 1.67 bits per message on the average. Since the binary code may be encoded into five primed messages, the six unprimed messages will be transmitted in five seconds (at a rate $R = 1.46 \times \frac{6}{5} = 1.75$ bits/second), whereas the theory still predicts a possible rate of two bits per second.

The reason that this particular code has not yielded a full realization of the channel capacity is that our code uses on the average 1.67 bits per message, while the actual information content is only 1.46 bits per message. A more efficient code may be found by encoding *pairs* of messages, rather than single messages into binary code. This is illustrated in Figure 7, where the assumption is made that the selection of any message is independent of the previous messages selected, so that the probability of a joint occurrence of two messages is the product of the probabilities of the two messages. The code of Figure 7 is used to encode a typical sequence of 72 messages, shown in Figure 8, into 53 primed messages (which require 53 seconds for transmission), and the rate is now

$$R = 1.46 \times \frac{72}{53} = 1.98 \text{ bits/second},$$

which is very close to the theoretical maximum value of 2 bits per second. Thus, in addition to the penalty of providing the storage media, we must in general also wait a certain prescribed interval of

AA	AB	BA	AC	CA	BC	BB	CB	CC
0.250	0.167	0.167	0.083	0.083	0.056	0.111	0.056	0.028
0	0	0	1	1	1	1	1	1
00	01	01	10	10	10	11	11	11
	010	011	100	101	101	110	111	111
				1010	1011		1110	1111

Figure 7.

AB BC AA CC CB AA AC BA BA BB AB AA
010 1011 00 1111 1110 00 100 011 011 110 010 00
01 01 01 10 01 11 11 11 00 01 00 01 10 11 11 00 10 00
B′ B′ B′ C′ B′ D′ D′ D′ A′ B′ A′ B′ C′ D′ D′ A′ C′ A′

BC CA AB CA CB AA AC BB AB AA CA AC
1011 1010 010 1010 1110 00 100 110 010 00 1010 100
10 11 10 10 01 01 01 01 11 00 01 00 11 00 10 00 10 10 10 00
C′ D′ C′ C′ B′ B′ B′ B′ D′ A′ B′ A′ D′ A′ C′ A′ C′ C′ C′ A′

BA BB BA AA AA AB BA AA AB AA BA BB
011 110 011 00 00 010 011 00 010 00 011 110
11 11 00 11 00 00 01 00 11 00 01 00 00 11 110
D′ D′ A′ D′ A′ A′ B′ A′ D′ A′ B′ A′ A′ D′ D′

Figure 8.

time to accumulate enough messages to make the encoding process efficient in order that the maximum rate of transmission can be reached.

DISCRETE CASE WITH NOISE

We now consider communication in a discrete channel in the presence of noise. That noise will affect the amount of information that can be transmitted through a channel is a fact that is exemplified by many familiar situations. Suppose, for instance, that before renting an apartment we ask a personal friend who has lived in this apartment what kinds of neighbors we will have, and he tells us that they are fine. We next ask the real estate agent who is going to rent us the property what kinds of neighbors we will have in our prospective new apartment, and he says we will have fine neighbors.

Which of these two messages conveys the most information? All will agree that the reply of our friend conveys more information than the reply of the real estate agent, the reason being that our friend is more reliable than the agent, who may have other motives than that of giving us accurate information. Thus, we could say that the reliability of the particular message would influence the amount of information conveyed. In a similar way, a person who receives a message over a channel which is unperturbed by any outside influence is assured of the fact that the message received is the one that was transmitted, while a person listening on a channel which is disturbed by outside influences cannot have such assurance. Thus, one may expect that the amount of information transmitted over a noiseless channel will be greater than that which could be transmitted over a similar channel in the presence of noise because of the greater reliability of the noiseless channel.

This raises the question of how one can express this difference quantitatively. Since the disturbing noise causes some of the transmitted messages to arrive at the receiver incorrectly, the simple measure of the degradation due to the noise would be to penalize the channel by an amount which is the rate at which information would have to be transmitted over another noiseless channel to correct for the mistakes made in the noisy channel.

The effect of the noise in the channel may be described by a set of *conditional probability distributions* $P_x(y)$, where $P_x(y)$ is the probability that if message x is transmitted, y is received. In the noise-free case considered in the previous section, the conditional distributions were of the form $P_x(y) = 1$ when $y = x$ and $P_x(y) = 0$ when $y \neq x$, since the message received is certain to be the transmitted message. However, in the presence of noise, the transmitted message may be incorrectly received so that $P_x(y) \neq 0$ when $y \neq x$. In this event, if message x is transmitted, we must transmit, on the average,

$$I(y) = -\sum_y P_x(y) \log P_x(y)$$

units of information to correct for the noise. By averaging this quantity over all possible transmitted messages, we obtain the average loss in information due to noise.

$$I_x(y) = -\sum_x P(x) \sum_y P_x(y) \log P_x(y)$$

The amount of information transmitted over a noisy channel may therefore be expressed as

$$\begin{aligned} I &= I(y) - I_x(y) \\ &= -\sum_y P(y) \log P(y) - \Big[-\sum_x P(x) \sum_y P_x(y) \log P_x(y)\Big] \end{aligned} \tag{4}$$

where $I(y)$ would be the information received if there were no noise and $I_x(y)$ is the average amount of information necessary to correct the error caused by the noise.

It is now possible to define the *capacity of a noisy channel* as the value of the product of the rate at which messages are being transmitted and the I in equation (4) when the transmitted message set is selected so as to maximize the received information. The following theorem can then be proved. *If the rate of transmission is less than the channel capacity, it is possible to encode a message for transmission over a noisy channel so that an arbitrarily small percentage of errors may be obtained at the receiver.*

To understand this remarkable theorem, consider the following example. Suppose that the noise in our channel is of such a nature that it influences, at most, one out of the block of seven binary symbols. Thus, in each group of seven binary digits there is at most one 0 which is converted to a 1 or one 1 which is converted to a 0. Assuming that we use a transmitted code in which the 1's and 0's are equally likely, we may calculate the channel capacity under the assumption that the binary digits are transmitted at a rate of one per second. Using equation (4) we have

$$\begin{aligned} I &= -[\tfrac{1}{2} \log \tfrac{1}{2} + \tfrac{1}{2} \log \tfrac{1}{2}] - \tfrac{1}{7}[-\tfrac{1}{2}(\tfrac{8}{8} \log \tfrac{1}{8}) - \tfrac{1}{2}(\tfrac{8}{8} \log \tfrac{1}{8})] \\ &= 1 - \tfrac{1}{7} \times 3 = \tfrac{4}{7} \text{ bits} \end{aligned}$$

Thus, the channel capacity is $\frac{4}{7}$ of a bit per second. The theory now tells us that it is possible to send information at this rate or less with essentially error-free transmission. This is illustrated by the use of the following code, invented by Hamming of the Bell Telephone Laboratories. In Figure 9 appears a list of sixteen different messages numbered 0 through 15, inclusive. Each one of these messages is encoded into a binary sequence employing seven bits. If we transmit these messages at the rate of one bit per second, each message will require seven seconds per transmission. Assuming equal likelihood of occurrence, each message will convey four bits of information. Provided these messages can be received unambiguously without error, we have succeeded in transmitting four bits of information in seven seconds, or at the rate of $\frac{4}{7}$ of a bit per second.

Error Correcting Code

Message	Digit Position 1	2	3	4	5	6	7
0	0	0	0	0	0	0	0
1	1	1	0	1	0	0	1
2	0	1	0	1	0	1	0
3	1	0	0	0	0	1	1
4	1	0	0	1	1	0	0
5	0	1	0	0	1	0	1
6	1	1	0	0	1	1	0
7	0	0	0	1	1	1	1
8	1	1	1	0	0	0	0
9	0	0	1	1	0	0	1
10	1	0	1	1	0	1	0
11	0	1	1	0	0	1	1
12	0	1	1	1	1	0	0
13	1	0	1	0	1	0	1
14	0	0	1	0	1	1	0
15	1	1	1	1	1	1	1

Check Positions: First digit 1, 3, 5, 7
Second digit 2, 3, 6, 7
Third digit 4, 5, 6, 7

Information Position 3, 5, 6, 7

Figure 9.

The code just described is called a *single-error-correcting code* because it has the property that the occurrence of a single error in any of the seven binary digit positions may be corrected as illustrated in Figure 10, where the tenth message is shown transmitted. Below this is shown the tenth message as received with an error in the third digit place. After the reception of this particular message, we now form a new binary number. The digits in this binary number are formed according to an even parity check of the digits as indicated in the bottom of Figure 9. To make an *even parity check* we add the digits in the places enumerated at the bottom of Figure 9, and if the sum of the number of 1's in the indicated positions is even, we interpret this as no error and write a 0, while if the sum of the number of 1's in the indicated positions is odd, we interpret this as an error and write a 1. The binary number so formed by this procedure is the number 011

Digit position number	—	1	2	3	4	5	6	7
Message as transmitted		1	0	1	1	0	1	0
Message as received		1	0	0	1	0	1	0

For the received message:

Sum of "1" 's in positions: 4, 5, 6, 7 is 2 (even) — 0
2, 3, 6, 7 is 1 (odd) — 1
1, 3, 5, 7 is 1 (odd) — 1

Therefore the check number is 011 which is decimal **3.**
Correcting the error in the third position

Corrected message 1 0 1 1 0 1 0

Information positions 3, 5, 6, 7 — 1 0 1 0
which is decimal 10 which is the message transmitted

Figure 10.

which is the binary representation of the decimal number 3 and indicates that the error was in the third position.

With this information, we now return to the received message, shown in Figure 10, and change the digit in the third position, thus obtaining the correct message as transmitted. We next decode this message to determine the original message that was transmitted by again referring to Figure 9, which indicates the information-bearing places of this seven digit code. These digits are written below the corrected received messages in Figure 10, and form a binary number 1010 which is the binary representation of the decimal number 10, which was the original message as transmitted. It is claimed that the particular code which is illustrated in Figure 9 has this error-correcting property for a single error in any position of any of the encoded messages. If no error has occurred in transmission, then the check binary number obtained in making the even parity check will be the binary number 000, which is then interpreted as "no error."

It is to be noted that the procedure used to encode a message for transmission over a noisy channel results in making the message longer so that it will have greater immunity to noise or errors, whereas in a noiseless channel the message was made artificially shorter so that the same information could be sent with the least number of symbols. The ratio of the number of bits of information transmitted to the number of bits used to transmit the information is called the *relative entropy* of the message; in our particular example this is $\frac{}{7}$. Unity minus the relative entropy is called the *redundancy*. In the above example, the redundancy is $\frac{3}{7}$, which expresses the fractional part of the message that does not convey information but which enables us to correct errors that may occur in the transmission of the information.

Studies made on written English indicate that the redundancy is approximately 50 per cent, or that on the average the same information could be packed into only half as many letters. What then is the purpose of the extra letters which are used? A probable answer is that they evolved in a natural manner to accommodate the inevitable illegibility that one finds in a hand-written message. Although these kinds of errors are relatively infrequent in machine-printed text, one must, of course, remember that languages and the written language evolved many centuries before the use of printing machines.

The redundancy of written English is illustrated by the garbled sentence, "The Mae ran tm the train." This could be corrected by almost anyone reasonably familiar with the English language to read, "The man ran to the train." This kind of correction is exactly the kind that was employed in the code of Figure 9, and can be attributed to the inherent redundancy of printed English.

CONTINUOUS CASE

Thus far the case has been considered in which our communications information source is a device which selects a message out of a finite list of messages. Next to be considered is a continuous information source in which the output of the source is an analogue quantity such as voltage, current, displacement, pressure, or a light intensity, etc.

An important example of the continuous case is the transmission of speech over a telephone line, in which one has a transducer, such as a microphone, which converts the varying intensity of air pressure, caused by the sound waves produced by the person speaking, into a voltage which is then amplified and transmitted over the telephone line. The message is conveyed by the continuous variations of the voltage sent over the telephone circuits.

One might naturally be led to attempt a treatment of the continuous case in a manner analogous to the handling of the discrete case. Here we could restrict our attention to the amplitude of the voltage waveform at discrete instances, spaced by equal time intervals as shown in Figure 11, and regard the numerical value of the message at

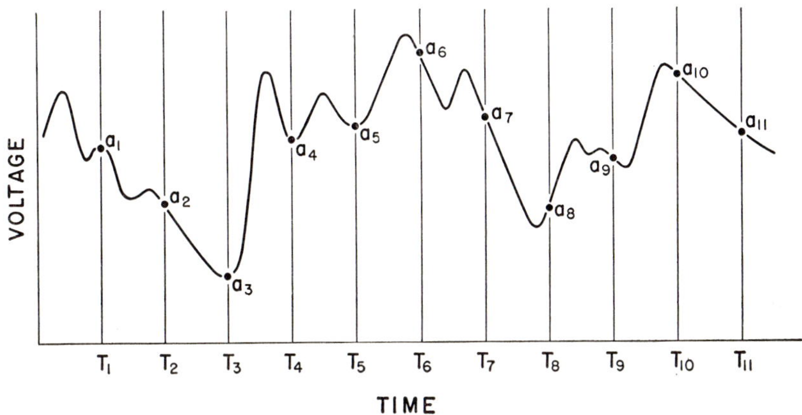

Figure 11. A continuous signal may be specified approximately by giving the numerical values, a_1, a_2, . . . , of the signal at successive instants in time, T_1, T_2, ...

each instant of time as the amplitude of the voltage at that instant. We are, however, immediately faced with two problems. First, how far apart should the sample points be in time? Because the voltage is a continuous function of time, to get an exact description of the voltage waveform would appear to require that we sample the waveform at infinitesimally small intervals of time. In other words, it would appear as though the basic rate at which the individual messages are selected is infinitely great. The second difficulty occurs when we attempt to evaluate the number of different messages that are available to the sender. Here we note that there might be some practical limitation as to the maximum amplitude of the voltage which appears between the two lines; but even so, with a finite upper bound for the maximum value of the voltage there is still an infinite number of different voltages which could be selected for our basic message. Thus, it appears as though the list of messages from which we are making our selection is infinitely long, and there is an unlimited amount of information associated with the selection of any one of these individual messages.

We will treat these two problems one at a time and in the order in which they were mentioned. That the first difficulty can be overcome is due to the fact that nearly all signals to be transmitted are *band-limited;* that is, they contain no components outside a given band of frequencies. For instance, it is known that in order to transmit intelligible speech, a band of frequencies, approximately 3,500 cycles per second wide is needed; to transmit music, a bandwidth of approximately 8,000 cycles per second is needed; and to transmit high-fidelity music, a bandwidth of approximately 15,000 cycles per second is needed.

Since this band-limited property seems to be inherent in many signals, we will build it into the theory by requiring that the information be conveyed by signals of limited bandwidth. This leads to a question: is it possible to represent the continuous band-limited signal in an interval of time "T" seconds long by a finite number of numbers? The answer to this problem was obtained many years ago by Nyquist, who found that, if the signal contained no frequencies greater than W cycles per second, the number of independent ordinates required to determine completely a given signal of duration T seconds is just $2WT$.

The proof of this is illustrated by Figure 12, in which we have

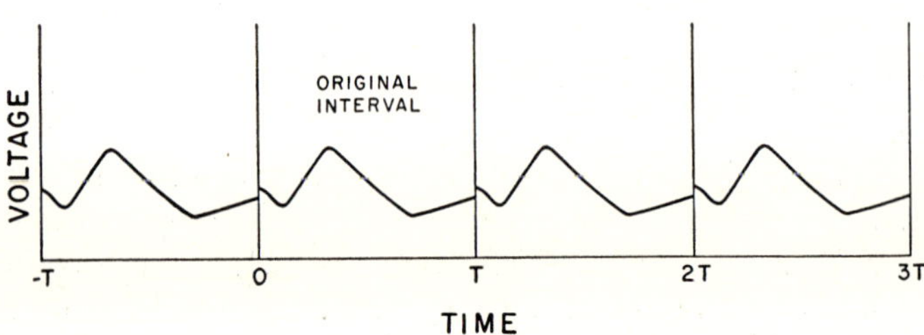

Figure 12. A signal of duration T may be considered as one period of a possible signal which may be represented by a Fourier Series.

selected an interval of the original signal T seconds long and have artificially constructed a repetitive signal of which this particular interval is one cycle. Since this signal is now repetitive, it may be represented for all time, and in particular for the time interval of T seconds, by a Fourier Series containing sine and cosine terms of fundamental frequency $f = 1/T$. Thus, to specify completely this waveform it is sufficient to give the amplitude of the sine and cosine terms in the Fourier Series expansion for the repetitive wave. Since the signal contains no frequencies greater than W cycles, and since all of the Fourier components are spaced f cycles apart, there are at most W/f sine terms, and W/f cosine terms or a total of $2W/f$ coefficients which must be specified to determine completely the original waveform. Since f is equal to $1/T$ the number of coefficients which need to be specified could also be written as $2WT$. Nyquist further reasoned that since $2WT$ numbers were needed to specify a waveform of duration T seconds, these numbers could be transmitted at an average rate of $2WT/T$ or $2W$ numbers per second. This is called the *Nyquist rate*—the average rate at which component messages are originated in the generation of the original band-limited waveform.

Although formally the problem is solved, there still remains a great deal of difficulty associated with the problem of actually determining these numbers which have to be transmitted at intervals of $\frac{1}{2}W$ seconds apart. The procedure, as we have outlined it according to Nyquist's proof, is to perform a Fourier analysis of the waveform which was generated during the previous T seconds and to transmit the amplitudes of the sine and cosine components. At the receiver, we may, in principle, reconstruct the original wave by adding together a group of sine and cosine waves of appropriate frequencies having the amplitudes that were transmitted. Thus, in principle, the first problem is solved, but in a way that is most impractical for engineering application. In practice, a neater solution may be based upon the *sampling theorem* in the time domain, next to be discussed.

If the original waveform contains no frequencies greater than W

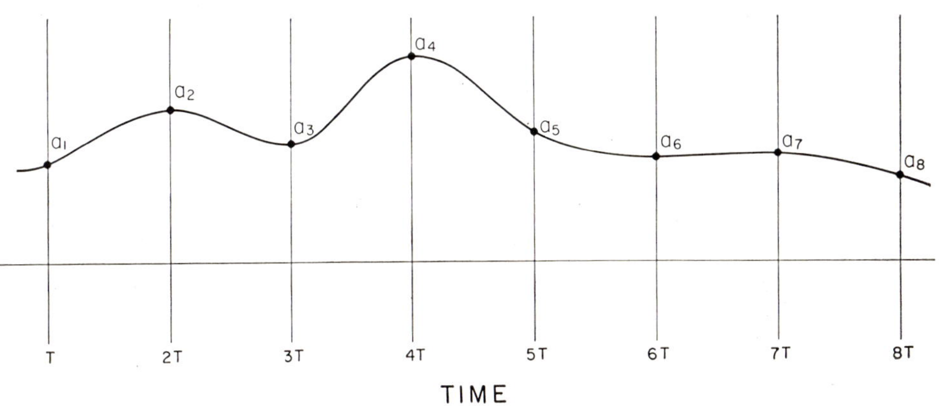

Figure 13(A). A band limited signal.

cycles per second, it can be reconstructed from the amplitudes of the original waveform measured at discrete intervals of time, notably just those intervals as prescribed by the Nyquist rate. This is shown in Figure 13(A). The continuous waveform need only be sampled at intervals of time $1/2W$ seconds apart, and from these amplitudes of the original waveform the entire original waveform may be constructed as shown in Figure 13(B). The process of reconstruction con-

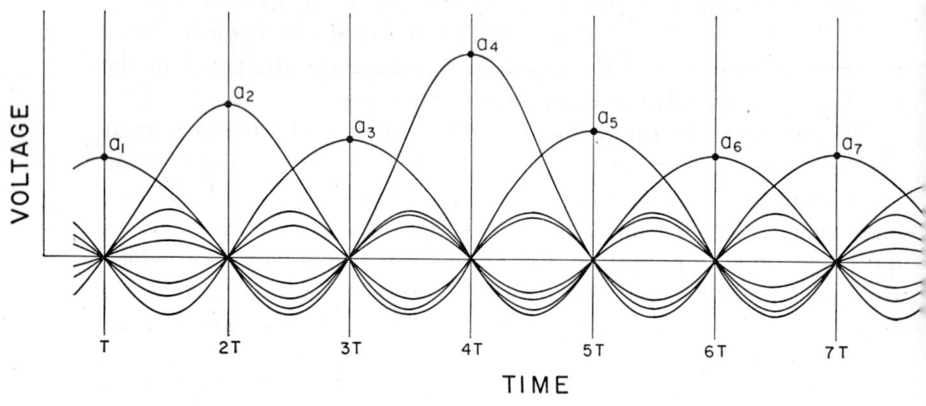

Figure 13(B). Its decomposition into $\frac{\sin 2\pi Wt}{2\pi Wt}$ components.

sists of taking a time function of the form $\frac{\sin 2\pi Wt}{2\pi Wt}$ as shown in Figure 13(C), centered at each of the sampling times and with an amplitude

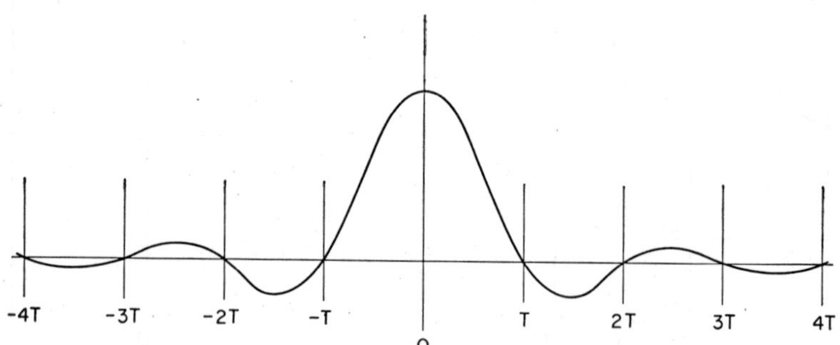

Figure 13(C). A single $\frac{\sin 2\pi Wt}{2\pi Wt}$ component.

equal to the amplitude of the original function at that time. Here, it is to be noted that, due to the regular spacing of the zero crossings of the $\sin x/x$ function, the original waveform is recovered exactly at each of the sampling instants, since only one of these functions contributes to the amplitude of the original function at each sampling instant. Thus, the amplitudes of the original function at time intervals $1/2W$ seconds apart are independent, as any one amplitude may be chosen arbitrarily without affecting the amplitudes at the other sample points.

To verify the fact that the original function is reconstructed exactly for all times in between the sample times, one needs to note first of all that the function $\frac{\sin 2\pi Wt}{2\pi Wt}$ contains no frequencies greater than W cycles per second. The sum of all of these functions of time used to reconstruct the original function will not introduce any frequency component greater than W cycles per second into the reconstructed waveform. In addition to this, for a function T seconds long we have specified $2WT$ independent numbers, and, since Nyquist's result assures us of the fact that, for a time function bandwidth limited to W cycles per second and duration T, there are only $2WT$ numbers needed to specify completely a unique function, the function which we have specified must be identical with the original function. We can thus view the transmission of information in the continuous-signal case, where the signal is bandwidth limited to W cycles per second, as reducible to the transmission of information in a discrete-signal case, where the discrete messages (i.e., the successive sampled amplitudes of the original function) are sent at the rate of $2W$ per second.

To complete our comparison between the transmission of information in the continuous and discrete cases, we examine the list of messages that is available at each of the sampling points, and again we note that if we place no restrictions, other than a maximum or an average maximum power limitation, on the transmitted waveform, there appears to be an infinite number of messages available for selection. We are forced to conclude for the noiseless case that either an infinite amount of information may be transmitted by a continuous waveform, or to ignore completely this situation and say that, since no communication channel is ever free of noise, we will not talk about the noiseless continuous case.

One recalls that in the noisy discrete case we penalized the communication channel for the errors that the noise introduced by subtracting from the capacity of the channel that rate at which information would have to be supplied to correct the errors introduced by noise. For the continuous signal, the selection of the amplitude of the signal at one of the sampling points could be imagined as the selection of one message out of a list of messages where the interval representing allowable signal amplitudes represents the list of messages and a small increment of this interval represents one of the messages that could be selected. Our previous argument was that, since the original waveform was continuous, an infinite number of these small increments was in reality possible. As we make the increments smaller and smaller, i.e., the messages more numerous, our ability to distinguish two different messages representing adjacent increments becomes more difficult in the presence of noise, so that the amount of information necessary to correct the errors increases. We note also that the amount of information necessary to correct the errors increases just as fast as the amount of information contained in the message because of an increased number of allowable messages. Thus, the difference between the two quantities, which is the amount of information per message or sample point, approaches a limit independent of the fineness of the subdivisions.

There is another way of looking at the same problem. It makes no sense to subdivide our allowable message amplitude into increments which are finer than the amplitude of the noise, as a subdivision finer than this would result in an almost certain corruption of one allowable message into another by the action of the noise. Thus, with noise in our channel, it does appear that we have a situation in the continuous case which is similar to that of the discrete case.

These same points are illustrated in the mathematical development of the theory. In this development we replace the discrete probabilities previously considered with probability densities, and summation by integration, thus obtaining for the quantity that we previously called I, the expression given by equation (5),

$$I = -\int_{-\infty}^{+\infty} p(x) \log p(x)\, dx. \tag{5}$$

The only difficulty with this quantity, I, is that it is not unique, but depends upon the particular co-ordinate system used to represent the variable. Thus, I will have one value if the waveform is expressed in volts, another value if expressed in millivolts, another value if expressed in microvolts, and so on. However, we are not talking about information as being analogous to the quantity I because we are talking only about the noisy situation, and we must look at an equation which is analogous to equation (4). This is equation (6).

$$\begin{aligned} I &= I(y) - I_x(y) \\ &= -\int_{-\infty}^{+\infty} p(y) \log p(y)\, dy \\ &\quad - \left[-\int_{-\infty}^{+\infty} p(x)\, dx \int_{-\infty}^{+\infty} p_x(y) \log p_x(y)\, dy \right] \end{aligned} \tag{6}$$

Thus, we are interested in evaluating the difference between two I's, one representing the signal and the other representing the effects of the noise. If we require that the variables associated with both of these I's be expressed in the same system of units (e.g., both in volts or amperes, etc.), then the choice of different units will produce changes in the I's which always exactly cancel when their difference is taken. Consequently, the *difference* given by equation (6) yields a unique value that is not dependent upon the co-ordinate representation.

The theory now states that it is possible to send the amount of information per sample point over a noisy channel as given by equation (6) with as small an error as desired. Equation (6) must be evaluated for the particular channel used, and here we again recall that the channel is not only restricted by the bandwidth available, but is also restricted by the power which is available for the signal waveform. Shannon was able to show that, for a channel having a bandwidth extending from zero to W cycles per second in which the average signal power is limited to P watts, and which is further disturbed by random gaussian noise having an average power of N watts, the channel capacity is given by equation (7).

$$C = W \log\left(1 + \frac{P}{N}\right) \tag{7}$$

It is indeed gratifying that the following heuristic derivation of what one might expect as a channel capacity comes so close to the actual formulation of equation (7). Suppose that a signal power of P watts, corresponding to a maximum signal amplitude of A volts, is available for transmission. Further suppose that the average noise power is N watts, corresponding to an rms noise amplitude of b volts. From our previous discussion, it appears useless to measure the available signal amplitude A on a scale having increments which are smaller than the noise voltage b, so there are A/b allowable messages in our signaling alphabet. The amount of information conveyed by each of these messages is given by equation (2) and is equal to the log A/b. Since the channel is bandwidth limited to W cycles per second, we may send $2W$ of these messages per second, so that the total information transmitted per second is $2W \log A/b$, and this therefore would represent the channel capacity. This formulation can be made to appear similar to equation (7) by taking the 2 from in front of the logarithm and incorporating it into the square of the argument of the logarithm. Thus, the capacity could be written $W \log (A/b)^2 = W \log P/N$, since the ratio of the squares of the amplitudes of the noise and of the signal is the same as the noise-to-signal power ratio.

An important interpretation of equation (7) indicates that it is possible to exchange *bandwidth* for *signal-to-noise ratio* in a communication channel. That is, if one does not have enough signal power available, so that the ratio of P/N is too small, it may be possible to transmit more information by increasing W appropriately. On the other hand, if the allowable band of transmission W is too small, by increasing the signal power sufficiently we may maintain the same rate of transmission. Examples of the former situation have been well known for a long time. Wideband frequency modulation is an example of just such a process where, by utilizing a wider band of frequencies, it is possible to achieve a reduction of the noise at the receiver. Examples of the latter case are relatively recent and were not explored until the pulse-modulation techniques were developed during the 1940's. It was mainly because of these new modulation techniques and the resulting flexibility of the communication channels to exchange bandwidth and signal-to-noise ratio that theoretical discussions of channel capacity and coding have been necessary in order to determine how efficiently these communication systems were operating. These applications have made Shannon's mathematical theory of communication a subject of considerable importance today.

REFERENCES

(1) Shannon, C. E., and Weaver, W. *The Mathematical Theory of Communication.* Urbana: University of Illinois Press, 1949.

(2) Davenport, W. B., Jr., and Root, W. L. *An Introduction to the Theory of Random Signals and Noise.* New York: McGraw-Hill, 1958.

(3) Woodward, P. M. *Probability and Information Theory with Applications to Radar.* New York: McGraw-Hill, 1953.

Twenty-One

FLOW-GRAPH REPRESENTATION OF SYSTEMS

WILLIAM H. HUGGINS

INTRODUCTION

The importance of mathematical models in systems engineering and operations research has been emphasized repeatedly in the other chapters of this book. These mathematical models often take the form of sets of algebraic and differential equations that use a mathematical notation which has evolved rather slowly over the last several hundred years. The algebraic part of Descartes' *Geometry*, published in 1637, reads not unlike a modern book on algebra, and the notation of the infinitesimal calculus, invented by Leibnitz and first published in 1684, has undergone little change within the last century. From the point of view of the mathematician, who is concerned with the development of mathematics as an idealized system of logical truths and not as a *tool* for representing real-world processes and relationships, there may be little need to revise or modify mathematical notation. However, those who are concerned with the use and application of mathematical models in the study of modern systems may have some justification in wondering whether the classical mathematical notation might not be augmented by a notation better suited for system representations.

The set of equations that suffice to describe the behavior of a system fails to portray in readily comprehended form the *structure* of the system-as-a-whole. Each equation reveals only one component of that structure, and the conventional notation does little to connect these pieces together into a coherent whole. Although the entire collection of equations does suffice mathematically to define the structure (in the same sense that a vector can be described by tabulating its components or projections upon some set of coordinates), it nevertheless fails to convey perceptually the structural (or topological) aspects of the system.

A common method of utilizing mathematics in science is to decompose the whole into a sum of its parts. As a limiting example of this method, many laws of the physical world have been expressed in terms of the relations between infinitesimal elements. The solution to any particular physical problem is then effected by assembling these infinitesimal elements (through integration of the appropriate differential equations) so as to satisfy the prescribed boundary conditions. The problems that are readily solved in physics usually pertain to systems that possess very simple structures with homogeneous properties, and the necessity for dealing with complex heterogeneous system structures has not been particularly great in the classical development of mathematical physics. But in the design of man-made systems, homogeneity is the exception rather than the rule. Each discrete element of the system may exhibit a behavior pattern that differs distinctly from that of all other elements—and we are still faced with the necessity for synthesizing these many elements into a composite system having prescribed overall characteristics.

To meet the requirements of a symbolism to designate this system structure, there has emerged in nearly all fields of modern technology various forms of flow-graphs called "flow charts," "block diagrams," etc. These diagrams are normally intended to portray not the things that comprise the system but rather the various operations that the system performs upon the "stuff" that it processes. Often, these flow-graphs are intended merely to provide a pictorial description of the system as a guide in mathematical analysis. However, it is the thesis of this chapter that these flow-graphs are evolving into a new type of mathematical notation that provides a concise, easily visualized description of system structure, and at the same time is capable of being manipulated and "solved" just as the conventional set of equations describing the system may be manipulated and solved.

Indeed, the rules for formulating and manipulating the flow-graph may be so prescribed that *one* flow-graph will be equivalent to *the entire set* of equations. Furthermore, each of the permissible operations on the graph will correspond to some permissible operation on the set of equations. Because of this one-to-one correspondence between the flow-graph and its described set of equations, either is capable of representing the same system. Is there anything, then, to be gained by using the flow-graph notation?

There is a growing awareness that the traditional symbolism of our conventional mathematics leaves something to be desired. Experience has repeatedly shown that advances in knowledge depend heavily upon the creation of a concise terminology and an efficient notation. The introduction of the Hindu-Arabic numerals led to major discoveries. Leibnitz attributed his own mathematical discoveries of the calculus to his earlier improvements in notation. Yet, experience has also shown that efforts to introduce new mathematical notations usually meet with great resistance. Some reasons for this may be quoted from a recent article by Karl Menger (12):

> Nothing is more distasteful to an active mathematician or scientist than discussions of symbolism and notation, and that dislike is perfectly understandable. After having overcome in his youth whatever difficulties the formal expression of ideas present, the mathematician finds that certain ways of writing have become his second nature and regards any suggestion of a change, even if he recognizes its merits, as nothing but a trivial nuisance.
>
> There are, however, situations in which a thorough discussion of such matters on the highest level is inevitable. They occur when, at turning points in the history of culture, it becomes imperative to make certain techniques and ideas of mathematics available to wider strata of the population. In the large groups to be initiated, many persons lack the ability to overcome the difficulties that the specialist overcame in his youth. Moreover, an immense collective benefit results if even persons with that ability are spared unnecessary complications.
>
> Such a turning point affected arithmetic when, during the Renaissance, mercantilism and experimental science were born. In banks and laboratories, the letters introduced by the Greeks and Romans as numerals proved to be utterly inefficient, even though they had served arithmeticians for over 2000 years. Unfortunately, medieval mathematicians misinterpreted the specialists' manipulative facility as intrinsic simplicity of the ancient numerals and regarded the Hindu-Arabic ideas as a pure nuisance. "Even in the 15th century," wrote G. Sarton, "there were still any number of learned doctors and professors who claimed that the Roman letters were much simpler than the Hindu numerals. Such prejudices confined the knowledge of arithmetic to a small elite and retarded its democratization as well as its progress. Eventually, however, as everyone knows, practical exigencies prevailed—incidentally, to the ultimate benefit of pure mathematics too."

Menger then goes on to illustrate the need for making a distinction between the notion of a variable, a parameter, and an operator, and for clarifying these distinctions by the use of new symbols. For instance, in the statement of the relationship between the values of the acceleration, a, produced by a force, f, acting upon a body of mass, m, it is conventional to write

$$f = ma.$$

Yet, the symbols a and f play a very different role from the symbol m, which is merely a constant of proportionality between the acceleration and force observables. This distinction in meaning is completely lost in the conventional mathematical equation. However, expression of this relationship in flow-graph form,

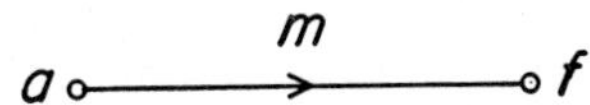

explicitly reveals that m is a measure of that which relates the *acceleration* and *force observables*.

The following sections of this chapter describe the flow-graph representation and show how it may be used to express and investigate the relations between the independent observables of a system. For linear and constant-parameter systems, the flow-graph can be manipulated so as, in effect, to solve the simultaneous equations that describe the system. For non-linear and time-varying systems, analytical solution of the corresponding equations becomes exceedingly difficult. However, the *form* of the flow-graph is unchanged and furthermore it is still capable of being manipulated (although not so generally as with linear, constant-parameter systems). This manipulation, which alters the structure of the graph, is of importance as a preliminary step to the obtaining of a quantitative solution, either by analysis or by computer. The flow-graph leads directly and simply to the wiring diagram of an analogue computer, or to the programming of a digital computer, upon which non-linear as well as time-varying systems may be simulated and studied without the often insurmountable difficulties that beset an attempt at analytical solution of such a system.

The flow-graph is thus a kind of mathematical "esperanto" that is well adopted to the needs of systems modeling for purposes of design and simulation, as well as analysis. Furthermore, it is a natural representation that utilizes the ability of the human eye to perceive at a glance the total pattern and the many relationships between the elements of the graph. Experience in teaching flow-graphs to students and others who are not sophisticated mathematically has suggested that this naturalness makes the flow-graph representation palatable to those who might have difficulty in swallowing the same material when expressed in conventional mathematical form.

FLOW-GRAPHS

Perhaps the simplest and most direct way to describe a flow-graph is by comparison with the set of algebraic equations that the graph represents. A typical set of algebraic equations and the corresponding flow-graph is shown in Figure 1.

$$
\begin{aligned}
x_0 &= x_0 \\
x_1 &= ax_0 + gx_2 + fx_3 \\
x_2 &= bx_1 + cx_2 \\
x_3 &= ex_1 + dx_2 \\
x'_3 &= x_3
\end{aligned}
$$

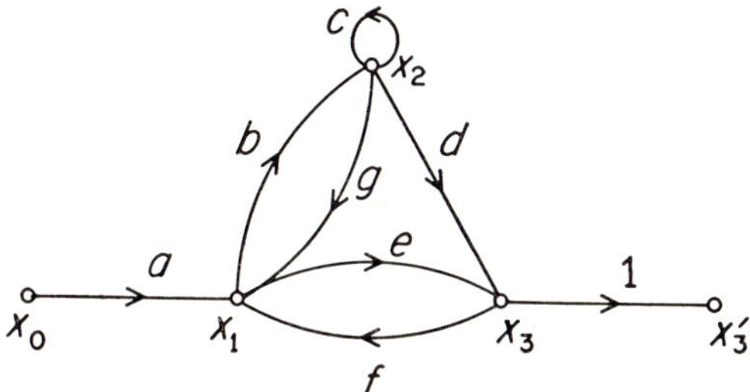

Figure 1. A set of equations and their representation by a flow-graph.

Definitions (References (2), (3), (4))

The flow-graph is composed of a network of *directed branches* that connect at points or *nodes*. The *branch jk* may be thought of as originating (or having its *input*) at node j, and *terminating* (or having its *output*) at node k. The assigned direction is conveniently indicated by an arrow pointing from node j to node k. Each branch jk has associated with it a *transmittance* g_{jk} which may be designated by a suitable letter written alongside the arrow of the branch, and each node j has an associated *node signal* x_j.

The signal transmitted over any branch jk is given by the product of the signal x_j, at the input end of the branch, multiplied by the branch transmittance g_{jk}. The value of the signal at any node k is given by the algebraic sum of the *incoming branch* signals at that node. Thus, with reference to Figure 1, there are three incoming branches terminating on node "1." That branch which originates at node "0" transmits a branch signal ax_0; that which originates at node "2" transmits a branch signal gx_2; and that which originates at node "3" transmits a signal fx_3. The value of the signal x_1 is therefore

$$x_1 = ax_0 + gx_2 + fx_3,$$

which is seen to be identical to the value given by the second equation. By writing similar expressions for the signal flow into nodes "2," "3," and "3′," one obtains the third, fourth, and fifth equations respectively.

The nodes considered in the preceding paragraph are *dependent nodes,* each having one or more incoming branches. We may also define a *source* as a node that has only *outgoing* branches (node "0" in Figure 1). A source corresponds to an *external* input whose value is assigned independently of all other node signals. A *sink* is a node having only incoming branches (e.g., node "3′" in Figure 1).

To describe the structure of the graph, we define a *path* as any continuous succession of branches, traversed in the indicated branch directions. Along an *open path,* no node appears more than once. For example, the two open paths between the source and sink nodes of Figure 1 are *abd1* and *ae1*. The set of open paths between a source node and some specified dependent node is uniquely determined by the graph.

A *loop* is a simple closed path along which each node is encountered once per traverse. In Figure 1, the loops are *bg, ef, bdf,* and *c,* the last being a *self loop* obtained when the input and output nodes of a branch coincide. A loop indicates the presence of *feedback.* The number of different paths through a graph containing a loop is infinite since a path may go around a loop any number of times. Notice, however, that in accord with our definition, a *non-simple* closed path, such as *bgef,* is not a loop. This path is comprised of two touching loops, *bg* and *ef.* A *non-touching loop set* is a set of loops no two of which have a common node. Of all possible pairs of loops in Figure 1, there is only one pair that is non-touching (*ef* and *c*). Of all possible triplets of loops, there are none that are non-touching.

Other terms that need definition include the *path transmittance,* which is the product of the branch transmittances in that path, and the *loop transmittance,* which is the product of the branch transmittances in that loop. The *graph transmittance* is the ratio of the signal at some specified *dependent node* to the signal applied at some specified *source node.* By using the principle of superposition, only one source and one dependent node need be considered at a time in order to "solve" the flow-graph.

Reduction and Transformation of Flow-Graphs

Graphs that do not contain loops are particularly easy to solve since, by starting at a source, each signal along an open path may be expressed explicitly in terms of the preceding signals. This permits one to simplify the original graph by various reductions, several of which are illustrated below.

CASCADE PATH:

PARALLEL PATHS:

CONTRACTION (OR NODE ABSORPTION):

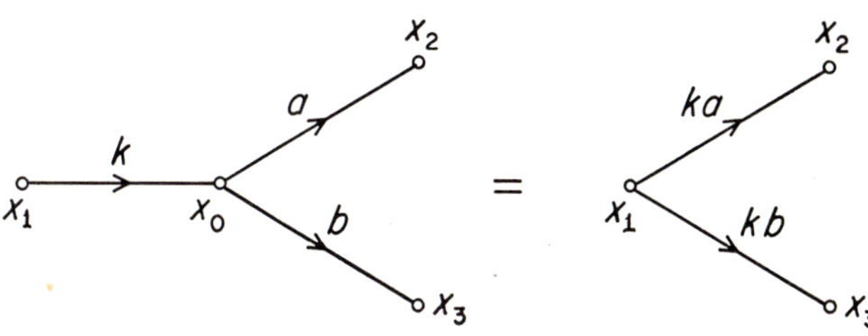

The equivalence between the graphs is with respect to the signal nodes that remain after transformation.

Another transformation that is of importance in preparing a graph for realization on an analogue computer is the change of sign of a node variable. For instance, to change x_0 to $-x_0$ in the preceding graph, the sign of each transmittance entering the node x_0 is changed as is also the sign of each transmittance leaving x_0.

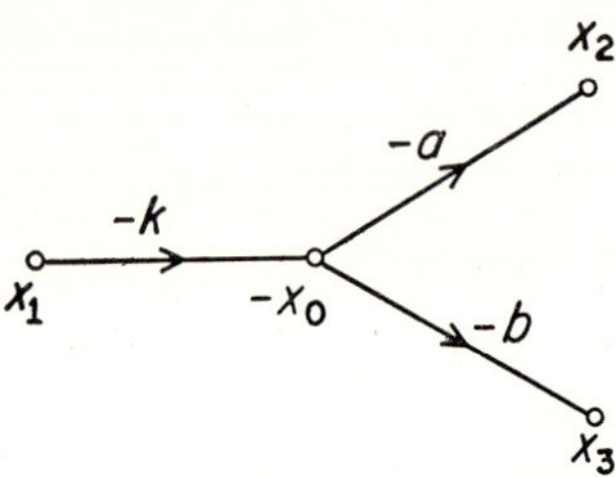

To visualize the reduction of a flow-graph that contains only *open paths,* consider the graph shown in Figure 2. The graph transmittance from "0" to "3" is given by the sum of the transmittances of all paths originating at x_0 and terminating at x_3. This is readily found to be $e + d(a + bc) + fb$.

When the graph contains one or more loops, not all nodes between a source or sink may be eliminated using the procedure just described. It is then useful to identify a set of *essential nodes*—the smallest col-

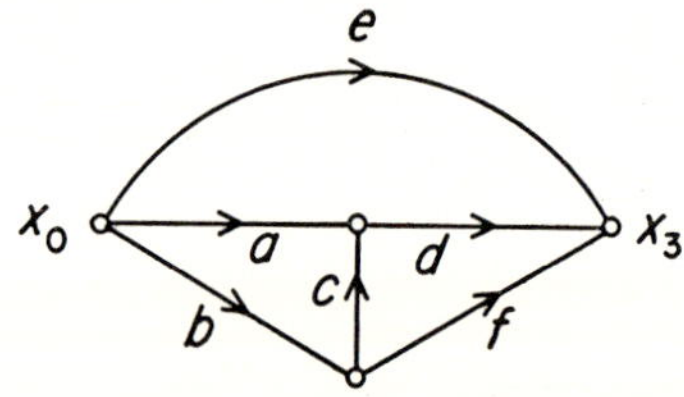

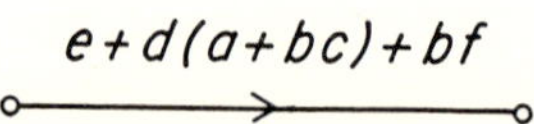

Figure 2. In an open graph, all nodes other than the source and sink are easily eliminated.

lection of nodes that when "killed" (i.e., made into sink's) will block all feedback loops in the graph. All nodes not belonging to an essential set lie on open paths that have essential nodes as sources and sinks. The non-essential nodes may therefore be eliminated from the graph by the methods already illustrated.

In the graph of Figure 1, there is a unique set of two essential nodes, x_1 and x_2. Interrupting signal flow through these two nodes opens all loops. Thus, the remaining node x_3 may be eliminated to obtain the reduced graph of Figure 3.

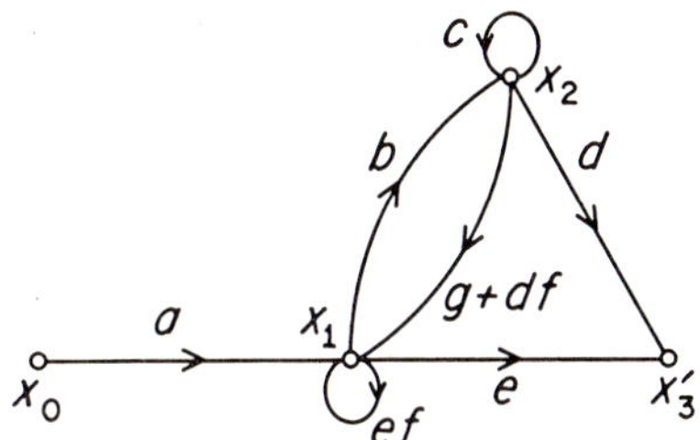

Figure 3. All non-essential nodes may be absorbed by contraction.

Further reduction of a loop graph, such as that shown in Figure 3, may be accomplished by eliminating the one or more self-loops that always appear. To see how this is done, consider the single self-loop of Figure 4.

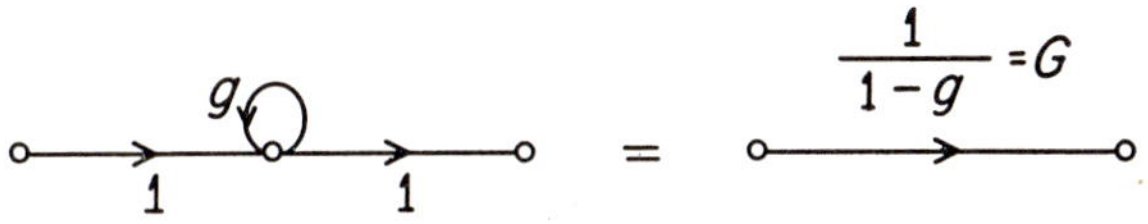

$$\frac{1}{1-g} = G$$

$$G = 1 + g + g^2 + g^3 + g^4 + \cdots$$
$$= \frac{1}{1-g}$$

Figure 4. The effect of a self-loop transmittance g on a node is to multiply all signal flow through that node by a factor $\dfrac{1}{1-g}$.

The paths from x_0 to x_2 include any number of repeated traverses of the loop. The path transmittance is increased by the factor g for each additional traverse of the loop. The sum of this infinite series of transmittances may be expressed in closed form as indicated in Figure 4. The result thus obtained is the basic equation of feedback theory. A slightly more general statement of a single-loop feedback graph and its reduction is given in Figure 5.

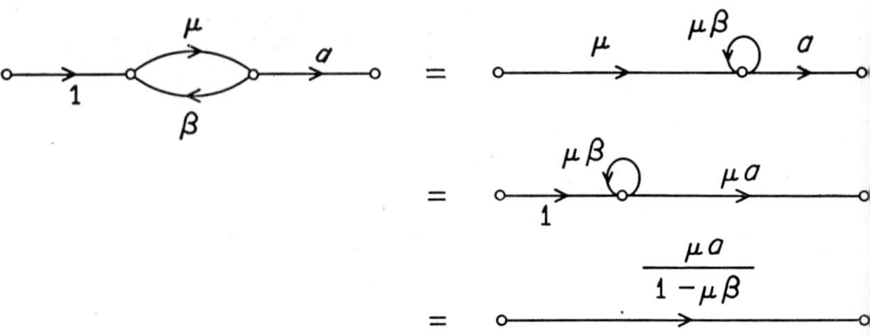

Figure 5. Illustrations of the reduction of a simple feedback graph.

Many other reduction procedures may be devised, see Reference (5), but those given above are adequate for reducing any flow-graph, as we shall illustrate by solving the graph of Figure 3 so as to express x'_3 directly in terms of x_0.

By a transformation, similar to that illustrated by Figure 4, the node x_2 with its self-loop, c, will, in effect, multiply all signal flow through it by the factor $\frac{1}{1-c}$. It may therefore be described as shown in Figure 6.

We see, upon clearing fractions, that

$$\frac{x'_3}{x_0} = \frac{a[e - ec + bd]}{1 - c - ef - bg - bdf + cef}.$$

MASON'S REDUCTION

Over one hundred years ago, Kirchhoff noted that a set of linear equations could be represented by flow-graphs and that the de-

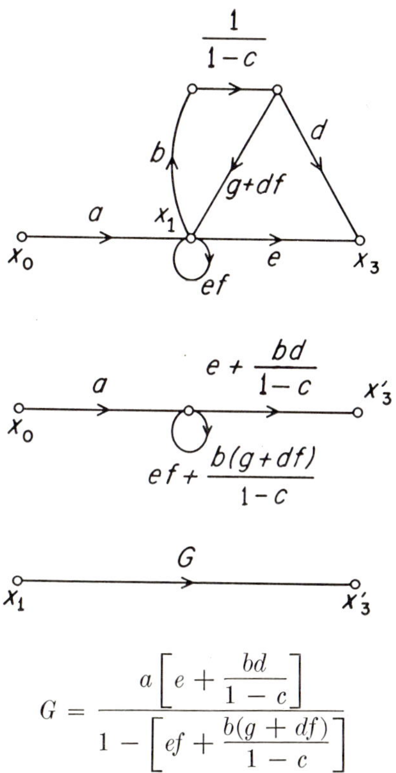

where,
$$G = \frac{a\left[e + \dfrac{bd}{1 - c}\right]}{1 - \left[ef + \dfrac{b(g + df)}{1 - c}\right]}$$

Figure 6. Reduction of Figure 3 to a single transmittance, G.

terminant and co-factors of the set of equations could be written down directly by inspection of the information provided by this graph (Reference (6)). These topological methods were little exploited until recent work by Tustin and Percival in England ((7), (8), (9)) and Mason (2) in the USA redirected attention to these methods and extended their applicability to modern control systems.

In this section, the method given by Mason for solving a flow-graph by inspection will be summarized using as an illustration the graph of Figure 1. We must determine the following:

Loop Transmittances—In this example, there are four loops and their transmittances:

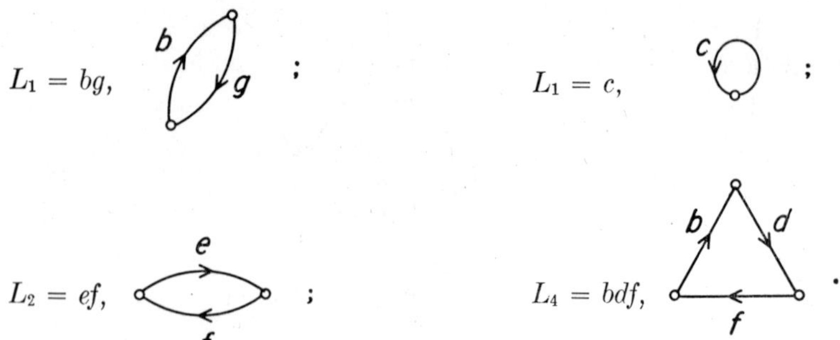

Of the possible sets of non-touching loops, there are four sets containing only one loop, and one set containing two loops. There are no sets containing three or more loops.

Open-Path Transmittances—In this example there are two open paths, and their transmittances:

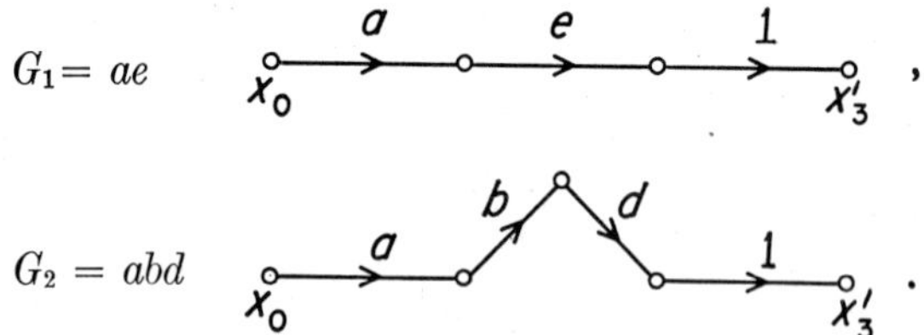

Graph Determinant—This important quantity is defined as unity plus the algebraic sum of the transmittances of *all* different non-touching loop sets contained in the graph. The algebraic sign is *positive* for an *even* number of loops in the set, and is *negative* when the number of loops is *odd*. If the set consists of a single loop, the loop-set transmittance is simply the transmittance of that loop. Thus, the graph determinant, Δ, is expressible as

$$\Delta = [1 - \Sigma L_i + \Sigma L_i L_j - \Sigma L_i L_j L_k + \cdots] \qquad (1)$$

where L_i represents all the different loops of the graph; $L_i L_j$ represents all of the different pairs of *non-touching* loops; $L_i L_j L_k$, are the different triplets, etc.

The graph determinant may be written alternatively as

$$\Delta = [(1 - L_1)(1 - L_2)(1 - L_3) \cdots (1 - L_n)]\dagger \qquad (2)$$

where L_1, L_2, $\ldots$, L_n are the loops transmittances of the n different loops in the graph and where the dagger, "†," indicates that in carrying out the multiplication within the bracket, a term will be dropped if it contains the transmittance product of touching loops.

In our particular example,

$$\Delta = 1 - (L_1 + L_2 + L_3 + L_4) + (L_2L_3).$$

Path Factor—This is the quantity analogous to the co-factor that arises in solving a set of equations by determinants. It is found by evaluating the graph determinant of that part of the graph not touching the specified path. Thus, a path factor is obtained from the graph determinant by striking out all terms containing transmittance products of loops which touch that path.

The path factor associated with the k'th open path will be denoted by Δ_k. In our particular example the path factors are

$$\Delta_1 = 1 - L_3$$
$$\Delta_2 = 1.$$

With these preliminary definitions we may finally solve for x_3 in terms of x_0. This involves finding the following:

Graph Transmittance—This is the signal at some specified *dependent* node per unit signal applied at some specified *source* node. The graph transmittance G is the weighted sum of the transmittances, G_k, of all the different open paths from the designated source node to the designated dependent node, where the weight for each path is the path factor Δ_k, divided by the graph determinant, Δ. That is

$$G = \frac{\Sigma \Delta_k G_k}{\Delta}. \tag{3}$$

In our particular example,

$$\frac{x'_3}{x_0} = \frac{\Delta_1 G_1 + \Delta_2 G_2}{\Delta} = G$$

$$\frac{x'_3}{x_0} = \frac{(1 - L_3)G_1 + G_2}{1 - (L_1 + L_2 + L_3 + L_4) + (L_2L_3)}$$

$$\frac{x'_3}{x_0} = \frac{(1 - c)ae + abd}{1 - bg - ef - c - bdf + cef}.$$

Illustrative Example

One of the most characteristic aspects of real-world problems is that they involve a multitude of factors which usually interact with one another to such an extent that it is often difficult to forecast the ultimate effect of changing any one factor. The obvious direct effect attributable to the one factor may be completely altered by the accumulation of the indirect effects from the changes induced in the other related factors. The flow-graph provides a visual structure, a universal language, a common ground upon which causal relationships among a number of variables may be laid out and compared. For systems designers and operations analysts, who must deal quantitatively with interacting variables, the flow graph should be particularly useful.

To determine how the flow-graph formulation may be used to study an engineering problem, let us suppose that in the much over-simplified design of a rocket, we consider only

x_1 the total weight of the rocket,
x_2 the propulsion thrust,
x_3 the payload.

Also, we assume that the state of the art is such that with all other factors held constant,

a. one unit increase in total weight requires one unit increase in propulsion thrust,
b. one unit increase in thrust requires $\frac{2}{3}$ units increase in total weight,
c. one unit of total weight costs 4 units of money, whereas one unit of thrust costs 10 units of money.

Determine:

1. The cost of increasing the payload one unit.
2. The cost of increasing the net thrust one unit.
3. The decrease in payload that must be accepted if the net thrust is to be increased by one unit without increasing the cost.

Designating the increase in the thrust not expended in the increase in weight by x_4, the pertinent interrelations between these factors may be shown by the flow-graph of Figure 7.

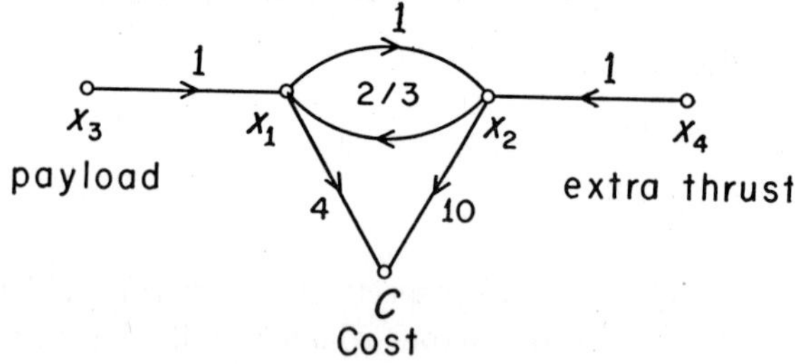

Figure 7. Flow-graph showing the interaction between weight, thrust, and cost of rocket.

The answers to the first and second questions can be obtained by evaluating the graph transmittances between nodes (x_3, C) and (x_4, C), respectively. By Mason's method, it is clear that there is one loop and two open paths in each case. Hence,

$$\frac{C}{x_3} = \frac{(1)(4) + (1)(10)}{1 - (1)(\frac{2}{3})} = \frac{14}{\frac{1}{3}} = 52$$

and

$$\frac{C}{x_4} = \frac{(\frac{2}{3})(4) + 10}{1 - (1)(\frac{2}{3})} = \frac{10\frac{8}{3}}{\frac{1}{3}} = 38.$$

Hence, the cost of increasing the payload one unit, while providing no net increase in thrust, is 52 units of money; whereas the cost of providing each unit extra thrust, while maintaining constant payload, is 38 units.

To answer the third question, we must treat the cost C as an independent variable and the payload x_3 as a dependent variable. This introduces the process of *path inversion,* a transformation whereby the roles of a source and a dependent node in a flow-graph may be interchanged.

Path Inversion

To interchange the role of an independent and a dependent variable, an open path from a source to the dependent node may be inverted, thus making the source into a dependent node and the dependent node into a source. It is also possible to invert a loop. Two methods appropriate for linear graphs will be presented. The first is given by Lorens (10) and the second by Mason (2). The inversion rules follow directly from the manipulation of the corresponding algebraic equations, and are self explanatory.

Original Graph: Path (1 – 3) to be inverted.

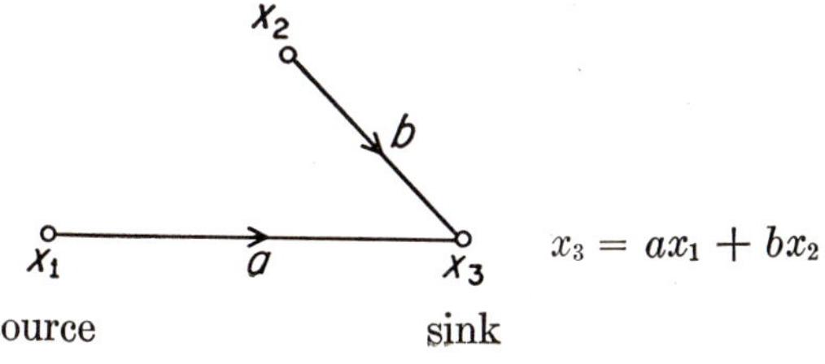

1st Method: Requires the bifurcation of node x_3 into a companion node. However, method is applicable to non-linear as well as linear graphs.

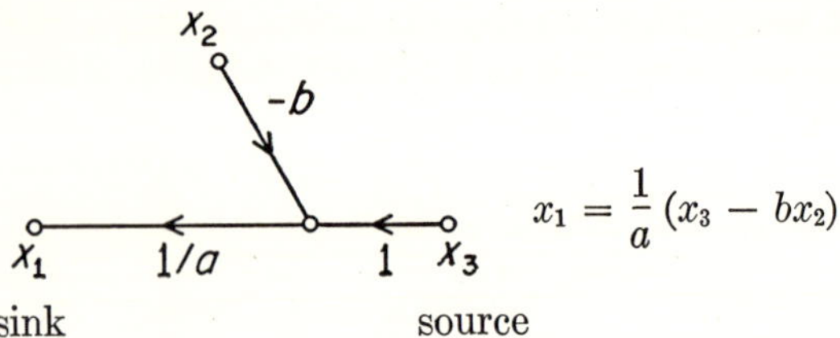

2nd Method: A further simplification of the 1st method that is possible *only* with linear graphs.

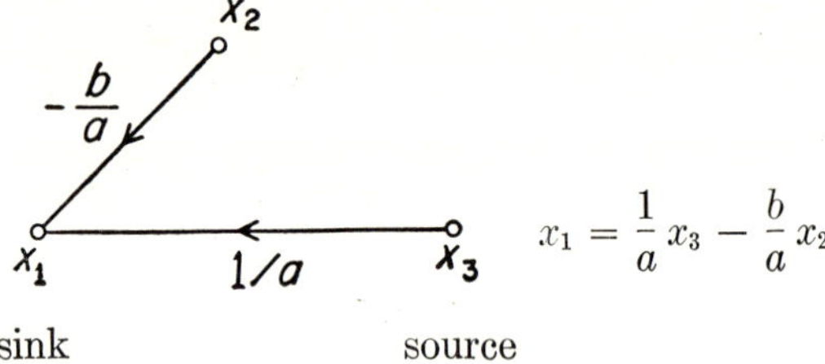

To invert an open path, begin with the branch leaving the original source, reverse the direction of the first branch and replace the transmittance by its reciprocal. In following the second method (which is the simpler for linear graphs), carry with the output end of the branch being inverted the output ends of all other branches that originally terminated in this same node, and divide the transmittance of each of these branches by the negative of the transmittance of the inverted branch. Repeat this process with the second branch of the open path and continue in this manner until all branches of the open path have been inverted. In following the first method, one proceeds in a similar fashion except that the output node of each branch being inverted is bifurcated into two nodes joined by a unit transmittance in the inverted direction. All outgoing branches are left with the original output node, whereas all incoming branches are moved to the new (interior) node and the signs of their transmittances are changed. This operation is difficult to describe in words but is actually very simple to perform, as the above diagrams show.

Path inversion may sometimes be used to reduce the number of feedback loops. This transformation is important in minimizing the difficulties with instability and the accumulative errors that sometimes arise in simulating a system on a computer. Furthermore, nonlinear "transmittances" are much more easily handled if the graph contains no feedback loops. For instance, the following graph represents a two-stage process with feedback around each stage.

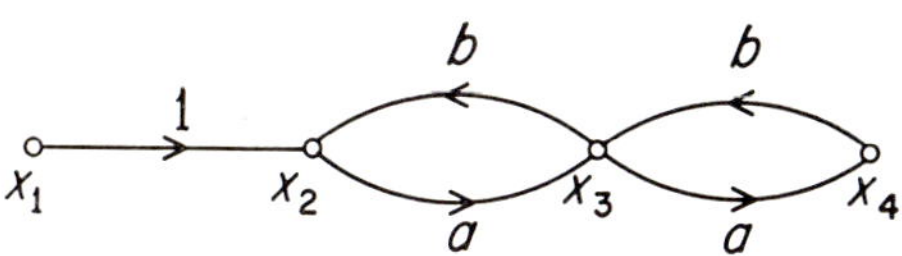

By Mason's rule, we may immediately establish the graph transmittance from x_1 to x_4. There are two loops, $L_1 = ab$ and $L_2 = ab$, and one open path, $G_1 = a^2$. Hence, $\Delta = 1 - 2ab$ and $\Delta_1 = 1$ and

$$\frac{x_4}{x_1} = \frac{a^2}{1 - 2ab}.$$

In order to transform this graph so as to eliminate the two feedback loops, the path from x_1 to x_4 may be inverted. That is, we regard x_4 as *specified* and inquire what x_1 *must have been.* (Notice that inversion does not imply a reversal of causality.) To invert the open path from x_1 to x_4 by the first method, we get

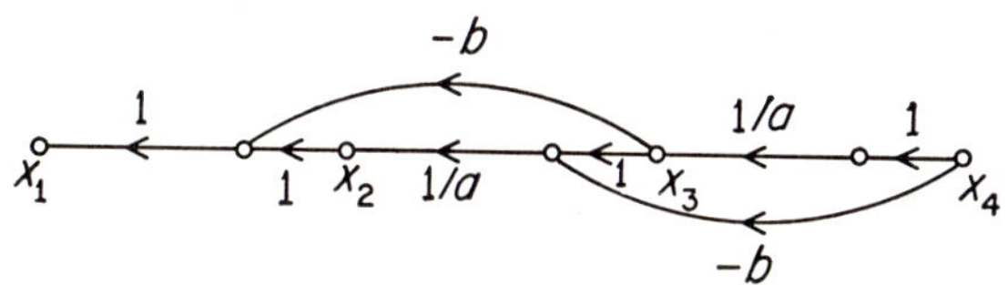

The total transmittance from x_4 to x_1 is thus easily found to be

$$\frac{x_1}{x_4} = \frac{1 - 2ab}{a^2}.$$

As this is a cascade graph, having no feedback loops, the signal at each node may be found explicitly and exactly once x_4 is given. (This would be so, even if the branch transmittances were known non-linear operators.)

Direct application of the second method, which is applicable only to linear graphs, yields the somewhat simpler result

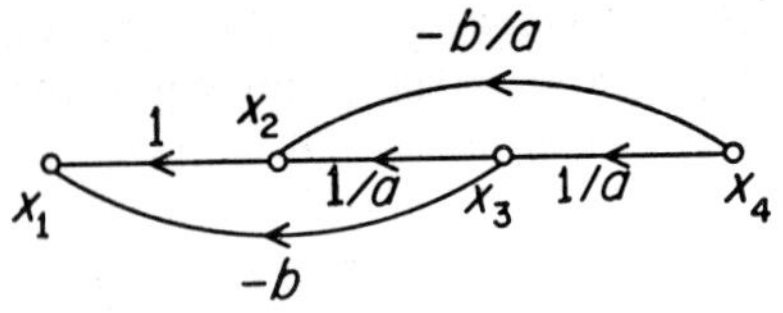

and evaluation of the total transmittance from x_4 to x_1 gives

$$\frac{x_1}{x_4} = -\frac{b}{a} + \frac{1}{a^2} - \frac{b}{a}$$

$$= \frac{1 - 2ab}{a^2}.$$

This inversion process is applicable to forward paths which start from a source and terminate on a sink, and also to feedback loops. Inversion of a loop often achieves considerable simplification in a graph. It is important in preparing the graph for realization on a computer since, if the transmittance of the original loop is large, the transmittance of the inverted loop will be small. Stability considerations are quite different in the two cases.

Illustrative Example Continued

To answer the third question of the illustrative example, the x_3 node may be made into a sink and the C node into a source by inverting the path (x_3, x_1, C) to obtain

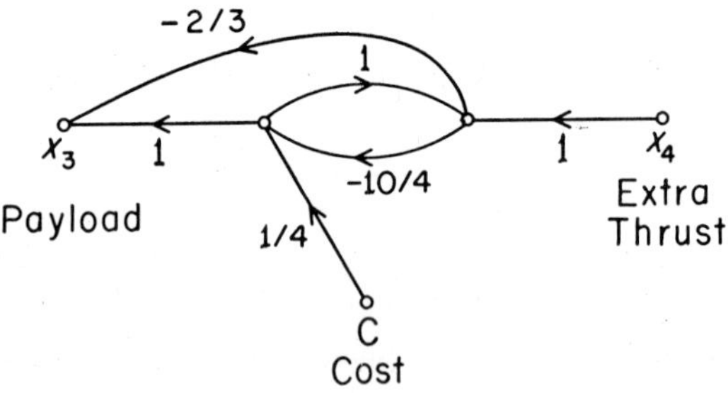

Thus, if the cost is held constant, the effect on the payload of providing one unit of extra thrust may be determined by evaluating the graph transmittance from x_4 to x_3. This yields

$$\frac{x_3}{x_4} = \frac{-\frac{2}{3} - \frac{10}{4}}{1 - (1)(-\frac{10}{4})} = -\frac{19}{21}.$$

Also, if the extra thrust is held constant, one unit of money will buy an increase in payload given by the transmittance from C to x_3,

$$\frac{x_3}{C} = \frac{\frac{1}{4}\,(1 - \frac{2}{3})}{1 - (1)(-\frac{10}{4})} = \frac{1}{42}.$$

In the graph shown above, the total transmittance around the loop is $-\frac{10}{4}$. Large loop transmittances sometimes pose difficulties if the system is to be simulated on an analogue computer. It is possible to invert the loop in such cases, thus yielding a loop transmittance that is smaller than unity. In this example, inversion of the loop yields the graph,

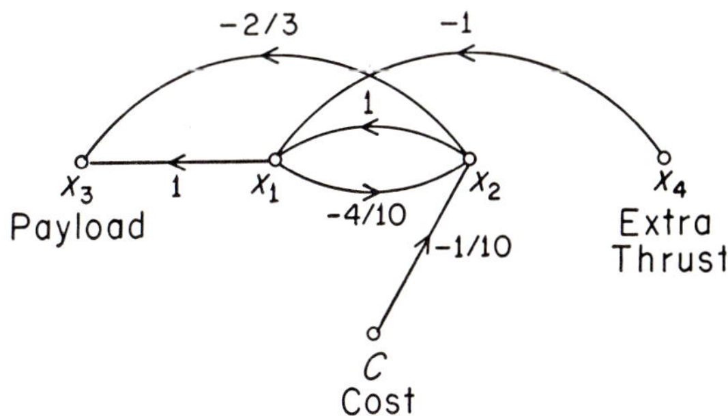

The graph transmittances are the same as those just determined; however, in this graph no branch transmittance is larger than unity.

Representation of Non-Linear Relationships

An interesting feature of the flow-graph is that when it is used to represent a system of linear algebraic equations, the flow-graph may be manipulated and solved in lieu of solving the equations. When the system is non-linear or time-varying, the algebraic solution of the equations breaks down, whereas the flow-graph with slight alterations may still be used to program a solution using numerical methods or a computer. Although some graph transformations can only be applied to linear graphs, there are some important transformations that are applicable even when the relations between the node signals are non-linear.

To represent a non-linear relationship, the branch transmittance is replaced by a functional form. Thus, the graph shown at the left is

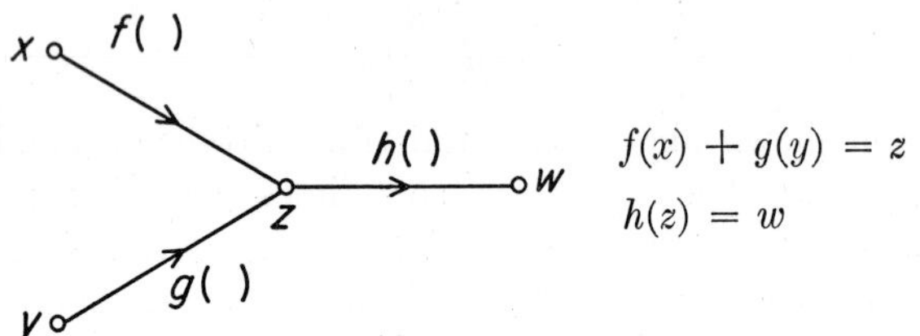

equivalent to the functional equations written at the right. In simulating this set of relationships on an analogue computer, the three non-linear functions could be realized with electronic function generators. If it were desired to express x as a function of w and y, the forward path from x to w could be inverted by one of several methods given by Lorens (10).

The first method is essentially the same as the first method already discussed for linear graphs. This method requires the realization of the inverse functions f^{-1} and h^{-1} (where $f^{-1}[f(x)] = x$, etc.).

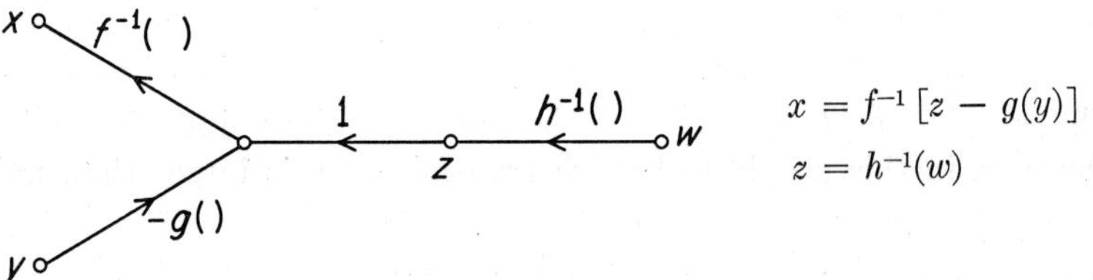

Thus, if the original non-linear relationship obeyed a *square* law, th inverse relationship would require the realization of a *square-roo* operator. Furthermore, the inverse function may not be so well be haved as the original function (e.g., it may be multiple valued, etc.).

Lorens avoids the use of inverse operations in his second method for path inversion. Since $w = h(z)$, the identity $z = z + \lambda[w - h(z)]$ must be valid. But this identity expresses z in terms of w, viz. if,

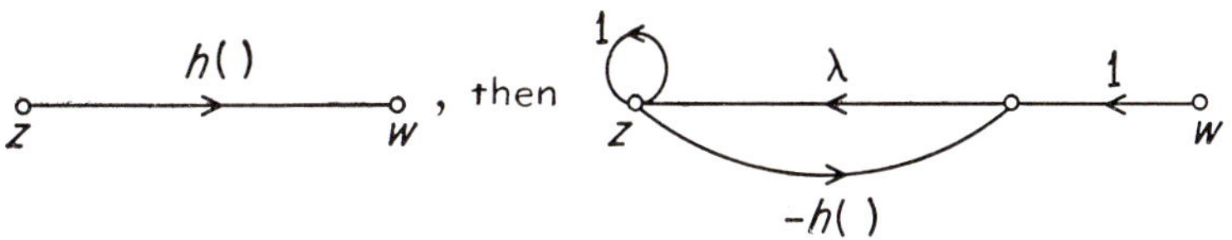

where λ is any transmittance chosen so as to obtain a stable solution. Using this transformation, the inversion of the path from w to x may also be written as

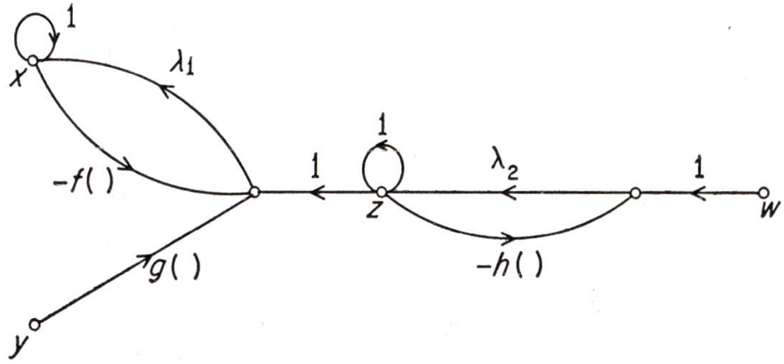

It will be observed that this method of simulating the inverse operator involves a kind of successive-approximation process. The transmittance λ must be chosen so that the approximation process will converge . . . beyond that, λ may be linear or non-linear and have quite general properties.

To illustrate the foregoing manipulations, consider the problem of multiplying together two signals, x and y. Here, it is convenient to use the logarithmic operator

x —ln( )→ ln(x)

and the exponential operator (which is the inverse of the logarithmic operator)

$$x \xrightarrow{e^{(\,)}} e^x \;.$$

Then, to represent $xy = z$,

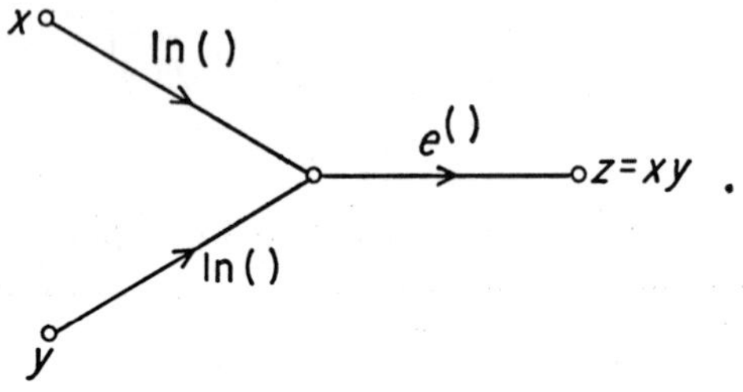

The exponential operator may be obtained from the logarithmic operator by inverting the forward path using the second method of Lorens, viz.

input $x \xrightarrow{\ln(\,)} \ln x = z$ output

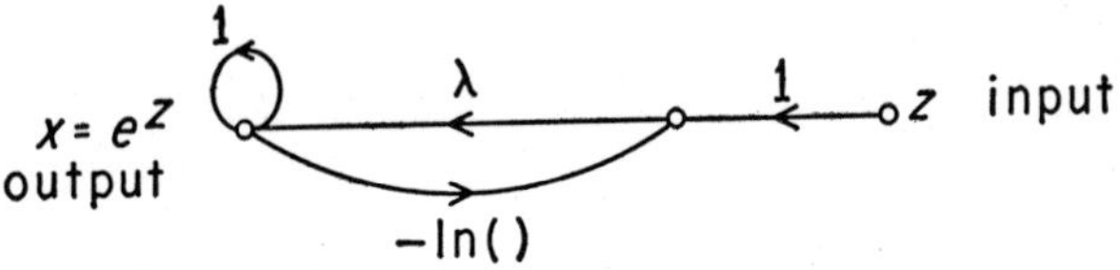

so that a multiplier using only logarithmic operators is,

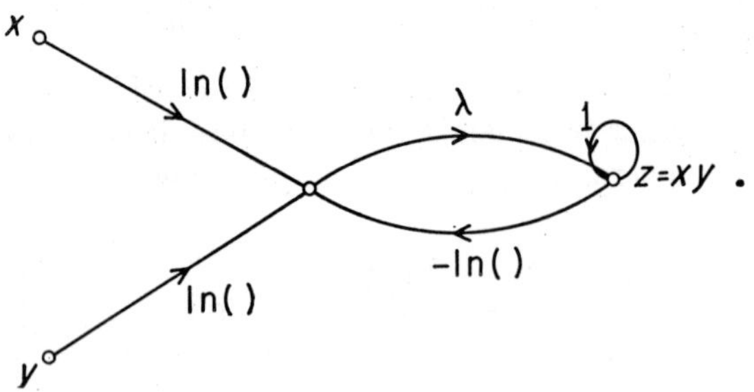

By simply changing the sign of the appropriate branch, it is possible to represent $x/y = z$, and by other scale-factor changes, x and y may be raised to any power desired.

Flow-Graph Representation of a Simple Control System

In this section, flow-graph techniques will be used to represent and study some of the properties of a closed-loop control system. Figure 8 shows a direct-current motor. It is known that the motor speed $\dot{\theta}$ is related to the applied voltage V and the load torque L by the motor equation

$$\dot{\theta} = aV - bL$$

where a and b are constants of the motor.

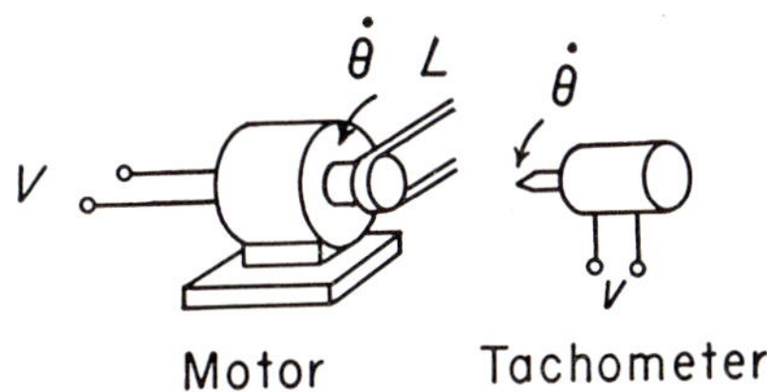

Figure 8. A DC motor and a tachometer.

If one attempted to control the speed of the motor by varying the applied voltage V, the speed will depend upon how heavily the motor is loaded as well as upon the motor "constants" a and b, both of which will vary with the temperature of the motor. These dependencies are illustrated by the flow-graph of Figure 9. Since the applied voltage is not influenced by the speed, this arrangement is often referred to as "open path (or loop)" control.

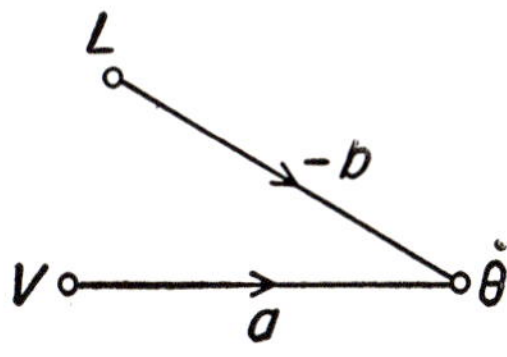

Figure 9. "Open path" speed control.

To control the speed more precisely, a tachometer, which generates a voltage $v = g\dot{\theta}$ proportional to its speed, may be driven by the motor. This tachometer signal may be fed back so as to control the voltage V applied to the motor, as illustrated in the flow-graph of Figure 10.

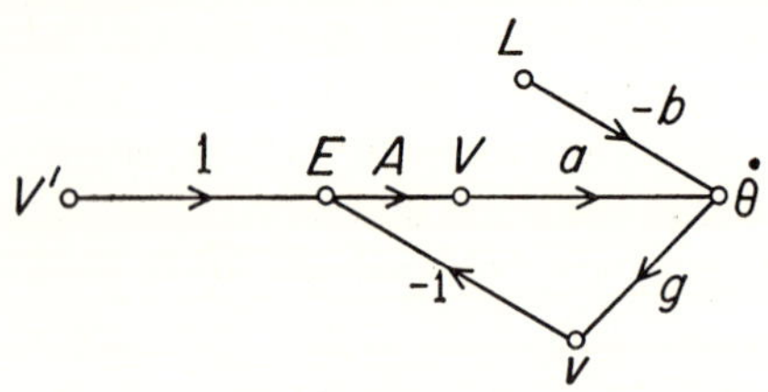

Figure 10. Closed-Loop speed control.

In the control system illustrated, the tachometer voltage is subtracted from a reference voltage V'. The difference $V' - v = E$ yields an *error signal* which is then amplified in a suitable device by a factor A and applied to the motor. It should be noted that the slower the motor turns, the larger will be the value of error signal E and this, in turn, will tend to cause the motor to speed up. However, as the speed increases, the error signal will decrease and, if the tachometer voltage v should become equal to the reference voltage V', the error signal would be zero, no voltage would be applied to the motor and it would slow down again. By making A quite large, the speed can be made to depend primarily upon the reference voltage V' and the tachometer constant, g, as will now be shown.

Starting with the flow-graph of Figure 10, the motor speed may be expressed in terms of the reference voltage V' and the motor load L, by computing the graph transmittances from the V' and L nodes to the $\dot{\theta}$ node. Because the graph is linear, the speed is given by the sum of these two contributions,

$$\dot{\theta} = \frac{Aa}{1 + gAa} V' - \frac{b}{1 + gAa} L.$$

As A becomes large, this expression approaches,

$$\dot{\theta} = \frac{1}{g} V' - \frac{b}{gAa} L$$

so that the speed is substantially independent of the motor parameter

a and depends instead upon the reference voltage V' and the tachometer constant g (which by careful engineering may be made truly constant). In addition, it will be noted that, since A occurs in the denominator of the coefficient of the load, the variations in motor speed with changes in motor load will also be greatly reduced.

It might appear that the best performance will be achieved by making A indefinitely large. This, however, ignores dynamic effects and questions of instability which will be discussed in Chapter 22 and in Chapter 23. In practice, the value of A cannot be made indefinitely large. Consequently, the speed will depend slightly upon the values of the transmittances A, a, etc. It is important to be able to estimate the sensitivity of the system performance to slight changes in these various system parameters.

Sensitivity of a System to Small Changes in Its Components

In any system, the components will usually deviate somewhat from their nominal values. A desirable objective in the system design may be to minimize the effect of these component variations upon the overall system performance. Feedback, as illustrated in the previous example, provides a method for reducing the effect of certain of the system parameters. However, it may be noted that even with unlimited amounts of feedback, the speed of the motor still depends directly upon the tachometer constant g. It is therefore clear that each system parameter may have a different effect upon the system performance, and that a particular feedback arrangement may decrease some of these effects and possibly increase others. Obviously, what is needed is a concise measure of the sensitivity of a specified aspect of system performance to changes in the value of a particular parameter.

The sensitivity of a transmittance G with respect to any arbitrary parameter b is defined as

$$S_b^G = \frac{\partial \ln G}{\partial \ln b} = \frac{\partial\, G/G}{\partial\, b/b}.$$

When the changes are small, *the sensitivity is the ratio of the percentage change in G to the percentage change in b.* It provides a useful estimate of the importance of the effects of errors or changes in any of the system parameters upon the graph transmittance between any specified source and dependent node.

A useful expression for the sensitivity, using quantities appearing in Mason's rule, has been given by W. A. Lynch (11). It is

$$S_b^G = \frac{\Delta^0}{\Delta} - \frac{\Sigma(G_k\Delta_k)^0}{\Sigma(G_k\Delta_k)}$$

where Δ^0 is the value of Δ when the transmittance of the specified branch b is set equal to zero; similarly for $(G_k\Delta_k)^0$.

As an example of the use of sensitivity factor, consider the closed-loop speed-control system of the previous section. In a typical DC motor, the parameter a relating the terminal voltage of the motor to its speed would depend upon the field excitation, which in turn would vary with changes in the resistance of the field windings as the motor heats up during operation. Let us see how a given variation in a would affect the transmittance form V' to $\dot{\theta}$. Here,

$$\begin{aligned} \Delta &= 1 + gAa \\ \Delta^0 &= 1 \\ G_1\Delta_1 &= Aa \\ (G_1\Delta_1)^0 &= 0. \end{aligned}$$

Therefore,

$$S_a^{V'\dot{\theta}} = \frac{1}{1 + gAa}.$$

Hence, if gAa is much larger than unity, the sensitivity will be much smaller than one and the dependency of motor speed upon the reference voltage V' will be virtually unaffected by changes in the parameter a. This possibility of obtaining very low sensitivity to changes in certain components is one of the very useful and desirable effects of feedback in a system.

On the other hand, the sensitivity with respect to the tachometer transmittance g is

$$S_g^{V'\dot{\theta}} = \frac{1}{1 + gAa} - \frac{Aa}{Aa} = -\frac{gAa}{1 + gAa}$$

which will be very nearly equal to -1 under the same conditions of large gAa. Hence, for precise control, the tachometer transmittance would have to be accurately maintained.

As a second illustration, let us return to the example of Figure 7. It was originally assumed that one unit increase in thrust would require $\frac{2}{3}$ unit increase in weight of the missile. Suppose that this ratio could be decreased by some percentage through the use of a different fuel. What would be the effect on the cost per unit payload? Here, we replace $\frac{2}{3}$ by b and we see that a given percentage change in weight/thrust ratio would effect twice that percentage change in the cost of the rocket.

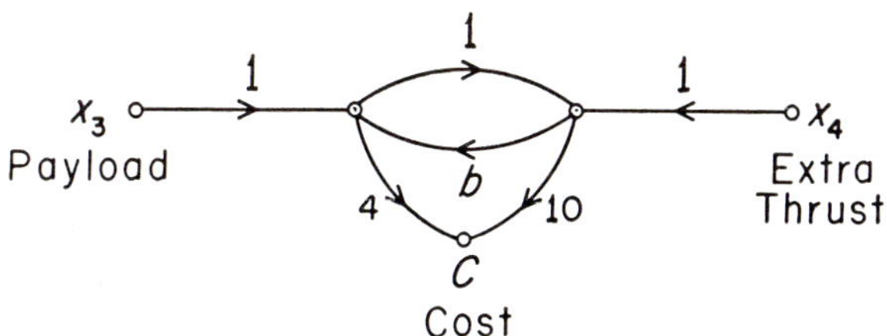

By inspection,

$$\Delta = 1 - b = \tfrac{1}{3} \text{ (for } b = \tfrac{2}{3})$$
$$\Delta^0 = 1.$$

Hence,

$$S_b^{x'_3C} = \frac{1}{\frac{1}{3}} - 1 = 2 \text{ (for } b = \tfrac{2}{3}).$$

On the other hand, the cost/unit-thrust ratio would be affected somewhat differently by the change in the fuel. Here, Δ and Δ^0 are the same as above, but

$$G_1\Delta_1 = 10$$
$$G_2\Delta_2 = 4b.$$

Hence,

$$S_b^{x'_4C} = 3 - \tfrac{15}{19} = 2\tfrac{4}{19} \text{ (for } b = \tfrac{2}{3}).$$

Thus, the effect of a decrease in the weight/thrust ratio of the fuel would produce a larger percentage decrease in the cost of each unit of net thrust than in each unit of payload.

It is hoped that these examples have given some indication of the use of the sensitivity value to estimate the significance of various parameters to particular relationships within the system, and of how the sensitivity may be computed directly from the flow-graph of the system.

REFERENCES

(1) Truxal, J. G. *Automatic Feedback Control System Synthesis*. New York: McGraw-Hill, 1955.

(2) Mason, S. J. "Feedback Theory: Some Properties of Signal Flow-Graphs," *Proceedings of the Institute of Radio Engineers,* Vol. 41 (September, 1953), 1144-56. "Feedback Theory: Further Properties of Signal Flow-Graphs," *Proceedings of the Institute of Radio Engineers,* Vol. 44 (July, 1956), 920-26.

(3) ——— and H. J. Zimmerman, *Electronic Circuits, Signals, and Systems,* John Wiley and Sons, Inc., 1960.

(4) Institute of Radio Engineers, Technical Committee 4. "Proposed Standard Definitions of Terms for Linear Signal Flow-Graphs," April, 1958.

(5) Happ, W. W. "Signal Flow-Graphs," *Proceedings of the Institute of Radio Engineers,* Vol. 45 (September, 1957), 1293.

(6) Ku, Y. H. "Resume of Maxwell's and Kirchhoff's Rules for Network Analysis," *Journal of the Franklin Institute,* Vol. 253, No. 1 (January, 1952), 211-24.

(7) Tustin, Arnold. *The Mechanism of Economic Systems.* Cambridge: Harvard University Press, 1953.

(8) ———. *Direct Current Machines for Control Systems.* New York: Macmillan Co., 1952.

(9) Percival, W. S. "The Graphs of Active Networks," *Proceedings of the Institute of Electrical Engineers,* Vol. 102, Part C (1955), 270.

(10) Lorens, C. S. *Theory and Application of Flow-Graphs,* Sc. D. Dissertation, Massachusetts Institute of Technology, July, 1956.

(11) Lynch, W. A. "A Formulation of the Sensitivity Function," *Institute of Radio Engineers Transactions, CT-4,* No. 3 (September, 1957), 289.

(12) Menger, Karl. "New Approach to Teaching Intermediate Mathematics," *Science,* Vol. 127 (6 June 1958), 1320-23.

Twenty-Two

SYSTEM DYNAMICS

WILLIAM H. HUGGINS

INTRODUCTION

When we survey the vast array of technological systems which incorporate many kinds of things—chemical, mechanical, and electrical, to name but a few—and range from the relative simplicity of the portable radio to the enormous complexity of the world-wide telephone system, we might have little hope of finding any significant aspect that is shared by them all. Yet, despite the infinite detail and endless variation in the physical appearance of the *things* that comprise these systems, there is one aspect, at least, in which they may resemble one another. This is in their *patterns of behavior*. Regardless of the kinds of things of which they are composed, it is found that some systems are passive and, like the run-down clock, remain motionless until activated by an external agent. Many systems are autonomous and self-regulatory and, like the wound-up clock, exhibit a sustained response once set into motion. A few systems are unstable and destroy themselves in extremes of activity.

Evidently, to study that which is most common to all systems we must ignore many of the physical attributes, such as size, shape, color, weight, etc., by which our senses most often identify any particular system, and instead focus our attention upon the *temporal patterns of change* of the system observables. All physical systems share a common attribute—that of *existence*—and the modes by which the state of existence may change from one moment to the next are likewise shared by many apparently unrelated systems. Indeed, these modes are so few in number and can be so explicitly stated (e.g., as physical laws) that it is often possible to *simulate* the behavior of many different systems on a single generalized system called a *computer*. Let us observe, however, that in this process, it is *not* the *things* (with their physical attributes of size, shape, color, weight, etc.) that are simulated, but rather the relationships between the *signals* (i.e., values of the *observables*) associated with each system.

This chapter will develop some concepts and techniques that are particularly helpful in studying the dynamic behavior of linear stationary systems, either analytically or by computer. The flow-graph representation introduced in the previous chapter is, thus far, limited to static systems exhibiting constant signals. For instance, an expression was obtained for the speed of an electric motor as a function of its load torque, when all quantities (or signals) were constants. However, should the load suddenly be removed from the motor, it is evident that the speed could not increase abruptly to its new value. More generally, the dynamics of a system must be considered whenever its observables are subject to change. Sometimes the changes in observables may be generated and sustained by the system itself, thus producing an unstable, oscillatory or "hunting" condition in which stationary equilibrium is never reached. The design problem of achieving system stability is of major importance in most control processes. Fortunately, in the case of linear stationary systems, a reasonably complete theory can be formulated using the methods described in this chapter and the next.

THE CONCEPT OF A "BLACK-BOX" (I.E., OPERATOR)

The notion that a system acts upon an input signal to produce a related output signal is made manifest in the concept of the "black box" illustrated in Figure 1. No information is conveyed regarding the details of what is going on inside the black box. In fact, there may be a wide variety of systems which are very different in internal structure but which nevertheless are equivalent to one another with respect to the relationship between some particular input and output signals. (Of course, these systems may not be equivalent in other aspects—for instance, one might utilize very few elements and be relatively in-

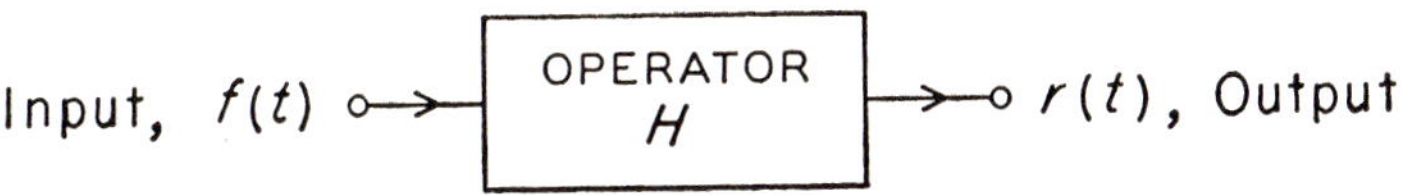

Figure 1. The relationship between the input and resulting output of a system may be represented by a "black-box" or system *operator,* H.

expensive whereas another might require many elements and be quite costly.)

The possibility of first specifying the desired input-output relations of a system and then proceeding by well-defined methods to *realize* the internal details of one or more systems that will possess these desired input-output relationships has been clearly demonstrated by the development of modern techniques for *network synthesis* (1). Unfortunately, the synthesis techniques that have been developed for electric circuits composed of simple *linear* R, L, C elements cannot be applied to systems composed of more complicated non-linear and time-varying elements. Nevertheless, the demonstration that such techniques may be developed for the special class of simple electrical systems is a challenging stimulus for the development of synthesis methods applicable to the much wider class of systems encountered in modern technology (2).

The notion of a "black-box" operator is basic to the development of a theory for systems. Three important problems associated with a single operator are illustrated in Figure 2.

Figure 2. Three problems associated with a single operator.

Problem	Given	Determine
Analysis	f, H	r
Instrumentation	r, H	f
Identification	f, r	H

The *analysis problem* arises whenever one wishes to determine the response of a known system to a given input. The design study of complicated control systems often involves repeated analysis of different configurations, using in each case a variety of representative inputs while modifying the parameters of the control system until an acceptable response is obtained.

The *instrumentation problem* is encountered whenever one attempts to infer the values of some signal $f(t)$ which, for some reason, cannot be observed directly. In principle, *any* measurement employs an operator H, and the signal $r(t)$ which is observed will usually differ from the signal $f(t)$ being measured. It is sometimes possible to design the measuring instrument so that its output will be a more-or-less accurate *time replica* of the input quantity. However, in many instances the output of the measuring instrument will differ from the input. The problem of inferring the input signal from these indirect observations is a subject of very considerable interest in most fields of human endeavor.

Finally, it should be noted that the *identification problem* is central to much of science. To probe a system with known stimuli and to determine from the observed responses to these stimuli, a *model* of the system is often the major objective of scientific experimentation.

Obviously, the notion of an operator as described in the preceding paragraphs is much too broad to allow concise quantitative development. We shall, therefore, restrict the class of operators to be considered in the sequel to those which are *linear* and *stationary*. (This restriction is to be assumed in all that follows unless specifically stated to the contrary.)

If an operator is *linear*, the *principle of superposition* may be applied. That is, if the input $f(t)$ to a linear operator is decomposed arbitrarily into the sum of two inputs, viz. $f(t) = f_1(t) + f_2(t)$, the corresponding output $r(t)$ may be expressed as $r(t) = r_1(t) + r_2(t)$, where $r_1(t)$ is the response of $f_1(t)$ only and $r_2(t)$ is the response to $f_2(t)$ only. Furthermore, if the input is multiplied by any constant factor a, the output will be multiplied by this same factor, viz. the response to $af(t)$ will be $ar(t)$.

An operator is *stationary* if it does not change its properties with time. More precisely, if, when any input signal $f(t)$ is replaced by a delayed signal $f(t - T)$, (where T is any *constant* time delay), the response $r(t)$ is likewise replaced by $r(t - T)$, then the operator is said to be stationary.

THE REPRESENTATION OF AN OPERATOR BY ITS TRANSMITTANCE

In general, the output $r(t)$ of an operation may bear little resemblance to the input. This is because many operators exhibit dynamic properties and the response depends upon how the input signal is changing with time. Fortunately, among all possible input signals, there will generally be a special class of input signals, any one of which has the property that the corresponding output resembles the input exactly except for a constant multiplying factor. Suppose that each member of this class of input signals is distinguished by an identifying subscript. Then, a particular input signal $f_k(t)$ will produce an output $H_k f_k(t)$ where H_k is a constant that depends both upon the properties of the operator H and the particular signal, f_k—hence, it is written as H_k. Any input $f_k(t)$ which produces the same output, except for a constant multiplier, is called a *characteristic signal* of the operator H.

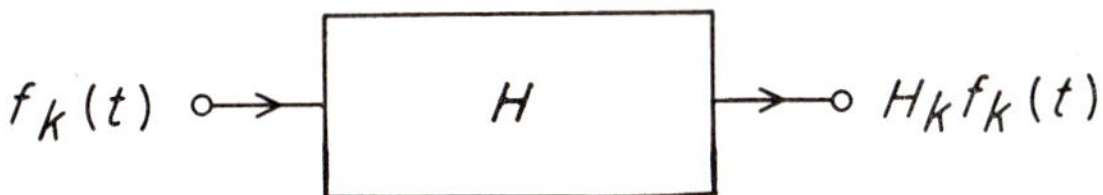

Figure 3. An input $f_k(t)$ is a characteristic signal of an operator H if it produces a response that is equal to a constant multiple, H_k, of the input.

The effect of the operator H upon any one of its characteristic input signals $f_k(t)$ may be represented exactly by specifying the value of the corresponding characteristic multiplier H_k, which is simply the ratio

$$\frac{\text{output}}{\text{input}} = \frac{H_k f_k(t)}{f_k(t)} = H_k. \tag{1}$$

This constant ratio will hereafter be referred to as the transmittance of the operator H for the k'th characteristic signal.

Now, as will be shown later, all stable linear stationary operators possess a common class of characteristic signals (i.e., the class of growing exponentials). Furthermore, it is possible to select from this class of growing exponentials a set which is complete. That is, any well-behaved signal can be synthesized by adding together suitable amounts of these various characteristic signals, viz.

$$f(t) = \sum_k F_k f_k(t) \tag{2}$$

where F_k is the amount of $f_k(t)$ in the decomposition of $f(t)$. If the transmittance of the operator is known for each characteristic signal appearing in (2), the response $r(t)$ of the operator to any input $f(t)$ is immediately expressible as the sum of the responses to each of the component inputs, viz.

$$\begin{aligned} r(t) &= \sum_k H_k F_k f_k(t) \\ &= \sum_k R_k f_k(t) \end{aligned} \tag{3}$$

where

$$R_k = H_k F_k \tag{4}$$

In other words, by expressing any signal as the sum of the characteristic signals of the operator, the calculation of the effect of the operator upon a given input signal is reduced to the simple arithmetical operations of multiplication and addition. Furthermore, specification of the set of transmittances H_k for each member of a complete set of characteristic signals provides a *representation* of the operator in the sense that this information is sufficient to determine exactly the output for any input signal. Hence, we may think of the symbol H as representing the specification of the transmittances H_k for all values of k.

A further remarkable advantage of this method of representing signals and operators is made evident when one considers *systems of operators*. Consider the result of *cascading* two operators, H and G, which share a common, complete set of characteristic signals. As indicated in Figure 4, the output $H_k f_k(t)$ serves as the input to the second operator. The result of cascading *two* linear, stationary operators is thus seen to be equivalent to a *single* operator characterized by the product $G_k H_k$ of the two cascaded transmittances G_k and H_k. It is evident that the order in which the operations are performed may be reversed without changing the final result (this is not generally true when the operators are non-linear or non-stationary, for then a complete set of characteristic signals that is common to both operators is not available).

When two linear stationary operators are connected in *parallel*, the

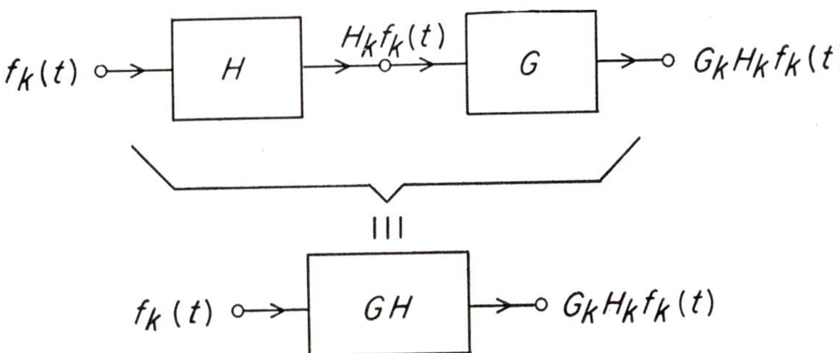

Figure 4. The cascading of two linear stationary operators is equivalent to a single operator having a transmittance which is the product of the transmittances of the cascaded operators.

same input is applied to each and the output of the combination is the sum of the two outputs as shown in Figure 5.

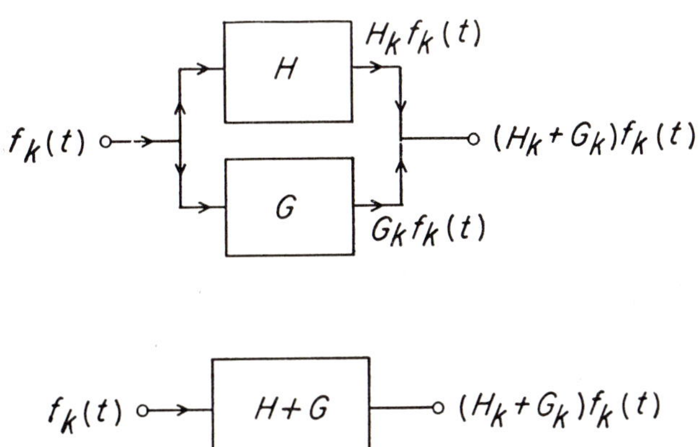

Figure 5. The paralleling of two operators is equivalent to a single operator having a transmittance which is the sum of the transmittances of the paralleled operators.

The composition rules illustrated by Figures 4 and 5 suggest that when any number of linear, stationary operators are interconnected, the relation between a specified input and output of the resulting system may be expressed by a single equivalent operator which is characterized by a set of transmittances that may be calculated in a straightforward manner from the transmittances of the component operators. The flow-graph representation of systems, discussed in the previous chapter, provides a convenient method for doing this, once each of the operators has been characterized by its transmittance. We therefore consider next the representation of some important linear stationary operators by their transmittances.

SOME IMPORTANT LINEAR STATIONARY OPERATORS

Four important operators that may be used to construct models of many physical systems are shown in Figure 6. Each of these operators

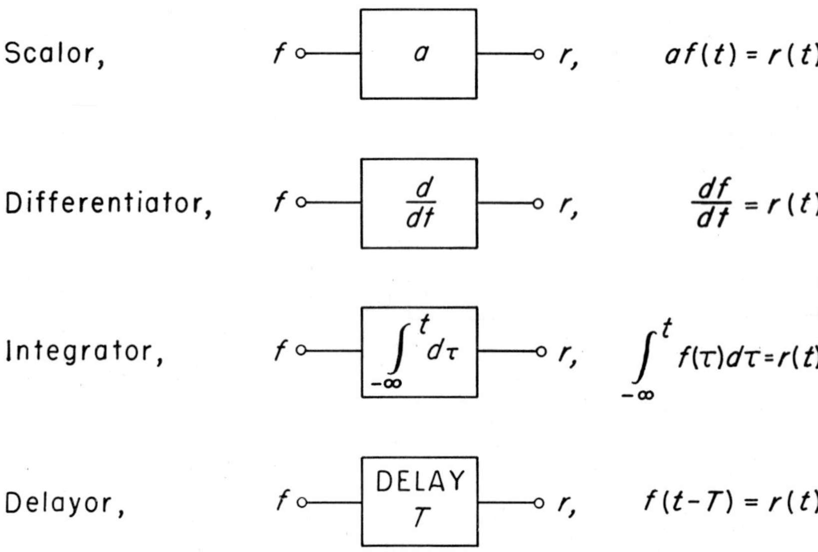

Figure 6. Four operators and their mathematical descriptions.

transforms any input signal $f(t)$ into another signal $r(t)$ in accord with the relationship indicated. The *differentiator* and *integrator* will generally alter drastically the wave shape of an input signal since the first gives an output which is at any instant equal to the rate of change of the input at that instant, whereas the second gives an output which at any instant is equal to the total accumulation of the input over all prior time. On the other hand, the *scalor* or *delayor* do not generally change the wave shape except, respectively, for multiplication by a constant amplitude scale factor or a constant shift (i.e., translation) in time.

Of these four operators, the scalor is the simplest. It may be represented in flow-graph notation by a single transmittance a which is

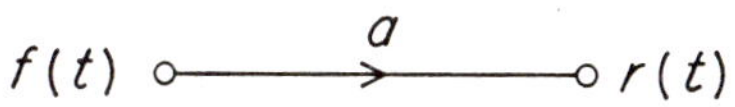

$$f(t) \circ \xrightarrow{a} \circ\, r(t)$$

Figure 7. Flow-graph representation of a scalor.

the same for all signals. Since the ratio of output to input is a constant for any signal $f(t)$, it is evident that *any signal is characteristic of a scalor.*

Consider next the differentiator shown in Figure 6. Here, for $f(t)$ to be a characteristic signal, it is necessary that the ratio

$$\frac{r(t)}{f(t)} = \frac{\dfrac{df}{dt}}{f(t)} \tag{5}$$

be a constant in time. That is, any signal $f(t)$ will be a characteristic signal of a differentiator if and only if

$$\frac{\dfrac{df}{dt}}{f(t)} = \frac{d}{dt} \ln f(t) = s, \tag{6}$$

where s is some constant. It follows that $\ln f(t) = st +$ constant, or that

$$f(t) = Ae^{st} = A \exp(st) \tag{7}$$

where A, being the value of $f(t)$ at $t = 0$, is called the *initial amplitude* (or often simply the amplitude) of the exponential signal, and s, being generally a complex constant, is called the *complex frequency* of the exponential signal $A \exp(st)$. Thus, *any* exponential signal is characteristic of a differentiator, which may then be replaced by a constant transmittance of value s.

Let us next determine the class of characteristic signals associated with an integrator. It is reasonable to inquire whether or not the class of exponential signals might also include the characteristic signals of an integrator. Assuming that $f(t) = A\exp(st)$, one finds that

$$r(t) = \int_{-\infty}^{t} Ae^{s\tau}\,d\tau \tag{8}$$

$$= \frac{A}{s}\left[e^{st} - e^{-\infty s}\right]. \tag{8a}$$

In order that the constant term (arising from the lower limit of the integral) will vanish, it is necessary that the *real part* σ *of the complex frequency* $s = \sigma + j\omega$, be positive. Then, there is no difficulty in forming the ratio

$$\frac{r(t)}{f(t)} = \frac{\frac{A}{s}e^{st}}{Ae^{st}} = \frac{1}{s}. \tag{9}$$

Thus, *any growing exponential,* with $\mathrm{Re}(s) > 0$, is *characteristic of an integrator.* For any member of this class of characteristic signals, the integrator may be replaced by the constant transmittance $1/s$.

It would be a triple blessing if the exponential signal should also happen to be a characteristic signal of the delay operator. Let us see if this can possibly be so. Here, for an input $f(t) = A\exp(st)$,

$$r(t) = Ae^{s(t-T)}$$

$$= [Ae^{-sT}]e^{st}$$

so that

$$\frac{r(t)}{f(t)} = e^{-Ts}. \tag{10}$$

Thus, *any exponential is characteristic of a delayor.* For any member of this class of characteristic signals, the delayor may be replaced by a constant transmittance $\exp(-Ts)$, which depends upon the complex frequency and the time delay T.

We have now shown that the exponential signal $\exp(st)$ (for s constant, with positive real part and $-\infty < t$) is a characteristic signal of all four operators shown in Figure 6.

If a single growing exponential, $A\exp(st)$, is applied to any one of these four operators, the effect of that operator may be completely represented by a transmittance (which depends upon the value of the complex frequency s) as shown in Figure 8.

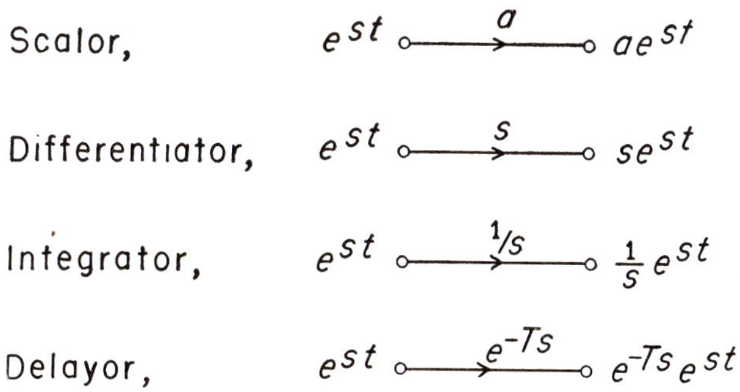

Scalor, e^{st} —a→ ae^{st}

Differentiator, e^{st} —s→ se^{st}

Integrator, e^{st} —$1/s$→ $\frac{1}{s}e^{st}$

Delayor, e^{st} —e^{-Ts}→ $e^{-Ts}e^{st}$

Figure 8. Four operators and their representation as transmittances for the growing exponential signal.

In Figure 8 the transmittance is shown as a function of the complex frequency s. It is evident that the differentiation operation yields a transmittance of value s for an exponential input signal, but is the converse true? Would an unidentified operator, which is found to have a transmittance equal to s for *any* growing exponential, necessarily be a differentiator? That is, does the specification of the *functional form* of the dependency of transmittance upon the complex frequency *completely specify* the operator?

Specification of the transmittance for *all* values of s does indeed completely determine the properties of the operator. There is a one-to-one correspondence between the conventional mathematical representations of Figure 6 and the transmittance representations of these same operators in Figure 8. We may therefore adopt the latter as a concise and useful notation of designating the operations of Figure 6.

A system formed by interconnecting any number of these operators will also possess a set of characteristic signals that is included within this same class of growing exponentials. This is so because differentiating, delaying, and integrating a growing exponential input produces outputs of exactly the same kind, and hence, the output of one operator when used as the input to another merely produces replications of itself that are identical except for a transmittance factor. For this reason it will be possible to reduce the model of a dynamic system to a flow-graph form involving constant branch transmittances (which depend, however, upon the value s of the complex frequency of the input to the system).

THE COMPLEX-FREQUENCY PLANE

From the discussion in the preceding section, it should be evident that it is the growing exponential signal, rather than the sinusoid so widely used in engineering measurements, that occupies the center of the stage in system theory. Fortunately, it is possible to decompose almost any signal encountered in engineering work into the sum of a number (perhaps infinite) of exponential components having different frequencies. Because we are dealing here with linear systems, the response to each exponential component may be found by simple flow-graph methods. Then, by superimposing the responses to these individual components, the response to the original signal is obtained.

Depending upon the value of its complex frequency s, the exponential signal takes on a wide variety of wave shapes so that many signals of practical importance can often be decomposed into just a few exponential components. The real part σ of the complex frequency $s = \sigma + j\omega$ will be referred to as the *neper frequency,* and the imaginary part, ω, will be referred to as the *radian frequency* so as to avoid confusion over what is meant by such expressions as "real" or "imaginary" frequency.

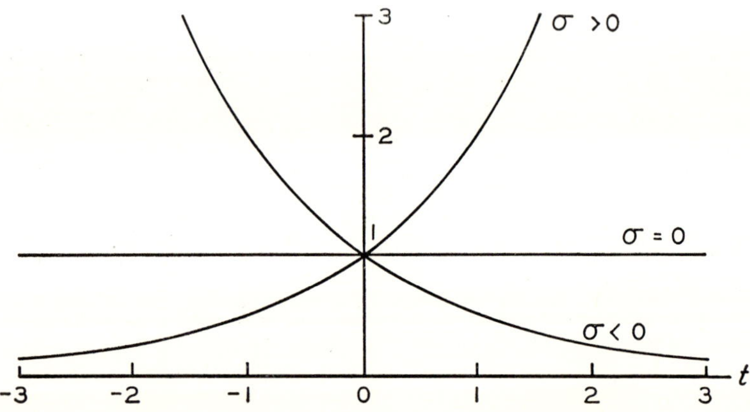

Figure 9. Neper frequency signals $\exp(\sigma t)$.

Figure 9 shows three signals whose radian frequencies are zero and whose neper frequencies are positive, zero, and negative, respectively.

In this particular plot, the amplitude A is assumed to be real and is equal to unity. In general, however, A may be any complex constant, in which event the exponential signal will have both real and imaginary parts, the amplitudes of which depend upon the phase angle of A. It should be noted that any constant signal, such as a direct voltage, is a special case of an exponential signal for which the complex frequency s is zero.

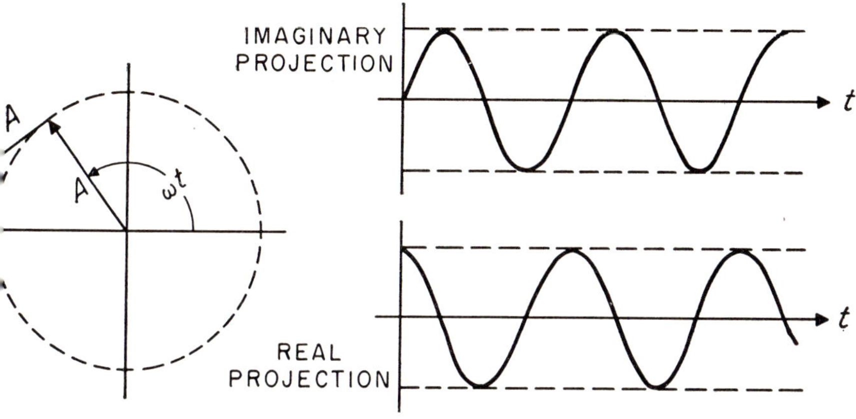

Figure 10. Radian-frequency signal, A exp(jωt).

Figure 10 illustrates a signal whose neper frequency is zero and whose radian frequency is positive. The value of the signal at any instant t is

$$Ae^{j\omega t} = A[\cos \omega t + j \sin \omega t] \tag{11}$$

which can be interpreted geometrically on an Argand diagram as a complex *phasor* of length A which is spinning in the counterclockwise direction with an angular velocity of ω radians/sec. The phase angle at any time is the constant argument of the *amplitude factor* A, plus the phase angle ωt of the time-varying factor $\exp(j\omega t)$. A *negative* radian frequency corresponds to a phasor that is spinning in the *clockwise* direction. Since

$$\sin \omega t = \frac{1}{2j}[e^{j\omega t} - e^{-j\omega t}] \tag{12}$$

$$\cos \omega t = \frac{1}{2}[e^{j\omega t} + e^{-j\omega t}], \tag{12a}$$

it is evident that a *cosine wave* consists of the *sum of two exponential signals,* as does also a sine wave (the latter differing from the former in that the initial amplitude factors are imaginary rather than real, thus accounting for the 90-degree difference in phase between the cosine and sine functions). Because the second exponential term is the complex conjugate of the first, the sum of the two terms is always real and it is only necessary to calculate the response of a circuit to a *single* input of the form $\exp(j\omega t)$—the real and imaginary parts of the complex signal thus obtained will be the responses to $\cos \omega t$ and $\sin \omega t$ input signals respectively.

Finally, when s is a complex number, we find that

$$Ae^{(\sigma+j\omega)t} = (Ae^{\sigma t})e^{j\omega t}. \tag{13}$$

This signal may be represented by a phasor, the length of which changes exponentially with a *neper frequency* σ while spinning with a radian frequency ω in the counterclockwise direction. The tip of the phasor traces out the logarithmic spiral illustrated in Figure 11.

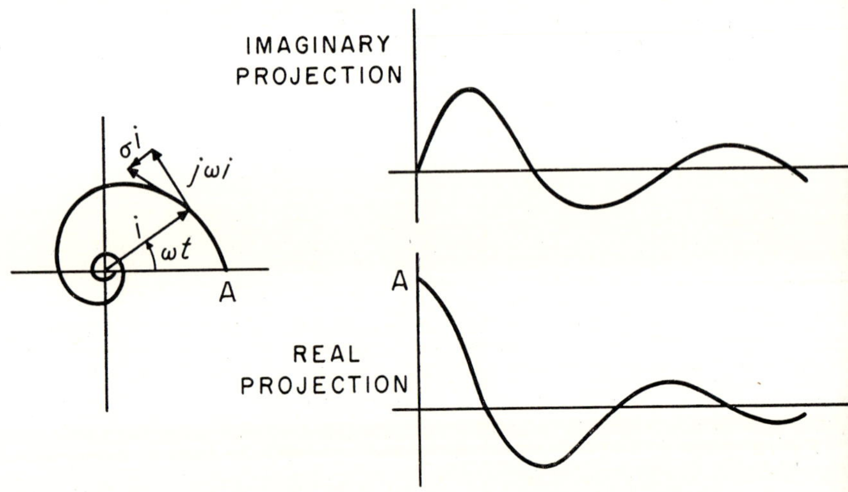

Figure 11. Complex-frequency signal, A $\exp(\sigma t + j\omega t)$.

To summarize, the waveform of an exponential signal is characterized by its complex frequency $\sigma + j\omega$, the real and imaginary parts of which define a point in the complex-frequency plane. The exponential signal may be interpreted geometrically as a phasor that changes with time in both length and direction angle at rates determined by σ and ω respectively. Figure 12 illustrates the typical motions of the phasors associated with different regions of the complex-frequency plane. It should be observed that the signals, whose complex frequencies fall in the *right half* of the s plane shown in Figure 12, are *unstable* in the sense that their instantaneous magnitudes are increasing exponentially without limit. The signals, whose frequencies lie on the j-axis, have real and imaginary parts which oscillate with constant peak amplitude. Hence, the frequencies on the j-axis correspond to the so-called *steady-state* signals. Finally, the signals whose frequencies fall in the *left half* of the s plane are *transient* in the sense that they decay to nothing with sufficient passage of time.

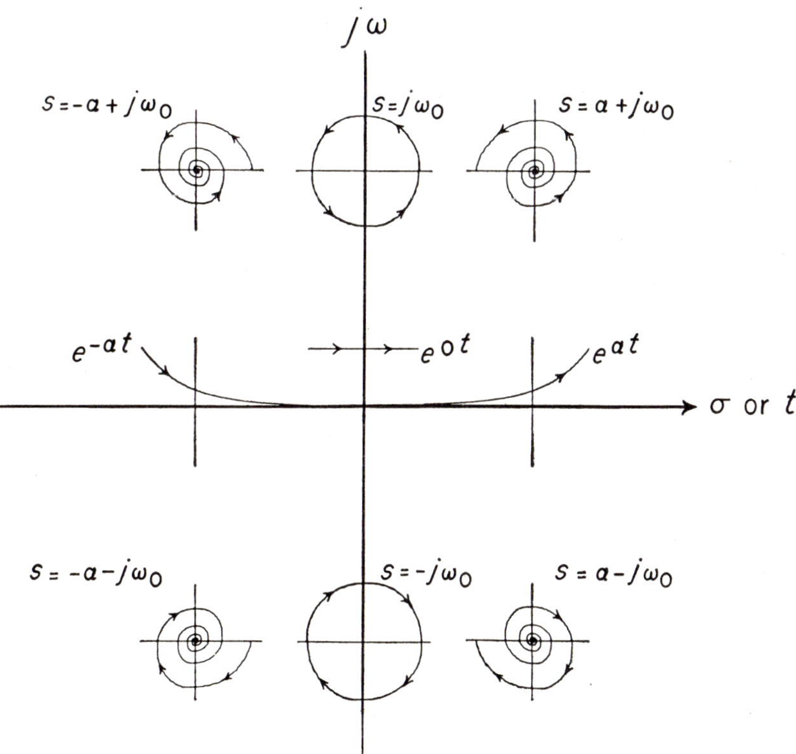

Figure 12. Illustrating the signals corresponding to nine different regions of the complex frequency plane.

THE CONCEPT OF A SIGNAL GENERATOR

In the previous discussion, we have been little concerned with how to present an input signal in quantitative manner suitable for computational purposes. The signals that arise in the real world are ordinarily considerably more complicated than the simple exponential signals considered thus far. Of course, one may always specify a signal $f(t)$ by tabulating its value at each of many instants in the time taken so closely together that all pertinent detail in the signal is preserved.[1] This, however, is of little use for analytical purposes, nor is it an economical method of representing a signal.

For interpretive and diagnostic purposes, it is most desirable to represent a signal by as few numbers as possible. Specification of a signal by tabulating its sampled values at many different instants in time will usually require that many more numbers be specified than if full use is made of the known structural properties which are inherent in most information-bearing signals (3). Perhaps the simplest and most meaningful way of describing a signal is to specify a generator of that signal.

A *signal generator* will be assumed to consist of a linear system of operators to which is applied a primitive input signal. This primitive input signal is subsequently shaped and processed by the linear system so as to produce at its output the desired signal.

What, then, is a good primitive signal that can serve as raw material from which all other signals might be made by performing suitable operations upon it? Before answering this question, let us observe that most signals have well-defined *epochs* or times-of-beginning. In addition to its epoch, a signal is characterized by its intensity, or amplitude. All remaining, distinguishable features of a signal, other than its epoch and intensity, relate to the structure or waveform of the signal.

Describing a signal in terms of its signal generator permits us to associate the *epoch* (or *temporal*) information with the primitive input signal, and the *wave shape* (or *structural*) information with the operator portion of the signal generator (4). A primitive signal, which is completely specified by its epoch and intensity, is the *unit impulse*,

[1] As mentioned in Chapter 20, this may always be done with a band-limited signal.

$\delta(t)$.[2] Although the unit impulse is a functional monstrosity, it is an acceptable and very useful signal when used as an input to any integrating operator (5). For instance, application of a unit impulse to an integrator generates a unit-step signal. If this unit step is itself applied to a second integrator, there results the *unit-ramp* signal. By successively integrating the unit impulse, the set of *power components* $u_{-n}(t) = \dfrac{t^{n-1}}{(n-1)!}$, $(t > 0)$ shown in Figure 13 is obtained. If we now tacitly agree always to use a unit impulse as the primitive original

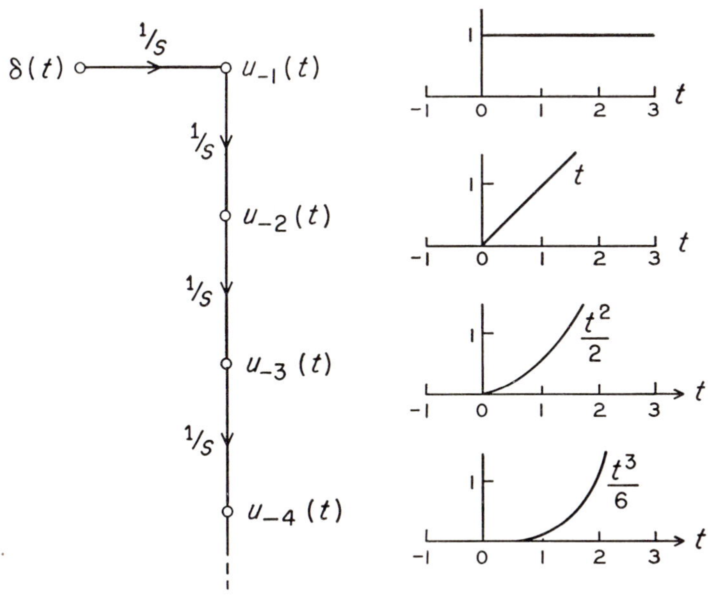

Figure 13. One-sided power components may be generated by repeatedly integrating a unit-impulse.

signal, it will only be necessary to specify the *operator portion* of the signal generator in order to describe concisely the signal that is generated. But, in the previous sections it was shown that the operator may

[2] The unit impulse $\delta(t)$ may be regarded as the limiting form, as T goes to zero, of any short positive pulse of duration T and of unit area centered around $t = 0$.

be specified concisely by its transmittance function. Thus, the unit step may be described either in the time domain as

$$u_{-1}(t) = \begin{cases} 1 \text{ for } t > 0 \\ 0 \text{ for } t < 0 \end{cases} \tag{14a}$$

or by the transmittance of its signal generator, viz.

$$U_{-1}(s) = \frac{1}{s}. \tag{14b}$$

Here we have adopted the usual convention of using a lower case letter to represent a signal in the time domain and an upper case letter to represent the *same* signal in the complex-frequency domain. In general, there is a correspondence between the "power" components in the time and frequency domains:

$$\frac{1}{s^n} \Leftrightarrow \begin{cases} \dfrac{t^{n-1}}{(n-1)!} & \text{for} \quad t > 0, \\ 0 & \text{for} \quad t < 0. \end{cases} \tag{15}$$

Either representation may be used to describe the *same* signal.

Since any linear stationary operator H may be characterized by its response $h(t)$ to a unit-impulse input applied at $t = 0$, it is possible to express the result of this same operation upon any input signal $f(t)$ by decomposing $f(t)$ into the sum of infinitesimal impulses,

$$f(t) = \int_{-\infty}^{\infty} \delta(t - \tau) f(\tau) \, d\tau. \tag{16}$$

Because the operator is assumed to be stationary, the response to the impluse $\delta(t - \tau)$ will be $h(t - \tau)$. The total response, summed over all inputs, will then be

$$r(t) = \int_{-\infty}^{\infty} h(t - \tau) f(\tau) \, d\tau \tag{17a}$$

which by a change of variables may be written in its equivalent form

$$r(t) = \int_{-\infty}^{\infty} h(\tau) f(t - \tau) \, d\tau. \tag{17b}$$

When $h(\tau)$ is known, this *convolution integral* may be evaluated for any reasonably well-behaved input signal $f(t)$ to obtain the corresponding response.

Thus, any linear stationary operator may be characterized by either its *frequency-domain* representation $H(s)$ or its *time-domain* representation $h(\tau)$. To establish the relationship between the frequency- and time-domain representations, let us determine the response of an operator, characterized by its impulse response $h(\tau)$, to a growing exponential input $\exp(st)$. By the convolution integral, (17b), we find that

$$r(t) = \int_{-\infty}^{\infty} h(\tau)\, e^{s(t-\tau)}\, d\tau$$

$$= \left[\int_{-\infty}^{\infty} h(\tau)\, e^{-s\tau}\, d\tau\right] e^{st}. \tag{18}$$

It will be noted that the response is also a growing exponential. The ratio of output/input thus yields a *constant* transmittance given by

$$H(s) = \int_{-\infty}^{\infty} h(\tau)\, e^{-s\tau}\, d\tau. \tag{19}$$

Equation (19) provides an analytic procedure for determining the transmittance H(s) *of the signal generator for* h(τ). By making the real part of s sufficiently positive, convergence of the integral defining $H(s)$ is always assured.

Incidentally, equation (19) demonstrates that an exponential signal, whose growth rate is sufficiently rapid to insure convergence of the integral expression for $H(s)$, will be a characteristic signal for a wider class of realizable linear stationary operators than considered hitherto. For instance, our results may be extended to continuous systems involving wave propagation, dispersion, etc., and may even include a class of *anticipatory* systems whose (non-physical) responses precede the input in time!

COMPARISON OF FREQUENCY-DOMAIN AND TIME-DOMAIN REPRESENTATIONS

What advantages does the frequency-domain description of a signal offer over its time-domain description? The first advantage is that the frequency-domain description is often much simpler, requiring the specification of a *single analytic function* of s, whereas the time-domain description may require the specification of several functions of t, each valid only over certain intervals of time (which must also be specified).

To illustrate, let us describe a rectangular pulse $f_1(t)$ of unit amplitude and duration T. A general procedure for constructing a signal generator is to differentiate the desired signal repeatedly until impulses appear (6). These may then be realized simply by means of scalors and delayors. Inversely, by integrating these impulses the same number of times as the original signal was differentiated, it is possible to recover the original signal. In the case of the rectangular pulse $f_1(t)$, a signal differentiation reveals two impulses and nothing else, as shown in Figure 14. Hence, to generate $f(t)$, we must produce at the

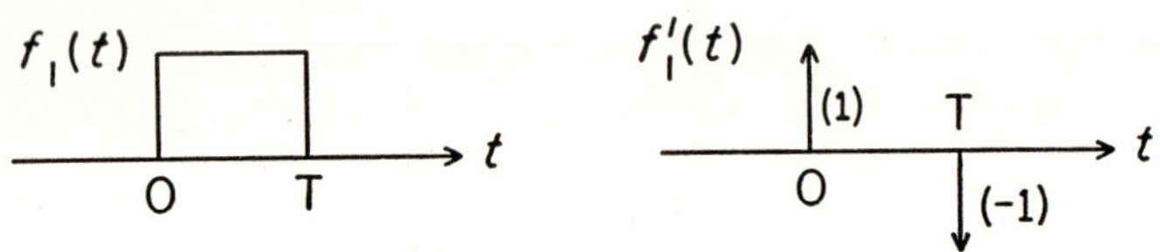

Figure 14. The derivative of $f_1(t)$ consists only of impulses.

input of an integrator an impulse at $t = 0$ of strength $+1$ and a delayed impulse at $t = T$ of strength -1. This is easily done with a scalor and a delayor, yielding as the signal generator for $f_1(t)$,

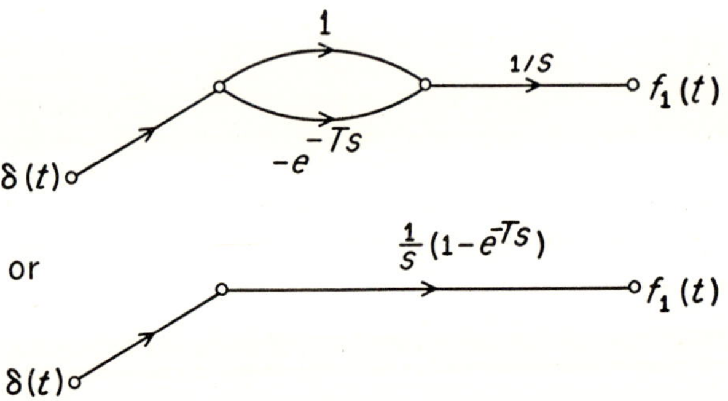

The transmittance of the signal generator is easily found by the methods of the previous chapter to be

$$F_1(s) = \frac{1}{s}(1 - e^{-Ts})$$

which is the frequency-domain description of the rectangular pulse.

A second advantage of the frequency-domain description is noted when the signal consists of the *sum* of a number of components. It is often possible to sum the resulting expressions for these components and obtain a simpler closed expression in the frequency domain. Consider, for instance, the representation of the periodic sequence of im-

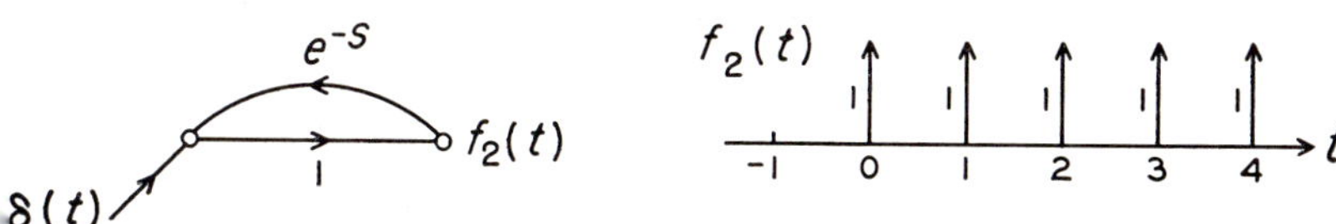

Figure 15. A periodic train of impulses may be generated by use of a recirculating delayor.

pulses shown in Figure 15. The time-domain description of this series of impulses involves an infinite series,

$$f_2(t) = \sum_{k=0}^{\infty} \delta(t - k) \tag{20a}$$

whereas, in the frequency domain, the operator transmittance may be written in closed form,

$$F_2(s) = \sum_{k=0}^{\infty} e^{-ks}$$

$$= \frac{1}{1 - e^{-s}}. \tag{20b}$$

To provide yet another illustration of the compactness of the frequency-domain description, let us find the signal generator for the single exponential signal

$$f_3(t) = \begin{cases} e^{-\alpha t} & t > 0 \\ 0 & t < 0. \end{cases} \tag{21a}$$

It is known that an exponential function may be expressed as the power series

$$f_3(t) = 1 - \alpha t + \frac{(\alpha t)^2}{2!} - \frac{(\alpha t)^3}{3!} + \cdots.$$

In the frequency domain, this expression corresponds to the geometric series,

$$F_3(s) = \frac{1}{s} - \frac{\alpha}{s^2} + \frac{\alpha^2}{s^3} - \frac{\alpha^3}{s^4} + \cdots$$
$$= \frac{1}{s}\left[\frac{1}{1+\dfrac{\alpha}{s}}\right] = \frac{1}{s+\alpha}, \tag{21b}$$

so that, again, a power series in time is equivalent to a simple algebraic expression in the frequency domain. It may be noted that the closed form involving $1/s$ may be realized by feedback around a single integrator,

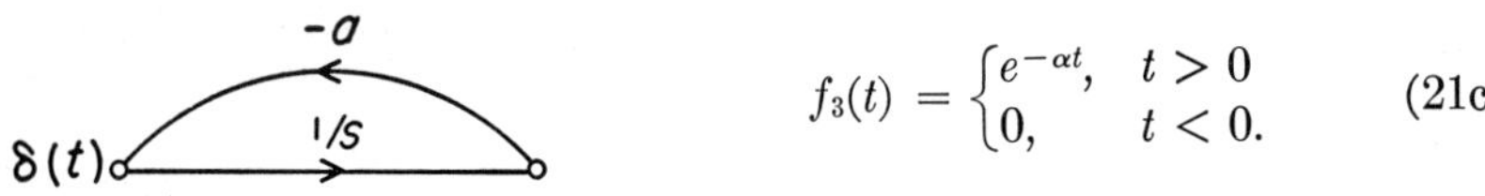

$$f_3(t) = \begin{cases} e^{-\alpha t}, & t > 0 \\ 0, & t < 0. \end{cases} \tag{21c}$$

The foregoing example has revealed a correspondence of fundamental importance: To generate *any* exponential

$$f_\nu(t) = \begin{cases} e^{s_\nu t}, & t > 0 \\ 0, & t < 0, \end{cases} \tag{22a}$$

a unit impulse may be applied to an operator characterized by the simple transmittance,

$$F_\nu(s) = \frac{1}{s - s_\nu}. \tag{22b}$$

Similarly, by repeatedly differentiating both (22a) and (22b) with respect to s_v, we may show that to generate the time-multiplied exponential

$$f_n(t) = \begin{cases} \dfrac{t^{n-1}}{(n-1)!}\, e^{s_\nu t}, & t > 0, \\ 0, & t < 0, \end{cases} \tag{22c}$$

a unit-impulse may be applied to an operator characterized by the transmittance

$$F_n(s) = \frac{1}{(s - s_\nu)^n}. \tag{22d}$$

The full importance of these correspondences will become evident later.

A third advantage of the frequency-domain description of signals is that it greatly simplifies the computation of the response of a linear stationary system to a prescribed input.

To find the response $r(t)$ to an input signal $f(t)$, of some system, H, (characterized by the transmittance $H(s)$), the input signal is first replaced by its signal generator, and the two systems are then *combined* into *a single equivalent signal generator* for $r(t)$, by setting $R(s) = F(s)\ H(s)$, as in Figure 16.

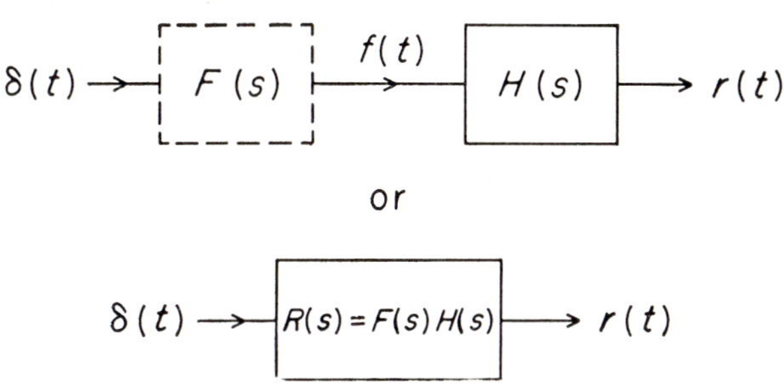

Figure 16. The input signal is replaced by its signal generator which, after combining with the system transmittance, yields the generator for the response r(t).

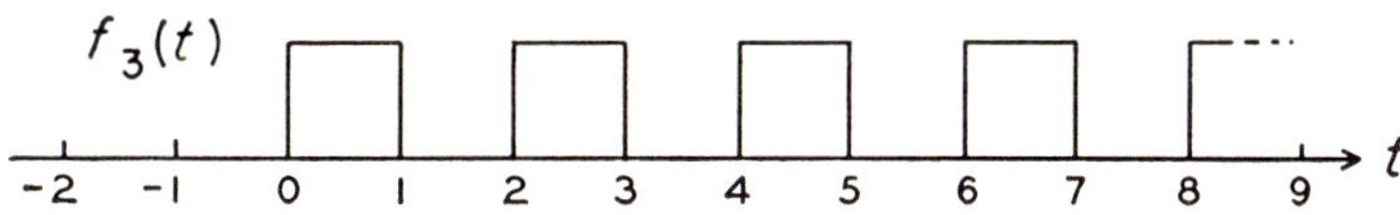

Figure 17. A square wave of two-second period, with its beginning epoch at $t = 0$.

The elegance of this result may be illustrated by finding the signal generator for the periodic square-wave signal shown in Figure 17. This signal may be considered to be composed of a number of identical rectangular pulses each of one-second duration, recurring regularly every two seconds. But to generate such a signal, we need only apply a periodic train of impulses to the input of the generator of the *single* rectangular pulse $f_1(t)$. Thus, a signal generator for $f_3(t)$ is

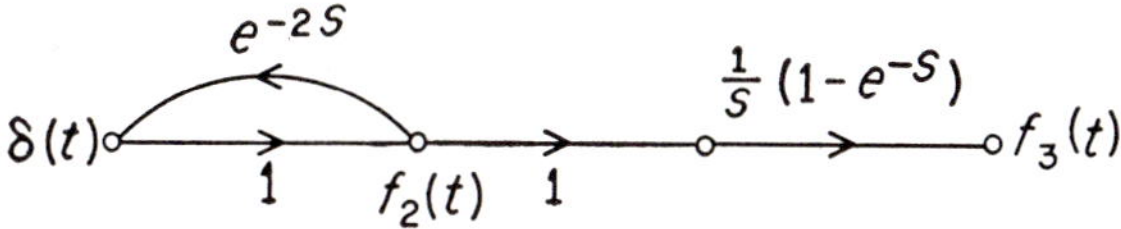

which, by conventional rules for the reduction of a flow-graph, immediately yields the graph transmittance

$$F_3(s) = \frac{1}{s}\frac{1 - e^{-s}}{1 + e^{-2s}} \tag{23a}$$

$$= \frac{1}{s}\frac{1}{1 + e^{-s}} \tag{23b}$$

$$= \frac{1}{s}[1 - e^{-s} + e^{-2s} - e^{-3s} + \cdots] \tag{23c}$$

In (23a) the common factor may be cancelled from numerator and denominator giving (23b). By expanding (23b) into a series by long division the infinite-series (23c) is obtained. Expressions (23b) and (23c) correspond to the transmittance of the much simpler signal generator which produces the signal of Figure 17 by integrating a periodic series of impulses of alternating polarity.

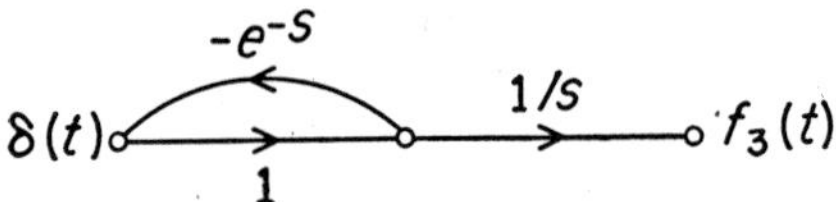

This example is of some interest because it shows how operators may be combined to yield an equivalent system of much simpler form. In fact, it is precisely because the elements of a system may be identified with various *algebraic* decompositions of a prescribed transmittance $H(s)$ that these methods are so useful in system design. Not only can several transmittances be combined into a single expression, but a single transmittance may be expanded into either the sum or product (or both) of several transmittances. The product decomposition may be associated with sub-systems connected in cascade, whereas the sum of two or more terms may be identified with sub-systems connected in parallel. For instance, if it were possible to separate $R(s)$ into the sum of n different components such that

$$R(s) = R_1(s) + R_2(s) + \cdots + R_n(s) \tag{24a}$$

we would then be able to expand the signal generator for $r(t)$ into a number of component signal generators. The task of computing $r(t)$ is thus reduced to that of finding the component signals $r_1(t)$, $r_2(t)$, ... $r_n(t)$, viz.

$$r(t) = r_1(t) + r_2(t) + \cdots + r_n(t). \tag{24b}$$

This expansion will be advantageous provided the output from each of the component signal generators is easily found. To see how thi

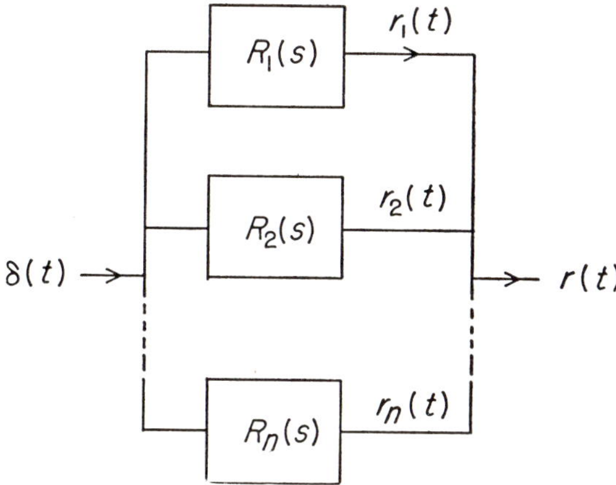

Figure 18. A signal generator may be expanded into several component generators.

may be done, let us first examine the algebraic properties of a typical system transmittance $H(s)$.

PROPERTIES OF A SYSTEM TRANSMITTANCE

It was shown in the previous chapter that the input-output relationship of any linear stationary system may be characterized by a graph transmittance

$$H = \frac{\Sigma G_k \Delta_k}{\Delta} = \frac{\text{output}}{\text{input}} \tag{25}$$

where, in general, the graph determinant Δ, the open-path transmittances G_k and their path factors Δ_k will all have values that depend upon the complex frequency s of the source.

Transmittances associated with different input and output nodes of the same system will have the *same graph determinant,* although the *path cofactors* and *open paths* will be different for different choices of input and output nodes. The graph determinant Δ is therefore a very significant quantity since, being common to all of these transmittances, it expresses a property of the over-all system that is independent of particular point of excitation or observation.

The *natural frequencies* of a system are those values of the complex frequency s for which the *graph-determinant* vanishes, i.e., $\Delta(s) = 0$.

By equation (25), the ratio of output-signal-amplitude to input-signal-amplitude *becomes infinite at each of the natural frequencies.* As a consequence, an exponential component characterized by any of these frequencies may be present in the output even though the amplitude of the same exponential component in the input is zero. These "natural components" account for the transient oscillations and other dynamic effects associated with a system.

If the system is to be stable, all of its natural frequencies must have negative real parts—otherwise the natural oscillations will grow in amplitude until they are limited by the non-linearities or destructive effects that always occur in physical systems. The graph determinant $\Delta(s)$ thus plays a central role in studies of system stability. Various criteria that may be used for testing a given $\Delta(s)$ will be discussed in the next chapter.

When the system is composed of a finite number of integrators, differentiators, and scalors, any of the graph transmittances may be expressed as a rational fraction (i.e., the ratio of two polynomials in s).

$$H(s) = \frac{N(s)}{D(s)} = \frac{a_m s^m + a_{m-1}s^{m-1} + \cdots + a_1 s + a_0}{s^n + b_{n-1}s^{n-1} + \cdots + b_1 s + b_0} \tag{26}$$

Instead of specifying the $1 + m + n$ coefficients of the numerator and denominator polynomials, it is possible to specify these same polynomials by giving a single scale factor and the m values of s for which the numerator becomes zero and the n values of s at which the denominator vanishes. The m values of s at which the numerator vanishes are called the *zeros* of $H(s)$ for obvious reasons. They will be designated by $z_1, z_2, \ldots z_m$. The n values of s at which the denominator vanishes and $H(s)$ becomes infinite, are called the *poles* of $H(s)$, and they will be designated by $p_1, p_2, \ldots, p_n$. Any transmittance of the form given by equation (26) may thus be expressed in terms of its poles, zeros, and one scale factor as

$$H(s) = H_0 \frac{(s - z_1)(s - z_2) \cdots (s - z_m)}{(s - p_1)(s - p_2) \cdots (s - p_n)}. \tag{27}$$

In the usual case, the degree m of the numerator is less than the degree n of the denominator, and $H(s)$ may be expanded into partial fractions

$$H(s) = \frac{k_1}{s - p_1} + \frac{k_2}{s - p_2} + \cdots + \frac{k_n}{s - p_n} \tag{28}$$

where $k_1, k_2, \ldots$ are the *residues* at the poles $p_1, p_2, \ldots, p_n$ which, for the moment, are assumed to be distinct (i.e., none of the zeros of $\Delta(s)$ are repeated). It is easy to show that the *residue* at any particular pole, p_ν, is found by evaluating the expression.

$$k_\nu = \lim_{s \to p\nu} [(s - p_\nu) H(s)]. \tag{29}$$

For the sake of completeness, it should be mentioned that when a pole is repeated r times, such that the nominator polynomial may be written in the form $D(s) = (s - p_r)^r D_1(s)$, then the r terms in the partial-fraction expansion arising from the multiple pole p_r will be of the form

$$\frac{k_{r,0}}{(s - p_r)^r} + \frac{k_{r,1}}{(s - p_r)^{r-1}} + \cdots + \frac{k_{r,r-1}}{s - p_r} \tag{30a}$$

where

$$k_{r,n} = \lim_{s \to p_r} \frac{1}{n!} \frac{d^n}{ds^n} [(s - p_r)^r H(s)]. \tag{30b}$$

It may be noted that equation (28) is a special case of equations (30) for $r = 1$ and $n = 0$.

We have thus shown that any system formed of integrators, scalors, and differentiators may be decomposed, as illustrated by Figure 18, into component generators, each having as a transmittance one of the partial-fraction terms of equations (28) or (30). By means of equations (20a and b), the signal produced by each of these component generators may easily be found. The sum of all these components then gives the desired response $r(t)$.

EXAMPLE OF CLOSED-LOOP SPEED-CONTROL SYSTEM

To illustrate the foregoing points, let us return to the example of the previous chapter in which a closed-loop speed-control system for a dc motor was analyzed. To make the problem dynamic, let us assume that the load torque, L, on the motor consists of two components, one proportional to the velocity, $\dot{\theta}$, and the other—a large inertial force—proportional to the acceleration $\ddot{\theta}$, viz.,

$$L = k\dot{\theta} + J\ddot{\theta}.$$

Also, let us assume that the controller A, which magnifies the error

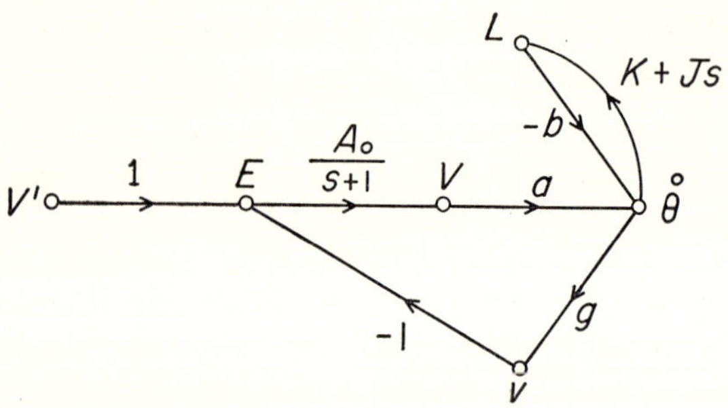

Figure 19. Closed-loop speed control with dynamic elements included.

voltage and supplies power to the motor, has a transmittance $A = \frac{A_0}{s+1}$. With these assumptions, the flow-graph of Figure 10 of the preceding chapter becomes that of Figure 19. Evaluation of the graph transmittance from V' to $\dot{\theta}$ yields

$$G^{V'\dot{\theta}} = \frac{\dot{\theta}}{V'} = \frac{\frac{aA_0}{s+1}}{1 + \frac{gaA_0}{s+1} + b(K + Js)}. \tag{31}$$

To simplify the example for purposes of numerical illustration, suppose that a, b, g, K and J all have the value of unity, thus giving for the graph determinant,

$$\Delta = 2 + s + \frac{A_0}{s+1}, \tag{32}$$

and for the one open-path transmittance

$$G_1 = \frac{A_0}{s+1}. \tag{33}$$

The *natural frequencies* of this system are those values of s for which the graph determinant is zero, or

$$(s + 1)(s + 2) + A_0 = 0. \tag{34}$$

If the amplification A_0 of the controller is made zero, the two natural frequencies are simply $p_1 = -1$ and $p_2 = -2$. The first natural frequency, corresponding to a transient component of the form $\exp(-t)$ is that associated with the controller alone, whereas the second natural frequency, corresponding to a transient component $\exp(-2t)$, is that associated with the dynamic load on the motor (i.e., if the power to the motor were suddenly interrupted, its speed would decrease exponentially with time as $\exp(-2t)$. For this rather useless condition, with $A_0 = 0$, the graph transmittance (31) is also zero and the motor speed due to any input voltage V' would be identically zero. However, as A_0 is increased from zero, the controller begins to interact with the load, and the natural frequencies of the system are modified thereby.

For any value of A_0, the natural frequencies of the system may be found by solving equation (34). This yields

$$p_{1,2} = -1.5 \pm \sqrt{0.25 - A_0}. \tag{34a}$$

When $A_0 = 0$, $p_1 = -1$ and $p_2 = -2$, as above. However, if A_0 were increased to the value 0.25, the two natural frequencies coincide and $p_{1,2} = -1.5$. For this critical value of A_0, the natural frequencies are still real. Any further increase in A_0 will cause the system to acquire an oscillatory natural behavior. For instance, when $A_0 = 10$, equation (31) becomes

$$G^{V'\dot{\theta}} = \frac{\dot{\theta}}{V'} = \frac{A_0}{(s+1)(s+2)+A_0}$$

$$= \frac{10}{s^2 + 3s + 12} \tag{35}$$

where the two zeros p_1 and p_2 of the denominator are found by the quadratic formula to be

$$p_1 = -1.5 + j\,3.12, \qquad p_2 = -1.5 - j\,3.12$$

so the system will exhibit a damped oscillatory natural response of the form $e^{-1.5t}\cos(3.12t + \phi)$. For an arbitrary, large value of A_0 the natural frequencies are given by $p_{1,2} = -1.5 \pm j\sqrt{A_0 - 0.25}$ and, as A_0 is increased, the frequency of the damped oscillation is quite closely given by $\sqrt{A_0}/2\pi$ oscillations per second, although the rate of decay remains constant at -1.5 nepers per second.

Figure 20 is a plot of the performance of this system as simulated on an analogue computer. With the system initially at rest, a constant value of $V' = 10$ is applied at $t = 0$ and the motor speed is measured as a function of time thereafter for each of several different values of A_0. The oscillatory nature of the response for $A_0 = 10$ is clearly shown in this figure. From the theoretical considerations of the previous paragraph, the "period" of this oscillation should be $2\pi/3.12 = 2.02$ sec-

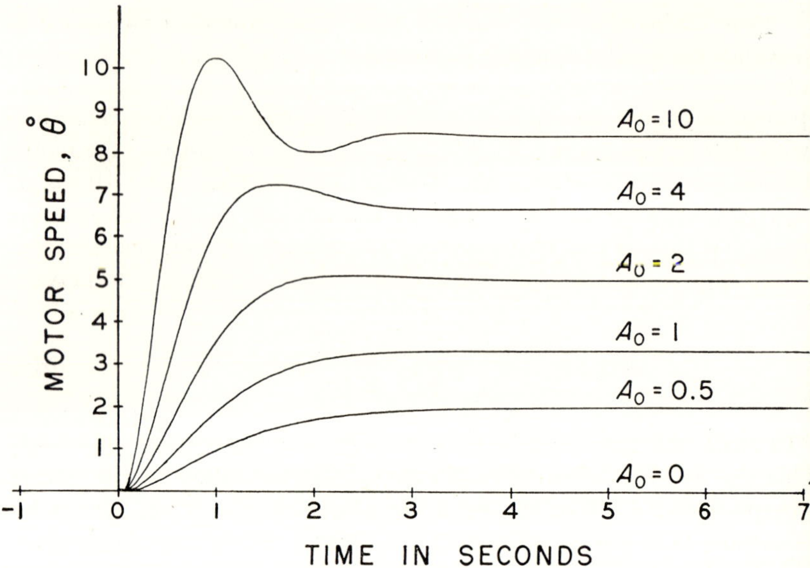

Figure 20. Change in motor speed $\dot{\theta}$ resulting from a step increase in the reference voltage V′ for several different values of controller amplification in A_o.

onds, which agrees closely with the observed time interval between the two successive minima.

An analytic expression for any of the response signals shown in Figure 20 can be obtained as follows. The constant input signal of $V' = 10$ for $t > 0$ can be generated by a single integrator. Hence, the motor speed $\dot{\theta}$ is represented by the following signal generator

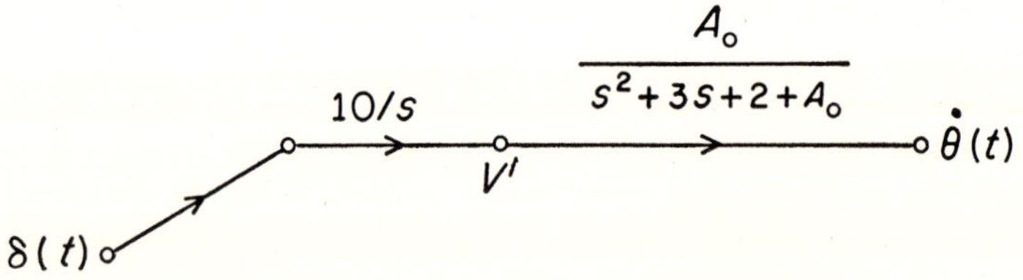

the over-all transmittance of which is,

$$\dot{\theta}(s) = \frac{10}{s} \cdot \frac{A_0}{s^2 + 3s + 2 + A_0}. \tag{36}$$

Next, decompose this transmittance into parallel components by making the partial-fraction expansion:

$$\dot{\theta}(s) = \left(\frac{10A_0}{2 + A_0}\right)\frac{1}{s} + \frac{10A_0}{p_1 - p_2}\left[\frac{1}{p_1}\left(\frac{1}{s - p_1}\right) - \frac{1}{p_2}\left(\frac{1}{s - p_2}\right)\right] \quad (37a)$$

where p_1 and p_2 are the values of the natural frequencies given by (34a). By use of equation (22), the exponential component due to each term is easily found, giving for $t > 0$,

$$\dot{\theta}(t) = \frac{10A_0}{2 + A_0} + \frac{10A_0}{p_1 - p_2}\left[\frac{1}{p_1} e^{p_1 t} - \frac{1}{p_2} e^{p_2 t}\right] \quad (37b)$$

To check this expression against the data of Figure 20, we may examine the constant or steady-state term (corresponding to the natural frequency at $s = 0$ of the input signal generator). When $A_0 = 2$, the constant term has a value of 5, and this is indeed the value that $\dot{\theta}$ approaches in Figure 20 as the transient terms die away. Similar agreement is obtained for the other values of A_0.

It will be recalled from the example of the previous chapter that the most accurate control of speed is obtained when A_0 is very large, since then the dependency of upon all system parameters (other than g) is reduced thereby. In the particular system considered here the natural frequencies expressed by equation (34a) will have the same negative real parts, whatever the value of A_0. Hence, the system will remain stable as A_0 is increased, although the frequency of the natural oscillations and the overshoot of the step response will increase. Examination of Figure 20 suggests that the motor will reach its *final* speed most quickly if A_0 is about equal to 2. To use a much larger value of A_0 could make the system susceptible to rapid fluctuations in speed.

The limitations on the choice of A_0, suggested by these considerations, would have been much more apparent had we considered only a slightly more complicated, third-order system. Then, both the real and imaginary parts of the natural frequencies would have been affected by increasing A_0. In particular, the real part of the natural frequencies would become less negative until for some critical value of A_0, the rate of decay of the natural oscillations would be zero and the system would continue to oscillate indefinitely at a constant amplitude. If A_0 were increased still further, the amplitude of the natural oscillation would grow from any arbitrarily small amplitude and the system would be *unstable.* (Considerable insight into the effect of A_0 upon the various natural frequencies of a system is provided by various "root-locus" techniques, a good account of which will be found in Chapter IV of reference 2.) The point to be emphasized here is that although increased feedback in a control system may lead to a more precise control in a static environment, unless intelligently applied it may change the dynamic behavior of the system so drastically as to degrade or destroy its useful performance.

PROBABILISTIC SYSTEMS

The system considered in the previous section is an example of a deterministic system in which the response at any instant may be found from knowledge of the input for all time prior to that instant. In the flow-graph representing this system, the nodes were associated with the various physical quantities, such as voltage, shaft-position, torques, etc., exhibited by the system. At any instant, each of these signals could be assigned a numerical value. The flow-graph branches were then associated with transmittances that expressed the relationships between these various signals.

Flow-graphs and transmittances may be used in yet another way to represent a much broader class of systems which exhibit non-metrical observables and random (i.e., non-deterministic) behavior patterns. We shall close this chapter by briefly outlining how these methods may be applied to this broader class of systems.

A fundamental notion is that of the *state* of a system. The prior inputs that a system has experienced will generally affect its *state* (or *condition of being*) at any instant. That is, the net effect of all past inputs may be summarized by specifying the state of the system at that instant. Two different *prior* input patterns that result in the same state of the system at a given instant may therefore be considered to be identical insofar as the future evolution of the system is concerned.

The change in the state of a system from any given instant to the next is, in principle, determined jointly by both the state of the system and the input at the given instant. It is very helpful to assume that the system is characterized by a finite number of discrete states. By making this assumption, we may represent each state by a node in a flow-graph. The signal at any particular node "j" now represents the *probability* p_j that the system is in state "j." The branches show how the system may pass from one state to the next. The transmittance of the branch from node "j" to node "k" represents the probability p_{jk} of transition from state "j" to "k." If all transitional probabilities p_{jk} are independent of the sequence of states through which the system passes, the system is referred to as a *linear* (Markov) process. If, furthermore, the transitional probabilities are constants that do not change with time, the system is *stationary,* and all of the methods already described for the study of linear stationary systems may be invoked in studying these probabilistic systems.

As a very simple and yet informative example, consider a coin-tossing machine which, by means of an optical sensor, can detect whether a "head" (i.e., H) or a "tail" (i.e., T) is on the upper face. Suppose that this machine starts to toss the coin once each second when a "start" button is pushed and continues tossing until some specified pattern of H's and T's occurs at which time it automatically stops. Certainly, the recognition of an H or a T would involve a different kind of measure than that employed for expressing the motor speed of the previous example. Furthermore, little is known about the detailed structure of the machine beyond the fact that it tosses the coin and is able to sense whether the coin has come up with an H or a T. How, then, is this system to be represented?

Let us suppose that the machine tosses the coin each time so that on any toss it is equally likely that either an H or T will come up, and that it continues to toss the coin until an H occurs for the first time. The behavior of this machine is then characterized by two states. In the first state, the machine has either just been started and is proceeding to make its first toss, or a T has just occurred and the machine is proceeding with its next toss. In the second state, an H has just occurred and the machine has stopped. A flow-graph representation of these two states will then have two nodes and will appear as

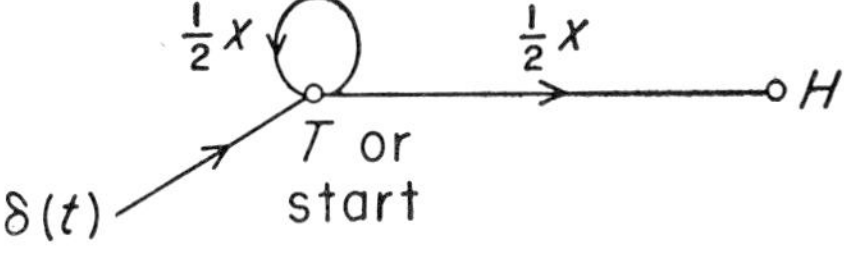

To start the machine, a unit impulse of probability is delivered to the "start" or T node at $t = 0$. If $x = e^{-s}$ represents the transmittance

of a unit delayor, at time $t = 1$ an impulse of strength $\frac{1}{2}$ will appear at the H node signifying that there is a probability of $\frac{1}{2}$ that an H will occur on the first toss. Also, at $t = 1$, an impulse of $\frac{1}{2}$ will appear at the T node indicating that there is probability of $\frac{1}{2}$ that a T will occur on the first toss. Thus, the two branches leaving the "start" node denote the two possibilities that can occur immediately subsequent to the system entering the "start" state. It will be observed that the sum of these two transitional probabilities is unity, which is to say that probability is conserved and the total measure of probabilities in the graph will remain at unity.

A typical sequence of states obtained on operating the machine might be $TTTH$. Note that this sequence corresponds to a path that circulates around the T loop thrice and then exits to the H node. The total transmittance around this path is $(\frac{1}{2}x)^3(\frac{1}{2}x) = (1/16)x^4$ which is to say that at $t = 4$, an impulse of strength 1/16 will appear at the H node. But this is precisely the probability of getting three "tails" followed by a "head"! Not only does the transmittance $(1/16)x^4$ inform us of the value of this probability, but the exponent "4" also tells us of the time at which this event can occur.

More generally, we see that the probability generator for the first occurrence of H is characterized by the graph transmittance

$$F_H(x) = \frac{\frac{1}{2}x}{1 - \frac{1}{2}x}. \tag{38}$$

By dividing the denominator into the numerator, (38) may be expanded into a power series in x, viz.,

$$F_H(x) = \tfrac{1}{2}x + \tfrac{1}{4}x^2 + \tfrac{1}{8}x^3 + \cdots. \tag{38a}$$

Thus, the probability of the system entering state H, after it is started with unit probability at $t = 0$, will be represented in the time domain by a series of impulses occurring at $t = 1, 2, 3$, etc. and having values of $\frac{1}{2}$, $\frac{1}{4}$, $\frac{1}{8}$, etc. Thus, (38) generates the probability distribution of the first occurrence of an H. In the statistical literature $F_H(x)$ is called the generating function of this probability distribution. However, in that literature, x is ordinarily not interpreted as representing a time-delay operator. (For an excellent treatment of the mathematical aspects of probability generating functions, the reader is referred to Feller (7). For a discussion of their relation to flow-graphs and other concepts familiar to the systems engineer, see (8, 9, and 10).

If the coin were biased such that the probability of an H were p and that of a T were q on any toss, the probability generator for the first occurrence of an H would be characterized by the transmittance

$$F_H(x) = \frac{px}{1 - qx}. \tag{39}$$

$$= px + pqx^2 + pq^2x^3 + \cdots . \tag{39a}$$

As previously remarked, the probability of first occurrence of an H on, say, the n'th toss is given by the coefficient, pq^{n-1}, of x^n. Other information of interest may be obtained directly from $F_H(x)$. For instance, consider the value of $F_H(1)$ obtained by setting $x = 1$. By (39a), this value is equal to the sum of the probabilities associated with the first, second, third, etc., tosses. That is, $F(1)$ is the probability that the event will *ultimately* occur without regard to when. Thus, if

$$F(1) = 1, \text{ the event is } \textit{certain} \tag{40a}$$

whereas if

$$F(1) < 1, \text{ the event is } \textit{uncertain.} \tag{40b}$$

And there is a probability $1 - F(1)$ that it will never occur. For the particular event described by (39), we have

$$F_H(1) = \frac{p}{1 - q} = \frac{p}{p} = 1$$

so an H is certain to occur if the coin is tossed for a sufficiently long time (assuming that $p > 0$).

Suppose that we are interested in the average time (i.e., number of tosses) to first occurrence of the event. If we consider *only those instances in which the event does ultimately occur,* it may be shown that the average (or mean) number of time-units until first occurrence is given by

$$\left.\frac{d}{dx} \ln F(x)\right|_{x=1} = \text{Mean occurrence time} \tag{41}$$

Thus, for $F_H(x)$ given by equation (39), we have

$$\ln F_H(x) = \ln px - \ln (1 - qx) \tag{42a}$$

$$\left.\frac{d}{dx} \ln F_H(x)\right|_{x=1} = \left[\frac{p}{px} - \frac{-q}{1 - qx}\right]_{x=1} = \frac{1}{p} \tag{42b}$$

so that, if $p = \frac{1}{2}$, the average number of tosses for the occurrence of an H will be 2.

On any one occasion, the terminating event may occur after any number of tosses. It may be shown that the variance in the time of occurrence is given by

$$\left[\frac{d}{dx}\ln F(x) + \frac{d^2}{dx^2}\ln F(x)\right]_{x=1} = \text{Variance of occurrence time} \qquad (43)$$

Thus, for the $F_H(x)$ of equation (39), we have by use of equation (42),

$$\frac{d^2}{dx_2}\ln F(x) = \frac{d}{dx}\left[\frac{d}{dx}\ln F(x)\right]$$
$$= -\frac{1}{x^2} + \frac{q^2}{(1-qx)^2}$$

so that evaluation of (43) in this case gives

$$\frac{1}{p} + \left[-1 + \frac{q^2}{(1-q)^2}\right] = \left(\frac{q}{p}\right)^2$$

as the variance. For $q = \frac{1}{2} = p$, this variance has the value of 1.

In order to Illustrate the power of the above formulation for dealing with more complicated systems, suppose that the machine continues to toss the coin until the third H has occurred. With the occurrence of this event (hereafter designated as $\mathcal{E}_2$) the machine will come to a halt. Let us find the probability distribution of the times of first occurrence of $\mathcal{E}_2$. To do this, we first construct the probability generator,

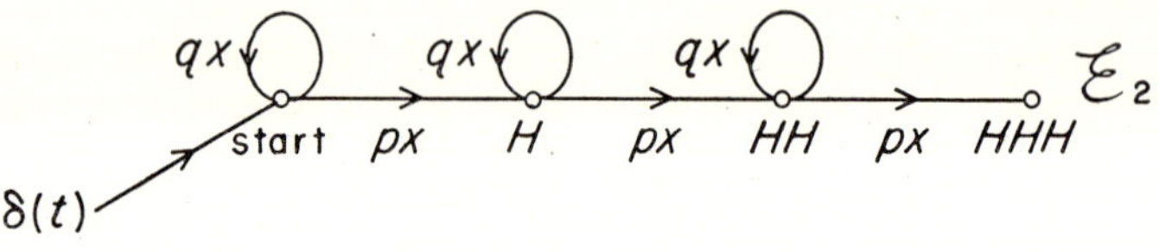

The probability of first occurrence of the event $\mathcal{E}_2$ is thus described by the transmittance

$$F_{\mathcal{E}_2}(x) = \left[\frac{px}{1-qx}\right]^3. \qquad (44)$$

By expanding (44) into a power series in x, the probability of $\mathcal{E}_2$ occurring on the n'th toss is found from the coefficient of x^n in this expansion. By setting $x = 1$ in (44), we find that $F_{\mathcal{E}_2}(1) = 1$, so $\mathcal{E}_2$ is a certain event. Also, since the logarithm of the cube of any quantity is three times the logarithm of the quantity itself, it is clear from the form of equations (41) and (43) that the mean and the variance of the occurrence time of $\mathcal{E}_2$ are both simply three times that for the occurrence of the first H already considered. (More generally, the mean and variance associated with the occurrence of the n'th H would be n/p and $n(q/p)^2$ respectively.)

To illustrate a still more complicated system, let us suppose that the coin-tossing machine operates until the occurrence of the event:

$\mathcal{E}_3$: *The first occurrence of a run,* HHH, *of three successive heads,*

at which time it stops. Evidently in order to recognize the event $\mathcal{E}_3$, the machine must contain a memory in which is stored the outcome of the previous two tosses. Thus, in addition to the "start" state, the system will have three other states depending upon whether a sequence of one, two, or three H's has just occurred. The flow-graph portraying the relationships between these states (for the special case of an unbiased coin) is

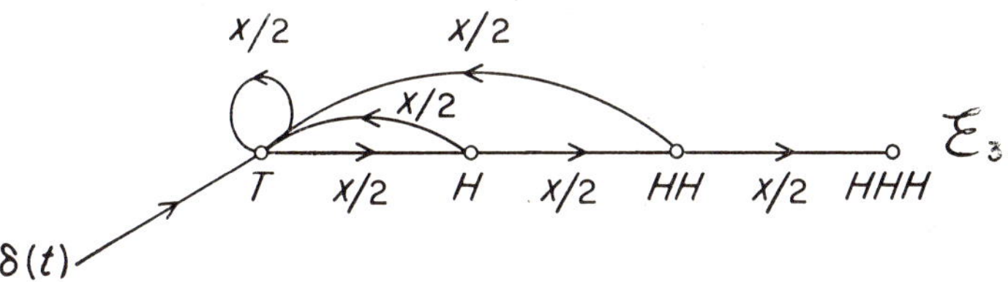

Thus, the probability generating function for the event $\mathcal{E}_3$ is

$$F_{\mathcal{E}_3}(x) = \frac{\left(\frac{x}{2}\right)^3}{1 - \frac{x}{2} - \left(\frac{x}{2}\right)^2 - \left(\frac{x}{2}\right)^3} \tag{45a}$$

$$= \frac{x^3}{8 - 4x - 2x^2 - x^3} \tag{45b}$$

$$= \frac{1}{8}x^3 + \frac{1}{16}x^4 + \frac{1}{16}x^5 + \frac{1}{16}x^6 + \frac{7}{118}x^7 + \cdots. \tag{45c}$$

The probability of a run of three H's occurring for the first time on the 7th toss is therefore 7/118, etc. Also, since $F_{\mathcal{E}_3}(1) = 1$, the event $\mathcal{E}_3$ is certain to occur. Furthermore, by evaluating equations (41) and (43) it is easy to show that the mean and variance of the occurrence time for $\mathcal{E}_3$ are 14 seconds and 142 (seconds)^2, respectively.

These examples serve to illustrate how fairly complicated random processes may be studied in a systematic manner. The structure outlined is considerably more powerful than the relatively simple examples given here are able to reveal. (References 8 and 9 considerably extend these results.) For instance, in the above example it is conceivable that the machine might be so constructed that it initially tosses the coin in such a way that either an H or T are equally likely to occur but that after the occurrence of an H, the tossing mechanism is modified so as to make $p = 0.4$ and after a run HH, so as to make $p = 0.2$. The graph corresponding to this system would be

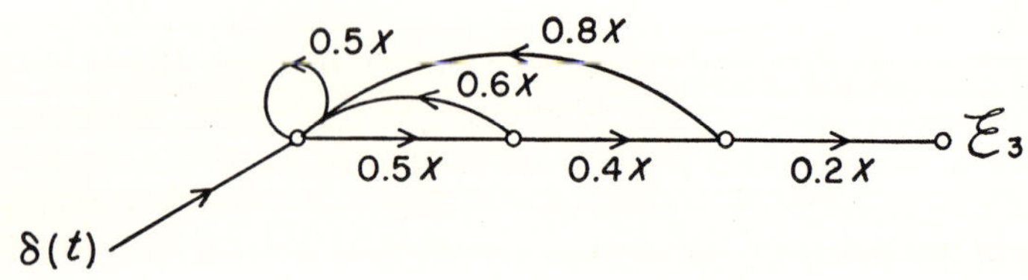

and the solution for the various probabilities would be carried through just as before. Also, although we have talked about coin-tossing machines it is clear that many significant real-world systems obey similar state relationships. It will be appropriate to mention one or two examples of this kind to conclude this chapter.

Renewal Processes

Suppose that the electric lights in an advertising sign are operated six days a week and on the seventh day those lights which are burned out are replaced. Suppose further that the light bulbs are statistically similar, each having a probability distribution $F(x)$ of time to failure which, for purposes of illustration, we shall assume to be given by (45a). That is, from our previous calculations we see that any bulb is certain to burn out ultimately; the average life of a bulb is fourteen weeks; and the standard deviation of the life until it burns out is $\sqrt{142} = 11.9$ weeks.

The graph representation of this renewal process is

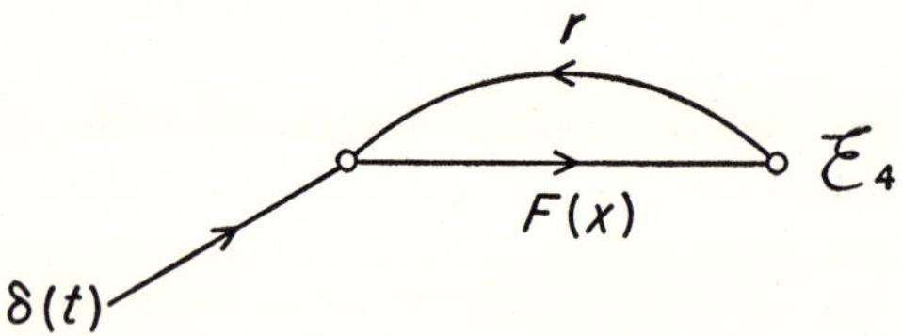

where the node $\mathcal{E}_4$ represents the likelihood that a bulb will have to be replaced at any time subsequent to the initial installation of the sign when all the bulbs were new. Without the branch labeled "r," the graph simply indicates the probability of failure $F(x)$. However, by returning this probability to the input, we have in effect renewed the process. The transmittance "r" is a *tagging variable that labels the number of replacements that have been made.* Thus, the likelihood of $\mathcal{E}_4$ is given by the graph transmittance

$$U_{\mathcal{E}_4}(x, r) = \frac{F(x)}{1 - rF(x)} \tag{46a}$$

$$= F(x) + rF^2(x) + r^2F^3(x) + \cdots \tag{46b}$$

which is composed of the sum of the probability distribution, $F(x)$, associated with the original bulbs, plus that of the first replacements, $rF^2(x)$, plus that of the second replacements, $r^2F^3(x)$, etc. By setting $r = 1$, we get

$$U_{\mathcal{E}_4}(x) = \frac{F(x)}{1 - F(x)} \tag{46c}$$

which is the *expectation of occurrence* of burn out, without regard to the generation of the bulb. Conversely, by expanding equation (46a) in a power series of x, viz.,

$$U_{\mathcal{E}_4}(x, r) = R_0(r) + R_1(r)\,x + R_2(r)\,x^2 + \cdots, \tag{46d}$$

the n'th coefficient thus obtained will give the probabilities over the various renewal generations of a bulb failure at the end of the n'th week. For instance, by expanding, say $R_8(r) = g_0 + g_1r + g_2r^2 + g_3r^3 + \ldots$ into a power series in r, the coefficient, g_3, of r^3 is the probability that a *third* (replacement) *generation* bulb will have failed during the *eighth week.*

The Machine-Repairman Problem

As the final example of this chapter, let us consider the dynamics of the machine-repairman problem discussed by Feller (7, pp. 416-420). The system consists of an automatic factory, containing a number of identical machines, and one or more repairmen who must keep the machines in operating condition. It is assumed that λdt is the probability of one of the operating machines failing in a small time interval dt and that μdt is the probability of a repairman completing his repairs and returning an inoperative machine to service in any time interval dt. These probabilities are assumed to be stationary and independent of the state of the system prior to the present instant.

Before setting up the flow-graph for the entire system, it will be informative to examine why these assumptions lead to *exponential distributions* by considering the probability that a single machine will have failed for the first time after it is placed into operation at $t = 0$. This process is represented by the flow-graph of Figure 21. The nodes

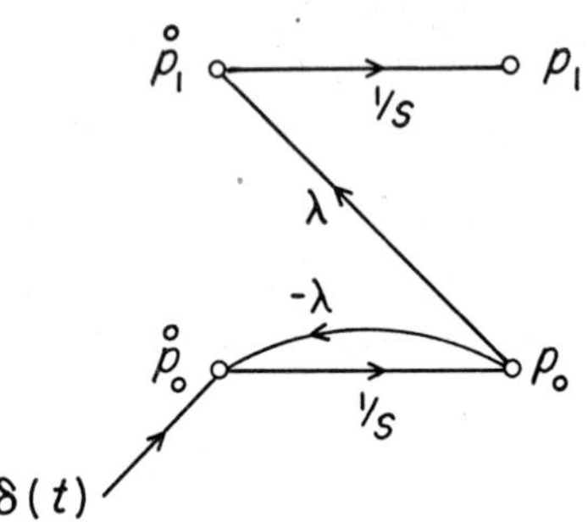

Figure 21. Flow-graph for the exponential failure of a one-machine system.

p_0 and p_1 on the right represent the probabilities of the machine being operative or inoperative, respectively. The nodes on the left represent the rates,

$$\dot{p}_0 = dp_0/dt \quad \text{and} \quad \dot{p}_1 = dp_1/dt,$$

at which the probabilities p_0 and p_1 are changing with time. These rates may be interpreted as the expectations of the system entering (or leaving) the states "0" or "1" in any small interval of time dt. Since the system must certainly be in one of these two states at any time, it is necessary that $p_0 + p_1 = 1$. Consequently $\dot{p}_0 + \dot{p}_1 = 0$, and any increase in p_1 must be balanced by a corresponding decrease in p_0. These transitions are represented by the branches labeled λ and $-\lambda$, respectively, in Figure 21.

At $t = 0$, a unit impulse of probability is applied to the node $\dot{p}_0$ and immediately thereupon the system is certainly in state "0" with unit probability. To find the probability p_0 of the machine being in the operative state "0" at any subsequent time, the transmittance of the probability generator may be written by inspection from Figure 21;

$$P_0(s) = \frac{\frac{1}{s}}{1 - \frac{-\lambda}{s}} = \frac{1}{s+\lambda} \tag{47a}$$

so that

$$p_0(t) = e^{-\lambda t}. \tag{47b}$$

Similarly, the probability of the machine being in the inoperative state "1" is given by the transmittances

$$P_1(s) = \frac{\lambda/s^2}{1 - \frac{-\lambda}{s}} = \frac{\lambda}{s(s+\lambda)} = \frac{1}{s} - \frac{1}{s+\lambda} \tag{48a}$$

so that

$$p_1(t) = 1 - e^{-\lambda t}. \tag{48b}$$

From this, it is clear that ultimate failure is certain.

Extending this simple example, we next represent the effect of the repairman by introducing a "renewal" path from the "1" state back to the "0" state as shown in Figure 22(A). By assumption, this transi-

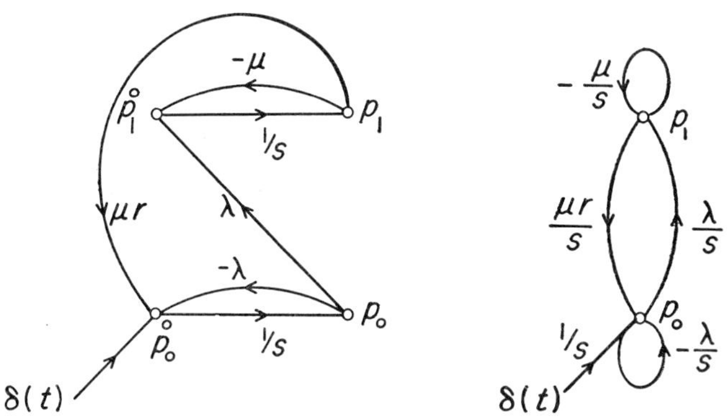

Figure 22(A). Flow-graph for a one-machine system subject to exponential failure and repair.

Figure 22(B). Reduced graph with unessential nodes "p_o" and "p_1" eliminated.

tion is to occur at an average rate μ, which is determined by the branch labeled "μr" between nodes "p_1" and "p_0." Here, again, to conserve probability a corresponding amount must be subtracted from the

p_1 node. (The "r" label is a tagging variable used to keep track of the number of times the machine has been repaired.)

The probabilities of the system being in the operative and inoperative states are immediately found from the graph transmittances

$$U_0(s, r) = \frac{\frac{1}{s}\left(1 + \frac{\mu}{s}\right)}{1 + \frac{\mu}{s} + \frac{\lambda}{s} - r\frac{\mu\lambda}{s^2} + \frac{\mu\lambda}{s^2}}$$

$$= \frac{s + \mu}{s^2 + (\mu + \lambda)\, s + (1 - r)\, \mu\lambda} \quad (49a)$$

and

$$U_1(s, r) = \frac{\lambda}{s^2 + (\mu + \lambda)\, s + (1 - r)\, \mu\lambda} \quad (49b)$$

By expanding $U_i(s, r) = \sum_{n=0}^{\infty} P_{i,n}(s)\, r^n$, we may obtain expressions for the probability $P_{i,n}(s)$ that a machine, which has been repaired n times will be in state "i." The *expectation* that the machine will be in state "i" *regardless of the number of repairs* is obtained by setting $r = 1$, viz.,

$$U_0(s) = \frac{s + \mu}{s(s + \mu + \lambda)} \quad (49c)$$

$$U_1(s) = \frac{\lambda}{s(s + \mu + \lambda)} \quad (49d)$$

These transmittances generate the time-expectations:

$$u_0(t) = \frac{\mu}{\mu + \lambda} + \frac{\lambda}{\mu + \lambda}\, e^{-(\mu+\lambda)t} \quad (49e)$$

$$u_1(t) = \frac{\lambda}{\mu + \lambda} - \frac{\lambda}{\mu + \lambda}\, e^{-(\mu+\lambda)t} \quad (49f)$$

which approach the steady-state probabilities

$$u_0(\infty) = \frac{\mu}{\mu + \lambda} \quad \text{and} \quad u_1(\infty) = \frac{\lambda}{\mu + \lambda} \quad (49g)$$

as $t \rightarrow \infty$.

The repair operator "r," appearing in equations (49a) and (49b), permit us to answer such questions as,

1) At any time t, what is the expected number of repairs that will have been made on an operative machine?

2) At any time t, what is the variance in the estimate of the preceding question?

Here, since

$$\frac{d}{dr} U_i(s, r)\bigg|_{r=1} = \sum_{n=0}^{\infty} nP_{i,n}(s) \tag{50a}$$

and

$$\frac{d}{dr}\left[r \frac{d}{dr} U_i(s, r)\right]_{r=1} = \sum_{n=0}^{\infty} n^2 P_{i,n}(s), \tag{50b}$$

we see that the first and second moments of the distribution with respect to the number of repairs may be found by forming the derivatives (50a) and (50b). Because of the linearity of these expressions, the corresponding time-domain expressions will also be subject to similar interpretation. Thus, if $u_i(t, r)$ is a time expectation corresponding to $U_i(s, r)$, we may show that

$$\frac{\dfrac{du_i(t, r)}{dr}}{u_i(t, r)}\Bigg|_{r=1} = \frac{d}{dr} \ln u_i(t, r)\bigg|_{r=1} = \bar{r}_i(t) \tag{51a}$$

is the *mean number* $\bar{r}_i(t)$ *of repairs* that have led to state "i" at time t. Similarly, for the variance, we may use (50b) to obtain

$$\frac{\dfrac{d}{dr}\left[r \dfrac{d}{dr} u_i(t, r)\right]}{u_i(t, r)}\Bigg|_{r=1} - (\bar{r}_i(t))^2 = \operatorname{var} r(t) \tag{51b}$$

But with some rearrangement equation (51b) may also be expressed in somewhat more elegant form as

$$\left[\frac{d^2}{dr^2} \ln u_i(t, r) + \frac{d}{dr} \ln u_i(t, r)\right]_{r=1} = \operatorname{var} r(t) \tag{51c}$$

which is identical in form to equation (43) for the variance of the occurrence time of a discrete process previously given.

To illustrate, let us suppose that $\lambda = 1 = \mu$. Then equation (49a) becomes

$$U_0(s, r) = \frac{s + 1}{s^2 + 2s + (1 - r)} \tag{52a}$$

$$= \frac{1}{s + 1 - \sqrt{r}} + \frac{1}{s + 1 + \sqrt{r}} \tag{52b}$$

so that

$$u_0(t, r) = e^{(-1+\sqrt{r})t} + e^{(-1-\sqrt{r})t}. \tag{52c}$$

By equation (51a), the mean number of repairs $\bar{r}_0(t)$ that have been made at any time upon an operative machine is

$$\left.\frac{d}{dr} \ln u_0(t, r)\right|_{r=1} = \left.\frac{\frac{1}{2} r^{-\frac{1}{2}} te^{-t} (e^{\sqrt{r}t} - e^{-\sqrt{r}t})}{e^{-t} (e^{\sqrt{r}t} + e^{-\sqrt{r}t})}\right|_{r=1}$$

$$= \left.\frac{t}{2\sqrt{r}} \tanh(\sqrt{r}t)\right|_{r=1} \tag{53a}$$

$$\bar{r}_0(t) = \frac{t}{2} \tanh t \tag{53b}$$

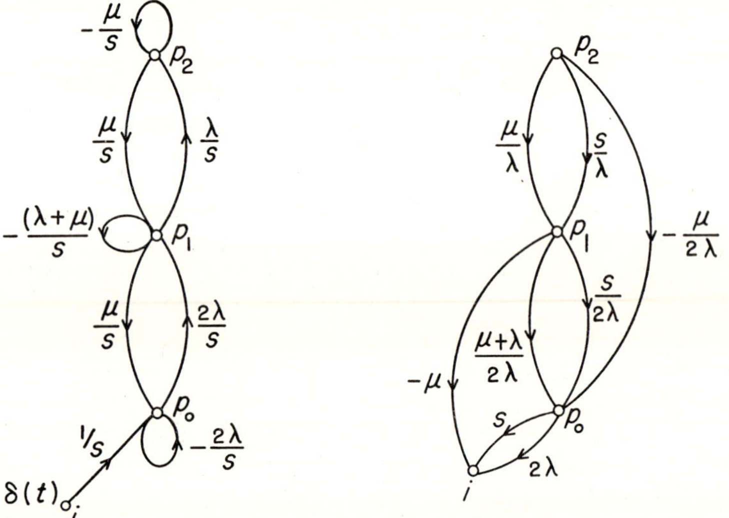

Figure 23(A). Reduced graph for a two-machine system with one repairman.

Figure 23(B). Modified graph formed by inverting the open path between "i" and "p_2" nodes.

which is a very plausible result since, for $\lambda = \mu = 1$ the machine will be operative only $\frac{1}{2}$ the time on the average. By evaluating equation (51c), the variance in the total number of repairs that have been made to place the machine in operation prior to time t is found to be

$$\text{var } r_0(t) = \frac{t}{4} \tanh t + \frac{t^2}{2} \operatorname{sech}^2 t \tag{53c}$$

It is easy in principle (although analytically tedious) to extend the above treatment to a system having multiple machines and one (or more) repairmen. In Figure 23(A) we show the essential graph of a system comprised of two machines and one repairman. There are now

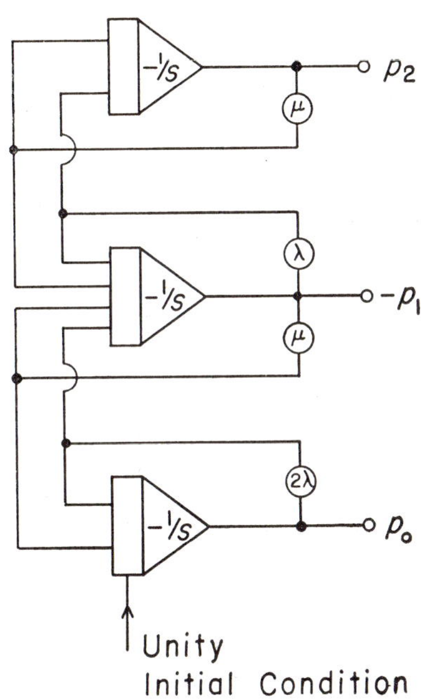

Figure 24. Realization of the graph of Figure (23a) on an Analogue Computer.

three states and the probability of transition from the state "p_0" to the state "p_1" has been doubled over that shown in Figure 22(B) since either machine may fail.[3] However, for the transition from "p_1" to "p_2," $\frac{\lambda}{s}$ is the correct transmittance since only one machine is

[3] The probability that both machines might fail simultaneously is negligible compared to the probability that either one may fail: It is this feature of the exponential transitions (namely, that the probability of simultaneous occurrence of two events may be neglected in comparison with the probability of the occurrence of one or the other) which permits us to obtain simple analytical solutions.

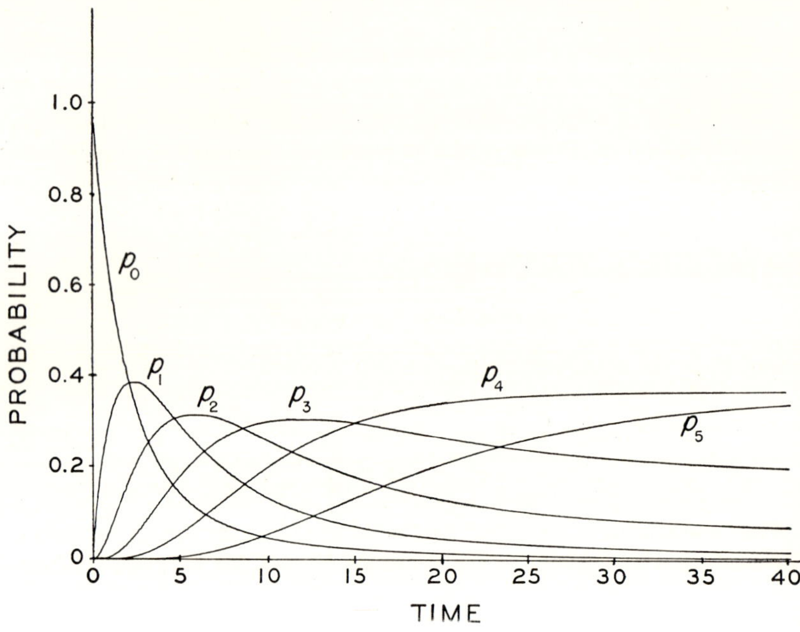

Figure 25. State probabilities as functions of time for 5-machines, one-repairman system with $\lambda = \mu = 0.1$. System initially in state zero.

operative in state "p_1." Transitions to lower states are effected by a machine being repaired. As there is only one repairman, these downward transitions are the same for both states "p_2" and "p_1." If, however, two repairmen were available, the probability of transition from "p_2" to "p_1" would be twice that from "p_1" to "p_0" (assuming that only one repairman can work on a given machine).

Suppose that the system is started in state "p_0." Let us find the probability that both machines will be out of order at any subsequent time. This requires finding the transmittance from "i" to "p_2" of Figure 23(A). The multiplicity of loops makes this computation rather messy. Considerable simplification is achieved by inverting the open path from "i" to "p_2," resulting in the graph of Figure 23(B) which is completely devoid of loops! The transmittance from "p_2" to "i" may then easily be found. It is equal to the reciprocal $1/P_2(s)$ of the desired graph transmittance. By normalizing our timescale so that $\lambda = 1$, we find that

$$P_2(s) = \frac{2}{s\left[s^2 + (3 + 2\,\mu)\,s + 2 + 2\mu + \mu^2\right]} \tag{54a}$$

$$= \frac{k_0}{s} + \frac{k_1}{s - p_1} + \frac{k_2}{s - p_2} \tag{54b}$$

where

$$p_{1,2} = -(\tfrac{3}{2} + \mu) \pm \sqrt{\tfrac{1}{4} + \mu}$$

The probability $p_2(t)$ is thus given by

$$p_2(t) = k_0 + k_1 e^{p_1 t} + k_2 e^{p_2 t} \tag{54c}$$

where k_0 is the *steady-state probability* given by

$$k_0 = \frac{1}{1 + \mu + \frac{\mu^2}{2}}. \tag{54d}$$

This problem may be solved easily by use of an analogue computer. Obvious transformations of the graph of Figure 23(A) yield the wir-

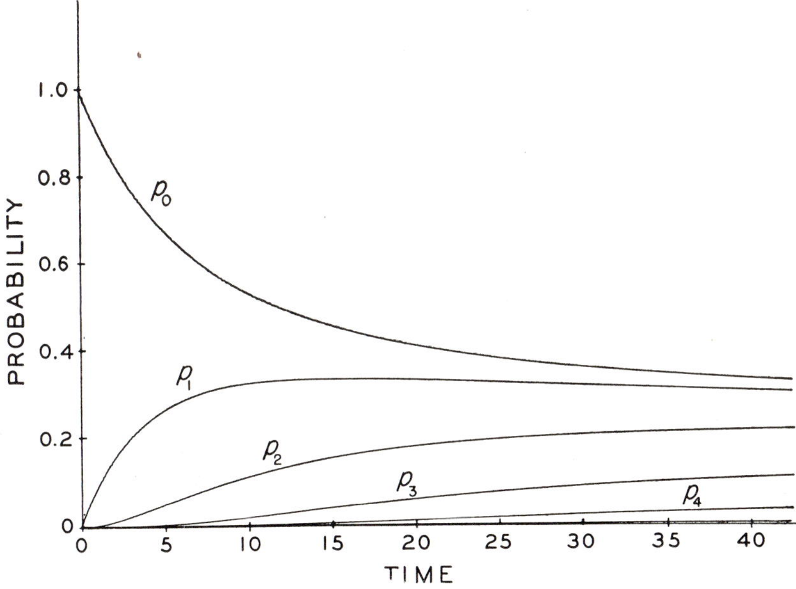

Figure 26. State probabilities as functions of time for 5-machines, one-repairman system with $\lambda = 0.02$ and $\mu = 0.1$. System initially in state zero.

ing schematic of Figure 24, where, for reasons of economy, the algebraic sign of P_1 has been reversed.

Figure 25 shows a typical set of data obtained by an analogue computer for a 5-machine system with one repairman. Time is measured in seconds and $\lambda = \mu = 0.1$. These data show the rapid degradation as the repairman struggles with a hopeless undertaking. In Figure 26 is seen the marked effect of decreasing λ from 0.1 to 0.02, leaving the other parameters unchanged. Only six integrators and one inverting amplifier were needed to solve this problem.

REFERENCES

(1) Guillemin, E. A. *Synthesis of Passive Networks.* New York: John Wiley and Sons, 1957.

(2) Truxal, J. G. *Automatic Feedback Control System Synthesis.* New York: McGraw-Hill, 1955.

(3) Huggins, W. H. "Signal Theory," *Institute of Radio Engineers Transactions, CT-3,* No. 4 (December, 1956), 210-16.

(4) ———. "A Theory of Hearing," *Communication Theory,* ed. W. Jackson. New York: Academic Press, 1953.

(5) Lighthill, M. J. *Introduction to Fourier Analysis and Generalized Functions.* Cambridge: Cambridge University Press, 1958.

(6) Janssen, J. M. L. "The Method of Discontinuities in Fourier Analysis," *Philips Research Reports,* Vol. 5 (December, 1950), 435-60.

(7) Feller, W. *An Introduction to Probability Theory and Its Applications.* 2nd ed.; New York: John Wiley and Sons, 1957 (particularly Chapters 11, 12, and 13).

(8) Sittler, R. W. "Systems Analysis of Discrete Markov Processes," *Institute of Radio Engineers Transactions, CT-3,* No. 4 (December, 1956), 257-66.

(9) Huggins, William H. "Signal-Flow Graphs and Random Signals," *Proceedings of the Institute of Radio Engineers,* Vol. 45, No. 1 (January, 1957).

(10) Mason, S. J. "About such Things as Varistors, Flow Graphs, Probability, Partial Factoring, and Matrices," *Institute of Radio Engineers Transactions, CT-4,* No. 3 (September, 1957), 90-97.

Twenty-Three

FEEDBACK AND STABILITY

NASLI H. CHOKSY

INTRODUCTION

This chapter is concerned with what is probably the most important characteristic of any system—its stability. The word "stability" has many interpretations, depending on the interpreter; a mathematician may define this term in a manner not readily understandable by an engineer, even though he may be referring to essentially the same property. We shall define in due course what is meant when we say that a system is "stable."

FEEDBACK

In Chapter 4 Gibson has indicated the important role that the principle of feedback control plays in the human system. Here we shall amplify somewhat the concept of feedback and the role it plays in the stabilization of systems. To control the behavior (i.e., the output) of a system so that it gives satisfactory performance, we have at our command two distinctly different methods: we may build the system such that it is 100 per cent perfect in all its parts and then, by applying a properly chosen, perfect stimulus (i.e., input), the desired perfect behavior is achieved. This is the method of *open-loop control.* If anything short of perfection is attained anywhere in the system, obviously it will not behave in the desired fashion; and, further, there is no possibility for the system to correct either itself or the behavior. The second method is to build the system in such a fashion that the stimulus applied to it takes cognizance of the fact that the behavior is not as it should be and uses the *difference* between the actual behavior and the desired behavior to control the system so as to make this difference as small as possible. This is the method of *closed-loop control.* It is called thus because, in order to adjust the stimulus for proper behavior, information about the actual behavior must be *fed back* so that the necessary difference in behavior may be sensed. The feedback path then closes a loop, as is shown in Figure 1a; or, in the signal-flow representation of Figure 1b (discussed in Chapter 21).

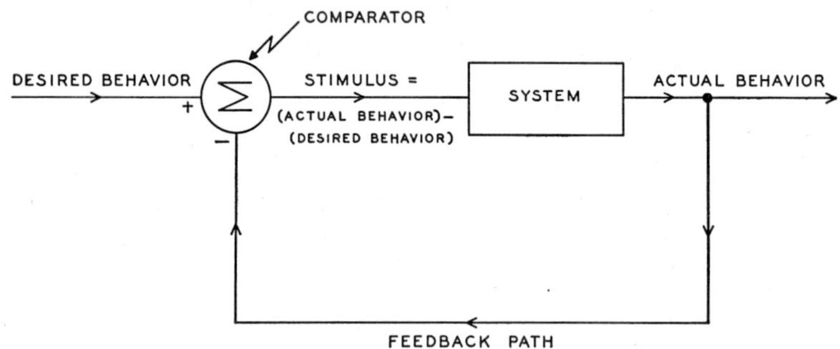

Figure 1(a). Representation of a closed-loop control system.

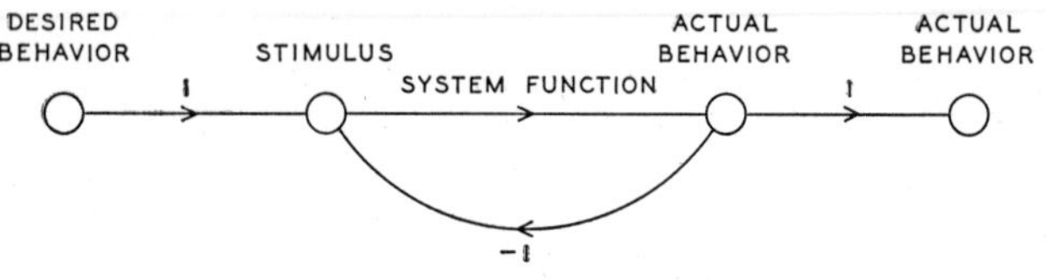

Figure 1(b). Signal-flow-graph representation of the closed-loop control system (above).

Examples

Examples of the use of feedback are abundant in nature and in the physical world around us. Gibson has given us a few such examples in Chapter 4. Let us consider here two more examples, one from the field of human actions, and one from the field of economic theory. For the former, we present the example made famous by Wiener (21)—that of learning to pick up a pencil. When the hand reaches out to pick up a pencil, the open-loop method calls for the brain to send commands to certain muscles in sequence and for the performance of the act by consciously willing these muscles to contract and relax in just such a fashion that the pencil is picked up. We are aware that this does not happen. Rather, we desire to pick up the pencil and reach for it, and then guide our hand to it by trying at all times to lessen the distance between the pencil and the hand. This is a closed-loop process—the difference between the position of the hand at any time and the position which is desired (namely, on the pencil) is used to control the further movement of the hand by feeding back this difference via the feedback path through our eyes. Trying to pick up the pencil with eyes closed will show the importance of this feedback path. However, it is known that when the pencil is in a familiar position (on our desk, say) it is quite easy to pick it up with the eyes closed. Does this mean that no feedback is involved? No! What happens is this. Initially, when we first learn to pick up the pencil from the position on the desk, we go through the process mentioned above. But, after a few repetitions, the exact movements of the hand (and the associated muscles) are remembered, and in future attempts to pick up the pencil the *memory* senses any deviations of the hand from the familiar path to the pencil and corrects them. Thus, feedback is still involved but is no longer visual but kinesthetic. We shall return to this example later.

For the economic example we choose Kalecki's Theory of the Trade Cycle (10). The equations developed in his theory are presented first as functions of time, and then as functions of the Laplace variable, s, that was introduced in Chapter 22. The equations are:

i) $$\frac{dK(t)}{dt} = L(t) - U(t) \qquad (1)$$

which states that the rate of increase in the total capital, $K(t)$, is less than the delivery rate of finished goods, $L(t)$, by the depreciation rate, $U(t)$.

ii) $$L(t) = D(t - T) \qquad (2)$$

which states that the delivery rate of finished goods depends directly on the rate of decision-to-invest, $D(t)$, which occurred an interval T units of time earlier.

iii) $$D(t) = a\,Y(t) - b\,K(t) \qquad (3)$$

which states that the rate of decision-to-invest is proportional to the difference between total income, $Y(t)$, and existing total capital, $K(t)$, each weighted by a suitable constant, a and b.

iv) $$Y(t) = I(t) + P(t) \tag{4}$$

which states that the total income is the sum of the rate of investment, $I(t)$, and the rate of production of consumption goods, $P(t)$.

v) $$I(t) = \frac{1}{T}\int_{t-T}^{t} D(t)\,dt \tag{5}$$

which states that the rate of investment is essentially the rate of decision-to-invest averaged over a time interval of T units preceding the time at which the investment is made.

The corresponding equations in terms of the Laplace variable are:

i) $$sK(s) = L(s) - U(s) \tag{6}$$

ii) $$L(s) = e^{-Ts} D(s) \tag{7}$$

iii) $$D(s) = aY(s) - bK(s) \tag{8}$$

iv) $$Y(s) = I(s) + P(s) \tag{9}$$

v) $$I(s) = \frac{1}{Ts}[1 - e^{-Ts}]\,D(s) \tag{10}$$

These relations can be represented by means of the signal-flow diagram of Figure 2. The mutual interdependence of the quantities K, D,

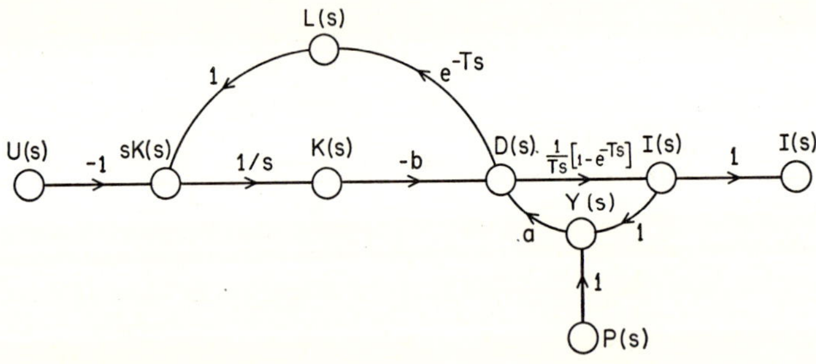

Figure 2. Signal-flow-graph representation of Kalecki's Theory of Trade Cycles.

and L, and of D, I, and Y illustrate the importance of the feedbacks involved in this system.

STABILITY

In order to study the relations and applications of feedback to the stability of a system, we now turn to the study of the question of stability itself. Webster defines stability thus:

> *stability:* that property of a body which causes it, when disturbed from a condition of equilibrium or steady motion, to develop forces or moments which tend to restore the body to its original condition.

In a general sense this is as good a definition of this term as any we may use. Let us then rephrase this definition to a form more suitable for our purposes. Suppose that we have a (linear) system which, under the influence of some stimulus, is in some (suitable) behavior state. For our purpose, a state of equilibrium or rest is to be considered as a special case of a general pattern of behavior. Now we assume that in some way the system suffers a disturbance which causes its behavior state to change from the desired one. We recognize that one of three possibilities will arise after the disturbance has ceased to act. After a sufficient length of time (called the "transient period") the system will (i) revert to its original behavior state, (ii) depart from the original behavior state in an ever increasing fashion, or, (iii) differ from the original behavior by some finite amount. If condition (i) is obtained we shall say that the system is *stable;* if condition (ii), *unstable;* if condition (iii), *limitedly stable.* For linear stationary systems it is known that it is enough to consider the system in an initially quiescent state and then to disturb it by means of an impulse. The resultant behavior will determine the stability, depending on which of the three conditions above are met. A linear stationary system which is stable in response to an impulsive excitation will then be stable for all excitations.

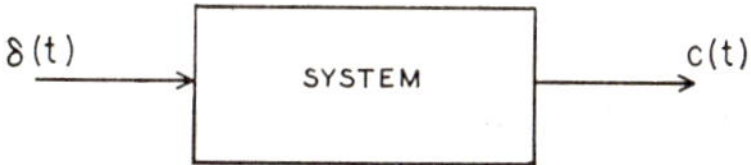

Figure 3. An initially quiescent linear system excited by an impulse function.

Let us express this mathematically. Consider Figure 3. This represents an initially quiescent linear system which is being disturbed by an impulse function $\delta(t)$, and consequently has the pattern of behavior denoted by $c(t)$. This behavior is obtainable as the solution of a linear differential equation with constant coefficients:

$$\frac{d^n c(t)}{dt^n} + a_1 \frac{d^{n-1}c(t)}{dt^{n-1}} + \cdots\cdots + a_{n-1}\frac{dc(t)}{dt} + a_n c(t) = \delta(t) \tag{11}$$

The solution of this equation, as has been shown in Chapter 22, is

$$c(t) = A_1 e^{s_1 t} + A_2 e^{s_2 t} + \cdots\cdots + A_n e^{s_n t} \tag{12}$$

where $A_1, A_2, \ldots, A_n$, are constants depending on the initial conditions, and $s_1, s_2, \ldots, s_n$, are the roots of the algebraic equation[1]

$$s^n + a_1 s^{n-1} + \cdots + a_{n-1}s + a_n = 0 \tag{13}$$

Let us now examine a general term on the right-hand side of equation (12): $A_k e^{s_k t}$. The root, s_k, of equation (13) will be, in general, a complex number, $s_k = \sigma_k + j\omega_k$. Figure 4 illustrates the form of the exponential

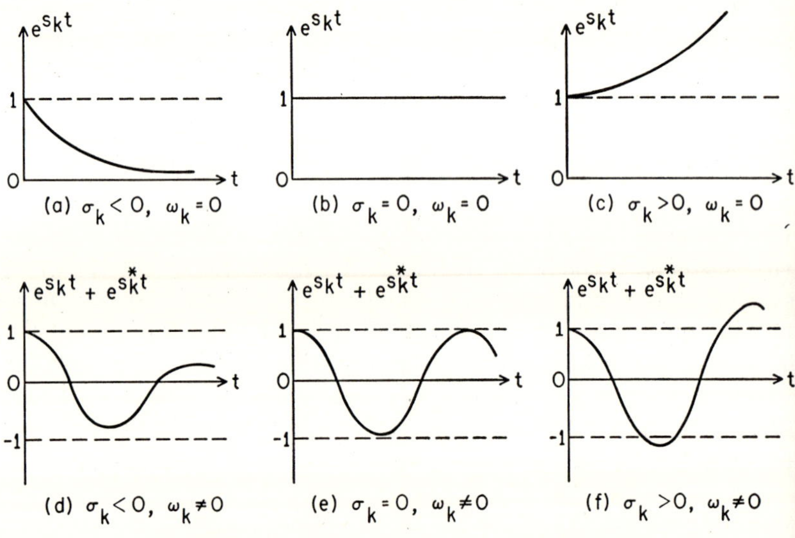

Figure 4. Typical curves for $e^{s_k t}$ where $s_k = \sigma_k + j\omega_k$.

function $e^{s_k t}$ for various values of σ_k and ω_k. Note that since the coefficients of equation (13) are real numbers, for each complex root $s_k =$

[1] This equation is obtained by equating to zero the graph determinant of the flow-graph representation of the system, as discussed in Chapter 22.

$\sigma_k + j\omega_k$ there must also be the complex conjugate root $s_k^* = \sigma_k - j\omega_k$. It is clear from Figure 4 that if we desire $c(t)$ to achieve the value zero as time goes on, i.e., to achieve the value it had before the disturbance was imposed, then *each* s_k *must be such that it is a negative real number* (Figure 4a) *or a complex number which has a negative real part* (Figure 4d). If any *one* s_k is such that it is a positive real number or a complex number with a positive real part, as time goes on, $c(t)$ will grow indefinitely large, as can be seen from Figures 4c and 4f respectively. We can now formulate a condition for the stability of a linear system:

> The system will be stable if and only if all the zeros of its characteristic function, $S(s) = s^n + a_1 s^{n-1} + \cdots + a_n$, have negative real parts.

Note that we have assumed that in general these zeros will be complex numbers, since a real number can always be considered to be a complex number whose imaginary part is zero.

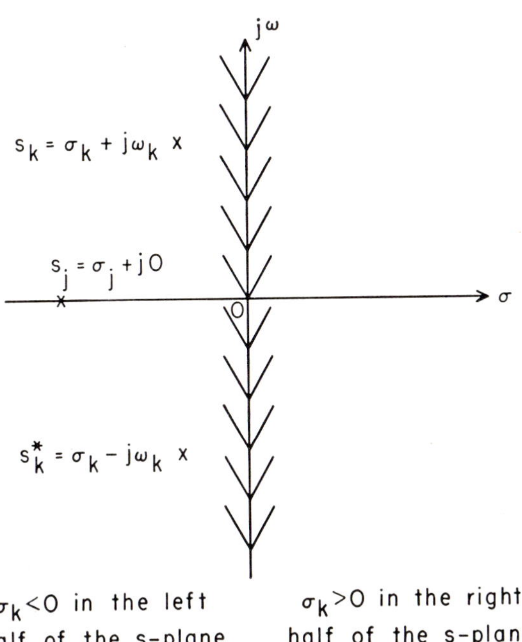

Figure 5. The s-plane (i.e., the complex plane on which is represented the complex variable $s = \sigma + j\omega$).

Figure 5 shows the complex plane associated with the complex variable, s. If the zeros of the characteristic function are to satisfy the condition stated above, it is easy to see that they must all be located in the left half of the s-plane. Figure 5 shows some typical zeros for a stable system. We can thus rephrase the condition for stability as follows:

> A linear system will be stable if and only if all the zeros of its characteristic function lie in the left-half of the s-plane.

Now, if we solve the differential equation (11) by the method of the Laplace transformation, the solution, $C(s)$, which corresponds to the time solution, $c(t)$, is given by

$$C(s) = \frac{1}{s^n + a_1 s^{n-1} + \cdots\cdots + a_{n-1}s + a_n} \tag{14}$$

If the system is excited by any other input, say $r(t)$, instead of the impulse function, the corresponding solution would be

$$\frac{C(s)}{R(s)} = \frac{1}{s^n + a_1 s^{n-1} + \cdots\cdots + a_{n-1}s + a_n} \tag{15}$$

where $R(s)$ is the Laplace transform of $r(t)$. The expression on the right-hand side of equation (15) is called the *transfer function* of the system. (In Chapter 22 this quantity is referred to as the *transmittance*). We note that the denominator of the transfer function is precisely the characteristic function, and that those values of s which are the zeros of the characteristic function will cause the transfer function to become infinite—they are the *poles* of the transfer function. But these are also the values of s which determine the stability of the system. We will now give the general form for the transfer function and then restate our stability condition. The general form is

$$\frac{C(s)}{R(s)} = G(s) = \frac{b_0 s^m + b_1 s^{m-1} + \cdots\cdots + b_{m-1}s + b_m}{s^n + a_1 s^{n-1} + \cdots\cdots + a_{n-1}s + a_n} \tag{16}$$

where, in general, the degree of the numerator is less than that of the denominator. If $G(s) = C(s)/R(s)$ is the *over-all* transfer function of a linear system, the condition for stability states:

> *The system will be stable if and only if the poles of the over-all transfer function all lie in the left-half of the s-plane.*

Now, if n, the degree of the polynomial in the denominator of equation (16), is larger than 4, it becomes a cumbersome task to factor

curve. Also, note in Figure 8 that we have marked the point $G_1(s) = -1 + j0$ (or, the "-1 point" as it will be called).

We are now in a position to state the criterion which enables us to determine the stability of the closed-loop system of Figure 6:

> Given the open-loop transfer function, $G_1(s)$, of a linear system, use it as a mapping function to map the mapping curve of Figure 7. Examine the mapped curve in the $G_1(s)$-plane and the corresponding mapped area. *The closed-loop system will, then, be stable if and only if the mapped area does not include the -1 point of the* $G_1(s)$*-plane.*

We see by this criterion that Figure 8 represents a stable system since the -1 point is not included in the shaded area. In Figure 9 we see an

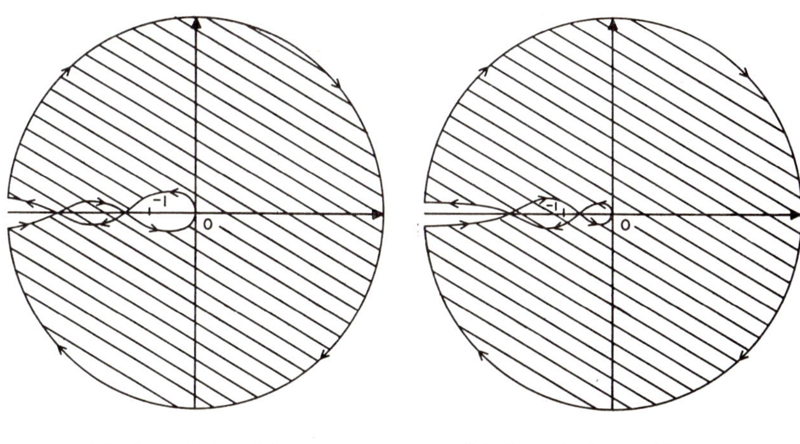

(a) A typical stable system. (b) A typical unstable system.

Figure 9. Typical stable and unstable systems.

example of a stable and of an unstable system. In this figure (both parts of which are drawn to the same scale) the only difference between the stable and the unstable system is that the former has a larger gain.

Example

In order to illustrate the use of feedback to stabilize an unstable system, let us consider the following hypothetical economic situation.

Let us assume that the total capital, $K(t)$, and the total income, $Y(t)$, are related by means of the differential equation:

$$\frac{d^2K(t)}{dt^2} + 2\frac{dK(t)}{dt} - K(t) = \frac{dY(t)}{dt} + 2Y(t)$$

(In actual practice such a relation would more likely be expressed by means of a graph, in which case either an equation would have to be obtained which fits the graph, or a graphical analysis would have to be made. Here we save some time and trouble by assuming a simple relationship.) This relationship expressed in terms of the Laplace variable becomes

$$K(s) = \frac{s + 2}{s^2 + 2s - 1}Y(s) \tag{23}$$

We show this relationship in Figure 10a as a signal-flow-graph of this open-loop system, where

$$\frac{K(s)}{Y(s)} = G_1(s) = \frac{s + 2}{s^2 + 2s - 1} \tag{24}$$

This open-loop transfer function has poles at $s = -2.414$ and $s = +0.414$, and hence is unstable since there is one pole in the right-half of the s-plane. Now, in this situation the existing capital is dependent on the total income, but the total income does not depend upon the

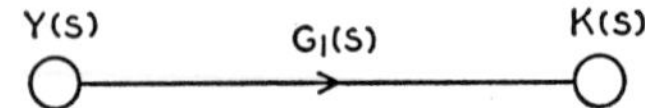

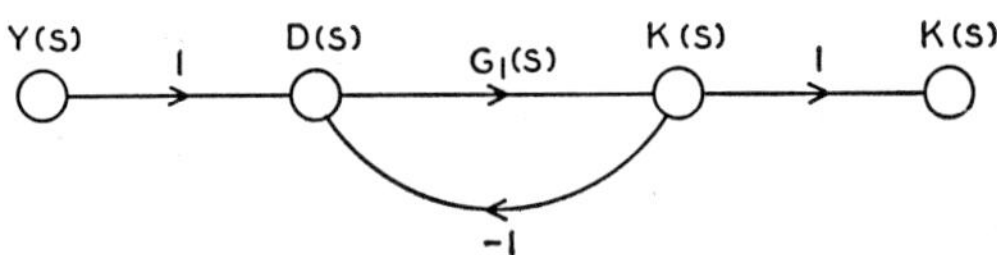

Figure 10. An example of the use of feedback to achieve stability (see text).

existing capital—which is not true in practice. Let us then feed back a signal which gives information about the total income, so that the system now looks as indicated in Figure 10b. The closed-loop transfer function is now given by

$$\frac{K(s)}{Y(s)} = G(s) = \frac{G_1(s)}{1 + G_1(s)} \tag{25}$$

To investigate the stability of this system, we use $G_1(s)$ as a mapping function as explained above. The resultant mapped curve and mapped region in the $G_1(s)$-plane are shown in Figure 11. Since the mapped

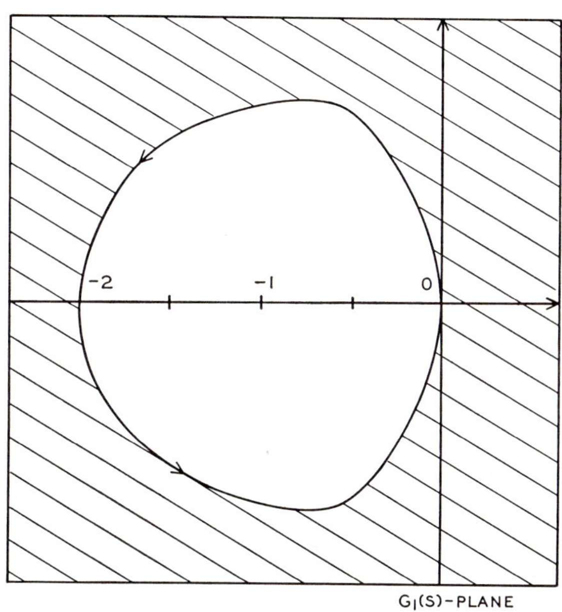

Figure 11. The mapped curve and mapped area for the example of Figure 10.

area does not include the -1 point, the closed-loop system is stable. We can check this by substituting for $G_1(s)$ in equation (25) from equation (24). The over-all transfer function, $G(s)$, is then

$$G(s) = \frac{s + 2}{s^2 + 3s + 1}$$

which has poles at $s = -2.618$ and $s = -0.382$, and since both lie in the left-half of the s-plane, the system is stable. The effect of the feedback was essentially to introduce a rate of decision-to-invest, $D(s)$, which was the difference between the total income and the existing capital, and then, using this rate of decision-to-invest, to control the existing capital through the unstable open-loop transfer function. We should recognize that this hypothetical example was presented to illustrate a point, and thus should not be taken as an example of practical economics.

STABILITY OF KALECKI'S ECONOMIC THEORY

Let us now apply our stability theory, as developed above, to the system which represents Kalecki's theory. However, the presence of the delay terms in the system (i.e., terms involving e^{-Ts}) complicates matters as regards the mapping procedure. But, in order to show that our stability criterion still holds for cases where such time lags exist, let us consider a simplified version of Kalecki's theory. We give then the simplified equations corresponding to equations (1)-(5), with explanations about the simplification which has been made in each equation:

i) $$\frac{dK(t)}{dt} = L(t) \tag{26}$$

where we have assumed that the depreciation rate is zero.

ii) $$L(t) = D(t-1) \tag{27}$$

where the delay between the delivery rate and the rate of decision-to-invest is taken to be one unit of time.

iii) $$D(t) = aY(t) - bK(t) \tag{28}$$

which is the same as equation (3).

iv) $$\frac{dY(t)}{dt} = I(t) \tag{29}$$

which states that the rate of increase of total income equals the rate of investment. (Compare with equation (4).)

v) $$I(t) = D(t) \tag{30}$$

where the investment is now made simultaneously with the decision-to-invest.

Combining these equations, we get the difference-differential equation

$$\frac{dI(t)}{dt} - aI(t) + bI(t-1) = 0 \tag{31}$$

The corresponding characteristic equation is

$$\begin{aligned} S(s) &= s - a + be^{-s} = 0 \\ &= 1 + G_1(s) \end{aligned} \tag{32}$$

where

$$G_1(s) = s - (a+1) + be^{-s} \tag{33}$$

In order to use this as a mapping function, we must select numerical values for a and b. Let $a = -2$ and $b = -1$. Then,

$$G_1(s) = s + 1 - e^{-s} \tag{34}$$

Use of this as a mapping function gives us Figure 12. Since the mapped

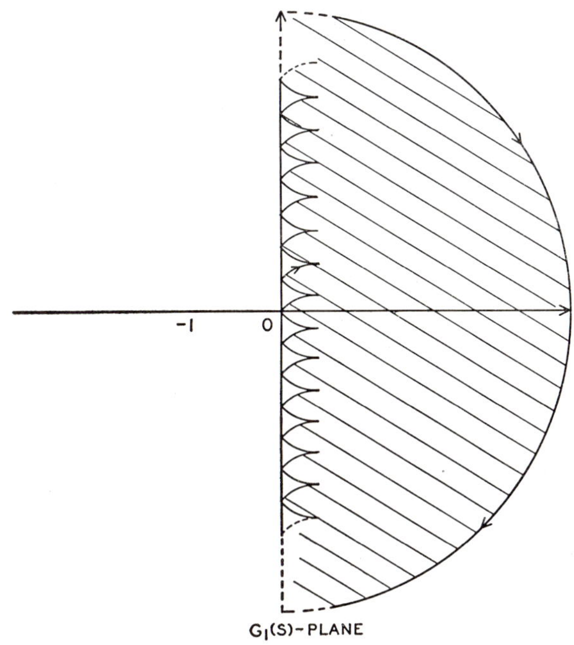

Figure 12. The mapped curve and the mapped area for the example of equation (34).

region does not include the -1 point, we have a stable system. In fact, it can be shown that we can determine the stability of such a system for any values of the coefficients a and b in equation (32) by means of Figure 13. Any combination of values of a and b that fall within the shaded region will give a stable system. Now, in actual practice, the investment cycle is usually a sinusoidally varying curve, corresponding to combinations of values of a and b which fall *on* the boundary of the curve shown in Figure 13. Thus, this curve, together

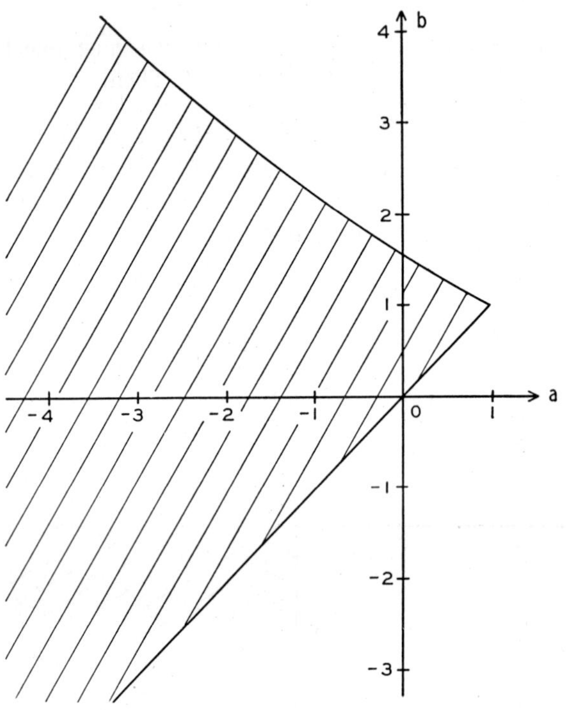

Figure 13. The characteristic equation $S(s) = s - a + be^{-s}$ corresponds to a stable system for all values of a and b falling within the shaded area.

with equations (28) and (30), tells us how to keep a balance between income and capital so that a steady investment cycle is maintained. Consider, for example, that $a = 1$ and $b = 1$. Equations (28) and (30) then give

$$I(t) = Y(t) - K(t)$$

and we have a condition in which a steady investment cycle is maintained. However, if without any change in the capital the income increases (i.e., a becomes greater than unity), Figure 13 shows that the system becomes unstable (i.e., the investment cycle starts to have an increasing amplitude which if not checked would entail ever-increasing investments). (In actuality the check might come by having no more money to invest—a nonlinear limiting of the variable—thus changing the system so that our equations are no longer valid.) In fact, for our assumed value of b, *any increase* in the value of a would lead to such an unstable condition. On the other hand, still keeping the capital at the same level, if the income level went down (i.e., a became less than unity), the investment cycle would start decreasing in amplitude, and this condition could be cured by either increasing or decreasing the total capital (i.e., adjusting the value of b) until steady conditions were obtained. A study of Figure 13 will enable many such practical interpretations of our simplified Kalecki's theory to be made.

CONDITIONAL STABILITY

Now, in the example of equation (23) we saw the usefulness of feedback to make an unstable system stable. It should be understood, however, that feedback does not automatically ensure stability, that is to say, a feedback system which is stable can be made unstable under certain conditions. One example of this was shown in Figure 9, where, as already mentioned, the only difference between the two systems represented by the two curves is one of gain. To understand this, let us write the open-loop transfer function, $G_1(s)$, in the more general form

$$G_1(s) = K_N \frac{s^m + b_1 s^{m-1} + \cdots + b_m}{s^N(s^n + a_1 s^{n-1} + \cdots + a_n)} \tag{35}$$

For example, in equation (21) we had: $K_N = \text{gain} = 1$; $N = 0$; $m = 1$; $b_1 = 2$; $n = 2$; $a_1 = 2$; $a_2 = -1$. These values resulted in the curve shown in Figure 11. If now, we had $K_N = \frac{1}{4}$ instead, the resultant curve would have been as shown in Figure 14, where now the -1 point is included in the mapped area and hence indicates that the closed-loop system is unstable. From these two curves (Figures 11 and 14) it is clear that K_N would have to be greater than $\frac{1}{2}$ if the system is to be stabilized by means of feedback. When $K_N = \frac{1}{2}$, the mapped curve passes right through the -1 point, and the closed-loop system is on

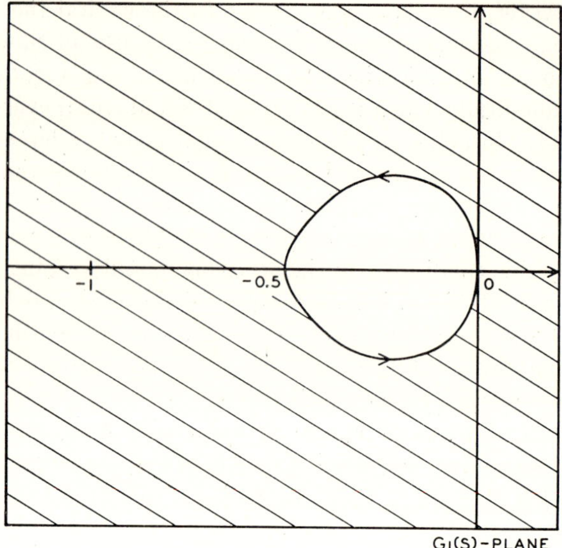

Figure 14. The mapped curve and the mapped area for the same system as shown in Figure 11, except that the gain has been reduced by one-fourth.

the verge of instability. This discussion shows that for some closed-loop systems, the corresponding open-loop gain may have to have values lying within a certain range (or ranges) for stability to be achieved; values lying outside this range (or ranges) lead to instability. Systems which are thus dependent on ranges of values of a certain parameter for their stability are called "conditionally stable" systems.

Example

Let us return to the example of picking up a pencil discussed at the beginning of this chapter. Sometimes damage to the brain or to the spinal column results in disturbances in the feedback paths and thus leads to instability. It is known that damage to a certain area of the brain will cause a person to suffer from what is called a "purpose tremor," and this results in that person not being able to pick up the pencil. His hand will come near the pencil and go past, it will start to return and go past in the opposite direction, and so on, the result being an oscillation of the hand about the position of the pencil (limitedly stable behavior). This system thus depends on a particular part of the system (brain area) working at a proper level (undamaged), and thus is an example of conditional stability.

CONCLUSION

In this chapter we have touched briefly on feedback and on the stability of systems. A criterion for the determination of the stability has been presented. Various examples illustrating the theory have been given. Within the limitations of space in this book, it is not feasible to explore all the various ramifications of feedback and stability. The list of books at the end of this chapter indicates where further exploration can be carried out.

ADDENDUM

The main body of this chapter was written so that a newcomer to the field of feedback theory and application would be in a position to apply what has been presented without bothering too much about some of the details. This addition is an attempt to fill in a few of these details.

Historically, the flyball governor, invented by James Watt (1736-1819) in 1788, is often cited as the first automatic control device using feedback. However, it was not until 1868 that a mathematical discussion of the principles involved in the working of governors was given. This was done by James Clerk Maxwell (1831-1879) in a paper read before the Royal Society of London on March 5, 1868 (11). Modern analysis of feedback systems stems from the papers by H. Nyquist (13) in 1932 and by H. L. Hazen (9) in 1934. The impetus given by World War II greatly increased the research, both theoretical and practical, in this field, and today there exists a vast amount of literature.

Referring to the main body of this chapter, we note that equation (14) gives us the response of a linear system to an impulse function, and since the expression on the right-hand side of equation (14) is later defined (in equation (15)) to be the transfer function of the system, the transfer function itself is often called the *impulse response* of the system.

Our method of determining the stability of a closed-loop system by using the corresponding open-loop transfer function as a mapping function was given by L. S. Dzung (8) and is thus often called the Dzung criterion. Now, H. Nyquist, in the paper referred to above, also formulated a criterion which necessitated the use of the open-loop transfer function as a mapping function, and hence, the mapped curve is often called the *Nyquist diagram* of the system. Further, we note from Figure 7 that the mapping curve consists largely of the variable s taking the value $s = j\omega$, for ω between $-\infty$ and $+\infty$. Since ω represents real frequency, the corresponding values of $G_1(s) = G_1(j\omega)$ represents the behavior of the open-loop transfer function at all frequencies, and thus the mapped curve is also known as the *frequency response curve* or the *harmonic response curve.*

Thus far we have been considering closed-loop systems in which the *total output* is fed back to the input without change (see Figures 1 and 6b, for example). Such systems are called *unity feedback systems.* However, it is quite possible to modify the output first, by passing it through another transfer function, as shown in Figure 15.

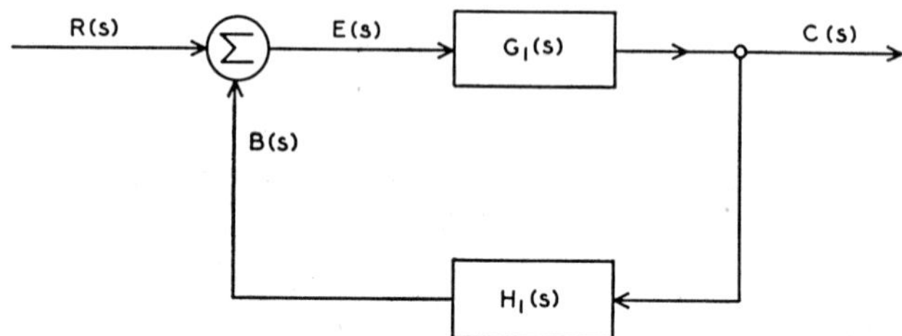

Figure 15. A closed-loop system with non-unity feedback.

In such a case, the open-loop transfer function is defined to be

$$\frac{B(s)}{E(s)} = G_2(s) = -G_1(s)\,H_1(s) \tag{36}$$

the corresponding closed-loop transfer function being

$$\frac{C(s)}{R(s)} = G(s) = \frac{G_1(s)}{1 - G_2(s)} = \frac{G_1(s)}{1 - G_1(s)\,H_1(s)} \tag{37}$$

Here $H_1(s)$ is called the *feedback transfer function.* For this system our stability criterion still holds except that we must now use $G_2(s) = -G_1(s)\ H_1(s)$ as the mapping function.

REFERENCES

(1) Allen, R. G. D. *Mathematical Economics.* New York and London: Macmillan and Co., 1957.
(2) Ashby, W. Ross. *Design for a Brain.* London: Chapman and Hall, 1952.
(3) Bower, J. L., and Schultheiss, P. M. *Introduction to the Design of Servomechanisms.* New York: John Wiley and Sons, 1958.
(4) Brown, G. S., and Campbell, D. P. *Principles of Servomechanisms.* New York: John Wiley and Sons, 1948.
(5) Bruns, R. A., and Saunders, R. M. *Analysis of Feedback Control Systems.* New York: McGraw-Hill, 1955.
(6) Chestnut, H., and Mayer, R. W. *Servomechanisms and Regulating System Design.* Vol. I. New York: John Wiley and Sons, 1951.
(7) Draper, C. S., McKay, W., and Lees, S. *Instrument Engineering.* Vol. II. New York: McGraw-Hill, 1953.
(8) Dzung, L. S. "The Stability Criterion," *Automatic and Manual Control.* Edited by A. Tustin. London: Butterworths Scientific Publications, 1952.
(9) Hazen, H. L. "Theory of Servomechanisms," *Journal of the Franklin Institute,* Vol. 218 (1934), 279-331.
(10) Kalecki, M. "A Macrodynamic Theory of Business Cycles," *Econometrica,* Vol. 3 (1935), 327-44.
(11) Maxwell, J. C. "On Governors," *Proceedings of the Royal Society of London,* Vol. 16 (1868), 270-83.
(12) Murphy, G. J. *Basic Automatic Control Theory.* New York: D. van Nostrand, 1957.
(13) Nyquist, H. "Regeneration Theory," *Bell System Technical Journal,* Vol. 11 (1932), 126-47.
(14) Porter, A. *Introduction to Servomechanisms.* New York: John Wiley and Sons, 1950.
(15) Savant, C. J., Jr. *Basic Feedback Control System Design.* New York: McGraw-Hill, 1958.
(16) Editors of *Scientific American. Automatic Control.* New York: Simon and Schuster, 1955.
(17) Sluckin, W. *Minds and Machines.* Baltimore and Harmondsworth, Eng.: Penguin Books, 1954.
(18) Thaler, G. J. *Elements of Servomechanism Theory.* New York: McGraw-Hill, 1954.
(19) Truxal, J. G. *Feedback Theory and Control System Synthesis.* New York: McGraw-Hill, 1955.
(20) Tustin, A. *The Mechanism of Economic Systems.* London: William Heinemann, 1957.
(21) Wiener, N. *Cybernetics.* New York: John Wiley and Sons, 1948.
(22) ———. *The Human Use of Human Beings.* Boston: Houghton, Mifflin and Co., 1950.

PART III

Case Studies

Twenty-Four

SIMULATION OF TACTICAL WAR GAMES [1]

RICHARD E. ZIMMERMAN

INTRODUCTION

War Games

The mission of the Tactical War Gaming Group at the Operations Research Office is to develop war gaming techniques for research purposes at all tactical levels. It is instructive to discuss briefly the connection between these tactical war games and previous efforts to use war gaming techniques for research purposes. The war game or map exercise has been used for many hundreds of years in an attempt to simulate military operations. Obviously, successful simulation techniques would assist in testing the value or effectiveness of new weapons, fighting techniques, or war plans. However, efforts to use the war game for these purposes have been severely hampered by a critical limitation of the means available to incorporate uncertainty or the play of chance, which is so prominent a feature of warfare. Thus, since the outcome of any given battle is uncertain, the significant results must be measured by the probability with which various alternative outcomes may be expected. A first requirement of the war gaming process is, therefore, that it permit repetition of the battle calculations while allowing nothing but the play of chance to vary so as to identify the

[1] This work was initiated by a suggestion of G. Gamow. The author is indebted to W. W. Nicholas, W. E. Cushen, and H. E. Adams for many of the ideas discussed and, in the case of H. E. Adams, for a great deal of the extended discussion out of which these ideas were developed. Of course, all errors and misstatements are the responsibility of the author alone.

spectrum of the possible outcomes and yield the associated frequency distribution.

A second requirement of the war gaming process is that a comparison be possible between the outcome of the battles using one weapon system and the outcome of similar battles using another weapon system, *all other things being held fixed.* In other words, we must be able to repeat the battle simulation many times while holding fixed all parameters except that one under investigation.

Attempts to repeat map exercises "played by hand," wherein numerous decisions are made throughout the battle by the human player, flounder on two points: (1) the intrinsic complexity of battle results in each game involving sometimes dozens of players and days, weeks, or even months of their time to complete a single game. Repetition of such games a hundred times or more, even if possible, would be prohibitively expensive; (2) the requirement that all parameters, except that one under investigation, be held fixed during these numerous games is frustrated by the dependence on the intuitive judgment of the human participants during the play of the game. For example, the players must resist the temptation to use in later games lessons learned in previous games, so long as such learned tactics themselves are not the variable being investigated. Any serious attempt to use operational simulation for war games as a research tool must therefore cope with these two problems.

This chapter discusses the evolution of a concept of war gaming and from it a model, referred to hereafter by its Operations Research Office code name, *Carmonette,* based upon the concept. *Carmonette* provides for the codification of all the decision processes occurring in the battle so that, once stated, those decisions may be accurately repeated as many times as is required.[2] Once this has been done, not only is it possible to repeat battles with all factors truly constant except the one under investigation, but also mechanical computing aids may be used to interpret the codified decision processes and speed up the over-all operation by factors of perhaps a thousand.

In this way we may hope to apply the war game process to prob-

[2] This does not *eliminate* judgment from the game, for that is impossible. Command decision—human decision—is the essential component of the operation of any organization. The effect is to require that judgment enter the play of the game in the form of rules formulated by experienced and responsible authority. When a war game is applied for training purposes, carefully controlled exceptions must be made to this requirement.

lems of military interest identical to those posed by professional military men for hundreds of years, but on a vastly larger scale. The fundamental limitation of this concept is that which has always restricted the usefulness of theoretical calculations to predict the future —theoretical analysis not associated with a substantial experimental program is usually of very little value.

Carmonette

Carmonette is a mathematical model of battle, of Monte Carlo type, which simulates in a simple straightforward manner the step-by-step progress of an isolated battle. It is designed so that all the calculations may be carried out on standard general purpose digital computers, except that a limited number of high-level decisions may be injected into the battle during the calculations. Some preliminary aspects of this work have already been published in reference (1).

Carmonette consists of two parts. The first, which details the manner of simulation of the combat activities of the individual participants in each battle, is in a high state of development. The second part, which deals with the integration of these separate combat elements into sensible and meaningful battles, is less well established. Both parts are discussed to provide a specific example of how very complicated processes involving decisions and vast amounts of information may, nevertheless, be simulated with some degree of success. Our discussion proceeds by considering in turn each of the various elements of the over-all problem.

The Problem

The specific problem is to determine the effectiveness of a radically new tank design on the combat effectiveness of small tactical units of company size—some hundreds of men and some dozens of weapons such as tanks.

Figure 1 lists the organization and equipment that might be associated with an armored unit using new tanks. It includes eighteen tanks, two platoons of infantry, some 4.2-inch mortars, machine guns and surface-to-surface guided missiles. (The number of combat elements was restricted to no more than thirty-six on each side, which was the capacity of simulation equipment used.)

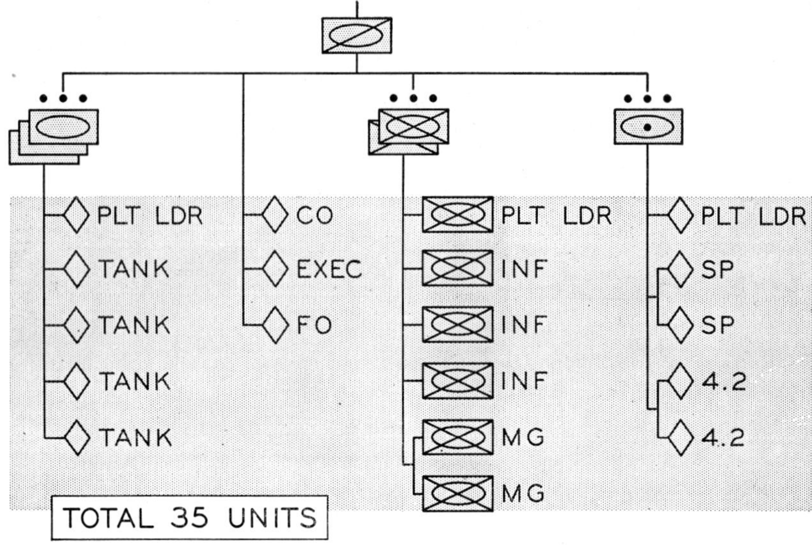

Figure 1. Schematic table of organization and equipment for a hypothetical armored unit. Under each unit symbol along the top is shown its composition. Key: CO, Commanding Officer; Exec, Executive Officer; FO, Forward Artillery Observer; Plt. Ldr, Platoon Leader; Plt. Hq, Platoon Headquarters; Inf., Infantry Squad; MG, Machine Gun Section; SP, Self-Propelled Missile Section; 4.2, 4.2-inch mortar section.

The Structure of the Battle

The major problem is to design a model of battle which can simulate typical or critical combat actions to test the effectiveness of the proposed tank company. Figure 2 indicates the distinct components of such battles which must be provided by the model. First, there are the *Actions* of the distinct combat elements—separate tanks and small groups of men like the squad or a gun crew. Second, there is the *Umpire* function, designed to monitor the exchange of information and the information gathering procedures of the distinct combat elements so as to limit these processes according to the restrictions imposed by the performance characteristics of the hardware and the cover and concealment associated with the terrain. The calculations of *Atomic Casualties* resulting from the use of atomic weapons is not a distinct component of battle but involves such special problems and massive calculations that it is here listed separately. The *Tactical*

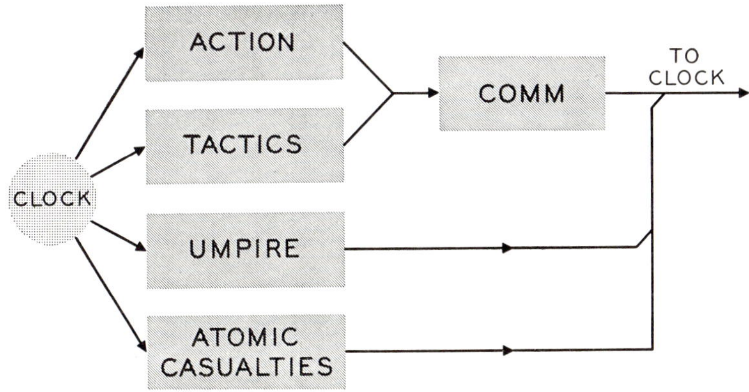

Figure 2. The computer must perform the various functions shown in a realistic time sequence determined by the clock.

Decision routine is responsible for integrating the separate combat actions of the individual combat elements into a sensible battle. Finally, the interactions between one combat element and another, especially the commanders of the various tactical units, make use of company *Communications*, which is therefore an important component of the battle to be simulated. These routines are assembled into a battle by the *Clocks* routine, which orders the different events in time.

In the next section, the simulation of the separate *Actions* of the combat elements will be described, and mention will be made of the *Communications* and *Umpire* routines.

Following this discussion of the simulation of the separate elements of the problems, a scheme will be described for simulating the major tactical decisions of the unit commanders who must integrate the individual actions into a sensible battle plan. Finally, some brief comments will be made concerning possible applications of the *model.*

SIMULATION OF SEPARATE COMBAT ELEMENTS

1101 Computer Battle Film

The idea of these combat actions is vividly conveyed by an animated cartoon motion picture that was prepared from consecutive still drawings, using trial results of some early test battles (1) programmed for the ERA 1101 computer to investigate the technical feasibility of *Carmonette*. The battle depicted for this cartoon is not presumed to have any particular military significance, although it has been put in a setting in Western Europe of some intrinsic interest. The battle takes place on a piece of ground a little over a mile square, about fifty miles south of the zonal boundary in Bavaria. Blue is deployed

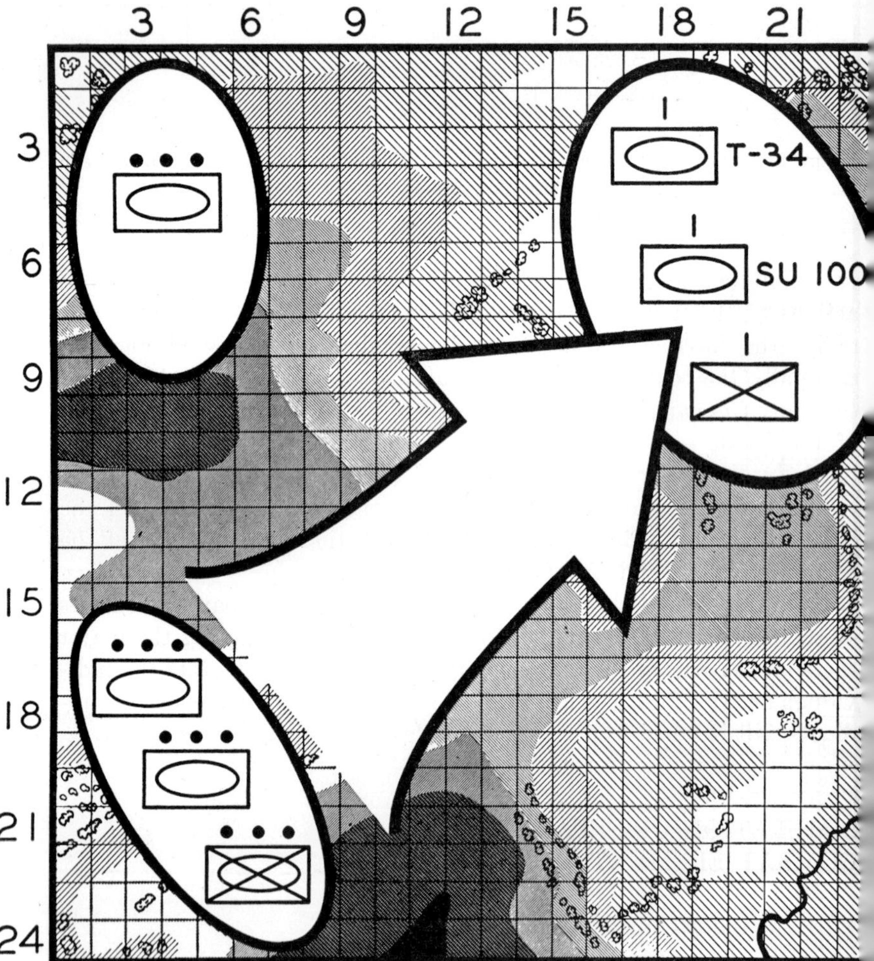

Figure 3. Blue forces deployed along west are to attack Red forces on northeast.

as shown on the west (Figure 3) and is ordered to attack Red on the northeast. Blue consists of a reinforced tank company of three tank platoons, a platoon of armored infantry, and a battery of 4.2-inch mortars (not shown on the map). This is a total of seventeen tanks, three squads of infantry, and twelve mortars. Red consists of a com-

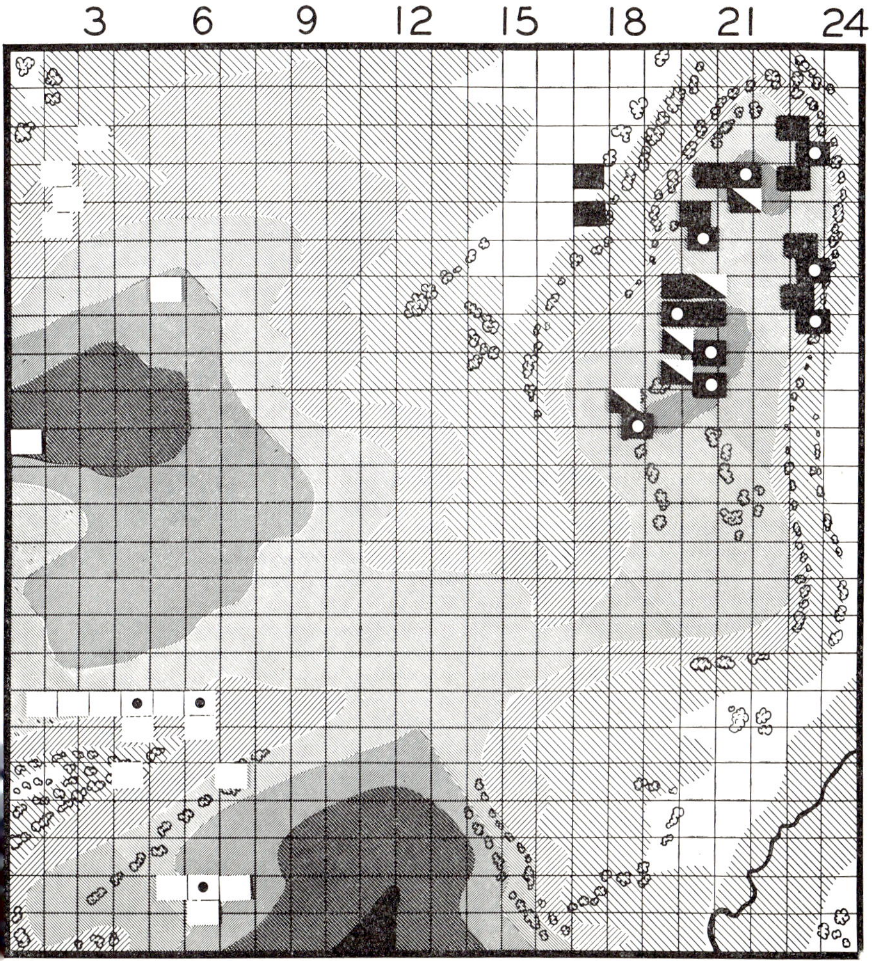

Figure 4. There are seventeen Blue tanks and three infantry squads deployed along west. Red forces on the northeast include ten T34 tanks (solid rectangles), five SU 100 guns (diagonally slashed rectangles), and nine squads of dismounted infantry. Time: Zero. Casualties: Blue: 0, Red: 0.

pany of ten T34 tanks, a company of five SU 100 self-propelled guns, and a company of nine squads of infantry. Figure 4 shows the initial position of the separate combat elements with reference to the 100 meter grid square system superimposed on the map. All the maneuver of the combat elements is related to these 100 meter grid squares.

The Blue tank platoon to the northwest is to remain in position firing on the Red forces while the remaining Blue units in the south-

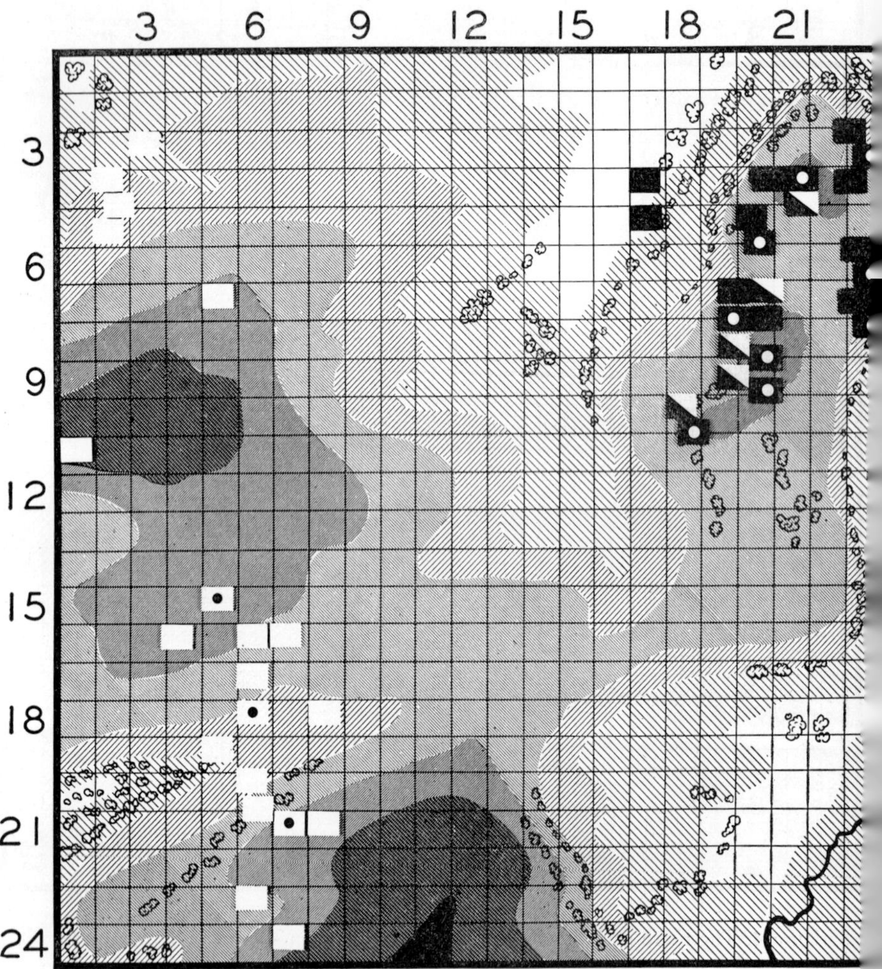

Figure 5. Time: 2 minutes. Casualties: Blue: 0, Red: 0.

west make a frontal attack up the hill into the Red position. As the attack starts (Figure 5), the assaulting Blue units start to move towards their terrain objective. Figures 6 through 11 depict the movement of the assaulting Blue forces during the first fifteen minutes of the battle simulation during which time (for these trial calculations) there is assumed to be no firing. Firing starts at fifteen minutes of battlefield time, and Figure 12 shows the state of the battle at sixteen

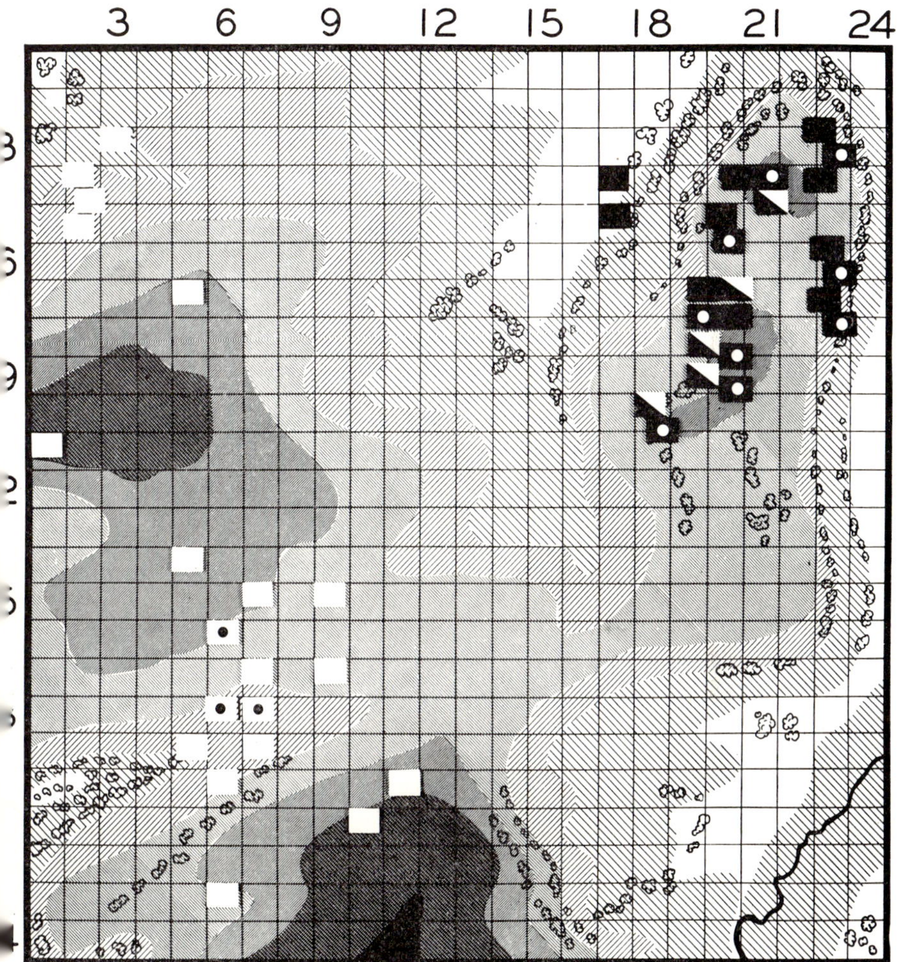

Figure 6. Time: 4 minutes. Casualties: Blue: 0, Red: 0.

minutes battlefield time, one minute after firing started. Notice that the casualties are already very heavy—Blue has lost six tanks and Red has lost three. Figures 13 through 19 give the further progress of the battle at two-minute intervals until at thirty minutes battlefield time the calculations were halted. While it is not appropriate to say the battle ends at this time, the casualties were so heavy—eleven Blue

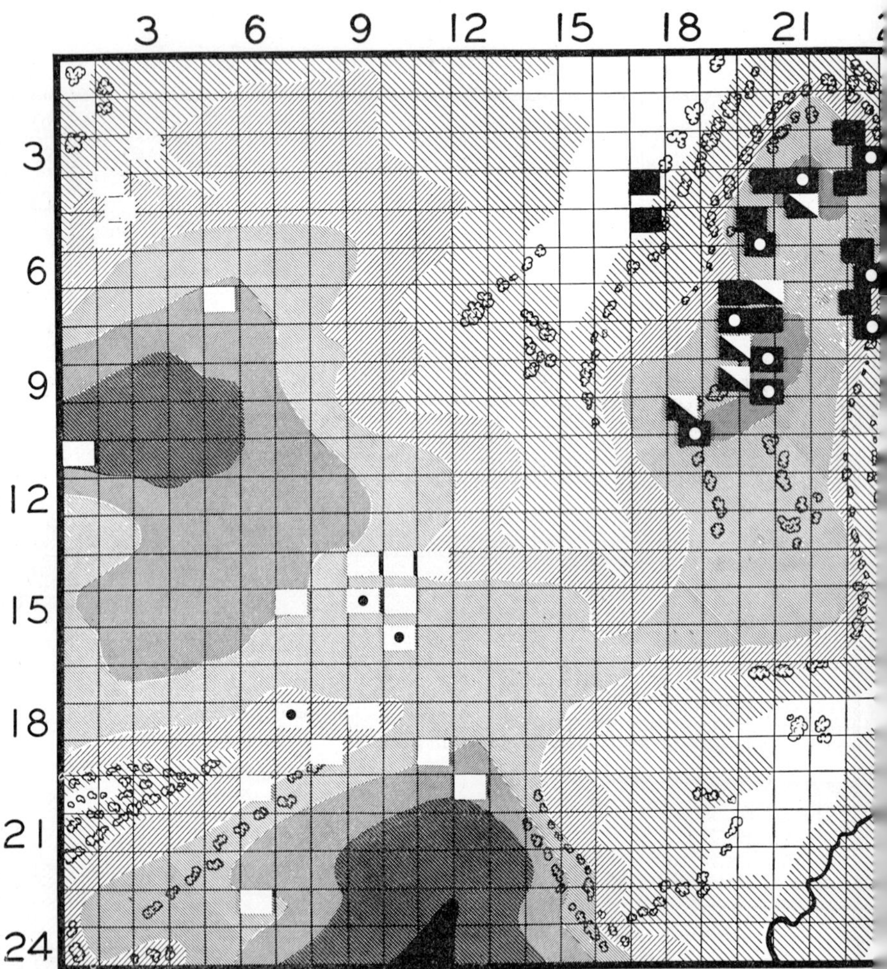

Figure 7. Time: 6 minutes. Casualties: Blue: 0, Red: 0.

versus eight Red tank casualties—that the battle may for all practical purposes be considered over. The basic combat actions of the combat elements in *Carmonette* are those demonstrated by the cartoon—actions of fire and maneuver and the associated decision processes. However, the detailed calculations in *Carmonette* are a great deal more extensive than those used for the cartoon battle.

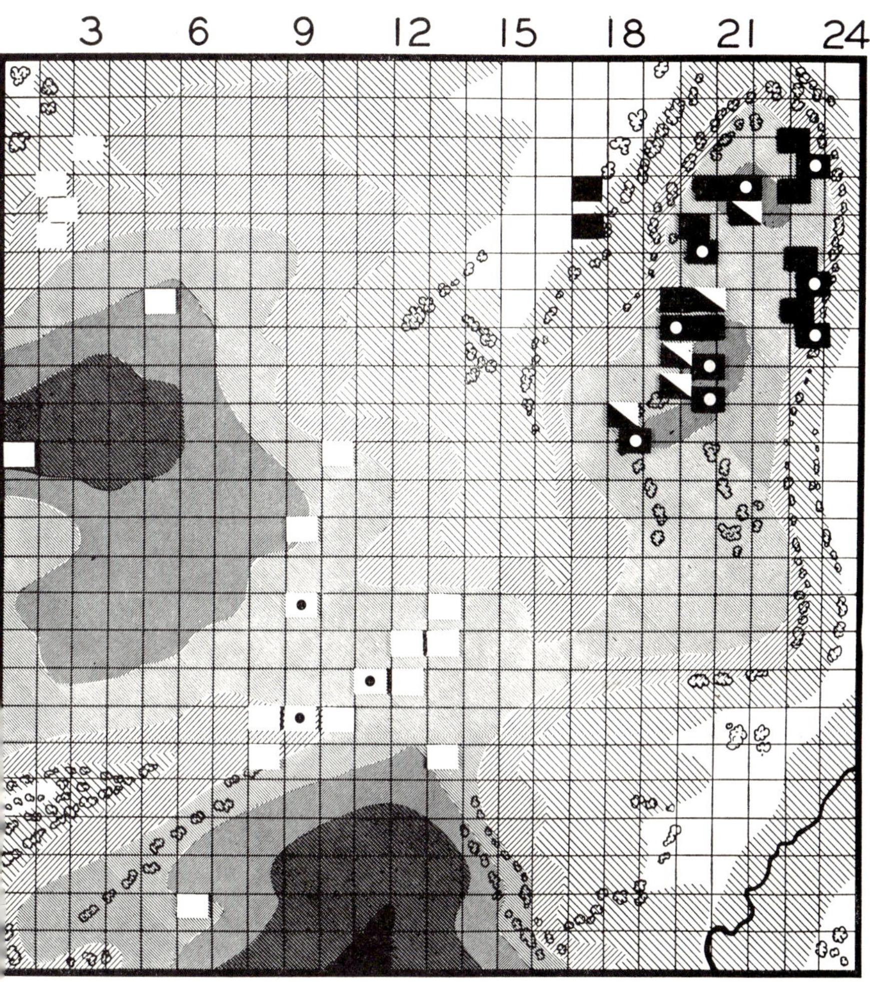

Figure 8. Time: 8 minutes. Casualties: Blue: 0, Red: 0.

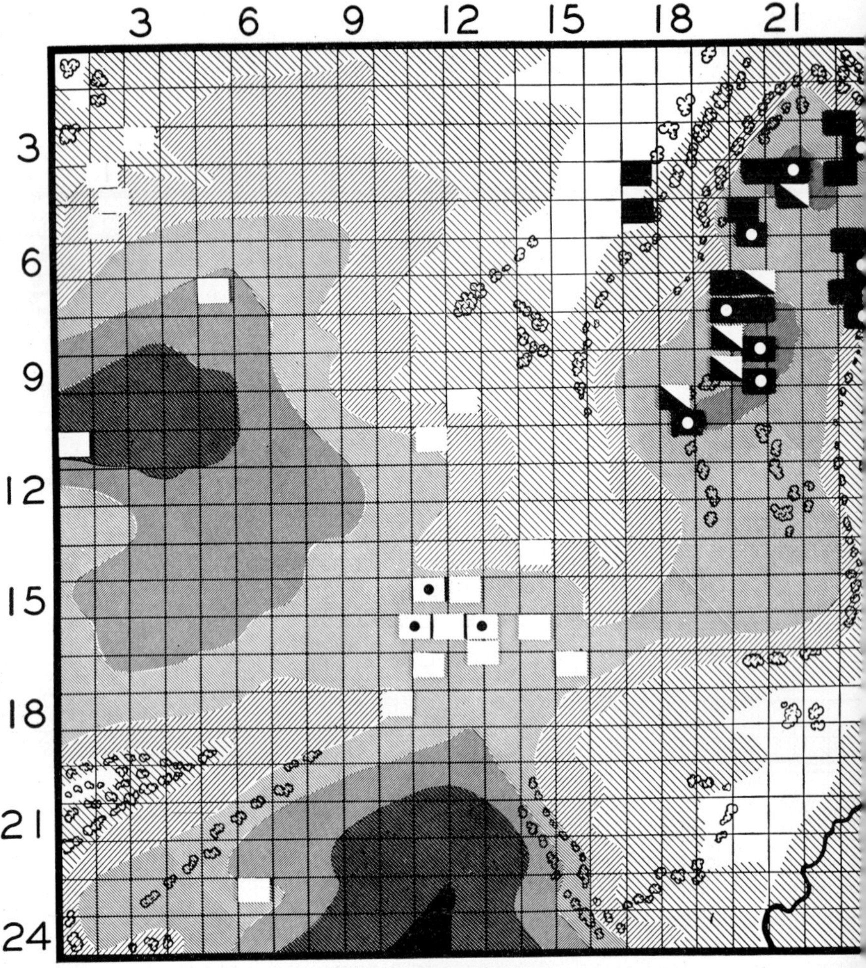

Figure 9. Time: 10 minutes. Casualties: Blue: 0, Red: 0.

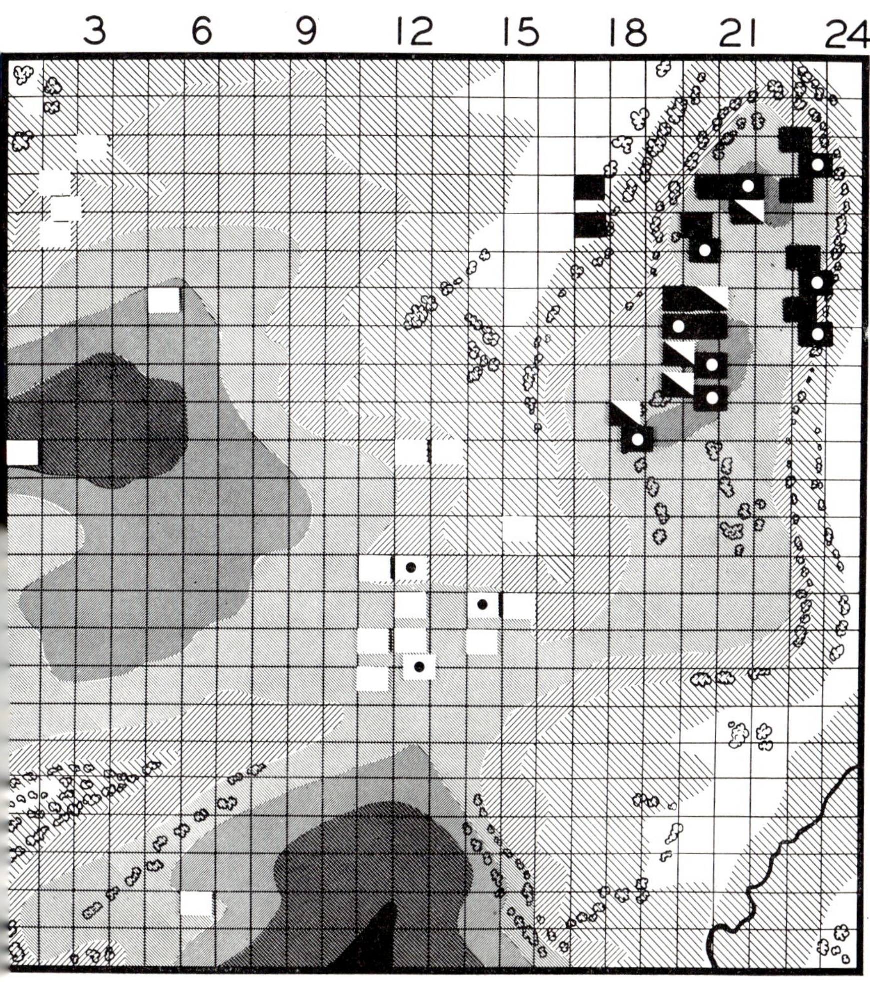

Figure 10. Time: 12 minutes. Casualties: Blue: 0, Red: 0.

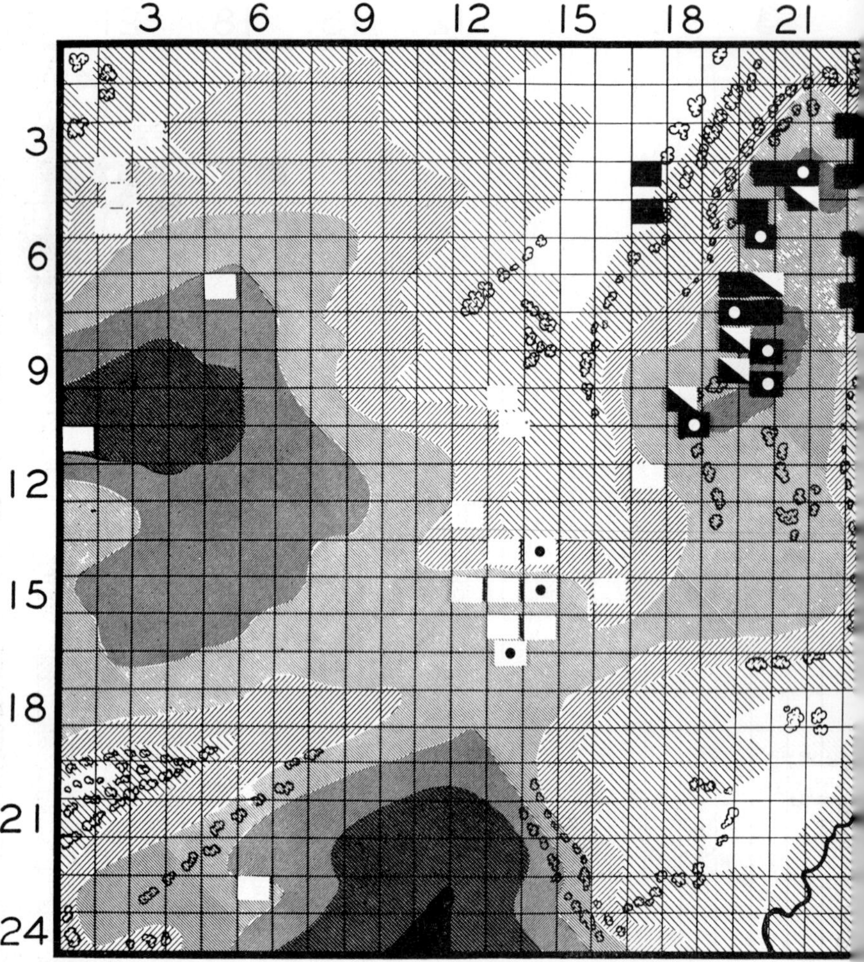

Figure 11. Time: 14 minutes. Casualties: Blue: 0, Red: 0.

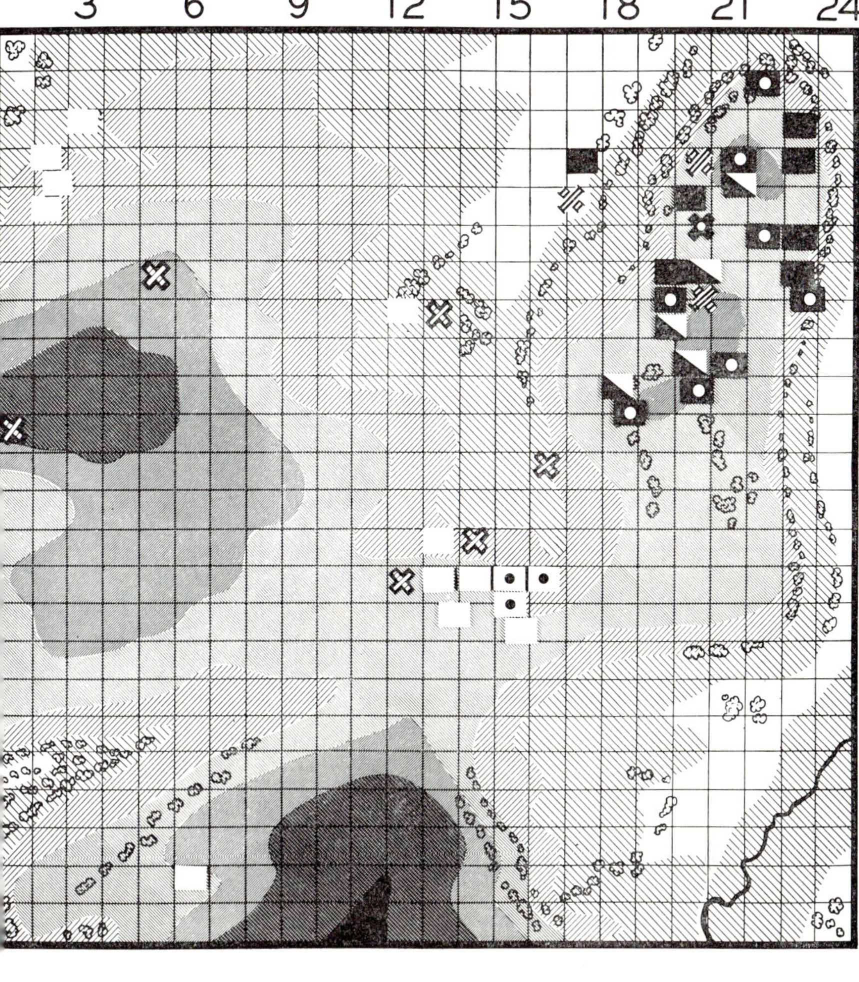

Figure 12. Time: 16 minutes. Casualties: Blue: 6, Red: 3.

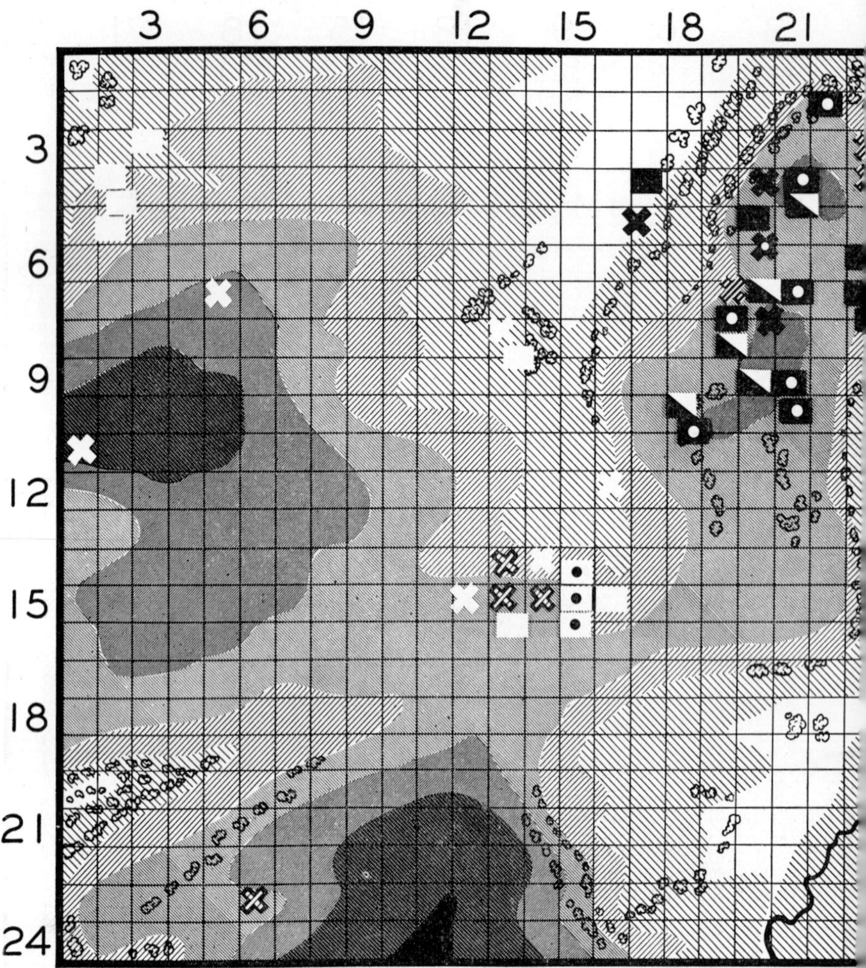

Figure 13. Time: 18 minutes. Casualties: Blue: 10, Red: 6.

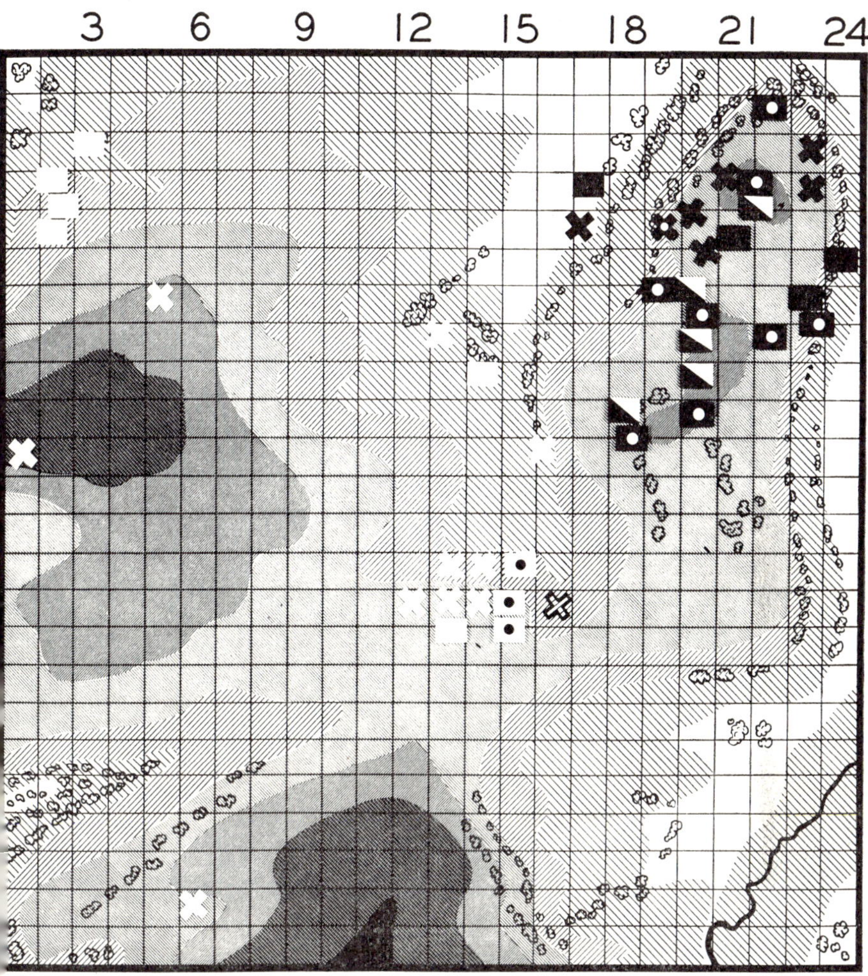

Figure 14. Time: 20 minutes. Casualties: Blue: 11, Red: 6.

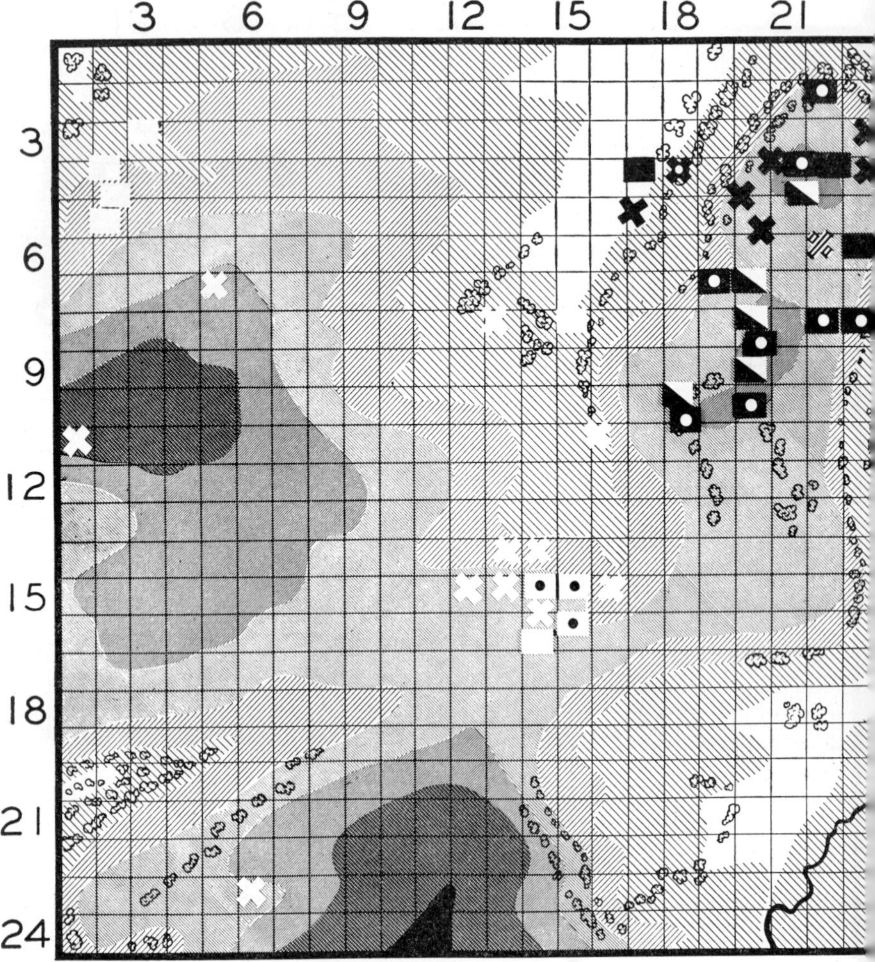

Figure 15. Time: 22 minutes. Casualties: Blue: 11, Red: 7.

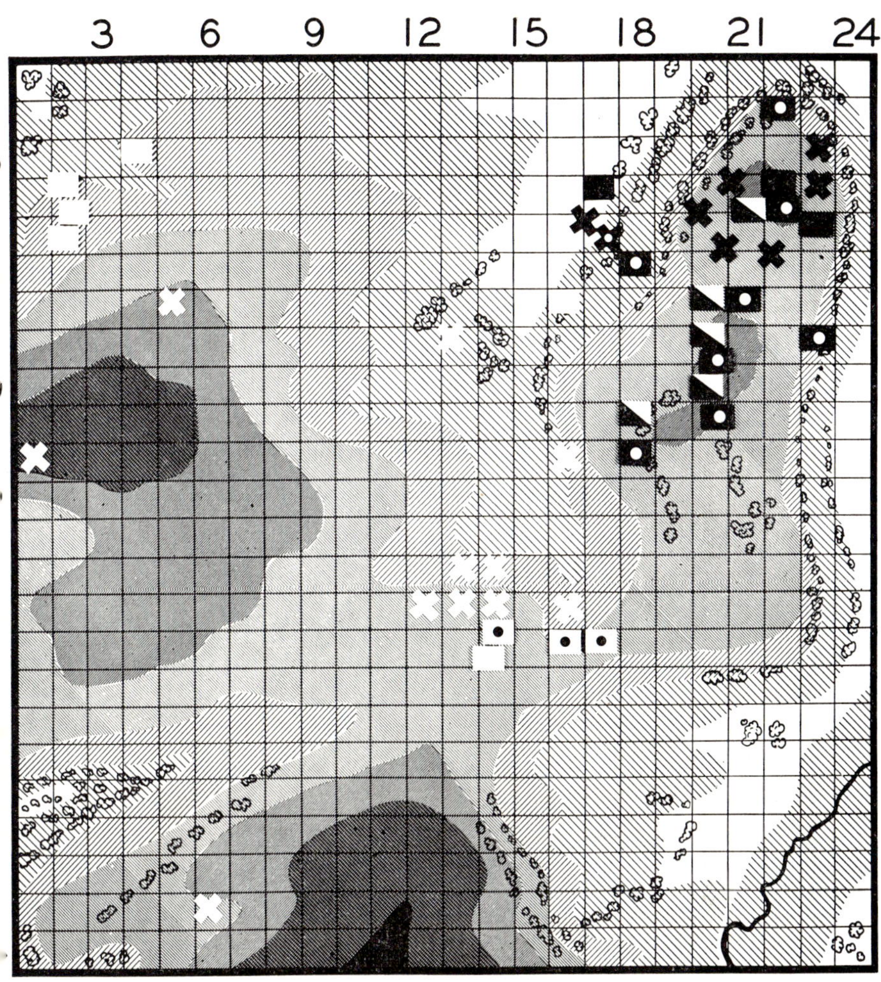

Figure 16. Time: 24 minutes. Casualties: Blue: 11, Red: 7.

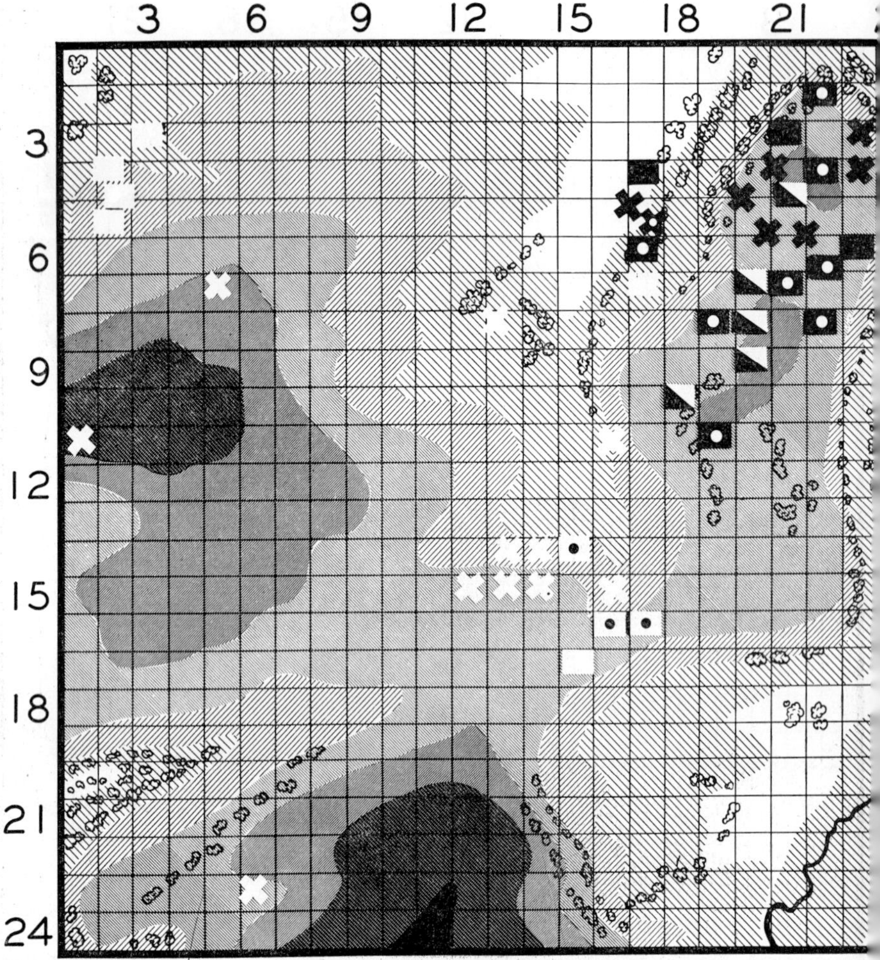

Figure 17. Time: 26 minutes. Casualties: Blue: 11, Red: 7.

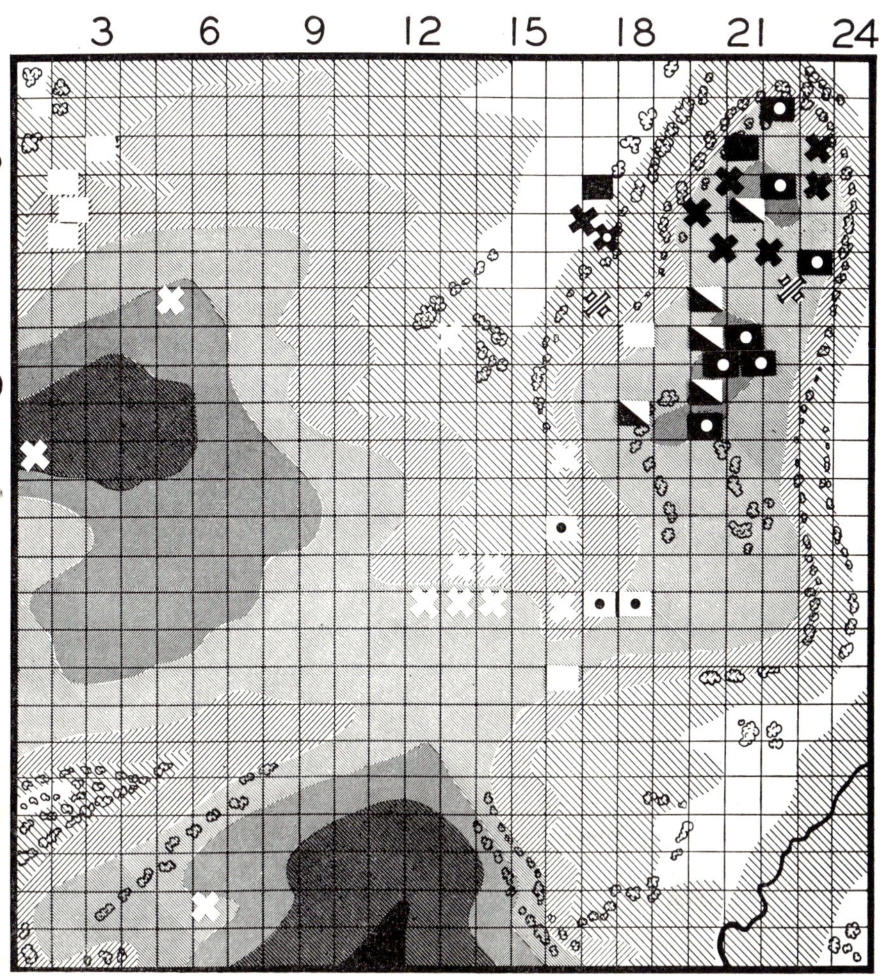

Figure 18. Time: 28 minutes. Casualties: Blue: 11, Red: 8.

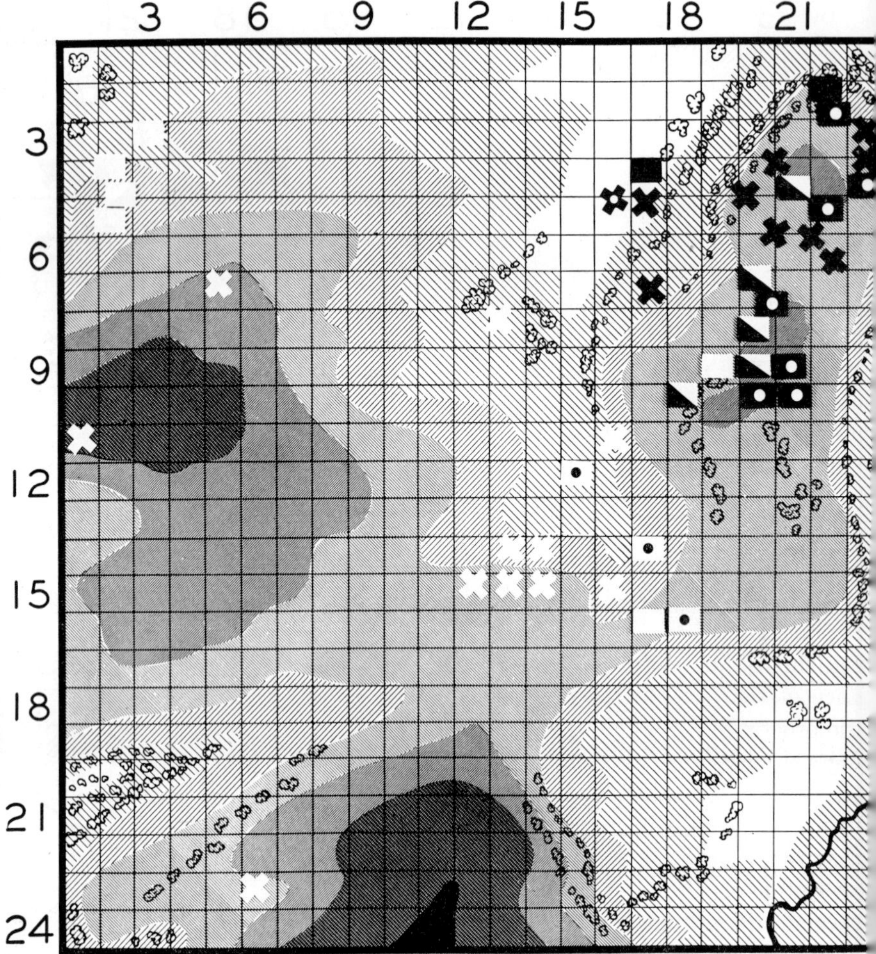

Figure 19. Time: 30 minutes. Casualties: Blue: 11, Red: 8.

Battlefield Time

Before a description of the simulation of individual events or actions it is appropriate to mention the way in which these are sequenced in time. Each unit has associated with it "alarm clocks" or numbers giving the time at which it will next be able to act. At the beginning of the battle (time zero) these clocks will all be set at a few seconds unless some units have been prohibited from moving or firing. The *Clocks* routine examines the clocks and finds the one with the smallest time. This corresponds to the first event that is scheduled to happen. This becomes the new battlefield time and the machine performs the calculations required to simulate the event. Suppose it is the firing of a tank. That tank's clock will be reset for the time at which it will be ready to fire again. Other clocks may be reset (for example, the clock of the enemy fired upon might be reset to allow it to return fire immediately). The clocks routine then determines the next event, the battlefield time is adjusted, and calculations continue.

Firing Actions

As was indicated by the cartoon, the units in *Carmonette* from time to time fire upon selected enemy units. For the cases of greatest interest, single-shot kill probabilities are sufficient to measure the performance of the gun. Of much greater interest, however, is simulation of the process of surveying the battlefield to discover enemy units, applying a priority system to select a target and the final decision to fire on the target. Figure 20 indicates the intimate connection between the firing decisions by combat elements and the decisions to move. Note that the upper part of the figure indicates that the unit will be given an option to fire only after it declines an option to move. Only in special circumstances is a unit first given an option to fire as in the lower part of the figure. In this special case, if the unit declines an option to fire, it is then offered an option to move. Thus, the actual decision process carried out by each combat element at frequent intervals throughout the battle may be considered as involving a selection of one of four alternatives: (1) to move or not move, (2) to fire, (3) prepare to fire, or (4) decline to fire. Clearly the move portion of this calculation is the more fundamental, since it ordinarily comes first.

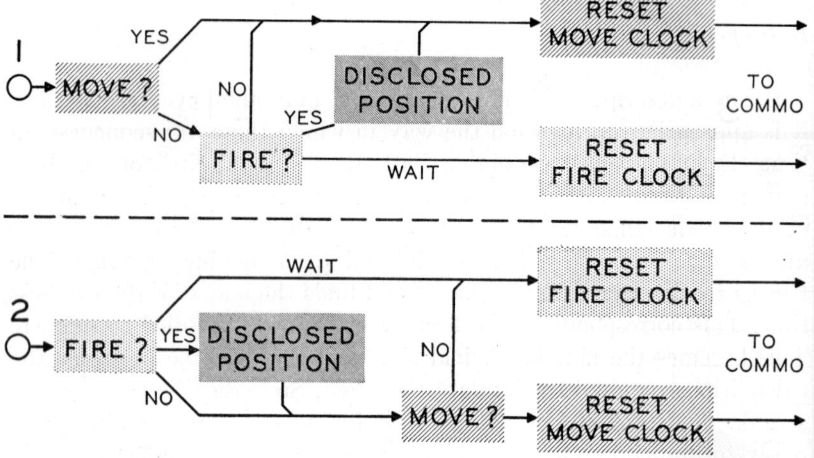

Figure 20. The calculation labeled "DISCLOSED POSITION" determines whether the act of firing has disclosed the firer's position to enemy units.

Terrain Quantification

Figure 21 shows the complete history of one of the Blue tanks throughout the cartoon battle. It is apparent that the progress of the Blue tank toward the terrain objectives resulted from a series of discrete moves from one square to an adjacent square. The decision process associated with selecting a particular adjacent square as the next position to be occupied in the course of the assault is the most fundamental decision process carried out by the individual combat elements. In order to effect sensible simulation of the activities of a real tank, these moving decisions must be intimately related to the terrain features. Therefore, the average value of important terrain characteristics for each square must be identified, stored in the computer, and allowed to influence the move choice. Figure 22 indicates the level of approximation associated with squares of this size for the cartoon battle and the types of terrain features of interest. On the cartoon battlefield there were 24 × 24 or 576 squares. For each square there was stored the degree of average concealment to be associated with the vegetation on the square in five steps, varying between open fields to dense forest. In addition, the elevation to the nearest meter was stored for each square. Finally, the presence of selected terrain features was noted, such as swamps or roads.

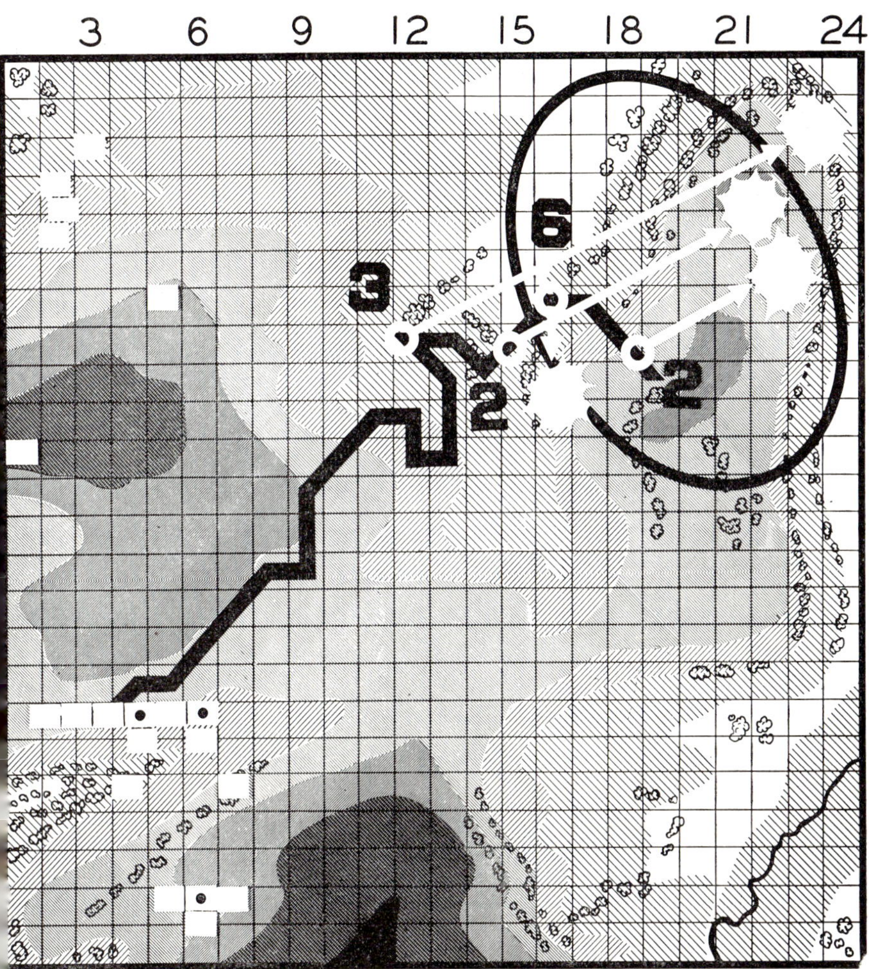

Figure 21. Activity of tank selected from "1101" battle. Numbers refer to number of shots fired from that position.

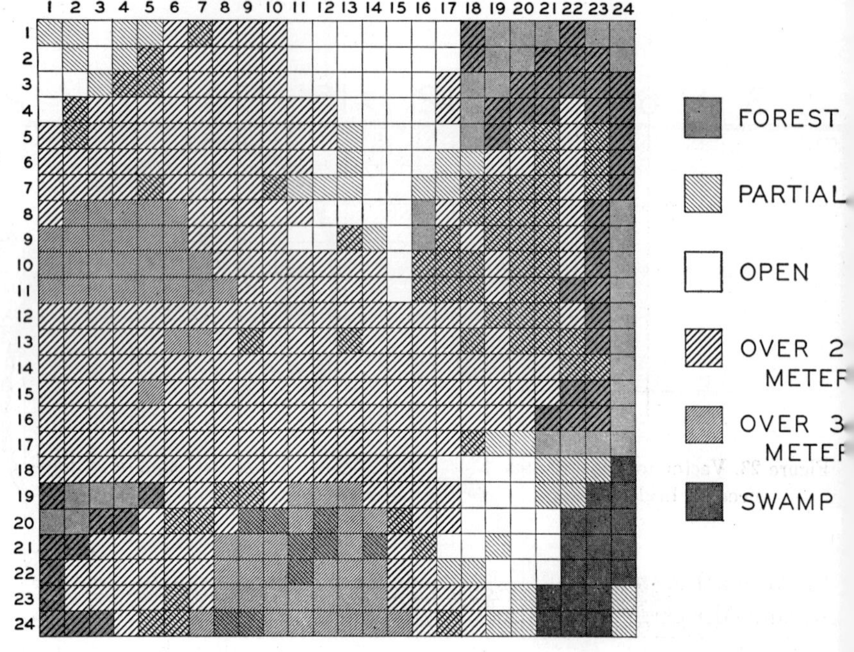

Figure 22. Schematic battlefield. Indicates representative terrain features associated with each grid square used in "1101" battle.

Carmonette provides an increase in the number of squares to 36 × 36, or 1296, and a considerable increase in the information stored fo such terrain features. The most significant additional terrain featur included in *Carmonette* involves terrain features we may term "vector" in nature. Thus, Figure 22 indicates only the "scalar" terrain char- acteristics of each square; that is, characteristics of the square alon which do not depend upon its neighbors. However, consider Figur 23, which is a schematic representation using the 100 meter gri squares of a typical combination of improved and unimproved roads a river, a bridge, and a river fording site. In this case it is clear tha one cannot characterize the terrain features to be associated with th central square except by identifying the appropriate adjacent square Thus, an improved road leads from the central square to the west an to the northeast. An unimproved road leads to the southeast. No roa leads to the northwest, the south, the southeast, or the east. Movemen

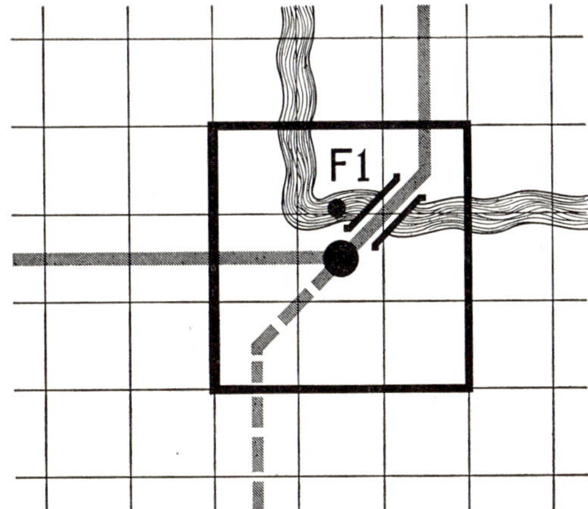

Figure 23. Vector terrain features. Mobility of element on central square is influenced by bridge, fording sight (F1), improved and unimproved roads.

to the north is only possible if the vehicle is capable of fording the stream. Movement to the northeast can be interrupted if the bridge is damaged. Thus, these terrain features have a direction and are properly termed vector terrain features.

Given such information about the terrain it is possible to cause a tank or other combat element to make each move dependent upon the terrain characteristics. As is indicated by Figure 24, the basic move decision involves a selection of one of the eight adjacent squares as the next position to be occupied (or a decision may be made to remain on the present position). The general nature of the factors we should expect to influence this choice are listed. The first, exposure to enemy positions, and the third, desirable terrain, depend upon the quantification of the terrain just described. The move is likely to have a preferred direction (associated with the terrain objective) and, further, to be influenced by exceptional circumstances, such as the knowledge that the tank is under fire or the inhibitions produced by the presence of friendly knocked-out tanks.

Without describing how these other features are to be handled in any detail, it is obvious that the computer move calculations can be made to depend on all terrain data stored for each of the adjacent squares in question. The computer is capable of examining the existence of a physical line-of-sight between any and all units using the

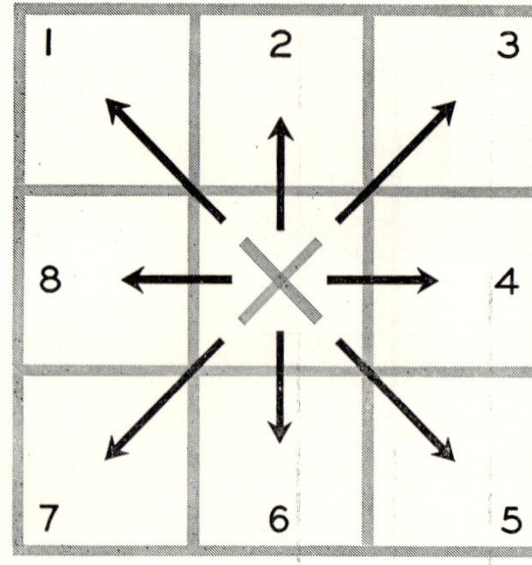

Figure 24. The elemental move decision is made repeatedly by each independent combat element in light of the battlefield situation.

elevation stored for each square; and, further, it can qualify the existence of such a line-of-sight by the influence of concealment afforded by vegetation, and thus obtain the degree of exposure to the enemy to be associated with each of the eight adjacent squares in turn.

The essential calculation to be made involves summing the relative desirability from the point of view of the tank commander, of each of these squares in turn, using the types of data available, and then selecting one of these squares on the basis of the weights so derived.

Before discussing in detail the specific procedure used in *Carmonette* for this weighting process (a matter of secondary interest), let us consider the more basic problem of what will be done with the weight—that is, the way in which one square will be selected. Figure 25 graphically illustrates this point. In fact, *Carmonette* proposes to use the weights developed by the scoring process, no matter how they are derived, as giving the relative probability that the tank will choose a particular square. This calculation will therefore be of the "Monte Carlo" type much used by applied mathematicians since World War II. This interpretation of the rating process has far-reaching consequences, since it introduces the play of chance into the battle calculations from the very beginning. Therefore, the results of any particular calculation have no general significance. Each battle must be repeated

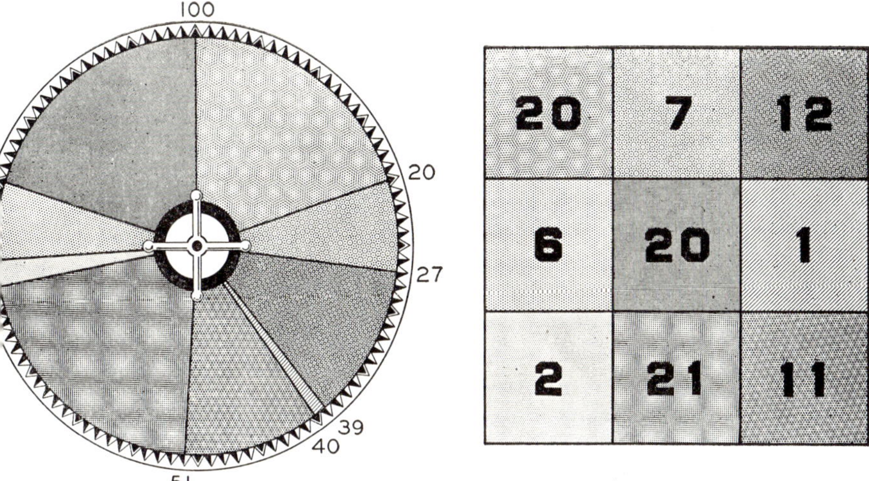

Figure 25. Sample weighting factors inscribed in each of nine squares are associated with probabilities by comparison with roulette wheel. A single spin of wheel then "selects" next square and thus determines movement of tank.

a number of times sufficient to generate the usual statistics associated with distributions.

Some Trial Battle Results

Figure 26 is a scatter diagram of the results (1) of fifty repetitions of the "cartoon" battle which differ only due to the play of chance. The diagram indicates no strong correlation between the Blue tank losses and the Red tank losses. Figure 27 shows the actual distributions separately—the average Blue losses being 10.4 tanks per battle and the average Red losses being 7.1 tanks per battle. Both distributions may be considered to be samples drawn from a Gaussian population within the usual confidence limits. Figure 28 indicates the convergence of the mean with increased numbers of battles. The differences between the mean Red tank losses and the Blue tank losses are statistically significant at a very high level. For this series of battles, therefore, fifty repetitions is an adequate sample size to determine

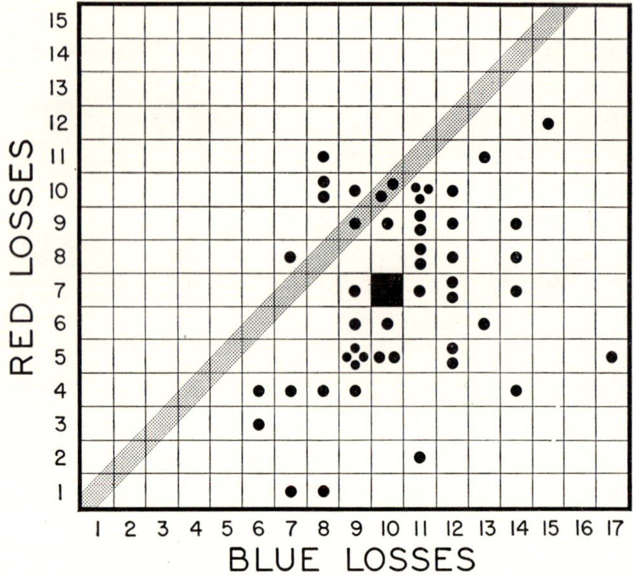

Figure 26. Scatter diagram of casualties from repeated calculations of casualties. Variation in outcome depended only on the operation of chance.

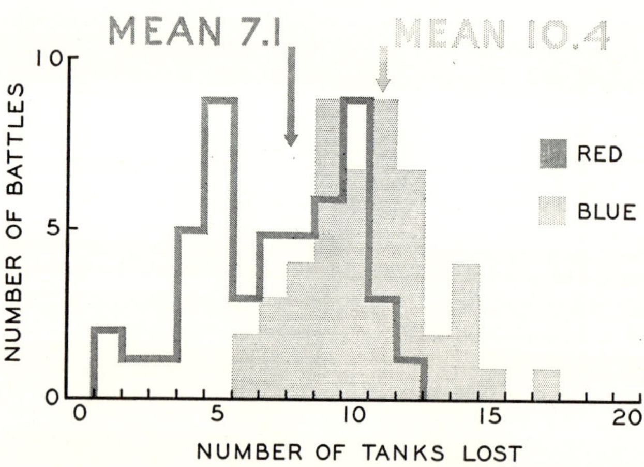

Figure 27. Blue casualty distribution indicated by solid area; Red casualty distribution indicated by outline.

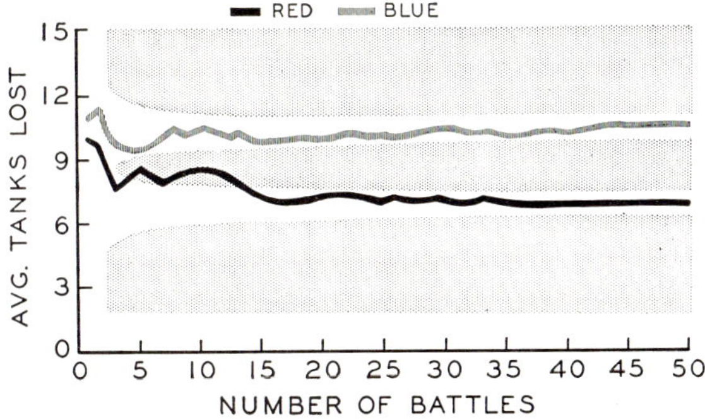

Figure 28. The upper curve indicates variation in Blue's cumulative mean casualties. The lower refers to Red. Also indicated is the standard deviation of the mean computed from the sigma of the distributions shown in Figure 28.

"winners" by comparing mean losses for all except the most marginal cases.

To demonstrate the sensitivity of the model to changes in the performance characteristics of a weapon, a second series of fifty battle calculations was made. In this case, the seventeen Blue tanks were replaced by seventeen hypothetical light tanks, with a doubled cross country speed and rate of fire, but substantially reduced armor thickness and gun power. All other factors were unchanged. Figure 29 gives

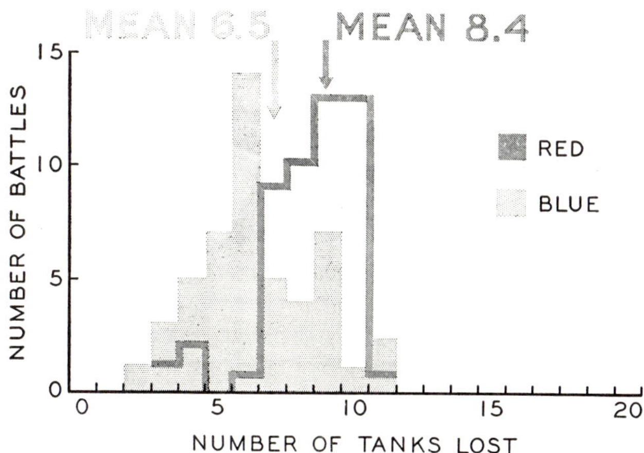

Figure 29. Blue light tank casualty distribution is shown by solid area; Red distribution by the outlined area.

the frequency distribution of tank losses for this second series of battles. Here the mean Blue tank losses were less than the mean Red tank losses at a very high level of confidence.

Significance of Probabilistic Model

It is evident that the above interpretation of the ratings as probabilities carries with it a high cost since it will then be necessary to repeat the calculations many times so as to determine the distributions. Three arguments may be offered in support of this choice as against the simpler alternative of selecting with certainty the most likely square.

Perhaps the most fundamental reason relates to the preference for a system which permits, at least in principle, direct verification by experiment. Thus, if it were desired to conduct a field experiment in which a representative group of tank commanders in identical circumstances were requested to identify the square to which they would move, we could be sure that the group would frequently demonstrate a variation in their choice. The experimental raw data would surely be a probability distribution, and the model must provide for its use in that form. Clearly, an experimental program designed to determine such information for a variety of battle situations and terrain types would be of tremendous size and cost. Yet the model must be compatible with the nature of the experimental data which may become available as a matter of principle and practice.

A second justification for the probabilistic interpretation is that it automatically compensates in a simple and straightforward way for the uncertainties bound to be associated with any particular rating process used. Thus, if the rating process used were to produce roughly equal scores for several squares, interpretation of these scores as relative probabilities avoids what must surely be an arbitrary and unsatisfactory selection of the one with a trivial superiority.

A third consideration which supports a probabilistic interpretation relates to the consequent possibility of investigating directly the sensitivity of a mechanical weapon system (with which there may be associated only small uncertainties) to variations in component performance resulting from the human factors which complicate the analysis of any real weapons system. In effect, the probabilistic interpretation is more consistent with the performance of man-weapon combinations and permits a more direct attack on the problems of such combinations than would any system which excludes stochastic factors.

Although the price is high, probabilistic interpretations and provisions for incorporating stochastic data are important, essential features of *Carmonette.*

The Terrain Feature Rating Process in Detail

Having discussed the probabilities character of the model, we will return to a description of the details of the weighting process. Each square is given a weight W that is composed of the sum of six distinct components

$$W = a_1L_1 + a_2L_2 + a_3L_3 + a_4L_4 + a_5L_5 + a_6L_6$$

where each component is the product of one factor, the L values associated with the battlefield, and a second factor, the a values associated with the value judgments or command decisions of the unit commanders. The a values provide the means by which individual actions may be assembled into a sensible battle, reflecting the command decisions as simulated by the *Tactics* routine to be discussed in the next section.

The L values represent six categories of "facts" concerning the battlefield:

L_1—Cover and concealment (defense)
L_2—Man-made objects
L_3—Trafficability (offense)
L_4—Fields of fire
L_5—Dispersion
L_6—Terrain objective

The intent is to use the terrain features of the battlefield in conjunction with each man's knowledge of the disposition of the enemy and the characteristics of his own unit in the construction of these L values. Notice that L_1 and L_2 comprise an assessment of the desirability of a square in terms of its defensive potential, while L_3 and L_4 provide the basis for an assessment of the desirability of each square from an offensive point of view (that is, the speed with which a unit can move across the square and the danger resulting from exposure to the enemy in the process). L_5 judges the square according to whether or not movement to the square will result in breaking formation. Each of the squares is also scored on the degree to which movement to that square is in the preferred direction. We may now observe the significance of the "a" coefficients. Their purpose is to adjust the influence of each of the first five components relative to the influence of the sixth component. Thus, we see that if a particular coefficient is set to zero, as the extreme case, the tank commander will no longer allow his movement to be influenced by that factor. For example, bold, even reckless, attacks will be associated with small values for the a_1 through a_4 coefficients, with the result that units will tend to take the most direct route to the terrain objective. Cautious, fast moving attacks will be associated with high values for a_3 but lower values for a_1, a_2, and a_4. Extremely cautious attacks would require high values for all the coefficients.

As a general rule the L values to be associated with a given combination of terrain features and other battle parameters are to be stored in the computer in the form of extensive function tables. Recourse to formulae will be made only after the memory capacity of the computer is substantially exhausted by the tables.

To illustrate the connection between various battle factors and these L values, the construction of L_4 value, "Fields of Fire" will be described. The L_4 value itself is related to a number of battle parameters,

$$L_4 = [C_1(L_1L_2) + C_2(L_3)][L_{41} + b_1L_{42} + b_2L_{43}]$$

where

L_{41} = Class 1 and 2 combat intelligence
L_{42} = Class 3 combat intelligence
L_{43} = Good firing positions
$C_1(L_1L_2)$ = Cross terms with cover and concealment factors
$C_2(L_3)$ = Cross terms with barrier
b_1,b_2 = Doctrine and/or training coefficient

This is the most complex of the L value components since it includes the essential cross terms with the other components. It can take on both positive and negative values corresponding to a desirable or undesirable combination of circumstances. The numerical value of the bracket farthest right is proportional to the threat associated with a particular square (by virtue of the existence of enemy units who can observe movement on that square) so far as is known or suspected by the tank commander. The three components of this bracket (L_{41}, L_{42}, L_{43}) correspond to different degrees of certainty in the mind of the tank commander as to the existence and type of these enemy units. The coefficients b_1 and b_2 degrade the influence of the less certain information. Function tables involving type of enemy unit, range, type of friendly unit, and similar factors yield the L_{41}, L_{42}, L_{43} values. Note that the L_{43} value is the most speculative of the three classes of combat intelligence since it refers to enemy units whose presence is only inferred from the excellent observation provided by the terrain feature if the enemy were to occupy it. Therefore, a large value for the coefficient b_2 corresponds to an exceedingly cautious tank commander.

The bracket farthest left on the right-hand side of the L_4 equation enhances or degrades the influence of the bracket farthest right according to the degree of cover, concealment, and trafficability afforded by the square. These are the cross terms with the other L value components. The intent is to qualify the threat implied by large values of the right-hand bracket by including a dependence on whether the local cover will permit enemy observation of movement within the square.

Combat Intelligence

The preceding discussion generated a requirement for classifying the tank commander's knowledge and opinion of the enemy's type and position. The general classes of such knowledge are:

1. Known—Identified enemy
2. Position known—Type estimated
   a. Machine gun (infantry)
   b. Anti-tank gun
   c. Tank
   d. Infantry
3. Position "known" incorrectly
   a to *d*, type estimated as above
   e Identified correctly at earlier time; unit has since made an undetected move.

This information ranges from precise and accurate knowledge of type and position through general knowledge of position and an estimated type (may be inaccurate) to a completely erroneous belief as to the position of an enemy unit. *Carmonette* provides for the generation of distinct knowledge and information for each separate unit on the battlefield.

There are three sources of information of an individual's knowledge about the enemy: (1) a direct and continuous survey of the battlefield making use of the *Umpire* routine, (2) a special calculation associated with the likelihood that the firing of one's weapons may reveal one's position to enemy forces, and (3) messages received through the communications system.

Umpire Routine

The umpire routine at regular intervals throughout the battle considers the mode of behavior of each combat element on the battlefield in turn. The result of such consideration is a list of those enemy units the combat element may be considered as having under observation. The calculations make use of function tables with the dimensions of range, type of unit being observed, its general activity, its previous exposure history, and the activities of the observer. This survey of the battlefield will be programmed to be completed at regular intervals (the order of 1-10 seconds) throughout the battle.

Firing; Disclosed Position Routine

Each individual's knowledge of the enemy may also be added to immediately after an enemy unit fires its weapons (Figure 20). Again the calculation makes use of function tables with substantially the same dimensions as those used by the umpire routine. Data available from field experiments may be used for these calculations.

Communications System

When a unit comes into the possession of new information (ordinarily information about enemy position and type) it may, depending on established communication procedures, transmit this information to selected friendly units. Figure 30 indicates in its upper half the general steps involved in the transmission of this information, and in the lower half gives a compact representation of the combat elements which may communicate directly with one another. Interpretation of this form, which is based on the binary number system used in the computer, is straightforward. Each column is associated with a particular combat element including the unit commanders. Each row indicates by 1's in the appropriate column that combination of combat elements which may communicate with one another. For example, the first column stands for the company commander; the second column

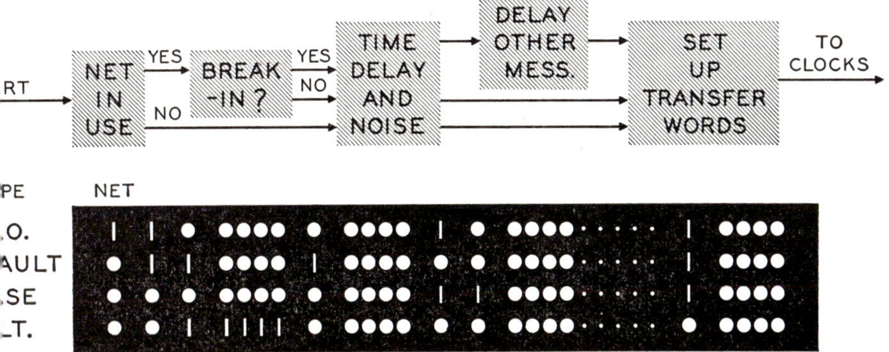

Figure 30. This illustrates the method of simulating the communication system. The upper flow diagram indicates the principal steps in the calculation. The lower half of the figure shows schematically by each row the members of a particular communications net.

stands for the commander of the assault group; the thirteenth column stands for the commander of the base of fire group; and the thirty-second column stands for the commander of the supporting artillery platoon. The first row, therefore, indicates that the above named individuals comprise one communications net. This particular means of representing information facilitates calculations by the computer.

Infantry Units

So far as their movement is concerned, the infantry squads (or, in special cases, platoons) move as a unit from square to square. However, the weapons system is complex, and the different types of weapons must be treated distinctly. In *Carmonette* each member of the fundamental infantry unit (usually the squad) is treated separately so far as weapons fire, ammunition stocks, and casualties are concerned. Figure 31 indicates the compact form used to store data indicating the types of squad members. Each column corresponds to a particular member of the squad as is indicated on the diagram. Again this scheme for storing information is designed to facilitate computer calculation, since it depends on a binary form which is used for all information retained by the computer.

INFANTRY UNIT

SQUAD (TYPE)

PLATOON

*CARRIERS

Figure 31. Forms for recording composition of infantry squad or platoon.

A TACTICS ROUTINE FOR COMMAND DECISION

We have described a model for the simulation of an ordered sequence of combat actions by the combat elements on each side. However, we have not yet described a mechanism that will insure a sensible sequence of actions.

It will be recalled that a number of parameters and coefficients have been introduced in the course of the discussion thus far which can profoundly alter the character of these calculations. These include (1) the "a" coefficients which have a very strong influence on the way in which combat units react to the situation around them in their moving deliberations; (2) the location of the terrain objective which, by its position relative to the enemy, may correspond to such extremes as attack or retreat; and (3) the priority system to be associated with the selection of targets which includes the option to decline to fire. Evidently, if the sequence of actions is to be arranged in a sensible way, we can most easily do so by appropriate adjustment of these general parameters. In other words, we have the *means* of implementing a sensible plan of battle if we can provide for the *generation* of a sensible plan of battle. The *Tactics* routine has this as its primary function.

In effect the *Tactics* routine simulates the deliberations of a commander as he generates a plan of action sensibly related to the course of battle up to that point.

A Hypothetical Battle

To clarify the logic of the decision process let us begin by considering the fortunes and misfortunes of a small armored formation as it strikes deep into enemy territory. Figure 32 indicates the first

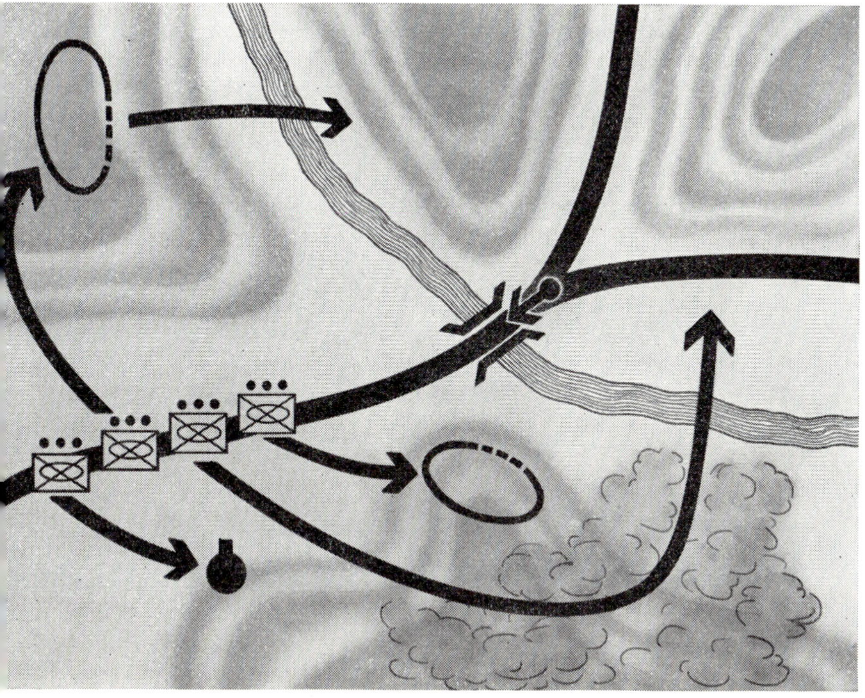

Figure 32. First assault plan implemented by commander.

phase of an engagement forced on a Blue company commander. He is proceeding in column along the road toward the northeast when he is brought to a halt by the fire of a powerful antitank gun in position just to the east of the bridge. We may imagine that the Blue commander surveys the situation and within a few seconds issues orders for an attack as indicated: the first and third platoons in the column to take up covering positions; the second platoon to flank the Red antitank gun on its left; the third platoon to send out a small patrol to investigate the enemy's right flank. The fourth platoon takes up a defiladed position to support the assault by indirect fire. The second phase of the battle develops when, as indicated by Figure 33, the as-

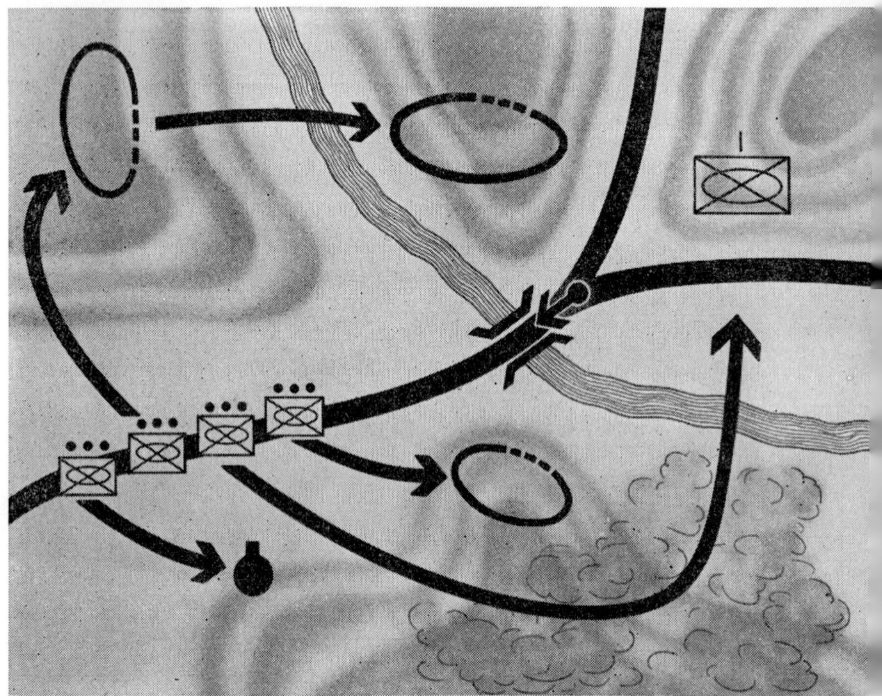

Figure 33. Contact intelligence generated by maneuver.

sault platoon comes under the fire of a strong enemy force believed to be in company strength. Simultaneously, the patrol crossing the river to the north discovers only light resistance in platoon strength. The reaction of the Blue commander is to place a small yield nuclear weapon on the Red position, Figure 34, and withdraw the forward

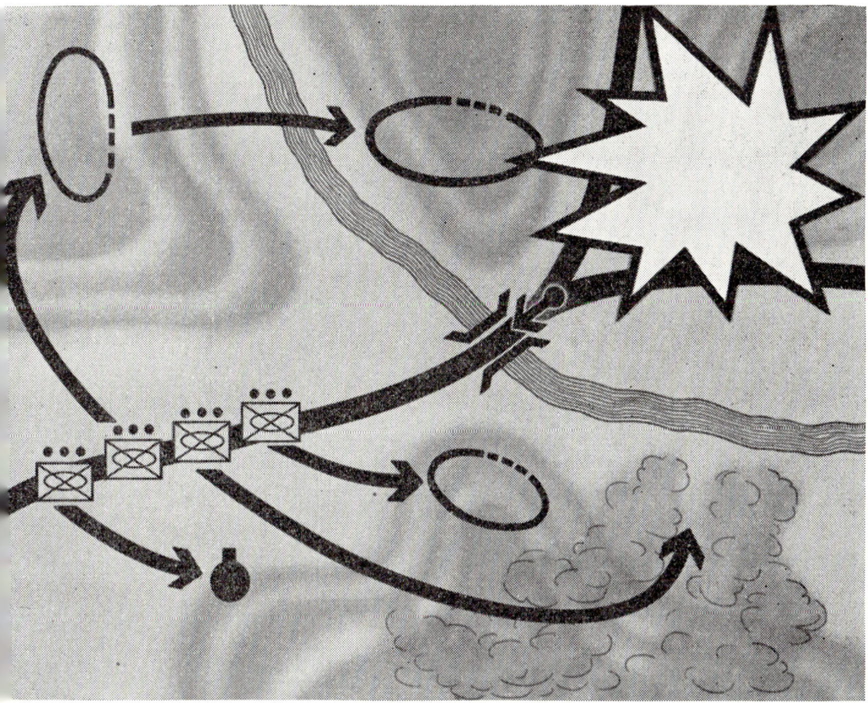

Figure 34. Nuclear fire is placed on enemy concentration.

elements of the assaulting platoon back across the river to take up a covering position. The left platoon then is designated the assaulting group, Figure 35.

During the hypothetical action just described, the Blue company commander twice made a major tactical decision drastically influencing the detailed combat actions of all of his subordinates: first, when the units deployed from their column formation to attack the enemy's left flank; second, when the Blue company commander halted this attack, caused the use of a nuclear weapon, and ordered a new attack on the enemy's right flank. The *Tactics* routine is required to simulate such major command decisions.

If we had inspected the above battle in more detail, we would have noticed a larger number of less drastic decisions by subordinate commanders, decisions having, however, the same general character and

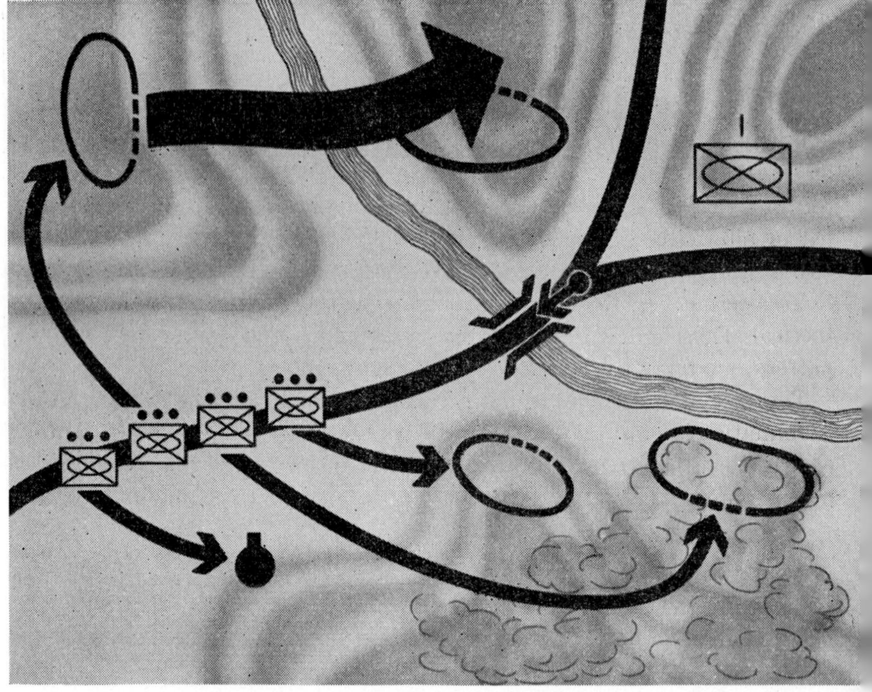

Figure 35. Revised assault plan.

consequences as the company commander's deliberations. The *Tactics* routine must simulate these decisions as well.

Clearly, either of the attack plans selected by the company commander in the preceding example could be implemented by the mechanisms already described for *Carmonette* by appropriate choice of the numbers. Further, the *Carmonette* mechanisms already described provide for explicit description of the known or suspected enemy positions which triggered the company commander's decisions.

The Logic of Decision Process

Consideration of the decision process is aided by listing the factors a commander is taught to consider by the military colleges.

1. Estimate of the Situation
   a. Mission

b. Own capabilities
c. Enemy capabilities
d. Terrain
2. Select and Transmit Decision

Further, the essential elements earlier found to apply to decisions by individual tank commanders were: (1) a quantification of the terrain; (2) the construction of a mutually exclusive and exhaustive list of alternatives (the eight squares plus the square occupied); and (3) a weighting system which described the relative desirability of these alternatives. It may be expected that the routine which effects a tactical decision for the company commander will follow this same pattern, modified as required for the change in scale.

The first requirement is for a scheme to quantify the terrain. An examination of military documents on tactical doctrine shows quite clearly that the quantification must be in terms of irregularly shaped areas, approximately homogeneous with respect to some important

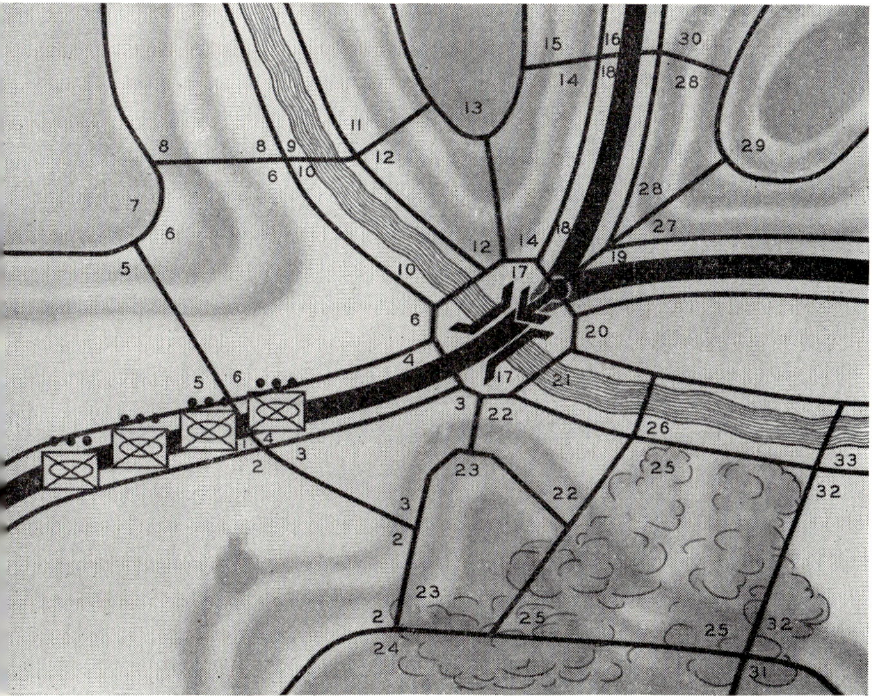

Figure 36. Identification of principal terrain features.

terrain feature such as hilltop, valley, forest, field, or village. Figure 36 indicates how the sample battlefield may be broken up into a small number of such areas. The second and third steps in the tactical decision process will likely involve a consideration of alternative positions and assault routes described in terms of the particular terrain feature areas they involve.

Now, to compose a mutually exclusive and exhaustive list of the alternative assault routes and overwatching positions available to the company commander, using all possible combinations of these tactical areas without restriction, would involve an astronomical number of alternatives. There are uncounted billions of possible combinations of troops and routes implied by the number of tactical areas. We must, therefore, seek an approximation of the list of alternatives which is sufficient for our purposes and which represents fairly the limited number of alternatives actually considered by the company commander.

Alternative Assault Routes

Certain elementary constraints may be placed on the formation of assault routes to help to reduce their number. For example, the constraints applied may be such that (1) each assault route will start at the initial position of the unit and terminate at the terrain objective of the unit, (2) no particular assault route will pass through the same area twice, and (3) no particular assault route will pass other than directly between two adjacent areas. Figure 37 shows about fifty-five possible assault routes which meet the above requirements. Evidently, additional restrictions must be applied to further reduce the number of alternatives.

It is desirable to avoid introducing sophisticated military judgments at this early stage in the *Tactics* routine. We have not yet been able to formulate a single additional rule which is sufficient to reduce further the number of alternative assault routes to the order of ten without in the process discarding the (intuitively) sensible routes. However, it seems likely that by a judicious combination of the requirement "all routes must have no areas in common, except their end points" with certain simple geometrical considerations, the desired reduction may be acquired.

For example, Figure 38 shows five routes which remain when the above argument is applied to the fifty-five assault routes shown in Figure 37. Thus, if one first selects that route which proceeds most

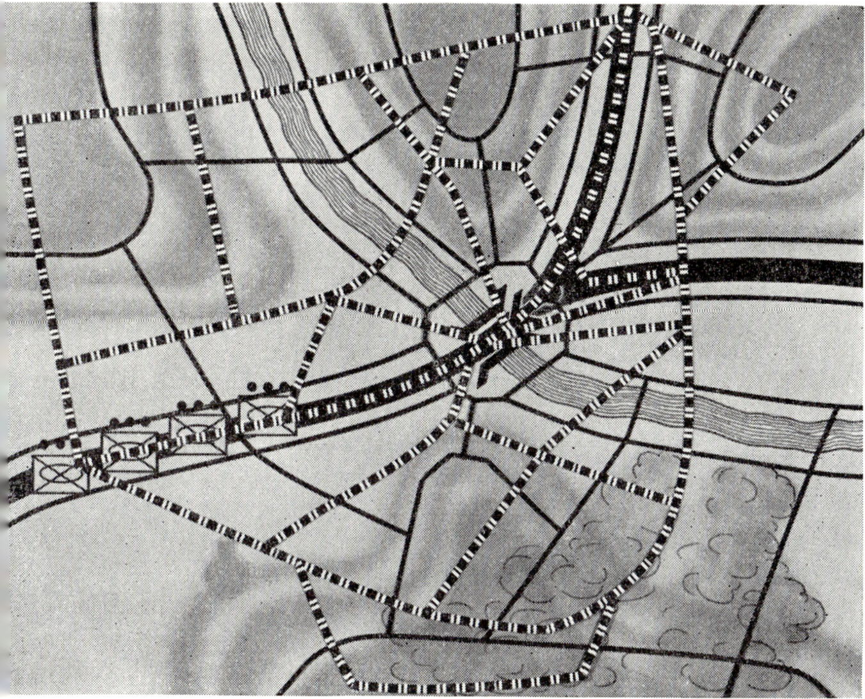

Figure 37. Dotted lines indicate location of all possible assault routes associated with given breakdown of terrain features.

directly from the present position to the terrain objective, and then determines the additional routes which are mutually exclusive in terms of the terrain areas they traverse, then only three assault routes remain. Figure 38 shows these three routes plus two additional routes which are distinct from the other three along most of their length.

Identification of impenetrable barriers and other undesirable terrain features will also be used to reduce the number of assault routes quickly.

It appears that appropriate combinations of the above rules may always be formed which are sufficient to reduce systematically the number of competing assault routes to the order of ten. Clearly the application of rules of the above type may be effected by a digital computer.

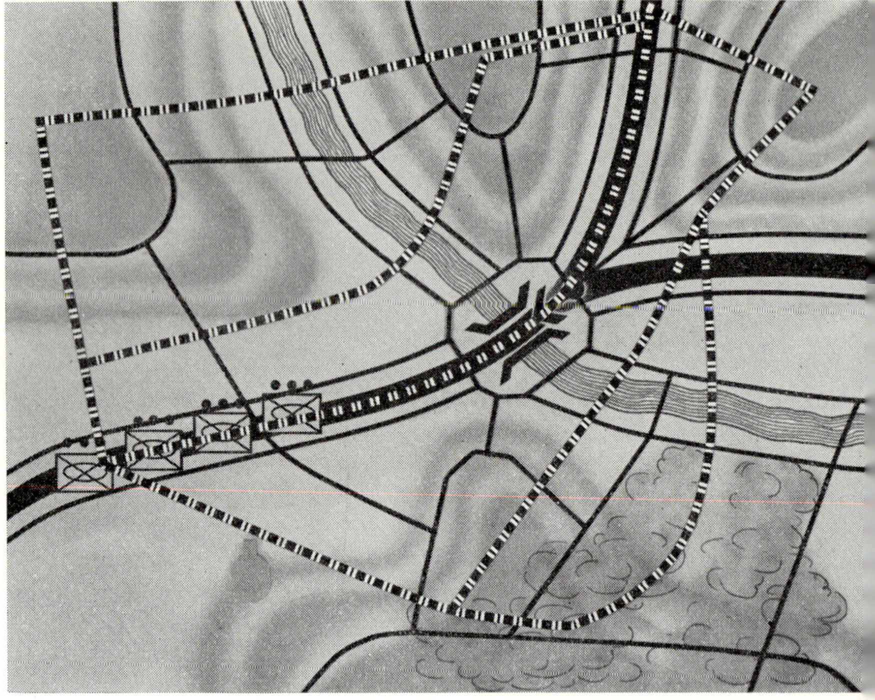

Figure 38. Dotted lines indicate principal assault routes.

Alternative Task Force Organizations

The alternative attack plans considered by the company commander include not only various assault routes but also various combinations of troops to lead the assault while others stay behind and cover the assault with fire. Suppose, for example, that the company commander has six platoons under his command; three tank platoons, two infantry platoons, and one indirect fire or heavy weapons platoon. Various combinations of these platoons could be assigned the several missions. Again the number of possible permutations of these groups when combined with the order of ten alternative assault routes gives rise to hundreds or thousands of alternatives—an unmanageable variety. But military tactical doctrines have much to say as to what are and what are not reasonable combinations of units. For example, a not improbable principle to be applied by the company commander is that tank units are not committed without accompanying infantry.

Further, the indirect fire platoon is always held behind in the support group and small unit commanders may often commit all their tanks and infantry to the attack. If these doctrines are applied to this example, the possible 3-group task forces are

Support	*Overwatch*	*Assault*
S	I	TTTI
S	TI	TTI
S	TTI	TI
S	TTTI	I

These four combinations exhaust the possible three-group task forces composed from the six platoons under the restriction that tank units are never assigned without infantry and may not be held in reserve. If the last rule is relaxed, five additional possibilities are obtained

Support	*Overwatch*	*Assault*
ST	I	TTI
ST	TI	TI
ST	TTI	I
STT	I	TI
STT	TI	I

It should be clear that the application of doctrinal statements in the manner just described illustrates the fundamental justification for the existence of the doctrine. The doctrines taught at military schools are very general and powerful rules, based on much experience, which the inexperienced can apply quickly to a problem so as to reduce it to manageable form. Therefore, the procedure just described is not an arbitrary or artificial one designed merely for matters of economy of computer time or to simplify the problem; rather it is suggested that this process faithfully simulates the general characteristics of one step in the actual command decision process.

Weighting of Alternative Attack Plans

The preceding steps have led to the creation of a limited list of alternative battle plans which are assumed to be approximately mutually exclusive and exhaustive. Therefore, only the third step remains —selection of one of these by a rating process. These ratings will be derived as follows: the computer will simulate the carrying out of each remaining assault plan in gross terms, the units being platoons instead of individual tanks; movement will be from one terrain area to another terrain area instead of from one small grid square to another; casualties will be the expected casualties as a certainty instead of being determined by sampling from the population distribution. In this way there will be determined for each attack plan the cost to a commander of achieving his objective in terms of equipment lost, casualties, and time consumed, together with the corresponding cost to the enemy.

The procedure for doing this was worked out in detail several years ago by Dr. W. E. Cushen (2) and is one of the earlier studies on which *Carmonette* is based.

"Value" of Battle Outcomes

At this point it will be helpful to propose a specific structure for the command decision process. The present proposal is to:

1. Form all possible combinations of routes and sub units
2. Discard those combinations that conflict with doctrine
3. "Simulate" on a gross scale each permissive battle plan
4. Record losses on both sides of time, troops, and equipment
5. Sum "value" of losses
6. Select course of action using some decision rule
7. Communicate operations order

The parts of the *Tactics* routine thus far described have accomplished the first four steps. But one important step remains. In step 5 the equipment and troop losses and time consumed must be combined into a single number which can then represent on a relative scale the "value" of that outcome. Clearly, no detailed calculations are required, but a profound and difficult judgment is inescapable. The company commander does indeed, by some obscure process, add *dollars* to *time* with *lives* at this stage in his deliberations. So must we. Without suggesting that there are available the comprehensive studies which will be required to effect such judgments, it is instructive to consider the initial orders our company commander may have received from his superior. He may have been told, "You must get to the top of that hill in two hours. Everything depends upon it." Or "Your objective is the top of the hill. But tomorrow we will be attacked by the enemy reserves. Therefore, you should try, if at all possible, to keep your losses to 10 per cent." Such statements unmistakably provide the basis on which our company commander will make this final "calculation." The general significance of such "value" concepts has been discussed elsewhere by N. Smith (3).

The remaining two steps are now simple and straightforward. In step 6 one of the remaining alternatives is selected either because it has the highest "value" to Blue or (particularly in the case of the junior commanders) the selection is made by treating the ratings as probabilities.

In step 7 the necessary translation is made between the form in which the tactical decision calculations were carried out and the form required to implement the combat element calculations. Here, for example, intermediate terrain objectives are assigned which will cause subordinate units to move generally along the curved assault route selected. Here also values of the "a" coefficients and similar parameters are selected which will cause appropriate elements to remain in position and provide covering fire. Note that this requires that alternative sets of such values be previously stored in the computer so that one may be selected at this stage.

Initiation of Command Decision

The above procedure for effecting a command decision is sufficient to start the battle. It is necessary, however, to provide the criteria for initiating additional command decisions. Simple measures suggest themselves. For example, a company commander may be required to initiate a command decision calculation if (1) his own casualties reach some threshold value, or (2) if the assault unit fails to meet time deadlines, or (3) if units start running out of ammunition, or (4) if stated numbers of previously undetected enemy units enter the battle. Thus, the *Tactics* routine must have supplied to it continuously throughout the battle summary statistics concerning the course of the battle, limited, of course, by the effectiveness of the communications system. A complete tactical decision is then initiated only when one of these thresholds is passed.

APPLICATION

It is estimated that the complete battle just described will initially take about an hour on the UNIVAC 1103A computer and may eventually be reduced to the order of ten minutes. We may, therefore, expect to be able to play from some hundreds to as many as 10,000 battles. Efficient utilization of this capacity will require, of course, careful attention to the statistics of experimental sampling procedures which have not been discussed at all.

Carmonette *vis-à-vis Field Experiments*

It is obvious that the application of *carmonette* requires the generation of an enormous quantity of input data. Further, the majority of the man-weapon performance data can only be accurately determined by costly field experiments which themselves are, at best, "reasonable" approximations of combat conditions. However, it would be an over-simplification to suggest, either, that a program of extensive field experiments is merely the servant of a series of tactical war games, or that nothing can be done with a tactical war game such as *Carmonette* without field experiments.

It is more meaningful to consider tactical war gaming and field experiments as equal and complementary components of a rational program for the investigation of military problems. Each has the capability of increasing the productiveness of the other program. The tactical war game provides a theoretical structure which sheds light on what constitutes desirable and fruitful experimental programs. The results of field experiments will provide improved input data for the theoretical models and provide the basis on which the theoretical models may be tested for error and corrected. Neither comes first. Both must be developed simultaneously to their mutual advantage.

Interpretation of Battle Outcomes

The interpretation of the battle results which may be obtained is itself a problem as difficult as the construction and playing of the company sized games that have been described. Essentially, each series of games will produce a measure of the cost of the operation(s) under study in terms of equipment, troops, and time. Selection of preferred outcomes and therefore the identification of the preferred weapons systems poses the same question of "value" as was posed during the command decision calculations. This suggests two bases on which the results of a computer battle may be interpreted. (1) The battle is considered to be typical of a long series of battles. In this case the question is merely one of bearable levels of attrition. (2) The battle may be a critical step in the implementation of some higher level and very important war plan. In this case the significance of the results is not fairly measured by the losses suffered during the battle but only by the contribution of the battle outcome to the success of the higher level war plans. In fact, we may expect that our primary concern will be battles of the latter type since the weapons systems under study will include radical doctrines and hardware with potentially drastic influences at every level. If this is the case, then the interpretation of the results of *company* sized war games will require the analysis of *battalion* sized war games. The interpretation of the *battalion* sized war games will require interpretation in terms of their influence on *still* higher level war games. There is, therefore, a requirement for tactical war games at all levels.

A Hierarchy of War Games

Figure 39 extends this notion to its inevitable conclusion. The problems leading to military operations at every level all involve the same

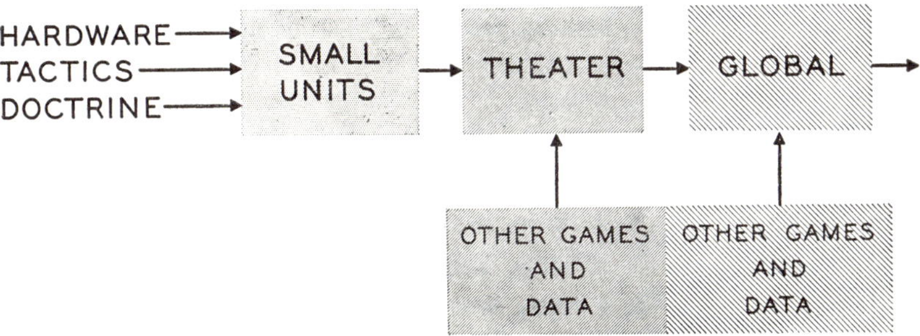

Figure 39. A hierarchy of war games.

degree of complications as impelled us to apply *Carmonette* to the company sized war game. This would lead to the construction of a hierarchy of games, each taking as input the results of analysis of a series of lower level games together with other pertinent information.

The description of *Carmonette,* as treating of a company sized action, is largely a matter of interpretation. If, instead of interpreting the individual combat elements as tanks and infantry squads, these elements were considered to be platoons of tanks and platoons of infantry with the grid squares being 300 yards on the side, then the *Carmonette* battle could encompass a battalion. Ultimately, sufficient data may be amassed to permit interpretation of the individual combat elements in *Carmonette* as divisions with each grid square 10 to 20 miles on a side. At this point, *Carmonette* might conceivably permit analysis of complete tactical operations within a theater.

It is in this more general sense that the logical structure of *Carmonette* is offered as a tool for the analysis of the effectiveness of new weapons systems in their operational context—a tactical war-gaming system applicable to all tactical levels.

REFERENCES

(1) Zimmerman, Richard E. "A Monte Carlo Model for Military Analysis," *Operations Research for Management.* Vol. II. Edited by Joseph F. McCloskey and John F. Coppinger. Baltimore: The Johns Hopkins Press, 1956.

(2) Cushen, Walter E. "Generalized Battle Games on a Digital Computer," *ORO-T-263* (19 February 1954). (Unclassified)

(3) Smith, N. "A Calculus for Ethics: A Theory of the Structure of Value," *Behavioral Science,* I, No. 2 and No. 3 (1956).

Twenty-Five

OPERATIONS RESEARCH IN A HOSPITAL*

CHARLES D. FLAGLE

INTRODUCTION

Operations research, as an essential part of the modern organization, is well established in the military and is experiencing rapid growth in industry. While not yet well publicized, there is an increasing awareness of the need and place of operations research in public service institutions. In these organizations, new analytical problems are posed, for such familiar forces as direct competition and the profit motive gave way to a set of more subtle, though nonetheless forceful, motivations and objectives.

The hospital is one of a class of organizations brought into being by the needs of the community. Relative to industrial or military organizations, individual hospitals seem small, for the total number of people involved in even a large hospital is on the order of several thousand. However, there are approximately 7,000 hospitals in the United States. Over a million people are employed in them and, with expenditures exceeding $5,000,000,000 each year, the total activity is comparable to our largest industries. The very size of the aggregate hospital activity makes it a topic of national importance, and since we are all subject to accident and illness, it is a matter of personal interest to everyone.

Each hospital is a remarkably self-sufficient community, with an intricate and intense set of relationships among several autonomous

* This work has been supported in part by U. S. Public Health Service Research Grants W-98 and W-167 from the Division of Hospital & Medical Facilities and GN-5537 from the Division of Nursing Resources.

personnel groups, the patients, and their families. The historical dominance of sociological and economic aspects of hospital organization is well illustrated in a paper by Edith Lenz[1] of the University of Minnesota.

The widespread growth of "third party" payment plans, the hospital and medical insurance agencies, has radically altered the bases for hospital income from a fee for service rendered to a recovery of cost. This throws new demands upon the financial officers of the institutions for more precise and rational determination and allocation of costs. These same plans have shifted the balance of type of service demanded toward more private facilities and less utilization of the public wards. There is a research need to develop within existing medical and financial statistics a set of sensitive measures of basic shifts in demand to indicate the appropriate direction and degree of management action.

Recently, the effect of new technological developments has begun to influence some of the traditional ways of doing things in hospitals. Through mass production and developments in materials and techniques, principally in plastics, an array of prepackaged, disposable, and in other ways labor-saving supplies, has been made available. A problem of research is to assess the potential benefits of the new technology and to seek to exploit these potentials in reshaping the organization to meet its contemporary challenges.

Today hospitals are plagued by serious economic problems publicly symptomized by continuing shortage of nursing personnel and ever increasing cost of operation. The reaction to the scarcity of hours of professional nursing service has been a proliferation of non-professional jobs, until, in some hospitals, employees now outnumber patients three to one. The Bureau of Labor Statistics reports the cost of hospital care in 1957 to be 149 per cent above the immediate post World War II level. This increase in cost of operation, the difficulty in maintaining stable staffing, and the increasing demands of modern medical care provide the incentive for a great deal of current research in organization and management of hospitals. To one familiar with industrial and military operations research, the difficulties faced in analysis of hospitals are more than offset by some of the unusual opportunities presented. The focus of attention and service upon patients as individuals makes for a structure of relatively small and in-

[1] Edith M. Lenz, "Hospital Administration—One of a Species," *Administrative Science Quarterly*, Vol. 1 (June, 1956), 444-63.

dependently functioning units, identified as clinics and wards. The pattern of organizational structure and procedure is basically similar on these wards, and so are their relationships with the medical and administrative staffs. Thus in study of one unit, one sees an activity repeated again and again throughout the hospital, and to a large extent, all hospitals. A great potential exists for widespread generalizations of results of local studies.

The United States Public Health Service, in its Division of Hospital and Medical Facilities and Division of Nursing Resources, has taken the lead in sponsoring and encouraging the participation of operations research people in hospital research. Under Public Health Research Grants, operations research groups at Ohio State and Johns Hopkins are now in relatively early analytical stages of defining problems, objectives, and measure of effectiveness. The construction of models to date has been principally for the purpose of explaining or understanding the operating complex rather than for optimization of some effectiveness measure.

An interesting example of pioneer research in hospitals and the potential contribution of community management talent to it is the study of Harper Hospital in Detroit, co-ordinated by Marion Wright.[2] Several operations research concepts can be recognized in this study, both long familiar to researchers in public health, the mixed team and the "total system" approach. Here a research group was created for a limited time period for a specific task.

The activity at Johns Hopkins, to be described here, carries the ideas a step further. It makes the research group and its work a permanent and integral part of the organization. As a case study, the early history of this activity may be of interest.

FORMATION OF OPERATIONS RESEARCH AT THE JOHNS HOPKINS HOSPITAL

A significant development in hospital management took place in the summer of 1956 with the formation of an Operations Research Division at The Johns Hopkins Hospital. Although established as an integral part of the Hospital organization, this division was charged not with operating responsibilities, but with an obligation to plan and

[2] Marion J. Wright, *The Improvement of Patient Care* (New York: G. P. Putnam's Sons, 1954).

perform a program of studies of the Hospital's operations and problems. The goal stated for the program was to strive to bring about improvements in operations by providing, through research and analysis, factual bases to aid management decisions, by helping to establish organizational mechanisms whereby essential changes could be effected, and by disseminating analytical techniques and attitudes throughout the organization.

In taking this action, the Hospital followed a lead established by all branches of the military since World War II and recently by many industries. Rapid technological, economic, and sociological change had increased the complexity of executive decisions and control to a point which demanded the development of new techniques and organizational aids. As a response to this need there emerged the profession of operations research and the formation of operations research groups as parts of the modern organizational structure.

The Johns Hopkins University has played a major role in this growth, and the history of this role is relevant to the recent development at the Hospital. The Operations Research Office, Johns Hopkins University, under contract with the United States Army, is one of the pioneer operations research groups in this country. With a large and professionally diverse staff, it brings to bear the gamut of scientific disciplines on problems of national defense. Since its incorporation into the University structure, there has been a kinship between this group and the University's Department of Industrial Engineering, whose chief concern is the study of human organization and the operations or systems of men and machines. On the Industrial Engineering staff, the fields of economics, statistics, accounting, and psychology are represented, along with engineering. This diversity, like that of the Operations Research Office, reflects the philosophy that the study of complex operations is inherently an interdisciplinary one—there are many facets to each problem and the solutions require the contributions of more than one science or art.

The laboratories for the faculty and students of Industrial Engineering have been, as a matter of policy, real organizations. Several years ago, upon the invitation of the Hospital's Director, Dr. Russell A. Nelson, to the Dean of Engineering, Robert H. Roy, students of the University began to make studies of the Hospital's activities. These studies were for dissertation purposes, and, under the traditions of those times, were limited to what one person could do in a one- or two-year period. This early experience demonstrated the potential contribution of people with training in research in organization and management—but not specifically hospital management. At the same time, it demonstrated the difficulties involved in bringing the results of research through to changed organization and procedures. As a step toward co-ordination of the research activity, the Director of the Hospital proposed the establishment of an Operations Research Division in the Hospital, responsible to the Director, headed by a member of the Industrial Engineering faculty.[3]

To the Hospital this meant continuity and co-ordination of research. To the University it presented a mechanism for sending faculty and students of various departments into the Hospital to do research in the field of human organization. To the individual research person it meant a home base within the area of study and knowledge that his project was part of a continuing program.

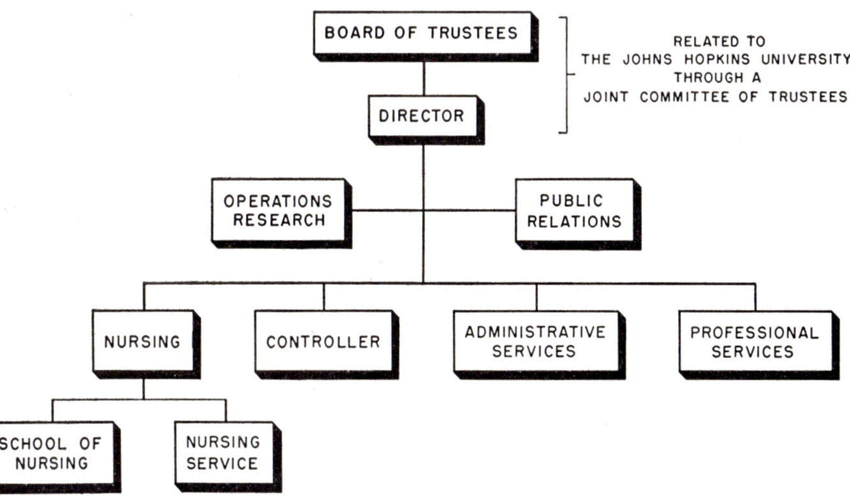

Figure 1. Organization of The Johns Hopkins Hospital.

The merits of the proposal and the potential mutual benefits to the Johns Hopkins institutions were apparent. The activity began officially

[3] The precedent for joint appointment to university faculty and to responsible hospital positions is well established at Johns Hopkins in the chiefs of medical services who are professors in the corresponding departments in the School of Medicine.

with the appointment, effective July 1, 1956, of a Director of Operations Research. By spring of 1958, the number of people involved in direct research had grown to twelve, some working individually, some in groups of four. The formal position of Operations Research within the organization had been established as shown on the organization chart, Figure 1. The Director of Operations Research stands, along with the Director of Public Relations, in staff relationship to the Director and Executive Vice President of the Hospital. The Director, the two staff officers, and four line officers shown in the chart form the Hospital's Administrative Council, which meets formally each week. This group, with the Dean of the School of Engineering, composes the Management Research Committee, chaired by the Director of Operations Research.

INITIATION OF THE RESEARCH PROGRAM

The first year was devoted to the organization of the operations research program—the interpretation of the diverse problems apparent or presented by the staff into a manageable, co-ordinated set of research studies; the attempt to determine the unifying objectives of the Hospital staff; and the establishment of personnel and organizational relationships which would permit prosecution of research and implementation of its results.

This was begun by a series of visits to Hospital areas and interviews with assistant directors and department heads. All had been asked to suggest problems they felt pressing or suitable for study—and these were forthcoming in great quantity and diversity. Should we adopt a newly developed method for measuring body temperature? How can we reduce delays in caring for pediatric outpatients? With whom should responsibility for transporting patients from hospital floors to radiology be placed? How can we reduce labor turnover in the auxiliary personnel group? Can medical records and statistics be adapted for managerial controls? What can be done to raise the practical upper limit of hospital census? What should be the information content of directors' meetings? What is the optimum interval between deliveries of supplies to their points of use?

Obviously the first problem of operations research was to evaluate the many suggestions for studies in light of its own capabilities and to abstract the basic problems. Next, the relative importance of the basic problems had to be assessed in terms of value to the Hospital. To determine "value to the Hospital" one must know Hospital objectives; but this is no simple thing, for the Hospital consists of several autonomous groups, each with long-established practices and traditions, and each has objectives which are not always compatible. Those simple unifying motivations, the profit motive and the threat of direct competition, which help simplify the problem of objectives for industry and the military, do not exist here.

Among the members of the medical staff, the practice and teaching of medicine, the learning of medicine, and medical research are the prime motivations; and the Hospital is seen by this group in the light of these objectives. Among those on the nursing staff, the giving of high-quality care to patients is the major concern, although like the doctors they, too, are concerned with teaching. In administration, the economic stability of the institution, the re-creation of the physical plant, are matters of principal concern.

Fortunately, for all the separate objectives of various groups, there is a central action toward which all effort focuses. This is the care of the patient. The care is given on each floor or ward of the hospital by a small group of nurses and doctors. It is remarkable that if one lists the objectives of any one of the ward groups, he finds them identical to the objectives of the institution as a whole. The care of the patient, the teaching and learning of medicine and nursing, medical research—all these are no more or less the goals of the ward staff than they are of the institution. About this central action of patient care the research program has been built. Almost all of the study problems suggested fitted neatly into research plans centered on patient care. One essential division of effort had to be made, and this related to the two distinct forms of care, inpatient and outpatient. The organizational problems arising from these two types of services contained basic differences. Different groups within the Hospital were responsible for these services.

In December, 1956, four graduate students of the School of Engineering were appointed Operations Research Assistants in the Hospital. This number has since been increased to nine participants from the University, including several from the Departments of Psychology and the Department of Biostatistics in the School of Hygiene and Public Health.

Another move of great importance to the development of research has been the participation of an Assistant Director for Methods Improvement, Nursing Department, in the operations research activity. No other single administrative step has done more to aid the initial efforts of the research group. The principal research efforts, described in the following sections, have been made by this group, assisted by the Hospital's medical, nursing, and administrative personnel.

THE STUDY OF INPATIENT CARE

The basic service of the Hospital is measured in patient days of care. Each patient day of care rendered by doctors and nurses in the inpatient unit is augmented by laboratories, pharmacy, kitchens, medical records, and an array of supply and administrative services. The continuous, hour-to-hour responsibility for patients rests in the hands of the nursing department. The staffing and organizational problems of the area are severe, for although many of the patient care activities can be anticipated and planned, there are some demands of patients which inherently cannot be scheduled. In addition to the variable demands arising from patient care, there are demands placed upon nursing personnel by the supporting services; and these are, more often than not, of a random rather than a scheduled nature. Thus, we have a centralization of responsibility for patient care on the nursing staff without concurrent authority to control the demands impinging from other services. The natural reaction of supervision under such circumstances is to provide a standby labor reserve to meet uncoordinated peak loads. This is generally true, and in most industrial or commercial situations in which load is imposed independently from a number of sources a reserve capacity of 30-40 per cent over normal is provided. This means, in consequence, idleness of facilities or personnel of 20-30 per cent of their scheduled time, which corresponds well with the degree of idleness among nursing auxiliary personnel reported by Cramer[4] and Speth[5] in their hospital studies.

There is a serious economic problem evident on the scene of patient care. In addition to the lost personnel time described above, there is some imbalance between expenditures for materials and personnel

[4] George Cramer, "An Analysis of Ward Nursing," Master of Science Essay, Department of Industrial Engineering, The Johns Hopkins University, 1955.

[5] Albert W. Speth, "A Recent Study at the Delaware Hospital," paper presented at the 16th Annual Conference, Maryland, D. C., Delaware Hospital Association.

time. The traditional daily patterns of patient care and other duties assumed on the hospital floor were established long ago when much volunteer or student labor was available and could be advantageously used to perform such tasks as servicing or cleaning equipment, thereby reducing costs of capital equipment and supplies. This permitted a high degree of self-sufficiency of the unit and was conducive to procedures quite compatible with the traditional, individualized relationship between the patient and his physician.

The impact of economic and social changes in recent years has deprived the nursing unit of much of its free help and has expanded the class of paid aides and orderlies. During the same time, the hospital supply industry has made important gains in development and efficient production of new products and services which are alternatives to work traditionally performed in the hospital. The effect of this relatively increased labor cost in the hospital is to shift the economic balance toward use of labor-saving or labor-replacing equipment and supplies, toward centralizing or grouping of many activities—and away from self-sufficiency of the unit.

To effect such a shift, thereby achieving economic balance and realizing the benefits of technological advance in hospital equipment, while at the same time retaining the essentially individualized nature of patient care, is a matter of reorganization of tasks on the nursing unit and throughout the hospital. The economic balance does not come about simply by piecemeal introduction of labor saving equipment, but can only come with a corresponding reallocation of personnel time.

The paradox of idleness among personnel in an activity which is chronically felt to be short of personnel is symptomatic of the previously described random and variable demands for which the staffing pattern must be designed. Typical of such situations, there is a serious morale problem, which is reflected in a high turnover rate of both professional nurses and auxiliaries. Typically also, the turnover rate is highest among the groups in which idleness is highest.

Four hundred of the nursing staff of 1,000 at The Johns Hopkins Hospital are in a sub-professional, auxiliary category—nursing aides and orderlies. Their duties include some direct patient care, with responsibility ranging from menial tasks up to elements of direct care such as the reading and recording of patient temperature, pulse, and respiration. Among this group, as in similar groups all over the country, annual turnover is 50 to 100 per cent with a drastically higher rate among new employees. The result is a constant fraction, one-third to one-half, of employees with less than four months' experience in the auxiliary group.

The three problems discussed—scheduling of staff, economic balance, and personnel turnover—are all interrelated, and their solution is a single one, the determination of the optimal system of staff, facilities, procedures, and co-ordination with other departments to meet the objectives of patient care under the present state of technical knowledge and economic conditions. The goal of the research program is to determine this optimal system; and although this effort centers principally in the Nursing Department, the Medical Staff, Administrative and Professional Services, and the office of the Controller have all necessarily participated. The attack on the problem is fourfold.

1) *Evaluation of Procedures and Advanced Equipment.* It is not sufficient to evaluate singly an item of equipment or procedure in which an increased first cost is justified by reduced labor costs. In the hospital equipment and procedures under consideration, labor is saved in small increments. The saving becomes effective only when new systems of equipment and procedures are considered which permit true realization of the benefits of time saved, either in payroll reduction or increased direct patient care. For this reason, the evaluation of equipment and procedures is being treated as a co-ordinated program, related to other changes. To date, substantial savings have been found inherent in the use of such prepackaged and disposable equipment as drainage tubing, enemas, and medication cups.

A great impetus has been given to this equipment phase of the problem by the recent creation, within Administrative Services, of a Central Supply Department. This group has taken from the hospital wards all of the burden of sterilizing equipment, and has done much to standardize and provide automatic resupply of sterile items and special equipment.

2) *Analysis of Random or Unpredictable Demand.* There are many duties performed by floor personnel on an unscheduled basis; for example, escorting patients to radiology or various clinics, and special trips to pharmacy or stores. In some cases, a procedure can be revised to eliminate the random nature of a demand. An example of this is a present study to develop a procedure which involves provision of a small inventory of emergency drugs to reduce unscheduled trips to the pharmacy. A second approach—when the process cannot be converted to a scheduled one—is to smooth the effects of randomness by pooling service facilities. This leads to consideration of centralization of some

of the escorting and messenger duties now performed on call by nursing personnel. In addition to taking advantage of the law of large numbers, centralizing such services near the origin of demands permits some rational scheduling, since the demand is more regular at its point of origin than it is at each of its points of focus.

3) *Systematic Pattern of Activities.* In addition to reserve labor required for random demands, an additional need for reserve arises from predictable peak loads. In an effort to assess the effect of the traditional sequencing and scheduling of patient care activities, a study has been made.[6] Definitive observations of time required to perform elements of direct patient care for a sample of approximately one hundred patients have been made for several categories of patient condition, several categories of personnel involved, and several clinics. All observations have been recorded in sequence against time of day. The direct analysis of these observations reveals quantitatively the dependence of the institution on its student and non-professional nursing personnel and the concentration of work load in parts of the day.

At the outset, the recognized categories of patients' conditions were numerous and based on many factors, such as mobility, mental state, ability to see, and need for isolation. As the study progressed, it became apparent that the total nursing effort required for a patient diminished sharply as the patient became able to leave his bed, to feed himself, and to walk. A few objectively discernible physical characteristics formed the basis for three different degrees of assistance needed.

The classification of patients by these categories of assistance needed—labelled total care, partial care, and self care—and the measurement of the fractions of patients in these categories has led to a research result of great interest and usefulness.

From over a year of observation and analysis of the organized professional and supporting activities which constitute hospital care, there emerges one characteristic of the activity which dominates all others in its effect on planning, staffing, and equipping in-patient facilities. This characteristic is the extreme variability of the needs of patients and their demands upon the time of the staff. The problem is basic. It stems from the fact that in any given hospital bed there may be a patient one day who needs constant observation and care, and the

[6] This study is the principal research effort of Mr. Robert Connor, a doctoral candidate. It will be published in total in his thesis and in part in the reports of the research group.

next day one who needs almost no attention at all. The resultant work load of the floor is the statistical sum of the basic variable demands centering on each bed. An example of the magnitude of this variation is demonstrated in the graph, Figure 2, which gives an estimate of the

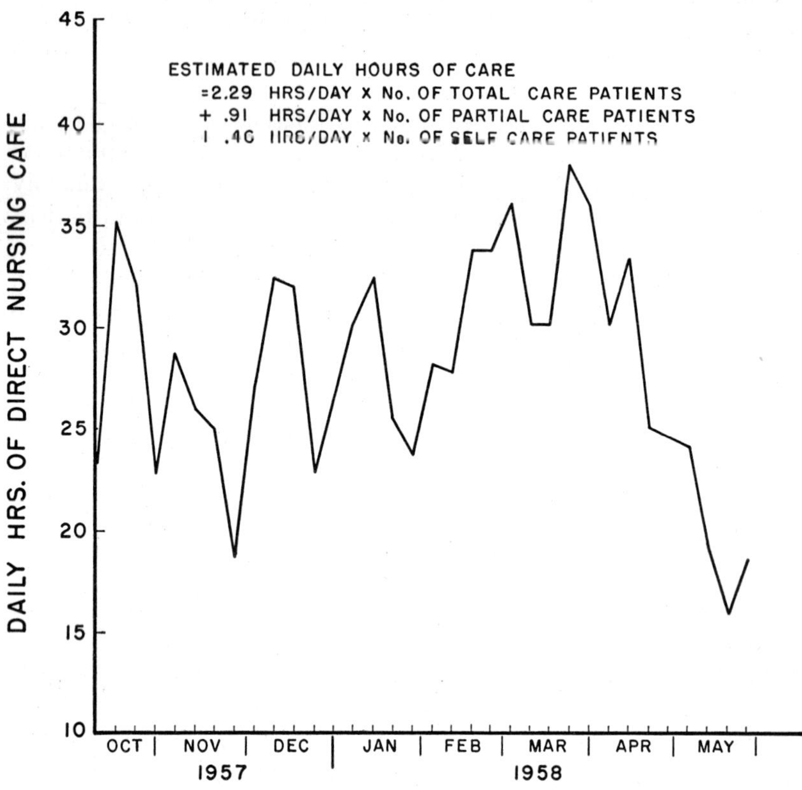

Figure 2. Daily hours of direct nursing care in a medical ward.

daily hours of direct care rendered patients in a 29-bed medical ward. The hours plotted are the aggregate of all time spent by nursing personnel, professional and auxiliary, at the patients' bedside, between the hours of 6:00 a.m. and 12:00 midnight. The work measurement studies previously described in this section have yielded average amounts of time devoted to patients in each of the three categories of total, partial, and self care. The total care rendered to all patients on any given day has been derived by classifying the patients on the ward and allocating to those in each class the average care hours observed for that class.

Note that the hours of care on a day in March were two and one-half times as great as those on another day late in May. While the graph plots only an estimate of hours of direct bedside nursing care, some indirect care, as well as the work of the medical staff and the needs for supplies, are strongly related to this measure. The picture presented is characteristic of hospital care. The illustration could have come from another ward in Johns Hopkins, or another period of time, or another ward in another hospital. Fortunately, the simultaneous observation in other wards has revealed that the peaks and valleys of care needed do not coincide, but occur independently from ward to ward. Thus, there appears the possibility of smoothing the activity in the hospital as a whole by a shifting of personnel, or patients, or both, on the basis of a continuous knowledge of the state of the patients. This implies a radical departure from conventional organization and administration, and the alternative ways in which this may be done will be treated a little later. Before doing so some further discussion of the nature of the patient care burden is necessary.

Note that the range, on Figure 2, within which the direct care varies on the medical floor studied is 15-37 hours a day. (The meaning of these numbers may be sharpened by consideration of the fact that they are only a fraction of the total scheduled nursing hours. For each hour of direct care rendered three nursing hours are scheduled.)

The range of 22 hours between the low of 15 and the high of 37 is large relative to the average of 26 hours of direct care given daily. It is also large relative to the total scheduled nursing hours in the day. The ratio of this range of work load fluctuation to total scheduled hours available may be thought of as a measure of instability of the activity, for the smaller the scheduled nurse hours or number of nurses over which the peak load must be distributed, the greater the apparent burden. This points up an apparent inconsistency in the philosophy of much modern hospital administration and research, including our own. It is unfortunate, but all the desirable administrative contributions in the form of improved architecture, supplies, and administrative technical support which would permit reduction in the number of scheduled nursing hours would now only amplify this instability measure. Obviously we are in need of some systematic recognition for control of the variability of demands for direct patient care.

Certainly there are ways in which the staff of a hospital ward weathers the days of unusual peak demand. Supplies may be borrowed from another floor, and if the situation becomes desperate, some additional personnel may be assigned temporarily. Usually these accommodations of people are met by unofficial and at times *sub rosa* actions on the part of staff. The actions themselves proceed from response to desperate need or some intuitive and often valid expectation of unusual need. The direction in which research is pointing is toward an official recognition and approval of many of these unofficial acts; toward the acceptance of unavoidable variability in work load and the development of rational and quantitative systems to meet it.

The greatest single contribution to the variability of work load is the fluctuation in the numbers of total care patients. As indicated in the legend of Figure 2, patients in this group receive an average aggregate direct care of $2\frac{1}{4}$ hours per day. This is five times as great as the average time spent with self care patients.

The importance of the total care patients is illustrated in Figure 3, in which their number as a percentage of ward capacity is plotted for the same ward and period of time as in Figure 2. As a matter of technical interest, the distribution of the number of total care patients

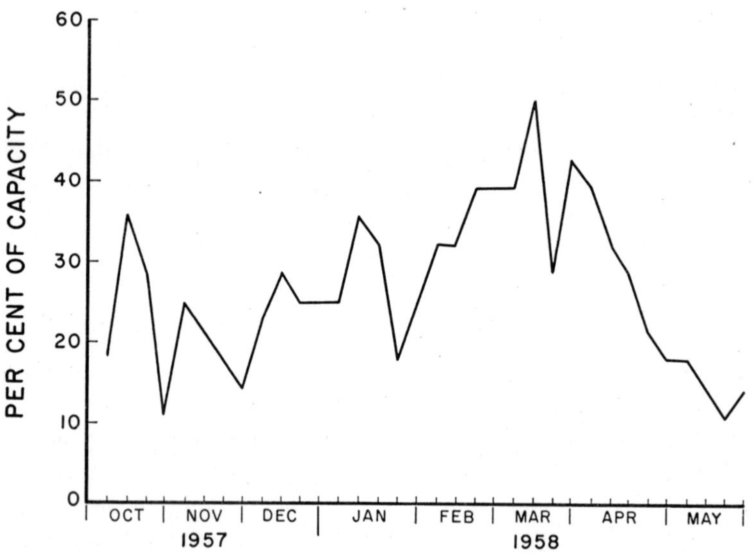

Figure 3. Fraction of intensive care patients in a medical ward.

is remarkably close to a Poisson distribution. This distribution of population is what one would expect theoretically[7] if total care patients appeared in the ward independently and randomly, either through admission or relapse from a less severe state, and if the lengths of stay in this condition are exponentially distributed, and if total care patients preempt the beds of the ward as a matter of policy. In actuality, all these assumptions hold very closely, and in our society there is very little that one could do or would want to do to change them. From the point of view of organization and management, the importance of this characteristic is that the variance of the population is equal to its average. We have as an inherent property of the hospital system a highly variable set of services to perform, from the point of view of the time required to perform them; and this holds even though the total patient population may be quite stable.

Let us consider some of the alternative forms of hospital organization as they appear in light of the variable nature of the job which must be done. The first form is a familiar one we can call "fixed, peak load staffing." This is a perfectly feasible form of organization, one which provides sufficient staff and equipment to cover historical or estimated future maximum demands. This is the simplest form of organization, since it requires no sensitivity to work load fluctuation. However, it has as its chief drawback the fact that it wastes one-fourth to one-third of staff time and inventory in a standby role under normal load.

Sensitivity to varying needs of patients may be transformed to several types of organization. We can fix staff in a given area and admit to that area patients whose demands will match the staff supply. This is the basis for the currently well-publicized "progressive patient care," in which the total, partial, and self care patients are housed in facilities to match their needs.

Conversely to moving patients to match staff, we can move staff to match patients. In a teaching hospital where it is desired to keep a patient in the same unit throughout his stay, a plan we might label "controlled variable staffing" is in order.

There are several modifications of the two alternatives described, each having the characteristic that a constant awareness of the patient

[7] In the chapter on Queueing Theory, page 412, an infinite parallel channel queueing system is shown to have a Poisson distribution of state probabilities when arrivals into the system are independent and random, and lengths of stay in the system are exponentially distributed.

need is an essential part of the control system. Each has the potential of a substantially full utilization of personnel time and equipment inventory. The development of these concepts into a workable system is now a principal effort of the research group at Johns Hopkins.

4) *Auxiliary Personnel Turnover.* The interest in personnel turnover began as part of an attempt to assess the cost of maintaining the staff of auxiliary employees. The problem appeared, however, to be one of such magnitude that it has been given the status of a project in itself. The initial results of this study show an extremely high casualty rate among new employees, as shown in Figures 4 and 5. In the sample of 359 employees, only one in four remained as long as six months. Among those who survived the first four to six months, however, turnover rate is much lower. Attempts have been made to correlate the tendency of employees to leave the job with a variety of objective factors, such as age, education, employment test scores, and domestic situation. Of these factors, only age appears significant—those over 26 years of age provide the stable core of employees. It is apparent that we are dealing here with a large-scale sociological problem, one which can be only partially solved within any one hospital.

Apart from the lost investment in recruiting and training, a high turnover rate has its most serious effect on organization in the constant need for indoctrination of inexperienced personnel. The analysis of this particular experience has led to some mathematical models of the relationship between organizational strength, measured in terms of experience level of its members, as a function of the rate of loss of new and old employees, and the length of time required for indoctrination. These analytical models are of general interest beyond their application to one particular study and are reproduced here.

Figures 4 and 5 are plots of the distribution of employment durations for a group of employees who terminated services during the year 1956. The numbers are real and typical of hospital auxiliary staffs. Figure 4 is a simple histogram; Figure 5 is a semi-logarithmic plot which reveals the negative exponential character of the distribution with two predominant rates. The implications here are that terminations are essentially accidental and independent, and that new employees are more, five times more, vulnerable than those who survive the first four to six months of employment. The representation of distribution of employment duration by a set of two (or more) connected negative exponential segments is of general interest. This

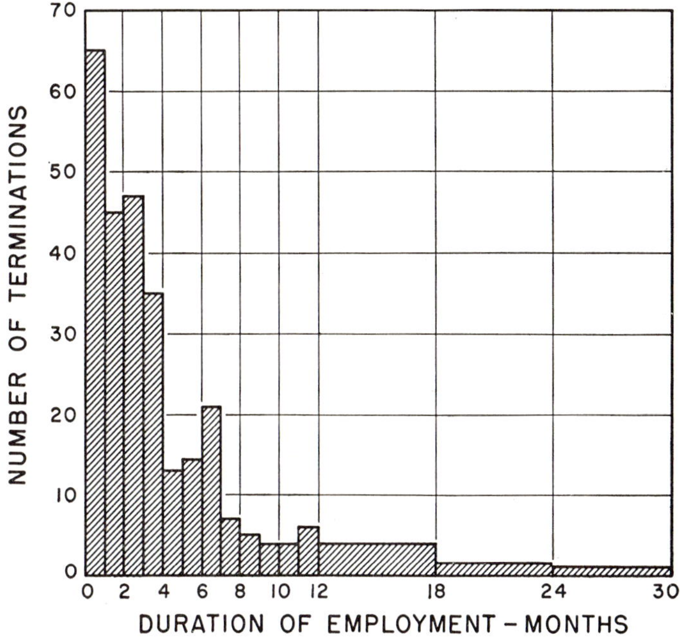

Figure 4. Employment duration of a group of 359 auxiliary nursing staff.

form has been found by other researchers to describe lifetimes of employees in industrial [8] and insurance[9] organizations.

It is the practice to replace as quickly as possible employees who leave, and a simple model of the process is as follows. Consider the work force in two parts, one the new, less experienced, highly vulnerable group. Members of this group who survive a probationary period form the second group, experienced and less vulnerable. Assuming all replacements to be new employees, we can treat the work force as two parts, as in Figure 6.

One measure of organization strength is the ratio of experienced employees. This ratio can be determined as a function of the individual

[8] K. F. Lane and J. E. Andrew, "A Method of Labor Turnover and Analysis," *Journal of the Royal Statistical Society,* Series A (general) 118, Part 3 (1955), 296-323.

[9] Louis L. Miller, "Three Examples of Operations Research in a Life Insurance Company." *Operations Research,* Vol. 5, No. 4 (1957).

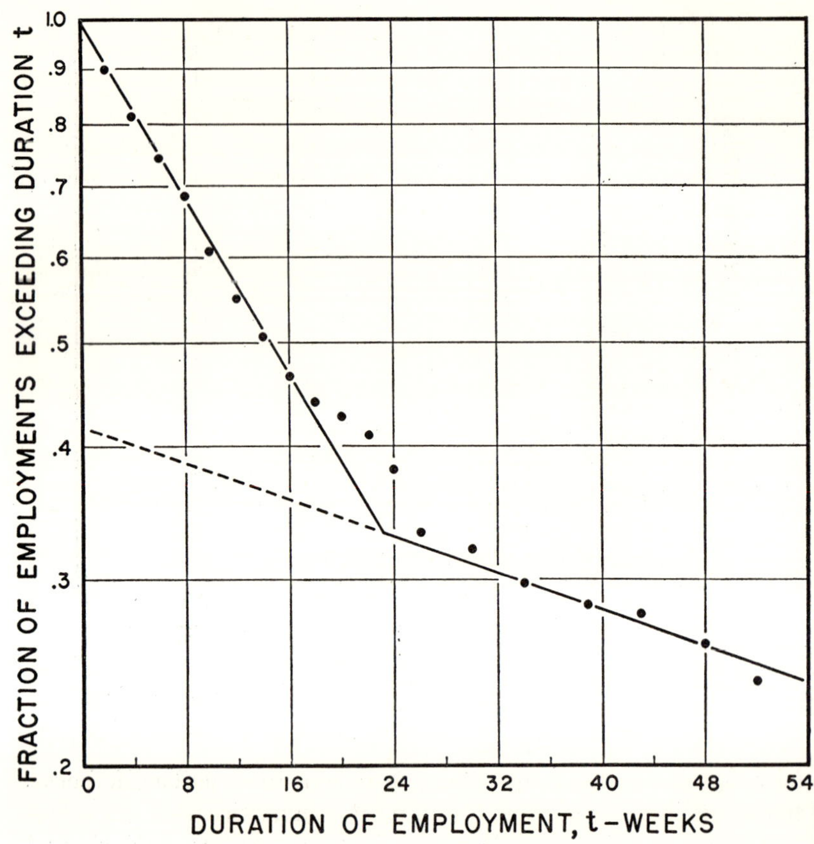

Figure 5. Cumulative employment duration.

termination rates M_1 and M_2 and the probationary period T, as follows.

The probability that a new employee will survive period T is e^{-M_1T} and the rate of survival is λe^{-M_1T}, which in equilibrium is also the termination rate in the experienced group N_2M_2. Thus

$$N_2M_2 = \lambda e^{-M_1T}. \tag{1}$$

Since the replacement rate is just equal to the total termination rate,

$$\lambda = N_1M_1 + N_2M_2 \tag{2}$$

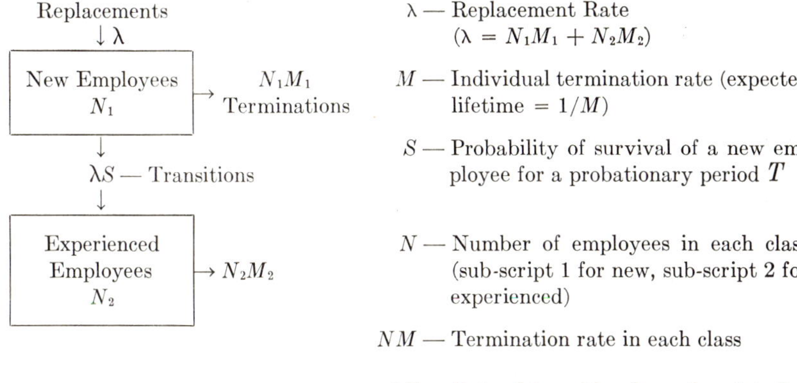

Figure 6. Schematic of personnel system.

Substituting in (1) for λ,

$$N_2M_2 = (N_1M_1 + N_2M_2)\, e^{-M_1T} \tag{3}$$

$$\frac{N_2}{N_1} = \frac{M_1}{M_2} \cdot \frac{e^{-M_1T}}{1 - e^{-M_1T}} \tag{3}$$

In the example at hand, $\frac{M_1}{M_2}$ is approximately 5, the expected lifetime of new employees is equal to the relative probation period (M_1T), and from equation (3), the ratio of old to new employees is a little less than 3.

The critical quantity in the relationship is M_1T. As both the initial casualty rate and the probation or transition period increase, the fraction of experienced employees falls rapidly. The nature of this behavior is shown in Figure 7, which reveals dramatically the necessity for matching long training periods, T, with low initial loss rates, M_1.

Since there is such obvious benefit to be derived from reduction in initial loss rate and transition period, corrective measures focus on training methods and early guidance and motivation. The determination of optimum recruiting and training policies requires the measurement of the effects of several different policies on these rates and comparison of changes in employment durations with the supervisory cost required to bring about these changes.

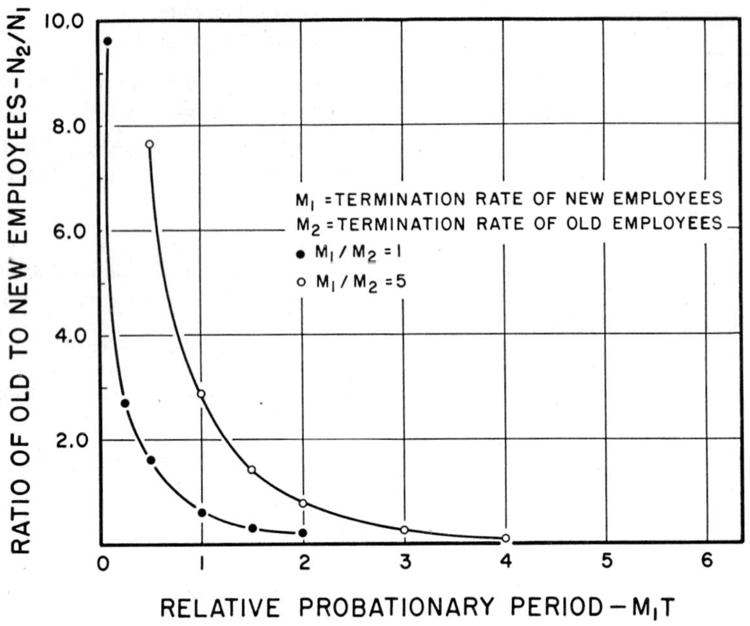

Figure 7. Effect of personnel turnover on employee experience level.

THE DIRECTION OF RESEARCH ON INPATIENT CARE

With a substantial number of items of labor-saving equipment now available and approved, with the major sources of random, external demands on inpatient units identified and plans for their alleviation drawn up, with quantitative knowledge of the personnel time needs of patient care as related to patient state recorded and analyzed, a test for a new organization on the inpatient unit is a logical next step. For this purpose the sixth floor of the Osler Medical Clinic has been designated as an experimental inpatient unit. It is operational in all respects, continuing to carry its normal patient load as revealed in Figures 2 and 3. It receives a greater degree of support from other services than in the past.

The interaction of nursing and other services on the unit are continuously studied to determine an appropriate division of the responsibility and means of communication. Indirect care activities, such as preparation of medications and general traffic of personnel, have been the subject of studies to guide architecture in the layout of nurse's station, patient rooms, and utility areas.

The principal changes which have been introduced in this area have dealt with the distribution of supplies, notably linen, and the assignment of personnel. It is interesting to note that these changes have been applied almost immediately to all of the wards of the Osler building, and at this writing plans are under way to extend them to other buildings of the Hospital. The significance of this is that the only systems or changes worth instituting have proved to be those which are incompatible with the concept of the ward as a self-sufficient entity. Because of the great variation in needs of any small, randomly selected group of patients, stability and effective use of resources can only be brought about by consideration of the largest statistical base possible. It is clear that many of the experimental procedures now in effect in one ward or building must be extended to the whole Hospital to achieve their fullest potential improvement in utilization of the Hospital's resources. It is not inconceivable that even fuller utilization of resources can be achieved by extending some of the systems of allocation of personnel and supply to the community of hospitals.

THE STUDY OF OUTPATIENT CARE

The outpatient may come to the Hospital for only a few minutes of medical attention. Yet the visit must be scheduled—he must pass through an admitting procedure on his first visit; he must pay his fee; he may need to be scheduled for reappointment; and he may need to have a prescription filled. There is often a wait between each of these functions, and the total visit may consume hours of the patient's time.

The aim of the outpatient study is to determine systems which will reduce delay to the patient, reduce congestion of Hospital space by waiting patients, and still assure the physicians an uninterrupted flow of patients.

In general the problems presented by outpatient services involve co-ordination of the flow of patients and the records associated with their cases. Because of the variability in arrival times of patients and the variability of times required to give the services and treatments needed, the analysis of the system is aided by queueing theory. Similar cases have shown the solution of congestion to be a reduction in the number of individual steps at which waiting occurs even at the expense of providing parallel lines of traffic flow. This necessarily involves higher equipment investment, broader training, and more versatility on the part of the staff. Because of the growing demands for outpatient care, there is serious need for practical means of increasing the capacity of clinics. The attempt here is to provide these valid practical solutions to problems while maintaining a research philosophy which seeks knowledge of general properties of the processes being observed.

From a research point of view, there is a basic problem to be solved in outpatient studies. That is the congestion reducing potential of providing several parallel channels of patient flow as opposed to single channels with many stops for services. The analytical aspects of this problem are treated in Chapter 14 in the section on a comparison of series and parallel channel queues, page 414.

Outpatient studies to date have included departments in pediatrics, ophthalmology, and emergency. Like most studies, they have begun with some stimulus of local interest, such as a reduction in delay to patients, but have very quickly produced areas of broader concern. The outpatient departments are a principal source of inpatients, particularly to the wards, and are thus of major importance to the program of medical education. The staffing of the outpatient departments by medical house staffs must be compatible with other demands on this group in the operating, class, and conference rooms. The interdependence of inpatient and outpatient services continually becomes more apparent, and the distinction in the research studies between the two types of services diminishes.

FUTURE OF OPERATIONS RESEARCH IN THE HOSPITAL

In scope, the program to the point described is essentially an administrative one. It investigates principally the organizational and support relationships between nursing and administrative departments and the hospital supply industry. It does not attempt to answer questions about the effectiveness of present patient care. It has some participation by the medical staff in the areas of approval of equipment and perhaps scheduling of some activities. But it does not yet deal with the reconciliation of the economics of hospital operation with the

PRELIMINARY STUDIES

To acquire familiarity with the manufacturing processes used and to acquaint member newspapers with research objectives, an initial survey was made during the summer of 1955.

Visits to a variety of newspapers[4] revealed similar methods of producing display advertising. Most newspapers have a "dispatch" department, into which all advertising material flows and from which all proofs are dispatched to advertisers. This department functions variously but in general serves to record times of arrival and departure, to examine copy, art work, and flat-cast stereotype material for completeness, and to send these components to the departments where they will next be used.

Copy or proofs received by the dispatch department will next be sent to "mark-up," a designation which describes the task of specifying on the copy and layout the type sizes, type faces, and measures in which the various lines are to be composed. Mark-up is thus essentially a planning function which can considerably influence the facility with which subsequent production operations can be performed.

Actual composition of the type follows mark-up. In every newspaper studied most of the lines are composed upon the Linotype or Intertype machine (there is no difference as far as this report is concerned), with the very large display lines being set upon another machine, the Ludlow, or by hand from type cases (again the difference is not significant here). With respect to machine composition, usual practice is to compose the lines in a given ad for which there are matrix magazines on that machine, and then to pass the copy successively from machine to machine, each performing its particular part. By this means magazine changes are minimized, however with much division of labor. Ludlow is composed in the same way, as though the Ludlow were simply another machine for large-size composition.

The slugs of type resulting from this procedure are assembled upon a galley and, when complete, assignment is made to an assembly or

[4] The newspapers visited were these: *Baltimore Sunpapers, Washington Star, Washington Post, Louisville Courier-Journal and Times, St. Petersburg Times* (Fotosetter), *South Bend Tribune* (Fotosetter), *Boston Herald-Traveler, Christian Science Monitor, Quincy Patriot-Ledger* (Photon), Fairchild Publications, *Allentown Call-Chronicle, Wilmington News* and *Journal-Every-Evening,* and *Philadelphia Inquirer.* Visits also were made to Photon, Inc. and to Mergenthaler Linotype Corp.

"make-up" compositor, who also receives or procures any engravings or flat-cast stereos which have been made while composition operations have been proceeding. The make-up man then proceeds to build the ad, arranging the component parts in proper juxtaposition, cutting, mitering, and inserting rules and decorative material, spacing out lines, etc., to the specifications indicated by the layout. Assembly involves from 40 to 60 per cent of all advertising composition time throughout the industry and is the most time-consuming single operation.

Following assembly, proofs are pulled. In many cases these are proofread against original copy and any errors found are corrected before proofs are sent to advertisers. Office corrections, as this stage is called, are not likely to involve the services of a mark-up man but will require machine composition and/or Ludlow, assembly, proofing and proof revision. Newspapers which do not make office corrections before sending proof to advertisers may proofread a duplicate while proofs are out, or they may rely entirely upon the customer to inspect his own ad.

Proofs sent to advertisers are in due course returned, either approved or, as is more likely, bearing corrections. Changes from original copy as opposed to errors made by the newspaper are called "store correc tions," "author's alterations," "reconstruction," or, more eloquently "tear down," and they represent a shockingly large part of advertising composition cost in some newspapers. Again, the correction procedur will involve the entire process of machine, Ludlow, assembly, proofing and proof revision, this time with mark-up almost certainly added This process of preparing successive revises will be repeated, usuall not more than once or twice, until the ad is approved by the custome It is then "released for page make-up," that is, for inclusion in th appropriate issue of the newspaper.

Production control procedures for this process are seemingly quit casual. Deadlines often specify that first proof will be submitted o the second morning (e.g., 9 a.m. Wednesday) for copy received b 6 p.m. (e.g., 6 p.m. Monday), that corrections and revises will be mad overnight, and that release for page make-up must be by 6 p.m. o the day before issue for a morning paper and by 9 a.m. of the date o issue for an evening paper. Internally, the dispatch room will sen art work to the engraving department, mats to the stereotype depar ment, and copy to the composing room with little apparent attempt t direct or to certify that these components will come together at th right time before assembly. That they do is a tribute to the experien and sagacity of composing room foremen and "head ad men" but also to the regularity and similarity of work patterns, and to the fact that such departments as engraving and stereo will promptly do all of the work before them, thereby operating with a minimum of in-process inventory.

As is well known, a given newspaper will vary greatly in size, not only seasonally but from day to day in any single week. Yet newspapers are produced to very rigid final deadlines because of distribution needs related to train, bus, and plane mail schedules, as well as to street sales and home deliveries. Because they are thus engaged in producing variable volume product to fixed terminal times, it is necessary that they (1) work against backlog during slack volume, (2) vary the work force in accordance with varying demand, (3) have stand-by labor during slack periods, or (4) employ some combination of these. We have had no opportunity to measure, but, through observation, have come to believe that stand-by labor is much more prevalent in the industry than is believed by newspaper executives. We have made numerous inquiries on this point but repeatedly have been told that stand-by labor is not a serious problem; yet the promptness with which engravings and flat-cast stereos are made and the absence of formal production control procedures are hard to reconcile on any other basis. Proposals for work sampling studies directed to the measurement of stand-by time have not yet received approval.

Some of the newspapers visited were found to employ job number systems by which each ad is identified and controlled, and a few used numbered job jackets or job envelopes as containers for all component proofs, cuts, etc. This practice appeared to facilitate sound control, has been recommended, and in some instances adopted.

RATES AND SERVICES

Newspapers conventionally sell advertising space at agate column line[5] or column inch rates. There are differences in price by classes of advertiser and discounts for quantity buying (Figure 1) but, within

[5] Most newspaper pages, excepting tabloids, are eight columns wide by about 300 agate lines long (2,400 column lines = 1 full page). There are 14 lines of type in the agate size in one inch, hence 14 column lines = 1 column inch. Since relatively few lines are composed in the agate size, it is chiefly important as a unit of measurement, analogous to the em in commercial printing.

RETAIL ADVERTISING RATES

These rates apply to local retail store display advertising only.

TRANSIENT RATES

1 inch to 99 inches	$3.25 per inch
Political Advertising [Cash in advance with copy]	3.85 per inch
Transient Amusement	3.85 per inch
Church Advertising, services of worship only, No admission charge	2.85 per inch
New and Used Automobile, locally placed, subject to retail regulations	3.25 per inch
General Advertising	3.92 per inch
General Advertising, locally placed, in excess of 1,400 lines, not commissionable	3.33 per inch
Reading Notices	60c per line

Flat Base Rate—$2.85 per inch

This rate is secured by all retail advertisers using a minimum of 100 inches in a calendar year and is subject to the following discounts based upon the volume used during any calendar month.

QUANTITY DISCOUNTS

Monthly Space	*Discount*	*Net*
40 to 125 inches	3%	$2.76
126 to 300 inches	5%	2.71
301 to 500 inches	7%	2.65
501 to 1,000 inches	9%	2.59
1,001 to 1,500 inches	11%	2.54
1,501 to 2,000 inches	13%	2.48
2,001 to 3,000 inches	15%	2.42
Over 3,000 inches	17%	2.37

Figure 1. Display advertising rate card. The prices shown are not current.

a given category, one advertiser will pay the same rate as another regardless of differences in demands for extra service. The extent of these services for which no specific charges are made was found to be quite shocking to newcomers to the industry.

In one newspaper, for example, commercial art and photography are available to advertisers at no extra cost. That newspaper employs a staff of twenty-five such persons who do little else besides draw or photograph merchandise for advertising illustrations. They will perform this service either by visiting the advertiser's own establishment or by having the goods brought to the newspaper office, where one can see suitcases, appliances, shoes, clothing, etc., awaiting such service

though we have no certain means of assessing the accuracy of the data reported, compositor co-operation in union and open shops alike has been all that one could ask.

To facilitate the data collection process, code numbers were devised for each operation and sample instruction forms prepared for preliminary presentation (Figure 2). Two newspapers, hereafter designated Newspapers 1 and 2, readily consented to the measurement procedure and, after introductory conferences at each composing room, data-taking began during the latter part of January, 1956. Arrangements also were made to secure from each newspaper complete copy-layout-proof sequences of each ad. By these means we would have a time record of the production requirements of each job accompanied by exact visual evidence from copy and proof of the operations required. Within the sample periods, data and proofs of from 2,000 to 2,500 local display ads were obtained from each of the two newspapers.

ANALYSIS

The log sheets collected from each of the two newspapers yielded roughly 40,000 to 60,000 entries requiring the subtraction of time intervals. This was done by students employed part-time, with verification of accuracy by a sampling inspection program. Time intervals, operation code numbers, job numbers, and other information were then punched into IBM cards and tabulated as indicated in Figure 3.

While data were being processed in this way, copy-layout-proof sequences were sorted by dates of issue, later by size ranges, and still later partially by complexity, in an effort to discover some meaningful functional relationship by which composition time could be predicted. Many display advertisements look exceedingly complex even in the printed newspaper; in the rougher arrangements of copy and layout they are very much worse, sometimes almost indecipherable.

Finally, this search for a meaningful relationship yielded the concept of an "assembly unit," which came to be defined as "any single type-high unit which must be handled individually by the assembly compositor." Knowledge of the process and of the capabilities of machines in any newspaper permits easy identification of assembly units either in the type form itself or from proof and layout. Examples of assembly units are shown in Figures 4 and 5.

LOCAL DISPLAY ADVERTISING DATA SHEET

DIRECTIONS

1. Each day you will find a data sheet with your time card. Keep this sheet with you and return it at the end of the day to the box near the time clock.

2. The various operations that you perform on the ad during its production are coded by number to facilitate your handling of the form.

3. Number 52 is "Delay" and should be used to show circumstances which prevent you from performing your job. When this code is used please explain the reason in the Remarks column. For example: breakdown; waiting for copy; etc.

4. You may sign your name to the data sheet if you wish but this is not required.

5. The last column will be used by us to compute the elapsed time of each operation.

6. The member of our research team who is working in your composing room will answer any questions you may have about the data sheet or about the project.

7. Sample data sheet.

Job No.	Operation No.	Remarks	Time Start	Time Stop
2057	11		8:00 AM	8:17
	52	Breakdown	8:17	8:32
1983	13		8:32	8:40
	52	Wait for copy	8:40	8:50
2034	11		8:50	9:13
2013	12		9:13	9:19
1995	14		9:19	9:24
2009	11		9:24	9:40

Figure 2. Instructions for self-recording by ad compositors.

JOB NUMBER	OPERATION CODE NUMBER	CARDS FOR OPERATION	MINUTES FOR OPERATION	MINUTES FOR JOB	SIZE OF AD IN COLUMN-LINES
17918	01	01	58		
	03	01	7		
	11	08	29		
	12	06	15		
	21	01	5		
	22	01	2		
	23	01	5		
	31	02	58		
	32	04	37		
	33	01	19		
	41	01	5		
	42	02	6	(215)	
	43	01	4		
	44	01	10		
	45	01	1	261	0315
17919	01	01	53		
	03	01	3		
	11	07	57		
	12	01	3		
	13	01	1		
	21	01	5		
	31	02	97		
	32	01	17		
	34	01	5	(242)	
	41	01	8		
	42	01	2		
	45	01	5	256	0420
17920	01	01	12		
	03	01	2		
	11	03	8		
	21	01	3		
	31	01	20		
	32	02	8		
	41	01	4	(57)	
	42	01	2		
	45	01	2	61	0250

Figure 3. Tabulation of composition times for individual ads. The figures in parentheses show composition time exclusive of store corrections.

Figure 4. Examples of assembly units, identified by arrows. The ad contains 32 assembly units.

Figure 5. Examples of assembly units, identified by numbers and arrows. The ad contains 120 assembly units.

When this concept was first explored, it was hoped that measurement could be achieved by a fixed and variable time relationship, represented by the familiar linear equation:

$$T = k + tU,$$

where T = total composition time, exclusive of store corrections,
k = a constant representing various get ready and put away elements,
t = time per assembly unit,
U = number of assembly units.

An effort to test this hypothesis accordingly was set in motion by counting the units of sample ads from Newspaper 1, now grouped partially by date of issue, partially by size range, and partially by complexity, rather heterogeneously, to be sure. A first attempt was made to plot units and times on rectangular co-ordinate cross-section paper but this yielded great congestion in the points plotted, because wide ranges in both time and numbers of units limited a small portion of the scales to the predominant number of small ads. On this account alone the plot was transferred to log-log paper with salutary effect.

An initial sample of about thirty-five ads, chosen without much plan from among various piles of copy-layout-proof sequences, gave such an obvious trend that a second, quite independent, sample was counted and plotted separately. When the two sheets were placed together and held against the light, registration of the points was very close and a combined plot was thereupon made (Figure 6). To insure an absence of possible bias in the counting of assembly units, all counts were made and recorded before taking time data from the IBM tabulation.

A first attempt to find a simple linear relationship between composition time and assembly units was abortive and led to the fitting of a straight line to the logarithms of these variables. This is the line shown in Figure 6, for which the equation is:

$$\text{Log } T = 0.874 + 0.790 \text{ Log } U$$

While this is not the final model recommended, it is a much better measure than the column line-time relationship conventionally used throughout the industry.

Although the attempt to find a linear relationship was abortive, stratification of the unit counts by ad size did reveal that composition time appears to be a function of size as well as of assembly units. This led finally to a decision to fit a plane in logarithms taking account of

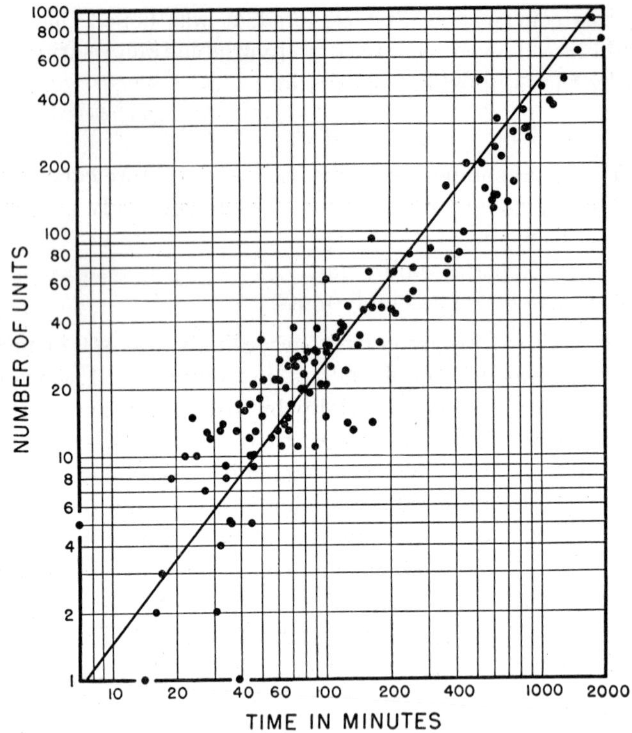

Figure 6. Log-log plot of time vs. number of assembly units.

both size and unit counts. This procedure yielded the following equation for Newspaper 1 (C = size in column lines):

$$\text{Log } T = 0.502 + 0.486 \text{ Log } U + 0.340 \text{ Log } C$$

Later, the same procedure gave this equation for Newspaper 2:

$$\text{Log } T = 0.616 + 0.557 \text{ Log } U + 0.228 \text{ Log } C$$

Apart from the fact that subsequent additional measurements in twenty-four newspapers have consistently confirmed the general form of the model:

$$\text{Log } T = k_1 + k_2 \log U + k_3 \log C,$$

it also rationalizes very well. In the composition of any ad, visible assembly units which print (that is, those which are "type-high") must be separated by spacing material which cannot be counted from proof. The amount of this will be a function of size, as well as a function of the number of assembly units because of factors related to complexity, spatial relationships, decipherability and legibility, weight, etc.

TESTS OF VALIDITY

As stated, the three-variable logarithmic model seemed reasonable in the light of composing room operations, and the values calculated for the constants did not seem out of line. The model produced a satisfactory coefficient of correlation in both cases, 0.87 at Newspaper 1 and 0.85 at Newspaper 2. The success of these preliminary tests suggested additional measures of the effectiveness of the model as a predictive device.

A measure of the relative effectiveness of the model in prediction is afforded by comparing it with the conventional, presently used column

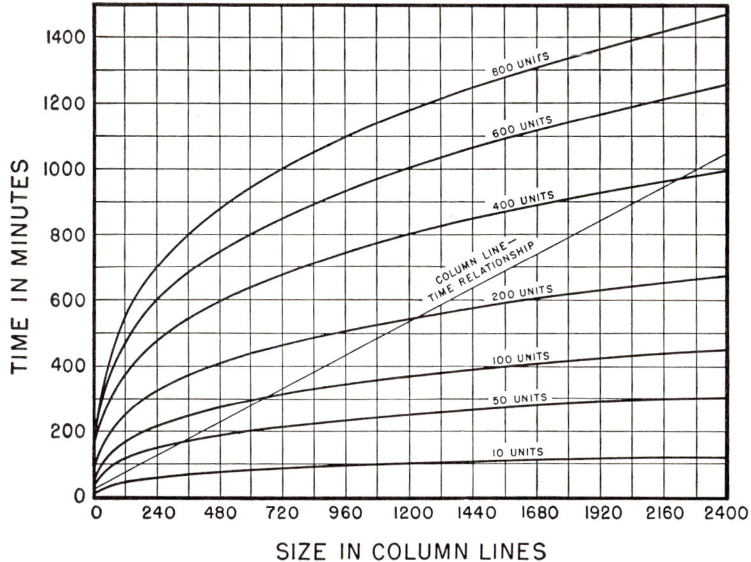

Figure 7. Comparison between the three-variable model and a column line-time model. Although the column line-time model has been assumed, it is clear that the three-variable model will measure more accurately than *any* column line-time model.

line-time model.[7] Figure 7 graphically represents the three-variable model in lines of constant units against an assumed column line-time relationship derived from the data, and Tables 1 and 2 show deviations from actual composition times of times predicted by each of the two models. Table 1 shows data from within the sample group used in deriving the three-variable model at Newspaper 1, and Table 2 shows data from a sample of thirty ads not used in the derivation. The aver-

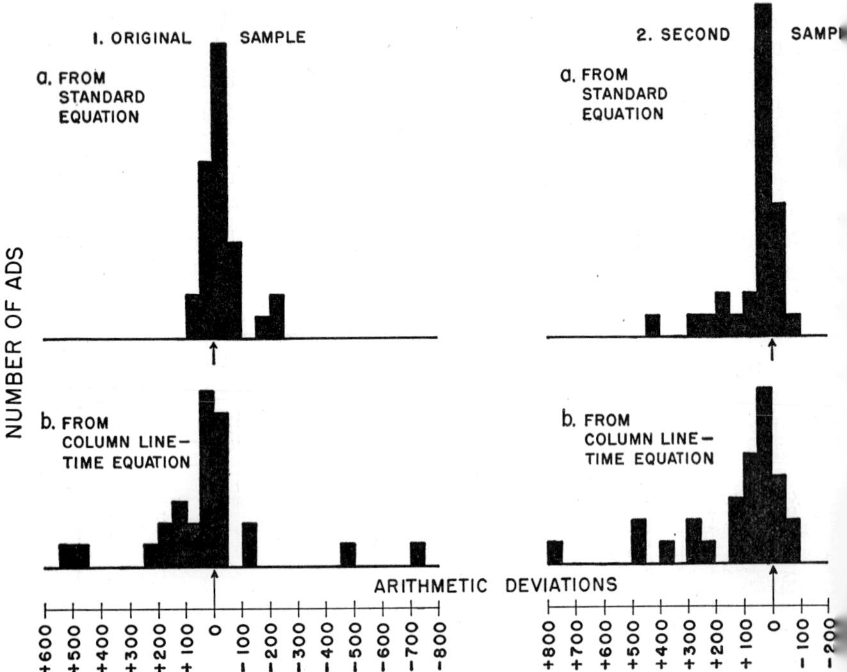

Figure 8. Histograms of the deviations shown in Tables 1 and 2.

age deviation in the column line-time relationship in Table 1 is 73.9 per cent; that for the three-variable model is 30.1 per cent. These figures in Table 2 are respectively 73.5 per cent and 34.1 per cent. This superiority also is revealed by Figure 8 which shows distribu-

[7] It is not quite correct to speak of a "presently used column line-time model," because no conscious use of *any* model is made. However, if one may assume that advertising rates are intended to be proportional to cost, then it is clear that the rate card shown in Figure 1 reveals a column line-time relationship.

Table 1. Deviations from Actual Composition Time

Comparing the Column Line–Time Relationship and the Three Variable Model

				Col. Line-Time			3-Variable Model		
Job Number	Size in Col. Lines	No. of Units	Actual Time* Minutes	Minutes	Deviation	% Deviation	Minutes	Deviation	% Deviation
52251	14	8	19	26	+7	36.8	22	+3	15.7
29827	300	46	182	149	−33	18.1	155	−27	14.8
25153	100	13	60	63	+3	5.0	54	−6	10.0
25301	360	39	119	175	+56	47.1	150	+31	26.1
54086	354	24	125	172	+47	37.6	114	−11	8.8
32633	168	12	128	93	−35	27.3	61	−67	52.3
09353	1150	146	605	513	−92	15.2	452	−153	25.3
29761	2160	99	441	946	+505	114.5	446	+5	1.1
37067	40	22	61	38	−23	37.7	54	−7	11.5
33317	640	20	79	295	+216	273.4	124	+45	57.0
25446	2400	260	880	1048	+168	19.1	788	−92	10.5
24097	2400	153	555	1048	+493	88.8	587	+32	5.8
44877	684	45	201	314	+113	56.2	199	−2	1.0
31867	130	15	102	76	−26	25.5	63	−39	38.2
31371	224	37	71	116	+45	63.4	125	+54	76.1
24011	276	1	39	139	+100	256.4	18	−21	53.8
30177	620	36	118	286	+168	142.4	170	+52	46.8
24101	258	11	89	131	−42	47.2	67	−22	24.7
30271	28	15	24	33	+9	37.5	39	+15	62.5
42063	1200	215	659	535	−124	18.8	567	−92	14.0
25293	630	32	179	290	+111	62.0	160	−19	10.6
24517	572	280	756	276	−480	63.5	518	−238	31.5
17981	2400	865	1776	1048	−728	41.0	1535	−241	13.6
29786	70	37	157	51	−106	67.5	86	−71	45.2
37666	70	12	83	51	−32	38.6	46	−37	44.6
25006	84	1	21	57	+36	171.4	12	−9	42.9
23967	98	4	32	63	+31	96.9	28	−4	12.5
16629	224	2	31	115	+84	271.0	73	+42	135.5
18729	16	13	28	27	−1	3.6	30	+2	7.1
29692	56	25	67	45	−22	32.8	64	−3	4.5

Column Line-Time Average Deviation = *73.9*
Three Variable Average Deviation = *30.1*
* The actual time given *excludes* store corrections.

Table 2. Deviations from Actual Composition Time

Comparing the Column Line–Time Relationship and the Three Variable Model

				Col. Line-Time			3-Variable Model		
Job Number	Size in Col. Lines	No. of Units	Actual Time* Minutes	Minutes	Deviation	% Deviation	Minutes	Deviation	% Deviation
31890	1512	217	604	668	+64	10.6	614	+10	1.7
36380	1800	291	502	791	+289	57.6	765	+263	52.4
17530	2400	115	295	1048	+753	255.3	501	+206	69.8
42090	170	8	33	93	+60	181.8	49	+16	48.5
38840	360	7	67	175	+108	161.2	58	−9	13.4
38830	1088	80	222	487	+265	119.4	318	+96	43.2
37740	480	48	134	226	+92	68.7	184	+50	37.3
37780	1620	68	219	714	+495	226.0	330	+111	50.7
24480	540	30	152	252	+100	65.8	147	−5	3.3
42160	200	8	43	106	+63	146.5	51	+8	18.6
25219	60	16	41	46	+5	12.2	51	+10	24.4
42280	1350	251	559	599	+40	7.2	642	+83	14.8
42099	420	65	264	200	−64	24.2	209	−55	20.8
36319	1500	193	174	663	+489	281.0	574	+400	229.9
38839	600	46	152	278	+126	82.9	193	+41	27.0
37749	112	13	78	69	−9	11.5	56	−22	28.2
29769	18	8	34	28	−6	17.6	24	−10	29.4
42279	85	21	55	57	+2	3.6	67	+12	21.8
37739	50	22	69	42	−27	39.1	58	−11	15.9
42489	1470	216	445	650	+205	46.1	607	+162	36.4
42154	2400	281	653	1048	+395	60.5	822	+169	25.9
31404	270	11	66	136	+70	106.1	68	+2	3.0
37734	100	13	45	63	+18	40.0	54	+9	20.0
41744	118	12	38	71	+33	86.8	54	+16	42.1
31634	67	25	41	49	+8	19.5	68	+27	65.9
31554	318	32	128	157	+29	22.7	129	+1	0.8
25344	165	37	148	91	−57	38.5	113	−35	23.6
36554	100	25	63	63	00	00.0	77	+14	22.2
31983	56	18	44	45	+1	2.3	54	+10	22.7
36553	168	34	100	39	−7	7.0	109	+9	9.0

Column Line-Time Average Deviation = *73.5%*
Three Variable Average Deviation = *34.1%*
*The actual time given *excludes* store corrections.

tions of the deviations for the two methods. The suggested model is clearly a better predictive device than the conventional column line-time assumption.

The comparison of actual times and predicted times for the individual ads of the 200 ad samples made it possible to estimate the "standard error of estimate" for individual ads. Using one-sigma limits, the lower limit was 66 per cent and the upper 153 per cent of predicted time at Newspaper 1. At Newspaper 2 the comparable figures were 63 per cent and 160 per cent. These were standard error of estimate limits for individual ads; for groups of ads or the sample as a whole, the likelihood of the predicted times being more nearly aligned with actual times is increased in proportion to the square root of the size of the sample.

Similar standard error of estimate limit analyses have been made at all newspapers tested. In every case the standard error fell somewhere between 64 per cent and 73 per cent at the lower limit and between 137 per cent and 155 per cent at the upper level. The striking agreement of these figures at all papers is an additional indication that the form of the model is generally applicable to newspaper composing room operations.

MEASUREMENT AT ADDITIONAL NEWSPAPERS

Development of the three-variable model at Newspapers 1 and 2 was followed by similar measurements at six additional newspapers of widely different size and location. Because the original operator-log technique of taking data yielded such large quantities of information and involved high cost for data-processing, it was decided to take additional, independent data by an alternative, less expensive means. This method, which consists of randomly selecting a sample of ads as they are received in the dispatch department, has come to be called the job-log method, because it involves time-recording, not by all operators for all ads, but only for the operations performed upon ads selected in the random sample. In this case the job-log form (Figure 9) travels with the ad in such a way that it represents the record for that ad alone.

This technique obviously eliminates the processing of vast quantities of data but at the cost of casting additional doubt upon accuracy of the record. Compositors keeping a log of their own activities

1001
JOB NUMBER

3 x 10
SIZE

Reed
ADVERTISER

2/14
INSERTION DATE

OPERATION CODE NUMBERS

MARK-UP	MACHINE	LUDLOW	AD ASSEMBLY	PROOFREADING
1 Original 2 Office Corrections 3 Store Corrections	11 Original 12 Office Corrections 13 Store Corrections	21 Original 22 Office Corrections 23 Store Corrections	31 Original 32 Office Corrections 33 Store Corrections	41 Original 42 Office Corrections 43 Store Corrections

Operation Code No.	Remarks	Date	Time Start	Time Stop	
1		Feb 10	8:01	8:12	
11		"	9:20	9:33	
11		"	9:52	10:07	
21		"	10:18	10:24	
31		"	1:05	1:48	
41	2 readers	"	2:02	2:07	
2		"	3:10	3:14	
12		"	4:19	4:21	
32		Feb 11	7:55	8:01	
42		"	9:13	9:15	

Figure 9. Job-log form.

throughout each day would perforce report all of their time, and it could be hoped that the "in between" operations of change-over, clean-up, etc., would be charged equitably against ad production. Besides, the operator-log method gave no special distinction to any ad; the compositor reported every ad and could have no way of knowing which would be used in analysis. The job-log method loses both of these advantages. Since wanted ads are differentiated by the presence of log sheets, it could be expected that this would enhance motivation and therefore attract special attention and handling with probable exclusion of delay elements.

Nevertheless, because of the far greater economy of the job-log method, it was tried at Newspapers 4 and 6, where operator-log data had also been taken. As was expected, the job-log method yielded lower constants than the operator-log data, but application of the various tests for validity showed slightly better consistency and led to a decision to adopt this approach in newspapers subsequently measured.

In each case, using a table of random numbers, 350 ads were sampled at each newspaper, data from these were carefully scrutinized, and a sample of 200 ads were used in deriving the equation. The job-log method has now been applied to a total of twenty-eight newspapers, making thirty-two papers in all which have been measured.[8] In every case for simplicity in computation and use, each equation is reduced to a factor table like that shown in Table 3.

[8] Measurements and factor tables have been completed or are in progress in these newspapers:

Birmingham News
Chicago Tribune
Cincinnati Enquirer
Decatur (Ill.) *Herald and Review*
Des Moines Register and Tribune
Detroit News
Duluth Herald, News-Tribune
East St. Louis Journal
Hackensack Bergen Evening Record
Hartford Courant
Lancaster Intelligencer-Journal; New Era, News
Louisville Courier-Journal and Times
Miami Herald
Milwaukee Journal
New Haven Register
New York Times
Norfolk Virginian-Pilot, Ledger-Star
Oklahoma City Times and *Daily Oklahoman*
Olympia (Wash.) *Daily Olympian*
Philadelphia Inquirer
Rock Hill (S.C.) *Evening Herald*
St. Paul Dispatch and Pioneer Press
San Diego Union & Evening Tribune
Santa Monica Evening Outlook
St. Petersburg Times
South Bend Tribune
Tallahassee Democrat
Toronto Star
Washington Post and Times Herald
Washington Star
Wilmington (Del.) *News* and *Journal-Every-Evening*
Wilmington (N.C.) *Star News*

Table 3. Newspaper 2—Factors for Calculating Standard Times

Units	Factor	Units	Factor	Units	Factor	Units	Factor	Col-Lines	Factor	Col-Lines	Factor
1	4	52	37	200	80	552	140	7	1.6	450	4.0
2	6	54	38	210	82	566	142	10	1.7	500	4.1
3	8	56	39	220	84	580	144	14	1.8	550	4.2
4	9	58	40	228	86	594	146	20	2.0	600	4.3
5	10	60	40	238	88	610	148	21	2.0	650	4.4
6	11	62	41	248	90	624	150	28	2.1	700	4.4
7	12	64	42	258	92	640	152	30	2.1	750	4.5
8	13	66	43	268	94	654	154	40	2.3	800	4.5
9	14	68	43	278	96	670	156	42	2.3	850	4.6
10	15	70	44	290	98	686	158	50	2.5	900	4.7
12	16	72	45	300	100	702	160	56	2.5	950	4.8
14	18	74	46	312	102	716	162	60	2.5	1050	4.9
16	19	76	46	322	104	732	164	70	2.6	1150	5.0
18	21	78	47	334	106	748	166	80	2.7	1250	5.1
20	22	80	48	346	108	766	168	90	2.8	1350	5.2
22	23	84	50	356	110	782	170	100	2.9	1450	5.3
24	24	92	52	368	112	800	172	110	2.9	1550	5.4
26	25	98	54	380	114	816	174	120	3.0	1700	5.5
28	26	106	56	394	116	832	176	130	3.0	1850	5.6
30	28	112	58	404	118	850	178	140	3.1	2000	5.7
32	29	120	60	418	120	868	180	150	3.1	2200	5.8
34	30	126	62	430	122	884	182	160	3.2	2380	5.9
36	31	134	64	444	124	902	184	170	3.2		
38	32	142	66	456	126	920	186	180	3.3		
40	33	150	68	470	128	938	188	190	3.3		
42	33	158	70	482	130	956	190	200	3.4		
44	34	166	72	496	132	974	192	250	3.6		
46	35	174	74	510	134	992	194	300	3.7		
48	36	182	76	524	136			350	3.8		
50	37	192	78	538	138			400	3.9		

To use this table select the next *lower* number of column-lines and units, and multiply the factors for each. For example, assume that a given ad contains 1400 column-lines and 752 assembly units. The respective factors are those shown for 1350 column-lines (5.2) and 748 units (166). The product of these factors, $5.2 \times 166 = 863$ minutes, the standard composition time for the ad.

REFINEMENTS IN THE MODEL

Subsequent to development of the three-variable model described above, refinements for improving accuracy and utility have been proposed. With the co-operation of Newspaper 4, where the suggestion originated, investigation was made of the practicality of developing a similar model for the assembly operation alone, in order that measurements could be made of assembly compositors as an important separate group in the composing room. This was done. The same model was found to apply, with smaller constants and lower values of T, of course, but with greater accuracy than for all operations together, as might be expected. The decision accordingly has been made to offer each newspaper measured two equations and two factor tables: one for the process as a whole and a second for the assembly operation alone.

As a corollary to this second study at Newspaper 4, it was decided to investigate the practicality of distinguishing "angle sections" from other assembly units, which are normally horizontal or vertical. In printing of all kinds, lines of type and other elements are "justified" to fill designated measures; the basic form taken by all assemblies in composition is therefore a rectangle, and the need to turn any typographic element at an angle other than 90 or 180 degrees poses out-of-the-ordinary problems to the compositor. An "angle section" was defined as "any typographical assembly which must be turned at an angle in such a way as to require the cutting and inserting of wedge-shaped spacing material." These are readily distinguishable in type (Figure 10).

Analysis of the extra time required by angle sections has been made in several newspapers by the simple calculation:

$$A = \frac{L - S}{n}$$

where A = time per angle section,
L = actual composition time from the job-log sheet,
S = standard time calculated from the factor table,
n = number of angle sections.

A final value or values for A has not yet been determined or announced because of the paucity of information from individual newspapers and the range of variability between them. However, it is quite clear that the angle section refinement is both practicable and mean-

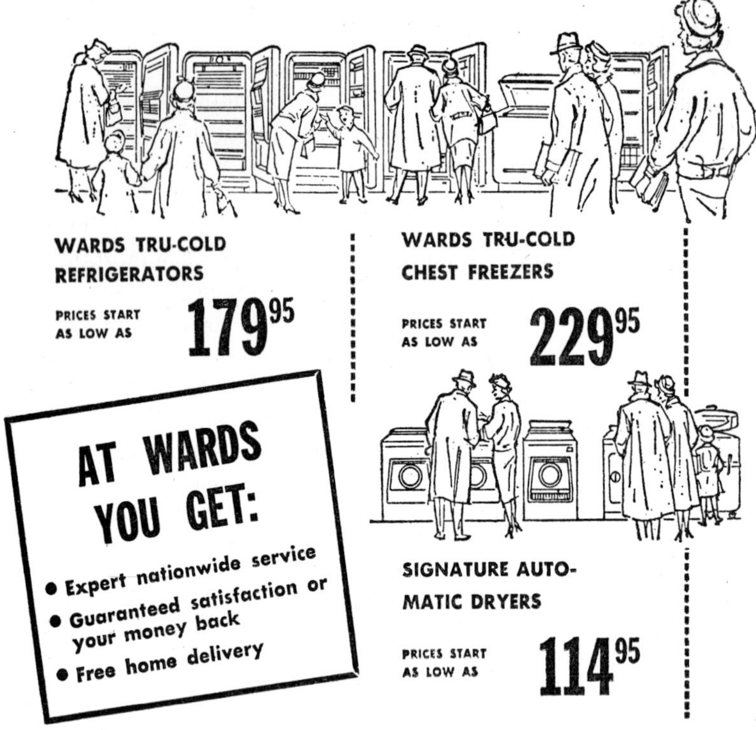

Figure 10. Example of an angle section.

ingful. In thirty-nine of forty-three ads showing angle sections, positive values (that is, additional times) have been found for A, with an average of about thirty minutes per angle section. Table 4 shows the marked improvement in accuracy resulting from adding angle section times to the factor table measurements of such ads.

It has been decided, therefore, to modify the model by introduction of the angle section concept and the general form is now:[9]

$$\text{Log}(T - nA) = k_1 + k_2\text{Log } U + k_3\text{Log } C$$

[9] Since this was written, a still further refinement in the model has been made as a consequence of special studies of ads composed from copy subjectively judged to be poor. In general, such ads are characterized by illegibility and poor spatial relationships caused by crowding a large number of assembly units into

Table 4. Improvement in Measurement of Ads Containing Angle Sections

Newspaper	Number of Ads	Standard Time per Angle Section in Minutes	Ad Sample Standard Times in Minutes: Without Angle Section Factor	Ad Sample Standard Times in Minutes: With Angle Section Factor	Standard Time Improvement in Per Cent
4*	43	20	3305	5689	72*
14	13	60	5237	8053	54
13	6	10	406	482	19
16	10	15	2091	2405	15
17	15	5	1369	1510	10

* Assembly operation only.

It is expected that the accumulation of additional data will yield a better value, or values, for A. Application of the refinement, it may be added, is quite simple; the factor table is used as before and then the extra time for angle sections, if any, is added to the product of the factors.

EFFICIENCY MEASUREMENTS

With the measurement technique just described, meaningful tests of newspaper efficiency in ad composition become possible for the first time. Internewspaper comparisons and comparisons of operations in different time periods for the same paper both are measures of efficiency. The internewspaper comparison is more glamorous and eye-

---

the available area. Measurement of these has been greatly improved by adding a fourth term to the model, so that it now reads:

$$\text{Log}\,(T - nA) = k_1 + k_2\text{Log}\,U + k_3\text{Log}\,C + k_4(\text{Log}\,U)^2$$

The value of k_4 is small (about 0.1), hence this term has but little effect upon T for small and medium values of U. For the large values of U found in complex ads, measurement has been substantially improved. Such ads comprise a very small proportion of the total number composed but are important sources of both trouble and loss.

catching, but the intranewspaper measures are likely to be of greater continuing importance.

The analysis at each newspaper results in a factor table like that shown above, but the values of the factors are different for each individual newspaper. Application of these factors to ads of representative size and complexity permits performance comparison among all of the newspapers measured in the manner shown by Table 5. Each newspaper may gauge its own operating effectiveness by comparison with the others.

Taking a sample ad of 600 column lines with 90 assembly units, Table 5 shows that the time required for composition varies from 92 minutes at Newspaper 6 to 297 minutes at Newspaper 9, or by a factor of more than three. The relative rank of the papers differs in a few cases for different size classes and complexity ranges, but there is a surprising degree of uniformity in relative position. It is clear that the larger papers require more time to compose ads than the smaller papers, although there are several in the class A linage group where performance is as good as in the class B papers. Small newspapers (the class C group has not been shown in Table 5) are known to have less sophisticated requirements in terms of diversity of type faces and type sizes than in the larger composing rooms. The problem of store corrections aud store revises is so much less apparent in the smaller newspapers that the factor tables for Newspapers 5, 6, 7, and 17 *include* store correction time, and this means that their performance superiority is greater than that indicated by Table 5. Some of the small newspapers measured do *not* proofread or correct their ads before proof is sent to customers and this absence of proofreading results in substantial economies. These variations in procedure help to explain but do not altogether account for the large differences in composition time.

Some explanation of a newspaper's relative standing is provided by Table 6. It shows how each paper performs in each of three categories of composition activity. Mark-up at Newspaper 9 was done by apprentices, who specify type size, face, and measure on an average of only 1.8 assembly units per minute, while this same operation at Newspaper 4 proceeds at a rate almost six times as great. Since specialists in mark-up are used at Newspaper 4, there is ground for surmising that mark-up there is not only done faster but is also done better for the performance of subsequent operations.

The range in assembly is not as great, from 2.5 units per minute at

Table 5. Standard Time Comparisons between Newspapers for Ads of Like Size and Complexity (not all measured newspapers are shown)

CLASS A NEWSPAPERS (over 20 million lines in 1957)

			Newspaper Number										
Units	Col. Lines	Col. Inches	1	8	9	14	15	18	23	25	26	27	28
5	14	1	18	20	18	18	12	8	14	10	11	7	11
10	28	2	31	31	30	31	20	14	25	20	22	13	20
20	56	4	55	49	58	57	37	27	42	35	39	24	40
25	100	7	72	61	80	74	47	37	54	49	54	32	55
40	250	18	124	95	143	123	77	66	86	79	91	51	95
50	450	32	168	120	189	155	101	86	110	109	128	68	132
90	600	43	246	167	297	237	152	142	163	168	202	110	205
150	900	64	367	228	460	364	231	227	230	265	317	169	311
200	1200	86	470	269	595	466	286	302	287	340	414	218	410
250	1800	129	593	331	786	588	360	384	355	433	525	270	536
300	2400	171	719	385	945	697	423	476	418	525	640	326	648
400	2400	171	818	431	1124	844	505	580	483	623	785	394	765
500	2400	171	917	470	1292	979	581	676	544	722	911	454	877

CLASS B NEWSPAPERS (10 to 20 million lines in 1957)

			Newspaper Number										
Units	Col. Lines	Col. Inches	2	4	6	7	10	12	16	19	20	21	22
5	14	1	18	10	5	8	10	8	11	10	6	25	26
10	28	2	32	16	11	15	17	17	23	20	12	42	42
20	56	4	55	31	22	25	24	34	39	36	24	70	70
25	100	7	70	38	27	36	32	48	53	48	34	87	91
40	250	18	119	60	43	64	58	85	86	81	59	140	145
50	450	32	148	80	59	87	81	121	113	109	77	181	189
90	600	43	215	120	92	125	119	201	168	176	132	271	269
150	900	64	320	174	138	179	171	315	249	280	205	377	382
200	1200	86	400	216	179	229	223	424	319	360	273	477	468
250	1800	129	495	264	230	294	283	566	403	458	358	586	581
300	2400	171	590	312	275	356	345	692	483	552	435	688	682
400	2400	171	684	369	329	397			571	676	538	800	778
500	2400	171	779	412	383	432			653	787	635	899	854

Table 6. Number of Assembly Units Handled per Minute, by Operation, Based on 200-Ad Sample

CLASS A	1	8	9	14	15	18	23	25	26	27	28							
Mark-up	3.5	9.1	1.8	2.1	5.0	5.4	3.8	3.6	4.3	13.4	3.4							
Assembly	0.8	1.0	0.9	0.6	2.1	1.2	0.9	0.9	0.9	1.9	0.8							
Proofreading	3.5	7.5	5.9	6.2	11.0	7.4	7.0	5.5	9.3	15.4	5.3							
CLASS B	2	4	6	7	10	12	16	19	20	21	22	24						
Mark-up	5.0	10.6	6.8	8.0	4.3	5.7	2.7	5.4	8.2	1.8	1.3	5.9						
Assembly	0.8	1.9	2.5	1.5	1.1	0.6	1.0	0.9	1.2	0.6	0.5	1.0						
Proofreading	7.0	3.6	66.5	10.6	19.8	10.3	8.0	5.8	45.4	6.5	5.7	11.4						
CLASS C	5	13	17															
Mark-up	2.5	8.9	12.7															
Assembly	1.1	1.1	1.5															
Proofreading	19.3	6.2	17.6															

Newspaper 6 to 0.8 at Newspapers 1 and 2. This is about 3 to 1 but, since composition time on assembly is greater than that for any other operation, the differences are very large and significant in absolute terms. Data for proofreading reflect not only differences in operator performance but also the economical absence of proofreading and correcting in some of the smaller newspapers.

The internewspaper differences are dramatic, but the continuing check on changes in effectiveness within a paper that is now made possible is probably of greater significance for control purposes. Efficiency analysis for successive periods becomes available since output can now be measured in meaningful quantitative terms. Actual labor time for production of a period is compared with the total of standard times for the period's output as calculated from the factor tables.

This is *not* a recommendation for conventional job cost accounting which requires compositors to record time start and stop for each ad worked on. Instead, the compositor need keep track of his time only by classes of work. At least three classes of work would be distinguished: productive time spent on local display ads, productive time spent on other kinds of work such as news, classified, etc., and indirect labor time. Since many compositors work full shifts without changing work classifications, time recording is likely to be kept to a practicable level. If store corrections are a significant factor, as they frequently are, they will have to be accounted for separately. Many papers now keep track of the time spent on store corrections by individual ads, and this procedure is recommended for general adoption even if efficiency measures are not used.

The intranewspaper efficiency measure can be stated in ratio form. If the following symbols are used:

E = efficiency for the period.
L = total composition minutes spent on local display advertising.
S = total store correction minutes.
T = total composition minutes estimated by the equation.
I = total indirect labor minutes.

Then the ratio is:

$$E = \frac{T + S}{L}$$

To secure control over indirect labor time, a second ratio may be desirable:

$$E = \frac{T + S}{L + I}$$

Changes in these ratios from period to period give a meaningful picture of the trend of productive efficiency.

COST-REVENUE COMPARISONS

The availability of a factor table for the computation of standard minutes for ad composition makes possible a comparison of cost and revenue for individual ads or groups of ads. By a series of allocations, average cost per productive minute in the composing room, average cost per column line in the other production departments, and selling cost per ad can be calculated. Since the size of each ad is known and the number of minutes required for its composition can be calculated by use of the measurement techniques described previously, the total cost of any ad or group of ads can be determined. Revenue for each ad is known from billing information, so a comparison of cost and revenue can be made. The comparison is especially meaningful when applied to groups of ads, such as all ads of a certain advertiser, or of a certain size range, or of groceries, department stores, or other types of businesses.

Table 7 is an illustration of the allocation of actual [10] costs at Newspaper 2 for an average month, so that unit costs per productive minute, per column line and per ad can be calculated. The accounting system of the newspaper studied classified all salaries and other expenses in five departments: advertising sales, production, circulation, editorial, and business. The salaries and expenses of administration, accounting, building maintenance, and general overhead were included in the business department total. Advertising sales costs are divided into four sub-departments: local display advertising, national advertising, classified advertising, and travel advertising; the production department

[10] All data in this figure and in the other calculations that follow have been multiplied by a figure known only to the research worker who gathered the data. All pertinent numerical relationships have been accurately preserved and all comparisons are thus based upon an actual situation. The computations for this newspaper are set forth in greater detail in "Newspaper Local Display Advertising—A Comparison of Cost and Revenue," by Joseph H. Shaffield, Sidney Davidson, and Irvin W. Kues. New York: ANPA Research Institute, 1958.

	Expense Item	Advertising Sales Department				Production Department			Circulation Depart-ment	Editorial Dept	Business Dept	Check Totals
		Local Adv Dept	Class Adv Dept	Nat'l Adv Dept	Travel Adv Dept	Composi-tion Room	Stereotype Depart-ment	Press Room				
	Salaries—Regular	$11,122	$5,346	$2,860	$1,943	$89,924	$19,328	$22,408	$33,154	$44,526	$9,965	$247,940
	premium	74	107	18		2,071	426	1,003	643	854	446	5,642
	sick					2,965	241	549	505			4,260
	vacation		157			4,953	921	660	758	819	361	8,629
5	General Salaries					995	208	245				1,448
	Payroll Taxes										6,388	6,388
	Other Taxes										1,798	1,798
	Insurance—company										862	862
	employees										3,821	3,821
10	Legal and Auditing										1,196	1,196
	Stationery and Supplies	188	138	27	73						888	1,314
	Postage								1,616		516	2,132
	Telephone and Telegraph									9,233	1,682	10,915
	Promotion	85	32	1,598	294						1,767	3,776
15	Employee Welfare, etc.										55	55
	Donations										740	740
	Depreciation—building										3,352	3,352
	equipment					2,345	468	3,053				5,866
	furn. & fixtures										846	846
20	Bad Debts Provision										1,548	1,548
	Pension Provision										13,536	13,536
	Heat										1,043	1,043
	Water										101	101
	Repairs, bldg.										2,183	2,183
25	Newsprint							122,391				122,391
	Ink							2,148				2,148
	Metal					188		188				376
	Power					1,058	529	1,059				2,646
	Repairs, equipment					604	302	100				1,006
30	Travel	519	176	163	289					578		1,725
	Photos, cuts and mats	444		27	174					4,031		4,676
	Correspondence									4,212		4,212
	Collection Fees										306	306
	Features									2,868		2,868
35	Car Allowances								4,391			4,391
	Transportation								17,198			17,198
	Extraordinary Handling								2,019			2,019
	Supplies					255	1,108	636	2,045		900	4,944
	Miscellaneous		79	131	9	25	24	24	406	176	446	1,320
40	TOTALS	$12,432	$6,035	$4,824	$2,782	$105,383	$23,743	$154,276	$63,735	$67,297	$62,810	$502,317
	Bus. Dept. Allocation											
	Basis: space	1,206	363	301	147	3,834	756	7,325	1,748	2,678	(18,358)	0
	Basis: salary	1,992	998	512	346	17,959	3,759	4,425	6,239	8,222	(44,452)	0
	TOTALS	$15,630	$7,396	$5,637	$3,275	$127,176	$28,258	$166,026	$70,722	$78,197	0	$502,31
45	Circ. & Ed.								$148,919			
	Circulation Revenue								(103,642)			(103,642)
	Deficit								$ 45,295			$398,693
	Allocation—Circ. & Ed. Deficit Basis: Total Expense	2,003	948	722	420	16,300	3,622	21,280	(45,295)			0
50	TOTALS	$17,633	$8,344	$6,359	$3,695	$143,476	$31,880	$187,306	0			$398,693
	Pages of Advertising	970.44	293.02	231.36	22.66							
	Dept'l Allocation to LDA	$17,633					$23,315	$119,782				
	Basis for conversion of LDA costs to unit costs	$390,000 of revenue				1,137,600 prod. min.	1,156,763 CL	1,156,763 CL				
55	UNIT COSTS	4.5% of revenue				$0.1261/ min.	$0.0202/ CL	$0.1035/ CL				

is divided into composing room, stereotype department, and press room.

All of the costs of running the newspaper for the period—slightly over $502,000—are divided among the several departments as shown on lines 1-40 of Table 7. Business department costs are then allocated among the other departments, partly on the basis of space occupied and partly on the basis of departmental salaries. Line 44 shows the total of newspaper costs for the month divided among the advertising sales department, the production department, the circulation department, and the editorial department.

One of the most controversial questions in the allocation process is the treatment to be accorded the circulation and editorial departments. In Table 7 these departments are treated as service departments. Their net cost after deducting circulation revenues, and without any of the expenses of the production departments being assigned to them, is allocated to the advertising sales and production departments on the basis of total expenses in those departments. The other chief alternative would involve treating the circulation and editorial departments as a revenue department and assigning it an appropriate share of production department costs, and not allocating its loss to other departments. This would result in lower unit costs per minute and per line and a very large loss—over $45,000—in the circulation and editorial departments.

With circulation and editorial department deficits allocated as shown in line 49 of Table 7, the total newspaper costs less circulation revenues are divided between the advertising sales and production departments. Unit cost calculations can then be made. All of the costs of the local advertising sales department, of course, relate to local display advertising. Ads are assumed to bear responsibility for the costs of this sales department in proportion to the revenue they produce. Department cost of $17,633 is divided by local display advertising revenue of approximately $390,000 to produce an allocation rate of 4.5 per cent of revenue.

Since none of the costs of the production departments were assigned to the editorial department, all of the production costs must be allocated to the four classes of advertising. Press room costs and stereotype costs were allocated to ads on the basis of space occupied. Rates of $0.1035 and $0.0202 per column line for press room and stereotype, respectively, or a combined space rate of $0.1237 per column line, were derived by dividing total department costs by advertising linage.

Table 8. Detailed Comparison of Revenue and Expense for 10 Sample Ads

	A	B	C	D	E*	F	G	H	I	J	K	L	M
Ad No.	Ad Size (Col. Lines)	Column Line Rate	Revenue: (A × B)	Assembly Units	Comp. Time (Std. Min.)	Store Corr. Time (Min.)	Total Time in Comp. Rm. (E + F)	Comp. Rm. Costs: (G × $0.1261)	Ad Size × $0.1237 (A × $0.1237)	Adv. Sales Costs: (C × .045)	Total Costs: (H + I + J)	Profit or Loss: (C − K)	Percent Profit or Loss: (L/C)
1	364	$.270	$ 98.28	80	157	3	160	$20.18	$ 45.03	$ 4.42	$ 69.63	$ 28.65	29
2	336	.280	94.08	36	120	46	166	20.93	41.56	4.23	66.72	27.36	29
3	63	.280	17.64	19	61	2	63	7.94	7.79	.79	16.52	1.12	6
4	14	.280	3.92	16	41	0	41	5.17	1.73	.18	7.08	−3.16	−81
5	28	.280	7.84	16	47	0	47	5.93	3.46	.35	9.74	−1.90	−24
6	322	.265	85.33	53	146	17	163	20.55	39.83	3.84	64.22	21.11	25
7	1690	.201	340.37	133	348	98	446	56.24	209.05	15.32	280.61	59.76	18
8	84	.300	25.20	29	80	0	80	10.09	10.39	1.13	21.61	3.59	14
9	2384	.265	631.76	336	625	66	691	87.14	294.90	28.43	410.47	221.29	35
10	2380	.257	611.78	160	415	0	415	52.33	294.41	27.53	374.27	237.51	39

* 6 min. added for make-up.

Composing room costs are almost all labor-time related, so a composing room cost per minute was calculated. Since no composing room costs were assigned to the editorial department, total composing room costs were divided by minutes spent in composing advertising in order to obtain the rate. A breakdown of composing room time between news and advertising was not available for the paper studied, but total composing time was allocated on the basis of relative news and ad space. A composing room cost rate of $0.1261 per minute then resulted.

The cost of any ad can then be calculated to be:

Selling cost of 4.5 per cent of revenue,
+ space cost of $0.1237 per column line,
+ time cost of $0.1261 per minute as predicted by the factor table.

Revenue data for individual ads were available, so profit or loss on any ad could be calculated. Table 8 is a detailed comparison of revenue and expenses for ten sample ads. Two hundred actual ads of this period were similarly analyzed. Profits on these ads ranged from a positive 61 to a negative 81 per cent of revenue. Losses were suffered on 13 per cent of the ads studied. Row A of Table 10 shows average percentage of profit or loss for various size classes, utilizing the cost allocations just described.

Alternative procedures of cost allocation are possible for many of the items shown in Table 7. However, in only two of the areas are there plausible alternative procedures that would produce substantially different results. The areas are those of local display advertising sales costs and editorial and circulation costs.

In the original allocation, the sales department allocation was made on the basis of ad revenue. A different assumption that is sometimes made is that all ads bear equal responsibility for the costs of this department. Under this assumption, department cost of $17,633 is divided by the number of ads, 3,588, to obtain a fixed selling cost of $4.90 per ad. Using this figure, percentage profit or loss was again calculated for each of the ten sample ads of Table 8. These profit or loss percentages are given in Row B of Table 9. In the broader 200-ad sample, 22 per cent of the ads produced losses. The relative profitability by ad size classes for this larger sample is indicated in Row B of Table 10.

If it were decided to treat editorial and circulation departments as a separate revenue department for which profit or loss was calculated independently, then that department should be assigned its "appropriate" share of production costs. Press room and stereotype costs

Table 9. Per Cent Profit or Loss by Alternative Cost Allocations for 10 Sample Ads

Method of Cost Allocation	Ad Number									
	1	2	3	4	5	6	7	8	9	10
A	29	29	6	−81	−24	25	18	14	35	39
B	29	28	−17	−201	−82	23	21	−1	39	43
C	55	55	41	−10	23	52	48	46	58	61
D	55	55	21	−116	−28	52	50	33	62	64

Method A—Circulation and Editorial Departments' deficit allocated and a fixed percent of revenue assigned as a selling cost.
Method B—Circulation and Editorial Departments' deficit allocated and a fixed amount of selling cost assigned per ad.
Method C—Circulation and Editorial Departments treated as revenue departments and a fixed percent of revenue assigned as a selling cost.
Method D—Circulation and Editorial Departments treated as revenue departments and a fixed amount of selling cost assigned per ad.

Table 10. Per Cent Profit or Loss by Size Ranges and by Alternative Cost Allocations for 200 Sample Ads

Method of Cost Allocation	Size Ranges in Column Lines													
	0–50	50–100	100–200	200–300	300–400	400–500	500–600	600–700	700–800	800–900	900–1000	1000–1500	1500–2000	2000–2500
A	−3	18	27	30	33	33	30	34	39	34	40	29	27	28
B	−75	−4	16	24	31	33	31	34	41	37	42	31	30	31
C	32	47	53	55	57	57	55	57	61	68	62	55	54	54
D	−24	28	45	51	56	57	56	58	62	60	63	57	56	54

would be divided by total linage, not just advertising linage, to get a space cost rate. Composing room costs would be divided by total productive minutes, not advertising minutes, to get the time cost rate. Under this assumption, the rates would become $0.0777 per column line and $0.0752 per minute instead of the former $0.1237 per column line and $0.1261 per minute rates.

Profit or loss on each ad could now be calculated on this assumption using the new rates. It would probably be more appropriate to speak of these amounts as contributions to the meeting of editorial department deficit and eventual profit, but for simplicity they are referred to as profit or loss. Row C of Table 9 shows the percentage profit or loss on each of the ten sample ads, assuming that selling costs are assigned as a percentage of revenue. Row D of Table 9 shows the profitability rates if a fixed selling cost per ad is used. Rows C and D of Table 10 show profitability of the 200-ad sample by ad size for each assumption.

Whatever the assumptions that an individual newspaper thinks appropriate, the important fact is that total cost per ad, or selected group of ads, can be calculated if assembly units are counted and factor tables used. With little, if any, expansion of existing accounting record keeping, profit or loss determination for individual ads, for advertising accounts, or for classes of advertising, becomes possible.

DEMAND ANALYSIS [11]

Data from Newspaper 1, having been taken by the operator-log method, covered all composition time by all compositors for all local display ads set during the four-week period. Because of this comprehensiveness, it became possible to study production input and output in considerable detail.

The local display advertising load upon the composing room (input) was measured by having each new ad time-stamped on arrival in the dispatch room, a procedure routinely followed by Newspaper 1, and by then charging to that arrival hour the aggregate composition time logged for each ad (Figure 3). More specifically, arrival times were punched into IBM cards, cards were sorted by hours of the day, and hourly input was tabulated as shown in Table 11. A plot of these inputs is shown for a central eight-day period in the lower portion of Figure 11, where the peak load effect of a 6 P.M. copy deadline is clearly revealed. Approximately 60 per cent of total advertising composition load is received in the dispatch room between 4:00 and 7:00 P.M.

The output of the "ad alley," as it is called, was measured in a

[11] Donald G. Warner, "Newspaper Local Display Advertising—Analysis of Composing Room Demand" (Master's essay, The Johns Hopkins University, 1957).

Table 11. Tabulation by Arrival Time

Copy in Date	Copy in Time	AM/PM	Job Number	Minutes per Job	Total Minutes per Hour
			31041	29	
			52311	133	1114
Jan 16	100	2	16629	31	
			16697	33	
			24469	108	
			32160	57	
			32448	50	
			52314	942	
			52315	1044	
			52383	55	
			52385	149	2469
	200	2	21441	43	
			21798	54	
			25154	234	
			25155	265	
			25463	74	
			30346	404	
			38493	118	
			39820	33	
			39821	27	1252
	300	2	37389	47	47
	400	2	16439	54	
			16440	34	
			16441	42	
			16442	22	
			21444	72	
			25301	119	
			29838	118	
			29899	392	
			29999	731	
			30000	132	
			32701	27	
			37142	506	
			37144	32	
			37902	355	
			38539	157	
			38797	60	
			42094	32	
			42095	69	
			42096	125	
			46078	95	3174

NOTE: 1 = AM, 2 = PM

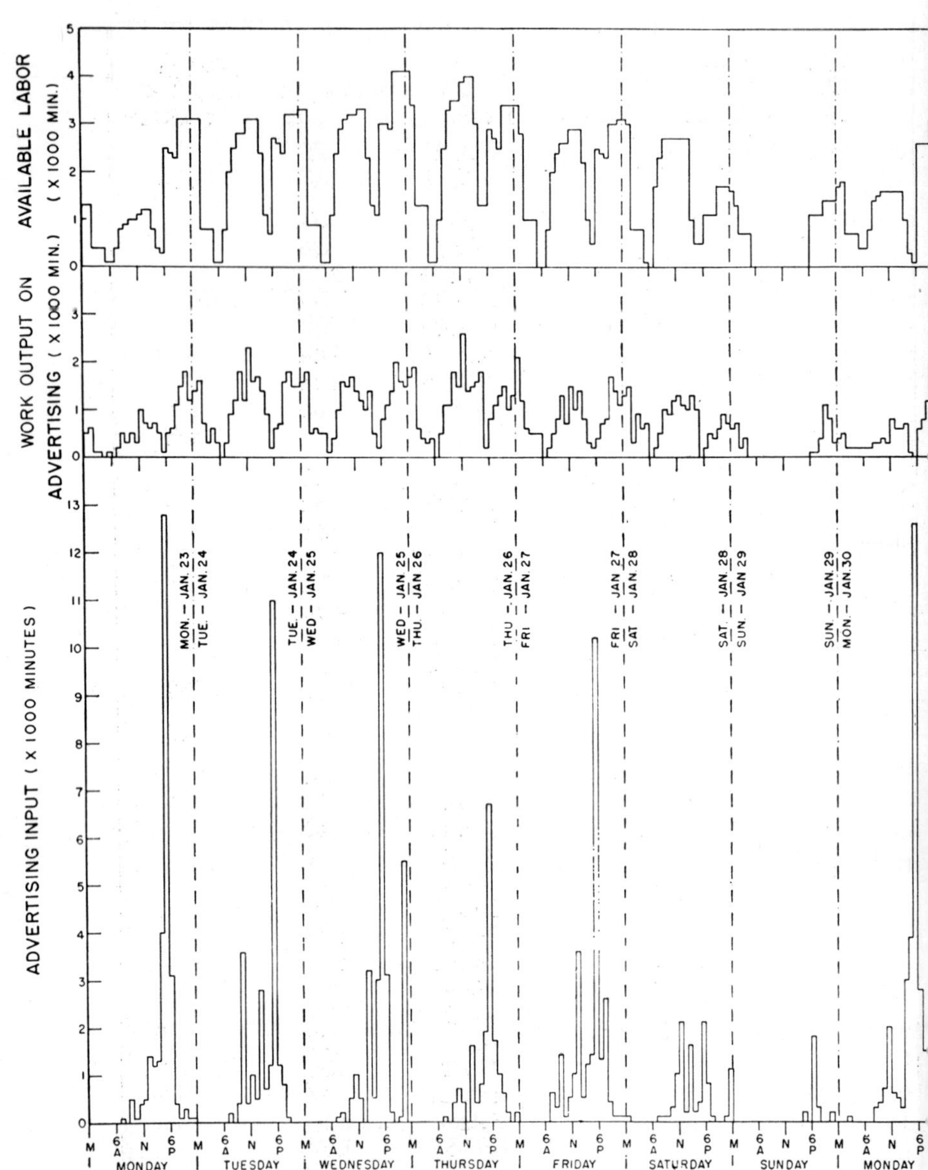

Figure 11. Hourly fluctuation of advertising input, work output on advertising, and available labor.

similar way, by grouping operation completion times by hours of the day and tabulating minutes of work done as shown in Table 12. Total output was plotted as shown in the middle portion of Figure 11, where it will be noticed that there are two daily peaks, one roughly centered about noon and the other about midnight.

These data were also plotted in detail, as shown in Figure 12. Here it can be seen that mark-up output is concentrated chiefly between 6:00 P.M. and midnight; the peak output on the linotype follows this by about two hours; assembly output has two peaks, one at night and one during the day; and proofreading output is spread out more during both day and night. The consequences of these peaks are discussed below.

As might be expected, available labor roughly corresponded with output. Man-minutes of labor available for each hour of the day were tabulated as shown in Table 13 and plotted as shown in the upper portion of Figure 11.[12] Again there is a peak during the day and another at night. These data reveal that 57 per cent of the ad composing room force receive premium pay for night-shift work. Wage scales at Newspaper 1 provide for a rate of $3.00 per hour for day work, $3.16 per hour for the evening shift, and $3.42 per hour (corrected for 35 hours per week) for the midnight or "lobster" shift. Thus it is clear that staffing to meet the evening input peak is costly.

A final plot of residual demand or backlog was made (Figure 13) by subtracting output from input by hours of the day. This is of interest chiefly because the minimum is seen to be 11,000 man-minutes. All of the data plotted, it may be added, were from the middle week of the sample period to avoid the complications of beginning and ending inventories.

To gauge the progress through the shop of individual jobs the tabulation shown in Table 14 was made. This gives the number of individual operations required by each job, the operation time, and the delay interval between completion of one operation and beginning of the next. These were summarized to show the averages and ranges of operation times and of intervals between operations as shown in Tables 15 and 16. The delay intervals were found to comprise 94 per cent of total shop time, against only 6 per cent for time spent in actual work.

[12] The considerable difference between total man-minutes of available labor and total output should not be ascribed to idleness or inefficiency. Ad compositors at Newspaper 1 also work on job work, composition for gravure magazines, and news, and such work is not included in the output data.

Table 12. Tabulation by Completion Time

Date	Completion Hour	AM PM	Operation Group	Minutes by Completion Hour by Operation Group	Total Time by Completion Hour
1 18	12 00	1	00	23	
			01	145	
			11	724	
			21	170	
			31	717	
			41	71	1850
	1 00	1	00	22	
			01	60	
			11	1072	
			21	103	
			31	638	
			42	83	1998
	2 00	1	00	8	
			01	104	
			11	87	
			31	470	
			41	38	707
	3 00	1	01	25	
			11	88	
			32	421	
			41	34	568
	4 00	1	01	112	
			11	93	
			32	819	
			41	32	1056
	5 00	1	01	20	
			11	3	
			33	139	
			42	70	232

NOTE: 1 = AM, 2 = PM

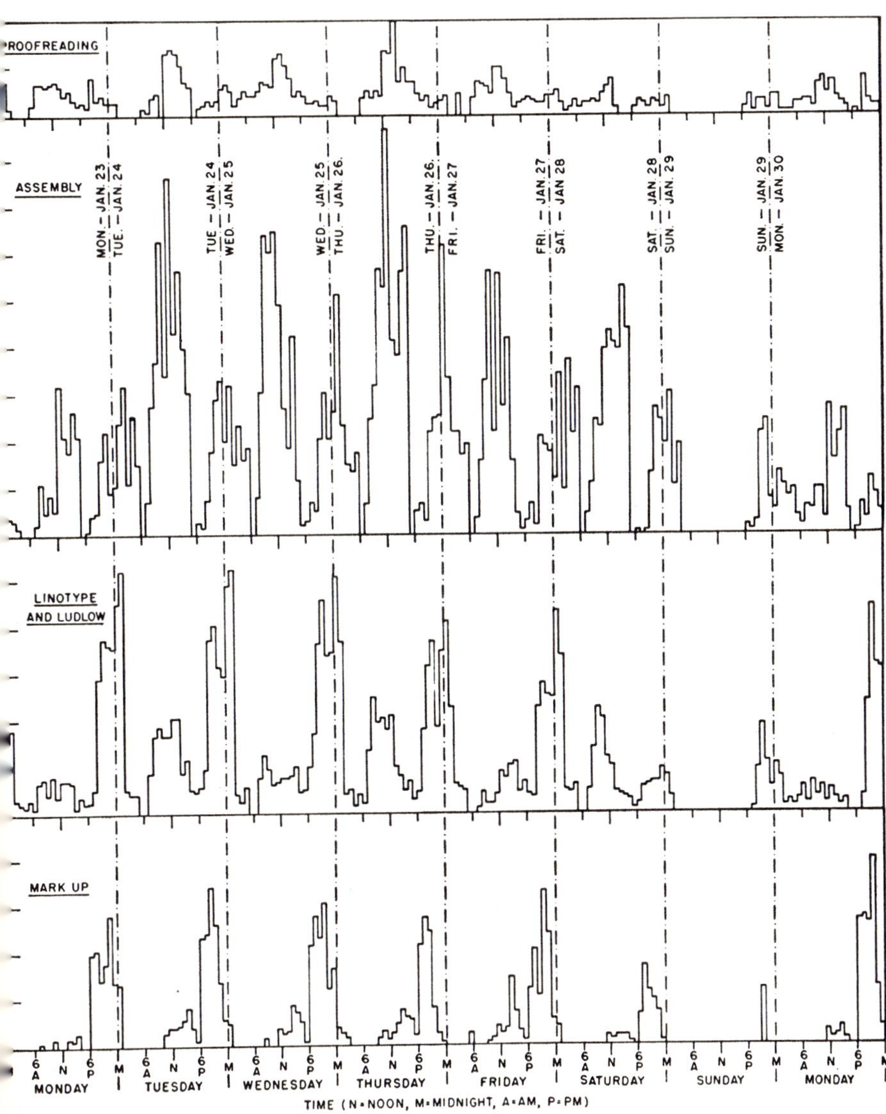

Figure 12. Hourly fluctuation of output by operations.

Table 13. Man-Minutes Available per Hour

Time	Man Hrs. Available	Minutes Available
Jan. 23		
12A-1A	22	1320
1-2A	22	1320
2-3A	7	420
3-4A	7	420
4-5A	7	420
5-6A	1	60
6-7A	1	60
7-8A	6.5	390
8-9A	13.5	810
9-10A	15	900
10-11A	16	960
11-12N	17	1020
12-1P	19	1140
1-2P	20	1200
2-3P	20	1200
3-4P	13.5	810
4-5P	6.5	390
5-6P	5	300
6-7P	41.5	2490
7-8P	40	2400
8-9P	38.5	2310
9-10P	51	3060
10-11P	51	3060
11-12M	51	3060

Moreover, the production of most ads was characterized by an extraordinary number of operations, many of which came from the practice of passing copy from one linotype to another so that each could set its particular portion. One ad in the sample, for example, was found to have moved between a total of 84 linotypes and the median for the ads sampled was found to be six. The average delay of 53 minutes between one linotype and another becomes strikingly significant in the light of the many moves made between machines.

One other detail is quite pertinent. At Newspaper 1 it was then the practice to meet each evening peak by the assignment of as many compositors as necessary to the initial mark-up operation. Once the peak in mark-up had been absorbed, most of these men would then

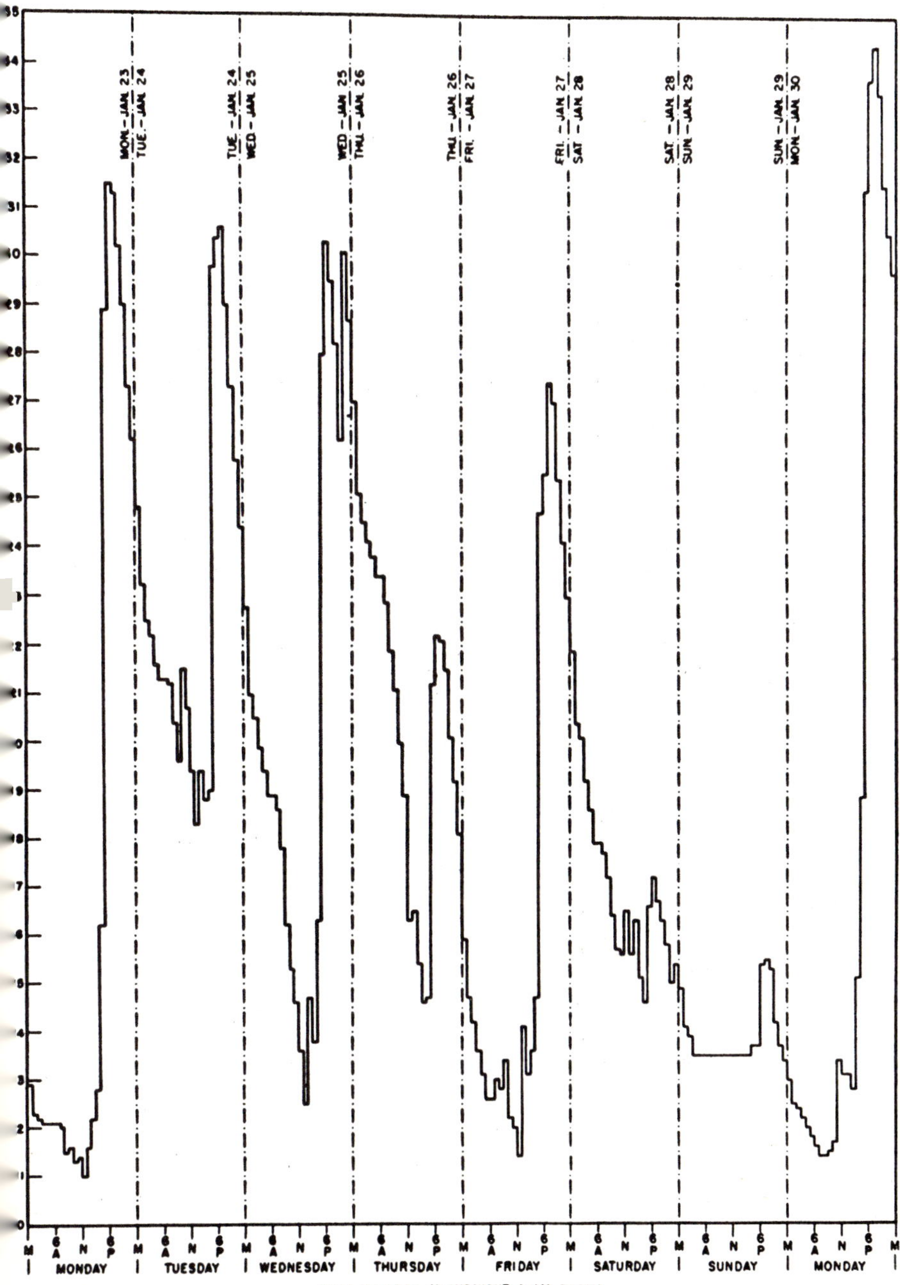

Figure 13. Hourly fluctuation of residual demand.

Table 14. Tabulation of Operations, Operation Times, and Delay Intervals

Job No.	Operation	Starting		Stopping		Minutes Elapsed
		Date	Time	Date	Time	
00000	Copy In	1-25	0530-2	1-25	0530-2	0
	Delay	1-25	0530-2	1-25	0815-2	165
	Mark-up (01)	1-25	0815 2	1-25	0826-2	11
	Delay	1-25	0826-2	1-25	1106-2	160
	Linotype (11)	1-25	1106-2	1-25	1110-2	4
	Delay	1-25	1110-2	1-26	1203-1	53
	Linotype (11)	1-26	1203-1	1-26	1207-1	4
	Delay	1-26	1207-1	1-26	0100-1	53
	Linotype (11)	1-26	0100-1	1-26	0104-1	4
	Delay	1-26	0104-1	1-26	0157-1	53
	Linotype (11)	1-26	0157-1	1-26	0201-1	4
	Delay	1-26	0201-1	1-26	0254-1	53
	Linotype (11)	1-26	0254-1	1-26	0258-1	4
	Delay	1-26	0258-1	1-26	0548-1	170
	Assembly (31)	1-26	0548-1	1-26	0610-1	22
	Delay	1-26	0610-1	1-26	1122-1	312
	Proofing (41)	1-26	1122-1	1-26	1126-1	4

move to linotype, assembly, or to proofreading for the remainder of their shifts. Out of 730 shifts worked, compositors on 126 worked on two or more tasks. Although it is traditional in the newspaper industry that journeymen compositors must be able to perform all composing room operations—the "run of the hook" as it is called—this means of assigning work involves important disadvantages:

1. Mark-up, as explained earlier, is a planning operation and the skill with which it is done largely determines the facility and economy with which subsequent operations are performed. By assigning many compositors variously to this most important task, the considerable advantages of specialization are lost.

2. Mark-up involves choices of type face and size and perception of spatial relationships, all of which have artistic attributes. Undoubtedly, there are very great differences between compositors in aptitudes for such work, regardless of skill in other composition tasks. Roving, temporary daily assignments lose the advantage of mark-up by those most capable of doing it.

Table 15. Averages and Ranges of Operation Times

Operation	Average Service Time	Range: Min. to Max.
Mark-up	11 min.	1-125 min.
Linotype	4 min.	1- 18 min.
Ludlow	7 min.	1- 51 min.
Assembly	22 min.	1-285 min.
Proofreading	4 min.	1- 44 min.

Table 16. Averages and Ranges of Delay Intervals

Between	Average Delay	Range: Min. to Max.
Copy in and First Operation	165 min.	3- 584 min.
Mark-up and Linotype or Ludlow	160 min.	0-1433 min.
Linotype and Linotype	53 min.	0- 751 min.
Linotype-Ludlow and Assembly	170 min.	0- 835 min.
Assembly and Proofreading	312 min.	9-2797 min.

3. Compositors who are assigned to mark-up for a few hours and then transferred to a second operation are not likely to make such a move without taking some sort of "break" at the time of transfer and they must, besides, go through a typical learning build-up at the beginning of each new operation. One may also speculate that motivation for really good mark-up may be diminished by temporary assignment to it.

4. Capital investment in mark-up desks, linotypes, and composing frames must be sufficient to meet operation staffing under peak conditions. For mark-up, which requires only a small desk, a type specimen book, and a few simple tools, the increased capital cost is trivial, but for composing frames and particularly for linotypes, the increased capital cost imposed by peak staffing is considerable.

From the foregoing it is clear that while the 6:00 P.M. deadlines which are fairly typical throughout the industry do not create high costs per se, they are conducive to high costs through staffing to meet attendant peak loads. It is equally clear, because of the magnitude of interoperation delays, that copy arrival need not be advanced to give more production time but may in part actually be retarded to the following day, thus not only benefiting the newspaper but the advertiser as well.

Newspaper 1 has taken dramatic action along these lines. Since from ten to fifteen advertisers buy about 50 per cent of local display advertising space, these key accounts have been offered a penny-a-line discount provided they will:

1. Submit a layout on ruled layout forms provided by the newspaper.
2. Submit all copy in typewritten form on separate 8½ x 11 inch paper. This permits copy separation by type sizes and faces and simultaneous machine composition, rather than the somewhat helter-skelter passing from machine to machine which formerly prevailed.
3. Pay for all store corrections.

These key accounts have been given next-day deadlines for same proof service and in some instances customers have agreed to accept first proofs of linotype and Ludlow slugs in galley form, rather than of the assembled ad as heretofore, a procedure which effects great economies when copy changes are required.

Newspaper 1 also has assigned carefully trained specialists to the mark-up function and has initiated an intensive study of store corrections, part of which involves customer liaison and education. Quantitative results from these innovations are not yet available but the net effect is expected to be salutary.

As reported above, this demand analysis at Newspaper 1 was made possible by possession of complete arrival and completion information for all local display ads set during the period, a circumstance which would seem to preclude similar analyses at newspapers measured by the job-log method. Fortunately, however, this is not so; possession of an equation and factor table and the time stamping of copy arrival in the dispatch room, a practice already common in the industry, plus the time-recording of store corrections, permits the calculation and plotting of input in standard minutes by hours of the day, just as in the case already described. While it is true that operation output cannot be detailed by this method, available labor can be ascertained from compositors' time cards, with aggregate results which should be just as useful for staffing decisions. Even without such measurement, the typical 6:00 P.M. copy deadlines found throughout the newspaper industry would appear to point to conditions like those found at Newspaper 1 and to the desirability of the kinds of action taken there.

STORE CORRECTIONS

Percentages of total ad composition time devoted to store corrections are shown in Table 17. Based upon the very conservative rate of $5.00 per hour, the approximate annual cost of store corrections is $68,000 for Newspaper 1, $24,000 for Newspaper 2, and $70,000 for Newspaper 4. These figures probably represent minimum out-of-pocket expenses and clearly indicate the fallacy of present practices with respect to store corrections and revises.

Most newspapers, as reported above, make no extra charges for correction services, but there are some that state a policy of charging for all store corrections in excess of some percentage of original composition time. Such statements of policy were accepted at face value for a time by the research team, until several ads from Newspaper 1 were noticed bearing blue-pencil entries such as, "8-2-2½-½-$2.50." Inquiry revealed that this meant estimates of eight hours for original composition and two and one-half hours for the store corrections called for by the proof. Two hours represented 25 per cent of estimated original composition time, leaving one-half hour excess to be charged for at $5.00 per hour. Log data for this and for other similar ads are shown in Table 18.

To illustrate further the effect of complex ads and heavy store corrections upon newspaper profits, a marginal cost rate of $5.04 was calculated for Newspaper 1. This was rounded to $5.00 per hour and used to compare composition cost and revenue for several ads of this kind. In one case a 112-column line ad cost $65.42 to compose and produced billing to the customer of only $44.80. Similar cases were found where composing room costs were 30 to 60 per cent of charges to customers. For example, a 1,080-column line ad billed at $302.40 cost $94.92 to compose, and a 70-column line ad billed at $32.20 had composition costs of $19.50. Considering the high cost of newsprint, presswork, etc., and the fact that these composition costs are stated in *incremental* terms, it is quite unlikely that these ads made any contribution to profits.

On the other hand, the composition costs of many ads were less than 3 per cent of billings for them and for some ads were less than 1 per cent of billings. For example, one full page ad had composition costs of $7.58 and revenues of $984.00, one 2,100-column line ad had composition costs of $1.25 and billings of $756.00, a 630-column line ad

Table 17. Percentage of Store Correction Composition Time to Original and First Office Correction Composition Time Based on 200-Ad Sample

CLASS A	1	8	9	14	15	18	23	25	26	27	28	
% Store Corrections	16.0%	16.6%	11.6%	39.3%	26.4%	9.2%	22.6%	24.9%	34.0%	20.5%	20.7%	
CLASS B	2	4	6[a]	7	10[a]	12	16	19	20	21	22	24
% Store Corrections	7.9%	13.6%	7.5%	5.1%	13.3%	3.7%	19.4%	37.2%	19.8%	32.7%	18.7%	15.4%
CLASS C	5	13	17[a]									
% Store Corrections	4.1%	11.5%	6.9%									

[a] For newspapers 6, 10, and 17, the percentages are of total correction minutes, office and store, divided by original composition minutes.

Table 18. Comparison of Store Correction Charges with Actual Store Correction Times, Newspaper 1

Job No.	Column Lines	Actual Comp. Min.		Estimated Comp. Min.	Estimated Store Corr. Min.	Excess Store Corr. Min.	Charge for Store Corr.
		Original Comp.	Store Corr.				
42268	2000	1218	366	480	210	90	$7.50
09418	1500	599	327	360	150	60	5.00
24116	2400	749	382	480	150	30	2.50
24096	2400	608	270	480	180	60	5.00
24097	2400	555	335	480	180	60	5.00

had composition costs of $1.75 and billings of $226.80. While the ads cited are not typical, they do emphasize the point that advertisers who submit clean, well laid-out copy are subsidizing advertisers who are careless or insist on complex, type-packed ads.

PREPARATION AND PROCESSING [13]

To gain insight into the nature and origin of store corrections, a study of copy preparation and layout, composition, proofreading, and correcting was carried out in two large department stores (designated here as Store X and Store Y), with the co-operation of Newspaper 3.

During this study a total of 77 run-of-paper ads and a 12-page tabloid advertising section were analyzed from Store X and, in a similar way, 19 ads were analyzed from Store Y. The method of analysis is indicated in Figure 14, where responsibility for ten classes of correction is allocated respectively to either store or newspaper.

The 77 ads from Store X ranged in size from 85 column lines to full page and comprised a total of 5,265 assembly units, of which no less than 1,796 or 34.1 per cent were changed in proof. Of this very large number of corrected units the following causes for correction were assigned:

[13] Jules Tewlow, "Newspaper Local Display Advertising: A Study of Its Preparation and Processing" (Master's essay, The Johns Hopkins University, 1957).

Worksheet For Changes and Corrections

Ad Description College Favorites

Insert Date 22 Aug 56

Ad Size 8 col x 295 lines

Proof No. 1 (x)/ No. 2 (x)

Paper: Morning ( ); Evening (x); Sunday ( )

Total Units 115

	N	S	Newspaper	Store
1. Typographical	9		13.6%	
2. Headlines:				
alignment	3	10	4.6	15.2
loading	--	2	--	3.0
changes				
errors	3	--	--	4.6
3. Body Copy:				
alignment				
leading	8	2	12.1	3.0
changes	--	8	--	12.1
errors	5	1	7.6	1.5
4. Transposition Errors	3		4.6	--
5. Prices & Merchandise:				
changes				
errors	--	1	--	1.5
6. Type Faces & Sizes:				
changes				
errors	--	1	--	1.5
7. Rules:				
changes				
errors				
8. Engravings:				
changes	--	2	--	3.0
errors	7	--	10.6	--
9. Store & Floor Lines:				
changes		1	--	1.5
errors				
10. Omissions				
	66 Units		100%	

Remarks:

57.4% of total units are changed or corrected $\left(\frac{66}{115} \times 100\%\right)$.

Note blouse section; heavy body copy changes.

Figure 14. Work sheet for changes and corrections.

1. Newspaper typographical and transposition errors .... 10.1 per cent
2. Newspaper's failure to follow instructions .... 21.6 per cent
3. Store buyers' changes .... 11.9 per cent
4. Artistic prerogative[14] .... 36.6 per cent
5. Store errors .... 19.8 per cent

Examples of some of these are shown in Figure 15.

These data indicate that almost 70 per cent of the changes originate with the store, but this figure is conservative, because many items assigned to Category 2 were doubtful, the instructions given by the store sometimes being obscure or ambiguous. It is therefore safe to say that *more than* 70 per cent of the 1,796 units corrected could properly be described as the responsibility of Store X. A smaller sample, aggregating 1,907 assembly units was analyzed from Store Y. In this instance 688 units or 36.1 per cent of all units were corrected and of these 79.4 per cent were the responsibility of the store.

These figures may be stated in a different and perhaps more meaningful way. If it is assumed that the correction of 35 per cent of all units compared is typical, and if 70 per cent of these corrections originate with advertisers, then the aggregate cost of store corrections and revises at Newspaper 3 is $.70 \times .35 = .245$, or 24.5 per cent of *total ad composition cost.* Because many other, smaller volume advertisers do not change their copy in this way, this figure undoubtedly is much too high and is more shocking than accurate. Nevertheless, it does reflect an element of the cost of newspaper operations which demands serious attention. The fact that some newspapers have greatly reduced this source of loss is indication that an attack upon the problem by newspapers is feasible.

THE EFFECT OF LONG DEADLINES

Initial surveys conducted in many newspapers during the early months of the research project led to the subjective opinion that deadlines giving a longer interval between submission of copy and date of issue were conducive to a greater volume of store corrections and

[14] Artistic prerogative is that type of change made to an ad which represents an artistic striving for perfection, when translating visual and aural imagery into graphic design. While observing the process of ad revision at Store X, the reason often given for altering a particular ad was, "This item or that word doesn't look or sound right."

Original Copy

12″ globe, $8.95

Colorful globe with 96-page booklet,
"See the World", will make history
and geography lessons come alive.

Buyer's Change

12″ globe, $8.95

Colorful globe with 96-page b
"See the World", will make
and geography lessons come a

Changes Due to Artistic Prerogative
(below)

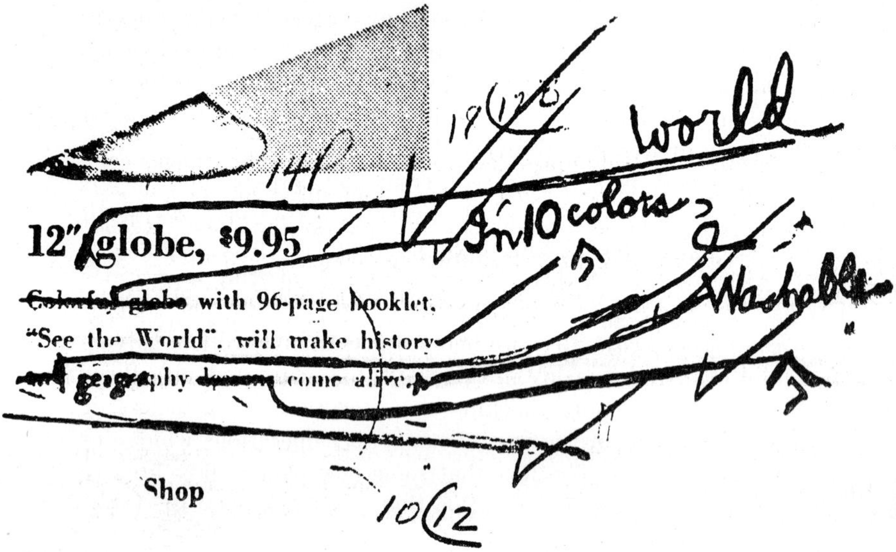

12″ globe, $9.95

~~Colorful globe~~ with 96-page booklet,
"See the World", will make history
~~and~~ geography ~~lessons~~ come alive.

Shop

Copy As It Appeared

In The Newspaper
(below)

12″ world globe, $9.95

In 10 colors, with 96-page booklet,
"See the World", will make history,
geography come alive. Washable.

Figure 15. Store corrections and revises.

revises. That opinion has been corroborated in substantial measure by study of a 12-page tabloid prepared by Store X and run by Newspaper 3.

This tabloid section was composed on a 10-day deadline and contained a total of 913 assembly units, of which 631 or 69.1 per cent were changed in corrections. This figure indicates that the longer the advertiser has to think about his ads, the more he will be dissatisfied with what he has submitted for publication. At the risk of over-dramatization, one must conclude that the newspaper which advances a deadline to gain more time for composition is inviting additional corrections.

UTILIZATION

The measurement techniques described in the preceding pages comprise an integrated system whereby each measured newspaper can:

1. Compare its own performance in the composition of local display advertising with that of all of the other newspapers measured.[15] This is done by the act of measurement itself, accompanied by internewspaper comparison tables provided to each newspaper measured. The provision of comparison data showing relatively poor performance could and should lead to managerial investigative and corrective action.

2. Compare production performance within its own composing room from one period to another. This may be done by the counting of all assembly units and column lines, followed by use of the factor table. The sum of such standard times, plus the recording of actual time spent on store corrections, yields a measure of advertising composition output for the accounting period chosen. Similarly, the recording of compositor productive time spent on advertising, plus associated indirect labor time, yields a measure of input. The output-input ratio derived from these two figures is a meaningful measure of composing

[15] Although each model is derived only from local display advertising composition data, the resulting factor tables will measure the composition of national and classified display advertising equally well. In national advertising the portion of an ad furnished in mat form is simply counted as one unit. Similarly, elements "picked up" as standing type from previously composed ads are counted as single units. Thus, the model is capable of measuring *all* advertising composition except straight classified, which is set to uniform measure in text type sizes and can therefore be measured in column lines.

room efficiency. Again, deviation of the ratio from previously established norms can either measure the extent of improvement or point to the need for corrective action.

3. Calculate the composition cost and total cost of producing any single ad, all or part of the ads of any customer or account, and all or any part of the ads in any particular class of advertising. By this means sources of inordinate loss or profit may be ascertained and corrective action taken.

4. Measure demand upon the composing room throughout the day and for each day of the week. Doing so can improve both scheduling and staffing, with attendant savings in overtime and night shift premium pay.

5. Compare the relative economy of hot metal composition with the emerging processes of photo composition.

As yet, no single newspaper of the many measured has adopted these proposals as an integrated system for enhanced managerial control.[16] This probably has been due to a variety of factors: inadequacy of communication from the research team to the industry and vice versa; the seemingly formidable three-variable logarithmic model; the mistaken belief that counting units and using the factor table is impracticable; reluctance to introduce time-recording and accounting procedures which are commonplace in other industries but lacking in this one; and, most of all, a lack of understanding of the value of quantitative information for managerial decision on the part of an industry whose criteria of excellence are dominated generally by editorial and public service attributes.

Nevertheless, despite this absence of total adoption, a good deal has been accomplished and much more is in prospect. Newspaper 1 has, as already reported, offered a discount and later deadlines to key accounts in return for copy and layout conformity and payment for store corrections. They have initiated altogether new procedures for machine assignments and proof submission and have abandoned the practice of roving assignments to mark-up, in favor of specialists.

[16] Because of this reluctance to adopt the measurement method as a continuing *system,* a sampling plan has been devised and preliminary indications are that it will be used. By measuring all ads of six columns to full page in size (about 1,800 to 2,400 column lines) and randomly sampling about 30 per cent of the remainder, total standard minutes, efficiency, and cost per standard minute can be calculated to confidence limits of less than ±4 per cent. Ads of six columns and larger comprise about 7 per cent of the total number of local display ads published.

They also carried out intensive studies of store corrections and used the log data as a basis of comparison for experiments in Photon composition, a method of ad production which is now operational. It is unfortunate that Newspaper 1 did not measure output by use of the model; had they done so, the extent of improvement could have been quantitatively determined.

Newspaper 2, at which cost-revenue comparisons originally were made, has revised its cost accounting procedures to take advantage of the method of measurement. Newspaper 21 has applied the average standard time per ad to all ads for comparison with labor input to get an approximate measure of efficiency. Newspaper 6 has used the model to arrive at a higher rate for a new advertiser of notoriously costly habits; they have counted units for a substantial period and have developed a nomograph of their particular model. Newspaper 18 plans to make extensive use of the system as a whole: for production control, for cost accounting, and ultimately for evaluation of advertising rates.

Several newspapers have utilized their factor tables for spot-checking performance in the composition of selected ads. These are logged through the shop and the actual times recorded are then compared with the standard times computed from the factor table. Utilization of a statistically determined model in this way may be questioned, but it is interesting to report that in several cases, where actual time logged was much less than standard, the log sheets showed no record of machine time at all, although such composition was shown to be necessary by the proof. Thus, in keeping with the experience of many jobbing shops, standard time is shown to be superior to so-called actual time.

CURRENT RESEARCH

As this is written, measurement of additional newspapers continues, now upon a fee basis. This service will be continued in the future by ANPA Research Institute, while university personnel engage in further research.

A detailed study of the mark-up function has been carried out by a former member of the research team, as an extension of this project, under the auspices of the Research Institute. A report of this study, which will contain practical recommendations for optimal procedures, has been published. A further study with the objective of determining and stating optimal methods for all advertising operations is now under way.

Development of an integrated data-processing system using punched cards and appropriate computer facilities is now under development and will be published in manual form as a guide to newspapers which have been measured.

Economic studies comparing the cost of composition by hot metal and Fotosetter have been carried out in three newspapers. It is hoped that this research eventually will provide a definitive answer to this important question.

A study of advertising rate structures and of the possible effect of the model upon them also is currently under way.

ACKNOWLEDGMENTS

Members of the research team have been the beneficiaries of unvarying support and co-operation from many newspaper executives and from a large number of compositors throughout the industry. Foremost among these has been Lisle Baker, Jr., Vice-President of the *Louisville Courier-Journal and Times* and past President of the ANPA Research Institute. The concept of operations research in the newspaper industry largely has been his and he has been a staunch supporter of the project throughout its course. Similar support has come from Franklin Schurz, publisher of the *South Bend Tribune* and past President of the ANPA Research Institute, and from Cyrus MacKinnon, former Managing Director of the Institute, who has successfully guided the project through difficult times. Brevity prevents more detailed and gracious acknowledgment to many others.

During the summer of 1955, Fred Wolke, member of the International Typographical Union, and several times President of the Louisville local, accompanied members of the research team on the initial survey. We are grateful for his insight and understanding and for his warm friendship.

Members of the University who participated in the project were: Dr. Alphonse Chapanis, Professor of Psychology and Industrial Engineering; Dr. Acheson J. Duncan, Associate Professor of Statistics; Dr. Charles D. Flagle, Associate Professor of Industrial Engineering and Director of Operations Research, The Johns Hopkins Hospital; and

Carville G. Bevans, Jr., Irvin W. Kues, Jules S. Tewlow, Donald G. Warner and Captain Joseph H. Shaffield, Research Assistants. Mr. Tewlow is now with the *New York Times,* and Mr. Warner is with the *Chicago Tribune.* All of us, including the authors, constituted a research *team* in the best operations research sense of the word.

Twenty-Seven

THE COST AND VALUE OF REPORTS—A CASE STUDY IN A TELEPHONE COMPANY [1]

MARVIN A. GRIFFIN

INTRODUCTION

Because the design of an optimal system of internal reports used for managerial control is indeed a complex problem, it may be conjectured that very few systems of reports have been optimally designed. The existing system of reports in an organization is more probably the result of many simple changes in the initial request of a decision-maker.

Perhaps some decision-maker decides that an analysis of a particular class of sales orders may be of some value in answering a specific question. He may then direct that the particular data be prepared. Should this request be repeated a few times, an enterprising office manager or chief clerk may observe that the preparation of the letters of transmittal could be eliminated by the design of an appropriate report form. This is subsequently prepared and the decision-maker continues to receive the report, thence to eternity.

Let us suppose that the report does have real value and that additional decision-makers hear of the report. The distribution list begins to grow. Through the years the report may have additional rows and columns of information added to meet the changing demand. Should

[1] The research reported in this chapter was performed under the general direction of Dr. Eliezer Naddor. Other members of the group were James Anderson, Marvin A. Griffin, John T. O'Conner, and Thomas Owens.

one decision-maker need the report more frequently and directly or indirectly effect this change, then every name on the distribution list is given a report with the increased frequency of issue. Should the report become too large for a standard size sheet of paper, there will probably be a separation into two reports with certain essential data duplicated on each report. This process of growth with attendant increase in cost may in fact be an optimization of the difference between the cost of the report and the value derived therefrom. Whether it is or not is seldom if ever known, although the magnitude of the cost associated with some report systems represents a considerable sum.

This evolutionary process of the report system would inevitably engender a certain amount of affection through tradition. The intertwined distribution lists and varied uses of reports eventually become so enmeshed in a large organization as to practically defy all attempts at change. Studying and understanding a complex system of reports is exasperatingly tedious and time-consuming. No cursory analysis can determine the obvious cost associated with such a system or begin to assess the value of the report to the users. Hence, it is indeed a problem area of sufficient magnitude and complexity to tax the resources of the current development of operations analysis.

A research project to determine the design parameters of a system of reports based on optimizing the difference between the total cost of the reports and the total value obtained from using them will be described in this chapter. The sponsor of the research is a telephone company with considerable interest in this field. Management decisions in this company are obviously made at several levels of the organization hierarchy. These decisions are generally based to some extent upon the analysis of significant enterprise transactions periodically issued to the decision-makers in the form of business reports.

Although the research has been problem centered, it is believed that fundamental laws or relationships which control optimal parameters in this particular enterprise would be applicable to the area of decision-making in general.

The Cost of Reports

The total cost of a system of reports has been estimated as "very high" many times in the literature; however, little searching cost analysis in the area is indicated. Why has so little effort been expended in determining a reasonably accurate cost of a report, as is done with many other internally consumed services or products? Most competitive as well as eleemosynary organizations have found it desirable, even necessary, to maintain records of all costs associated with the production of a particular product or service offered for sale. Many other goods or services which are consumed internally have in recent years been subjected to the scrutiny of the cost accountant and later the decision-maker's control. A very reasonable and worthwhile extension to the cost control philosophy seems to be the analysis of the reporting function itself. It seems somewhat ironic that the service of providing these reports has escaped searching analysis. The question of the desirability of establishing cost codes for labor, materials, equipment, and overhead items to determine the cost of a system of reports is beyond our present scope. However, in order to optimize the cost-value relationship we must have quantified facts. This, then, dictates an immediate need, if not a continuing one, for relevant, reasonably accurate costs pertaining to the reports.

ELEMENTS OF COST

Although the detailed determination of the elements of cost will be deferred for later treatment, it seems desirable to state the criteria which were used to include or reject the numerous expenses related to the production of the system of reports. It was decided to treat the reports as products which were to be offered for sale. Hence, rather than consider only the direct labor and materials, all costs were included. The cost elements are as follows:

1. Direct Labor. This element includes the actual time of each clerk employed in the process, as well as the actual time of the various machine operators.
2. Supplies and Materials. The expense supplies used by the clerks and the operators; the IBM cards, paper tapes, printing supplies, postage, and other similar items are included in this element.
3. Equipment. The normal carrying charges of all equipment used in the manufacture or production of the reports were included, plus the IBM rental cost.
4. Facilities. The cost of the space occupied by the clerk and equipment and the cost of heating, lighting, cleaning, and other services were included in this element.
5. Overhead. This element includes the cost of supervision, training, social security, vacations and holidays, sick and excused leave. No attempt was made to determine whether or not a particular

supervisor would be added or transferred as a result of changing the number of clerks in a particular section, nor was any attempt made to determine whether or not an additional piece of equipment would be added or released.

This approach can certainly produce serious errors in a practical situation. However, it is considered feasible as well as expedient to make the assumption that in the long run a supervisor will be released, or rather reassigned, if a significant number of his subordinates have been released through a series of cost reductions. The same reasoning applies to the equipment or to rented office space.

COST AREAS

Since many organizational units were engaged in the production of the system of reports, it was considered necessary to break the problem into fairly distinct functions. These functions were as follows:

1. Preparation. The cost of each element commencing with the receipt of the basic source data through the completion of one copy of the report was included in the preparation area.
2. Reproduction. This area included the cost of each element commencing with one completed copy of the report and ending with the required number of copies completely collated and stapled or otherwise secured.
3. Distribution. This area included all packaging, addressing, and mailing or delivery costs associated with the system of reports.
4. Storage. This area included all elemental costs incurred in providing vault and filing cabinet storage for the system of reports.
5. Use. Initially it was considered desirable to evaluate the cost of using the information given on a particular report. Hence, if each decision-maker was required to run subtotals before using the information, the cost of this would be added to the other costs of producing the reports. Subsequent investigation revealed that the uses of the reported information were so varied that it would be impossible to assess accurately the costs associated with using the reports.

It appears that two pertinent observations should be made before dispensing with this area of cost. First, there seems to be no justification for a report preparation group to fail to provide the necessary arithmetic required by several users. (In this particular case there are about 200 users.) However, it requires very little imagination to show that the combinations and permutations of a few thousand numbers can soon become astronomical. Hence, it is economical for the report preparation group to continue to compute information so long as their costs are less than the sum of the costs incurred by the users.

Secondly, the presentation of the reported information should be made in such a manner that it may be accurately and readily understood. The field of human engineering (3) might well provide some much needed design parameters to guide those people responsible for the form design function.

The Value of Reports

The cost problem with all the attendant complexities appears to be quite simple when compared to the value measurement problem. Communication-information theory (5, 10, 11) is very useful in the design of effective transmission-receiving systems; however, the question of what should or should not be transmitted is given little or no consideration. The design of an efficient telephone system or other communication network necessarily assumes that the messages must be handled. However, in the design of a system of reports one might well ask the question: "What messages are of sufficient value to be transmitted?" This problem is somewhat external to the theory of communications.

In order to evaluate a particular piece of information we must determine the effect it has on a decision. We can gain some insight into the problem by constructing some meaningful models with a finite amount of information contained therein. By varying the amount of information given, we are then in a position to perceive the effect on the solution or decision of a particular piece of information. The theory of games offers systematic solutions to a problem for purposes of perceiving the effect of varying the amount of information given (7, 13, 14). In the example of the practical application of game theory applied to business problems cited by Weart (13), we see that only two of about six variables have any real effect on the game theory solution. Hence, we might conclude that any resources further expended in measuring the effect of the other variables would produce no positive value.

The cost of collecting data to evaluate a decision is frequently neglected, with the result that some factors are measured which have no significant effect. The relevancy, the sensitivity, and the range of the variables might well be considered before launching a "special cost study" to investigate some economic opportunity.

The question considered in this chapter is somewhat analogous to the problem of the "special cost study" in making a major decision.

However, the emphasis is no longer on a particular investment or decision, but rather the short- and long-term decisions which continually confront a decision-maker. It is quite obvious that a particular report containing the basic sales information of a large enterprise can be presented in countless ways. There is practically no end to the computed and summarized statements that can be presented from even a few thousand transactions. Most of the statements will be of little value, however, and only a few will remain which have real meaning.

The meaningful facts can usually be identified by first determining the objectives of the decision-makers who receive the report and, second, by determining which information in the report is relevant to the objectives.

The next link in this model relates to the decision-maker. It is obvious that he cannot receive more meaningful information than is contained in the report. It is suspected that he actually receives somewhat less. The chance of a misinterpretation is also ever present. After the decision-maker has selected whatever facts he believes pertinent, he is then in a position to make a decision, execute a report, or to take some other appropriate action.

At this point we must ask the question, "What are the consequences of the action taken?" Let us suppose that the action taken was the best of the available alternatives. The next question may then be, "What would have been the action in the absence of the reported facts?" It appears that the value of the report can finally be discussed. The quantified difference between the probable action taken with the reported facts versus the probable action taken without the reported facts seems to be an appropriate measure of the value of the report. In a large organization we will usually have a sizable distribution list, and the total value of the report will be the sum of the values to the individual users.

ORIGIN OF THE CASE STUDY

At a board meeting a few years ago, one of the directors commented upon the growth of reports and other documents the company used to conduct its business. The subject aroused some interest, and it was subsequently decided that one copy of each report used in the telephone company should be collected and displayed on the conference table for the next meeting. The display was quite dramatic: a pile of documents, no two of which were duplicates, about five feet long and two feet in height.

Although the company is quite large in terms of employees, gross income, and area of operation, it was felt that the amount of reports required was excessive. Hence, a decision was reached to look into the matter to determine if something could be done to decrease the number and the growth of reports. Specifically, an operations research project was organized to investigate the matter. In March, 1957, a contract was signed between the telephone company and The Johns Hopkins University to conduct an operations research study relating to the initiation, distribution, usage, modification, and storage of certain telephone company reports, records, documents and other papers.

The initial question confronting the research team was the determination of a point of entry into the problem. No doubt this is a critical question for a new team, and some general recommendations (4) have been formulated. The team recognized that it would be physically impossible to do more than attempt to develop a methodology within the period of the contract. However, a methodology, to be of proved value, must be developed from the solution of specific problems. Hence, it was decided to begin with one report which had wide distribution, high frequency of issue, and involved considerable cost, an approach often used by methods engineers in manufacturing plants (1, 8, 12).

At the suggestion of the comptroller of the company, a report known as the *Station Movement Report* was selected as a point of entry into the problem. This report, illustrated in Figure 1, is issued on a monthly basis to approximately 200 people. It shows the number of telephones (stations) installed, disconnected, up-graded, down-graded, or otherwise changed by each central office in the state. Certain computations are also shown on the report. The central offices are summarized by districts, divisions, and state. Also the net gain, the difference between newly installed telephones and disconnections, is shown. This report has been in use for some thirty-odd years and has met the qualifications stated above. The word "movement" in the title of the report refers to change. There are seven "movements" shown as column headings and some three hundred rows or locations shown on the report.

The Research Team

The research team was composed of five people, including the director, three full-time graduate students, and an employee of the tele-

	COMPANY STATIONS										COMPANY CIRCUITS				SERVICE STATIONS	
					OTHER CHANGES		NET GAIN*				NET GAIN*				In Service End of Month	
CENTRAL OFFICE (a)	Connec- tions (b)	Change of Address "In" (c)	Total Inward Move- ment (d)	Total Outward Move- ment (e)#	"In" (f)	"Out" (g)	This Month (h)	Year To Date (i)	Adjust- ments (j)	In Service End of Month (k)	This Month (l)	Year To Date (m)	Ad- just- ments (n)	In Service End of Month (o)	Stas. (p)	Ccts. (q)
e (Columbia) P	10		10	4	1	1	6	28		277		6	1	51		
tal	65	21	86	42	21	21	44	461		3761	20	206	1	1468		
idge	28	19	47	33	2	2	14	320		2799	1	128		1281		
rloo P	7	1	8	6			2	40		413	2	12	1	78		
tal	35	20	55	39	2	2	16	360		3212	3	140	1	1359		
Burnie								-7717				-4058				
nfield 6 (G.B.)	224	84	308	180	271	268	131	9370	-2	9439	164	5393	278	5715		5
n								-541				-85				
land 9 (Crain)	18	2	20	18	3	6	-1	599		602	2	102		105		
tal	242	86	328	198	274	274	130	1711	-2	10041	166	1352	278	5820		5
icum	223	23	246	37	46	46	209	767		4217	36	303		1720		
r 6 (Pikesvl.)	111	90	201	96	29	29	105	1154		9881	47	569		5098		
ield 3 (Roslyn) P	47	50	97	31	12	12	66	685		3965	19	343		2152		
tal	158	140	298	127	41	41	171	1839		13846	66	912		7250		
pect P	61	19	80	51	23	23	29	357		3848	11	86	-29	2293		
erstown	45	24	69	31	6	6	38	339		3109	9	106		1531		
ington	5	1	6	5			1	23		329		6		79		
tal	50	25	75	36	6	6	39	362		3438	9	112		1610		
sor 4	50	42	92	33	31	31	59	504		5058	52	351		3063		
stock								-203				-89				
s 8 (Woodstock)	1		1	5			-4	239		231	-2	130		132		
. MONTH	1272	550	1822	860	526	526	962		-3		464		251			
. YEAR TO DATE	10657	6496	17153	8374	14624	14624		8779	73	70700		4199	151	36679		5
n District																
field 5 (Boulvd.)	78	83	161	86	60	65	70	173		10950	53	301	-8	7202		
field 8 (Boulvd.)	80	75	155	51	29	25	108	1951		3154	40	1054	-9	1651		
al	158	158	316	137	89	90	178	2124		14104	93	1355	-17	8853		
ater 5 (Chase)	13	17	30	24	1	1	6	131		1737	1	18		481		
ater 8 (Chase)	1	2	3	6			-3	34		157		1		98		
al	14	19	33	30	1	1	3	165		1894	1	19		579		
ysville	30	14	44	13	11	11	31	249		1935	9	82		834		
P	6	3	9	5	3	3	4	41		764		7		119		
al	36	17	53	18	14	14	35	290		2699	9	89		953		
er 4 (Dundalk)	143	69	212	89	12	20	115	326		11141	30	397		6185		
er 5 (Dundalk)	100	31	131	65	7	7	66	1977		5031	15	1220		2585		
er 8 (Dundalk)	128	33	161	30	12	4	139	3042		3040	47	1467		1467		
al	371	133	504	184	31	31	320	5345		19212	92	3084		10237		
ck 6 (Essex)	125	39	164	97	14	16	65	34		7799	37	-30		4177		
ck 7 (Essex)	514	33	547	463	9	7	86	956	1	10186	17	16		3462		
brook 2 P	83	12	95	12	3	2	84	515		515	51	377		377		
al	722	84	806	572	26	25	235	1505	1	18500	105	363		8016		
	20	13	33	8	2	2	25	202		1972		15		592		
ows Point	66	20	86	60	17	17	26	232		4465	18	150		1612		
y 3 (Towson)	279	142	421	179	97	98	241	-1034		15448	117	-419		7885		
y 5 (Towson)	126	91	217	145	124	123	73	636	-1	12184	78	810		6015		
al	405	233	638	324	221	221	314	-398	-1	27632	195	391		13900		
l 7 P	45	25	70	43	50	50	27	3821		3821	17	1899	-4	1895		
MONTH	1837	702	2539	1376	451	451	1163				530		-21			
YEAR TO DATE	18035	7384	25419	12133	11315	11319		13286	-34	94299		7365	-285	46637		
MORE DIVISION																
MONTH	9362	3371	12733	8025	2675	2675	4708		-10		2033		24			
YEAR TO DATE	87780	37220	125000	81666	48919	48919		43234	-17	558121		20019	-2010	291660	1024	49
olis District																
ial 3 (Annap.)	123	45	168	116	6	4	54	647		6976	2	269		2966		
ial 8 (Annap.)	66	23	89	54	3	5	33	451		7436	3	205		2887		
al	189	68	257	170	9	9	87	1098		14412	5	474		5853		
er	55	30	85	57	10	10	28	448		2801	6	93		1036		
t P	27	12	39	21	8	8	18	263		1747	8	94		926		
al	82	42	124	78	18	18	46	711		4548	14	187		1962		
sta	6		6	4			2	14		382		18		240		
ew 2 (Brandywine)	18	3	21	25	1	1	-4	139		1505	-2	43		383		
n Island	2	2	4	8			-4	76		627	-2	40		312		
Mills Dial								-3430				-1097				
teer 2(Gt.M.Dial)	81	26	107	84	5	5	23	1590		1590	2	731		726		
teer 3(Gt.M.Dial)	56	4	60	33	3	3	27	2296		2296	1	382		379		
Mills (Manual)	36	8	44	31	1	1	13	189		1486	3	43		328		
al	173	38	211	148	9	9	63	645		5372	6	59		1433		
side 3(Ind.Head)	30	5	35	19	2	2	16	26		1651	3	-7		436		

* Excludes Purchases, Sales, and Adjustments.

\# To Obtain Disconnections, Subtract Change of Address "IN" From Total Outward Movement.

Figure 1. Illustration of a portion of the *Station Movement Report*.

phone company. The director, a member of the faculty, devoted approximately one-fourth of his time to the project; the graduate students devoted approximately one-half of their time to the project; the company employee was the only full-time member of the team. The team also utilized certain professional consultants on specific areas of the research.

Progress Reports

The team presented verbal progress reports to the committee responsible for the research on a bi-monthly basis. The committee was composed of the company president, the comptroller, and other top level managers. It was believed that the verbal reports, with outline charts, was a better method of presentation than the written report method. These meetings provided a forum whereby management could appraise the quality of the research in formulating eventual implementation plans. The team was able to benefit from the meetings in several ways. First and foremost was the determination of company objectives and restrictions. By questioning executives about their probable action in a given situation the team was well aware of what type of change could or could not be installed. Hence, the research was co-ordinated along the lines of company policy and the final approval of the solution required very little redesign.

Many other meetings were held with various operating managers for the purpose of obtaining competent assistance in answering specific problems under investigation.

THE ORIENTATION STUDY

The first phase of the research work may be described as an orientation study. This was quite necessary since none of the team members had expert knowledge about the report problem in general or the telephone movement reports in particular. The purpose of the orientation study was for the team to become acquainted with the problem area; to determine the feasibility of the area selected as a meaningful problem; and to conclude with a statement of the problem.

The investigation of the movement reports commenced with the source documents, proceeded through the preparation, reproduction, and distribution areas, and terminated with interviews with the users.

In order to make a decision as to whether or not the team was working on a meaningful problem, it was decided to use as a criterion the total cost of the report. Hence, it became necessary to measure the costs incurred in producing the report. Operation process charts were used to record all steps in the production of the report. Supervisors provided estimates of the costs of the various operations under their direction.

The process charts soon showed the fallacy of an investigation of the *Station Movement Report* as a separate entity. The IBM cards which were used in the preparation of the *Station Movement Report* were also used to prepare some nine other related reports. Hence, it became apparent that the one report initially selected must be studied as part of the larger group. The group of related reports is referred to as a "system of reports."

Formulation of the Problem

The primary purpose of this phase of the study was to maximize the effectiveness of the efforts of the team. Maintaining realistic and practical aims required continual revision of the statement of the problem, since the desire for grandiose accomplishment was ever present. As the budgeted time and other resources were consumed without commensurate achievement more realistic aims were indicated. Certain objectives were revised or eliminated purely from practical necessity in order to achieve some measure of accomplishment within the restrictions of the contract.

OBJECTIVES

The decision-makers in this case were the comptroller, the operating departmental supervisors, and the company president. Although the reports were made in the accounting department under the direction of the comptroller, they were used by the operating supervisors as well as by the president. Hence any change in the report would produce some effect of a company-wide nature.

The objectives were as stated below:

1. To reduce the total cost of the system of reports.
2. To maximize the value derived from using the reports to provide improved customer service; to maintain a stable labor force; and to produce a reasonable rate of return on the invested capital.

3. To improve the "index" relating to *Station Movement Reports.* (This is one of the work-volume-cost standards used for measuring the efficiency of the company.)

MEASUREMENT OF OBJECTIVES

The total cost of the system of reports is to be measured by summing the elements of cost in each area previously discussed.

The total value of the system of reports is to be measured by determining the effect of the reported information on the various decisions relating to the company objectives.

The index measurement plan currently in effect will continue to provide a measurement of the third objective from established cost accounting routines.

ALTERNATIVE SYSTEMS OF REPORTS

There is a considerable amount of variation in which source data may be presented in final summary type reports. The ten reports comprising the system were issued periodically, ranging from a daily basis to a quarterly basis. Some of the reports presented significant movement information on a central office or telephone exchange location, whereas other reports presented the information summarized for the state. By considering all of the details on the source documents and the many and varied users in different departments and organizational levels, it is obvious that a very large number of alternative systems of reports could be developed. This is one of the parameters with which the research team had considerable opportunity for redesign.

ALTERNATIVE METHODS OF PREPARATION

Another design parameter is the preparation method. For example, a report may be hand tallied or IBM cards may be key-punched and sorted by machine. A totaling operation may be carried out mentally or by an adding machine. The present system of reports was prepared by pulling pre-punched IBM cards from a tub file which matched the transactions reported on the source documents illustrated in Figure 2

Figure 2, opposite. Illustration of a service order showing *location, movement* and *class of service.* The location is shown by OD, the initials of the central office; the movement shown is an OUT, a disconnect; and the class of service is indicated by 2 MR MES, a two party, measured rate, residence telephone.

S T A S	No. on Prem.	Un-chgd.	NO.	MOVED	CHANGED

INSTALLATION WORK

IW

LI CARD FILE

CHECKED ☐

DATE COMP.

INSTALLER

DATE S 3-17-58 APPL. ORD 0-683407*-T

BUS OFC BW DUE 3-19 RO

LST

CLHD

ALST

S E & R

IN

LST WOLCOTT WM H MAJ USA/
3 POOKS HILL RD BETH

OD 2-6371

CLHD

ALST

S E & R
2 MR MES 4 60
1 SPRG CD 00

OUT

SCC AP DEP

BILL RMKS
FB % STUDENT DETACHMENT ASSOCIATE SIGNAL
OFFICERS ADVANCE COURSE THE SIGNAL SCHOOL
FT MONMOUTH N J/
LOC APT 811/

CL. SER. DIR A C S ADVTG LI B

SER RECAP

DIST TER BLK

UG 17 TER Base Pooks Hill TZ PROT

BRG 3409 R PTY

LINE	DIST PR	Bnd'g Pst	UG PR	LCH-FR	PAN-GR-BAY COL	JK-TER-LS SW-V	DROP YES NO NOTES
			442	OD	014	54	2832

	INSTALLER	DATE	WIRE CHIEF
DISC. & LEFT IN			
FINAL COMPLETION OR REMOVAL	Johnson	3-19-58	

These cards were processed on the IBM 407 accounting machine. The preparation methods which were considered feasible were the present method (P_1), the IBM 650 (P_2), a hand tally method (P_3), the IBM 650 plus sampling techniques (P_4), and a hand tally method plus sampling techniques (P_5). These alternative methods of preparation, as already indicated, were designated as P_1, P_2, P_3, P_4, and P_5 for convenient reference.

Although the primary consequence of choosing a method of preparation affects cost, it may also affect the time to prepare the report as well as the accuracy of the reported facts.

RELATION BETWEEN THE ALTERNATIVES AND OBJECTIVES

The models relating the cost of a system and the value of a system to the method of preparation (P_j) and the specifications for the system (S_i) are given below,

$$\text{Cost} = f(S_i, P_j)$$
$$\text{Value} = g(S_i, P_j)$$

These relationships state that the cost as well as the value of a system of reports are functions of the system itself—namely, the information contained in the reports, the frequency of issue, the amount of summarization, etc.—and are also functions of the method of preparation. The problem, then, is to find S^* and P^*, the optimal system of reports and the optimal method of preparation. If we define the measure of optimization as utility (U), we may formally state the model as follows:

$$\text{Utility} = \text{Value} - \text{Cost}$$

ILLUSTRATION OF A MOVEMENT REPORT

Let us suppose that one of the eight customers shown at C in Figure 3 contacts the business office and requests that a telephone be installed in his residence. The service representative records the customer's name, his address, and other specific information concerning the type of telephone he wishes to have installed. Subsequently, the business office prepares a tape for the purpose of transmitting all of the necessary information for installing the telephone over a teletype system to

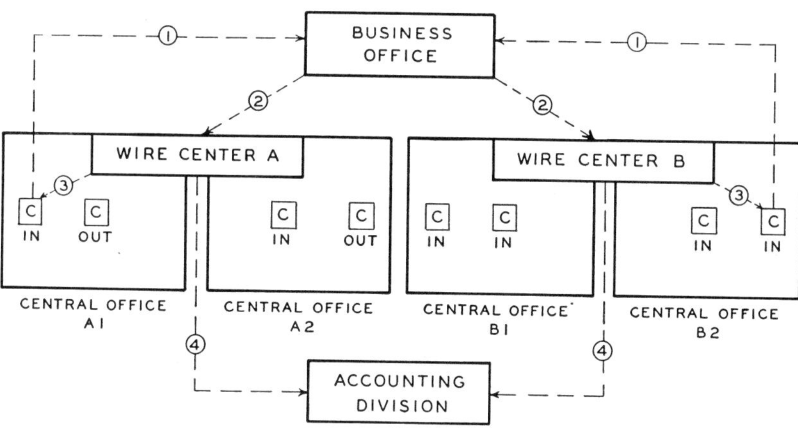

Figure 3. An illustration of the origin of basic movement and the Movement Report reflecting these changes. The typical sequence of events is shown by number: (1) Customer-Business Office; (2) Business Office-Wire Center; (3) Installer-Customer; (4) Service order-Accounting.

MOVEMENT REPORT—JUNE 10, 1958

Location	In	Out	Net Gain
Central Office A1	1	1	0
Central Office A2	1	1	0
Central Office B1	2	0	2
Central Office B2	2	0	2
District—AB	6	2	4

Wire Center A which is located in the vicinity of the customer's residence.

Although several copies of the service order as shown in Figure 2 are produced simultaneously by the teletype machine in Wire Center A, we are concerned with only two copies. One of the copies is delivered to a telephone installer who proceeds to the customer's residence and installs the telephone. Upon completion of the installation, the installer calls a clerk located in Wire Center A who also has an identical copy of the service order. The installer informs the clerk of any significant changes in the service requested by the customer. The clerk corrects her copy of the service order to show these changes and notes that the work has been completed.

At the close of each day the clerk in Wire Center A collects all of the service orders, sorts them by Central Office A1 and Central Office A2, and forwards them to the accounting department responsible for preparing the *Station Movement Reports*. The transactions contained in the completed service orders constitute the basic information considered in this problem.

The *In* and *Out* shown beneath each customer's residence represent respectively the installation of a new telephone and the removal of an installed telephone. Note that Wire Center A contains two central offices, A1 and A2. Likewise, Wire Center B also contains two central offices, B1 and B2. Using this information and making the assumption that Wire Center A and B constitute a District AB, we are in a position to prepare the movement report shown in Figure 3 for the conditions illustrated.

ILLUSTRATION OF A SYSTEM OF REPORTS

A system of reports is illustrated in Figure 4. This system contains three reports. Report No. 1 presents the weekly transactions for a given month for two movements. Report No. 2 presents the same information for a month without a weekly summary; however, we now have three locations and three classes of service. Report No. 3 presents the information by central offices (locations), summarized for the districts. Two types of movements are shown. The net gain in each case is 103 stations.

This illustration indicates the various ways in which certain transactions may be reported. It also indicates the computations required to be performed on the basic source data to produce the reports. Sorting, totaling, resorting, totaling, and subtracting is required to produce each of these reports.

It is quite impossible to compare various alternative report systems containing a large variety of classes of information without some convenient means of coding the type of data contained therein. Although the team spent a great deal of time attempting to relate the present

NO. 1

Movement / Frequency of Preparation	In	Out	Net Gain
1- 7	52	23	29
8-15	79	19	60
16-22	26	22	4
22-30	45	35	10
Month	202	99	103

NO. 2

Location / Class of Service	AB	CD	EF	Total
Residence	7	37	4	48
Commercial	2	26	4	32
Industrial	6	19	−2	23
Month	15	82	6	103

NO. 3

Movement / Location	In	Out	Net Gain
A	7	2	5
B	18	8	10
A + B	25	10	15
C	43	10	33
D	120	71	49
D + C	163	81	82
E	6	2	4
F	8	6	2
E + F	14	8	6
Month	202	99	103

Figure 4. Illustration of a system of reports with two dimensions fixed and two dimensions variable.

system to various alternative systems, little or no progress was made until a coding system had been developed. The coding system was a natural development from recognition of the fact that the information under consideration was a function of the type of movement, the class of service, the location, and the frequency of preparation of the report.

The present problem involved seven movements, 39 classes of service, 300 locations, and 260 frequencies of preparation.

A further restriction was imposed on the coding system for practical reasons. The various summaries and computed totals deceptively confused the analysis as did the various dimensions. Hence it was necessary to consider only basic data regarding the dimensions. Basic data refers to the source documents. For example the basic data regarding movement in Figure 3 is of two types, *In* and *Out*. We do not consider the *Net Gain* information as basic data, since it is simply a computation and could be determined from the other data presented if it did not appear on the report.

Figure 4 provides an illustration of the various dimensions or transaction classifications used in the analysis of the alternatives (9). Each report contains only two dimensions which vary. The other two dimensions are fixed for each report. The location dimension and the class of service dimension would be shown in the title of report No. 11; otherwise the report would have no meaning. The inability of a sheet of a report to handle the four variable dimensions necessarily dictates several reports concerned with presenting these basic data. Discovering

Report Number	Dimension			
	Movement	Class of Service	Location	Frequency of Preparation
1	2	1	1	4
2	1	3	3	1
3	2	1	6	1
Basic	2	3	6	4

Figure 5. Identification of the basic data contained in the system of reports illustrated in Figure 4.

the most useful combinations of the two variables can lead to significant cost reductions while maintaining or increasing the value of the report system.

The system of reports illustrated in Figure 4 can be meaningfully summarized as shown in Figure 5. The bottom line of Figure 5, the amount of basic data contained in the system of reports, offers a convenient specification for analyzing or designing a system of reports.

This method of analysis was used to describe the present system of reports. The results of this analysis are presented in Figure 6. The *Station Movement Report* shown in Figure 1 is referred to in Figure 6 as number 8. This report gives seven basic movements, one class of

Report Number	Dimension			
	Movement	Class of Service	Location	Frequency of Preparation
1	7	4	14	260
2	3	3	1	260
3	3	2	1	260
4	3	1	1	260
5	3	1	12	52
6	3	2	14	52
7	7	8	1	12
8	7	1	300	12
9	1	39	300	12
10	3	4	2	52
Basic	7	39	300	260

Figure 6. Identification of the basic data in the present system of reports. The numbers in the columns indicate the sub-classifications given on each of the ten reports for each dimension. The frequency of preparation sub-classification is based on a year, with daily (260), weekly (52), and monthly (12) reporting periods indicated.

service (main stations only), 300 row headings or locations, and is prepared twelve times per year.

DETERMINATION OF FEASIBLE ALTERNATIVE SYSTEMS

Once the specifications or dimensions of the present system of reports were fully understood, it was fairly easy to design individual reports. This was accomplished by specifying two fixed and two variable dimensions. The variable dimensions could be less but not greater than the basic dimensions shown in Figure 6. The problem of designing a system and selecting the individual reports in the system was somewhat more complex.

Throughout the orientation study certain restrictions were mentioned by the users of the movement reports. Some of the data were used to prepare corporate reports, since some accounting entries for depreciation and standard cost determinations were based on currently reported data. These facts were cataloged along with many others for the purpose of designing practicable alternatives. The team prepared a system of reports based on these minimum restrictions; these were referred to as *Simplified Movement Reports.* These reports were believed to contain the least amount of information acceptable to the majority of the users of the present system.

In order to verify the findings and to discover any other restrictions before actually designing additional alternatives, it was necessary to contact each person receiving one or more of the reports in the present system. A questionnaire was chosen for this purpose. The *Simplified Movement Reports* were used in conjunction with the questionnaire. It was believed that the items on the questionnaire could be more realistically comprehended and answered in terms of what was wrong with the new reports than from any other point of view. At least many of the variables in the dimensions were eliminated from consideration, thereby calling attention to those which remained.

At the suggestion of the team the comptroller prepared a letter of transmittal to be attached to the questionnaire and the *Simplified Movement Reports.* The comptroller's signature on the letter seemed desirable since he is responsible for the preparation and issuance of all the reports in the present system. The questionnaire is shown in Figure 7. Since it was impossible for the team to foresee all possible answers

STATION MOVEMENT QUESTIONNAIRE

The attached Simplified Station Movement Report is a proposed replacement for the Station Movement Reports listed below. These reports would no longer be available should the new report be adopted.

2704 Summary of Station Movement
2705 Station and Circuit Report

1. A check has been placed by the above reports you now receive. Could you effectively perform your job if you no longer received these reports?

2. Would you accept the attached Simplified Station Movement Report as a satisfactory replacement for the above reports?

3. Please list the additional station movement information (rows, columns, locations, and frequency of issue) that could be added to the attached report which you consider to be *absolutely* essential for you to perform your job. Be specific. You may use a sketch to show the additional data you require.

4. State the specific use you will make of each item requested in 3. Give examples of past decisions, recommendations or reports which would be affected if you did not receive the additional data.

5. What would probably happen, in terms of your decisions, recommendations, or reports, if you were not given the additional data requested in 3? You are again urged to be specific; to give an answer for each additional item you would require. It would be helpful if you could estimate, in dollars, the effect of not having the additional information.

6. What additional analysis would you have to make in your office to allow you to make decisions, recommendations, or reports, if you were not given the additional data requested in 3? Give an estimate, in dollars, of the cost of taking these steps.

SUPPLEMENT

The attached *Informal* Simplified Station Movement Report is a proposed replacement for the Station Movement Reports listed below. These reports would no longer be available should the new report be adopted. This informal report would normally be issued every week. Under unusual circumstances it would be issued more frequently.

R-284 Daily Summary of Station Movement
R-334 Maryland Suburban Station Data
R-335 Daily Station Movement, Total State
R-336 Projection of Station Movement
TWX Teletyped Station Movement Report

If you receive any of these reports please answer questions 1 through 6 as above, this time as they pertain to the informal reports. The same detailed answers are requested.

Figure 7. Questionnaire used to determine restrictions on the system of reports and the uses of station movement information.

to the questionnaire, it was decided that a forced set of answers would greatly inhibit the responses of the users.

Analysis of the Questionnaire

The summarized analysis of the questionnaire is shown in Figure 8. In order to prepare Figure 8 it was necessary to edit carefully the answers to each question and to code the questionnaire. The coding scheme contained a key for the organizational level, the department, and a specification for each of the four dimensions. Figure 8 was readily prepared from the coded questionnaires. In some cases it was necessary to contact the user of the reports to determine exactly what was wanted before the code could be affixed to the questionnaire. A number of users stated that they no longer needed the report and suggested that their names be dropped from the distribution list.

Certain users indicated that their needs could only be met by the present system of reports. Interviews were arranged and these requirements were investigated in detail for specific uses. The summation of the questionnaire and the interviews provided a fairly complete picture of the significant uses of the information contained in the present system. Generally the categories of use were informational, to prepare other reports, and to make decisions.

It became apparent to the team during the analysis of the questionnaire that the problem of placing a dollar value on the report would be next to impossible. It seems that the dollar value of a report which is used to make decisions, to prepare other reports, and to provide general information to executives can only be measured by executive judgment, except perhaps for the case of theoretical and trivial examples.

The stated requirements of the executives given in Figure 8 provide a measure or quantification of value which can be used in the solution to this problem. At least we have a finite listing of restrictions and an optimal system of reports can be designed within these restrictions. Certainly it is desirable to be able quantitatively to measure restrictions, i.e., to be able to state that a given system of reports is worth a certain amount of money. However, when no means of direct quantitative measurement appears feasible, some indirect method must be sought.

Analysis of the data shown in Figure 8 indicated that approximately twelve alternative systems of reports could be considered feasible.

Department	Mgt. Level	Location Code				Frequency Code				Movement Code		Class of Service Code					
		1	2	3	4	1	2	3	4	1	2	1	2	3	4	5	6
Executive	5				1			1		1		1					
	4				1			1		1		1					
	3	1			1			2		2		2					
	2	1						1		1		1					
	1																
Accounting	5																
	4				3		1	2		3		3					
	3	2			2			3	1	4		2	2				
	2			1	4	1		4		5		3		1			1
	1				1			1		1		1					
Engineering	5																
	4	1						1		1			1				
	3																
	2	1						1		1		1					
	1																
Commercial	5																
	4	4				1		3		4			3			1	
	3	8			5			11	2	13		7	4			1	1
	2	27			10		1	34	2	37		15	19		1	2	
	1	2															
Plant	5				1			1		1		1					
	4	3		1	4			7		7		3	1		2		1
	3	8			3			10	1	10	1	6	4	1			
	2																
	1																
Traffic	5				1			1		1		1					
	4	2			1			3		3		1	1		1		
	3	1			11			12		12		12					
	2																
	1				1			1		1		1					

Location Code	Frequency Code	Movement Code	Class of Service Code
1. Cent. Office	1. Day	1. As on Questionnaire	1. As on group #1
2. Traff. Cent.	2. Week	2. Other	2. As on group #2
3. Wire Cent.	3. Month		3. As on group #3
4. District	4. Quarter		4. As on group #4
			5. As on group #5
			6. Other

Figure 8. Summarized results of movement report questionnaire, by dimensions desired, management level, and department. For explanation of dimension codes see bottom of figure. Class of service codes refer to convenient groupings not illustrated. The number in each cell represents the total number of questionnaires falling in that particular category.

Some of the specifications of these systems were quite arbitrary but were generally based on restrictions and qualitative statements obtained from the questionnaire. The determination of the cost of the various alternative systems, using the various methods of preparation previously mentioned, and the qualitative evaluation of the requirements on the questionnaires will provide the basis for selecting the best alternative system.

MEASURING THE COST OF THE FEASIBLE ALTERNATIVES

While the cost problem seems to be conceptually simple, difficulty arises from a lack of factual information concerning the increments of cost and from the tremendous amount of effort required to compute the cost of each alternative. The cost of a particular system of reports involves the determination of the elemental cost for each of the cost areas previously defined. Hence, the problem is to prepare a cost sheet, as illustrated in Figure 9, for each of the twelve alternative systems. Since each system could be prepared feasibly by any of the five methods of preparation, a total of 60 cost sheets would be required. Only 12 were prepared in great detail. It was possible to eliminate the other 48 without developing detailed costs, because of certain restrictions on methods of preparation involving sampling, restrictions on systems which provided minimal information, and judgment.

The particular problem was further complicated by the fact that the present system of reports and probably the selected alternative system would involve two separate locations. The state is organizationally divided into two divisions, and each division prepares the reports for its organization. The summarized state totals are, of course, prepared by only one of these divisions.

The obvious starting point in the problem is to calculate the cost of the present system. However, the tremendous amount of work to be done with the various methods of preparation and alternatives clearly dictates careful planning. The decision was made to develop usable standard costs wherever possible in order to simplify the problem. Hence, the problem confronting the team was to complete a cost sheet as shown in Figure 9 for the present system of reports, at the same

Cost Elements	Cost Areas			
	Preparation	Reproduction	Distribution	Storage
Direct Labor				
Materials				
Equipment				
Facilities Overhead Supervision				
Vacations and Holidays				
Sick and Excused Leave				
Relief and Pensions				
Social Security				
Training				

Figure 9. Cost information required for each alternative by each of five methods of preparation.

time preserving any measurements that would facilitate calculations for possible alternatives.

Direct Labor Costs

The preparation of the present system primarily involved four groups of employees: two groups of clerks who pulled the IBM cards for each piece of needed information on the service order, and two groups of IBM operators who prepared the intermediate and final tabulations. The clerks in the clerical operations section performed many tasks in addition to those associated with the present system of movement reports. There were no adequate measurements available to indicate the amount of time spent on the various jobs performed. This problem when analyzed indicated that the most practicable means of measuring would be by use of the technique of work sampling (2, 6). An all-day time study was made of one of the clerks for the purpose of listing and measuring the various specific jobs performed by the clerks. A master list of possible jobs was compiled from the time study and edited by the first and second level supervisors. The final work sample sheet used is shown in Figure 10. All of the initial observations were made by a member of the team, but toward the end of the study the first level supervisor also made a number of observations. The use of this method of measuring was explained to the clerks, who gave excellent co-operation throughout the study. The presence of the observer was scarcely noticed after the first week of observations. Since there were five or six clerks engaged in this operation, a great number of observations were obtained in a fairly short period of time.

A control chart was maintained throughout the study to determine relevant statistical properties of the system. When the degree of probable error of the computed time spent on each job reached the arbitrarily assigned permissible error, the study was discontinued. These measurement methods were explained to the decision-makers responsible for approving final recommendations, and the final results were accepted without question.

The next step in determining the cost of the direct labor of the clerks applicable to the present system of reports was the classification of the jobs performed. The observer became quite familiar with the work and was able to make a reasonable assignment of all of the jobs pertaining to movement reports, all not pertaining to movement reports, and a classification pertaining to both of the previous classes. This classification was checked by the first level supervisor and a determination was made to split the combination class of jobs between movement and other reports.

The measurement of the direct labor used in the preparation of the movement reports in the two IBM sections was accomplished in essentially the same manner. The work sample observation sheet shown in Figure 11 was used to record the observations in the IBM sections. Note that Figure 11 contains the equipment used in the preparation of the movement reports as well as the operators. The cost of the equipment and facilities used in the clerical operations section was

WORK SAMPLE DATA SHEET
CLERICAL OPERATIONS SECTION

Date Feb. 21, 1958 Observer L. A. Herzog Location ________

Activity Key

1. Open Mail, Stamp Date, Check Invoice
2. Note "C" Orders and PBX
3. Pull IBM Cards
4. Verify Cards
5. Progressing
6. Prepare Cards for IBM
7. Write Up Adjustments
8. Work 284, Check IBM Runs
9. Set Up (File, Get Materials)
10. Out of Office—Deliver Work
11. Phone and Check Plant for Information
12. Talk—Not Work, Visit
13. Make Up Work Volume
14. Break
15. Reassociations
16. Dispose of I.W's.
17. Tally Aux. Service Rpt.
18. Tally Colored Sets
19. Tally Labor Units
20. Tally Service Units
21. Tally ITMC Orders
22. Sort WA's—Billing
23. Work Supplementals
24. Miscellaneous Audit
25. Audit Dist. WA's—IW's
26. Division IN—OUT Rept.
27. Order Invoices
28. Other Assignment

	Time Observations Began							
Operator Observed	8:52	9:42	10:14	11:54	1:17	2:18	3:24	4:51
Wolf	23	23	19	20	5	25	22	8
Inkenich	3	21	4	1	16	23	5	25
Lewis	8	8	5	5	2	19	11	4
Mayr	19	1	3	5	6	13	5	21
Moessinger	5	11	6	3	10	25	8	5

Figure 10. Work sample data sheet-clerical operations section.

WORK SAMPLE DATA SHEET
IBM SECTION

Date 5-5-58 Observer R. Klien Location

Machine Activity Key

1. Available—Not in Use
2. Not Available—Maintenance
3. Operating—Training
4. Operating—Not Move. Rpts.
5. Set Up—Not Move. Rpts.
6. Operating—Movement Rpts.
7. Summarizing—Movement Rpts.
8. Set Up—Movement Rpts.

Machine Observed	#	Time Observations Began							
		8:58	9:06	10:09	11:40	1:25	2:25	3:42	4:19
Sorter	1	4	4	4	2	4	4	1	4
Sorter	2	6	4	4	4	4	4	4	4
Sorter	3	6	4	6	1	1	6	4	6
Sorter	4	4	1	4	4	4	4	4	4
Sorter	5	4	4	4	4	4	2	2	4
Sorter	6	1	1	4	4	4	4	4	4
Collator	1	1	1	1	7	5	4	4	4
Reproducer	1	4	4	1	6	6	6	1	1
Tab 402	1	8	6	6	1	5	4	4	4
Tab 402	2	5	5	4	4	4	1	1	4
Tab 407	1	7	7	7	4	5	4	7	7
Interpreter	1	1	1	1	4	1	1	1	1
Interpreter	2	4	4	1	1	1	4	4	4
Key Punch	1	1	1	1	1	1	1	1	1
Key Punch	2	4	4	4	4	4	4	4	4
Key Punch	3	4	6	4	4	4	4	4	4
Key Verifier	1	4	4	4	4	4	4	4	4

Operator Activity Key

1. Set Up—Movement Rpts.
2. Running—Movement Rpts.
3. Checking—Movement Rpts.
4. Set Up—Not Movement Rpts.
5. Running—Not Movement Rpts.
6. Checking—Not Movement Rpts.
7. Out of Office
8. Waiting for Work
9. Other Assignment

	Time Observations Began							
Operator Observed	8:58	9:06	10:09	11:40	1:25	2:25	3:42	4:19
Operator #1	2	2	1	2	2	2	3	2
Operator #2	1	2	2	5	6	2	2	3

Figure 11. Work sample data sheet–IBM section.

divided between movement and not-movement on the basis of a ratio between these two broad classes of work. The probable error of this allocation of cost seemed to be small from a cursory analysis of the situation. The effect of a considerable error in this division would have little effect on the final computation since the order of magnitude of these items was very low. However, the allocation of IBM equipment rental cost could not be treated so simply. In the first place, the hourly rental cost of the IBM 407 exceeds the yearly carrying cost of a clerk's desk. Hence, the measurement problem in the IBM sections required the determination of the operators' time devoted to the movement reports, as well as the operating time of the IBM machines actually used.

The direct labor of the clerks in the clerical operation section and the IBM operators were taken from the summarized observation sheets and entered on the cost sheet. The average labor rate for each group was used to determine the direct labor cost. There were several other clerical groups who performed some direct labor on the present system. The amount of time spent by these clerks on the present system of movement reports was measured, for the most part, by the first level supervisors. These supervisors were asked to make a careful estimate of the time of each clerk in operating the multilith machine, transcribing IBM computations to a report, collating individual sheets of the report, mailing the finished report, telephoning information, and a host of similar jobs. The supervisors timed the clerks over a reasonable period and subsequently produced useful indications of the time normally spent in the performance of the jobs associated with the present system of movement reports. The various clerks were identified by name, and the labor rate could be determined. The product of the labor rate and hours, summed for all jobs in each cost area, produced additional direct labor costs for the present system.

Material Costs

The material used in the present system was determined by beginning and ending inventories for certain periods, by counting the number of sheets in the reports and multiplying by the distribution list, by estimating the average IBM cards and tapes actually used, and estimating the quantities of other incidental supplies known to be used in producing the present system of reports. These items were priced by the accounting department from actual billing records and summarized under each cost area. Estimates were pròvided from supervisors, clerks, and members of the team. One fairly significant item of cost in the distribution area was postage, since many of the reports were quite large. A number of the reports were distributed by company mail. Rather than attempt to measure the cost of delivery by company mail it was decided to use U.S. postal rates and to cost all reports as if delivered by the Post Office.

Equipment Costs

Fortunately, every piece of capital equipment in the company has an equipment tag affixed to it. A complete inventory of the equipment used in connection with the present system of reports was taken and delivered to the appropriate accounting section for the purpose of having the carrying charge of each item determined. The maintenance and carrying charge for most of the equipment was approximately $4\frac{1}{2}$ per cent of the purchase price.

The rental cost of the IBM equipment was also furnished by the accounting department based on recent contractual arrangements. The actual hours normally required for each piece of IBM equipment was determined by the work sample previously mentioned. A simultaneous work sample was made to determine the utilization of each piece of equipment used in manufacturing the present system of movement reports. The hourly cost of the IBM equipment was determined by dividing the actual monthly rental costs by the number of hours actually used, thereby taking into account the actual use of the machine rather than the time for which the machine was available for use. In some cases this is a significant difference, i.e., low utilization.

FACILITY COST

The space in number of square feet used in manufacturing the present system of reports was inventoried. This, of course, included the office area occupied by the clerks as well as the floor space assigned by the various pieces of equipment used in the process.

Costing a square foot of floor space, including lighting, heating, cleaning, carrying and other associated items, can be a formidable task indeed. The problem was solved quite simply in this case, however. The company actually rents a very large amount of office space to house its activities. Some of this space is rented at approximately $5.00 per square foot per year for assignable office use. This price includes the heating, lighting, cleaning, carrying, and other associated charges. Hence the facility cost was determined by taking the product of the number of square feet of space used in the present system, under each area of cost, and the above rental charge per square foot.

Overhead Cost

All of the elements of overhead cost were readily determined from cost accounting records, except the supervision and training costs. The supervision costs were related to the hourly cost of the clerks and operators as a per cent of the direct labor cost per hour. This was determined by summing all first and second level supervisory salaries, summing all clerk and operator salaries, and expressing the relationship as a percentage. Although this is somewhat arbitrary, it does provide a method of charging the products produced (reports) with the significant supervisory expense.

The training cost per clerk or operator was determined from company records. In prior years the company had made a study for these and other job classes to determine the time required to train new employees to perform at 100 per cent effectiveness. The data indicated that the effective output of a new employee during the first month on a particular job would be 50 per cent. This rate would increase during succeeding months by some stated percentage and possibly reach 100 per cent at the end of nine months. The data were based on measured performance for protracted periods and were used to determine force requirements.

The average length of time employees spent on these jobs was determined from personnel records. The training time was computed in hours by computing the area between 100 per cent performance for the period of time spent on the job and the actual output determined by the per cent effectiveness curve. The cost of this time was computed by using the average hourly rate for the training period. This cost constituted approximately 10 per cent of the total employees' earnings based on average turnover.

Summary-Cost of Present System

The actual time required to cost the present system involved several hundred man-hours. However, the results were useful and the degree of accuracy initially decided upon was maintained. The standard times and various other determinations were invaluable in costing the other alternative systems and methods of preparation. The costs of the alternative methods of preparations were to a large degree synthesized from the initial cost work. In some cases, however, experiments had to be designed to evaluate new methods. The general approach was essentially the same as described above.

The final cost of the present system was considerably in excess of the initial estimate of $40,000. The initial estimate served its purpose, however, in that it indicated the order of magnitude of the problem, and its determination was quickly and economically accomplished. Figure 12 shows the completed cost sheet for the present system. The elements of cost are expressed as a percentage of total cost in order to draw significant conclusions. First and foremost, it seems that direct labor costs and the associated overhead are the major expenses, representing approximately 76 per cent of the total cost. Secondly, approximately 96 per cent of the total cost is incurred in the preparation

Cost Elements	Cost Areas			
	Preparation	Reproduction	Distribution	Storage
Direct Labor	49.00%	1.00%	0.25%	0.15%
Materials	1.50%	1.00%	1.00%	0.01%
Equipment	13.00%	0.01%	0.10%	0.01%
Facilities	7.00%	0.10%	0.03%	0.04%
Overhead				
Supervision Vacations and Holidays Sick and Excused Leave Relief and Pensions Social Security Training	25.35%	0.25%	0.10%	0.10%

Figure 12. Cost of each element in each area in the present system of reports expressed as a percentage of the total cost.

area. Hence, any significant reductions must necessarily be made in these areas.

It seems probable that a cost equation could be determined, after a number of reports were similarly costed, to evaluate the cost of other systems of reports. The equation might take into account the number of pages, the number of rows, columns, the method of preparation, the amount of computation, and other related variables. Its use might be along the line of the per-square-foot-cost idea used by the construction industry.

Cost of Alternative Systems

It was not too difficult to cost the alternative systems, once all of the cost elements for the present system had been measured and checked. A few remaining problems, such as measuring the hand tally time required to process a service order, and the set-up and run time on the IBM 650, did require considerable effort, however. In both of these cases as well as some others the team was able to experiment with conditions which would be encountered under actual operations. Considerable care was exercised in costing the alternatives as had been done and described above for the present system.

Figure 13 shows the relative cost of each of the twelve systems and each of the five methods of preparation. This figure clearly indicates

	S_1	S_2	S_3	S_4	S_5	S_6	S_7	S_8	S_9	S_{10}	S_{11}	S_{12}
P_1	100	76	81	83	86	79	84	86	89	90	91	92
P_2	120	40	45	53	57	45	59	90	103	50	68	105
P_3	60	23	28	30	33	41	46	48	51	44	45	46
P_4	50	40	41	43	48	45	47	48	49	46	47	48
P_5	30	20	21	23	28	23	25	27	29	28	29	30

Figure 13. Summarized cost for each alternative system (S_i) and each method of preparation (P_j). Numbers represent relative cost.

that the cost of a report system is indeed a function of the method of preparation as well as the specifications of the system. Although the relative costs shown in Figure 13 did not agree with the preconceived notions of the team, assignable causes of variation were not too difficult to determine.

It should be recalled that the solution to the problem cannot be determined from an analysis of the relative costs only. We must also consider relative value.

SOLUTION TO THE PROBLEM

The final solution recommended by the team and accepted for implementation by the managers of the telephone company regrettably omitted the degree of quantification of value initially expected. The team did measure the value of the various systems in terms of executive judgement and was quite confident that the report systems which were acceptable to the majority of users had value comparable with the present system. Hence, in terms of the model, Utility equals Value minus Cost, by holding Value essentially constant, the team was able to produce a scale for ordering each of the twelve alternatives indicated on the questionnaires.

Top-level managers from all divisions concerned with movement reports were present at the final meeting of the project. A particular method of preparation and a particular system of reports was recommended by the team and an attempt was made to reconcile all objections. The reasoning behind the recommendation was fully explained and in most cases verified by measurements. The recommendation of the team was essentially accepted by the company. The changes desired were discussed in terms of their probable cost and the decision was made to implement the new system as soon as practicable.

Briefly, the new system replaces the IBM cards and machines with a hand tally method of preparation which is very flexible and more economical. Some of the reports were reduced in size and redundant content. It was also found that certain information on some of the reports could be more economically obtained by using different source documents than were used in the original system.

IMPLEMENTATION

A tentative schedule of implementation was approved at the final meeting. The schedule included a listing of significant jobs to be accomplished prior to the cut-in point for the new system. The program was assigned to a responsible accounting supervisor with authority for making the score of decisions normally required in the installation of a complex system. The company employee who worked as a member of the team and one graduate student were assigned the implementation job under the direction of the accounting supervisor. A detailed account of the implementation of this project will be published at a later date.

The new system produces a reduction in operating cost of approximately $15,000 per year. This is the actual current reduction, which disregards such step functions as cost of floor space, supervision, and other items of overhead. The actual saving comes about by lowering the cost of the preparation method, elimination of certain redundant, high cost operations, and a change in the frequency of preparation of some of the data.

CONCLUSIONS

The approach to the problem of recognizing value and cost in a system of reports sheds new light in this little explored area, at least to the users of reports in this company.

The recognition of the specifications of a report in terms of the dimensions and basic data proved to be of great value in redesigning the system and in understanding feasible alternatives. The maximum requirement of basic data under each dimension seems to affect the cost of the system of reports to a significant degree. Careful attention is indicated in the preparation area and particularly to direct labor cost in the design of a system of reports. The first copy of the reports represented 96 per cent of the cost of the present system. When only a few hundred copies of a report are required, there seems to be no appreciable effect on the total cost of the system by increasing or decreasing this number over fairly wide ranges.

The value problem seems to be an area which offers very high potential rewards for further research. High-speed data processing systems will very likely increase the need for value determinations.

It is believed that intangible as well as tangible results are attributable to this case study. The various executives contacted in connection with the study have been motivated to give more careful consideration to the general paper work problem. Many of these managers are now more aware of the tremendous costs involved in producing reports, and subsequent decisions in this area are likely to reflect this newly gained knowledge. Not to be forgotten, of course, is the fact that the new system of reports provides as much useful information as the old—at a substantial reduction in annual cost.

REFERENCES

(1) Barnes, Ralph M. *Motion and Time Study*. New York: John Wiley and Sons, 1958.
(2) ———. *Work Sampling*. New York: John Wiley and Sons, 1957.

(3) Chapanis, Alphonse, Garner, Wendel R., and Morgan, Clifford T. *Applied Experimental Psychology*. New York: John Wiley and Sons, 1949.

(4) Churchman, C. West, Ackoff, Russell L., and Arnoff, E. Leonard. *Introduction to Operations Research*. New York: John Wiley and Sons, 1957.

(5) Goldman, Stanford. *Information Theory*. New York: Prentice-Hall, 1953.

(6) Heiland, Robert E., and Richardson, Wallace J. *Work Sampling*. New York: McGraw-Hill, 1957.

(7) McKinsey, J. C. C. *Introduction to the Theory of Games*. New York: McGraw-Hill, 1952.

(8) Morrow, Robert Lee. *Motion Economy and Work Measurement*. New York: Ronald Press, 1957.

(9) Naddor, Eliezer. "What Data to Process," *Operations Research,* Vol. 6, No. 4 (July-August, 1958), 629-30; and private communications.

(10) Rosenstein, Allen B. "The Industrial Engineering Application of Communication-Information Theory," *Journal of Industrial Engineering,* Vol. 6, No. 5 (September-October, 1955).

(11) Shannon, Claude, and Weaver, Warren. *The Mathematical Theory of Communications*. Urbana: The University of Illinois Press, 1949.

(12) Taylor, Frederick W., *Scientific Management*. 3rd ed. New York: Harper and Brothers, 1947.

(13) Weart, Spencer A. "Practical Application of the Theory of Games to Complex Managerial Decisions," *Journal of Industrial Engineering,* Vol. 8, No. 4 (July-August, 1957).

(14) Williams, J. D. *The Compleat Strategyst*. New York: McGraw-Hill, 1952.

INDEX